Measurement Techniques Sensors and Platforms

Editor-in-Chief

John H. Steele

Marine Policy Center, Woods Hole Oceanographic Institution, Woods Hole,
Massachusetts, USA

Editors

Steve A. Thorpe

National Oceanography Centre, University of Southampton,
Southampton, UK
School of Ocean Sciences, Bangor University, Menai Bridge, Anglesey, UK

Karl K. Turekian

Yale University, Department of Geology and Geophysics, New Haven,
Connecticut, USA

Subject Area Volumes from the Second Edition

Climate & Oceans edited by Karl K. Turekian
Elements of Physical Oceanography edited by Steve A. Thorpe
Marine Biology edited by John H. Steele
Marine Chemistry & Geochemistry edited by Karl K. Turekian
Marine Ecological Processes edited by John H. Steele
Marine Geology & Geophysics edited by Karl K. Turekian
Marine Policy & Economics guest edited by Porter Hoagland, Marine Policy Center,
Woods Hole Oceanographic Institution, Woods Hole, Massachusetts
Measurement Techniques, Sensors & Platforms edited by Steve A. Thorpe
Ocean Currents edited by Steve A. Thorpe

ENCYLOPEDIA
OF
OCEAN SCIENCES: MEASUREMENT TECHNIQUES, SENSORS AND PLATFORMS

Editor

STEVE A. THORPE

BOSTON • HEIDELBERG • LONDON • NEW YORK • OXFORD
PARIS • SAN DIEGO • SAN FRANCISCO • SINGAPORE • SYDNEY • TOKYO
Academic Press is an imprint of Elsevier

ELSEVIER

ACADEMIC PRESS

Academic Press is an imprint of Elsevier
32 Jamestown Road, London NW1 7BY, UK
30 Corporate Drive, Suite 400, Burlington, MA 01803, USA
525 B Street, Suite 1900, San Diego, CA 92101-4495, USA

Material in the work originally appeared in *Encyclopedia of Ocean Sciences* (Elsevier Ltd., 2001) and *Encyclopedia of Ocean Sciences*, 2nd Edition (Elsevier Ltd., 2009), edited by John H. Steele, Steve A. Thorpe and Karl K. Turekian.

The following articles are US government works in the public domain and are not subject to copyright:

Satellite Oceanography, History and Introductory Concepts
Satellite Passive-Microwave Measurement of Sea Ice

Notice
No responsibility is assumed by the publisher for any injury and/or damage to persons or property as a matter of products liability, negligence or otherwise, or from any use or operation of any methods, products, instructions or ideas contained in the material herein, Because of rapid advances in the medical sciences, in particular, independent verification of diagnoses and drug dosages should be made

British Library Cataloguing in Publication Data
A catalogue record for this book is available from the British Library

Library of Congress Control Number: 2009907113

ISBN: 978-0-08-096487-4

For information on all Elsevier publications
visit our website at www.elsevierdirect.com

CONTENTS

SENSORS: TURBULENCE

SENSORS: OPTICAL

SENSORS: CHEMICAL

ACOUSTIC METHODS

SATELLITE, AIRCRAFT, OR SHIP-BORNE REMOTE SENSING

APPENDICES

INDEX

MEASUREMENT TECHNIQUES, SENSORS AND PLATFORMS: INTRODUCTION

Measurement is the foundation of any branch of science, and no less so in oceanography. New instruments have led to greatly improved understanding of the ocean, sometimes to radical changes in knowledge of the way in which it works. The challenges of extremely high pressures — at the mean ocean depth of 3750 meters the pressure is almost 370 times that at the surface — corrosion, and the severe motions induced on instruments lowered from the undulating sea surface into a sometimes violently moving sea, have resulted in many failures and have been overcome only by instruments designed by some of the most ingenious and talented engineers. The history of devices to measure the ocean is a major part of the history of the science itself, matching, as time has advanced, developments in basic technology and other sciences, so that whereas the early instruments those prior to the first World War, were largely mechanical with moving parts, those made in recent years have taken advantage of the immense and rapid developments in electronics and particularly in electronic data storage. Compare, for example, the 1904 Ekman propeller current meter lowered on a wire, its data storage limited to fewer than about 50 measurements and direction measurement relying on the falling of small balls into cups on a compass-mounted disc with 36 segments (a 10° resolution), with a present-day free-fall turbulence probe with eight (or more) different sensors to measure current shear, pressure, temperature, conductivity, and so density, each sampled to high precision more than 100 times a second and with a capacity to continue sampling for many hours to depths of 5000 meters.

The two World Wars stimulated notable advances in instrumentations, particularly underwater sonar to detect submarines during WW I. The ability to track neutrally buoyant floats using acoustic methods that eventually followed in the 1950s, led to the discovery that in most parts of the ocean the kinetic energy is dominated by mesoscale eddies with scales of many tens of kilometres and periods of several months. The thermistor chain developed in WWII has led to much better understanding of these eddies and of the structure of the ocean and its shorter-term variations such as internal waves.

Whilst only salinity was once sampled to a useful accuracy (by collection of water in water bottles, closed mechanically through the impact of a 'messenger' sliding down the supporting wire followed, on recovery, by analysis on board ship), a range of chemicals can now be routinely sampled *in situ*. The articles in this volume provide an extensive coverage of the instruments, sensors and platforms used today. Many instruments designed before the electronic revolution of the early 1960s are no longer in use, having been made re-dundant by recent advances and, although very important in their time, are not described here. Nor are some essential details that hindered or delayed improvements in instruments, such as reliable underwater seals, O-rings, water tight electrical connections and dependable devices to release buoyant instruments from their holding weights on the seabed.

Ocean measurement provides a story of the search for improved accuracy, precision and resolution, higher sampling rates and data storage capacity, the endless need for quality control and precision manufacture. New instruments and ways to measure, sample or 'image' the ocean often lead to the development of understanding of the ocean's dynamics and properties, more measurements stimulating advances in under-standing that demand further accuracy and the ability to broaden the area coverage of measurements. In this respect, satellite observation developed since the early 1970s has radically extended knowledge of the ocean, both of the surface and also of some aspects of its motion at depth. Much has been learnt about the distribution of plankton in the upper ocean, of the seasonal and area variations in sea-surface properties, temperature, wave amplitude, mesoscale eddies and surface currents, which could never have been obtained by other means.

Included in this volume are some articles on the development of subsurface vehicles capable of carrying scientists into the depth of the ocean. Whilst operations in mid-water have resulted in discoveries of or-ganisms so delicate and often so large that they cannot be caught without damage in nets or trawls, the most remarkable and important studies have been those of the seabed, especially in the observations of hydro-thermal vents and the prolific life that surrounds them. Much greater use is now being made of autonomous vehicles that can remain submerged for many days, pre-programmed to follow a set course or under com-mand from a mother vessel at the surface, sampling the water through which they pass and enabling access to regions previously little explored, such as those below the ice shelves around Antarctica. Ships are still essential, but many of the observations once only made from ships can now be more efficiently and economically made from autonomous underwater vehicles or floats that periodically communicate their data back to shore.

It is regretted that some important instruments or measurement techniques of ocean properties that are useful in prediction are not included. Most notably are those relating to waves, tides and sea level, the first (waves) important in ship routing and required in the second World War to provide means of predicting conditions during military land invasions from the sea, the last (sea level) now vital in estimating the effects of climate change. Tidal elevation is one of the oldest measurements made of the ocean. Continuously recording coastal tide gauges had been invented by the 1830s but ways to measure tides in deep water were not devised until the 1960s. (Cartwright, 1999, gives a comprehensive description of the history of tidal theory and measurement.) Although research vessels are the subject of an article, there is no reference to FLIP, the Floating Instrument Platform operated by the Scripps Institution of Oceanography in La Jolla, California, that can be towed into deep water, and which then 'flips' to become a vertical and very stable spar, accommodating scientists and from which the properties of the upper ocean can be measured. GLORIA, the Geological Long Range Inclined Asdic, that has contributed much to the mapping and discovery of features of scientific and economic importance on the seabed around the continents and the oceanic island chains, is another major instrument that is not described. (Knowledge of ocean bathymetry is fundamental to geological and physical oceanography, and instruments for measuring depth, or of 'sounding' the ocean, were some of the earliest oceanographic instruments to be developed. Robert Hooke's depth sounding device was described in 1667 but, like some of its successors, was not necessarily successful — wooden floats became waterlogged at depth, losing their buoyancy and, in deep water, the instrument failed to return to the surface!)

Much reference is naturally made to the properties of seawater on which the sensing methods depend for their response. Where it has seemed useful to do so, for example where the method and complexity of measurement depend critically on the seawater properties, articles describing these properties are included as well as articles on the measurement sensors themselves. The volume might therefore be regarded and used as an introduction to the physical, biological, chemical and geological properties of the ocean, as well an account of the methods of measurement.

Each author of the articles is an expert in the subject addressed. All the authors are distinguished researchers who have given time to write concisely and lucidly about their subjects, and the Editors are indebted to them all for the time given and care taken in preparing these accounts.

The articles in this volume would not have been produced without the very helpful advice and encouragement of the several members of the Encyclopedia's Editorial Advisory Board, particularly Gwyn Griffiths, Ken MacDonald and Robert Spindel. Each provided suggestions about the content and authorship of particular subject areas. In addition to thanking the authors of the articles in this volume, the Editors wish to thank the members of the Editorial Board for the time they gave to identify and encourage authors, to read and comment on (and sometimes suggest improvements to) the written articles, and to make this venture possible.

Steve A. Thorpe
Editor

REFERENCES

Cartwright DE (1999) Tides: A Scientific History, 292 pp. Cambridge: Cambridge University Press.

RESEARCH VESSELS

SHIPS

R. P. Dinsmore, Woods Hole Oceanographic
Institution, Woods Hole, MA, USA

Introduction

Oceanographic research vessels are shipboard platforms which support the conduct of scientific research at sea. Such research may include mapping and charting, marine biology, fisheries, geology and geophysics, physical processes, marine meteorology, chemical oceanography, marine acoustics, underwater archaeology, ocean engineering, and related fields.

Unlike other types of vessels (i.e., passenger, cargo, tankers, tugs, etc.) oceanographic research vessels (RVs) are a highly varied group owing to the diverse disciplines in which they engage. However, characteristics common to most RVs are relatively small size (usually 25–100 m length overall); heavy outfit of winches, cranes, and frames for overboard work; spacious working decks; multiple laboratories; and state-of-the-art instrumentation for navigation, data acquisition and processing.

Categories of RVs may vary according to the geographic areas of operations as well as the nature and sponsorship of work. Examples are:

1. By Region:
   Coastal, usually smaller vessels of limited endurance and capability
   Ocean going, larger vessels usually with multipurpose capabilities, ocean-wide or global range
   Polar, ice reinforced with high endurance and multipurpose capability
2. By discipline:
   Mapping and charting, emphasis on bathymetry
   Fisheries, stock surveys, gear research, environmental studies
   Geophysics, seismic and magnetic surveys
   Support, submersibles, buoys, autonomous vehicles, diving
   Multiple purpose, biological, chemical, and physical oceanography; marine geology; acoustics; ocean engineering; student training; may also include support services.
3. By sponsor:
   Federal, usually mission oriented and applied research

Military, defense, mapping and charting, acoustics
Academic, basic research, student practicum
Commercial, exploration, petroleum, mining and fisheries

According to information available from the International Research Ship Operators Meeting (ISOM), in 2000 there were approximately 420 RVs (over 25 m length overall) operated by 49 coastal nations.

History

The history of RVs is linked closely to the cruises and expeditions of early record. Most of these were voyages seeking new lands or trade routes. Oceanic studies at best were limited to the extent and boundaries of the seas. Little is known of the ships themselves except it can be assumed that they were typical naval or trading vessels of the era. A known example is an expedition sent by Queen Hatshepsut of Egypt in 1500 BC. Sailing from Suez to the Land of Punt (Somaliland) to seek the source of myrrh, the fleet cruised the west coast of Africa. Pictures of the ships can be seen on reliefs in the temple of Deir-el-Bahri.

These ships represent the high-water mark of the Egyptian shipbuilder and although handsome vessels, they had serious weaknesses and did not play a role in the development of naval architecture.

The earliest known voyage of maritime exploration comes from Greek legend and sailors' tales. It is that of Jason dispatched from Iolcus on the northeast coast of Greece to the Black Sea *c*.1100 BC with the ship *Argo* and a celebrated crew of Argonauts in quest of the Golden Fleece. Although embellished by Greek mythology, this may have a factual basis; by sixth century BC there were Greek settlements along the shores explored by Jason, and the inhabitants traditionally collected gold dust from rams fleeces lain in river beds.

Better recorded is an expedition commissioned by Pharaoh Necho II of Egypt in 609 BC. A Phoenician fleet sailed south from Suez and circumnavigated Africa on a four-year voyage. Information from this and other Phoenician voyages over the next 300 years (including Hanno to the west coast of Africa in 500 BC, and Pytheas to the British Isles in 310 BC), although not specifically oceanographic research, contributed to a database from which philosophers and scholars would hypothesize the shape and extent of the world's oceans. In today's terminology these vessels would be characterized as ships of opportunity.

Under the Roman Empire, sea-trade routes were forged throughout the known world from Britain to the Orient. Data on ocean winds and currents were compiled in early sailing directions which added to the growing fund of ocean knowledge. Aristotle, although not a seafarer, directed much of his study to the sea both in physical processes and marine animals. From samples and reports taken from ships he named and described 180 species of fish and invertebrates Pliny the Elder cataloged marine animals into 176 species, and searched available ocean soundings proclaiming an average ocean depth of 15 stadia (2700 m). Ptolemy (AD 90–168), a Greek mathematician and astronomer at Alexandria employed ship reports in compiling world maps which formed the basis of mapping new discoveries through the sixteenth century.

With the fall of the Roman Empire came the dark ages of cartography and marine exploration slowed to a standstill except for Viking voyages to Greenland and Newfoundland, and Arab trade routes in the Indian Ocean and as far as China. Arabian sea tales such as 'Sinbad the Sailor' or the 'Wonders of India' by Ibn Shahryar (905) rival those of Homer and Herodotus. The art of nautical surveying began with the Arabs during medieval times. They had the compass and astrolabe and made charts of the coastlines which they visited.

In the West the city-state of Venice became in the ninth century the most important maritime power in the Mediterranean. It was from Venice that Marco Polo began his famous journeys. In the fifteenth century, maritime exploration resumed, much of it under the inspiration and patronage of Prince Henry of Portugal (1394–1460), known as the Navigator, who established an academy at Sagres, Portugal, and attracted mathematicians, astronomers, and cartographers. This led to increasingly distant voyages leading up to those of Columbus, Vasco da Gama, and Magellan. With the results of these voyages, the shape of the world ocean was beginning to emerge, but little else was known. From the tracks of ships and their logs, information on prevailing winds and ocean currents were made into sailing directions. Magellan is reported to have made attempts at measuring depths, but with only 360 m of sounding line, he achieved little.

By the late seventeenth and early eighteenth centuries instruments were being devised for deep soundings, water samples, and even subsurface temperatures. However, any ship so engaged either a naval or trading vessel remained on an opportunity basis. In the mid to late eighteenth century the arts of navigation and cartography were amply demonstrated in the surveys and voyages of Captains James Cook and George Vancouver. The latter carried out surveys to especially high standards, and his vessel HMS *Discovery* might well be considered one of the first oceanographic research vessels.

Hydrographic departments were set up by seafaring nations: France in 1720, Britain in 1795, Spain in 1800, the US in 1807. Most were established as a navy activity, and most remain so to this day; exceptions were the French Corps of Hydrographic Engineers, and the US Coast Survey (now the National Ocean Survey). A date significant to this history is 1809 when the British Admiralty assigned a vessel permanently dedicated to survey service. Others followed suit; the first such US ship was the Coast Survey Schooner *Experiment* (1831). By now hydrographic survey vessels were a recognized type of vessel.

In the early to mid-nineteenth century expeditions were setting to sea with missions to include oceanographic investigations. Scientists (often termed 'naturalists') were senior members of the ships' complement. These include the vessels: *Astrolobe*, 1826–29 (Dumont D'Urville); *Beagle*, 1831–36 (Charles Darwin); US Exploring Expedition, 1838–42 (James Wilkes and James Dana); *Erebus & Terror*, 1839–43 (Sir James Ross); *Beacon*, 1841 (Edward Forbes); *Rattlesnake*, 1848–50 (Thomas Huxley). Most of the vessels participating in these expeditions were navy ships, but the nature of the work, their outfitting, and accomplishments mark them as oceanographic vessels of their time.

A new era in marine sciences commenced with the voyage of the HMS *Challenger*, 1872–76, a 69 m British Navy steam corvette. Equipped with a capable depth sounding machine and other instruments for water and bottom sampling, the *Challenger* under the scientific direction of Sir Charles Wyville Thomson obtained data from 362 stations worldwide. More than 4700 new species of marine life were discovered, and a sounding of 8180 m was made in the Marianas Trench. Modern oceanography is said to have begun with the Challenger Expedition. This spurred interest in oceanographic research, and many nations began to field worldwide voyages. These included the German *Gazelle* (1874–76); Russian *Vitiaz* (1886–89); Austrian *Pola*; USS *Blake* (1887–1880); and the Arctic cruise of the Norwegian *Fram* (1893–96).

The late 1800s and early 1900s saw a growing interest in marine sciences and the founding of both government- and university-sponsored oceanographic institutions. These included: Stazione Zoologia,

Naples; Marine Biological Laboratory, Woods Hole, USA; Geophysical Institute, Bergen, Norway; Deutsche Seewarte, Hamburg; Scripps Institution of Oceanography, La Jolla, USA; Oceanographic Museum, Monaco; Plymouth Laboratory of Marine Biology, UK. These laboratories acquired and out-fitted ships, which although mostly still conversions of naval or commercial vessels, became recognized oceanographic research vessels. The first vessel designed especially for marine research was the US Fish Commission Steamer *Albatross* built in 1882 for the new laboratory at Woods Hole. This was a 72 m iron-hulled, twin-screw steamer. It also was the first vessel equipped with electric generators (for lowering arc lamps to attract fish and organisms at night stations). Based at the Woods Hole Laboratory, the *Albatross* made notable deep sea voyages in both the Atlantic and Pacific Oceans. As with the *Challenger*, the *Albatross* became a legend in its own time and continued working until 1923.

In the early 1900s research voyages were aimed at strategic regions. Nautical charting and resources exploration were concentrated in areas of political significance: the Southern Ocean, South-east Asia, and the Caribbean. Many of the ships were designed and constructed as research vessels, notably the *Discovery* commanded by R. F. Scott on his first Antarctic voyage 1901–04. Another was the non-magnetic brig *Carnegie* built in 1909 which carried out investigations worldwide. The *Carnegie* was the first research vessel to carry a salinometer, an electrical conductivity instrument for determining the chlorinity of sea water thus avoiding arduous chemical titrations. Research voyages by now were highly systematic on the pattern set by the *Challenger* expedition.

Another era of oceanic investigations began in 1925 with the German Atlantic Expedition voyage of the *Meteor*, a converted 66 m gunboat, that made transects of the South Atlantic Ocean using acoustic echo sounders, modern sampling bottles, bottom corers, current meters, deep-sea anchoring, and meteorological kites and balloons. Unlike the random cruise tracks of most earlier expeditions, the *Meteor* worked on precise grid tracklines. Ocean currents, temperatures, and salinities, and bathymetry of the Mid-Atlantic Ridge were mapped with great accuracy.

After World War II interest in oceanography and the marine sciences increased dramatically. As echo sounding after World War I was a milestone in oceanography, the advent of electronic navigation in the 1950s was another. Loran (LOng RAnge Navigation), a hyperbolic system using two radio stations transmitting simultaneously, provided ships with continuous position fixing. Universities and government agencies both embarked on new marine investigations, the former concentrating on basic science and the latter on applied science. In addition, commercial interests were becoming active in resource exploration. The ships employed were mostly ex-wartime vessels: tugs, minesweepers, patrol boats, salvage vessels, etc. A 1950 survey of active research vessels showed 155 ships operated by 34 nations.

The first International Geophysical Year, 1957–58, brought together research ships of many nations working on cooperative projects. This decade also saw the formation of international bodies including: the International Oceanographic Congress; Inter-governmental Oceanographic Commission (IOC); the Special Committee on Oceanic Research (SCOR) of the International Council of Scientific Unions, all of which added to the growing pace of oceanic investigations.

The decade of the 1960s is often referred to as the 'golden years of oceanography.' Both public and scientific awareness of the oceans increased at an unprecedented rate. Funding for marine science both by government and private sources was generous, and the numbers of scientists followed suit. New shipboard instrumentation included the Chlorinity–Temperature–Depth sounder which replaced the old method of lowering water bottles and reversing thermometers. Computers were available for shipboard data processing. Advances in instrumentation and the new projects they generated were making the existing research ships obsolete; and the need for new and more capable ships became a pressing issue. As a result, shipbuilding programs were started by most of the larger nations heavily involved in oceanographic research. One of the most ambitious was the construction of fourteen AGOR-3 Class ships by the US Navy. These ships, especially designed for research and surveys, were 70 m in length, 1370 tonnes, and incorporated quiet ship operation, centerwells, multiple echo sounders, and a full array of scientific instrumentation. Of the fourteen in the class, 11 were retained in the US and three were transferred to other nations. During this period, the Soviet Union also embarked on a major building program which resulted by the mid-1970s in probably the world's largest fleet – both by vessel size and numbers. In 1979 there was an estimated total of 720 research vessels being operated by 72 nations, the USSR (194 ships), USA (115), and Japan (94) being the leading three.

By the mid-1980s, the shipbuilding boom which started in the late 1950s had dwindled, but many of those vessels themselves were becoming obsolete. New ships were planned to meet the growing needs of shipboard investigators. This resulted in larger ships with improved maneuverability, seakeeping, and data-acquisition capabilities. The new ships built to meet these requirements plus improvements to selected older vessels constitute today's oceanographic research vessels. The worldwide fleet is now smaller in terms of numbers than 25 years ago, but the overall tonnage is greater and the capability vastly superior.

The Nature of Research Vessels

The term 'oceanographic research vessel' is relatively new; earlier ships with limited roles were 'hydrographic survey vessels' or 'fisheries vessels.' As marine science evolved to include biological, chemical, geological, and physical processes of the ocean and its floor and the air–sea interface above, the term 'oceanography' and 'oceanographic research vessel' has come to include all of these disciplines.

When research vessels became larger and more numerous it was inevitable that regulations governing their construction and operation would come into force. Traditionally, ships were either commercial, warships, or yachts. Research ships with scientific personnel fit into none of these, and it became necessary to recognize the uniqueness of such ships in order to preclude burdensome and inapplicable laws. Most nations have now established a definition of an oceanographic research vessel. The United Nations International Maritime Organization (IMO) has established a category of 'Special Purpose Ship' which includes 'ships engaged in research, expeditions, and survey'. Scientific personnel are defined as '… all persons who are not passengers or members of the crew … who are carried on board in connection with the special purpose …'

United States law is more specific; it states

> The term oceanographic research vessel means a vessel that the Secretary finds is being employed only in instruction in oceanography or limnology, or both, or only in oceanographic or limnological research, including those studies about the sea such as seismic, gravity meter, and magnetic exploration and other marine geophysical or geological surveys, atmospheric research, and biological research.

The same law defines scientific personnel as those persons who are aboard an oceanographic research vessel solely for the purpose of engaging in scientific research, or instructing or receiving instruction in oceanography, and shall not be considered seamen.

The specific purposes of research vessels include: hydrographic survey (mapping and charting); geophysical or seismic survey; fisheries; general purpose (multidiscipline); and support vessels. Despite their differences, there are commonalities that distinguish a research vessel from other ships. These are defined in the science mission requirements which set forth the operational capabilities, working environment, science accommodations and outfit to meet the science role for which the ship type is intended. The science mission requirements are the dominant factors governing the planning for a new vessel or the conversion of an existing one. The requirements can vary according to the size, area of operations, and type of service, but the composition of the requirements is a product of long usage.

A typical set of scientific mission requirements for a large high-endurance general purpose oceanographic research ship is as follows.

1. General: the ship is to serve as a large general purpose multidiscipline oceanographic research vessel. The primary requirement is for a high endurance ship capable of worldwide cruising (except in close pack ice) and able to provide both overside and laboratory work to proceed in high sea states. Other general requirements are flexibility, vibration and noise free, cleanliness, and economy of operation and construction.

2. Size: size is ultimately determined by requirements which probably will result in a vessel larger than existing ships. However, the length-over-all (LOA) should not exceed 100 m.

3. Endurance: sixty days; providing the ability to transit to remote areas and work 3–4 weeks on station; 15 000 mile range at cruising speed.

4. Accommodations: 30–35 scientific personnel in two-person staterooms; expandable to 40 through the use of vans. There should be a science-library lounge with conference capability and a science office.

5. Speed: 15 knots cruising; sustainable through sea state 4 (1.25–2.5 m); speed control ± 0.1 knot in the 0–6 knot range, and ± 0.2 knot in the range of 6–15 knots.[1]

6. Seakeeping: the ship should be able to maintain science operations in the following speeds and sea states:

- 15 knots cruising through sea state 4 (1.25–2.5 m);
- 13 knots cruising through sea state 5 (2.5–4 m);
- 8 knots cruising through sea state 6 (4–6 m);
- 6 knots cruising through sea state 7 (6–9 m).

7. Station keeping: the ship should be able to maintain science operations and work in sea states through 5, with limited work in sea state 7. There

---

[1] 1 (Nautical) mile = 1.853 km, 1 knot = 1.853 km h^{-1} = 0.515 m s^{-1}.

should be dynamic positioning, both relative and absolute, at best heading in 35-knot wind, sea state 5, and 3-knot current in depths to 6000 m, using satellite and/or bottom transponders; $\pm 5°$ heading and 50 m maximum excursion. It should be able to maintain a precision trackline (including towing) at speeds as low as 2 knots with a $45°$ maximum heading deviation from the trackline under controlled conditions (satellite or acoustic navigation) in depths to 6000 m, in 35 knot wind; and 3-knot current. Speed control along track should be within 0.1 knot with 50 m of maximum excursion from the trackline.

8. Ice strengthening: ice classification sufficient to transit loose pack ice. It is not intended for icebreaking or close pack work.

9. Deck working area: spacious stern quarter area – 300 m^2 minimum with contiguous work area along one side 4×15 m minimum. There should be deck loading up to 7000 $kg\,m^{-2}$ and there should be overside holddowns on 0.5 m centers. The area should be highly flexible to accommodate large, heavy, and portable equipment, with a dry working deck but not more than 2–3 m above the waterline. There should be a usable clear foredeck area to accommodate specialized towers and booms extending beyond the bow wave. All working decks should be accessible for power, water, air, and data and voice communication ports.

10. Cranes: a suite of modern cranes:

(a) to reach all working deck areas and offload vans and heavy equipment up to 9000 kg;
(b) articulated to work close to deck and water surface;
(c) to handle overside loads up to 2500 kg, 10 m from side and up to 4500 kg closer to side;
(d) overside cranes to have servo controls and motion compensation;
(e) usable as overside cable fairleads at sea.

The ship should be capable of carrying portable cranes for specialized purposes such as deploying and towing scanning sonars, photo and video devices, remotely operated vehicles (ROVs), and paravaned seismic air gun arrays.

11. Winches: oceanographic winch systems with fine control (0.5 m min^{-1}; constant tensioning and constant parameter; wire monitoring systems with inputs to laboratory panels and shipboard data systems; local and remote controls including laboratory auto control.

Permanently installed general purpose winches should include:

• two winches capable of handling 10 000 m of wire rope or electromechanical conducting cables having diameters from 0.6 mm to 1.0 cm.

• a winch complex capable of handling 12 000 m of 1.5 cm trawling or coring wire, and 10 000 m of 1.75 cm electromechanical conducting cable (up to 10 kVA power transmission and fiberoptics); this can be two separate winches or one winch with two storage drums.

Additional special purpose winches may be installed temporarily at various locations along working decks. Winch sizes may range up to 40 mtons and have power demands up to 250 kW. (See also multichannel seismics.) Winch control station(s) should be located for optimum operator visibility with communications to laboratories and ship control stations.

12. Overside handling: various frames and other handling gear able to accommodate wire, cable, and free-launched arrays; matched to work with winch and crane locations but able to be relocated as necessary. The stern A-frame must have 6 m minimum horizontal and 10 m vertical clearances, 5 m inboard and outboard reaches, and safe static working load up to 60 mtons. It must be able to handle, deploy and retrieve very long, large-diameter piston cores up to 50 m length, 15 mtons weight and 60 mtons pullout tension. There should be provision to carry additional overside handling rigs along working decks from bow to stern. (See also multichannel seismics).

13. Towing: capable of towing large scientific packages up to 4500 kg tension at 6 knots, and 10 000 kg at 2.5 knots in sea state 5; 35 knots of wind and 3 knot current.

14. Laboratories: approximately 400 m^2 of laboratory space including: main lab (200 m^2) flexible for subdivision providing smaller specialized labs; hydro lab (30 m^2) and wet lab (40 m^2) both located contiguous to sampling areas; bio-chem analytical lab (30 m^2); electronics/computer lab and associated users space (60 m^2); darkroom (10 m^2); climate-controlled chamber (15 m^2); and freezer(s) (15 m^2).

Labs should be arranged so that none serve as general passageways. Access between labs should be convenient. Labs, offices, and storage should be served by a man-rated lift having clear inside dimensions not less than 3×4 m.

Labs should be fabricated of uncontaminated and 'clean' materials. Furnishings, doors, hatches, ventilation, cable runs, and fittings should be planned for maximum lab cleanliness. Fume hoods should be installed permanently in wet and analytical labs. Main lab should have provision for temporary installation of fume hoods.

Cabinetry should be of high-grade laboratory quality with flexibility for arrangements through the use of bulkhead, deck and overhead holddown fittings.

Heating, ventilation, and air conditioning (HVAC) should be appropriate to labs, vans and other science spaces being served. Laboratories should be able to maintain a temperature of 20–23°C, 50% relative humidity, and 9–11 air changes per hour. Filtered air should be provided to the analytical lab. Each lab should have a separate electric circuit on a clean bus and continuous delivery capability of at least 250 VA m^{-2} of lab area. Total estimated laboratory power demand is 100 kVA. There should be an uncontaminated seawater supply to most laboratories, vans, and several key deck areas.

15. Vans: carry four standardized 2.5 × 6 m portable vans which may have laboratory, berthing, storage, or other specialized use. With hook-up provision for power, HVAC, fresh water, uncontaminated sea water, compressed air, drains, communications, data and shipboard monitoring systems. There should be direct van access to ship interior at key locations. There should be provision to carry up to four additional vans on working and upper decks with supporting connections at several locations. The ship should be capable of loading and off-loading vans with its own cranes.

16. Workboats: at least one and preferably two inflatable boats located for ease of launching and recovery. There should be a scientific workboat 8–10 m LOA fitted out for supplemental operation at sea including collecting, instrumentation and wide-angle signal measurement. It should have 12 h endurance, with both manned and automated operation, be of 'clean' construction and carried in place of one of the four van options above.

17. Science storage: total of 600 m^3 of scientific storage accessible to labs by lift and weatherdeck hatch(es). Half should have suitable shelving, racks and tiedowns, and the remainder open hold. A significant portion of storage should be in close proximity to science spaces (preferably on the same deck).

18. Acoustical systems: ship to be as acoustically quiet as practicable in the choice of all shipboard systems and their installation. The design target is operationally quiet noise levels at 12 knots cruising in sea state 5 at the following frequency ranges:

- 4 Hz–500 Hz seismic bottom profiling
- 3 kHz–500 kHz echo sounding and acoustic navigation
- 75 kHz–300 kHz Doppler current profiling

The ship should have 12 kHz precision and 3.5 kHz sub-bottom echo sounding systems, and provision for additional systems. There should be a phased array, wide multibeam precision echo sounding system; transducer wells (0.5 m diam) one located forward and two midships; pressurized sea chest (1.5 × 3 m) located at the optimum acoustic location for afloat installation and servicing of transducers and transponders.

19. Multichannel seismics: all vessels shall have the capability to carry out multichannel seismic profiling (MCS) surveys using large source arrays and long streamers. Selected vessels should carry an MCS system equivalent to current exploration industry standards.

20. Navigation/positioning: There shall be a Global Positioning System (GPS) with appropriate interfaces to data systems and ship control processors; a short baseline acoustic navigation system; a dynamic positioning system with both absolute and relative positioning parameters.

21. Internal communications: system to provide high quality voice communications throughout all science spaces and working areas. Data transmission, monitoring and recording systems should be available throughout science spaces including vans and key working areas. There should be closed circuit television monitoring and recording of all working areas including subsurface performance of equipment and its handling. Monitors for all ship control, environmental parameters, science and overside equipment performance should be available in most science spaces.

22. Exterior communications: reliable voice channels for continuous communications to shore stations (including home laboratories), other ships, boats and aircraft. This includes satellite, VHF and UHF. There should be: facsimile communications to transmit high-speed graphics and hard copy on regular schedules; high-speed data communications links to shore labs and other ships on a continuous basis.

23. Satellite monitoring: transponding and receiving equipment including antennae to interrogate and receive satellite readouts of environmental remote sensing.

24. Ship control: the chief requirement is maximum visibility of deck work areas during science operations and especially during deployment and retrieval of equipment. This would envision a bridge-pilot house very nearly amidships with unobstructed stem visibility. The functions, communications, and layout of the ship control station should be designed to enhance the interaction of ship and science operations; ship course, speed, attitude, and positioning will often be integrated with science work requiring control to be exercised from a laboratory area.

These science mission requirements are typical of a large general-purpose research vessel. Requirements for smaller vessels and specialized research ships can be expected to differ according to the intended capability and service.

Design Characteristics of Oceanographic Vessels

In general the design of an oceanographic ship is driven by the science mission requirements described above. In any statement of requirements an ordering of priorities is important for the guidance of the design and construction of the ship. In the case of research vessels the following factors have been ranked by groups of practicing investigators from all disciplines.

1. Seakeeping: station keeping
2. Work environment: lab spaces and arrangements; deck working area; overside handling (winches and wire); flexibility
3. Endurance: range; days at sea
4. Science complement
5. Operating economy
6. Acoustical characteristics
7. Speed: ship control
8. Pay load: science storage; weight handling

These priorities are not necessarily rank ordered although there is general agreement among oceanographers that seakeeping, particularly on station, and work environment are the two top priorities. The remaining are ranked so closely together that they are of equal importance. The science mission requirements set for each of these areas become threshold levels, and any characteristic which falls below the threshold becomes a dominant priority.

General Purpose Vessels

Ships of this type (also termed multidiscipline) constitute the classic oceanographic research vessels and are the dominant class in terms of numbers today. They have outfitting and laboratories to support any of the physical, chemical, biological, and geological ocean science studies plus ocean engineering. The science mission requirements given above describe a large general purpose ship. Smaller vessels can be expected to have commensurately reduced requirements.

Current and future multidiscipline oceanographic ships are characterized as requiring significant open deck area and laboratory space. Accommodations for scientific personnel are greater than for single purpose vessels due to the larger science parties carried. Flexibility is an essential feature in a general purpose research vessel. A biological cruise may be followed by geology investigations which can require the reconfiguration of laboratory and deck equipment within a short space of time.

In addition to larger scientific complements, the complexity and size of instrumentation now being deployed at sea has increased dramatically over the past half century. As a result, the size of general purpose vessels has increased significantly. A research vessel of 60 m in length was considered to be large in 1950; the same consideration today has grown to 100 m and new vessels are being built to that standard. Even existing vessels have been lengthened to meet the growing needs.

The majority of oceanographic research vessels, however, are smaller vessels, 25–50 m in length, and limited to coastal service.

Mapping and Charting Vessels

Mapping and charting ships were probably the earliest oceanographic vessels, usually in conjunction with an exploration voyage. Incident to the establishment of marine trade routes, nautical charting of coastal regions became routine and the vessels so engaged were usually termed hydrographic survey ships. Surveys were (and still are) carried out using wire sounding, drags, and launches. Survey vessels are characterized by the number of boats and launches carried and less deck working space than general purpose vessels. Modern survey vessels, however, are often expected to carry out other scientific disciplines, and winches, cranes and frames can be observed on these ships.

Recent developments have affected the role (and therefore design) of this class of vessel. As a result of the International Law of the Sea Conferences, coastal states began to exercise control over their continental shelves and economic zones 200 miles from shore. This brought about interest in the resources (fishery, bottom and sub-bottom) of these newly acquired areas, and research vessels were tasked to explore and map these resources. The usual nautical charting procedures are not applicable in the open ocean regions, and modern electronic echo-sounding instruments have supplanted the older wire measurements. This involves large hull-mounted arrays of acoustic projectors and hydrophones which can map a swath of ocean floor with great precision up to five miles in width, and at cruising speeds. The design of vessels to carry this equipment requires a hull form to optimize acoustic transmission and reception, and to minimize hull noise from propulsion and auxiliary machinery. Further, such new ships also may be outfitted to perform other oceanographic tasks incident to surveys. Their appearance, therefore, may come closer to a general purpose vessel.

Fisheries Research Vessels

Fisheries research generally includes three fields of study: (1) environmental investigations, (2) stock assessment, and (3) gear testing and development. The first of these are surveys and analyses of sea surface and water column parameters; both synoptic and serial. These are biological, physical, and chemical investigations (as well as geological if bottom fisheries are considered); and can be accomplished from a general purpose oceanographic research vessel.

Ships engaged in fish stock assessment and exploratory fishing, or development work in fishing methods and gear, fish handling, processing, and preservation of fish quality on board, are specialized types of vessels closely related to actual fishing vessels. Design characteristics include a stern ramp and long fish deck for bringing nets aboard, trawling winches, and wet labs for analyses of fish sampling. Newer designs also include instrumentation and laboratories for environmental investigations, and extensive electronic instrumentation for acoustic fish finding, biomass evaluation, and fish identification and population count.

As with mapping and charting ships, most fishery research vessels are operated by government agencies.

Geophysical Research Vessels

The purpose of marine geophysical research vessels is to investigate the sea floor and sub-bottom, oceanic crust, margins, and lithosphere. The demanding design aspect for these ships is the requirement for a MCS system used to profile the deep geologic structure beneath the seafloor. The missions range from basic research of the Earth's crust (plate tectonics) to resources exploration.

The primary components of an MCS system are the large air compressors needed to 'fire' multiple towed airgun sound source arrays and a long towed hydrophone streamer which may reach up to 10 km in length. The supporting outfit for the handling and deployment of the system includes large reels and winches for the streamer, and paravanes to spread the sound source arrays athwart the ship's track. This latter results in the need for a large stern working deck close to the water with tracked guide rails and swingout booms. Electronic and mechanical workshops are located close to the working deck. The design incorporates a large electronics room for processing the reflected signals from the hydrophones and integrating the imagery with magnetics, gravity and navigation data.

The highly specialized design requirements for a full-scale marine geophysics ship usually precludes work in other oceanographic disciplines. On the other hand, large general purpose research ships often carry compressors and portable streamer reels sufficient for limited seismic profiling.

Polar Research Vessels

Whereas most oceanographic research vessels are classed by the discipline in which they engage, polar research vessels are defined by their area of operations. Earlier terminology distinguished between polar research vessels and icebreakers with the former having limited icebreaking capability, and the latter with limited or no research capability. The more current trend is to combine full research capability into new icebreaker construction.

Arctic and Antarctic research ships in the nineteenth and early twentieth centuries were primarily ice reinforced sealing vessels with little or no icebreaking capability. World War II and subsequent Arctic logistics, and the International Antarctic Treaty (1959) brought about increased interest in polar regions which was furthered by petroleum exploration in the Arctic in the late twentieth century. Icebreakers with limited research capability early in this period became full-fledged research vessels by the end of the century.

The special requirements defining a polar research vessel include increased endurance, usually set at 90 days, helicopter support, special provisions for cold weather work, such as enclosed winch rooms and heated decks, and icebreaking capability. Other science mission requirements continue the same as for a large general purpose RVs. Of special concern is seakeeping in open seas. Past icebreaker hull shapes necessarily resulted in notoriously poor seakindliness. Newer designs employing ice reamers into the hull form offer improved seakeeping.

Ice capability is usually defined as the ability to break a given thickness of level ice at 3 knots continuous speed, and transit ice ridges by ramming. Current requirements for polar research vessels have varied from 0.75 m to 1.25 m ice thickness in the continuous mode, and 2.0 to 3.0 m ridge heights in the ramming mode. These correspond to Polar Class 10 of Det Norske Veritas or Ice Class A3-A4 of the American Bureau of Shipping.

Support Vessels

Ships that carry, house, maintain, launch and retrieve other platforms and vehicles have evolved into a class worthy of note. These include vessels that support submersibles, ROVs, buoys, underwater habitats, and scientific diving. Earlier ships of this class were mostly converted merchant or fishing vessels whose only function was to launch and retrieve and supply hotel services. Recent vessels, especially those dedicated to major programs such as submersible support, are large ships and fully out-fitted for general purpose work.

Other Classes of Oceanographic Research Vessels

In addition to the above types, there are research ships which serve other purposes. These include ocean drilling and geotechnical ships, weather ships, underwater archaeology, and training and education vessels. The total number of these ships is relatively small, and many of them merge in and out of the category and serve for a limited stretch of time.

Often ships will take on identification as a research vessel for commercial expediency or other fashion not truly related to oceanographic research. Such roles may include treasure hunting, salvage, whale watching, recreational diving, ecology tours, etc. These vessels may increase the popular awareness of oceanography but are not bona fide oceanographic research vessels.

Research Vessel Operations

Oceanographic cruises are usually the culmination of several years of scientific and logistics planning. Coastal vessels, typically 25–50 m length, will usu-ally remain in a home, or adjacent regions on cruises of 1–3 weeks' duration. Ocean-going vessels, 50–100 m length, may undertake voyages of 1–2 years away from the home port, with cruise segments of 25–35 days working out of ports of opportunity. New scientific parties may join the ship at a port call and the nature of the following cruise leg can change from, for instance biological to physical ocean-ography. This involves complex logistics and careful planning and coordination, within a typical 4–5 day turnaround.

From the time of the *Meteor* expedition of 1925, cruise plans are usually highly systematic with work concentrating along preset track lines and grids, or confined to a small area of intense investigation. Work can take place continuously while underway using hull-mounted and/or towed instrumentation, or the ship will stop at a station and lower instru-ments for water column or bottom sampling. Typical stations are at 15–60-mile intervals and can last 1–4 h. Measurements or observations that are com-monly made are shown in **Table 1**. In addition, work may include towed vehicles along a precise trackline at very slow controlled speeds making many of the observations, chiefly acoustic, photographic, and video.

Table 1   Common measurements and observations made from RVs

Underway	On station
Single channel echo sounding	Echo sounding
Multichannel echo sounding	Sub-bottom profiling
Sub-bottom profiling (3.5 kHz)	Acoustic Doppler profiling
Acoustic Doppler profiling	Surface to any depth
Sea surface:	Temperature
Temperature	Salinity
Salinity	Sound velocity
Fluorometry	Dissolved oxygen
Dissolved oxygen	Water sample
Towed magnetometer	Bottom sampling
Gravimeter	Bottom coring
Meteorological	Bottom photography and video
Wind speed and direction	Bottom dredging
Barometric pressure	Geothermal bottom probe
Humidity	Biological net tows and trawls
Solar radiation	Biological net tows
Towed plankton recorder	

Cooperative projects among research vessels in-cluding different nations have become commonplace. These share a common scientific goal, and cruise tracks, times, methodology, data reduction, and archiving are assigned by joint planning groups.

A significant factor affecting oceanographic research cruises today is the permission required by a research vessel to operate in another nation's 200-mile eco-nomic zone. As a result of the United Nations Law of the Sea Conventions and the treaties resulting there-from (1958–1982), coastal nations were given juris-diction over the conduct of marine scientific research extending 200 nautical miles out from their coast (including island possessions). This area is termed ex-clusive economic zone (EEZ). As of 2000 there were approximately 151 coastal states (this number varies according to the world political makeup), and 36% of the world ocean falls within their economic zones. Most coastal nations have prescribed laws governing research in their zones. The rules include requests for permission, observers, port calls, sharing data, and penalties. This often poses a burden on the operator of a research vessel when permission requirements are arbitrary and untimely. It is not, however, an un-workable burden if done in an orderly manner, and does have desirable features for international cooper-ation. Problems arise when requests are not submitted within the time specified, or resulting data are not forthcoming. These are complicated by unrealistic re-quirements on the part of the coastal state, delay in acknowledging or acting on a request, or ignoring it totally. These can have a profound effect on scientific research and need to be addressed in future Law of the Sea Conventions.

Table 2 Countries operating 10 or more oceanographic research ships

Russia	86
United States	84
Japan	66
China	17
Ukraine	14
Korea	11
Germany	11
United Kingdom	10
Canada	10
All others (39)	111
Total	420

World Oceanographic Fleet

The precise number of oceanographic research vessels worldwide is difficult to ascertain. Few nations maintain lists specific to research vessels, and numbers are available chiefly by declarations on the part of the operator. Some ships move in and out of a research status from another classification, e.g., fishing, passenger, yacht, etc. Also, some operators keep hydrographic survey, seismic exploration, and even fisheries research ships as categories separate from oceanographic research; here they are included within the general heading of oceanographic research vessels.

Based on the best available information, 48 nations or international agencies operated 420 oceanographic research vessels of size greater than 25 m LOA in 2000. Of these, 310 ships were from nine nations operating 10 or more ships each (Table 2).

A significant step in international cooperation affecting oceanographic research vessels and the marine sciences they support has been the International Ship Operators Meeting, an intergovernmental association, founded in 1986, comprising representatives from various ship operating agencies that meets periodically to exchange information on ship operations and schedules, and work on common problems affecting research vessels. In 1999, 21 ship-operating nations were represented and extended membership by other states is ongoing.

Future Oceanographic Ships

The interest in, and growth of, marine science over the past half century shows little or no indication of diminishing. The trend in oceanographic investigations has been to carry larger and more complex instrumentation to sea. The size and capability of research vessels in support of developing projects has also increased.

Future oceanographic research ships can be expected to become somewhat larger than their counterparts today. This will result from demands for more sizeable scientific complements and laboratory spaces. Workdeck and shops will be needed for larger equipment systems such as buoy arrays, bottom stations, towed and autonomous vehicles. Larger overside handling systems incorporating motion compensation will make demands for more deck space.

There will be fewer differences between basic science, fisheries, and hydrographic surveying vessels so that one vessel can serve several purposes. This may result in fewer vessels, but overall tonnage and capacity can be expected to increase.

New types of craft may take a place alongside conventional ships. These include submarines, 'flip'-type vessels which transit horizontally and flip vertically on station, and small waterplane-area twin hull ships (SWATH). SWATH, or semisubmerged ships, are a relatively recent development in ship design. Although patents employing this concept show up in 1905, 1932, and 1946, it was not until 1972 that the US Navy built a 28 m, 220 ton prototype model. The principle of a SWATH ship is that submerged hulls do not follow surface wave motion, and thin struts supporting an above water platform which have a small cross-section (waterplane) are nearly transparent to surface waves, and have longer natural periods and reduced buoyancy force changes than a conventional hull. The result is that SWATH ships, both in theory and performance, demonstrate a remarkably stable environment and platform configuration which is highly attractive for science and engineering operations at sea.

LOWERED, EXPENDABLE, OR TOWED INSTRUMENTS

CONTINUOUS PLANKTON RECORDERS

A. John, Sir Alister Hardy Foundation for Ocean Science, Plymouth, UK
P. C. Reid, SAHFOS, Plymouth, UK

Introduction

The Continuous Plankton Recorder (CPR) survey is a synoptic survey of upper-layer plankton covering much of the northern North Atlantic and North Sea. It is the longest running and the most geographically extensive of any routine biological survey of the oceans. Over 4 000 000 miles of towing have resulted in the analysis of nearly 200 000 samples and the routine identification of over 400 species/groups of plankton. Data from the survey have been used to study biogeography, biodiversity, seasonal and inter-annual variation, long-term trends, and exceptional events. The value of such an extensive time-series increases as each year's data are accumulated. Some recognition of the importance of the CPR survey was achieved in 1999 when it was adopted as an integral part of the Initial Observing System of the Global Ocean Observing System (GOOS).

History

The CPR prototype was designed by Alister Hardy for operation on the 1925–27 Discovery Expedition to the Antarctic, as a means of overcoming the problem of patchiness in plankton. It consisted of a hollow cylindrical body tapered at each end, weighted at the front and with a diving plane, horizontal tail fins, and a vertical tail fin with a buoyancy chamber on top (**Figure 1A**). Hardy designed a more compact version with a smaller sampling aperture for use on merchant ships and this was first deployed on a commercial ship in the North Sea in September 1931 (**Figure 1B**). During the 1980s the design was modified further to include a box-shaped double tail-fin that provides better stability when deployed on the faster merchant ships of today (**Figure 1C**). The space within this tail-fin is used in some machines to accommodate physical sensors and flowmeters. The normal maximum tow distance for a CPR is approximately 450 nautical miles (834 km).

By the late 1930s there were seven CPR routes in the North Sea and one in the north-east Atlantic; in 1938 CPRs were towed for over 30 000 miles. After a break for the Second World War, the survey restarted in 1946 and expanded into the eastern North Atlantic. Extension of sampling into the western North Atlantic took place in 1958. The survey reached its greatest extent from 1962 to 1972 when CPRs were towed for at least 120 000 nautical miles annually. Sampling in the western Atlantic, which had been suspended due to funding problems in 1986, recommenced in 1991 and is still ongoing. **Figure 2A** shows the extent of the survey in 1999.

Initially based at the University College of Hull, the survey moved to Leith, Edinburgh in 1950 under the management of the Scottish Marine Biological Association (now the Scottish Association for Marine Science). In 1977 it finally moved to Plymouth as part of the Institute for Marine Environmental Research (now Plymouth Marine Laboratory). After a short period of uncertainty in the late 1980s, when the continuation of the survey was threatened, the Sir Alister Hardy Foundation for Ocean Science (SAHFOS) was formed in November 1990 to operate the survey. Since 1931 more than 200 merchant ships, ocean weather ships, and coastguard cutters – known as 'ships of opportunity' – from many nations have towed CPRs in a voluntary capacity to maintain the survey. The Foundation is greatly indebted to the captains and crews of all these towing ships and their shipping and management companies, without whom the survey could not continue.

During the 1990s CPRs were towed by SAHFOS in several other areas, including the Mediterranean (1998–99), the Gulf of Guinea (1995–99), the Baltic (1998–99), and the Indian Ocean (1999). A separate survey by the National Oceanic and Atmospheric Administration/National Marine Fisheries Service using CPRs along the east coast of the USA off Narragansett has been running since 1974; CPRs are currently towed on two routes in the Middle Atlantic Bight. Following a successful 2000 mile trial tow in the north-east Pacific from Alaska to California in July–August 1997, a 2-year survey by SAHFOS using CPRs in the north-east Pacific started in March 2000. In addition to five tows per year on the Alaska–California route, there is one 3000 mile tow annually east–west from Vancouver to the north-west Pacific (**Figure 2B**). A 'sister' survey, situated in the Southern Ocean south of Australia between 60°E and 160°E, is operated by the Australian Antarctic Division. In this survey CPRs have been deployed since the early 1990s on voyages between Tasmania and stations in the Antarctic.

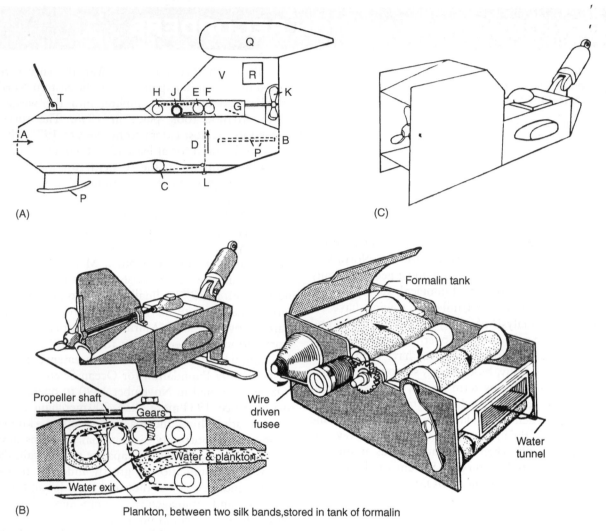

Figure 1 (A) Diagram of the first Continuous Plankton Recorder used on 'Discovery'. (Reproduced with permission from Hardy, 1967). (B) The 'old' CPR, used up to around 1983, showing the internal filtering mechanism. (C) The CPR in current use, with the 'box' tail-fin.

As the operator of a long-term international survey, which has sampled in most of the world's oceans, SAHFOS regularly trains its own staff in plankton identification. In recent years SAHFOS has also trained scientists from the following 10 countries: Benin, Cameroon, Côte d'Ivoire, France, Finland, Ghana, Italy, Nigeria, Thailand, and the USA.

The Database and Open Access Data Policy

The CPR database is housed on an IBM-compatible PC and stored in a relational Microsoft Access DATABASE system. Spatial and temporal data are stored for every sample analysed by the CPR survey since 1948. This amounts to >175 000 samples, with around 400 more samples added per month. There are more than two million plankton data points in the database, which also contains supporting information, including sample locations, dates and times of samples, a taxon catalog, and analyst details. In the near future it will also hold additional conductivity, temperature, and depth (CTD) data. Routine processing procedures ensure that, despite various operational difficulties, the previous year's data are usually available in the database within 9 months.

In 1999 SAHFOS adopted a new open access data policy, i.e. data are freely available to all users worldwide, although a reasonable payment may be incurred for time taken to extract a large amount of data. The only stipulation is that users have to sign a SAHFOS Data Licence Agreement. Details of the database can be found on the web site: http://www.npm.ac.uk/sahfos/. This site advertises the availability of data and allows requests for data to be made easily.

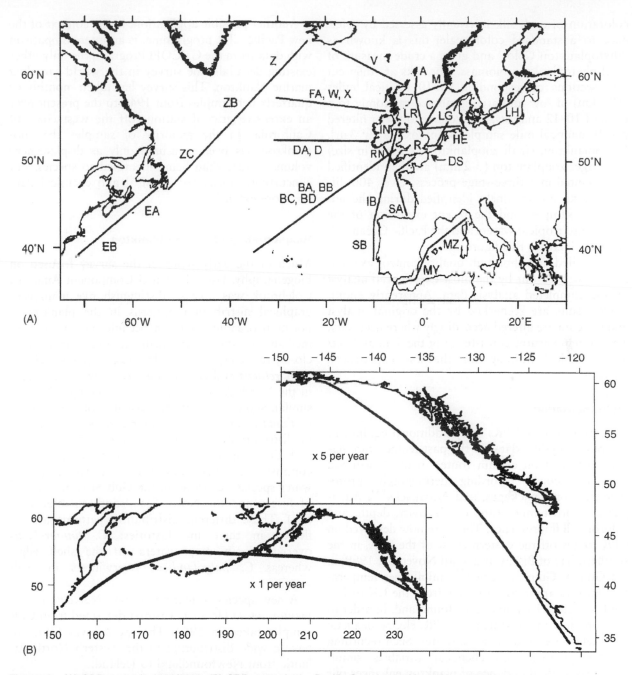

Figure 2 (A) CPR routes in 1999/2000. (B) CPR routes in the Pacific Ocean towed in 2000 and 2001.

The CPR Bibliography, which is available on the SAHFOS web site, lists over 500 references using results from the survey. During the early years many of the papers based on CPR data were published in the 'in-house' journal *Hull Bulletins of Marine Ecology*, which from 1953 onwards became the *Bulletins of Marine Ecology*; this was last published in 1980.

Methods

Merchant ships of many nations tow CPRs each month along 20–25 standard routes (**Figure 2A**) at a depth of 6–10 m. Water enters the CPR through a 12.7 mm square aperture and travels down a tunnel that expands to a cross-section of 50×100 mm, where it passes through a silk filtering mesh with a mesh size of approximately 280 μm. The movement of the CPR through the water turns a propeller that drives a set of rollers and moves the silk across the tunnel. At the top of the tunnel the filtering silk is joined by a covering silk and both are wound onto a spool located in a storage chamber containing formaldehyde solution. The CPRs are then returned to SAHFOS in Plymouth for examination. The green coloration of each silk is visually assessed by reference to a standard color scale; this is known as 'Phytoplankton Color' and gives a crude measure of total phytoplankton biomass. The silks are then cut into sections corresponding to 10 nautical miles (18.5 km) of tow and are distributed randomly to a team of 10–12 analysts. The volume of water filtered per 10-nautical-mile sample is approximately 3 m³. Phytoplankton, small zooplankton (<2 mm in size) and larger zooplankton (>2 mm) are then identified and counted in a three-stage process. Over 400 different taxa are routinely identified during the analysis of samples and the recent expansion of the survey into tropical waters and the Pacific Ocean will certainly increase this figure.

A detailed and thorough quality control examination is carried out by the most experienced analyst on the completed analysis data. Apparently anomalous results are rechecked by the original analyst and the data are altered accordingly where necessary. This system ensures consistency of the data and acts as 'in-service' training for the less experienced analysts.

Instrumentation

On certain routes CPRs carry additional equipment to obtain physical data. In the past temperature has been recorded on certain routes in the North Sea using Braincon™ recording thermographs, prototype electronic packages, and Aquapacks™. Aquapacks record temperature, conductivity, depth, and chlorophyll fluorescence. These are now deployed on CPR routes off the eastern coast of the USA, in the southern Bay of Biscay and, until November 1999, in the Gulf of Guinea. Vemco™ minilogger temperature sensors are used on routes from the UK to Iceland, and from Iceland to Newfoundland. In order to measure flow rate through the CPR, electromagnetic flowmeters are used on some routes. Such recording of key physical and chemical variables simultaneously with abundance of plankton enhances our ability to interpret observed changes in the plankton.

Results and Applications of the Data

The long-term time-series of CPR data acts as a baseline against which to measure natural and anthropogenic changes in biogeography, biodiversity, seasonal variation, inter-annual variation, long-term trends, and exceptional events. The results have applications to studies of eutrophication and are increasingly being applied in statistical analysis of plankton populations and modeling. Some examples are given below.

Another possible application, in the context of the new Pacific CPR programme, is an inter-comparison with data from the CalCOFI Program, the only other existing decadal-scale survey in the world sampling marine plankton. This survey has taken monthly or quarterly net samples from 1949 to the present over an extensive grid of stations off the west coast of California. In the majority of samples the zooplankton has been measured only as displacement volume, rather than being identified to species, but concurrently measured physical and chemical data are more extensive.

Biogeography of Marine Plankton

Much of the early work of the survey focused on biogeography. Using Principal Component Analysis, Colebrook was able to distinguish five main geographical distribution patterns in the plankton – northern oceanic, southern oceanic, northern intermediate, southern intermediate, and neritic. Two closely related species of calanoid copepod – *Calanus finmarchicus* and *C. helgolandicus* – which co-occur in the North Atlantic and are morphologically very similar, show very different distributions (**Figure 3**). *C. finmarchicus* is a cold-water species whose center of distribution lies in the north-west Atlantic gyre and the Norwegian Sea ('northern oceanic'). In contrast, *C. helgolandicus* is a warm–temperate water species occurring in the Gulf Stream, the Bay of Biscay and the North Sea ('southern intermediate'). These different distribution patterns are reflected in their life histories; *C. finmarchicus* overwinters in deep waters off the shelf edge, whereas *C. helgolandicus* overwinters in shelf waters.

A new species of marine diatom, *Navicula planamembranacea* Hendey, was first described from CPR samples taken in 1962. The species was found to have a wide distribution in the western North Atlantic from Newfoundland to Iceland.

An atlas of distribution of 255 species or groups (taxa) of plankton recorded by the CPR survey between 1958 and 1968 was published by the Edinburgh Oceanographic Laboratory in 1973. An updated version of this atlas, covering more than 40 years of CPR data and over 400 taxa, is in preparation.

Phytoplankton, Zooplankton, Herring, Kittiwake Breeding Data, and Weather

A study in the north-eastern North Sea found that patterns of four time-series of marine data and weather showed similar long-term trends. Covering

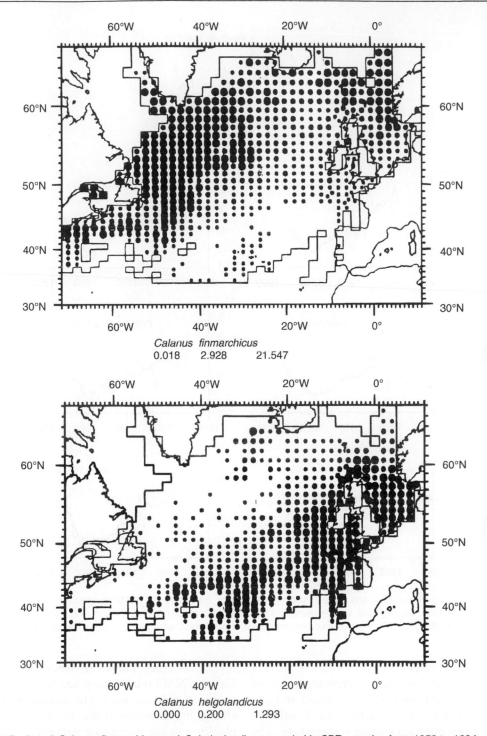

Calanus finmarchicus
0.018   2.928   21.547

Calanus helgolandicus
0.000   0.200   1.293

Figure 3 Distribution of *Calanus finmarchicus* and *C. helgolandicus* recorded in CPR samples from 1958 to 1994.

the period 1955–87, these trends were found in the abundance of phytoplankton and zooplankton (as measured by the CPR), herring in the northern North Sea, kittiwake breeding success (laying date, clutch size, and number of chicks fledged per pair) at a colony on the north-east coast of England, and the frequency of westerly weather (**Figure 4**).

The mechanisms behind the parallelism in these data over the 33-year period are still not fully understood.

Calanus and the North Atlantic Oscillation

The North Atlantic Oscillation (NAO) is a large-scale alternation of atmospheric mass between

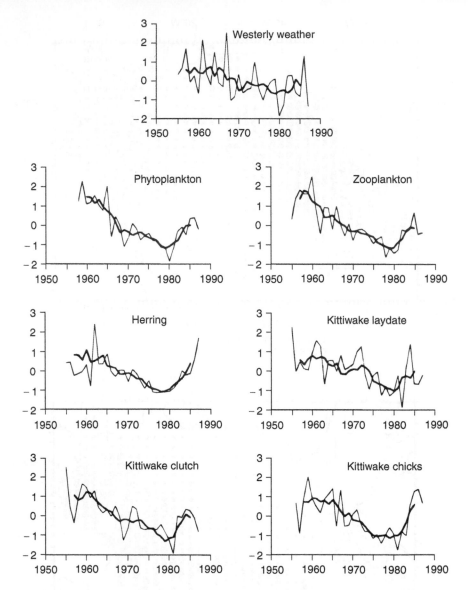

Figure 4 Standardized time-series and 5-year running means for frequency of westerly weather, and for abundances of phytoplankton, zooplankton, herring, and three parameters of kittiwake breeding (laying date, clutch size, and number of chicks fledged per pair), from 1955 to 1987. (Reproduced with permission from Aebischer NJ *et al.* (1990) *Nature* 347: 753–755.)

subtropical high surface pressure, centred on the Azores, and subpolar low surface pressures, centred on Iceland. The NAO determines the speed and direction of the westerly winds across the North Atlantic, as well as winter sea surface temperature. The NAO index is the difference in normalized sea level pressures between Ponta Delgadas (Azores) and Akureyri (Iceland). There is a close association between the abundance of *Calanus finmarchicus* and *C. helgolandicus* in the north-east Atlantic and this index (**Figure 5**). At times of heightened pressure difference between the Azores and Iceland, i.e. a high, positive NAO index, there is low abundance of *C. finmarchicus* and high abundance of *C. helgolandicus*; during a low, negative NAO index the

reverse is true. However, since 1995 this strong *Calanus*/NAO relationship has broken down and the causes of this are presently unknown. It suggests a change in the nature of the link between climate and plankton in the north-east Atlantic.

North Sea Ecosystem Regime Shift

Recent studies have shown changes in CPR Phytoplankton Color, a visual assessment of chlorophyll, for the north-east Atlantic and the North Sea. In the central North Sea and the central north-east Atlantic an increased season length was strikingly evident after the mid-1980s. In contrast, in the north-east Atlantic north of 59°N Phytoplankton Color

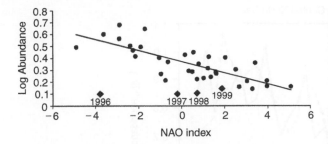

Figure 5 Annual log abundance of *Calanus finmarchicus* in the north-east Atlantic Ocean against the NAO winter index for the period 1962–99. (Adapted with permission from Fromentin JM, and Planque B (1996) *Marine Ecology Progress Series* 134: 111–118.)

declined after the mid-1980s (**Figure** 6). These changes in part appear to be linked to the recent high positive phase of the NAO index and reflect changes in mixing, current flow, and sea surface temperature. The increase in Phytoplankton Color and phytoplankton season length after 1987 coincided with a large increase in catches of the western stock of horse mackerel *Trachurus trachurus* in the northern North Sea, apparently connected with the increased transport of Atlantic water into the North Sea. From 1988 onwards the NAO index increased to the highest positive level observed in the twentieth century. Positive NAO anomalies are associated with stronger

and more southerly tracks of the westerly winds and higher temperatures in western Europe. These changes coincided with a series of other changes that affected the whole North Sea ecosystem, affecting many trophic levels and indicating a regime shift.

North Wall of the Gulf Stream and Copepod Numbers

Zooplankton populations in the eastern North Atlantic and the North Sea show similar trends to variations in the latitude of the north wall of the Gulf Stream, as measured by the Gulf Stream North Wall (GSNW) index, which is statistically related to the NAO 2 years previously. **Figure** 7 shows the close correlation between total copepods in the central North Sea and the GSNW index. This relationship is also evident in zooplankton in freshwater lakes and in the productivity of terrestrial environments, indicating a possible climatic control.

Biodiversity

Analyses of long-term trends in biodiversity of zooplankton in CPR samples indicate increases in diversity in the northern North Sea. This may be related to distributions altering in response to climatic change as geographical variation in biodiversity of the plankton shows generally higher diversity at low latitudes than

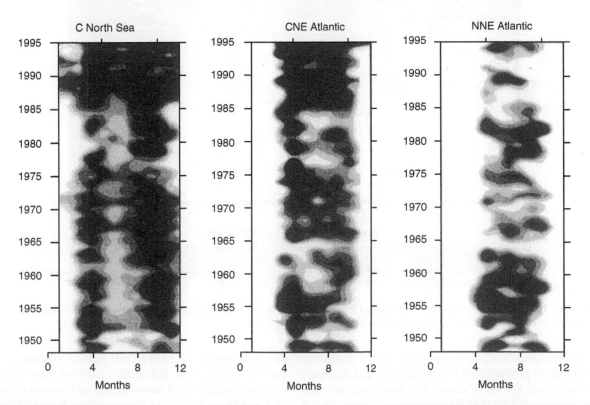

Figure 6 Contour plots of mean monthly Phytoplankton Color during 1948–95 for the central North Sea, and for the central and northern north-east Atlantic. (Reproduced with permission from Reid PC *et al.* (1998) *Nature* 391: 546.)

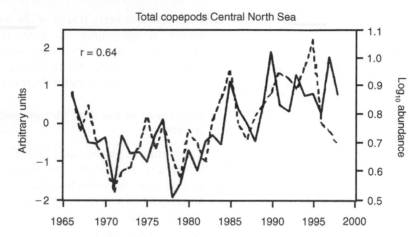

Figure 7 The latitude of the Gulf Stream (the GSNW index 'arbitrary units', broken line) compared with the abundance of total copepods in the central North Sea (solid line). Adapted with permission from Taylor AH *et al.* (1992) *Journal of Mar. Biol. Ass.* UK 72: 919–921.

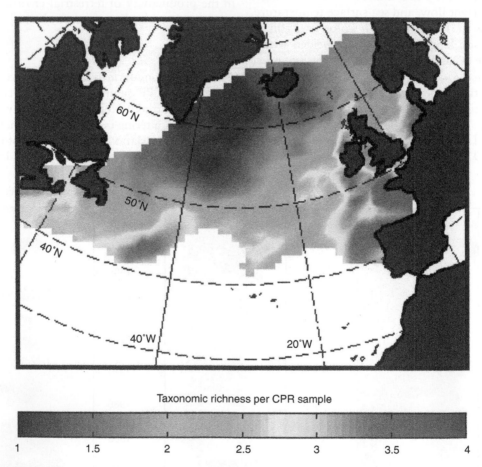

Figure 8 The biodiversity (taxonomic richness) of calanoid copepods in the CPR sampling area. (Adapted with permission from Beaugrand *et al.* (2000) *Marine Ecology Progress Series* 204: 299–303.)

at high latitudes. Calanoid copepods are the dominant zooplankton group in the North Atlantic and the large data set from the CPR survey has been used to map their diversity. This has demonstrated a pronounced local spatial variability in biodiversity. Higher diversity was found in the Gulf Stream extension, the Bay of Biscay, and along the southern part of the European shelf. Cold water south of Greenland, east of Canada, and west of Norway was found to have the lowest diversity (**Figure 8**).

Monitoring for Nonindigenous and Harmful Algal Blooms

The regularity of sampling by the CPR enables it to detect changes in plankton communities. Few case histories exist that describe the initial appearance and subsequent geographical spread of non-indigenous species. In 1977 the large diatom *Coscinodiscus wailesii* was recorded for the first time off Plymouth, when mucilage containing this species was found to be clogging fishing nets. *C. wailesii* was previously known only from southern California, the Red Sea, and the South China Sea and it is believed that it arrived in European waters via ships' ballast water. Since then the species has spread throughout north-west European waters and has become an important contributor to North Sea phytoplankton biomass, particularly in autumn and winter. Such introduced species can, on occasions, have considerable ecological and economic effects on regional ecosystems.

There has been an apparent worldwide increase in the number of recorded harmful algal blooms and the CPR survey is ideally placed to monitor such events. The serious outbreak of paralytic shellfish poisoning that occurred in 1968 on the north-east coast of England was shown by CPR sampling to have been caused by the dinoflagellate *Alexandrium tamarense*.

Increased nutrient inputs into the North Sea since the 1950s have been linked with an apparent increase in the haptophycean alga *Phaeocystis*, particularly in Continental coastal regions where it produces large accumulations of foam on beaches. In contrast, long-term records (1946–87) from the CPR survey, which samples away from coastal areas, show that *Phaeocystis* has declined considerably in the open-sea areas of the north-east Atlantic and the North Sea (**Figure 9**). It is notable that the decline occurred both in areas not subject to anthropogenic nutrient inputs (Areas 1 and 2, west of the UK) and in the most affected area (Area 4, the southern North Sea). This decrease in *Phaeocystis* up to 1980 is also shown by many other species of plankton, suggesting a common causal relationship.

Exceptional Events

Doliolids are indicators of oceanic water and in CPR samples are normally found to the west and south-west of the British Isles; they occur only sporadically in the North Sea and are rarely recorded in the central or southern North Sea. On two occasions in recent years, in October–December 1989 and September–October 1997, the doliolid *Doliolum nationalis* was recorded in CPR samples taken in the

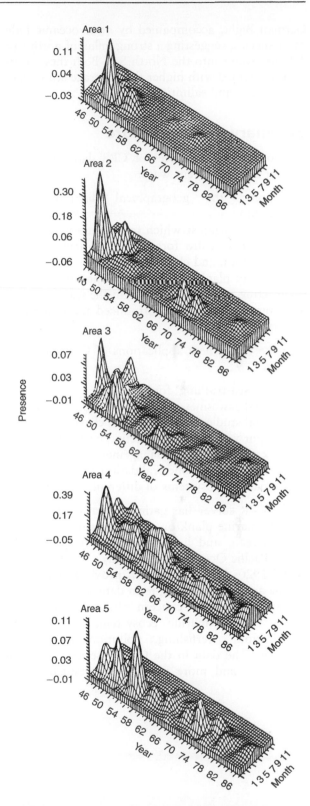

Figure 9 Presence of *Phaeocystis* in five areas of the northeast Atlantic Ocean and the North Sea. Data are plotted for each month for 1946–87 inclusive. (Reproduced with permission from Owens NJP *et al.* (1989) *Journal of Mar. Biol. Ass. UK* 69: 813–821.)

German Bight, accompanied by other oceanic indicator species, suggesting a strong influx of north-east Atlantic water into the North Sea. Both these occasions coincided with higher than average sea surface temperature and salinities.

Summary and the Future

The long-term time-series of CPR data have been used in many different ways:

- mapping the geographical distribution of plankton
- a baseline against which to measure natural and anthropogenically forced change, including eutrophication and climate change
- linking of plankton and environmental forcing
- detecting exceptional events in the sea
- monitoring for newly introduced and potentially harmful species.

In the future new applications of CPR data may include:

- use as 'sea-truthing' for satellites
- regional assessment of plankton biodiversity
- regional studies of responses to climate change
- as input variables to predictive modeling for fish stock and ecosystem management
- for construction and validation of new models comparing ecosystems of different regional seas.

The CPR survey has gathered nearly 70 years of data on marine plankton throughout the North Atlantic Ocean, and has recently extended into the North Pacific Ocean. Alister Hardy's simple concept in the 1920s has succeeded in providing us with a unique and valuable long-term data set. There is increasing worldwide concern about anthropogenic effects on the marine ecosystem, including eutrophication, overfishing, pollution, and global warming. The data in the CPR time-series is being used more and more widely to investigate these problems and now plays a significant role in our understanding of global ocean and climate change.

See also

Satellite Remote Sensing of Sea Surface Temperatures.

Further Reading

Colebrook JM (1960) Continuous Plankton Records: methods of analysis, 1950–59. *Bulletins of Marine Ecology* 5: 51–64.

Gamble JC (1994) Long-term planktonic time series as monitors of marine environmental change. In: Leigh RA and Johnston AE (eds.) *Long-term Experiments in Agricultural and Ecological Sciences*, pp. 365–386. Wallingford: CAB International.

Glover RS (1967) The continuous plankton recorder survey of the North Atlantic. *Symp. Zoological Society of London* 19: 189–210.

Hardy AC (1939) Ecological investigations with the Continuous Plankton Recorder: object, plan and methods. *Hull Bulletins of Marine Ecology* 1: 1–57.

Hardy AC (1956) *The Open Sea: Its Natural History. Part 1: The World of Plankton.* London: Collins.

Hardy AC (1967) *Great Waters.* London: Collins.

IOC and SAHFOS (1991) *Monitoring the Health of the Ocean: Defining the Role of the Continuous Plankton Recorder in Global Ecosystem Studies.* Paris: UNESCO.

Oceanographic Laboratory, Edinburgh (1973) Continuous plankton records: a plankton atlas of the North Atlantic and the North Sea. *Bulletins of Marine Ecology* 7: 1–174.

Reid PC, Planque B, and Edwards M (1998) Is observed variability in the observed long-term results of the Continuous Plankton Recorder survey a response to climate change? *Fisheries Oceanography* 7: 282–288.

Warner AJ and Hays GC (1994) Sampling by the Continuous Plankton Recorder survey. *Progress in Oceanography* 34: 237–256.

CTD (CONDUCTIVITY, TEMPERATURE, DEPTH) PROFILER

A. J. Williams, III, Woods Hole Oceanographic
Institution, Woods Hole, MA, USA

Introduction: The *In Situ* Measurement of Salinity, Temperature, and Density in the Ocean

One of the most useful instruments developed for determining seawater properties during the last four decades has been the CTD (conductivity, temperature, depth). This device has supplanted the traditional hydrocast using Nansen bottles and reversing thermometers that was standard physical oceanographic practice from about 1910 to 1970. The CTD, although an electronic instrument, has its origin in the older technology. The computations of properties such as depth, salinity, density, speed of sound, and potential temperature have been greatly facilitated by having the measurements of conductivity, temperature, and pressure in digital format for direct entry into standard formulas, originally in FORTRAN but now in Matlab and other computational engines.

Temperature

The temperature of seawater is directly important because many physical properties depend upon temperature and indirectly important because calculations of salinity from conductivity measurements are dominated by the temperature dependence of conductivity. Temperature is a conservative property of seawater. It is generally modified only when the fluid is at the surface where it can exchange heat with the atmosphere or rarely when it is in contact with another body of water where exchanges of heat may occur by mixing. Temperature has been sampled in hydrocasts (hydrographic stations) with reversing thermometers since the late nineteenth century. In a reversing thermometer, mercury in a glass bulb expands or contracts filling a capillary tube to a greater or less extent much as occurs in a normal fever thermometer. However, when the reversing thermometer has equilibrated with the seawater at its depth along the hydrowire, a messenger falling from the

surface along the wire releases a latch, which causes the thermometer to invert, breaking the column of mercury in the capillary tube, thus capturing the volume of mercury expanded into the capillary tube (**Figure 1**). When the reversing thermometer is returned to the surface, this length of mercury in the capillary tube can be measured, the change in length due to the change in temperature from the sample depth to the surface corrected for, and the *in situ* temperature of the seawater at the sampled depth computed. The technique is sensitive and reliable with well-characterized reversing thermometers

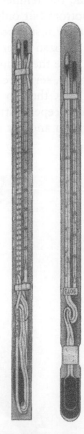

Figure 1 The reversing thermometer has a constriction in the capillary tubing that causes the mercury column to break when the thermometer is returned to its upright orientation after being deployed upside down. This allows the temperature to be measured at depth with the thermometer in the normal, connected column, but then when the thermometer is inverted by the agency of a messenger sent down the hydrographic wire, the mercury beyond the constriction is captured and can be read upon recovery to the surface. An auxiliary thermometer allows the surface reading temperature to be applied to correct for the expansion of the trapped mercury column. General Oceanics, Inc.

where a long history of their calibration has been kept. Resolution of temperature in deep expanded-range thermometers is typically 1 or 2 millidegree. However, hydrostatic pressure at depth would compress the glass and cause the mercury to move along the capillary a greater distance than if the pressure were kept at 1 atm, so *in situ* temperature is measured with a pressure-protected reversing thermometer inside a pressure-resisting glass tube.

The depth of the measurement can be determined even if the hydrowire is not a straight vertical line the length of which can be measured. Current shear in the water column bends the hydrowire into a curve and causes the sample depth to be less than that determined from the meter wheel reading: the distance that was paid out from that when the reversing thermometer entered the water until the lowering was stopped. So for more than a century it has been possible to determine the temperature profile from the surface to the bottom but only at discrete intervals of depth. A heavily instrumented hydrographic cast may have had a dozen reversing thermometers on it but these only gave the temperature at a dozen depths. In some stations, several casts are made with instruments only in, say, the top 1000 m for higher resolution on one and deeper instruments with more widely spaced depths on another (**Figure 2**).

Figure 2 The photograph shows a hydrographic cast being taken with a scientist on the hero platform attaching a bottle to the hydrographic wire. Woods Hole Oceanographic Institution.

A continuous profile of temperature became possible first with the bathythermograph (BT) and subsequently by the STD (salinity, temperature, depth) profiler. The BT, developed by Athelstan Spillhaus in the 1930s, was a valuable tool during World War II for US destroyers and submarines in determining the acoustic refraction in the upper water column with consequences for acoustic submarine detection. Sharp thermal interfaces were revealed at places where sound was refracted by the difference in speed of sound of the water on the two sides of the interface. In the BT, a glass slide plated with carbon (smoked) or a thin gold layer was scratched by a stylus that was moved along one arc by expansion of a fluid-filled coiled tube responding to temperature and along another arc by a second coiled tube (Bourdon tube) empty of fluid and acted upon by pressure. The fine scraped line on the slide could be read in a viewer that was calibrated for the BT with which it was used. In practice, temperatures at regular depths were read off and transcribed to a paper chart, thus missing the promise of a continuous profile of temperature. In fact, there were frequently apparent jiggles in the traces that were initially ignored and only much later were discovered to be real indications of physical phenomena and termed fine structure.

Electronic instrumentation, accepted grudgingly at first, made the measurement of temperature with a platinum resistance thermometer a practical option at sea. This sensor in which a temperature-sensitive element of fine platinum wire is placed inside a pressure-protecting metal sleeve was stable and could be calibrated to a few millidegrees or even to sub-millidegree precision. The near absence of work hardening of the platinum wire and its freedom from corrosion made the platinum resistance thermometer a standard in temperature measurement in the lab and in the sea. It did, however, require some assumptions to define the specific resistance at temperatures other than the critical points of boiling water (100 °C) at standard pressure (1013 millibar) and normal isotopic composition, and the melting point of ice (0 °C) under the same conditions. The ice melting point is sensitive to trace contamination with electrolytes and to the state of the water in thermal equilibrium with the ice, so a more reliable critical point has been used to pin down the low-temperature end of the platinum resistance thermometer scale, the triple point of water. When standard isotopic composition water is in equilibrium with all three phases, liquid (water), vapor (steam), and solid (ice), and there are no other fluids or vapors present, the temperature is 0.010 °C, a very well characterized temperature. Sealed triple point cells have been

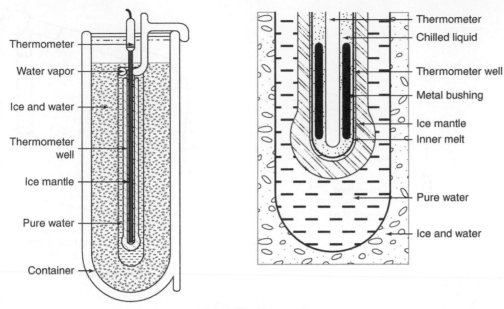

Triple point of water cell

Figure 3 The triple point cell is made with a film of ice in a sealed finger of glass containing only pure degassed water. At the point where the ice surrounding the internal cavity starts to melt, the temperature is 0.010 °C. Sea-Bird Electronics, Inc.

constructed to establish ice/water/vapor equilibrium for precise temperature calibration. A triple point cell has an internal cavity into which a sensor to be tested can be placed (**Figure 3**). Before its use, an ice layer is grown on the outside of the cavity in contact with the water by dropping dry ice into the cavity and freezing the water to a thickness somewhat less than the clearance of the cavity from the outer wall of the triple point cell. Then a little warming is permitted until the ice covering can be spun around the cavity without sticking. This is the point where the cavity wall is at the triple point, 0.010 °C.

Two points may define a linear relation between resistance and temperature but in reality, this relationship is not exactly linear. The temperature scale is in practice defined thermodynamically as that which an ideal gas would obey: $PV = nRT$, where P is absolute pressure, V is specific volume, n is the number of moles of gas, R is the Boltzman constant, and T is the absolute temperature. Even though there is no ideal gas that obeys this relation due to van der Waals forces between molecules of gas at low temperatures, helium approximates it better than any other gas. A practical temperature scale based upon resistance of a platinum resistance thermometer was defined in 1978 but careful work led to an improved temperature scale being established in 1983 that retained the critical endpoints but adjusted the nonlinear shape between the endpoints with a difference of several millidegrees (about 5 millidegree) near 35 °C. This is a serious error to introduce into long-term measurements of, for example, global warming, so the exact temperature scale used for calibration and interpretation is critical.

Thermistors are semiconductors where the resistance changes much more radically with a change in temperature than the platinum resistance thermometer but they are not as stable. Thermistors are heterogeneous ceramic beads exhibiting a negative or a positive temperature coefficient of conductivity; the latter are generally used in this application as they become more conductive as they get warmer. The beads, however, frequently contain defects that grow or heal and change the thermistor resistance at a fixed temperature, so ultrastability is not a characteristic of thermistors unlike platinum resistance thermometers. Thermistors are useful as a secondary sensor of temperature but are calibrated against a platinum resistance thermometer which in turn is calibrated against a triple point cell. The XBT or expendable bathythermograph uses a thermistor to profile temperature from a sinking body that is temporarily connected to the launcher by a fine electrical wire (**Figure 4**).

There is one other enhancement of the calibration of temperature in the oceanographic temperature range. A gallium triple point cell can be used to establish a critical point, 29.764 6 °C, closer to the oceanic temperature range than the boiling point of water. In any case, the calibration of thermometers is critical for long-duration monitoring of oceanic water masses to discern warming trends but is less important for gradient measurements.

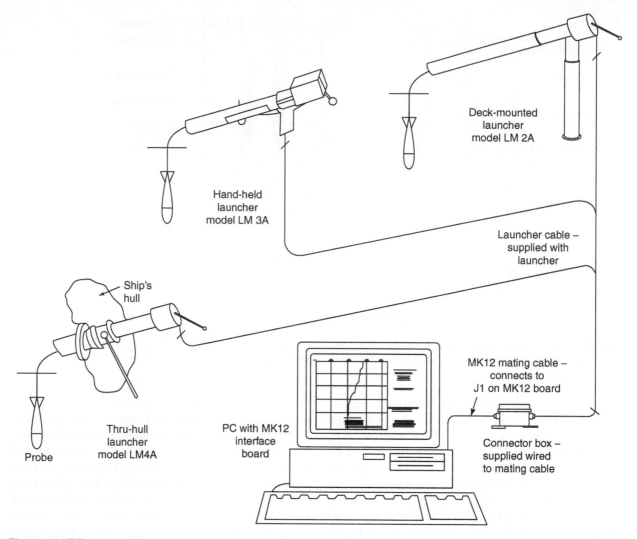

Figure 4 XBT or expendable bathythermograph is a continuous recording temperature profiler that can be dropped from a moving ship or an aircraft. A thermistor in the nose of the probe is sensed through a fine wire that is streamed by both the falling probe and the moving vessel. Depth is assumed from time elapsed after probe contact with the water and is calibrated for fall rate. Lockheed Martin Sippican, Inc.

As electronic temperature measurements were installed upon profilers, in particular the STD profiler (Bissett-Berman, c. 1968), observations of fine structure in temperature came to be accepted as real, which led to an understanding of internal thermal mixing and stratification. Microstructure, as the layering at scales of meters first became known, turned out to result from double-diffusive convection in many cases and produced layers on the order of tens of meters thick, which became known as fine structure and which are now known to be nearly ubiquitous in the ocean (**Figure 5**).

Salinity

The equation of state of seawater, which relates the density to temperature, salinity, and pressure, has been determined with great care by laboratory methods. Of almost as great interest as the density (and more for water watchers) is the salinity. Salinity is, if anything, more conservative than temperature and only changes upon exposure of the water to the atmosphere or another body of water at a different salinity where mixing may occur. Hydrothermal vents on the seafloor have been suspected of also changing the salinity of deep water masses by injection of dissolved materials. The present theory even suggests that hydrothermal vent circulation establishes the overall salinity of the sea, not river runoff followed by evaporation. So determination of salinity is important, not just for the equation of state where salinity affects density.

The direct determination of salinity is awkward. It is defined as the weight of solids in grams in 1 kg of

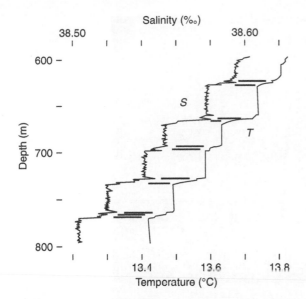

Figure 5 The profile of sheets and layers from an instrument named self-contained imaging microprofiler (SCIMP) shows at very high resolution the microstructure in the Tyrrhenian Sea. The sensor in SCIMP was the first CTD to use internal recording. Freedom from a cable allowed very smooth sinking and high spatial resolution of temperature and salinity. Molcard R and Williams AJ, III (1975) Deep stepped structure in the Tyrrhenian Sea. *Memoires Societe Royale des Sciences de Liege* VII: 191–210.

seawater when the water has been evaporated and all the carbonates have been converted to oxides, bromine and iodine converted to chlorine, and all organic matter completely oxidized. Direct evaporation does not work because chlorides are lost. But a simpler indirect measure can be based on the almost constant composition of seawater (same ratios of major ions nearly everywhere, only the water content varies). This involves titrating the chloride (and other halogens) with silver nitrate and indicating with potassium chromate. The relation is

$$\text{salinity} = 0.03 + 1.805 \times \text{chlorinity}$$

Titrating is slow and awkward, so an attempt was made to determine the salinity by electrical conductivity measurements. The comparison was made between conductivity of diluted standard seawater and full-strength standard seawater at a common temperature. The relation was fairly linear even though seawater is more than a very dilute solution. Once the relations were worked out from laboratory measurements, it became possible to measure salinity by putting the unknown sample in a temperature bath and measuring the ratio of its conductivity to that of a known sample in the same temperature bath. Schleicher and Bradshaw at Woods Hole Oceanographic Institution did this work in the 1960s.

The resulting instrument was the lab salinometer and it allowed faster, more precise determinations of salinity than the older titration method. But it did require a supply of standard seawater for comparison.

Standard seawater is a commodity required by the case for oceanographic cruises where many salinity samples are expected. On a typical hydrographic cast, each reversing thermometer was attached to a salinity sample bottle. Fridtjof Nansen perfected metal sample bottles used on a hydrowire at the turn of the twentieth century and these, which were triggered with the same messenger that caused the reversing thermometers to reverse, closed valves at the top and bottom capturing a water sample that could later be titrated or compared to standard seawater in a salinometer (**Figure 6**). Reversing thermometers were commonly attached to Nansen bottles so that a single triggering action might capture a temperature and a salt sample at one instant. Standard seawater is collected at an open ocean site of which several were used initially. Currently the IAPSO standard seawater is of North Atlantic origin. After being filtered through a 200-nm filter, this water is evaporated and diluted to a specific conductivity to serve as a standard, the principle being that for many thousands of samples the ionic composition is essentially the same. By bringing its conductivity to a standard value with the addition or subtraction of pure water, all samples are assumed to be interchangeable. An earlier standard, Copenhagen standard seawater, was unfortunately not representative of all seawater of the world being somewhat anomalous in having a different ionic composition, so that the constant 0.03 had to be added to the chlorinity determined by titration. However, as a conductivity standard rather than as a chlorinity standard, Copenhagen water served well.

The temperature bath-controlled salinometer was a boon to laboratory analyses of salinity measurements from bottles but it was not sufficient for an ocean-going salinity-profiling instrument. Two more steps were required. The first was to determine the temperature coefficient of conductivity and this permitted correction of the conductivity measurement without a thermostatic bath. The principal variable responsible for conductivity changes in seawater is temperature, not salinity, so the temperature had to be measured very accurately and the lab work done very carefully. Actually, the conductivity ratio could still be used as long as the standard seawater was at the same temperature as the unknown sample. But, it was also possible to just calibrate the salinometer occasionally with standard seawater and to calculate the difference in conductivity expected from the temperature of the sample, which was different from the standard seawater calibration temperature.

Figure 6 A case of standard seawater flasks is shown as carried on research cruises where many samples will be taken and analyzed at sea. The standard seawater is used to calibrate the salinometer.

Salinometers permitted salinities to be run at sea from Nansen bottles, which improved accuracy somewhat because saltwater samples can sometimes get spoilt if kept too long. But the observations were made from only a few points in the profile. Then Neil Brown in 1961 combined pressure measurements with conductivity and temperature measurements to make an *in situ* sampler, the STD profiler. Schleicher and Bradshaw determined the pressure effect on conductivity (by now a three-variable problem) and Brown and Allentoft extended the conductivity ratio measurements.

The STD opened a new window on the ocean and immediately presented problems for physical oceanographers by showing fine structure in a way that could no longer be ignored. The STD converted the conductivity measurement to salinity with *in situ* analog circuitry using temperature and pressure. However, only a few years later, computers began to go to sea and Brown realized a better algorithm could be applied to raw digital conductivity, temperature, and pressure measurements by shipboard-based computers than by using the analog conversions in the *in situ* instrumentation. Furthermore, if recorded digitally, the original data could always be reprocessed at a later time upon the improvement of the algorithm. Finally, the precision and accuracy of the measurement could be improved and the size of the sensors reduced to push the scale of observations from the meter to the centimeter scale. It was the latter that drew Brown to Woods Hole Oceanographic Institution in 1969 to develop the microprofiler. This was the first CTD.

CTD Sensors

Temperature can be measured to about 2 millidegree with reversing thermometers and salinity can be relied upon to a few parts per million. To improve on this, Brown aimed for resolution of salinity to 1 ppm which required resolution of temperature to 0.5 millidegree Celsius. Stability had to be very good to make calibrations to this standard meaningful. For standards work, the platinum thermometer is used and Brown chose that for the CTD. To minimize size and retain high stability with the conductivity measurement, Brown designed a ceramic, platinum, and glass conductivity cell. For pressure, he used a strain gauge bridge on a hollow cylinder. Using temperature, conductivity, and pressure, salinity can be calculated and from temperature, salinity, and pressure, density can be calculated. Depth can then be determined from pressure, the integral of density to the surface, and a local value for gravity. The correction from pressure to depth is small and for many purposes inconsequential so that profiles of temperature and salinity against pressure are used in place of profiles against calculated depth in most cases.

Pressure

Originally Brown planned to build each sensor himself but technology in the commercial world provided him with an adequate pressure sensor initially in a pressure transducer produced by Paine Instruments, Inc., subsequently improved with temperature

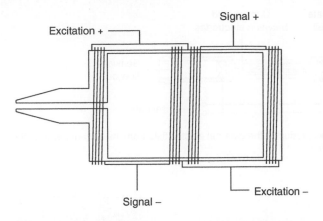

Figure 7 This drawing of a strain gauge pressure sensor shows four sets of wires around the body of the sensor. Internal pressure from the port at the left stretches the center of the cylinder more than the heavy ends but the thermal expansion is similar in all four coils. Driven as a resistance bridge, the sensing leads experience a doubling of the effect of a single coil but cancel the thermal effect on stretching of the wires.

correction for reducing hysteresis during a profile. Typical lowering speeds for deep CTD profiles are between 30 and 100 m min^{-1}, limited by the need to prevent the cable from going slack and jumping the sheave at the top or getting a loop in the cable near the bottom. But this speed causes the temperature in the instrument housing to vary rapidly, especially while transiting the thermocline, and temperature gradients inside the instrument are a problem for sensors that were designed for constant temperature. The hysteresis in pressure has remained a problem up to the present time and has only been tolerated because the errors are not serious for the computation of salinity. They are principally of concern when trying to measure motion of a surface of constant salinity or temperature between profiles or between the down profile (clean because the instrument is at the leading edge of the insertion) and the up profile a few minutes to an hour later (**Figure 7**).

Conductivity

Direct measurement of conductivity presents problems because of polarization of seawater at the electrodes, so the salinity measurements of the STD made electronically used an inductive cell without electrodes. In this cell, made with dimensionally stable materials to fix the geometry, a toroidal transformer was constructed in which seawater formed a single shorted secondary turn through the hole. Electric current was induced in this shorted turn and the conductivity of the seawater in this path measured as a transformed conductivity in the toroidal primary winding of the cell. While

reasonably stable, this technique did not offer the high spatial resolution desired in the CTD nor was it the only route to removing the difficulties with electrodes, so Brown elected to design a miniature stable conductivity cell for the CTD.

The conductivity cell of Brown's CTD had four electrodes, two for current and two to measure voltage, to minimize electrode effects with a symmetry that made it insensitive to local contamination of the electrodes. It was only 3 mm in diameter and 8-mm long, so it was hoped that this small size would be able to resolve centimeter-scale structure. Measurements were made at 10 kHz to circumvent polarization at the current electrodes. Other geometries for electrode-type conductivity cells have been used, a three-electrode cell from Sea-Bird Electronics, Inc., for example (**Figure 8**). Conductivity cell design is a present occupation of sensor technologists. For example, long-duration, fast cells are being developed by Ray Schmitt of Woods Hole Oceanographic Institution for deployment on gliders (autonomous underwater vehicles that profile along oblique paths by gliding between two depths).

Original plans by Brown to make his own thermometer, in a helium-filled ceramic capillary tube, were discarded when it was discovered how hard it was to work with ceramics. Endless difficulties in glass to ceramic and glass to metal seals developed and overcoming these in the conductivity cell which had no voids was hard enough. A commercial platinum resistance thermometer was chosen from Rosemont Inc., with a time constant of 300 ms and a guaranteed stability of 10 millidegree in a year but in practice somewhat better.

The 300-ms response time of the thermometer meant that for 1-cm resolution, his original target, descent rates of 0.3 m min^{-1} would be the limit. This was a bitter result; however, Brown added a fast response thermistor to correct the temperature measurement at faster descent rates. The correction technique added the derivative of the thermistor temperature measurement in an analog circuit to the stable platinum resistance thermometer temperature measurement to replace the high-frequency variations that were lost, without affecting the overall accuracy of the platinum resistance temperature measurement in less dynamic regions. Later, he increased the size of the conductivity cell to facilitate manufacture. This potentially degraded spatial resolution. Flushing of the conductivity cell has been an issue and the original cell, although only 8-mm long, had a flushing length at speeds above 10 cm s^{-1} of about 3.5 cm. The new cell flushed in about 8 cm of descent during lowering on a hydrographic station. With a thermistor response time of 30 ms, a 10-cm vertical resolution was

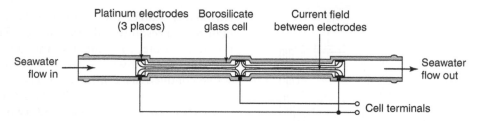

Figure 8   This drawing of a conductivity cell shows three electrodes, reducing the external electric field from the current between the outer electrodes and the central electrode. Sea-Bird Electronics, Inc.

possible at descent speeds of $50\,cm\,s^{-1}$ or $30\,m\,min^{-1}$, a reasonable winch speed. The requirement of resolving structure to 10 cm at a descent rate of $30\,m\,min^{-1}$ meant a sample rate of 10 Hz. (The original resolution target was higher and the first microprofiler had three channels running at 32 ms each in parallel.) The subsequent Neil Brown Instrument Systems Mark III CTD successively digitized conductivity, pressure, and temperature at 32 ms in each channel so that it obtained a complete sample every 96 ms, which was fast enough to resolve 10-cm thick features at a lowering speed of $30\,m\,min^{-1}$.

Practical solutions to the requirements of very high accuracy and reasonable speed are hallmarks of the very best instruments and the Mark III CTD established a standard. The range in temperature is about 32 °C, from freezing to nearly the warmest surface water. For packing efficiency, straight binary integers were used and the value of 2^{15} is 32 768. Thus a 16-bit measurement of temperature gives 0.5 millidegree resolution and 0–32.8 °C range. (For some work, a -2 °C lower end is needed and this was later incorporated.) Conductivity varies over the same range because it tracks temperature. Sixteen bits generally permit salinities up to 38‰ to be measured to 0.001‰ precision. The depth range needed for most of the ocean is 6500 m (or a pressure range of 6500 decibar) and, with a digitizer capable of 16-bit resolution, 10-cm depth resolution is permitted, again right on target for the resolution of the sensors. But to make a measurement to a part in 65 536 (2^{16}) and have it remain accurate and stable is not easy. Furthermore, the conductivity measurement must be made at about 10 kHz to minimize electrode polarization. Neil Brown's solution to these problems was to make all of the digitizations with transformer windings, weighted in a binary sequence and added electronically. These were driven at 10 kHz so that polarization effects were minimal. Precision in the measurements was ensured by the turns ratio in the transformer.

While the Neil Brown CTD was the first of the new profiling instruments, others soon followed. Guildline, Inc., produced an excellent lab salinometer and that technology similarly permitted a

Figure 9   A CTD with rosette sampler of large Nisken bottles and reversing thermometers is being lowered off the side of a ship. Woods Hole Oceanographic Institution.

profiler to be developed. Sea-Bird Electronics, Inc., produced sensors for temperature and conductivity based upon a Wein bridge oscillator that were precise, compact, and easy to incorporate into instruments and soon Sea-Bird produced its own CTD based upon these sensors. Sea-Bird's conductivity cell is a three-electrode cell with a slow natural flushing time but it is generally flushed with an external pump to establish a constant flushing time irrespective of lowering rate, thus improving the spatial resolution. This device is now widely used on oceanographic research vessels. Ocean Sensors, Inc., has produced a small CTD with internal recording for inclusion on

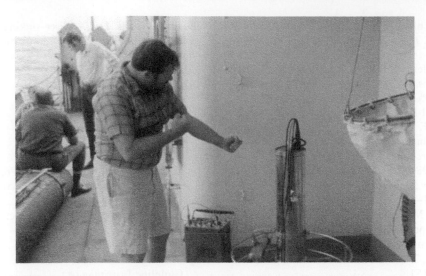

Figure 10 Neil Brown (1927–2005), a long-time contributor to precise continuous measurements of physical properties of the ocean, shown here confronting the CTD he invented, had a great sense of humor. His use of transformer-based analog-to-digital converters provided the precision that permitted electronic measurements of temperature, conductivity, and pressure to be made in profilers.

autonomous instruments. Idronaut, S.r.l., has a CTD which can accommodate additional sensors for oxygen, carbon dioxide, ambient light, pH, and optical backscatter, for example. In fact, the ability to add sensors to a digital instrument has been an advantage not lost on instrument builders, so that a typical profiler package on a ship may contain a CTD with a suite of these additional sensors, some sensors duplicated for redundancy, and generally includes a rosette sampler of Niskin bottles to capture water on trigger from the surface.

Extended Deployments of the CTD

Battery-supplied power for the CTD and the development in the early 1970s of digital magnetic tape recorders freed the CTD from cable connection to the ship (**Figure 9**). The salinity and temperature profiles shown in **Figure 5** were recorded on a Sea-Data cassette tape from a free vehicle. Now with massive solid state memory such as compact flash, replacing hard disks of the 1980s and 1990s, digital data storage is not a problem and CTDs are found on autonomous underwater vehicles, moored profilers, and on gliders and floats.

The CTD is now a standard oceanographic instrument and has replaced the Nansen cast as a hydrographic tool (**Figure 10**). The data are sent up conducting cable as a frequency-shift-keyed signal. Multiconductor or fiber-optic armored cable has replaced the hydrographic wire for most hydrographic surveys. Recording as an acoustic signal on tape has been replaced by direct storage of digital data on a PC or dedicated server but the decoding and computation of salinity is still generally done on deck. The

algorithm has been clumsy in that it sticks close to the actual relations derived from the laboratory data sets. These were derived by going from salinity to conductivity, not the other way around. Dynamically one wants to know density and this is now computed as a second step when it could be done directly. But computational complexity is a negligible cost with even the smallest PCs. Many CTDs export salinity directly, having done the conversion internally. However, conductivity recording still permits enhanced processing if raw values are retained. A direct measurement of density might be the next step in sensor development. However, the demand for sensors of light, chemistry, and properties other than the classical ones of temperature, salinity, and pressure has seemed to be more significant and the modern CTD is often just the central element of a complex suite of sensors.

See also

Gliders. Satellite Remote Sensing: Salinity Measurements. Satellite Remote Sensing of Sea Surface Temperatures.

Further Reading

Bacon S, Culkin F, Higgs N, and Ridout P (2004) IAPSO standard seawater: Definition of the uncertainty in the calibration procedure, and stability of recent batches. *Journal of Atmospheric and Oceanic Technology* 24: 1785–1799.

Bradshaw A and Schleicher KE (1970) Direct measurement of thermal expansion of seawater under pressure. *Deep-Sea Research* 17: 691–706.

Bradshaw A and Schleicher KE (1980) Electrical conductivity of seawater. *IEEE Journal of Oceanic Engineering* 5: 50–56.

Brown NL and Allentoft B (1966) *Salinity, Conductivity and Temperature. Relationships of Seawater over the Range of 0–50‰*. US ONR Contract Nr-4290(00), MJO 2003 Final Report, Washington, DC.

Cox RA, Culkin F, and Riley JP (1967) The electrical conductivity/chlorinity relationship in natural seawater. *Deep-Sea Research* 14: 203–220.

Dauphinee TM (1980) Introduction. *Special Issue on the Practical Salinity Scale 1978. IEEE Journal of Oceanic Engineering* 2: 1–2.

Forch C, Knudsen M, and Sorensen SPL (1902) Berichte uber die Konstantenbestimmungen zur Aufstellung der hydrographischen Tabellen. Kgl. Danske Videnskab. Selskabs Skrifter, 7 Rackke, Naturvidensk. Og Mathem. Afd. 12, pp. 1–151.

Lewis EL and Perkin RG (1978) Salinity: Its definition and calculation. *Journal of Geophysical Research* 83: 466–478.

Miyake M, Emery WJ, and Lovett J (1981) An evaluation of expendable salinity–temperature profilers in the eastern North Pacific. *Journal of Physical Oceanography* 11: 1159–1165.

Molcard R and Williams AJ, III (1975) Deep stepped structure in the Tyrrhenian Sea. *Memoires Societe Royale des Sciences de Liege* VII: 191–210.

Ridout P and Higgs N (online) An Overview of the IAPSO Standard Seawater Service. Havant: OSIL. http://www.ptb.de/de/org/3/31/313/230ptbsem/230ptbsem_osil_ridout.pdf (accessed Mar. 2008).

UNESCO (1981) International Oceanographic Tables, Vol. 3. *UNESCO Technical Papers in Marine Science* 39, pp. 1–111. Paris: UNESCO.

Williams AJ, III (1974) Free-sinking temperature and salinity profiler for ocean microstructure studies. *IEEE International Conference on Engineering in the Ocean Environment* 2: 279–283.

Relevant Websites

http://www.paineelectronics.com
   – Downhole and Differential Pressure Transducer, Sensor, and Transmitter, Paine Electronics.

http://www.sippican.com
   – Expendable Probes, Sippican, Inc.

http://www.generaloceanics.com
   – General Oceanics, Inc.

http://www.guildline.com
   – Guildline Instruments.

http://www.idronaut.it
   – Idronaut, S.r.l.

http://www.agu.org
   – News article 'Athelstan Spilhaus dies at 86' (30 March 1998), Science and Society (American Geophysical Union).

http://www.mnc.net
   – Nordic Visitors Norway.

http://www.oceansensors.com
   – Ocean Sensors, Inc.

http://www.emersonprocess.com
   – Rosemount Temperature Products, Emerson Process Management.

http://www.seabird.com
   – Sea-Bird Electronics, Inc.

http://www.whoi.edu
   – Woods Hole Oceanographic Institution.

DEEP-SEA DRILLING METHODOLOGY

K. Moran, University of Rhode Island, Narragansett, RI, USA

Introduction

The technology developed and used in past scientific drilling programs, the Deep Sea Drilling Project and the Ocean Drilling Program, has now been expanded in the current Integrated Ocean Drilling Program (IODP). These technologies include innovative drilling methods, sampling tools and procedures, *in situ* measurement tools, and seafloor observatories. This new IODP technology will be used to drill deeper into the seafloor than was possible in the previous scientific programs. The first drilling target of the IODP using the new technology is the seismogenic zone offshore Japan, a location deep in the Earth (7–14 km) where earthquakes are generated.

Drilling Technology

The Deep Sea Drilling Project and the Ocean Drilling Program used the same basic drilling technology, the open hole method. Today, the IODP has extended this capability to include closed hole methods, known as riser drilling.

Open Hole or Nonriser Drilling

Drilling is the process of establishing a borehole. The open hole method uses a single drill pipe that hangs from the drill ship's derrick, a tall framework positioned over the drill hole used to support the drill pipe. The drill pipe is rotated using drilling systems, specifically a hydraulically powered top drive located above the drill floor of the ship. Surface sea water is flushed through the center of the pipe to lubricate the rotating bit that cuts the rock and then flushes sediment and rock cuttings away to the seafloor (**Figure 1**). Open hole refers to the resulting borehole which remains open to the ocean during drilling. This method is also called a riserless drilling system. Important parts of the deep-water drilling system are a drilling derrick that is large and strong enough to hang a long length of drill pipe reaching deep ocean and subseafloor depths (up to 8 km); a system that rotates the drill pipe; a motion compensator that isolates the

ship's motion from the drill pipe; and a pump that flushes sea water through the drill pipe.

Open hole methods are successfully used in all of the Earth's oceans (**Figure 2**). Scientific ocean-drilling achievements include drilling in very deep water (6 km) and to >2 km below the seafloor (**Table 1**). Although there have been many achievements using these methods, there are also limitations. Although the exact depth limit of open-hole drilling method is not yet known, it is likely limited to 2–4 km below the seafloor. This limitation exists because the drill fluid must be modified to a lower density so that the deep cuttings can be lifted from the bit and flushed out of the hole when drilling deep into the seafloor. Another

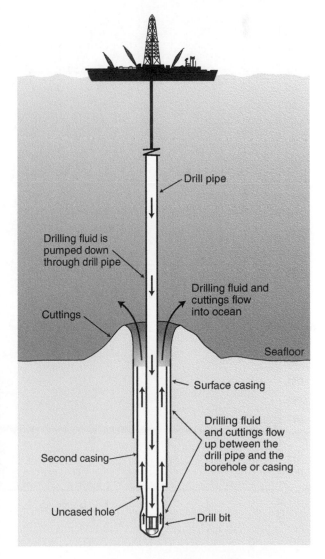

Figure 1 Diagram of a nonriser drilling system.

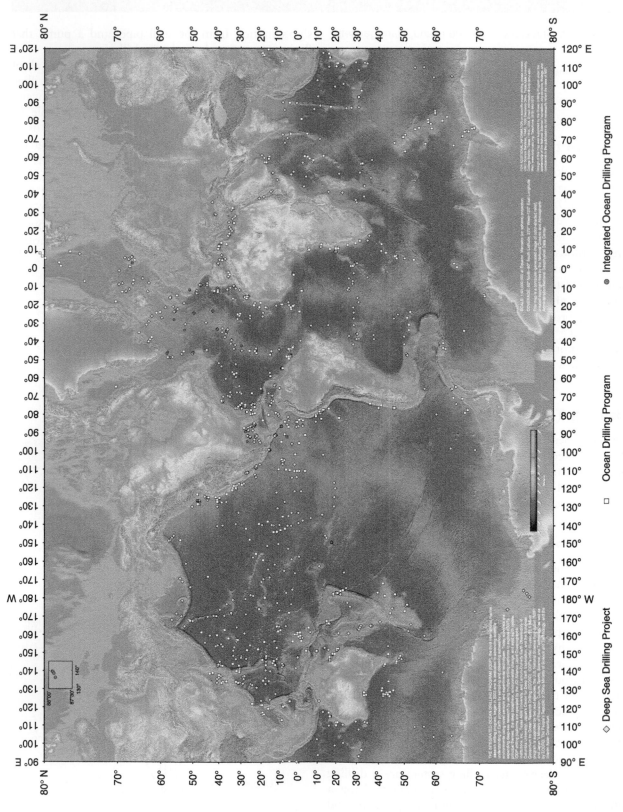

Figure 2 Map of all sites drilled by the Deep Sea Drilling Project (DSDP), the Ocean Drilling Program (ODP), and the IODP.

◇ Deep Sea Drilling Project   □ Ocean Drilling Program   ● Integrated Ocean Drilling Program

Table 1 Fact sheet about the *JOIDES Resolution,* the research vessel used by the Ocean Drilling Program

Total number of days in port		445 days
Total number of days at sea		4751 days
Total distance traveled		507 420 km
Total number of holes drilled		1445 holes
Deepest water level drilled	Leg 129	5980 m
Deepest hole drilled	Leg 148, Hole 504B (South-eastern Pacific Ocean, off coast of Ecuador)	2111 m
Total amount of core		180 880 m
Most core recovered on single	Leg 175 – Benguela (15 Aug.–10 Oct. 1997)	8003 m
Northern-most site drilled	Leg 113, site 911	Latitude 80.4744° N Longitude 8.2273° E
Southern-most site drilled	Leg 151, site 693	Latitude 70.8315° S Longitude 14.5735° W
Year and place of constitution	1978	Halifax, Nova Scotia, Canada
Laboratories and other scientific equipment installed	1984	Pascagoula, Mississippi
Gross tonnage		9719 tons
Net tonnage		2915 tons
Engines/generators		Seven 16 cyl index Diesel 5@2100 kW (2815 hp) 2@15500 kW (2010 hp)
Length		143 m
Beam		21 m
Derrick		62 m
Speed		11 knots
Crusing range		120 days
Scientific and technical party		50 people
Ship's Crew		65 people
Laboratory space		1115 m^2
Drill string		8838 m

limitation of the open hole method is that drilling must be restricted to locations where hydrocarbons are unlikely to be encountered. In an open hole, there is no way to control the drilling fluid pressure. In locations where oil and gas may exist, the formations are frequently overpressured (similar to a champagne bottle). If these formations would be punctured with an open hole system, the drill pipe would act like a straw that connects this overpressured zone in the rock to the ocean and the ship. This type of puncture is called a 'blow-out' and is a serious drilling hazard. The explosion as gases are vented through the straw to the ship's drill floor could cause serious damage, or worse yet, the change in the density of the seawater as the gas bubbles are released into the overlying ocean could cause the ship to sink. Without a system to control the pressure in the borehole, there is no way to prevent a blow-out.

Closed Hole or Riser Drilling

The new deep-sea drilling technology used in IODP is a closed system, also known as riser drilling. This technology has been used, in shallow to intermediate water depths, by the offshore oil industry to explore for, and produce oil and gas.

Riser drilling uses two pipes: a drill pipe similar to that used for open hole drilling and a wider diameter riser pipe that surrounds the drill pipe and is cemented into the seafloor (**Figure 3**). The system is closed because drill fluid (seawater and additives) is pumped down the drillpipe (to lubricate the bit and flush rock cuttings away from the bit) and then returned to the ship via the riser. With riser drilling, the drill fluid density can be varied and borehole pressure can be monitored and controlled, thus overcoming the two limitations of open hole drilling. The single limitation of riser drilling is water depth. The current water depth limit of riser technology is approximately 3 km. This water depth limit occurs because the riser, filled with drill fluid and cuttings, puts a large amount of pressure on the rock. This pressure is greater than the strength of the rock; thus, the rock breaks apart under the riser pressure, causing the drilling system to fail.

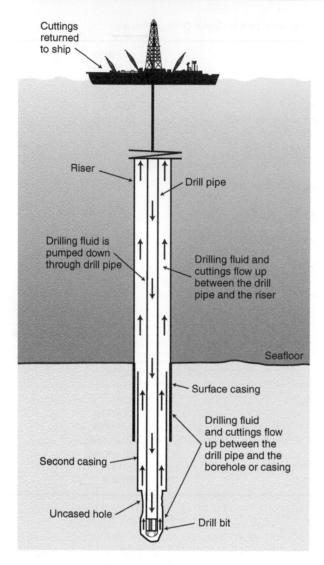

Figure 3   Diagram of a riser drilling system.

Deep-Sea Sampling

Scientific ocean drilling not only requires a borehole, but – more importantly – the recovery of high-quality core samples taken as continuously as possible. Recovering sediment and rock samples from below the seafloor in deep water requires the use of wireline tools. Wireline tools are pumped down the center of the drill pipe to the bottom (called the bottom hole assembly) where they are mechanically latched into place near the drill bit in preparation for sampling. Different types of tools are advanced into the geological formation and take a core sample in different ways, depending on the type of sediment or rock. After the tool samples the rock formation, it is unlatched from the bottom hole assembly with a mechanical device, called an overshot, that is sent down the pipe on a wire. The overshot is used to unlatch the wireline tool, return it to the ship, and recover the core sample.

In scientific ocean drilling, three standard wireline core sampling tools are used: the advanced piston corer, the extended core barrel, and the rotary core barrel.

The piston corer is advanced into the sediment ahead of the drill bit using pressure applied by shipboard pumps through the drill pipe (**Figure 4(a)**). The drill fluid pressure in the pipe is increased until the corer shoots into the sediment. After the corer is shot 10 m ahead of the bit, it is recovered using the wireline overshot. The drill pipe and bit are then advanced by rotary drilling another 10 m, in preparation for taking another core sample. The piston corer is designed to recover undisturbed core samples of soft ooze and sediments up to 250 m below the seafloor. In soft to hard sediments, the piston corer can achieve 100% recovery. However, because the cores are taken sequentially, sediment between consecutive cores may not be recovered. To ensure that a continuous sedimentary section is recovered, particularly for paleoclimate studies, the IODP drills a minimum of three boreholes at one site. The positions of the breaks between consecutive core samples are staggered in each borehole so that if sediment is not sampled in one borehole at a core break depth, it will be recovered in the second or third borehole.

The extended core barrel is a modification of the oil industry's rotary corer and is designed to recover core samples of sedimentary rock formation (**Figure 4(b)**). Typically, the extended core barrel is deployed at depths below the seafloor at which the sediment is too hard for sampling by the piston corer. The extended corer uses the rotation of the drill string to deepen the borehole and cut the core sample. The cutting action is done with a small bit attached to the core barrel. An innovation of this tool is an internal spring that allows the core barrel's smaller bit to extend ahead of the drill bit in softer formations. In hard formations, where greater cutting action is needed, the spring is compressed and the small core bit rotates with the main drill bit.

The rotary core barrel is a direct descendant of the rotary coring system used in the oil industry and is similar to the extended core barrel in its retracted mode. The rotary corer is designed to recover core samples from medium to very hard formations, including igneous rock. The corer uses the rotation of the drill string and the main drill bit to deepen the hole and cut the core sample (**Figure 4(c)**).

In all drilling operations, there are times when the drill pipe must be recovered to the ship, for example, to change a worn bit or to install different bottom hole equipment. Before pulling the pipe out of the

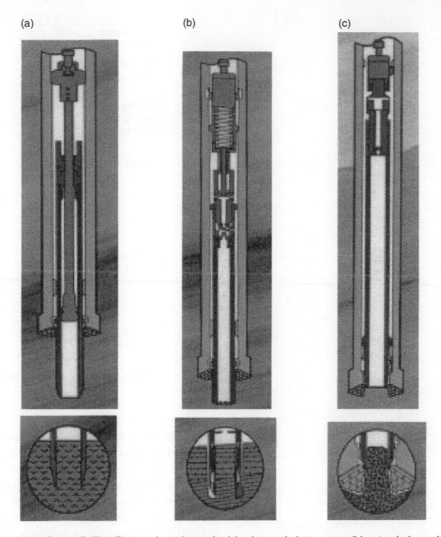

Figure 4 Diagrams of the Ocean Drilling Program's coring tools: (a) advanced piston corer; (b) extended core barrel; and (c) rotary core barrel.

borehole, a re-entry cone (**Figure 5**) is dropped down the outside of the drill pipe where it free-falls to the seafloor. The cone is used as a guide to re-enter the borehole with a new bit or equipment. The cone is a very small target in deep water (1–6 km) and ancillary tools are needed to locate it for re-entry into the borehole. The cone is first located acoustically, using seafloor transponders. To precisely pinpoint the cone, a video camera is lowered with the drill pipe to visually pinpoint the location and drop the drill pipe back into the borehole.

Special corers are used to sample unusual and difficult formations. For example, gas hydrates, ice-like material that is stable under high pressure and low temperature, commonly occur in deep water below the seafloor and require special samplers. Gas hydrates have generated much public interest since they can contain methane gas trapped within their structure, which is thought to be a potential future energy

source. However, when gas hydrate is sampled, it must be kept at *in situ* pressure conditions to maintain the integrity of the core. Thus, a pressure core sampler is used to sample hydrates. The sampler is similar to the extended corer in that it has its own bit, but it has an internal valve that closes before the sampler is removed from the formation. The closed valve maintains the sample at *in situ* pressure conditions.

Drilling Measurements

Once sampling is completed, logging tools are lowered into the borehole to measure the *in situ* geophysical and chemical properties of the formation. In IODP, logging tools, developed for use in the oil industry, are most commonly leased from Schlumberger. The origins of logging go back to 1911 when the science of geophysics was new and was just beginning to be used

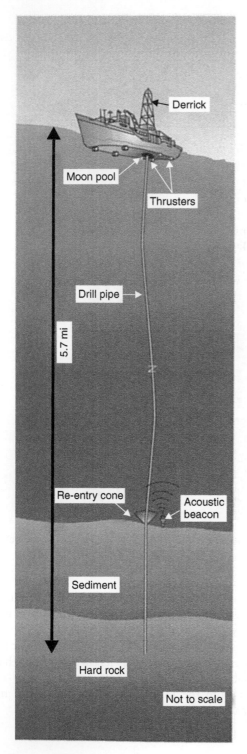

Figure 5 Diagram of the ship, drill pipe, and re-entry cone used in the drilling process.

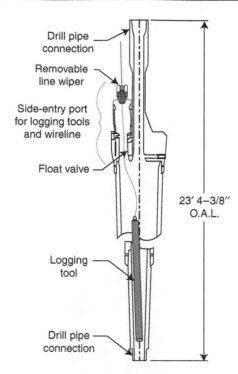

Figure 6 Diagram of the side-entry sub technology.

to explore the internal structure of the Earth. Conrad and Marcel Schlumberger, the founders of Schlumberger, conceived the idea that electrical measurements could be used to detect ore (precious minerals). Working at first alone and then with a number of

associates, they extended the electrical prospecting technique from the surface to the oil well. Now, the use of electric prospecting, called logging, is widely accepted as a standard method in oil exploration.

Logging tools are lowered into the borehole using a cable that also transmits the data, in real time, to the ship. The term logging refers to the type of data collected. For example, a borehole log is a record or ledger of the sediment and rock encountered while drilling. Logs are geophysical and chemical records of the borehole. The logging tools typically comprise transmitters and sensors or a sensor alone encased in a robust stainless steel tube. Examples of tool measurements include electrical resistivity, gamma ray attenuation, natural gamma, acoustic velocity, and magnetic susceptibility. Log data provide an almost continuous record of the sediment and rock formation along length of the borehole.

Log data are of high quality when collected in the open hole, outside of the drill pipe. The most common method for deploying logging tools is to deploy a device that releases the drill bit from the drill pipe once all drilling and sampling operations are completed. The drill bit falls to the bottom of the borehole and is left there. The drill pipe is retracted and only 75–100 m of pipe is left at the top of the borehole to keep the upper, loose part of the borehole stable. Logging tools are lowered through the drill pipe to the bottom of the borehole. Log data are acquired by slowly raising the tool up the borehole at a constant speed. When boreholes are unstable and the walls are collapsing into the open hole, another method is used, unique to scientific ocean drilling. In unstable conditions, the drill pipe cannot be retracted to within 75 m of the seafloor without borehole walls collapsing and blocking or bridging the hole. In these situations, after the bit is released, the logging tools are lowered inside the drill pipe to the bottom of the hole. The drill pipe is retracted only enough to expose the logging tools to the open hole, while protecting the remainder of the borehole walls. Then, drill pipe is retracted at the same speed as logging tools are pulled up through the borehole. The technology developed that allows for this unique operation is called the side-entry sub (**Figure 6**). When inserted as a part of the drill string, the cable, to which the tools are attached, exits the drill pipe at

the side-entry sub, positioned well below the ship. In the way, the logging cable does not interfere with removal of drill pipe.

Data in the borehole are also collected using wireline-deployed tools. Sediment temperature is measured with a temperature sensor mounted inside the cutting edge of the piston corer. In addition, other tools are deployed that do not recover a sample. They are pushed into the sediment ahead of the drill bit, left in place for 10–15 min to record temperature, and then pulled back to the ship, where the data are downloaded and analyzed. Special wireline tools have also been used to measure pressure and fluid flow properties of sediment and rock.

IODP also uses an oil industry-developed method for logging sediment and igneous rocks – logging-while-drilling or LWD. In this method, some of the

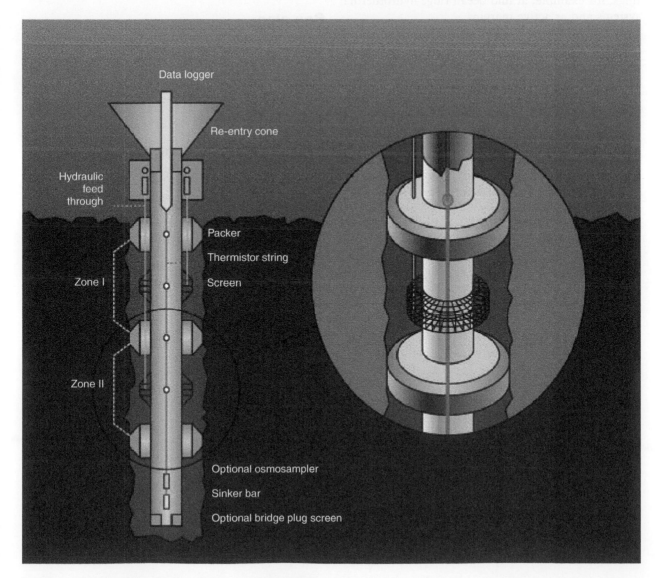

Figure 7 Diagram of the advanced CORK system used to seal instruments in the borehole.

same wireline logging sensors are repackaged and attached to the drill string, immediately behind the drill bit. The sensors log the geophysical properties of the sediment and rock as drilling proceeds, thus eliminating problems associated with borehole collapse.

Deep Seafloor Observatories

The circulation obviation retrofit kit (CORK) is a seafloor observatory that measures pressure, temperature, and fluid composition – important parameters for the study of the dynamics of deep-sea hydrologic systems. CORKs are installed by the IODP for measurements over long periods of time (months to years). Since 1991, observatories have been installed on the deep seafloor in different settings, for example, at mid-ocean ridge hydrothermal systems and at active margins.

The CORKs are installed by the drill ship. After a borehole is drilled, a CORK is installed to seal instruments in the borehole away from the overlying ocean (**Figure 7**). The CORK has two major parts: the CORK body that provides the seal and an instrument cable, for measuring fluid pressures, and temperatures that hangs from the CORK into the borehole. A data recorder is included with the instrument cable. The data recorders have sufficient battery power and memory for up to 5 years of operation. Data are recovered from CORKs using manned submersibles or remotely operated vehicles. The instruments in the CORK measure pressure and temperature spaced along a cable that extends into the sealed borehole. The CORK also includes a valve above the seal where borehole fluids can be sampled. Advanced CORKs are also used to isolate specific and measure the properties of different sediment or rock zones.

The IODP installs another type of long-term seafloor observatory for earthquake studies. Seismic monitoring instruments are installed in deep boreholes located in seismically active regions, for example, off the coast of Japan. These data are used to help established predictive measures to prevent loss of life and damage to cities during large earthquakes.

Deep-sea seismic observatories contain a strainmeter, two seismometers, a tilmeter, and a temperature sensor. The observatories have replaceable data-recording devices and batteries like CORKs, and are serviced by remotely operated vehicles. Eventually, real-time power supply and data retrieval will be possible when some of the observatories are connected to nearby deep-sea fiber-optic cables.

Summary

Deep-sea scientific drilling applies innovative sampling, instrument, and observatory technologies to the study of Earth system science. These range from the study of Earth's past ocean and climate conditions using high-quality sediment cores, to the study of earthquakes and tectonic processes using logging tools and seafloor observatories, to exploring gas hydrates (a potential future energy source) using specialized sampling tools.

In 2004, the IODP succeeded two earlier scientific programs, Deep Sea Drilling and Ocean Drilling. The IODP operates two ships: the D/V JOIDES Resolution and the D/V Chikyu and leases ships for special operations in shallow water and ice-covered seas. The IODP is supported by the US, Japan, and Europe.

See also

Deep   Submergence: Science   of.   Manned Submersibles, Deep Water. Remotely Operated Vehicles (ROVs).

Further Reading

DSDP (1969–86) Initial Reports of the Deep Sea Drilling Project. Washington, DC: US Government Printing Office. http://www.deepseadrilling.org/i_reports.htm (accessed Mar. 2008).

Integrated Ocean Drilling Program (2001) Earth, Oceans and Life: Scientific Investigation of the Earth System Using Multiple Drilling Platforms and New Technologies, Integrated Ocean Drilling Program, Initial Science Plan, 2003–2013.

Joint Oceanographic Institutions (1996) *Understanding Our Dynamic Earth: Ocean Drilling Program Long Range Plan.* Washington, DC: Joint Oceanographic Institutions.

Oceanography Society (2006) *Special Issue: The Impact of the Ocean Drilling Program. Oceanography* 19(4).

ODP (1985–present) Proceedings of the Ocean Drilling Program, Initial Reports, vols. 101–210. College Station, TX: ODP. http://www-odp.tamu.edu/publications (accessed Mar. 2008).

ODP (1985–present) Proceedings of the Ocean Drilling Program, Scientific Results, vols. 101–210. College Station, TX: ODP. http://www-odp.tamu.edu/publications (accessed Mar. 2008).

ODP (2004–present) Proceedings of the Integrated Ocean Drilling Program, vols. 300–. College Station, TX: Integrated Ocean Drilling Program.

WHOI (1993–94) *Oceanus: 25 Years of Ocean Drilling*, vol. 36, no. 4. Woods Hole, MA: Woods Hole Oceanographic Institution.

EXPENDABLE SENSORS

J. Scott, DERA Winfrith, Dorchester, Dorset, UK

Introduction

Expendable sensors represent an approach to ocean measurement in which some degree of measurement precision may be sacrificed in the interests of lower costs and operational expediency. Two requirements of physical oceanography have driven their development: the problem of achieving adequate spatial sampling of the ocean on timescales commensurate with temporal variability; and the requirement by naval forces for under-way assessments of sonar propagation conditions – the first (and still the dominant) application of *operational oceanography*.

The naval requirement first arose in the area of physical oceanography, in the need to know the depth variation of water temperature. In practice, of the three parameters that determine sound speed – temperature, salinity, and pressure – it is temperature that predominates. Pressure is normally deducible with adequate precision from depth, and salinity is normally sufficiently constant to be neglected or simply 'modeled' using an archived (T,S) relation. However, salinity may be important near ice, in fiords, and estuaries, and in regions of freshwater influence (ROFIs). The naval requirement is normally for the vertical sound speed profile, and it is the *shape* of this profile that is important, rather than its mean value.

The expendable measurement facility was quickly taken up by the civilian oceanographic community. It gives a means of tackling the problem of how to make synoptic ocean structure measurements where features are likely to move significantly during a survey. A survey with spatial scales small enough to capture interesting features is seriously degraded by their movement and development. Particularly at mid- to high latitudes, a survey using conventional profiling instruments – such as the conductivity-temperature-depth (CTD) probe – cannot be carried out in a time that is small compared with the timescales of motion and development of features such as frontal boundaries and eddies.

Surveys are severely limited by deployments that require a vessel to be regularly stationary for casts with a profiling speed of $\sim 1\,\mathrm{m\,s^{-1}}$. Expendable probes allow use at ship speeds up to 20–30 knots,

and air-dropped expendables can clearly outstrip even this.

The technique also provides standard results, and the expendable bathythermograph, or XBT, is now considered a central component of global climate monitoring programs such as the Global Ocean Observation System (GOOS). Near-real-time transfer of these data from ships under way is now also an important input to global meteorological forecasting.

The XBT was originally intended to improve on the (nonexpendable) *mechanical* bathythermograph (MBT) which was the principal operational naval device up to the mid-1970s. The MBT was lowered into the water from a vessel, and it *inscribed* a temperature-depth trace, with a sharp stylus on a small coated glass slide. The temperature-sensing element was a xylene-filled copper tube, whose (temperature-dependent) pressure moved the stylus across the slide via a Bourdon tube. Stylus movement along the slide was determined by a copper bellows, compressed by the increasing water pressure. Data were read from the trace using an optical projector and scale.

The XBT was a major advance, allowing operation while under way and dispensing with the intricate measurement routine of the MBT, with the need for a deployment/recovery winch and with the need for calibration. It uses a pre-calibrated thermistor measurement, read onboard in real time. Inference of its depth uses knowledge of its rate of fall through the water. There are now several manufacturers, although the originators. Sippican Inc. (Marion, MA, USA), still lead in the number of available probe types.

Current expendable probe capabilities include, in addition to temperature, the measurement of sound speed, conductivity, ocean current, optical properties, and (recently) seabed properties. This review of expendables summarizes the variety of measurable parameters and then outlines the available deployment options. A number of examples are then given of their use in oceanographic research. Expendables specific to naval activities, such as noise-measuring (sonobuoy) systems, will not be covered here, although in some cases they have a limited ocean measurement capability. Air-deployed drifting systems are also omitted.

Expendable Sensor Types

Expendable Bathythermograph (XBT)

The purpose of the XBT is to provide a vertical profile of temperature, from the surface to as great a depth as required, if possible to the seabed. A thermistor measures temperature as the probe descends, and the depth for each measurement is deduced from time of descent using an empirical equation. A number of variants are available with different depth capabilities, the deepest (the T5) reaching 1830 m, which may necessitate data extrapolation in deeper water. Apart from expense, which increases with depth capability, operational constraints become more restrictive for the deeper probes. Whereas the T-7 (760 m) can operate at platform speeds up to 15 knots, a T-5 is limited to 6 knots.[1,2] Other probe types are (or have been) available,[3] but T-7 and T-5 types are in most regular use.

The stated accuracy of all probe types is $\pm$ 0.15°C and $\pm$2% of indicated depth, with a depth resolution of 0.65 m.

Operational effects of finite depth Various approaches have been adopted to overcome the limited depth capability of XBTs. The best means of doing this is generally accepted to be extrapolation with the help of relevant (same survey) full-depth CTD casts. This has the added advantage of allowing a check on the XBT depth data. In naval operational terms, however,[4] this is rarely an option. In conditions where the measured temperature has stabilized at the maximum depth, extrapolation to the seabed using the data trend may be reasonable. A second approach uses extrapolation using archived data, although if these are mismatched, this may be a problem.

Probe design The standard XBT has two main parts: a protective 'shell' which remains on the vessel after launch, and the probe itself, which falls through the water and passes data along the

connecting wire. Electrical contact with the probe is achieved when the unit is loaded into the launcher, allowing initialization of the onboard electronics before the probe is released by withdrawal of a 'firing pin' from its tail section.

Throughout the operation of the device the probe remains connected to its shell, and thence to the onboard electronics, by two-strand wire. This wire is arranged in two coils, one within the shell, dispensing wire horizontally as the vessel moves away from the launch location, and one within the probe body, which dispenses wire upwards as the probe falls. Data are collected until wire breakage, either when the probe reaches the seabed or (in deeper water) when the wire has been expended without the bottom being reached. If the deploying vessel travels faster than the design speed, the upper coil of wire may be exhausted first.

The success of XBT is the result of a number of critical design features. One of these is the small compartment in the shell that contains the electrical contacts, which is designed to avoid the problem of making good instant electrical contacts between a probe, which may have spent many months awaiting use, and a launcher normally sited on an exposed ship deck. In the compartment, the probe contacts are embedded within a thick gelatinous insulator, which is penetrated by the pointed launcher contacts as the breech closes. The material cleans the launcher contacts as they penetrate, and maintains their clean state by the practice of leaving the spent shell in the launcher between probe launches.

The free-falling probe itself involves three principal components: the thermistor element, making the temperature measurement; the two-strand wire that connects the thermistor circuit to the onboard electronics; and a weighted, hydrodynamically shaped body. Each of these components plays a vital part in the remarkable success of the instrument as a whole. The thermistor, of course, is indispensable, this small fragment of temperature-sensitive semiconductor providing the measurement capability of the unit. This is positioned in an aperture at the probe tip.

The connecting wire may represent a technical achievement at least as great as any of the other XBT components. Two thin strands of copper wire are covered by a thin insulating lacquer which binds them securely but which is sufficiently non-sticky to avoid the problem of self-attachment within the coils, in which hundreds of meters are compactly wound.

The probe body consists of a weighty metal nose cone attached to a lightweight faired hollow plastic tail, equipped with fins to ensure vertical travel,

---

[1] The 450m T-4 (460 m, 30 knots), with a rated ship speed of 30 knots, used to be the routine choice for operational use, but this generally gave way to the deeper T-7 at the end of the 1980s.

[2] The specified maximum may be exceeded, but premature-breakage of the wire will then limit the depth of data collected.

[3] The T-7, T-5, and T-4 are complemented by the T-6 (460 m, 15 knots), T-10 (200 m, 10 knots), and T-11 (460 m, 6 knots), the latter giving a 0.18 m vertical resolution. A T-7 variation called 'Deep Blue' (760 m, 20 knots) was developed for (faster moving) Volunteer Observing Ships.

[4] The naval application differs significantly from purely oceanographic applications. At great depth the sound speed increases slowly with depth (pressure), providing weak upward refraction of sound and decreased seabed interaction. Even small temperature gradients may negate this effect, and small errors in the extrapolation may have disproportionate effects.

and an even metering of wire from the contained coil. The metal cone acts as a 'sea electrode' to complete the measurement circuit with the (effectively grounded) vessel. The final part of the instrument is a soft plastic cap, removed before use, which restricts the movement (and possible damage) of the probe in its packaged state.

Probe operation Operation of the XBT involves five stages (1) removal of the shell of the previous probe from the launcher; (2) unpacking and insertion of the fresh probe into the launcher, completing the electrical circuit by closing the breech; (3) initialization of the onboard electronics to recognize the probe and to prepare for data acquisition; (4) launch, by withdrawing the 'firing pin'; and (5) following data acquisition, completion of the data file closure procedure. Data acquisition begins when the probe completes the earth–loop circuit on reaching the water, and continues until the wire breaks. Acquisition may be ended before this if, for example, the probe is known to have reached the seabed. This process is common to all ship-launched probes.

Expendable Sound Velocity Probe (XSV)

In the XSV the active sensor, instead of the XBT's thermistor, is a small sound speed sensor using the 'sing-around' principle. In this, an ultrasonic transmitter/receiver pair are arranged with a fixed separation in an electrical circuit with strong feedback. The circuit's 'resonance' ('sing-around') frequency is determined mainly by the time acoustic path, and its measurement allows inference of the sound speed.

XSVs are almost solely limited to military use,[5] principally by operational submarines, and both air-launched and submarine-launched variants are available for this reason. For naval operations, they offer an improvement over the XBT in regions such as Arctic, Mediterranean, and coastal waters where salinity variations may cause significant sound speed changes.

The probes have a specified precision of ± 0.25 m s^{-1}, and they are available with two main depth options: the XSV-01 (850 m, 15 knots) and the XSV-02 (2000 m, 8 knots). Both give depth resolution of 0.32 m. A higher resolution, slower-falling XSV-03 (850 m, 5 knots) is available, giving 0.1 m depth

resolution. Depth precision is quoted as ± 4.6 m or $\pm 2\%$ of indicated depth, whichever is greater.

Expendable Conductivity-Temperature-Depth Probe (XCTD)

The XCTD is one of the newer expendable probes, measuring conductivity as well as temperature. This is not a simple extension of capability, since precision conductivity measurement is acknowledged to be particularly difficult, and first-time operation must be assured, even following a lengthy unattended shelf-life. A four-electrode configuration is used in a resistive measurement of conductivity, with a thermistor for the temperature measurement. As with other probes, depth is inferred from the fall rate of the probe. In the XCTD the measuring system converts basic C, T data into frequency-modulated signals for transmission along the standard two-wire connection. Each unit is calibrated in a three-salt-water-bath procedure following manufacture, with the resulting calibration coefficients stored in nonvolatile memory in the probe canister.

Expendable Current Profiler (XCP)

The Sippican XCP is the first expendable technique for ocean current measurement. Its measurement of current velocity utilizes measurements of the weak[6] electrical current generated by the motion of conducting sea water through the Earth's magnetic field, which is directly proportional to the velocity at any given depth. The falling probe interrupts this current and measures the electrical potential thus produced. The probe spins at a prescribed rate, converting the potential seen by separated electrodes into a sinusoidal signal whose orientation is deduced from a co-rotating 'compass' coil. This allows resolution of the measured current into north and east components; the third measurement made is that of temperature, by the thermistor mounted in the usual probe nose position. The data are passed along the wire as three frequency-modulated signals.

The XCP is available only as a stand-alone instrument using a radiofrequency (rf) link, similar to that used for air-launched versions of the XBT and XSV. It may therefore be used from either ship or aircraft, and reception need not be from the deploying platform. An aircraft allows greater reception range, particularly in high sea states. A time interval is allowed between launch and probe release, to allow a deploying ship to move away, reducing its

---

[5] An important exception is their potential use in precision hydrographic surveying, to assess the mean sound speed of the water column.

[6] In the measurement, a water velocity of 1 m s^{-1} corresponds to about 5 μV at mid-latitudes.

electromagnetic disturbance. Operation involves only the removal of a number of a few items of protective packaging before the unit is dropped into the sea. Energizing of the seawater battery, deployment of the rf antenna, operation of the probe, and eventual scuttling all take place without intervention. The probes have a 1500 m capability, and the 16 Hz (rotation rate determined) sampling rate allows 0.3 m vertical resolution, with a specified velocity resolution of ± 10 mm s^{-1} rms and $\pm 3\%$ rms horizontal shear current accuracy. The specified temperature resolution is $\pm 0.2°C$.

Expendable Optical Irradiance Probe (XKT)

The XKT, another relatively recent innovation, is used to measure the vertical profile of light penetrating the upper layer of the ocean, allowing an estimate of the optical diffuse attenuation, K, in addition to temperature. The upward-looking probe has a cosine spatial response and is sensitive to the wavelength band 490 ± 10 nm, operating down to 200 m with about 0.15 m s^{-1} vertical resolution. Irradiance is measured over a dynamic range of 10^5 within 5% log conformity and 10^6 within 10%. An air-dropped version is available.

A second variety of optical properties probe, the XOTD/AXOTD, measuring suspended particle concentrations in addition to temperature, has been reported as 'under development'. This operates using the scattering of light from an included source and is intended for the particle concentration range 5 μg l^{-1} to 3 g l^{-1} in the depth range to 500 m.

Expendable Bottom Penetrometer Probe (XBP)

The XBP is the most recent addition to the armoury of expendables, and it is currently still at the development stage, available by special arrangement for evaluation. The requirement which it fulfils is, once again, driven by military operations, and relates to the need to know certain seabed properties, particularly in shallow water (<200 m). Sonar behavior in shallow water is often determined by the geoacoustic properties of the seabed, and aspects of mine counter-measures, particularly relating to the probability of mine burial, are sensitively dependent on the properties of seafloor sediments.

The sensor carried by the XBP in place of the XBT's thermistor is an accelerometer, whose purpose is to monitor the deceleration of the probe on impact with the seabed. Hard or rocky seabeds involve rapid deceleration, whilst sediments allow a smoother, longer period of retardation.

Normal (Surface Ship) Deployment

The standard deployment of all probes is from a surface vessel, normally from the stern, so that the wire emerges freely behind the vessel as it moves away from the launch point. Two types of launcher are normally available: a robust (heavy) military standard deck launcher, usually mounted permanently on deck, and a much smaller hand-held unit which may be used even from small craft. Care is required with either type in strong wind conditions, as the thin wire may be 'caught' by the wind and may become entangled with parts of the vessel's superstructure. (Through-hull launchers are an option used by some military vessels, allowing operation in adverse weather conditions.)

Since vessel heading is normally set by survey or operational demands, rather than by wind direction, it is often found that having a pair of launchers, mounted on the port and starboard quarters of the vessel, allows greater flexibility in cross-winds. A hand-held launcher is sometimes used to complement a single deck-mounted launcher for this reason. A less conventional way of addressing this problem is to apply mild restraint to the emerging wire – such as loosely guiding the wire through the fingers. This can reduce the effect of the wind on pulling excess wire from the dispenser within the shell, although care must be taken not to impede its normal flow.

Successful use of expendables may also be limited if the vessel is towing equipment, since the XBT wire streaming aft can become fouled by tow cables.

The only processing of the data normally needed involves the removal of values from the top 3–5 m, influenced by transient effects as the probe adjusts to the water temperature, and the removal of values obtained after the probe has reached the seabed, an event frequently denoted by a sharp spike in the data.

Deployment Variations

The XBT and XSV are available with additional deployment options – air-launched and submarine-launched, – developed mainly for military use. As was noted above, the XCP is suited to both ship and airborne deployment.

The Air-Launched XBT and XSV (AXBT, AXSV)

The measurement parts of the AXBT and AXSV are substantially the same as for the standard probes, including the sensor, the wire attaching the falling probe to the surface unit, and the weighted probe body. In this case, however, the only wire coil used is that within the probe, connecting the sensor to a buoyant electronics package which conditions the temperature signal and communicates it via an rf transmitter link to the launch aircraft. AXBTs and AXSVs are packaged in the standard-size canister used for sonobuoys – the 'A-size', 914 mm × 124 mm – and they may be launched at air speeds up to 370 knots and altitudes up to more than 9000 feet. Descent is controlled by a parachute, deployed when the buoy leaves the aircraft, and operation begins after a short delay in which the probe reaches temperature stability and the seawater battery (for the rf transmitter) becomes energized.

Each unit has a user-selectable rf frequency, which allows simultaneous monitoring of a number of probes. Although as many as 99 channels may be available, the number of probes being deployed is also limited by the number of channels available for simultaneous monitoring by the aircraft. Although transmissions cease when the probe has reached its maximum depth, and the scuttling mechanism is initiated, another probe using the same frequency cannot be launched until this has occurred. The receiver used for AXBTs is a standard unit normally fitted only to military (Maritime Patrol) aircraft.

Two types of AXBT are available, designed for maximum depths of 302 m and 760 m. Their spatial resolution is rather better than that of the standard XBT, at about 0.15 m. Specified depth accuracy is 2% of indicated depth, and temperature accuracy is $\pm 0.18°C$. The standard AXSV operators to 760 m, with vertical resolution of about 0.15 m, 2% accuracy, and a specified sound speed accuracy of $\pm 0.25 \, m \, s^{-1}$.

In practice, although AXBTs allow rapid deployment over substantial horizontal scales, it is difficult to simultaneously achieve high spatial sampling, because of the combined effect of the finite reload time and the finite number of available receiving channels. To achieve a probe spacing smaller than about 30 km along a single track it is normally necessary to make at least two passes along the track, the second interleaving dropping probes between the stations covered on the first. Despite the operational difficulties, and the relative inaccessibility of such activities to nonmilitary agencies, AXBTs are the only means of executing large-area surveys (hundreds of kilometers) of dynamic regions for which the synoptic requirement requires a time-spread of <1 day.

The Submarine-Launched XBT and ASV (SSXBT and SSXSV)

These variants of the XBT and XSV satisfy the uniquely military requirement for a submerged submarine to assess the sonar propagation characteristics of its environment. Without it, a submarine would need either to surface for a conventional deployment or to move vertically, making measurements with onboard sensors.

The technique is related to that of the AXBT, in that the temperature profile is measured by a probe falling from a buoyant package which floats at the surface. In this case, however, the package rises to the surface under its own buoyancy following its submerged launch from a signal ejector probe, and remains connected by wire to the submarine. As might be imagined from its requirement to pass probes through the pressure hull of a submarine, this technique involves expensive technology, and is unlikely to be used for purely oceanographic, as opposed to operational, purposes.

An equivalent version of the XCTD is understood to be imminent.

Data Recording and Handling

Current practice for the recording of data normally involves a PC, with a dedicated electronic interface unit which checks the continuity and integrity of the individual probe electronics before launch and then transfers data only after it has detected the probe reaching the water. Data display options exist using either dedicated display programs or standard data handling routines. As in many oceanographic applications, PCs have dramatically simplified the data recording process. In the years between the emergence of the XBT and the eventual prominence of the PC there was a period in which a dedicated inboard electronic unit was necessary for acquisition and data display using a paper trace. Digital recording on magnetic media (long-since obsolete) was also an option, although the practice of digitization from the paper trace persisted operationally into the 1980s. This inboard equipment frequently tended to be temperamental – a feature that is difficult to forgive when linked to the use of expendable probes.

The operational context of XBT data (for meteorological and military purposes) has led to the establishment, by the World Meteorological Organization (WMO) of a standard data exchange format, known as the JJYY format (formerly JJXX the format was officially changed from JJXX to JJYY by the WMO on November 8, 1995), following the alphabetical code used to prefix each 'bathy report'. Details such as date, time, location, probe type, and recorder type are included in these reports. This format involves reduction of the XBT data to a small number of 'inflection points', or 'break points', which are intended to capture the main features of the temperature profile whilst minimizing the data transfer requirement.

Measurement Precision

Assessment of the precision of an expendable device is necessarily limited by the expected loss of the probe following use. However, practical experience normally indicates performance within specified tolerances for the measured variables. Experience is considerably greater for the XBT than for the other probe types, as this is used much more widely for routine purposes.

The growing importance of XBTs in the population of global databases has led to some detailed consideration of their precision in the scientific literature. This has centered principally on the derivation of probe depth using the fall-rate equations provided by the manufacturers. Direct verification for individual probe types is not possible because of the large vertical distances involved, and the inappropriateness of attaching verification sensors. The best available means of checking the depth data of expendables, of any kind, is to carry out parallel measurements with a temperature-measuring instrument, such as a CTD, which has a direct pressure-measuring capability. Distinctive individual temperature features in the water column may then be used to indicate a depth correction for the region of the survey.

A number of surveys, several of them involving dense sampling of XBT and CTD casts, have found significant systematic errors greater than the quoted tolerances, the manufacturer's equation always underestimating the fall rate. A number of alternative fall-rate equations have been proposed, and it appears that a consensus is steadily being reached on the optimum equation.

The calibration issue is particularly significant for assessing trends in climate change, and it is acknowledged to be particularly important to follow an agreed standard procedure for handling the known depth errors in databases. Use of a variety of fall-rate equations would lead to major confusion in interpretation of ensembles of data. It is now accepted that only data using the manufacturer's equation should be used for archived data, and that it should be left to subsequent analysis to make whatever adjustments are felt necessary.

It is generally found that temperature errors are within the bounds indicated by the manufacturer. However, it has been indicated that performance may be improved by calibrating each probe against a secondary standard before use. Although this method can realistically assess only one temperature, giving an error to be applied as an offset to the subsequent launch, it is a reasonably practical means of improving confidence in the data.

Use for Ocean Surveying

Although XBT use for purely oceanographic survey purposes is probably still dominated by naval requirements this is an effective way of enhancing a traditional survey using CTD casts. Without impact on survey time it is possible to increase the spatial sampling rate by a factor of two or more by interpolating XBT launches between CTDs. Another common use is in the support of surveys undertaken with towed undulators or instrument chains, XBTs allowing a degree of downward extrapolation of the detailed upper-layer data that these collect. A third common use of XBTs is as a 'fall-back' option for use when weather conditions are unsuitable for other equipment.

AXBTs, and their deployment, are considerably more expensive, and tend to be used only when rapidity is essential. For example, the use of repeated large-scale AXBT surveys of the highly dynamic Iceland-Faroes frontal zone has been reported. AXBTs have also been used to give near-synoptic sections using aircraft underflights of satellite altimeter tracks, to validate the use of residual height data for ocean monitoring.

The viability of these techniques depends on the proportion of budget available for expendable items, and the expense of the probes must be seen in the context of the high basic cost of trials.

The other main contribution made to ocean science by expendables relates to deductions made from data archives. These are frequently dominated by XBT data and they often allow spatial and temporal coverage of regions that would otherwise have insufficient coverage for reliable deductions to be made.

None of the other probe types described here comes close to the XBT in its contribution to the science. Perhaps the most detailed and intensive use of the other expendables has involved the XCP, whose data (drawn from a number of different ocean regions) have been used to draw conclusions about the horizontal shear environment of the ocean.

See also

Acoustics, Shallow Water. Inherent Optical Properties and Irradiance. Ships. Sonar Systems.

Further Reading

Black PG, Elsberry RL, and Shay LK (1988) Airborne surveys of ocean current and temperature perturbations induced by hurricanes. *Advances in Underwater Technology, Ocean Science and Offshore Engineering* 16: 51–58.

Budéus G and Krause G (1993) On-cruise calibration of XBT probes. *Deep-Sea Research* 40: 1359–1363.

Carnes MR and Mitchell JL (1990) Synthetic temperature profiles derived from Geosat altimetry: Comparison with air-dropped expendable bathythermograph profiles. *Journal of Geophysical Research* 95: 17979–17992.

Smart JH (1984) Spatial variability of major frontal systems in the North Atlantic Norwegian Sea area: 1980–1981. *Journal of Physical Oceanography* 14: 185–192.

Smart JH (1988) Comparison of modelled and observed dependence of shear on stratification in the upper ocean. *Dynamics of Atmospheres and Oceans* 12: 127–142.

Stoll RD and Akal T (1999) XBP – Tool for rapid assessment of seabed sediment properties. *Sea Technology* February: 47–51.

Thadathil P, Ghosh AK, and Muraleedharan PM (1998) An evaluation of XBT depth equations for the Indian Ocean. *Deep-Sea Research* 45: 819–827.

GRABS FOR SHELF BENTHIC SAMPLING

P. F. Kingston, Heriot-Watt University, Edinburgh, UK

Introduction

The sedimentary environment is theoretically one of the easiest to sample quantitatively and one of the most convenient ways to secure such samples is by means of grabs. Grab samplers are used for both faunal samples, when the grab contents are retained in their entirety and then sieved to remove the biota from the sediment, and for chemical/physical samples when a subsample is usually taken from the surface of the sediment obtained. In both cases, the sampling program is reliant on the grab sampler taking consistent and relatively undisturbed sediment samples.

Conventional Grab Samplers

The forerunner of the grab samplers used today is the Petersen grab, designed by C.G.J. Petersen to conduct benthic faunal investigations in Danish fiords in the early part of the twentieth century. It consisted of two quadrant buckets that were held in an open position and lowered to the seabed (**Figure 1**). On the bottom, the relaxing of the tension on the lowering warp released the buckets and subsequent hauling caused them to close before they left the bottom. The instrument is still used today but is seriously limited in its range of usefulness, working efficiently only in very soft mud.

Petersen's grab formed the basis for the design of many that came after. One enduring example is the van Veen grab, a sampler that is in common use today (**Figure 2**). The main improvement over Petersen's design is the provision of long arms attached to the buckets to provide additional leverage to the closing action. The arms also provided a means by which the complex closing mechanism of the Petersen grab could be simplified with the hauling warp being attached to chains on the ends of the arms. The mechanical advantage of the long arms can be improved further by using an endless warp rig; this has the added advantage of helping to prevent the grab being jerked off the bottom if the ship rolls as the grab is closing. The van Veen grab was designed in 1933 and is still widely used in benthic infaunal studies owing to its simple design, robustness, and digging efficiency. The van Veen grab

Figure 1 Petersen grab.

Figure 2 van Veen grab.

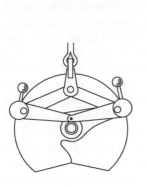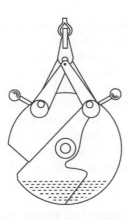

Figure 3 Diagram of a Hunter grab. Reproduced from Hunter B and Simpson AE (1976) A benthic grab designed for easy operation and durability. *Journal of the Marine Biological Association* 56: 951–957.

Figure 4   Smith–McIntyre grab.

Figure 5   Day grab.

typically covers a surface area of $0.1 \, m^2$, although instruments of twice this size are sometimes used.

A more recent design of frameless grab is the Hunter grab (**Figure 3**). This is of a more compact design than the van Veen. The jaws are closed by levers attached to the buckets in a parallelogram configuration, giving the mechanism a good overall mechanical advantage. The closing action requires no chains or pulleys and the instrument can be operated by one person. Its disadvantage is that the bucket design does not encourage good initial penetration of the sediment, which is important in hard-packed sediments.

A disadvantage of the grab samplers discussed so far is that there is little latitude for horizontal movement of the ship while the sample is being secured: the smallest amount of drift and the sampler is likely to be pulled over. The Smith–McIntyre grab was designed to reduce this problem by mounting the grab buckets in a stabilizing frame (**Figure 4**). Initial penetration of the leading edge of the buckets is assisted by the use of powerful springs and the buckets are closed by cables pulling on attached short arms in a similar way to that on the van Veen grab. The driving springs are released by two trigger plates, one on either side of the supporting frame to ensure that the sampler is resting flat on the seabed before the sample is taken. In firm sand the Smith–McIntyre grab penetrates to about the same depth of sediment as the van Veen grab. Its main disadvantage is the need to cock the spring mechanism on deck before deploying the sampler, a process that can be quite hazardous in rough weather.

The Day grab is a simplified form of the Smith–McIntyre instrument in which the trigger and closing mechanism remains the same, but without spring assistance for initial penetration of the buckets (**Figure 5**). The Day grab is widely used, particularly for monitoring work, despite its poor performance in hard-packed sandy sediments.

Most of the grabs thus far discussed have been designed to take samples with a surface area of 0.1 or $0.2 \, m^2$. The Baird grab, however, takes samples of $0.5 \, m^2$ by means of two inclined digging plates that are pulled together by tension on the warp (**Figure 6**). The grab is useful where a relatively large surface area needs to be covered, but has the disadvantage of taking a shallow bite and having the surface of the sample exposed while it is being hauled in.

Warp Activation

All the grabs described above use the warp acting against the weight of the sampler to close the jaws. However, direct contact with the vessel on the surface during the closure of the grab mechanism poses several problems.

Warp Heave

As tension is taken up by the warp to close the jaws, there is a tendency for the grab to be pulled up off the bottom, resulting in a shallower bite than might be expected from the geometry of the sampler. This tendency is related to the total weight of the sampler and the speed of hauling and is exacerbated by firm sediments. For example, the theoretical maximum depth of bite of a 120-kg Day grab is 13 cm (based

Figure 6 Baird grab.

on direct measurements of the sampler); however, in medium sand, the digging performance is reduced to a maximum depth of only 8 cm (**Figure 7(e)**). The influence of warp action on the digging efficiency of a grab sampler can also depend on the way in which the sampler is rigged. This is particularly true of the van Veen grab. **Figure 7(b)** shows the bite profile of the chain-rigged sampler in which the end of each arm is directly connected to the warp by a chain. The vertical sides of the profile represent the initial penetration of the grab while the central rise shows the upward movement of the grab as the jaws close. **Figure 7(c)** shows the bite profile of a van Veen of similar size and weight (30 kg) rigged with an endless warp in which the arms are closed by a loop of wire passing through a block at the end of each arm (as in **Figure 2**). The vertical profile of the initial penetration is again apparent; however, in this case, the overall depth of the sampler in the sediment is maintained as the jaws close. The endless warp rig increases the mechanical advantage of the pull of the warp while decreasing the speed at which the jaws are closed. The result is that the sampler is 'insulated' from surface conditions to a greater extent than when chain-rigged, giving a better digging efficiency.

Grab 'Bounce'

In calm sea conditions it is relatively easy to control the rate of warp heave and obtain at least some consistency in the volume of sediment secured. However, such conditions are seldom experienced in the open sea where it is more usual to encounter wave action. Few ships used in offshore benthic

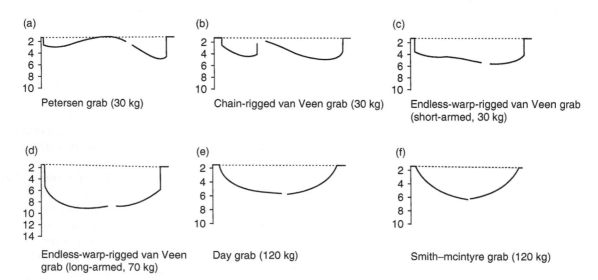

Figure 7 Digging profiles of a range of commonly used benthic grab samplers obtained in a test tank using a fine sand substratum. (a) Peterson grab (30 kg); (b) chain-rigged van Veen grab (30 kg); (c) endless-warp-rigged van Veen grab (short-armed, 30 kg); (d) endless-warp-rigged van Veen grab (long-armed, 70 kg); (e) Day grab (120 kg); (f) Smith–McIntyre grab (120 kg).

studies are fitted with winches with heave compensators so that the effect of ship's roll is to introduce an erratic motion to the warp. This may result in the grab 'bouncing' off the bottom where the ship rises just as bottom contact is made, or in the grab being snatched off the bottom where the ship rises just as hauling commences. In the former instance, it is unlikely that any sediment is secured; in the latter, the amount of material and its integrity as a sample will vary considerably, depending on the exact circumstances of its retrieval.

The intensity of this effect will depend on the severity of the weather conditions. **Figure 8** shows the relationship between wind speed and grab failure rate, which is over 60% of hauls at wind force 8. What is of more concern to the scientist attempting to obtain quantitative samples is the dramatic increase in variability with increase in wind speed with a coefficient of variation between 20 and 30 at force 7. The high cost of ship-time places considerable pressure on operators to work in as severe weather conditions as possible and it is not unusual for sampling to continue in wind force 7 conditions with all its disadvantages.

Drift

For a warp-activated grab sampler to operate efficiently it should be hauled with the warp positioned vertically above. Where there is a strong wind or current, these conditions may be difficult to achieve. The result is that the grab samplers are pulled on to their sides. This is a particular problem with samplers, such as the van Veen grab, that do not have stabilizing frames. Diver observations have shown, however, that at least in shallow water, where the drift effect is at its greatest on the bottom, even the framed heavily weighted Day and Smith–McIntyre grabs can be toppled.

Initial Penetration

It is clear that the weight of the sampler is an important element in determining the volume of the sample secured. Much of the improved digging efficiency of the van Veen grab shown in **Figure 7(d)** can be attributed to the addition of an extra 40 kg of weight which increased the initial penetration of the sampler on contact with the sediment surface.

Initial penetration is one of the most important factors in the sequence of events in grab operation, determining the final volume of sediment secured. **Figure 9** shows the relationship between initial penetration and final sample volume obtained for a van Veen grab. Over 70% of the final volume is determined by the initial penetration. Subsequent digging of the sampler is hampered, as already shown, by the pull of the warp.

For most benthic faunal studies it is important for the sampler to penetrate at least 5 cm into the sediment (for a 0.1 m² surface area sample this gives 5 l of sample). In terms of number of species and individuals, over 90% of benthic macrofauna are found in the top 4–5 cm of sediment. **Figure 10** shows how the number of individuals relates to average sample

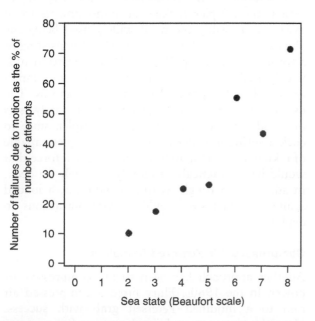

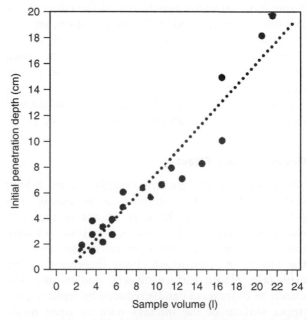

Figure 8 Relationship between wind speed and grab failure rate.

Figure 9 Relationship between initial penetration of a van Veen grab sampler and volume of sediment secured.

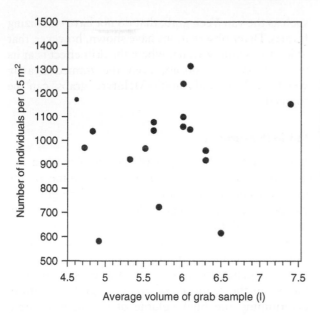

Figure 10 Relationship between number of benthic fauna individuals captured and sample volume for a boreal offshore sand substratum.

Figure 11 Shipek grab.

volume for 18 stations in the southern North Sea (each liter recorded represents 1 cm of penetration). Although there is a considerable variation in the numbers of individuals between stations, there is no significant trend linking increased abundance with increased sample volume penetration. No sample volumes of less than 4.5 l were taken, indicating that at that level of penetration most of the fauna were being captured.

Samplers in which the jaws are held rigidly in a frame have no initial penetration if the edges of the jaw buckets, when held in the open position, are on a level with the base of the frame. The lack of any initial penetration in such instruments has the added disadvantage in benthic fauna work of under-sampling at the edges of the bite profile (see **Figures 7(e)** and **7(f)**), although the addition of weight will usually increase the sample volume obtained.

Pressure Wave Effect

The descent of the grab necessarily creates a bow wave. Under field conditions, it is usually impracticable to lower the grab at a rate that will eliminate a preceding bow wave, even if the sea were flat calm. There have been several investigations of the effects of 'downwash' both theoretical, using artificially placed surface objects, and *in situ*. The effects of downwash can be reduced by replacing the upper surface of the buckets with an open mesh. Although there is still a considerable effect on the surface flock layer (rendering the samples of dubious

value for chemical contamination studies), the effect on the numbers of benthic fauna is generally very small.

Self-Activated Bottom Samplers

There can be little doubt that one of the most important factors responsible for sampler failure or sample variability in heavy seas is the reliance of most presently used instruments on warp-activated closure. The most immediate and obvious answer to this problem is to make the closing action independent of the warp by incorporating a self-powering mechanism.

Spring-Powered Samplers

One solution to the problem is to use a spring to actuate the sampler buckets. Such instruments are in existence, possibly the most widely used being the Shipek grab, a small sampler ($0.04 \, \text{m}^2$) consisting of a spring-loaded scoop (**Figure 11**). This instrument is widely used where small superficial sediment samples are required for physical or chemical analysis. The use of a pretensioned spring unfortunately sets practical limits on the size of the sampler, since to cock a spring in order to operate a sampler capable of taking a $0.1 \, \text{m}^2$ sample would require a force that would be impracticable to apply routinely on deck. In addition, in rough weather conditions, a loaded sampler of this size would be very hazardous to deploy.

Compressed-Air-Powered Samplers

Another approach has been to use compressed air power. In the 1960s, Flury fitted a compressed air ram to a modified Petersen grab with success. However, the restricted depth range of the instrument and the inconvenience of having to recharge the air reservoir for each haul limited its potential for routine offshore work.

Hydraulically Powered Samplers

Hydraulically powered grabs are commonly used for large-scale sediment shifting operations such as seabed dredging. The Bedford Institute of Oceanography, Nova Scotia, successfully scaled down this technology to that of a practical benthic sampler. Their instrument is relatively large, standing 2.5 m high and weighing some 1136 kg. It covers a surface area of $0.5\,m^2$ and samples to a maximum sediment depth of 25 cm. At full penetration, the sediment volume taken is about 100 l. The buckets are driven closed by hydraulic rams powered from the surface. The grab is also fitted with an underwater television camera which allows the operator to visually select the precise sampling area on the seabed, close and open the bucket remotely, and verify that the bucket closed properly prior to recovery. The top of the buckets remain open during descent to minimize the effect of downwash and close on retrieval to reduce washout of the sample on ascent. The current operating depth of the instrument is 500 m. The instrument has been successfully used on several major offshore studies, but does require the use of a substantial vessel for its deployment.

Hydrostatically Powered Samplers

Hydrostatically powered samplers use the potential energy of the difference in hydrostatic pressure at the sea surface and the seabed. The idea of using this power source is not new. In the early part of the twentieth century, a 'hydraulic engine' was in use by marine geologists that harnessed hydrostatic pressure to drive a rock drill. Hydrostatic power has also been used to drive corers largely for geological studies. However, these instruments were principally concerned with deep sediment corers and were not designed to collect macrofauna or material at the sediment–water interface.

A more recent development has been that of a grab built by Heriot-Watt University, Edinburgh. The sampler uses water pressure difference to operate a hydraulic ram that is activated when the grab reaches the seabed. **Figure 12** shows the general layout of the instrument. Water enters the upper chamber of the cylinder when the sampler is on the seabed, forcing down a piston that is connected to a system of levers that close the jaws. The actuating valve is held shut by the weight of the sampler and there is a delay mechanism to prevent premature closure of the jaws resulting from 'bounce'. Back on the ship, the jaws are held shut by an overcenter locking mechanism

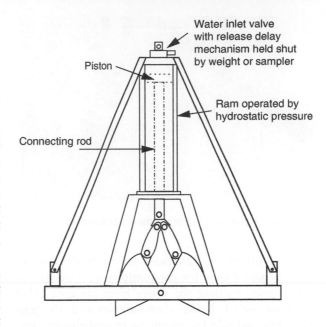

Figure 12 Diagram of a grab sampler using hydrostatic pressure to close the jaws.

and, on release, are drawn open by reversal of the piston motion from air pressure built up on the underside of the piston during its initial power stroke. Since the powering of the grab jaws is independent of the warp, the sampler may be used successfully in a much wider range of surface weather conditions than conventional grabs.

Alternatives to Grab Samplers

Ideally a benthic sediment sample for faunal studies should be straight-sided to the maximum depth of its excavation and should retain the original stratification of the sediment. Grab samplers by the very nature of their action will never achieve this end.

Suction Samplers

One answer to this problem is to employ some sort of corer designed to take samples of sufficient surface area to satisfy the present approaches to benthic studies. The Knudsen sampler is such a device and is theoretically capable of taking the perfect benthic sample. It uses a suction technique to drive a core tube of $0.1\,m^2$ cross-sectional area 30 cm into the sediment. Water is pumped out of the core tube on the seabed by a pump that is powered by unwinding a cable from a drum. The sample is retrieved by pulling the core out sideways using a wishbone arrangement and returning it to the surface bottom-side up (**Figures 13** and **14**). Under ideal conditions, the device will take a straight-sided sample to a depth

Figure 13 Knudsen sampler in descent position.

of 30 cm. However, conditions have to be flat calm in order to allow time for the pump to operate on the seabed and evacuate the water from the core. This limits the use of the Knudsen sampler and it is generally not suitable for sampling in unsheltered conditions offshore. Mounting the sampler in a stabilizing frame can improve its success rate and it is used regularly for inshore monitoring work where it is necessary to capture deep burrowing species.

Spade Box Samplers

Another approach to the problem is to drive an open-ended box into the sediment, using the weight of the sampler, and arrange for a shutter to close off the bottom end. The most widespread design of such an instrument is that of the spade box sampler, first described by Reineck in the 1950s and later subjected to various modifications. The sampler consists of a removable steel box open at both ends and driven into the sediment by its own weight. The lower end of the box is closed by a shutter supported on an arm pivoted in such a way as to cause it to slide through the sediment and across the mouth of the box (**Figure 15**). As with the grab samplers previously described, the shutter is driven by the act of hauling on the warp with all the attendant disadvantages. Nevertheless, box corers are very successful and are used widely for obtaining relatively undisturbed samples of up to $0.25 \, m^2$ surface area (**Figure 16**). One big advantage of the box sampler is that the box can usually be removed with the sample and its overlying water left intact, allowing detailed studies of the sediment surface. Furthermore, it is possible to subsample using small-diameter corers for studies of chemical and physical characteristics. Despite their potential of securing the 'ideal' sediment sample, box corers are rarely used for routine benthic monitoring work. This is largely because of their size (a box corer capable of taking a $0.1 \, m^2$

Figure 14 Knudsen sampler in ascent position.

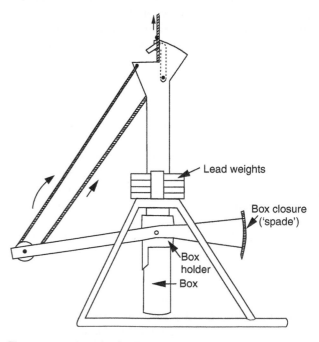

Figure 15 Diagram of a Reineck spade box sampler.

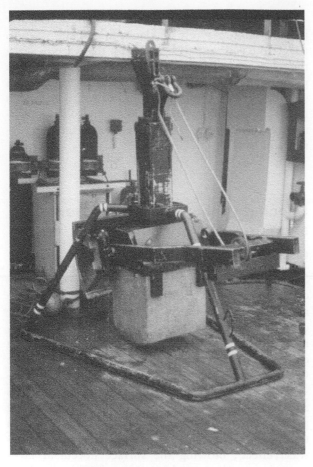

Figure 16 A 0.25 m² spade box sampler.

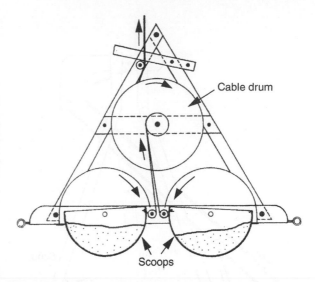

Figure 17 Diagram of a Holme scoop. Reproduced from Holme NA (1953) The biomass of the bottom fauna in the English Channel off Plymouth. *Journal of the Marine Biological Association* 32: 1–49.

macrobenthos sampling. At present, there is no instrument that fulfils the requirements for a quick turnaround precision multiple corer for offshore sampling.

Sampling Difficult Sediments

Most of the samplers so far discussed operate reasonably well in mud or sand substrata. Few operate satisfactorily in gravel or stony mixed ground either because the bottom is too hard for the sampler to penetrate the substratum or because of the increased likelihood of a stone holding the jaws open when they are drawn together. To get around this problem various types of scoops have been devised. The Holme grab has a double scoop action with two buckets rotating in opposite directions to minimize any lateral movement during digging. The scoops are closed by means of a cable and pulley arrangement (**Figure 17**) and simultaneously take two samples of 0.05 m² surface area.

The Hamon grab, which has proved to be very effective in coarse, loose sediments, takes a single rectangular scoop of the substratum covering a surface area of about 0.29 m². The scoop is forced into the sediment by a long lever driven by pulleys that are powered by the pull of the warp (**Figure 18**). Although the samples may not always be as consistent as those from a more conventional grab sampler, the Hamon grab has found widespread use where regular sampling on rough ground is impossible by any other means.

Precision Corers

For chemical monitoring, it is important that the sediment–water interface is maintained intact, for it is the surface flock layer that will contain the most recently deposited material. Unfortunately, such undisturbed samples are rarely obtained using grab samplers or box corers. Precision corers are capable of securing undisturbed surface sediment cores; however, they are unsuitable for routine offshore work because of the time taken to secure a sample on the seabed and dependence on warp-activated closure. Additionally, the cross-sectional area of the core (0.002–0.004 m²) would necessitate the taking of large numbers of replicate samples in order to capture sufficient numbers of benthic macrofauna to be useful. This would be impracticable given the time taken to take a single sample. Large multiple precision corers have been constructed; these are usually too large and difficult to deploy for routine

sample weighs over 750 kg and stands 2 m high) and the difficulty in deployment and recovery in heavy seas.

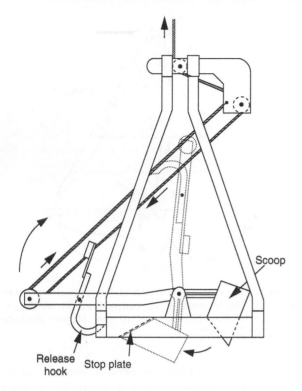

Scoop

Release
hook   Stop plate

Figure 18   Diagram of a Hamon scoop.

Present State of Technology

It is perhaps surprising that given the high state of technology of survey vessels, position fixing, and analytical equipment, the most commonly used samplers are relatively primitive (being designed some 40 or more years ago). Yet the quality of the sample is of fundamental importance to any research or monitoring work. Currently the most popular instruments are grab samplers, probably because of their wide operational weather window and apparent reliability. Samplers such as the Day, van Veen, and Smith–McIntyre grab samplers are still routinely used for sampling sediment for chemical and biological analysis despite their well-documented shortcomings. As discussed earlier, the most important of these are the substantial downwash that precedes the sampler as it descends and the disturbance of the trapped sediment layers by the closing action of the jaws. Both chemical and meiofaunal studies are particularly vulnerable to these.

Although box corers go some way to reducing disturbance of the sediment strata, the all-important surface flocculent layer is invariably washed away. A big disadvantage of the box corer for routine offshore work is that it is sensitive to weather conditions; in addition, the larger instruments do not perform well on sand substrata.

A generic disadvantage of most samplers presently in use is that they rely on slackening of the warp to trigger the action and the heave of the warp to drive the closing mechanism. In calm conditions, this presents no great problem, but with increasing sea state the vertical movement of the warp decreases reliability dramatically until the variability and finally failure rate of hauls make further sampling effort fruitless.

Specific Problems, Requirements, and Future Developments

Chemical Studies

Studies involving sediment chemistry require precision sampling if undisturbed samples at the sediment/water interface are to be obtained. This can be critical, particularly when the results of recent sedimentation are of interest. The impracticality of using existing hydraulically damped corers such as the Craib corer for offshore work has led to the widespread use of less 'weather sensitive' devices such as spade box corers and grabs for routine monitoring purposes. However, studies carried out have shown that these samplers produce a considerable 'downwash effect', blowing the surface flocculent layer away before the sample is secured. This can have serious consequences if any meaningful estimation of the surface chemistry of the sediment is desired. Repetitive and accurate sampling is also a prerequisite for determining spatial and temporal change in sediment chemistry.

Meiofauna Studies

Meiofauna has increasingly been shown to have potential as an important tool in benthic monitoring. One of the major factors limiting its wider adoption is the lack of a suitable sampler. Although instruments such as the Craib corer and its multicorer derivatives are capable of sampling the critically important superficial sediment layer, these designs provide a poor level of success on harder sediments and in anything but near-perfect weather conditions. They also have a slow turnaround time. Box corers are widely used as an alternative; however, they are known to be unreliable in their sampling of meiofauna. As with the macrobenthos, meiobenthic patchiness results in low levels of precision of abundance estimates unless large numbers of samples are taken.

Macrobenthic Studies

The measurement and prediction of spatial and temporal variation in natural populations are of great importance to population biologists, both for fundamental research into population dynamics and productivity and in the characterization of benthic communities for determining change induced by environmental impact. Though always an important consideration, cost-effectiveness of sampling and sample processing is not so crucial in fundamental research, since time and funding may be tailored to fit objectives. This is rarely the case in routine monitoring work where often resolution and timescales have to fit the resources available.

Benthic fauna are contagiously distributed and to sample such communities with a precision that will enable distinction between temporal variation and incipient change resulting from pollution effects, it is generally accepted that five replicate $0.1\,m^2$ hauls from each station are necessary (giving a precision at which the standard error is no more than 20% of the mean). This frequency of sampling requires approximately 10–15 man-days of sediment faunal analysis per sample station. While this may be acceptable in community structure studies in which time and manpower (and thus cost) are not a primary consideration, this high cost of analyzing samples is of importance in routine monitoring studies and has led monitoring agencies and offshore operators to reduce sampling frequency on cost grounds to as few as two replicates per station. This reduces the precision with which faunal abundance can be estimated to a level at which only gross change can be demonstrated. However, sampling to an acceptable precision may be achieved from an area equivalent to $0.1\,m^2$ if a smaller sampling unit is used. For example, 50 5-cm core samples (with a similar total surface area) have been shown to give a similar precision to that of 5 to 12 $0.1\text{-}m^2$ grab samples. Thus a similar degree of precision may be obtained for around one-fifth to one-twelfth the analytical costs using a conventional approach.

The problem is to be able to secure the 50 core samples per site that would be needed for the macrofaunal monitoring in a timescale that would be realistic offshore. At present, there is no instrument capable of supporting such a sampling demand and operating in the range of sediment types that widescale monitoring studies demand. Clearly a single core sampler would be impracticable, and one must look to the future development of a multiple corer that is capable of flexibility in its operation, which allows a quick on-deck turnaround between hauls.

Further Reading

Ankar S (1977) Digging profile and penetration of the van Veen grab in different sediment types. *Contributions from the Askö Laboratory, University of Stockholm, Sweden* 16: 1–12.

Beukema JJ (1974) The efficiency of the van Veen grab compared with the Reineck box sampler. *Journal du Conseil Permanent International pour l'Exploration de la Mer* 35: 319–327.

Eleftheriou A and McIntyre AD (eds.) (2005) *Methods for the Study of the Marine Benthos.* Oxford, UK: Blackwell.

Flury JA (1967) Modified Petersen grab. *Journal of the Fisheries Research Board of Canada* 20: 1549–1550.

Holme NA (1953) The biomass of the bottom fauna in the English Channel off Plymouth. *Journal of the Marine Biological Association* 32: 1–49.

Hunter B and Simpson AE (1976) A benthic grab designed for easy operation and durability. *Journal of the Marine Biological Association* 56: 951–957.

Riddle MJ (1988) Bite profiles of some benthic grab samplers. *Estuarine, Coastal and Shelf Science* 29(3): 285–292.

Thorsen G (1957) Sampling the benthos. In: Hedgepeth JW (ed.) *Treatise on Marine Ecology and Paleoecology, Vol. 1: Ecology.* Washington, DC: The Geological Society of America.

TOWED VEHICLES

I. Helmond, CSIRO Marine Research, Tasmania, Australia

Introduction

As a platform for marine instruments the towed vehicle (often termed a 'towfish' or 'towed body') combines the advantages of a ship-mounted instrument that gathers surface data while under way, and an instrument lowered from a stationary ship to gather data at depth. This article discusses the types of vehicles, the significance of the hydrodynamic drag of the tow cable and methods to reduce it. It also outlines the basic hydrodynamics of towed vehicles and presents the results of a model of the vehicle/cable system, indicating depths and cable tensions for a typical system.

Types of Towed Vehicles

A towed vehicle system has three main components: the vehicle, the tow cable and a winch. The vehicles fall into two broad categories: those with active depth control and those without.

Vehicles With Active Depth Control

Depth-controlled towed vehicles (or 'undulators') can move vertically in the water column while being towed horizontally by the ship. The main advantage they have over the lowered instrument is that they can quickly and conveniently measure vertical profiles of ocean properties with high horizontal spatial resolution. The main disadvantage is that it is difficult to reach depths greater than 1000 m while being towed at useful speeds; most available systems are limited to 500 m at best. The reason for the depth limitation is that the hydrodynamic drag of the tow cable must be overcome by a downward force on the vehicle, produced either by weight or a downward hydrodynamic force from hydrofoils or wings. However the cable's strength limits the allowable size of these forces.

The following are the common types of depth-controlled vehicles.

- A vehicle with controllable wings towed with an electromechanical cable that connects it to a controlling computer on board the vessel. The electromechanical cable allows the data to be transmitted to the ship and displayed and processed in real time. This can be an advantage when following a feature such as an ocean front; the ship's course can be adjusted to optimize the data collection. It also has the advantage of enabling fast, real-time response, an important consideration for bottom avoidance when operating in shallow water. As with most towed vehicles, cable drag dominates performance, so cable fairing is commonly used to reduce drag. Examples of this type of towed vehicle are the Batfish (Guildline Instruments, Canada), SeaSoar (Chelsea Instruments, UK) and the Scanfish (MacArtney A/S, Denmark).

- A vehicle with controllable wings that is totally self-contained. It is preprogrammed for maximum and minimum depths and records the data internally. As such a vehicle can be towed on a simple wire rope, it is convenient to use on ships not equipped for research. The lack of real-time control and data can be a disadvantage. An example of this type is the Aquashuttle (Chelsea Instruments, UK).

- A passive vehicle, often with fixed wings, where changes in depth are made by winching the tow cable in and out. This necessitates a high-speed, computer-controlled winch to produce the depth variation, but the vehicle can be simple. A recent development of this type is the Moving Vessel Profiler (Brooke Ocean Technology Ltd, Canada). This system has a winch that spools out the cable fast enough to allow the vehicle to free fall while the ship is under way, and then retrieves it after it has reached its maximum depth, which may be as deep as 800 m. Because the profiler free-falls on a slack cable, the usual cable drag constraints are not as relevant. This allows good depths to be achieved without the complication of cable fairing.

Vehicles Without Active Depth Control

The vehicle without active depth control produces the depressing force by means of its weight, fixed wings or both. It maintains a constant depth for a given tow-cable length and tow speed. The vehicle can be towed with either an electromechanical cable to provide real-time data to the ship (as used for underwater survey instruments such as the side-scan sonar) or with a wire rope (as often used for plankton recorders).

Tow Cable Drag

Tow cables are either wire rope or double-armored electromechanical cables. In the latter, two layers of armor provide mechanical strength, and because the layers are wound in opposite directions they are approximately torque balanced, i.e., the cable has little tendency to rotate when loaded. The electrical conductors handle data and power and optical fibers are used for high data rates. Except for systems with short cables used for shallow operations the tow cable is usually the dominant part of a towed vehicle system. Even a modest cable of 8 mm diameter and 500 m length has a mass of about 150 kg and a longitudinal cross-sectional area of 4 m^2. This large cross-sectional area means that the cable drag dominates the performance of the system.

Drag Caused by Flow Normal to the Cable

The normal drag on a moving body in a fluid is given by

$$D_N = C_{DN}\tfrac{1}{2}\rho u_N^2 A$$

For a cable, the cross-sectional area A is the product of the cable's diameter (d) and length (l) (see list of symbols at end of article). This drag is the sum of drag due to the shape of the body (the form or pressure drag) and drag due to surface friction (additionally a shape that produces lift also generates induced drag). The value of the drag coefficient C_{DN} depends on the Reynolds number R_e (the ratio of inertial to viscous forces), which is defined as $R_e = ud/v$. For a long smooth cylinder with normal flow at Reynolds numbers less than about 3×10^5 the flow is laminar and $C_{DN} \approx 1.2$. At higher Reynolds numbers, the flow becomes turbulent and C_{DN} drops to about 0.35. This change in drag coefficient is due to the large area of separated flow in the wake of the cylinder in laminar flow decreasing when the flow becomes turbulent. For most cables used for towed vehicles, the value of the Reynolds number is less than 10^5. To exceed this value a 10 mm-diameter cable moving through water requires a normal velocity greater than 10 m s^{-1}. For cables with a rough surface, such as the usual stranded cable, $C_{DN} \approx 1.5$ when $10^3 < R_e < 10^5$ (**Figure 1**).

There is a mechanism that increases the drag coefficient of a cable above the value of a rigid cylinder. This is vortex-induced oscillation, commonly referred to as 'strumming'. Vortices shed from the region of flow separation alter the local pressure distribution and the cable experiences a time-varying force at the frequency of the vortex shedding. This frequency, f, is a function of the normal flow velocity u_N and the

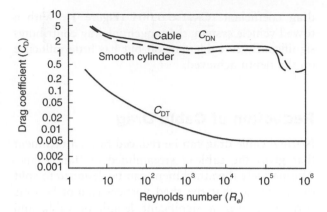

Figure 1 Normal and tangential drag coefficients for a typical cable.

cable diameter d. The Strouhal number S_n is defined as $S_n = fd/u_N$, and $S_n \approx 0.2$ over the range of Reynolds numbers $10^2 < R_e < 10^5$. If this frequency is close to a natural frequency of the cable, an amplified oscillation occurs. A tow cable has many natural frequencies and there are many modes excited by vortex shedding; the result is a continuous oscillation of the cable. These vortex-induced oscillations, which can have an amplitude of up to three cable diameters, increase cable fatigue and drag. The increase in drag can be as much as 200%, resulting in drag coefficients as high as 3. Values around 2 are common.

The values of the drag coefficient for towed cables cited in the literature differ widely because each case has its own set of conditions: cable curvature, tension, Reynolds number and surface roughness vary from case to case. A good starting point for estimating the normal cable drag is $C_{DN} \approx 1.5$ for cables that are not strumming and $C_{DN} \approx 2$ for strumming cables.

For the towed vehicle to be able to dive, the normal cable drag force must be overcome by downward or depressing forces: the cable weight, the towed vehicle weight and the hydrodynamic forces produced by the vehicle. The normal drag does not contribute directly to cable tension but by influencing the angle of the cable in the water it determines the component of cable weight that contributes to tension.

Drag Caused by Flow Tangential to the Cable

The tangential drag on a long cylinder is due to surface friction. It is given by:

$$D_T = C_{DT}\tfrac{1}{2}\rho u_T^2 \pi A$$

For tow cables with Reynolds numbers greater than about 10^3, a typical value for the tangential drag coefficient $(C_{DT}) \approx 0.005$ (**Figure 1**). With a towed vehicle system, the tangential drag contributes significantly to cable tension but has little influence on the depth achieved.

Reduction of Cable Drag

Normal cable drag can be reduced by an attachment that gives the cable a streamlined or 'fair' shape. Alternatively the attachment can be designed to split the wake, so that the shed vortices cannot become correlated over a significant length of cable and strumming is prevented. These attachments are usually called 'fairing'.

Rigid Airfoil-shaped Fairing

'Wrap-round' fairing is the most effective method of reducing normal drag. An example is shown in **Figure 2**. A good airfoil shape can have a normal drag coefficient of 0.05, but the practicalities of having a rigid moulded plastic shape that can wrap around a cable, be passed over sheaves and spooled onto a winch often results in a drag coefficient of about 0.2. The greater drag is primarily due to the circular nose of the fairing (to accommodate the cable) and gaps between fairing sections. Because of the large surface area of the fairing, the tangential drag coefficient (based on cable surface area) is about 0.05. This means that, although the normal drag coefficient of a faired cable is only a tenth of a bare cable, the tangential drag coefficient is about ten times greater. A consequence of this is large cable tensions. For a typical system with 500 m of faired cable, at least 50% of the cable tension can be due to the fairing.

Another consequence of the high tangential drag is that this loading must be transferred from the fairing sections to the cable. Every 2–3 m, a 'stop ring' is swaged or clamped to the cable to take the load. If this force accumulates over too great a length the fairing sections will not rotate freely, and in the extreme they can break under the high compressive load.

Another form of rigid fairing is the 'clip-on' fairing (see **Figure 2**). Unlike the 'wrap-round' type this does not totally enclose the cable, but is essentially an after-body attached to the cable with clips. Because of the gap between the cable and the fairing the drag coefficient is higher. Typical values are $C_{DN} \approx 0.4$.

A problem that can occur with rigid fairing is the phenomenon of 'tow-off'. If the fairing sections are not free to align accurately with the flow, they can generate a considerable lift force, which can cause the cable to tow off to the side and decrease the depth it achieves.

Although the rigid airfoil fairing is the most effective method of decreasing drag, it comes at a high cost. Not only is it expensive but it also requires special winches and handling gear. However, if a system is set up well it gives reliable performance.

Flexible Ribbon and 'Hair' Fairing

Ribbon fairing is made of a flexible material in the form of trailing ribbons (**Figure 2**). Fibers or 'hairs' attached to the cable are also effective. These fairing devices do not usually produce a fair shape, but achieve their effect by splitting the wake. Their main effect is, therefore, to reduce strumming and reduce the normal drag coefficient to the bare cable value of

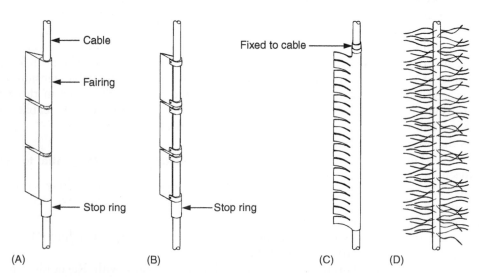

Figure 2 (A) 'Wrap-round' fairing; (B) 'clip-on' fairing; (C) 'ribbon' fairing; (D) 'hair' fairing.

about 1.5. In some cases ribbon fairing can streamline the cable, reducing the normal drag coefficient to about 0.7 (due to the reduction of form drag). Hair fairing reduces strumming but can increase the normal drag coefficient so that $C_{DN} \approx 2$.

These devices increase the surface area over a bare cable so the tangential drag is increased, resulting in greater cable tensions. Flexible fairing does not need special handling gear and can be wound onto a regular winch. However, the fairing deteriorates rather quickly requiring regular repair and replacement.

Basic Aerodynamics of the Towed Vehicle

Vehicles that Generate Lift

Most towed vehicles use wings (hydrofoils) to generate the force required to pull the tow cable down. **Figure 3** shows an example of a winged vehicle. As shown in **Figure 4**, a wing moving through a fluid experiences a force perpendicular to the direction of flow (the lift), a force directly opposing the motion (the drag), and a moment tending to rotate the wing (the pitching moment). The pitching moment is usually referenced about a point termed the aerodynamic center, chosen so that the moment coefficient is constant with angle of attack.

The lift force is given by:

$$L = C_L \tfrac{1}{2}\rho u^2 S$$

The drag force is given by:

$$D = C_D \tfrac{1}{2}\rho u^2 S$$

The pitching moment is given by:

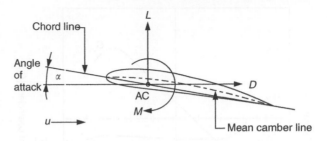

Figure 4 Forces and moment on an airfoil.

$$M = C_M \tfrac{1}{2}\rho u^2 Sc$$

The lift coefficient C_L is proportional to the angle of attack. The theoretical relationship for a thin, symmetrical airfoil gives the slope of the curve of lift coefficient against angle of attack

$$dC_L/d\alpha = a_0 = 2\pi/\text{radian} = 0.11/\text{degree}$$

The theoretical aerodynamic center is at the quarter chord point $(c/4)$ and $C_M = 0$. If the angle of attack is increased beyond a certain value the flow separates from the low-pressure side of the wing rapidly causing the lift to decrease and the drag to increase. This is termed 'stall' (**Figure 5**).

An asymmetrical or cambered airfoil, where the camber line (the line drawn halfway between the upper and lower surfaces) deviates from the chord line (**Figure 4**), has the same theoretical slope of the lift curve, 2π, but has a nonzero value of the lift coefficient when $\alpha = 0$. The aerodynamic center is also at the quarter chord point but $C_M \neq 0$ (**Figure 6**).

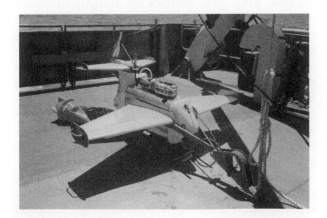

Figure 3 CSIRO (Australia) modified SeaSoar with a Sea-Bird Electronics Inc. conductivity, temperature and depth (CTD) instrument.

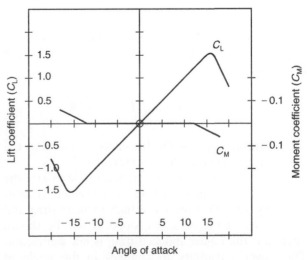

Figure 5 Typical section characteristics for symmetrical airfoil type NACA0012.

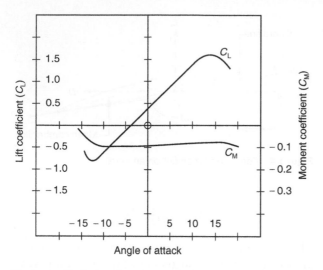

Figure 6 Typical section characteristics for asymmetrical airfoil type NACA4412.

These properties describe airfoil sections that are two-dimensional. In a real wing, the span is finite and there is spanwise flow. The effect of this is a 'leakage' around the wing tips from the high-pressure side to the low-pressure side. This generates wing-tip vortices, which in turn produce a downward flow around the wing – the downwash. The angle of this local flow relative to the wing subtracts from the angle of attack so that the wing actually experiences a smaller effective angle of attack. Since the lift vector is normal to the local relative flow it becomes inclined behind the vertical and so has a rearward component – the induced drag. This drag can be the dominant drag on a towed vehicle. The induced drag coefficient is given by:

$$C_{Di} = C_L^2/(\pi AR)$$

The reduction in the effective angle of attack of a wing with finite span also reduces the slope of the lift curve, a.

$$dC_L/d\alpha = a = a_0/(1 + a_0/\pi AR)$$

The reduced slope of the lift curve means that the wing has a higher angle of attack at stall.

To achieve the necessary mechanical strength, the aspect ratio of wings used on towed vehicles is usually very low. This results in high induced drag and low values of the lift curve slope. It will be shown later that the higher induced drag is not significant. The lower sensitivity to changes in the angle of attack (and the higher angle at stall) can be an advantage: it makes the vehicle more tolerant of flight

perturbations such as those experienced when towing in rough seas.

The delta wing (a wing with a triangular planform) is a popular form for vehicles without active depth control. Flow over a delta wing is dominated by large leading-edge vortices and cross-flow that enable the wing to operate at large angles of attack without stalling. A delta wing has a typical lift curve slope of about 0.05/degree (about half that of a conventional wing), but can operate with an angle of attack up to 30° before stalling. Delta wings make robust depressors and are often given large dihedral angles to increase roll stability. The dihedral is the angle of inclination of the wings in relation to the lateral axis.

To control the depth of a towed vehicle, the magnitude and direction of the wing's lift force are usually varied by:

- the use of control surfaces (flaps) on the trailing edge;
- an independent control surface, usually at the tail;
- rotating the entire wing about its aerodynamic center to vary its angle of attack.

The first two methods cause the whole vehicle to adopt an angle of attack and so the body of the vehicle also generates lift. Some towed vehicles, such as the Aquashuttle (Chelsea Instruments, UK), operate without wings, generating all the lift from the body. Others, such as the Scanfish (MacArtney A/S, Denmark) and the V-Fin (Endeco Inc, USA), are effectively flying wings. The third method, used by the Batfish (Guildline Instruments, Canada) and SeaSoar (Chelsea Instruments, UK), maintains the body aligned with the flow, which is an advantage for some types of sensors that need to be aligned to the flow. Both these vehicles use a highly cambered wing section (NACA6412) that has a large moment coefficient ($C_M \approx -0.13$). Thus a large torque is needed to rotate the wing. If a small operating torque is required the wings are typically controlled by an electric servomotor. If large forces are needed a hydraulic system is used. A symmetrical wing section pivoted at the quarter chord point requires only a small torque, but the control system needs to be robust enough to survive rough handling on the ship's deck.

To gain the maximum benefit of any lift force, the vehicle must fly so that the force is directed near to vertical, that is it should fly without a significant roll angle. A towed vehicle often needs to be able to direct its lift force both down and up, to maximize its depth range. A consequence of this is that dihedral, a

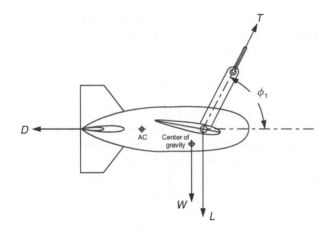

Figure 7 Basic forces on a towed vehicle.

servomotor systems. The other basic requirement is for pitch and yaw stability. This is achieved by ensuring that the tow point is ahead of the aerodynamic center (AC) of the vehicle (**Figure 7**). The position of the vehicle's aerodynamic center is controlled by providing a suitably sized tailfin.

Vehicles That Do Not Generate Lift

These passive towed vehicles use their weight to produce the required downward force. The drag of the cable is overcome by the combined weight of cable and vehicle. Depth is controlled by varying the cable length. The depth is also very dependent on the tow speed. This is because the cable drag is proportional to the square of the tow speed and the depressing force is fixed by the cable and vehicle weight. This contrasts with the vehicle that generates lift; its depth is less speed-dependent because both lift and drag vary with speed in the same manner. A heavy passive vehicle can have good pitch and roll stability if the position of the tow point, center of mass, center of gravity and the aerodynamic center are carefully chosen. The lack of flow separation over lifting surfaces also makes them acoustically quiet. This stability and quietness make them useful vehicles for underwater acoustic work (**Figure 8**).

common method of providing roll stability, cannot be used. What would be stable for lift in one direction would be unstable in the other. By having the tow cable attachment point above the center of gravity of the vehicle, the vehicle's weight contributes to roll stability (**Figure 7**). But to stabilize the large lift forces needed for good depth performance, additional aerodynamic control by means of ailerons or similar devices is needed. These can be simple systems driven directly with gravity by using a heavy pendulum device or more sophisticated electric

Figure 8 The CSIRO (Australia) multifrequency towed vehicle used for fish stock measurements. Note the ribbon-faired cable.

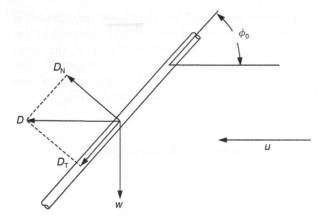

Figure 9 Forces on the tow cable.

Performance of the Vehicle/Cable System

When a cable is towed through water, it assumes an equilibrium angle where the drag force (D) is balanced by the cable weight (w). If the cable properties are uniform, this angle is constant along the cable length

By referring to **Figure 9**

$$\tan \phi_0 = w/D$$

Or expressed in terms of the normal drag coefficient

$$\cos \phi_0 = -1 \pm (1 + 4B^2)^{1/2}/2B$$

where $B = \rho u^2 dC_{DN}/(2w)$

The equilibrium depth is ($l \sin \phi_0$). When a vehicle is attached to the end of the cable it perturbs this equilibrium depth by the extent that its weight and lift force either drag the cable deeper when diving, or lift the cable when climbing. This defines the depth range of the vehicle/cable system (**Figure 10**). As

shown in **Figure 7** the angle of the cable at the vehicle, ϕ_1 is determined by the weight (W), lift (L) and drag (D) of the vehicle.

$$\tan \phi_1 = (W + L)/D$$

The cable profile starts at an angle of ϕ_1 at the vehicle and gradually approaches ϕ_0 up the cable.

The vehicle drag is the sum of the form drag, the friction drag and the induced drag. With vehicles that generate lift the induced drag is the main component. Even with a poorly streamlined vehicle it dominates, providing perhaps 75% of the drag. The rather poor performance of the typical low aspect ratio wing used on towed vehicles gives the vehicle a lift to drag ratio of about 3. This makes the cable angle $\phi_1 \approx 70°$. Further improvement in the lift to drag ratio does not gain much in cable angle or depth. **Table 1** compares the equilibrium depths for bare and faired cables; **Table 2** compares the performance of bare and faired cable when towing a vehicle with controllable wings. These data were produced by a computer model of the vehicle/cable system.

Effects of Wave-Induced Ship Motion on the Towed Vehicle

A problem with towing in rough seas is that the wave-induced motion of the towing vessel is propagated down the tow cable to the vehicle. Motion normal to the cable is rapidly damped, but there is surprisingly little attenuation of the tangential motion of the cable. The amplitude of the perturbation of the vehicle is approximately the same as the cable at the ship. This causes changes in the vehicle's pitch angle and depth which can be very significant for towed acoustic systems and vehicles such as camera units operating very close to the seafloor.

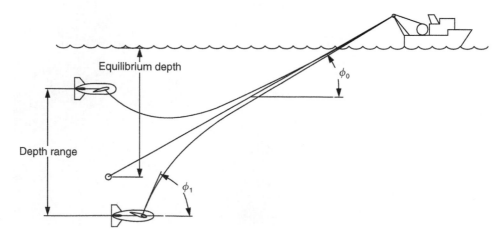

Figure 10 Towed cable profile.

Table 1 Comparison of equilibrium depths for 500 m of 8.2 mm diameter cable with a weight of 2.5 N m^{-1} towed at 4 m s^{-1} for faired and bare cable

	Equilibrium depth (m)	Tension (kN)	ϕ_0 (degrees)
Faired $C_{DN} = 0.2$	142	3.6	17
Bare $C_{DN} = 2.0$	61	0.7	7

The following methods are used to minimize the effects of ship motion.

- The vehicle can be tuned to minimize the pitching effect by carefully adjusting the position of the vehicle's aerodynamic center and center of mass in relation to the tow point.
- A constant-tension winch reduces the cable displacement at the ship by spooling cable in and out as the ship surges.
- A device called an accumulator produces an effect similar to the constant-tension winch. The cable runs over a pair of sheaves that are mounted on a sprung or pneumatic arrangement that allows them to take up and pay out cable as needed.
- In the 'two-body system' the instrumented vehicle is passive and near-neutrally buoyant. It is towed behind the depressor or depth-controlled vehicle on a cable that is approximately horizontal. This geometry decouples ship's motion from the second vehicle.
- A system that has a cable angle close to horizontal at the ship is insensitive to ship pitch and heave as these displacements are almost normal to the cable. A system that uses a long cable, a cable without fairing or a high tow speed has this characteristic.

Flight Control

Tow speeds vary from as low as 1 ms^{-1} for seafloor survey instruments to 10 ms^{-1} for high-speed systems. The fast systems are limited to shallow depths.

Speeds of 3–5 ms^{-1} are common for oceanographic surveys and depths of up to 1000 m can be achieved. Depth-controlled vehicles operate with vertical velocities up to about 1 ms^{-1}.

Depth-controlled vehicles are commonly operated in an undulating mode with maximum and minimum depths set to specific values to give a triangular flight path. The depth is measured by the water pressure at the vehicle and the wings or control surfaces are adjusted to make the vehicle follow the defined path by a servo system. The servo-loop parameters are usually controlled by the shipboard computer; however the actual servo-loop system may reside in the towed vehicle or combine ship- and vehicle-based components. The control algorithm needs to be carefully tuned to achieve smooth flight.

Sensors

Some of the earliest towed vehicles were used to collect plankton (in fact a trawl net is a form of towed vehicle). These plankton collectors are often separate nets and depressors but can also be single units. An early system, the Hardy Continuous Plankton Recorder, dates from the 1930s. Several commercially available vehicles are a development of this type, for example the Aquashuttle (Chelsea Instruments, UK) and the U-Tow (Valeport Ltd, UK).

Depth-controlled vehicles are commonly equipped with a conductivity, temperature and depth (CTD) instrument. They are also suitable platforms for

Table 2 Comparison of depths, cable tensions and cable angles (ϕ_2 cable angle at ship, ϕ_1 cable angle at vehicle) for the same cable as **Table 1** towing a typical vehicle with the following characteristics: weight in water, 1500 N; cross-sectional area, 0.2 m^2; wing area, 0.5 m^2; wing aspect ratio, 1; tow speed, 4 m s^{-1}. Positive lift coefficients indicate lift force upwards

	Wing C_L	Depth (m)	Tension (kN)	ϕ_2 (degrees)	ϕ_1 (degrees)
Faired $C_{DN} = 0.2$	−1.0	360	11.2	36	72
	+0.5	0	4.3	7	−36
Bare $C_{DN} = 2.0$	−1.0	140	6.5	8	72
	+0.5	31	1.6	7	−36
	+1.0	0	3.2	6	−55

many other types of sensors, such as fluorometers, radiometers, nutrient analysers and transmissometers. In the case of a CTD it is recommended that the sensors be duplicated. When a CTD is lowered from a stationary ship in the usual manner the calibration of the conductivity sensor is checked by collecting water samples for laboratory analysis. This option is not usually available on a towed instrument so a check of sensor stability can be obtained by using dual sensors. The conductivity cells can also be blocked by marine organisms especially when towing near the surface. Dual cells allow recognition of this problem. **Figure 11** shows the data from a CTD section across an ocean front demonstrating the high spatial resolution realized with a towed system.

Passive towed vehicles are often used for acoustic survey work. Examples are side-scan sonars, towed multibeam systems for seafloor mapping, and vehicles for estimating fish stocks. The towed vehicle

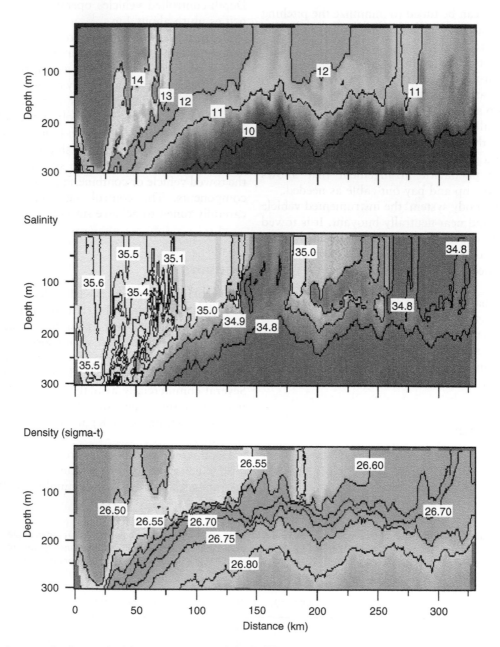

Figure 11 An example of a conductivity, temperature and depth (CTD) section across the Sub-Tropical Front south of Australia using a SeaSoar towed vehicle equipped with a Seabird CTD instrument. (From Tomczak M and Pender L (1999) *Density compensation in the Sub-Tropical Front in the Indian Ocean South of Australia.*http://www.es.flinders.edu.au/ ~ mattom/STF/ fr1098.html.

Figure 12 A faired cable winch and SeaSoar towed vehicle. This winch holds 5000 m of cable, 400 m faired and 4600 m of bare cable. A combination of faired and bare cable can be a cost-effective method of achieving greater depths.

offers advantages over hull-mounted transducers by deploying the acoustic transducer away from the high noise environment and bubble layer near the ship and closer to the object of interest. A well-designed system can also be a more stable platform for the acoustic transducers than a ship in rough seas.

The Winch

Towed-vehicle systems using electromechanical cables usually require a special, purpose-built winch with accurate spooling gear and slip-rings to make the electrical connection to the rotating drum. Cable lengths can vary from 100 m to 5000 m. If a cable with rigid fairing is used special blocks and fairing guiding devices are needed to handle the cable. The faired cable has a large bending radius and can only be wound onto the winch drum in a single layer. As illustrated in **Figure 12**, this type of winch is quite large. If the towed vehicle system uses wire rope for the tow cable, then the winch can be a standard type.

Nomenclature

A Area
AC   Aerodynamic center
AR   Wing aspect ratio $= b^2/S$
C_{Di} Induced drag coefficient
C_{DN} Normal drag coefficient
C_{DT} Tangential drag coefficient
C_L Lift coefficient
C_M Moment coefficient

D Drag force
D_N Normal drag force
D_T Tangential drag force
L Lift force
M Pitching moment
R_e Reynolds number $= ud/v$
S Wing planform area
S_n Strouhal number $= fd/u$
T Cable tension
W Tow vehicle weight
a Slope of the lift curve for a wing $= dC_L/d\alpha$
a_0 Slope of the lift curve for an airfoil section $= dC_L/d\alpha$
b Wing span
c Chord length
cg Center of gravity
d Cable diameter
f Frequency
l Cable length
u Velocity
u_N Normal velocity
u_T Tangential velocity
w Cable weight per unit length
α Angle of attack
ϕ Angle of the cable to the horizontal
ϕ_0 Cable equilibrium angle
ϕ_1 Cable angle at the towed vehicle
ϕ_2 Cable angle at the ship
μ Dynamic viscosity ($\approx 1 \times 10^{-3}$ kg m^{-1} s for water)
v Kinematic viscosity $= \mu/\rho$ ($\approx 1 \times 10^{-6}$ m^{2}s^{-1} for water)
ρ Density (≈ 1000 kg m^{-3} for water)

See also

Acoustic Scattering by Marine Organisms. Ships. Sonar Systems.

Further Reading

Abbott IH and von Doenhoff AE (1959) *Theory of Wing Sections.* New York: Dover Publications.

Anderson JD (1991) *Fundamentals of Aerodynamics,* 2nd edn. New York: McGraw-Hill.

Wingham PJ (1983) Comparative steady state deep towing performance of bare and faired cable systems. *Ocean Engineering* 10(1): 1–32.

ZOOPLANKTON SAMPLING WITH NETS AND TRAWLS

P. H. Wiebe, Woods Hole Oceanographic
Institution, Woods Hole, MA, USA
M. C. Benfield, Louisiana State University,
Baton Rouge, LA, USA

Introduction

In the late 1800s and early 1900s, quantitative ocean plankton sampling began with non-opening/closing nets, opening/closing nets (mostly messenger-based), high-speed samplers, and planktobenthos net systems. Technology gains inelectrical/electronic systems enabled investigators to advance beyondsimple vertically or obliquely towed nets to multiple cod-end systems and multiple net systems in the 1950s and 1960s. Recent technological innovation has enabled net systems to be complemented or replaced by optical and acoustics-based systems. Multi-sensor zooplankton collection systems are now the norm and in the future, we can anticipate seeing the development of real-time four-dimensional plankton sampling and concurrent environmental measurements systems, and ocean-basin scale sampling with autonomous vehicles.

From the beginning of modern biological oceanography in the late 1800s, remotely operated instruments have been fundamental to observing and collecting zooplankton. For most of the past century, biological sampling of the deep ocean has depended upon winches and steel cables to deploy a variety of instruments. The development of quantitative zooplankton collecting systems began with Victor Hensen in the 1880s (**Figure 1A**). His methods covered the whole scope of plankton sampling from the building and handling of nets to the final counting of organisms in the laboratory.

Three kinds of samplers developed in parallel: waterbottle samplers that take discrete samples of a small volume of water (a few liters), pumping systems that sample intermediate volumes of water (tens of liters to tens of cubic meters), and nets of many different shapes and sizes that are towed vertically, horizontally, or obliquely and sample much larger volumes of water (tens to thousands of cubic meters) (**Table 1**). Net systems dominated the equipment normally used to sample zooplankton until recent technological developments enabled the use of high-frequency acoustics and optical systems as well.

Net Systems

A variety of net systems have been developed over the past 100 + years and versions of all of these devices are still in use today. They can be categorized into eight groups: non-opening/closing nets, simple opening/closing nets, high-speed samplers, neuston samplers, planktobenthos plankton nets, closing cod-end samplers, multiple-net systems, and moored plankton collection systems.

Non-opening/Closing Nets

Numerous variants of the simple non-opening/closing plankton net have been developed, which are principally hauled vertically. Most are simple ring-nets with mouth openings ranging from 25 to 113 cm in diameter and conical or cylinder-cone nets 300–500 cm in length. Among the ring-nets that have been widely used are the Juday net (**Figure 1B**), International Standard Net, the British N-series nets, the Norpac net, the Indian Ocean Standard net (**Figure 1C**), the ICITA net, the WP2 net, the Cal-COFI net, and the MARMAP Bongo net (**Figure 1D**). Early nets were made from silk, but today nets are made from a square mesh nylon netting. Typical meshes used on zooplankton nets range from 150 μm to 505 μm, although larger and smaller mesh sizes are available. Most of these nets are designed to be hauled vertically. They are lowered to depth cod-end first and then pulled back to the surface with animals being caught on the way up. Others, such as the CalCOFI net and the Bongo net are designed to be towed obliquely from the surface down to a maximum depth of tow and then back to the surface. The Reeve net was a simple ring-net with a very large cod-end bucket designed to capture zooplankton alive. The Isaacs-Kidd midwater trawl (IKMT) has been used to collect samples of the larger macro-zooplankton and micronekton. It has a pentagonal mouth opening and a dihedral depressor vane as part of the mouth opening. Four sizes of IKMTs, 3 foot (91 cm), 6 foot (183 cm), 10 foot (304 cm), and 15 foot (457 cm) are often cited.

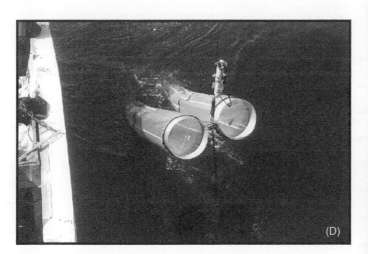

Figure 1 Some commonly used non-opening/closed nets. (A) The Hensen net. (Reproduced with permission from Winpenny, 1937.) (B) The Juday net; note the use of messenger release on this version of the net. (Reproduced with permission from Juday, 1916.) (C) The Indian Ocean Standard net. (Reproduced with permission from Currie, 1963.) (D) The Bongo net with CTD (*c.* 1999). (Photograph courtesy of P. Wiebe.) (E) The Tucker trawl. (Reproduced with permission from Tucker, 1951.)

Non-opening/closing nets with rectangular mouth openings were not widely used until the Tucker trawl was first described in 1951 (**Figure 1E**). This simple trawl design with a 180 cm × 180 cm mouth opening gave rise to a substantial number of opening/closing net systems described below.

Simple Opening/Closing Nets

The development of nets that could obtain depth-specific samples evolved from those of very simple design (a simple ring net) at an early stage. In the late 1800s and early 1900s, there was considerable effort

Table 1   Summary of zooplankton sampling gear types

Sampling gear	Type of sampling	Size fraction	Resolving scale		Typical operating range	
			Vertical	Horizontal	Vertical	Horizontal
Conventional methods						
Waterbottles	Discrete samples	Micro/meso	0.1–1 m	—	4000 m	—
Small nets	Vertically integrating	Micro/meso	5–100 m	—	500 m	—
Large nets	Vertical, obliquely Horizontally integrating	Meso/macro	5–1000 m	50–5000 m	1000 m	10 km
High-speed samplers	Obliquely, horizontally integrating	Meso/macro	5–200 m	500–5000 m	200 m	10 km
Pumps	Discrete samples	Micro/meso	0.1–100 m	—	200 m	—
Multiple net systems						
Continuous plankton recorder	Horizontally integrating	Meso	10–100 m	10–100 m	100 m	1000 km
Longhurst-Hardy plankton recorder	Obliquely, horizontally integrating	Meso	5–20 m	15–100 m	1000 m	10 km
MOCNESS	Obliquely, horizontally integrating	Meso/macro	1–200 m	100–2000 m	5000 m	20 km
BIONESS	Obliquely, horizontally integrating	Meso/macro	1–200 m	100–2000 m	5000 m	20 km
RMT	Obliquely, horizontally integrating	Meso/macro	1–200 m	100–2000 m	5000 m	20 km
Multinet	Vertically Obliquely, horizontally	Meso/macro	2–1000 m	100–2000 m	5000 m	5 km
Electronic optical or acoustical systems						
Electronic plankton-counter	High resolution in the horizontal/vertical plane	Meso	0.5–1 m	5–1000 m	300 m	100s of km
In situ silhouette camera net system	High resolution in the horizontal/vertical plane	Meso	0.5–1 m	5–1000 m	1000 m	10 km
Optical plankton counter	High resolution in the horizontal/vertical plane	Meso	0.5–1 m	5–1000 m	300 m	100s of km
Video plankton recorder	High resolution in the horizontal/vertical plane	Meso	0.01–1 m	5–1000 m	200 m	100s of km
Ichthyoplankton recorder	High resolution in the horizontal/vertical plane	Meso	0.1–1 m	5–1000 m	200 m	10 km
Multifrequency acoustic profiler system	High resolution in the horizontal/vertical plane	Meso/macro	0.5–1 m	5–1000 m	100 m	10 km
Dual-beam acoustic profiler	High resolution in the horizontal/vertical plane	Meso/macro	0.5–1 m	1–1000 m	800 m	100s of km
Split-beam acoustic profiler	High resolution in the horizontal/vertical plane	Meso/macro	0.5–1 m	1–1000 m	1000 m	100s of km
ADCP	High resolution in the horizontal/vertical plane	Meso/macro	10 m	5–500 m	500 m	100s of km

Most vertical nets are hauled at a speed of 0.5–1 m s^{-1}. Normal speed for horizontal tows are ~2 knots (1 m s^{-1}) and for high-speed samplers ~5 knots (2.6 m s^{-1}). For further categorization of pumping systems which are used by a number of investigators, reference is made to the review paper by Miller and Judkins (1981). (Reproduced with permission from Sameoto D, Wiebe P, Runge S, *et al.* (2000) Collecting zooplankton. In: Harris R, Wiebe P, Lenz J, Skjoldal HR, and Huntley M (eds.) *ICES Zooplankton Methodology Manual*, pp. 55–81. New York: Academic Press.)

to develop devices that closed or opened and closed nets at depth. Most employed mechanical release devices which were attached to the towing wire and activated by messengers traveling down the towing wire. The single-messenger Nansen closing mechanism and its variants were very popular during most of early to mid-twentieth century (**Figure 2**). Double-messenger systems that opened and then closed a net

quickly followed. In the mid-1930s, the Leavitt net system became popular and variants of this system are still being used today (**Figure 2B**). Another popular system still in use today is the Clarke and Bumpus sampler, a two-messenger zooplankton collection system that can be deployed as multiple units on the wire and has a positive means of opening and closing the mouth of the net (**Figure 2C**).

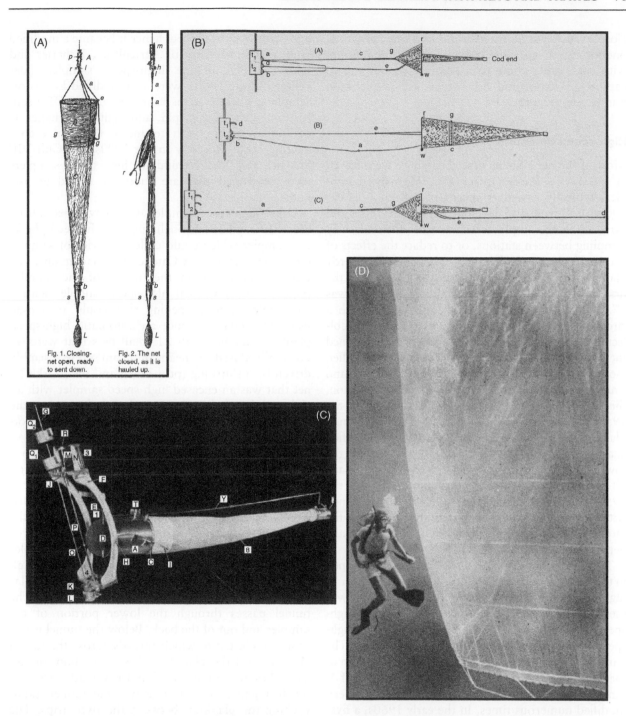

Figure 2 Some commonly used simple opening/closing nets. (A) The single-messenger Nansen closing net. (Reproduced with permission from Nansen, 1915.) (B) The two-messenger Leavitt net. (Reproduced with permission from Leavitt, 1935.) (C) The two-messenger Clarke-Bumpus net. (Reproduced with permission from Clarke and Bumpus, 1939.) The plankton purse seine (D) represents an unusual way to collect plankton from a specific region. (Reproduced with permission from Murphy and Clutter, 1972.)

Mechanical tripping mechanisms activated by pressure, by combinations of messengers and flowmeter revolutions, or clocks have also been devised.

Nontraditional approaches to collecting plankton include designs to catch plankton on the downward

fall of the net rather than the reverse – so-called pop-down nets; to sample under sea ice using the English umbrella net; to sample plankton from several depths simultaneously, using a combination of nets and a pumping system; to sample plankton from the nuclear submarine, *SSN Seadragon*; to open and close a Tucker-style trawl using two towing cables, one for the top spreader bar and one for the bottom, with each cable going to a separate winch; and to capture plankton and fish larvae with a plankton purse seine (**Figure 2D**).

High-speed Samplers

Most of the net systems described above were towed at speeds <3 knots (150 cm s^{-1}). High-speed samplers typically towed at speeds of 3–8 knots (150–400 cm s^{-1}) were also developed in the late 1800s and early 1900s to sample in bad weather, for underway sampling between stations, or to reduce the effects of net avoidance by the larger zooplankton. The Hardy plankton indicator, developed in the 1920s, was the first widely used device. The original version was 17.8 cm in diameter and 91.4 cm in length with a circular filtering disk on which plankton were collected. It was subsequently modified (and renamed the standard plankton indicator) to make it smaller, more streamlined, and equipped with a depressor and stabilizing fins (**Figure 3**). An even smaller version, the Small Plankton Sampler, was developed. In the 1950s, it was further modified and named the Small Plankton Indicator, and in the 1960s, it was modified again so that multiple units could be used on the towing wire at speeds of 7–8 knots with a multiplane kit otter depressor at the end of the wire. Until the 1950s, only one high-speed collector was designed with a double-messenger system that enabled the mouth to be opened and closed; most could not make depth-specific collections.

The 'Gulf' series of high-speed samplers developed in the 1950s and early 1960s gave rise to a number of high-speed samplers still in use today. The first was the Gulf I-A which looked similar to earlier high-speed samplers. The Gulf III was a much larger high-speed sampler that was enclosed in a metal case. The Gulf V was an unencased and scaled-down version of the Gulf III (**Figure 3B**). The Gulf III and Gulf V samplers have been very popular, and have been modified numerous times. In the early 1960s, a five-bucket cod-end sampling device was added to the Gulf III that was electrically activated from a deck unit through two-conductor cable. HAI (shark) was the German version of the Gulf III built in the mid-1960s. A hemispherical nose cone and an opening/closing lid were added to the HAI. This German system evolved further when 'Nackthai' (naked shark), a modified Gulf V sampler, was developed in the late 1960s. Also in the 1960s, the British modified the Gulf III sampler, which was subsequently called the Lowestoft sampler (**Figure 3C**). Subsequently, the Lowestoft sampler was scaled down

and made opened bodied; hence it became a modified Gulf V. The Ministry of Agriculture, Fisheries and Food MAFF/Guildline high-speed samplers, developed in the 1980s, were also modified Lowestoft samplers. These systems have a Guildline CTD sensor unit with oxygen, pH, and digital flowmeter as additional probes with telemetry through a conducting cable. Recently in the 1990s, the Gulf VII/Pro net and MAFF/Guildline high-speed samplers were developed that are routinely towed at 5–7 knots.

Other high-speed samplers were developed during the 1950s and 1960s, including a high-speed plankton sampler which could collect a series of samples during a tow; the 'Bary Catcher' that had an opening/closing mechanism in the mouth of the sampler (**Figure 3D**); a vertical high-speed sampler with a rectangular mouth opening that could be closed using the Juday method; an automatic high-speed plankton sampler with 21 small nets that were sequentially closed by means of a cam/screw assembly driven by a ships log (propellor); and the Clarke Jet net that was an encased high-speed sampler with an elaborate internal passageway designed to reduce the flow speed of water within the sampler to that normally experienced by a slowly towed net.

The continuous plankton recorder (CPR) is in a class by itself when it comes to high-speed plankton samplers, because it can take many samples and can be towed from commercial ships (**Figure 3E**). Originally built in the 1920s, it has evolved over the years to become the mainstay in a plankton survey program in the North Atlantic. This encased sampler weight 87 kg and is about 50 cm wide by 50 cm tall by 100 cm long. The 1.27 cm × 1.27 cm rectangular aperture expands into a larger tunnel opening. The tunnel passes through the lower portion of the sampler and out of the back. Below the tunnel is one spool of silk gauze which threads across the tunnel and captures the plankton. A second spool of silk gauze lies above the tunnel and is threaded to meet the first gauze strip as it leaves the tunnel, sandwiching the plankton between the two strips. The gauze strips are wound up on a take-up spool which resides in a formalin-filled tank above the flow-through tunnel, preserving the plankton. The take-up spool is driven by a propeller on the back of the sampler behind the tail fins. This sampler is usually towed at 20 knots from commercial transport vessels at a fixed depth of about 10 m below the surface, thus it only samples the surface layer of the ocean. The undulating oceanographic recorder (UOR) was developed in the 1970s to extend the vertical sampling capability of high-speed plankton collection systems. The UOR carries sensors to measure

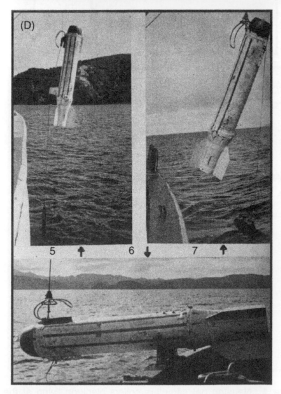

Figure 3 Some examples of high-speed plankton samplers. (A) The standard plankton indicator. (Reproduced with permission from Hardy, 1936.) (B) The encased Gulf III sampler. (Reproduced with permission from Gehringer, 1952.) (C) The open-bodied Lowestoft sampler (Gulf V type). (Reproduced with permission from Lockwood, 1974.) (D) The Bary catcher. (Reproduced with permission from Bary, 1958.) (E) The continuous plankton recorder (CPR). (Reproduced with permission from Hardy, 1936.)

temperature, salinity, and pressure; data are logged internally at 30 observations per minute. A propellor drives the rollers winding up the gauze and provides the power for the electronics.

Neuston Samplers

Nets to collect neuston, the zooplankton that live within a few centimeters of the sea surface, by-and-large are non-opening/closing. The first net specifically designed to sample zooplankton neuston was built in about 1960. A rectangular mouth opening design is typical of most of the systems. Neuston nets come either with a single net which collects animals right at the water surface or vertically stacked sets of two to six nets extending from the surface to about 100 cm depth (**Figure 4**). Normally they are towed from a vessel, but a 'push-net' was developed in the 1970s with a pair of rectangular nets positioned side-by-side in a framework and mounted in front of a small catamaran boat that pushed the frame through the water at ~2.6 knots.

Planktobenthos Plankton Nets

The ocean bottom is also special habitat structure for zooplankton, and gear to sample zooplankton living here ('planktobenthos') was developed early. The first nets were designed in the 1890s specifically to sample plankton living very near the bottom. Non-opening/closing systems were succeeded by samplers with mechanically operated opening/closing doors or with a self-closing device (**Figure 5A**).

An entirely different strategy has been to employ manned submersibles or deep-towed vehicles to collect deep-sea planktobenthos. A pair of nets mounted on the front of DSRV *Alvin* was used for making net collections at depths >1000 m in the 1970s; the pilot opened and closed the net (**Figure 5B**). A multiple net system was used on the Deep-Tow towed body. This system was attached to the bottom of the Deep-Tow and used for sampling within a few tens of meters above the deep-sea floor in the 1980s (**Figure 5C**). This net system was later adapted for use on DSRV *Alvin* for near-bottom studies of plankton in the vicinity of hydrothermal vent sites in the 1990s.

On other benthic habitats, such as coral reefs, fixed or stationary net systems which orient to the current's flow and filter out zooplankton drifting by, nets pushed by divers, and traps have been used to capture plankton close to the bottom. The Horizontal Plankton Sampler (HOPLASA) creates its own current to collect zooplankton on or near the bottom in coral reef areas with variable or little current flow (**Figure 5D**).

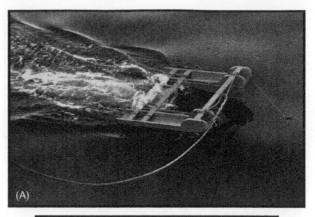

(A)

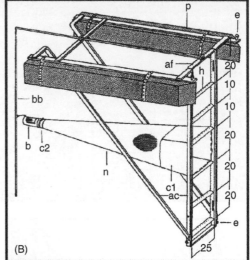

(B)

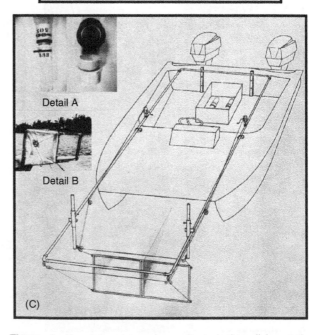

(C)

Figure 4 Neuston net samplers collect plankton living at the sea surface. (A) A single net system. (Reproduced with permission from David, 1965.) (B) A multinet system. (Reproduced with permission from Ellertsen, 1977.) (C) A push net. (Reproduced with permission from Miller, 1973.)

Figure 5 Some planktobenthos samplers. (A) Early system with opening/closing doors. (Reproduced with permission from Wickstead, 1953.) (B) DSR Alvin opening/closing system. (Reproduced with permission from Grice, 1972.) (C) The Deep-Tow multiple net system. (Reproduced with permission from Wishner, 1980.) (D) A system for coral reef sampling (HOPLASA). (Reproduced with permission from Rutzler, 1980.)

Closing Cod-end Systems

In the late 1950s and 1960s, conducting cables and transistorized electronics were beginning to be adapted for oceanographic use and sophisticated net systems began to do more than collect animals at specific depth intervals. Single nets equipped with closing cod-end devices preceded multiple net systems by only a few years. One of the first systems used a 1950s version of a serial device in the high-speed sampler that was mechanically driven by a propellor. Another had a pressure-actuated catch-dividing bucket (CDB) attached to the back of an IKMT (**Figure 6A**). The Mark III Discrete Depth Plankton Sampler (DDPS) also developed for use with an IKMT or a 1 m diameter net, had four catch chambers separated by solenoid-activated damper doors (**Figure 6B**). This latter system was one of the first to carry underwater electronics to sample depth and temperature, and to telemeter the data up a single conductor cable for display at the surface. The multiple plankton sampler (MPS, described below) was turned into a cod-end sampler for an IKMT

and later modified by adding environmental sensors and an electronically controlled opening/closing mechanism.

The Longhurst-Hardy plankton recorder (LHPR), a modification of the CPR, was developed in the 1960s (**Figure 6D**). The recorder box was attached to the back end of a net and gauze strips in the box were advanced in discrete steps (15 s to 60 s) by an electronics package on the tow frame; data on pressure, temperature, and flow were logged on an internal recorder; power was supplied by a NICAD battery pack. The LHPR was redesigned in the 1970s to reduce problems with hang-ups and stalling of animals in the net which caused smearing of the

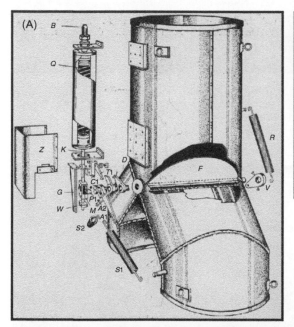

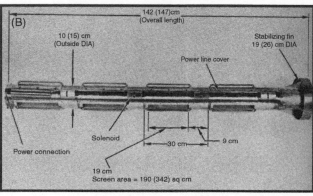

Figure 6 Some discrete-depth samplers using a closing cod-end. (A) The catch dividing cod-end. (Reproduced with permission from Foxton, 1963.) (B) The Mark III multiple cod-end bucket. (Reproduced with permission from Aron *et al.* 1964.) (C) ARIES. (Reproduced with permission from Dunn *et al.* 1993.) (D) A version of the LHPR. (Photograph courtesy of J. Smith, 1966.)

distributions of animals and losses of animals from the recorder box. The modified LHPR was used without a net on the conning-tower of the US Navy research submarine *Dolphin* in the 1980s. Another modification of the LHPR was made by the British in 1980s. They used an unenclosed Lowestoft sampler to mount a pair of recorder boxes to collect meso- and micro-zooplankton. The system acoustically telemetered depth, flow, and temperature. It also carried a chlorophyll sensor with a recorder system. The LHPR was further modified for use in catching Antarctic krill. A descendant of the LHPR developed in the 1990s is the Autosampling and Recording Instrumental Environmental Sampler (ARIES) (**Figure 6C**). This cod-end plankton sampling device is a stretched version of the Lowestoft-modified Gull III frame. It has a multiple cod-end system, water sampler, data logger, and an acoustical telemetry system.

Multiple Net Systems

The development of multiple net systems began with the simple non-opening/closing Tucker trawl system. In the mid-1960s, timing clocks were used to open and close the Tucker trawl mouth. Then late in the 1960s, the British rectangular mouth opening trawl (RMT), which was opened and closed acoustically, was developed. The RMT was expanded into the NIO Combination Net (RMT $1+8$), which carries nets with $1\,m^2$ and $8\,m^2$ mouth openings (**Figure 7A**). This was expanded into a multiple net system with three sets of 1 m and 8 m nets controlled acoustically. The acoustic command and telemetry system for the RMT $1+8$ was replaced in the 1990s by a micro-computer-controlled unit connected by conducting cable to an underwater electronics unit.

In a parallel development in the 1970s, a five-net and a nine-net Tucker Multiple Net Trawl was developed on the West Coast of the USA. The system was powered electrically through conducting wire and controlled from the surface. A modified Tucker trawl system, the Multiple Opening/Closing Net and Environmental Sensing System (MOCNESS), with nine nets and a rigid mouth opening was built soon after on the US east coast (**Figure 7E**). The current versions of the MOCNESS are computer-controlled (**Table 2**). Sensors include pressure, temperature, conductivity, fluorometer, transmissometer, oxygen, and light.

The design of the Bé multiple plankton sampler (MPS) (**Figure 7B**), initially messenger operated in the late 1950s and then pressure-actuated in the 1960s, was the basis for the Bedford Institute of Oceanography Net and Environmental Sensing System (BIONESS), with 10 nets, developed in the 1980s (**Figure 7D**). A modified version of the MPS was developed in Germany at about the same time and named the Multinet; it carried five nets, which were opened and closed electronically via conducting cable (**Figure 7C**). A scaled-up version of BIONESS built in the 1990s was the Large Opening Closing High Speed Net and Environmental Sampling System (LOCHNESS). Another variant of the MPS was the Ocean Research Institute's (Japan) vertical multiple plankton sampler developed in the 1990s in which the nets are opened/closed by surface commands transmitted via conducting cable to an underwater unit.

Moored Plankton Collection Systems

Only a few instrument systems have been developed that autonomously collect time-series samples of plankton from moorings. Most were patterned after the CPR or LHPR (e.g. the O'Hara automatic plankton sampler built in the 1980s; a modified version of the O'Hara system built in the 1990s; the moored, automated, serial zooplankton pump (MASZP) built in the late 1980s) (**Figure 8**). The lack of such systems may be due to the difficulty of powering them for long periods underwater.

Optical Systems

Optical survey instruments can be divided into two categories, based on whether the systems produce an image of their zooplankton targets (e.g. video, photographic, and digital camera systems) or use the interruption of a light source to detect and estimate the size of particles (e.g. the optical plankton counter). The first attempts to quantify plankton optically appear to have been made in the 1950s using a beam of light projected into the chamber from a 300 W mercury vapor lamp and a Focabell camera (Orion Camera, Tokyo).

Image-forming Systems Mounted on Non-opening/Closing Nets

In the 1980s, a 35 mm still camera with a high-capacity film magazine in front of the cod-end of a plankton net attached to a rigid frame was used to take *in situ* silhouette photographs of zooplankton as they passed into the cod-end. This was a field application of the laboratory-based silhouette photography system developed in the late 1970s. The camera provided a series of photographic images at points along the trajectory of the net separated by $<1\,m$. In the development of the ichthyoplankton recorder, the still camera was replaced with a video

Figure 7 Some examples of multiple net plankton sampling systems. (A) The RMT 1 + 8. (Reproduced with permission from Baker, 1973.) (B) The Bé net. (Reproduced with permission from Bé, 1959.) (C) The Multinet. (Photograph courtesy of B. Niehof.) (D) The BIONESS. (Photograph courtesy of P. Wiebe, 1993.) (E) The 1 m² MOCNESS. (Photograph courtesy of Wiebe, 1998.)

camera, which was located in front of the cod-end of a high-speed Gulf V-type net (*Nackthai*). It had an estimated horizontal spatial resolution of 3 cm. One consequence of going from camera film to video tape was a loss of image resolution.

Stand-alone Image-forming Systems

The video plankton recorder (VPR) was developed in the early 1990s as a towed instrument capable of

imaging zooplankton within a defined volume of water (**Figure 9A**). The original VPR had four video cameras; each camera imaged concentrically located volumes of water ranging from 1 ml to 1000 ml, but it has been modified to a one- or two-camera system. It has been possible to image undisturbed animals in their natural orientations. The current VPR image processing system is capable of digitizing each video field in real time and scanning the fields for targets using user-defined search criteria for

Table 2 MOCNESS system dimensions and weights

System	Number of nets	Width of frame (m)	Height of frame (m)	Net width (m)	Mouth area at 45° towing angle (m)	Length of net (m)	Approx. weight in air (kg)	Rec. wire diameter (mm)
MOCNESS-1/4	9	0.838	1.430	0.50	0.5	6.00	70	6.4
MOCNESS-1/4-Double	18/20	1.430	1.430	0.50	0.5	6.00	155	7.4
MOCNESS-1	9	1.240	2.870	1.00	1.0	6.00	150	7.4
MOCNESS-1-Double	18/20	2.560	2.870	1.00	1.0	6.00	320	12.1
MOCNESS-2	9	1.650	3.150	1.41	2.0	6.00	210	11.8
MOCNESS-4	6	2.140	4.080	2.00	4.0	8.44	460	11.8
MOCNESS-10	6	3.410	4.690	3.17	10.0	18.25	640	11.8
MOCNESS-20	6	5.500	7.300	4.47	20.0	14.50	940	17.3

The MOCNESS systems are denoted by the mouth area when being towed. Thus a MOCNESS-1/4 has a 0.25 m² mouth opening. The 'Double' systems have two sets of nets side-by-side in a single rigid framework. Nets can be opened and closed on one side and then opened and closed on the other.

brightness, focus, and size. The targets are identified using a zooplankton identification program to provide near-real-time maps of the zooplankton distributions.

A number of VPR-based systems are currently in operation or under development: a single-camera system is mounted on the BIOMAPER II vehicle (described below); an internally recording VPR has been constructed and used to quantify radiolarians and foraminiferans; and one has been mounted on a 1 m² MOCNESS net system to map the fine-scale distributions of the larval cod prey items. A moored system called the Autonomous Vertically Profiling Plankton Observatory (AVPPO) utilizes an internally recording, two-camera VPR, and has been deployed in coastal waters off New England.

Image resolution constraints inherent in the use of standard video formats have driven the development

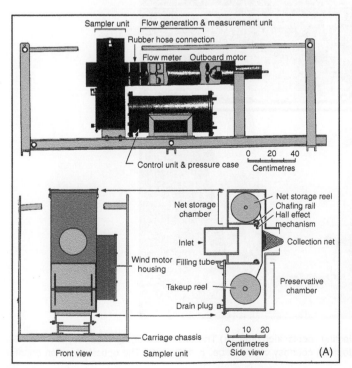
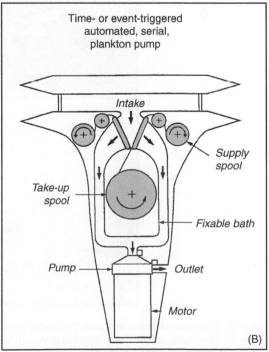

Figure 8 Two examples of moored plankton collecting systems. (A) A modified version of the O'Hara sampler. (Reproduced with permission from Lewis and Heckl, 1991.) (B) MASZP. (Reproduced with permission from Doherty *et al.* 1993.)

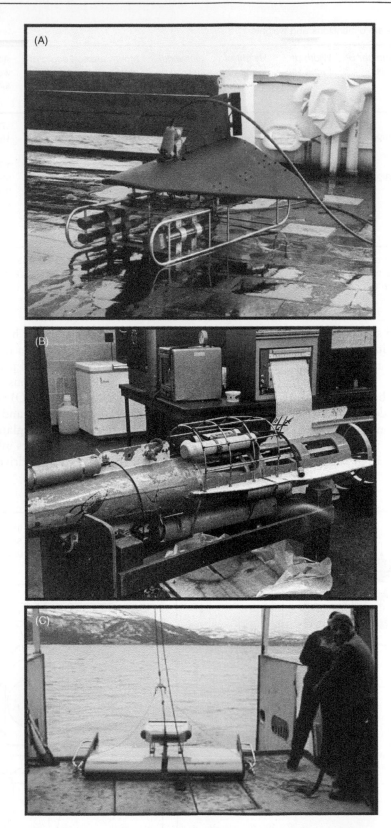

Figure 9 Examples of optical or electrical systems for collecting zooplankton data. (A) The VPR. (Photograph courtesy of P. Alatalo, 1999.) (B) The *in-situ* zooplankton detecting device. (Photograph courtesy of P. Wiebe, *c.* 1972.) (C) The optical plankton counter (OPC). (Photograph courtesy of M. Zhou, 2000.)

of optical systems that utilizes higher-resolution formats. A modification of the continuous underway fish egg sampler (CUFES, described below) utilizes a line-scanning digital camera to quantify the abundances of fish eggs. The shadowed image particle profiling and evaluation recorder (SIPPER) utilizes high-resolution digital line-scanning cameras to quantify zooplankton passing through a laser light sheet. The SIPPER has been mounted either on a towed vehicle called the high-resolution sampler (HRS) or an AUV.

The need for systems to quantify the abundance of 'marine snow' prompted development of profiling systems based on both still and video cameras. In the 1980s, a profiling system called the large amorphous aggregates (LAA) camera was constructed which employed a photographic camera and a pair of strobes to photograph marine aggregates. A video profiling instrument called the underwater video profiler (UVP) has been used to quantify the vertical distribution and size frequency of marine snow, and to examine the distributions of macrozooplankton. The UVP consists of a Hi-8 video camera imaging a collimated light sheet coupled with a CTD, data logger, and batteries. A profiling system called ZOOVIS recently has been developed around a high resolution (2048×2048 pixel) digital camera and CTD linked to a surface workstation via a fiber-optic cable. A color video camera has been mounted on the front of a Sea Owl II remotely operated vehicle (ROV) and used to quantify the vertical distribution of gelatinous zooplankton off the west coast of Sweden.

Still holographic imaging of plankton in a laboratory was first reported in 1966. It was refined in the 1970s to record movies of live plankton in the laboratory. In the 1990s, a submersible internally recording in-line holographic camera that records up to 300 holograms on a film emulsion was developed.

Many zooplankton produce or induce the production of bioluminescent light that can be detected with sensitive CCD cameras. One system is mounted on the Johnson SeaLink manned submersible and consists of an intensified silicon-intensified target (ISIT) video camera mounted on and aimed forward at a 1 m diameter transect screen to quantify the distribution, abundance, and identities of bioluminescent zooplankton.

Particle Detection Systems

Particle detection systems refer to non-image-forming devices that utilize interruption of an electrical current or a light beam to detect and estimate the size of a passing particle. The first *in situ* particle

counting and sizing system appeared in the late 1960s and was referred to as the *in situ* zooplankton detecting device (**Figure 9B**). A shipboard version of the device was connected to a continuously pumped stream of water and employed to analyze spatial heterogeneity of zooplankton in surface waters in relation to chlorophyll fluorescence and temperature. A version of this conductive zooplankton counter was deployed aboard a Batfish towed vehicle in the 1980s.

A second group of particle detectors utilized photodetectors rather than changes in voltage. The Opto-Electronic Plankton Sizer was a laboratory-based system designed in the 1970s to automate the measurement of preserved plankton samples. The HIAC particle size analyzer was modified at the Lowestoft Laboratory during the late 1970s for plankton counting. The optical plankton counter (OPC) was developed during the mid-1980s (**Figure 9C**). This instrument measures changes in the intensity of a light beam that occur when a particle crosses the beam. The OPC has been mounted on a variety of towed platforms or in shore-based or shipboard applications. The OPC has also been incorporated into a shipboard device called the continuous underway fish egg sampling system (CUFES) which enumerates the distribution and abundance of fish eggs in surface waters. In spite of the prevalence of OPC systems in current use, interpretation of OPC data remains a subject of some controversy.

Optical Instruments for Nonquantitative Studies

The ecoSCOPE is an optical video-endoscope that enables direct observation of predator–prey interactions between juvenile fish and zooplankton. The ecoSCOPE has been operated from an ROV, from the keel of a sailing vessel, and in towed and moored modes, but the best recordings of predator/prey interactions have come from free-drifting deployments, when the instrument was hovering within schools of feeding juvenile herring. A software package called dynIMAGE animates sequential images keeping the fish and its prey in the middle of the viewing field.

Optical sensors can provide valuable ground-truthing for acoustical sensors. In the 1990s, a megapixel digital still camera was mounted on a FishTV sonar array and the resulting system was named the Optical-Acoustical Submersible Imaging System (OASIS). In this system, high acoustic returns are used to trigger the camera taking a picture of the acoustical target. An analog video camera aimed at the focal point of an acoustic array mounted on the front of a MAXRover ROV has been used to take pictures of individual zooplankton passing through the acoustic beam.

High-frequency Acoustics

High-frequency acoustics (≥ 38–1000 kHz) provide the foundation for another class of tools to study zooplankton. The utility of the acoustic systems derives from their ability to operate with high ping rates and precision range-gating. Mapping planktonic distributions on a wide range of space and timescales is becoming possible because of the continued development of acoustics systems and appropriate ground-truthing methods. There are two fundamental measurements: volume backscattering (integration of the energy return from all individuals in a given ensonified volume, i.e. echo integration) and target strength (echo strength from an individual). Statistical procedures have been developed to estimate animal assemblage size distribution using the data from single-beam transducers. In some cases, it is possible to extract estimates of animal target strength distribution in addition to volume backscattering from a series of single-beam transducers operating at different frequencies. Multibeam acoustical systems provide a direct means of determining individual target strength (TS). The two current designs, dual-beam and split-beam, both provide a hardware solution to the problem of TS determination.

The Current State of Plankton Sampling Systems

The diversity of zooplankton samplers in use today reflects the fact that no single collection system adequately samples all zooplankton. Non-opening/closing nets, such as the WP2, the modified Juday net, and the Bongo net, are used in large ocean surveys. Simple, double-messenger opening/closing nets similar to those developed in the first half of the last century are still manufactured and used. The Multinet, RMT 1 + 8, BIONESS, and MOCNESS are widely used multiple-net systems that also carry additional sensors to measure other water properties. Plankton pumps are also being used, especially to collect micro-zooplankton.

The advent of high-speed computers and towing cables with optical fibers and electrical conductors have enabled development of multi-sensor towed systems which provide real-time data while the instrument package is deployed. The MOCNESS has been equipped with a high-frequency acoustic system for forward or sideways range-gated viewing (Figure 10A). An EG&G Edgerton model 205 camera and a flash light were mounted on the top of a modified MOCNESS and on the top of BIONESS to take black and white photographs about 2 m in front of the net mouth. The BIONESS has also been equipped with an OPC and video lighting system, and used in conjunction with an echosounder.

The BIo-Optical Multi-frequency Acoustical and Physical Environmental Recorder – BIOMAPER II – was developed to conduct high-speed, large-area surveys of zooplankton and environmental property distributions to depths of 500 m (Figure 10B). Mounted inside are a multi-frequency sonar (upwards-looking and downwards-looking pairs of transducers operating at five frequencies: 43, 120, 200, 420, and 1000 kHz), an environmental sensor package (CTD, fluorometer, transmissometer), and several other bio-optical sensors (down- and upwelling spectral radiometers, spectrally matched attenuation, and absorption meters). A single-camera video plankton recorder (VPR) system is mounted above and just forward of the nose piece. The lower four acoustical frequencies involve split-beam technology and are able to make target strength and echo integration measurements.

A variety of vehicles have been built that actively change their vertical position without changing the towing wire length. Examples for surveying zooplankton include the undulating oceanographic recorder and SeaSoar equipped with optical (VPR and OPC) and/or acoustical (the Tracor Acoustical Profiling System, TAPS). Remotely operated vehicles (ROVs) have also been equipped with acoustical and video systems to study zooplankton. A SeaRover ROV was equipped with the same dual-beam acoustic system and environmental sensors. A VPR rigged to provide 3-D images of plankton and an environmental sensor package (temperature, conductivity, pressure, fluorescence) were mounted on the front of the ROV JASON and on the SeaRover ROV (Figure 10C). FishTV (FTV) has been used on a Phantom IV ROV and a combination of acoustics and video has been used on the front of a MAXRover ROV. Dual-beam acoustics (420 and 1000 kHz) have also been deployed on the DSRV Johnson SeaLink.

Future Developments

The future promises vastly increased application of remote sensing techniques and sensor development, and real-time data telemetry, processing, and display. Three-dimensional (space) and four-dimensional visualization (space and time) of biological and

Figure 10 Examples of multi-sensor plankton sampling systems. (A) MOCNESS with a dual-beam acoustic system. (Photograph courtesy of P. Wiebe, 1994.) (B) BIOMAPER-II. (Photograph courtesy of P. Wiebe, 1999.) (C) The JASON-ROV with 3-D VPR system. (Photograph courtesy of P.Alatalo, 1995.)

acoustic data are also an increasingly important aspect of data processing. For a number of research programs today, the development of an image of the spatial arrangement of organisms is but the first step in efforts to study and understand their relationships to each other and to their environment. Thus, there is need for real-time 3-D and 4-D images.

Autonomous self-propelled vehicles (AUVs) have only recently begun to be used widely to gather-oceanographic data. The remote environmental measuring units (REMUS) are a new class of small AUVs which can carry an impressive array of environmental sensors including a VPR. Another class of autonomous vehicles is epitomized by the autonomous benthic explorer (ABE), which is equipped with precise navigation and control systems that enable it to descend to a worksite, navigate preset tracklines or terrain-follow, and find a docking station. A much larger AUV which has been employed

for biological studies is the Autosub-1 that carries a gyrocompass, ADCP, an echosounder, and acoustic telemetry and surface radio electronics. It can be programmed to run a geographically based course using GPS surface positions and dead reckoning.

The autonomous Lagrangian circulation explorer (ALACE) and the more recently developed profiling version (PALACE) floats that carry temperature and conductivity probes are vertically migrating neutrally buoyant drifters. They track the movements of water at depths between the surface and 1000–2000 m depth. Hundreds to thousands of the PALACE floats will be deployed over the next few years and it is expected that they will become a mainstay in the Global Ocean Observing System (GOOS). The next generation of neutrally buoyant floats is an autonomous glider named SPRAY. SPRAY will be able to sail along specific preprogrammed tracklines. A further step in their development is to provide biological instrumentation to complement the physical sensors.

High-resolution optical systems, suchas the VPR, combined with computer-based identification programs can now provide higher level taxa identifications in near-real time. Classification of species using acoustic signatures is less well developed and it now seems unlikely that the technology to develop species-specific acoustic signatures will be developed soon. Molecularly based species identification is likely to make significant strides in the next decade. It is now conceivable that this information will enable simultaneous analysis, identification, and quantification of all species occurring in a zooplankton sample.

See also

Acoustic Scattering by Marine Organisms. Continuous Plankton Recorders. Grabs for Shelf Benthic Sampling. Platforms: Autonomous Underwater Vehicles; Satellite Remote Sensing SAR.

Further Reading

Harris RP, Wiebe PH, Lenz J, Skjoldal HR, and Huntley M (eds.) (2000) *ICES Zooplankton Methodology Manual.* New York: Academic Press.

Kofoid CA (1991) On a self-closing plankton net for horizontal towing. *University of California Publications in Zoology* 8: 312–340.

Miller CB and Judkins DC (1981) Design of pumping systems for sampling zooplankton with descriptions of two high-capacity samplers for coastal studies. *Biol Oceanogr* 1: 29–56.

Omori M and Ikeda Y (1976) *Methods in Marine Zooplankton Ecology.* New York: John Wiley & Sons.

Schulze PC, Strickler JR, Bergström BI, *et al.* (1992) Video systems for in situ studies of zooplankton. *Arch Hydrobiol Beih Ergebn Limnol* 36: 1–21.

Sprules WG, Bergström B, Cyr H, *et al.* (1992) Non-video optical instruments for studying zooplankton distribution and abundance. *Arch Hydrobiol Beih Ergebn Limnol* 36: 45–58.

Tranter DJ (ed.) (1968) *Part I. Reviews on Zooplankton Sampling Methods. Monographs on Oceanographic Methodology, Zooplankton Sampling.* UNESCO

MOORINGS, FLOATS, AND CURRENT METERS

DRIFTERS AND FLOATS

P. L. Richardson, Woods Hole Oceanographic
Institution, Woods Hole, MA, USA

Introduction

Starting in the 1970s, surface drifters and subsurface
neutrally buoyant floats have been developed, improved, and tracked in large numbers in the ocean.
For the first time we have now obtained worldwide
maps of the surface and subsurface velocity at a few
depths. New profiling floats are measuring and reporting in real time the evolving temperature and
salinity structure of the upper 2 km of the ocean in
ways that were impossible a decade ago. These
measurements are documenting variations of the
world ocean's temperature and salinity structure.
The new data are revealing insights about ocean
circulation and its time variability that were not
available without drifters and floats.

Surface drifters and subsurface floats measure
ocean trajectories that show where water parcels go,
how fast they go there, and how vigorously they are
mixed by eddies. Ocean trajectories, which are called
a Lagrangian description of the flow, are useful both
for visualizing ocean motion and for determining its
velocity characteristics. The superposition of numerous trajectories reveals that very different kinds
of circulation patterns occur in different regions.
Time variability is illustrated by the tangle of crossing trajectories. Trajectories often show the complicated relationship between currents and nearby
seafloor topography and coastlines. Drifters and
floats have been used to follow discrete eddies like
Gulf Stream rings and 'meddies' (Mediterranean
water eddies) continuously for years. When a drifter
becomes trapped in the rotating swirl flow around an
eddy's center, the path of the eddy and its swirl velocity can be inferred from the drifter trajectory. Thus
the movement of a single drifter represents the huge
mass of water being advected by the eddy. Trajectories of drifters launched in clusters have provided
important information about dispersion, eddy diffusivity, and stirring in the ocean.

When a sufficient number of drifters are in a region,
velocity measurements along trajectories can be
grouped into variously sized geographical bins and
calculations made of velocity statistics such as mean
velocity, seasonal variations, and eddy energy. Gridded

values of these statistics can be plotted and contoured
to reveal, for example, patterns of ocean circulation
and the sources and sinks of eddy energy. Maps of
velocity fields can be combined with measurements
of hydrography to give the three-dimensional velocity
field of the ocean. Oceanographers are using the newly
acquired drifter data in these ways and also incorporating them into models of ocean circulation.

Care must be used in interpreting drifter measurements because they are often imperfect current
followers. For example, surface drifters have a small
downwind slip relative to the surrounding water.
They also tend to be concentrated by currents into
near-surface converge regions. The surface water
that converges can descend below the surface, but
drifters are constrained to remain at the sea surface
in the convergence region. Surface currents sometimes converge drifters into swift ocean jets like the
Gulf Stream. This can result in oversampling these
features and in gridded mean velocities that are different from averages of moored current meter
measurements, which are called Eulerian measurements of the flow. Bin averages of drifter velocities
can give misleading results if the drifters are very
unevenly distributed in space. For example, drifters
launched in a cluster in a region of zero mean velocity tend to diffuse away from the cluster center by
eddy motions, implying a divergent flow regime. The
dispersal of such a cluster gives important information about how tracers or pollutants might also
disperse in the ocean. Drifters tend to diffuse faster
toward regions of higher eddy energy, resulting in a
mean velocity toward the direction of higher energy.
This is because drifters located in a region of high
eddy kinetic energy drift faster than those in a region
of low kinetic energy. Errors concerning array biases
need to be estimated and considered along with
gridded maps of velocity.

Surface Drifters

Surface drifter measurements of currents have been
made for as long as people have been going to sea.
The earliest measurements were visual sightings of
natural and man-made floating objects within sight
of land or from an anchored ship that served as a
reference. Starting at least 400 years ago, mariners
reported using subsurface drogues of different shapes
and sizes tethered to surface floats to measure currents (**Figure 1**). The drogues were designed to have a
large area of drag relative to the surface float so that

Figure 1 Schematic of an early drifter and drogue from the *Challenger* expedition (1872–76). Adapted from Niiler PP, Davis RE, and White HJ (1987) Water following characteristics of a mixed layer drifter. *Deep-Sea Research* 24: 1867–1881.

the drifter would be advected primarily with the water at the drogue depth and not be strongly biased by wind, waves, and the vertical shear of near-surface currents. Over the years many kinds of drogues, tethers, and surface floats have been tried, including drogues in the form of crossed vanes, fishing nets, parachutes, window shades, and cylinders.

Ship Drifts

Probably the most successful historical drifter is a ship; the drift of ships underway as they crossed oceans provided millions of ocean current measurements. A ship drift measurement is obtained by subtracting the velocity between two measured position fixes from the estimated dead reckoning velocity of the ship through the water over the same time interval. The difference in velocity is considered to be a measure of the surface current. This technique depends on good navigation, which became common by the end of the nineteenth century. Most of what we have learned about the large-scale patterns of ocean currents until very recently came from compilations of historical ship drift measurements. Pilot charts used by most mariners today are still based on historical ship drifts. Problems with the ship drift technique are the fairly large random errors of each velocity measurement ($\sim 20\,\mathrm{cm\,s^{-1}}$) and the suspected systematic downwind leeway or slip of a ship through the water due to wind and wave forces. New velocity maps based on satellite-tracked drifters are providing a much more accurate and higher-resolution replacement of ship drift maps.

Drifting derelict ships gave an early measurement of ocean trajectories during the nineteenth century. Wooden vessels that had been damaged in storms were often abandoned at sea and left to drift for months to years. Repeated sightings of individual vessels reported in the US pilot charts provided trajectories.

Other Drifters

Bottles with notes and other floating objects have been a popular form of surface drifter over the years. The vectors between launch and recovery on some distant shore provided some interesting maps but ones that were difficult to interpret. The technique was improved and exploited in the North Atlantic by Prince Albert I of Monaco during the late 1800s. More recently, 61 000 Nike shoes and 29 000 plastic toy animals were accidentally released from damaged containers lost overboard from ships in storms in the North Pacific. The recovery of thousands of these drifters along the west coast of North America has given some interesting results about mean currents and dispersion.

Bottom drifters are very slightly negatively buoyant and drift along the seafloor until they come ashore and are recovered. The vectors between launch and recovery show long-term mean currents near the seafloor.

Tracking

Early measurements of drifters were visual sightings using telescopes, compasses, and sextants to measure bearings and locations. Later during the 1950s, radio direction finding and radar were used to track drifters over longer ranges and times from shore, ship, and airplane. Some drifter trajectories in the 1960s were obtained by Fritz Fuglister and Charlie Parker in the Gulf Stream and its rings using radar. These early experiments did not obtain very many detailed and long trajectories but did reveal interesting features of the circulation. It was clearly apparent that a remote, accurate, relatively inexpensive, long-term tracking system was needed. This was soon provided in the 1970s by satellite tracking, which revolutionized the tracking of drifters in the ocean.

The first satellite tracking of drifters occurred in 1970 using the Interrogation Recording and Location System (IRLS) system flown on *Nimbus 3* and *4* satellites. This system measured the slant range and bearing of a radio transmitter on a drifter. The IRLS drifters were very expensive, too expensive at $50 000 to be used in large numbers (but cheap compared to the cost of the satellite). During 1972–73, several drifters were tracked with the (Corporative Application Satellite) EOLE system, which incorporated Doppler measurements of the drifter radio transmissions to determine position. In the mid-1970s, NASA developed the Random Access Measurement System (RAMS), which used Doppler measurements and was flown on the *Nimbus 6* satellite. The radio transmitters were relatively inexpensive at $1300, and tracking was provided free by NASA (as proof of concept), which enabled many oceanographers to begin satellite tracking of surface drifters. The modern Doppler-based satellite tracking used today, the French Argos system flown on polar orbiting satellites, is similar to the early RAMS but provides improved position performance. The modern drifter transmitter emits a 0.5-W signal at 402 MHz approximately every minute. Positions are obtained by Service Argos about six times per day in the equatorial region increasing to 15 times per day near 60° N. Position errors are around 300 m. The cost per day of satellite tracking is around $10, which becomes quite expensive for continuous year-long trajectories. To reduce costs, some drifters have been programmed to transmit only one day out of three or for one-third of each day. This causes gaps in the trajectories that need to be interpolated.

Recently, drifters with Global Positioning System (GPS) receivers have been deployed to obtain more accurate (~ 10 m) and virtually continuous fixes. GPS position and sensor data need to be transmitted to shore via the Argos system or another satellite that can relay data. Experiments are underway using new satellite systems to relay information both ways – to the drifters and to the shore – in order to increase bandwidth, to decrease costs, and to modify sampling.

WOCE drifter The development of satellite tracking in the 1970s quickly revealed the weakness of available drifters – most performed poorly and most did not survive long at sea. Many problems needed to be overcome: fishbite, chafe, shockloading, biofouling, corrosion, etc. Early drogues tended to fall off fairly quickly and, since drogue sensors were not used or did not work well, no one knew how

long drogues remained attached. Over the years many people tried various approaches to solve these problems, but it was mainly due to the impetus of two large experiments, Tropical Ocean and Global Atmosphere (TOGA) and World Ocean Circulation Experiment (WOCE), and with the persistent efforts of Peter Niiler and colleagues that a good surface drifter was finally developed, standardized, and deployed in large numbers. The so-called WOCE drifters have good water-following characteristics and the slip of the drogue has been calibrated in different conditions. The WOCE drifter works fairly reliably and often survives longer than a year at sea. As of March 2007 there were around 1300 drifters being tracked in the oceans as part of the Global Drifter Program. Data assembly and quality control is performed by the Drifter Data Assembly Center at the National Oceanic and Atmospheric Administration (NOAA) in Miami, Florida. Recent analyses include mapping surface velocity over broad regions and the generation of maps of mean sea level pressure based on drifter measurements.

The WOCE drifter consists of a spherical surface float 35 cm in diameter, a 0.56-cm diameter plastic-impregnated wire tether, with a 20-cm diameter subsurface float located at 275 cm below the surface and a drogue in the shape of a 644-cm-long cloth cylinder 92 cm in diameter with circular holes in its sides (**Figure 2**). The fiberglass surface float contains a radio transmitter, batteries, antenna, and sensors including a thermometer and a submergence sensor that indicates if the drogue is attached. Additional sensors can be added to measure conductivity, atmospheric pressure, light, sound, etc. The basic WOCE drifter costs around $2500, ready for deployment.

The WOCE drifter's drogue is centered at a depth of 15 m below the sea surface. The ratio of the drag area of the drogue to the drag area of tether and float is around 41:1, which results in the drogue's slip through the water being less than $1 \, \mathrm{cm \, s^{-1}}$ in winds of $10 \, \mathrm{m \, s^{-1}}$. The slip was measured to be proportional to wind speed and inversely proportional to the drag area ratio. From this information, the slip can be estimated and subtracted from the drifter velocity.

Drogue Depth

Drogues have been placed at many different depths to suit particular experiments. The drogues of Coastal Ocean Dynamics Experiment (CODE) drifters developed by Russ Davis were located in the upper meter of the water column to measure the surface velocity. WOCE drogues are placed at 15 m

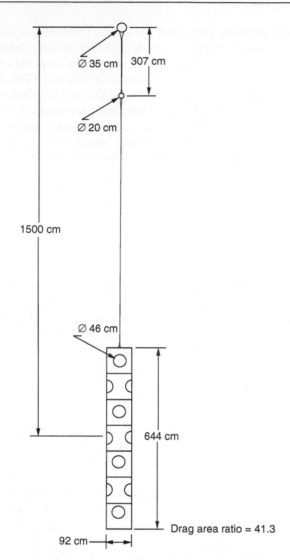

Figure 2 Schematic of WOCE surface drifter. Adapted from Sybrandy AL and Niiler PP (1991) *WOCE/TOGA Lagrangian Drifter Construction Manual*, SIO Reference 91/6, WOCE Report No. 63. La Jolla, CA: Scripps Institution of Oceanography.

to measure a representative velocity in the Ekman layer but below the fastest surface currents. Many scientists have deployed drogues at around 100 m to measure the geostrophic velocity below the Ekman layer. An argument for the 100-m depth is that it is better to place the drogue below the complicated velocity structures in the Ekman layer, Langmuir circulations, and near-surface convergence regions. An argument against the 100-m depth is that the drag of the long tether and surface float in the relatively fast Ekman layer could create excessive slip of the drogue and bias the drifter measurement of geostrophic velocity. The controversy continues. The 15-m depth is widely used today, but many earlier drifters had deeper drogues at around 100 m. Some drogues have been placed as deep as 500–1000 m

often to track subsurface coherent eddies such as meddies.

Subsurface Floats

Many kinds of freely drifting subsurface floats are being used to measure ocean currents, although most floats are usually either autonomous or acoustic. An autonomous float measures a series of subsurface displacements and velocities between periodic surface satellite fixes. An acoustic float measures continuous subsurface trajectories and velocities using acoustic tracking. Acoustic floats provide high-resolution ocean trajectories but require an acoustic tracking array and the effort to calculate subsurface positions, both of which add cost. Thousands of autonomous and acoustic floats have been deployed to measure the general circulation in the world ocean at various depths but concentrated near 800 m. Historical and WOCE era float data can be seen and obtained on the WOCE float website along with references to a series of detailed scientific papers, and newer float data on the Argo website.

WOCE Autonomous Float

The autonomous WOCE float was developed in the 1990s by Russ Davis and Doug Webb. The float typically drifts submerged for a few weeks at a time and periodically rises to the sea surface where it transmits data and is positioned by the Argos satellite system. After around a day drifting on the surface, the float resubmerges to its mission depth, typically somewhere in the upper kilometer of the ocean, and continues to drift for another few weeks. Around 100 round trips are possible over a lifetime up to 6 years.

The float consists of an aluminum pressure hull 1 m in length and 0.17 m in diameter (**Figure 3**). A hydraulic pump moves oil between internal and external bladders, forcing changes of volume and buoyancy and enabling the float to ascend and descend. An antenna transmits to Argos and a damping plate keeps the float from submerging while it is floating in waves on the sea surface. Some floats are drogued to follow a pressure surface; others can be programmed with active ballasting to follow a particular temperature or density surface; more complicated sampling schemes are possible.

Autonomous floats have been equipped with temperature and conductivity sensors to measure vertical profiles as the floats rise to the surface. Electric potential sensors have been added to some floats by Tom Sanford and Doug Webb in order to measure vertical profiles of horizontal velocity. These

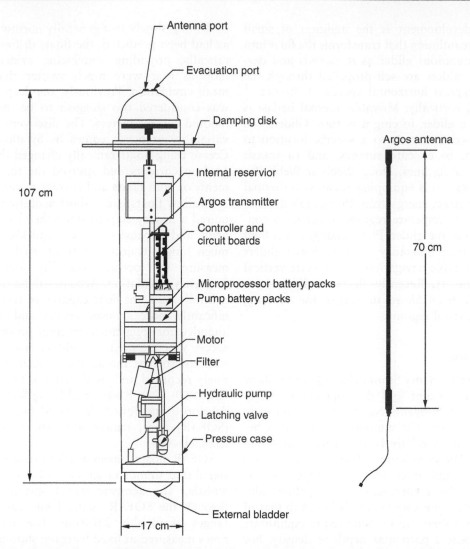

Figure 3 Schematic of Autonomous Lagrangian Circulation Explorer (ALACE) float. For ascent, the hydraulic pump moves oil down from an internal reservoir to an external bladder. For descent, the latching valve is opened, allowing oil to flow back into the internal reservoir. The antenna shown at the right is mounted on the top hemispherical end cap. Adapted from Davis RE, Webb DC, Regier LA, and Dufour J (1992) The Autonomous Lagrangian Circulation Explorer (ALACE). *Journal of Atmospheric and Oceanic Technology* 9: 264–285.

floats were recently used to measure the ocean response of hurricanes.

Starting in 2000, an array of profiling floats began to be launched as part of an international program called Argo. Plans are to build up the float array reaching 3000 profiling floats by 2007 and to replace them as they are lost. As of March 2007, there are around 2800 Argo floats operational. The floats profile temperature and salinity to a depth of 2000 m and measure velocity at the drift depth near 1000 m. Profiles of temperature and salinity are being used to map large areas of the ocean including velocity at the drift depth and are being incorporated into predictive numerical models. The profiles are being combined with earlier and sparser hydrographic profiles to document oceanic climate changes.

The basic drift data from an autonomous float are subsurface displacements or velocity vectors between surface satellite fixes or between extrapolated positions at the times of descent and ascent. Errors in position are estimated to be around 3 km. The surface drifts cause gaps in the series of subsurface displacements, so the displacements cannot be connected into a continuous subsurface trajectory. Subsurface displacements are typically measured over several weeks, which attenuates the higher-frequency motions of ocean eddies. The main benefit of these floats is that they can be used to map the low-frequency ocean circulation worldwide relatively inexpensively. The cost of a WOCE profiling autonomous float is around $16 000 (cheaper than a day of an ocean-going ship).

A recent development is the addition of small wings plus streamlining that transforms the float into a simple autonomous glider as it ascends and descends. These gliders are self-propelled through the ocean with typical horizontal speeds of $30 \, \text{cm s}^{-1}$ while moving vertically. Movable internal ballast is used to bank a glider, forcing it to turn. Gliders can be programmed to return to a specific location to hold position, to execute surveys, and to transit ocean basins along lines. Doug Webb at Webb Research Corporation is equipping some with thermal engines that extract energy from the ocean's thermal stratification in temperate regions in order to continuously power the glider. Phase changes of a fluid are used to force buoyancy changes. Some gliders incorporate suitable navigation and measure vertical profiles of velocity. Recently, fleets of gliders have been directed from shore to survey the evolving structure of coastal regions.

Acoustic Floats

In the mid-1950s, Henry Stommel and John Swallow pioneered the concept and development of freely drifting neutrally buoyant acoustic floats to measure subsurface currents. The method uses acoustics because the ocean is relatively transparent to sound propagation. The deep sound channel centered at a depth around 1000 m enables long-range acoustic propagation. The compressibility of hollow aluminum and glass pressure vessels is less than that of water, so that a float can be ballasted to equilibrate and remain near a particular depth or density. For example, if the float is displaced too deep, it compresses less than water and becomes relatively buoyant, rising back to its equilibrium level, which is consequently stable. Once neutrally buoyant, a float can drift with the currents at that depth for long times.

In 1955, Swallow built the first successful floats (since called Swallow floats) and tracked them for a few days by means of hydrophones lowered from a ship. A moored buoy provided a reference point for the ship positioning. The first pressure hulls were made out of surplus aluminum scaffolding tubes; Swallow thinned the walls with caustic soda to adjust compressibility and buoyancy. Although several floats failed, two worked successfully, which led to further experiments. In 1957, Swallow tracked deep floats as they drifted rapidly southward offshore of South Carolina, providing the first convincing proof of a swift, narrow southward flowing deep western boundary current previously predicted by Stommel. A second experiment in 1959 tracked deep Swallow floats in the Sargasso Sea west of Bermuda. Instead of drifting slowly in a generally northward direction as had been predicted, the floats drifted fast and erratically, providing convincing evidence of eddy motions that were much swifter than long-term mean circulation. Previously, the deep interior flow was considered too sluggish to be measured with moored current meters. The discovery of mesoscale variability or ocean eddies by Swallow and James Crease using floats radically changed the perception of deep currents and spurred the further developments of both floats and current meters.

Swallow floats had a short acoustic range and required a nearby ship to track them, which was difficult and expensive. It was quickly realized that much longer trajectories were needed in order to measure the ocean variability and the lower-frequency circulation. Accomplishing this required a neutrally buoyant float capable of transmitting significantly more acoustic energy and operating unattended for long times at great pressures. Second, access was required to military undersea listening stations, so that the acoustic signals could be routinely recorded and used to track the floats. In the late 1960s, Tom Rossby and Doug Webb successfully developed and tested the sound fixing and ranging (SOFAR) float, named after the SOFAR acoustic channel.

SOFAR floats transmit a low-frequency (250 Hz) signal that sounds in air somewhat like a faint boat whistle. The acoustic signal spreads horizontally through the SOFAR channel and can be heard at ranges of roughly 2500 km. The acoustic arrival times measured at fixed listening stations are used to calculate distances to the float and to triangulate its position. The first success with a SOFAR float drift of four months in 1969 led to further developments and the first large deployment of floats in 1973 as part of Mid-Ocean Dynamics Experiment (MODE). Very interesting scientific results using the float data led to many more experiments and wider use of floats. Later improvements included swept-frequency coherent signaling in 1974, active depth control in 1976, higher power for longer range in 1980, and microprocessors and better electronics in 1983. Moored autonomous undersea listening stations were developed in 1980, freeing experiments from military stations and enabling floats to be tracked in the Gulf Stream and other regions for the first time. SOFAR floats are large ($\sim$5-m long) and heavy ($\sim$430 kg), which makes them difficult to use in large numbers. In 1984, Rossby developed the RAFOS (SOFAR spelled backward) float, a much smaller, cheaper float that listens to moored sound sources and at the end of its mission surfaces and reports back data via satellite. This float made it much easier and cheaper to conduct larger experiments; this style of float was improved and tracked in large numbers in the North and South Atlantic as part of WOCE. Various float groups have collaborated in tracking floats at different depths and in maintaining moored tracking arrays.

WOCE RAFOS Float

The modern acoustic RAFOS float consists of a glass hull 8.5 cm in diameter and 150–200-cm long, enclosing an electronic package, Argos beacon, and temperature and pressure sensors (**Figure 4**). An acoustic transducer and external drop weight are attached to an aluminum end cap on the bottom. RAFOS floats are capable of operating at depths from just below the sea surface to around 4000 m. Usually several times per day they listen and record the times of arrival of 80 s 250-Hz acoustic signals transmitted

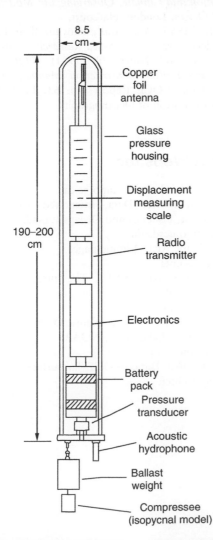

Figure 4 Schematic of RAFOS acoustic float. Adapted from Rossby HT, Dorson D, and Fontaine J (1986) The RAFOS system. *Journal of Atmospheric and Oceanic Technology* 3: 672–679.

from an array of moored undersea sound sources. At the end of the mission, a few months to a few years in length, the float drops an external weight, rises to the sea surface, and transmits recorded times of arrival, temperatures, and pressures to the Argos system. The float remains drifting on the surface for roughly a month before all the data are received and relayed ashore by satellite. A typical float costs $4000–5000 and is considered expendable because it is difficult and expensive to retrieve. The times of arrival are used with the known transmit times of sources and the estimated speed of sound to triangulate the float's position.

A drifting RAFOS float closely follows a pressure surface. A compressee consisting of a spring and piston in a cylinder is sometimes suspended below a RAFOS float, so that it matches the compressibility of seawater. If the compressibilities are the same, the float will remain on or close to a constant density surface and more closely follow water parcels. Some floats have active ballasting and can track a column of water by cycling between two density surfaces.

To ballast a RAFOS float, it is weighed in air and water, which gives its volume. Its compressibility is measured by weighing the float at different pressures in a water-filled tank. The amount of weight to be added to make the float neutrally buoyant at the target depth (or density) is calculated using the compressibility and thermal expansion of the float and the temperature and density of the water in the tank and at the target depth. Floats usually equilibrate within 50 m of their target depths or density.

Some floats combine acoustic tracking with the active buoyancy of the autonomous float, so that the float can periodically surface and relay data to shore at intervals of a few months. This avoids the long wait for multiyear RAFOS floats to surface and avoids the loss of all data should a float fail during its mission. French MARVOR floats developed by Michel Ollitrault report data back every 3 months and typically survive for 5 years.

Drogues have been added to neutrally buoyant floats by Eric d'Asaro to enable them to better measure three-dimensional trajectories. Vertical velocities from these floats are especially interesting in the upper ocean and in the deep convective regions like the Labrador Sea in winter. Another technique used to measure vertical water velocity is the addition of tilted vanes attached to the outside of a float. Water moving vertically past the vanes forces the float to spin and this is measured and recorded.

At least two moored sound sources are required to position a RAFOS float. Often three or more are used to improve accuracy. The sources transmit an 80-s swept-frequency 250-Hz signal a few times per day for up to 5 years. Sources are similar to the old SOFAR floats and cost around $33 000. Mooring costs of wire rope, flotation, acoustic release, and other recovery aids can double this figure. Recently, louder, more efficient, and more expensive sound sources have increased tracking ranges up to 4000 km.

Errors of acoustic positioning are difficult to estimate and vary depending on the size and shape of the tracking array, the accuracy of float and source clocks, how well the speed of sound is known, etc. Estimates of absolute position errors range from a few kilometers up to 10 km (or more). Fix-to-fix relative errors are usually less than this because some errors cancel and others such as clock errors vary slowly in time. Corrections are made for the Doppler shift caused by a float's movement toward or away from a source. The typical correction amounts to around 1.3 km for a speed of $10\,\mathrm{cm\,s}^{-1}$. Tides and inertial oscillations add high-frequency noise to positions and velocities, but since a float integrates these motions, it provides an accurate measure of lower-frequency motions.

Further Reading

Burns LG (2007) *Tracking Trash: Flotsam, Jetsam, and the Science of Ocean Motion*, 56pp. Boston, MA: Houghton Mifflin.

D'Asaro EA, Farmer DM, Osse JT, and Dairiki GT (2000) A Lagrangian float. *Journal of Atmospheric and Oceanic Technology* 13: 1230–1246.

Davis RE (2005) Intermediate-depth circulation of the Indian and South Pacific Oceans measured by autonomous floats. *Journal of Physical Oceanography* 35: 683–707.

Davis RE, Sherman JT, and Dufour J (2001) Profiling ALACEs and other advances in autonomous subsurface floats. *Journal of Atmospheric and Oceanic Technology* 18: 982–993.

Davis RE, Webb DC, Regier LA, and Dufour J (1992) The Autonomous Lagrangian Circulation Explorer (ALACE). *Journal of Atmospheric and Oceanic Technology* 9: 264–285.

Gould WJ (2005) From swallow floats to Argo – the development of neutrally buoyant floats. *Deep-Sea Research* 52: 529–543.

Griffa A, Kirwan AD, Mariano AJ, Rossby T, and Ozgokmen TM (eds.) (2007) *Lagrangian Analysis and Prediction of Coastal and Ocean Dynamics.* Cambridge, MA: Cambridge University Press.

Niiler PP, Davis RE, and White HJ (1987) Water following characteristics of a mixed layer drifter. *Deep-Sea Research* 24: 1867–1881.

Pazan SE and Niiler P (2004) New global drifter data set available. *EOS, Transactions of the American Geophysical Union* 85(2): 17.

Richardson PL (1997) Drifting in the wind: Leeway error in shipdrift data. *Deep-Sea Research I* 44: 1877–1903.

Rossby HT, Dorson D, and Fontaine J (1986) The RAFOS system. *Journal of Atmospheric and Oceanic Technology* 3: 672–679.

Sanford TB, Price JF, Webb DC, and Girton JB (2007) Highly resolved observations and simulations of the ocean response to a hurricane. *Geophysical Research Letters* 34: L13604.

Siedler G, Church J, and Gould J (eds.) (2001) *Ocean Circulation and Climate, Observing and Modelling the Global Ocean.* London: Harcourt.

Swallow JC (1955) A neutrally-buoyant float for measuring deep currents. *Deep-Sea Research* 3: 74–81.

Sybrandy AL and Niiler PP (1991) *WOCE/TOGA Lagrangian Drifter Construction Manual*, SIO Reference 91/6, WOCE Report No. 63. La Jolla, CA: Scripps Institution of Oceanography.

Relevant Websites

http://www.meds-sdmm.dfo-mpo.gc.ca
– Archived Drifter Data, Integrated Science Data Management, Fisheries and Oceans Canada.

http://www.argo.ucsd.edu
– Argo Floats, ARGO home page.

http://www.argo.net
– Argo Floats, ARGO.NET, The International Argo Project Home Page.

http://www.aoml.noaa.gov
– Global Drifter Program, Atlantic Oceanographic and Meteorological Laboratory, NOAA.

http://wfdac.whoi.edu
– WOCE Float Data, WOCE Subsurface Float Data Assembly Center.

MOORINGS

R. P. Trask and R. A. Weller, Woods Hole
Oceanographic Institution, Woods Hole, MA, USA

Introduction

The need to measure ocean currents throughout the water column for extended periods in order to better understand ocean dynamics was a driving force that led to the development of oceanographic moorings. Today's moorings are used as 'platforms' from which a variety of measurements can be made. These include not only the speed and direction of currents, but also other physical parameters, such as conductivity (salinity), temperature, and sea state, as well as surface meteorology, bio-optical parameters, sedimentation rates, and chemical properties.

Moorings typically have three basic components: an anchor, some type of chain or line to which instrumentation can be attached, and flotation devices that keep the line and instrumentation from falling to the seafloor. Shackles and links are typically used to connect mooring components and to secure instruments in line. The choice of hardware, line, and flotation for a particular application, as well as the size and design of the anchor, depends on the type of mooring and the environment in which it is deployed.

Most moorings fall into two broad categories – surface and subsurface. The main difference between the two is that the surface mooring has a buoy floating on the ocean surface, whereas the subsurface mooring does not. Although the two mooring types have similar components, the capabilities of the two are very different. With a surface buoy, it is possible to measure surface meteorology, telemeter data, and make very near-surface measurements in the upper ocean. The surface mooring, however, is exposed to ocean storms with high wind and wave conditions and therefore must be constructed to withstand the forces associated with those environmental conditions. In addition, the wave action may transmit some unwanted motion to subsurface instruments if care is not taken. The subsurface mooring, on the other hand, is away from the surface forcing and can be fabricated from smaller, lighter components, which are less expensive and easier to handle. However, it is difficult to make near-surface measurements from a subsurface mooring.

Early attempts at mooring work in the 1960s began with surface moorings. Problems with the mooring materials and the dynamic conditions encountered at the ocean surface resulted in poor performance by these early designs, and attention turned to developing subsurface moorings. The introduction of wire rope as a material for fabricating mooring lines and the advent of a remotely triggered mechanism to release the mooring's anchor were significant milestones that helped make the subsurface mooring a viable option. It has since proved to be a very successful oceanographic tool. Interest in the upper ocean and the air–sea interface prompted a reexamination of the surface mooring design. The evolution of the subsurface mooring as a standard platform for oceanographic observations and the more recent development of reliable surface moorings are summarized here.

Subsurface Mooring Evolution

Early moorings consisted of a surface float, surplus railroad car wheels for an anchor, and lightweight synthetic line, such as polypropylene or nylon, to connect the surface float to the anchor. Several kilometers of line are required for a full-depth ocean mooring, and weight of the line itself, even in water, is not negligible. Instrumentation was connected to the synthetic line along its length. The anchor was connected to the mooring line by means of a corrosible weak link. The initial method of recovering the moorings was to connect to the surface float and pull it with the hope that the tension would break the weak link leaving the anchor behind. Unfortunately, at the time of recovery, the mooring line was often weaker than the weak link and would break, allowing the line and instrumentation below the break to fall to the seafloor.

Studies showed that the synthetic ropes were being damaged by fish. Analysis of many failed lines revealed tooth fragments and bite patterns that were used to identify the type of fish responsible for the damage. Statistics concerning the number of fishbites, their depths, and their locations were collected, and it was found that the majority of fishbites occurred in the upper 1500–2000 m of the water column. Prevention of mooring failure due to fish attack required lines that could resist fishbite.

Ropes made of high-strength carbon steel wires were an obvious candidate. Wire rope would not only provide protection from fish attack, but also would have minimal stretch, unlike the synthetic ropes, and would provide high strength with relatively low drag. Many types of wire rope construction and sizes were tested, in addition to methods for terminating the wire rope; terminations are the fittings attached to the ends of wire sections. In constructing a mooring whose components can be shipped separately and handled safely on the deck of a ship at sea, the practice is to cut the wire into sections of specific lengths (shots) that allow connection to other wire shots or to instrumentation in series (end to end). A desirable termination is one that is as strong as the wire rope itself. If the technique used to terminate a rope imposes stress concentrations, which significantly reduce the strength of the wire rope, then the whole system is weakened. Methods of terminating wire include the formation of eyes into which shackles can be attached either from swaged fittings or from zinc- or resin-poured sockets. Swaged terminations utilize a fitting that is slid onto the end of the wire and pressed or swaged onto the wire with a hydraulic press. In the case of a poured-socket termination, the wire is inserted into the socket and the individual wires are splayed outward or 'broomed out'. Once the wires are properly cleaned and positioned, a filler material (molten zinc or uncured epoxy resin) is poured into the socket and allowed to harden. A strain relief boot is often used in conjunction with a swaged fitting termination, as well as with the poured sockets. The boots are often an injection-molded urethane material designed to extend from the fitting out over a short section of the wire to minimize the bending fatigue that can occur between the flexible wire and rigid fitting.

At present, galvanized 3×19 wire rope is widely used for oceanographic applications. The designation '3×19' denotes three strands or groups, each with 19 individual wires: the 19 wires are twisted together to form a strand. Three strands are then wound together to form the rope. The rotation characteristics of wire rope are critically important in certain oceanographic applications. If the rope has the tendency to spin or rotate excessively when placed under tension, there is a tendency for that wire to develop loops when the tension is reduced quickly. If the load is quickly applied again to the line, the loops are pulled tight into kinks, which can severely weaken the wire rope. Wire rope with minimal rotation characteristics is preferred for mooring applications, particularly surface mooring work. Wire ropes are available with varying degrees of rotation resistance. Swivels are sometimes placed in series with the wire to minimize the chances of kink formation. In addition to using galvanized wires when fabricating the rope to provide protection

against corrosion, some wire ropes have a plastic jacket extruded over the wire. Types of plastics used for jacketing materials include polyvinyl chloride, polypropylene, and high-density polyethylene.

In the early years, mooring recoveries that were initiated by pulling on deteriorated mooring lines often resulted in line breakage and instrument loss. A preferable approach was to detach the mooring from its anchor prior to hauling on the mooring line. This would reduce the load on the mooring line since the line would never 'feel' the weight of the anchor nor the tension required to pull the anchor out of the bottom sediments. This approach became possible with the development of an acoustically commanded anchor release. The acoustic release is deployed inline on the mooring and is typically positioned below all instrumentation and close to the anchor. To activate the release mechanism, a coded acoustic signal is sent from the recovery vessel. The acoustic release detects the signal and disconnects from the anchor.

When mooring work was in its infancy, the surface buoy was a vital, visible link to the mooring below. Without it, the exact location of the mooring was unknown. The introduction of acoustic releases not only provided a way to disconnect the line and instruments from the anchor, but also provided a way to locate the exact position of the mooring by acoustic direction finding. This eliminated the need for a surface float, which at that time, before the development of meteorological sensors for buoys, was used solely for recovery purposes. Instead of having a mooring that stretched all the way from the ocean bottom to the surface, the mooring was shortened so that the top of the mooring was positioned below the surface of the water. Sufficient buoyancy was placed at the top of the mooring to keep all of the mooring components as vertical as possible throughout the water column. With this design, which became known as the subsurface mooring, the mooring would ascend to the surface once it was acoustically released from its anchor. At the surface, it could be pulled out of the water by the waiting recovery vessel. A great advantage of the subsurface mooring is having the hardware below the ocean surface, which is the most dynamic part of the water column. As a result, there is a considerable reduction in component fatigue due to surface-wave action. In addition, the mooring is no longer visible to surface vessels and is less vulnerable to vandalism.

The buoyancy used on subsurface moorings is usually in the shape of a sphere, because of its low drag coefficient. Other shapes have been used depending on the specific application. Various materials are used, including steel spheres, glass spheres with protective plastic covers, ceramic spheres, and syntactic foam spheres. Syntactic foams consist of small pressure-resistant glass microspheres (2–300 μm in diameter), as well as larger glass fiber-reinforced spheres (0.15–10 cm in diameter) embedded in a thermosetting plastic binder. An advantage of the syntactic foam is that it can be molded to form custom shapes. Unlike the steel spheres, whose use is depth-limited (maximum working depth is approximately 1000 m), the syntactic foam can be engineered to withstand full ocean-depth pressures. In addition, it can be designed to provide the same buoyancy as a string of glass balls (commonly 43 cm in diameter) with considerably less drag. This makes syntactic foam spheres attractive for use in high-current regimes, since the drag on the mooring will be less; consequently, less buoyancy will be needed to keep the mooring near-vertical.

Subsurface moorings with a single element of buoyancy at the top are still at some risk. Should the buoyant element be lost or damaged, the mooring would fall to the bottom, leaving no secondary means of bringing it back to the surface when the acoustic release mechanism is activated. To provide a higher degree of reliability, buoyancy is often provided in the form of glass balls attached along the length of the mooring. In addition, buoyancy is often added to the bottom of the mooring, just above the acoustic release, to provide what is sometimes referred to as the 'backup recovery'. With this design feature (**Figure 1**), should a mooring component fail and the upper part of the mooring be lost, no matter where the failure occurs, there should be sufficient buoyancy below that point to bring the remaining section of the mooring back to the surface. Instrumentation that would otherwise have been lost if deployed with a single buoyant element is recoverable with this configuration. Equally important, the recovery provides the opportunity to identify the failed component and correct the problem. As a recovery aid, pressure-activated submersible satellite transmitters are frequently installed on the upper buoyancy sphere. In the event of a mooring component failure that causes the top of the mooring to surface, it can be tracked via satellite. This allows for possible recovery of whatever instrumentation hangs below.

The number of measurements made from a subsurface mooring depends on several factors including the load the mooring is designed to support and the resources available for instrumenting the mooring. Depending on the resources, a mooring may have on the order of 10 instruments located at discrete depths. If one or more of the instruments malfunction during their deployment, the data from that particular depth are missing, but there may be other

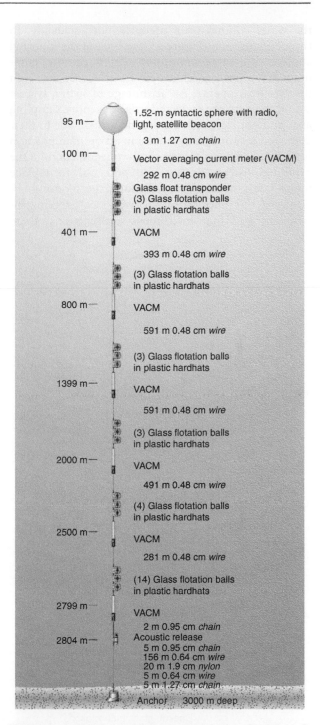

Figure 1 A typical subsurface mooring design. Design by S. Worrilow.

instruments possibly above and below the failed unit that continue to collect data. Such a design requires numerous lengths of terminated wire, which must be deployed in a particular order so the individual instruments end up at the desired depths.

An alternative approach is to have a single instrument that uses the mooring wire as a guide along which an instrument moves up and down profiling the water column on a predetermined schedule. In this configuration (**Figure 2**), the bulk of the mooring consists of a single continuous length of wire, eliminating the need for multiple lengths used between discrete instruments. An advantage of such a system is that the data are collected from the entire sampling area and not from specific depths, which, due to their deployed location, may not be in an optimal position for observing an interesting phenomenon. One profiling instrument has the potential to replace many instruments deployed along the mooring. If, however, the mooring only carries one profiling instrument and that instrument malfunctions, there

is a chance that the mooring may not collect any data. For that reason, it is not uncommon for a mooring to have a combination of instrument types, which include profiling as well as several discrete instruments.

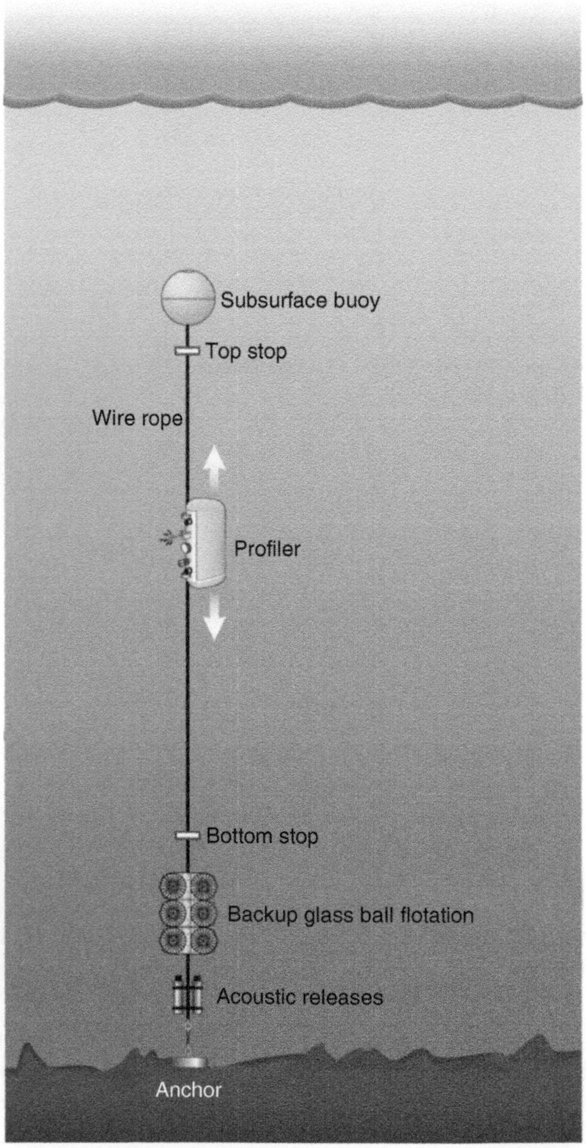

Duplicate measurements by different instruments can improve the chances of collecting a full data set. Since redundant moorings are seldom an option, a mooring failure can have a catastrophic impact on the total data return. With both instruments and mooring components, attention to detail is critical. The care taken in preparing and testing instrumentation, in the selection of quality mooring components, and in the fabrication techniques utilized is often a deciding factor. Other factors that can impact a mooring's success include the quality of the information that went into the design of the mooring, that is, how similar were the predicted environmental conditions to the actual conditions encountered? Is the design unique or is it a variation of a design that has historically worked well? Attention to detail during the actual deployment of the mooring, as well as uncontrollable outside influences such as fishing activities in the area, can also be contributing factors. Despite all the variables, it is possible to routinely achieve success rates that are greater than 90%.

Advances in Surface Mooring Technology

Growing interest in understanding interactions between the ocean and the atmosphere has rekindled interest in using surface moorings. The surface mooring is a unique structure. It extends from above the surface to the ocean bottom, providing a platform from which both meteorological and oceanographic measurements can be made in waters that range from shallow to 5 km in depth. Surface-mooring designs must consider the effects of surface waves, ocean currents, biofouling, and other factors that can vary with the time of year, location, and regional climate and weather patterns. The success of a surface-mooring deployment often depends on the abilities both to accurately estimate the range of conditions that the mooring may encounter while deployed and to design a structure that will survive those conditions. The primary goal of any mooring deployment is to keep the mooring on location and making accurate measurements. Adverse environmental conditions not only influence the longevity itself but also impact the instruments that the mooring supports. It is often very difficult to keep the instruments working under such conditions for long periods.

Surface moorings are used to support submerged oceanographic instrumentation from very close to

Figure 2 A subsurface mooring schematic with a profiling instrument that runs up and down the mooring line on a predetermined schedule. Illustration by J. Doucette/WHOI Graphics.

the surface (sometimes floating at the surface) to near the bottom, which is typically 5 km in depth. Measurements of physical properties, such as temperature, velocity, and conductivity (salinity), as well as of biological parameters, such as photosynthetically available radiation (PAR), beam transmission, chlorophyll fluorescence, and dissolved oxygen, are routinely made from surface moorings. The surface buoy also provides a platform from which meteorological measurements can be made and a structure from which both surface- and subsurface-collected data can be telemetered via satellite. Meteorological sensors typically deployed on a surface buoy measure wind speed, wind direction, air temperature, relative humidity, barometric pressure, precipitation, and long-wave and short-wave radiation. The meteorological data are stored in memory and telemetered via satellite to a receiving station ashore. The telemetered data often play an important part in real-time analysis and reaction to conditions on site. The data can also be passed to weather centers for forecasting purposes.

There are a number of different types of surface buoys. Some shapes have been in use since the early days of mooring work, and others are relatively new. Buoy shapes include the toroid or 'donut', the discus, and the hemispherical hull. The toroid hull in various configurations is widely used throughout the scientific community. Where a significant amount of instrumentation must be supported, a discus-shaped buoy with as much as 6800 kg of buoyancy may be used for both deep- and shallow-water applications. Some buoys are designed with modular buoyancy elements that can be added to maximize the available buoyancy. The discus buoy design is widely used by the US National Data Buoy Center in Mississippi in coastal waters, at the Great Lakes stations, and for directional wave measurements. The 3-m discus-shaped hull was also adopted by the Atmospheric Environment Service in Canada for its coastal buoys. Smaller discus-shaped hulls are used for shallow-water applications.

Buoy hulls are made of aluminum, steel, fiberglass over foam, and various closed-cell foams. Several closed-cell foams are extremely resistant to wear and have low maintenance. Ionomer foam and polyethylene foam are common materials for buoy and fender applications. Depending on the material, various outer skin treatments are used to increase the hull's resilience to wear. These include the application of heat and pressure, as well as bonding a different material, such as urethane, to the exterior of the hull.

The mooring materials used on surface moorings resemble those used on the subsurface moorings. Component sizes are usually increased to compensate for the larger forces and the increased wear. Materials include chain, plastic-jacketed wire rope, and synthetic line. Chain is used directly beneath the buoy for strength, ease of handling, and, because of its additional mass, stabilization of the buoy during its deployment. If the water is sufficiently deep and the design permits, the wire rope is usually extended to a depth of at least 1500 m and often as deep as 2000 m for fishbite protection.

The surface mooring needs some form of built-in 'compliance' (ability to stretch) to compensate for large vertical excursions that the buoy may experience during the change of tides and with passing waves and swell. The compliance also compensates for the buoy being displaced laterally on the surface by the drag forces associated with ocean currents and prevents the buoy from being pulled under when such forces are applied. In deep-water applications, compliance is provided through the use of synthetic materials, such as nylon. The synthetic line acts like a large rubber band that stretches as necessary to maintain the connections between the surface-following buoy and the anchor on the bottom.

A challenge in the design process, particularly in shallow water, is to achieve an appropriate mix of compliant materials and fishbite-resistant materials, which tend to be unstretchable. The 'scope' of the mooring – the ratio of the total unstretched length of the mooring components to the water depth – can be one of the sensitive design factors. A mooring with a scope of less than 1.0 relies on the stretch of the nylon for the anchor to reach the bottom. Such a taut mooring remains fairly vertical with a relatively small watch circle (the diameter of the area on the ocean surface where the buoy can move about while still anchored to the ocean bottom), but it carries a penalty: such a vertical mooring is under considerable tension, or 'preloaded', at the time of deployment. Currents and waves impose additional loads beyond the initial preloaded condition. Moorings with scopes between 1.0 and c. 1.1 are generally referred to as 'semi-taut' designs. The mooring shown in **Figure 3** is typical of a semi-taut design.

Early surface moorings were designed using only a static analysis program, which used steady-state current profiles as input to predict mooring performance. However, experience has shown that it is necessary to consider the combined effects of strong currents and surface waves. An investigation of the dynamic effects of surface forcing on the performance of surface moorings found that semi-taut moorings could have a resonant response to forcing in the range of surface wave periods, causing high dynamic loads. These high tensions limit the

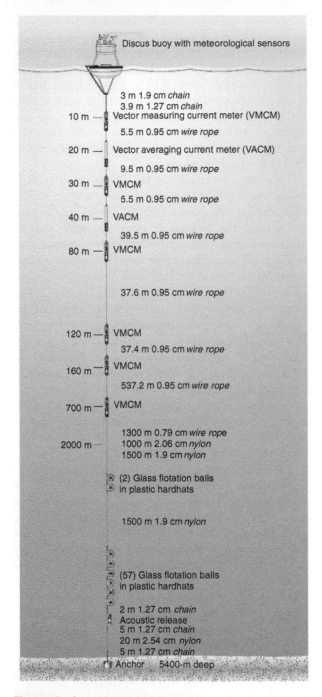

Figure 3 A semi-taut surface mooring design. Design by P. Clay.

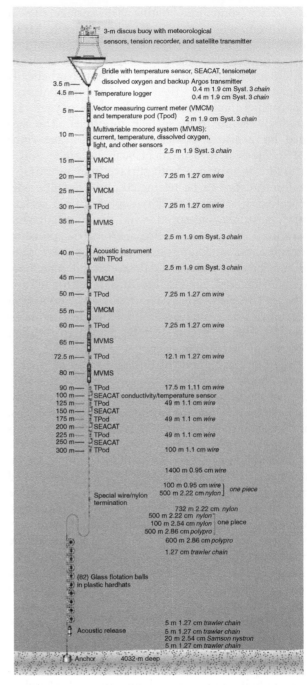

Figure 4 An inverse catenary mooring design. Design by G. Tupper.

instrument-carrying capacity of the mooring and can lead to failure of mooring components.

An alternative design fashioned after the US National Data Buoy Center 'inverse catenary' mooring has evolved in response to difficulties encountered using taut surface mooring designs. With wire rope in the upper part of the mooring and with nylon line spliced to a buoyant synthetic line such as polypropylene below, the inverse catenary design (**Figure 4**)

offers larger scope (typically 1.2) for high-current periods, yet performs well in lesser currents. In low currents, the positively buoyant synthetic line keeps the slightly negatively buoyant nylon from tangling with the rest of the mooring below it. Thus, the inverse catenary design can tolerate a wider range of environmental conditions. The inverse catenary design lowers the static mooring tension, as shown in **Table 1**. The dynamic tension contribution to the total

Table 1 A comparison of semi-taut surface mooring and an inverse catenary design (subjected to the same ocean current forcing)

	Semi-taut	Inverse catenary	Difference
Mooring scope	1.109	1.285	
Tension at the buoy (kg)	2065	1602	463
Anchor tension (kg)	2292	1783	509
Horizontal excursion (m)	1208	1735	527

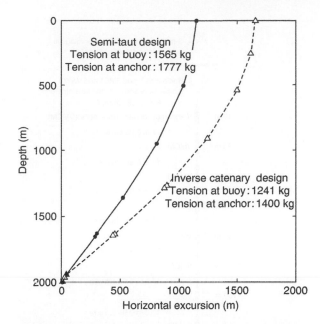

Figure 5 A comparison of the shape of a semi-taut design with that of an inverse catenary design when subjected to the same ocean current forcing. Note the differences in buoy and anchor tensions as well as the horizontal excursions at the surface.

tension, however, is unchanged, and care must still be taken in the design process to prevent the mooring from having a resonant response to forcing in the range of surface wave periods.

In some regions of the world's oceans, the dynamic loading due to high wind and sea state conditions may be so severe that ultimate strength considerations are superseded by the fatigue properties of the standard hardware components. In these cases, in addition to appropriate mooring design, attention must be paid to the choice and preparation of mooring hardware. Cyclic fatigue tests revealed that, in certain applications, mooring hardware that had been used reliably in the past lost a significant part of its service life owing to fatigue and either failed or showed evidence of cracks. Where possible, different hardware components that are less susceptible to fatigue failure in the range of expected tensions are now substituted.

In situations where there is no replacement hardware available, the fatigue performance is improved by shot peening. Shot peening is a process whereby a component is blasted with small spherical media, called shot, in a manner similar to the process of sand blasting. The medium used in shot peening is more rounded rather than angular and sharp, as in sandblasting. Each piece of shot acts like a small ball-peen hammer and tends to dimple the surface that it strikes. At each dimple site, the surface structure of the material is placed in tension. Immediately below the surface of each dimple, the material is highly stressed in compression so as to counteract the tensile stress at the surface. A shot-peened part with its many overlapping dimples, therefore, has a surface layer with residual compressive stress. Cracks do not tend to initiate or propagate in a compressive stress zone. Since cracks usually start at the surface, a shot-peened component will take longer to develop a crack, thereby increasing the fatigue life of the part.

With both the semi-taut and the inverse catenary surface mooring designs, it is difficult to make deep-current measurements because the mooring line at these depths is sometimes inclined more than 15° from vertical. This is a problem for two reasons: first, some instruments fitted with compasses do not work well if the compass is inclined more than 15°; and second, some velocity sensors require the instrument to be nearly vertical. An inverse catenary mooring, with its greater scope, has inclination problems at shallower depths as compared to the semi-taut design. **Figure 5** compares the mooring shape of a semi-taut design with that of an inverse catenary mooring subjected to the same environment conditions.

In addition to the inclination problem, there is also a depth-variability problem. Compliant members on a surface mooring are usually synthetics, which must be placed below the fishbite zone (nominally 2000-m depth). The deep instruments are, therefore, in line in the synthetics; their depth can vary by several hundred meters depending on the stretch of the material. A pressure sensor on the instrument can be used to record the instrument depth, but if a particular depth is desired, it is not possible with the conventional design. Hence, the trade-off for being able to withstand a wider range of environmental conditions is a reduction in the depth range for making certain kinds of measurements.

A partial solution to the problem of deep measurements on a surface mooring is illustrated in the mooring design shown in **Figure 6**, which combines features of both the subsurface- and inverse

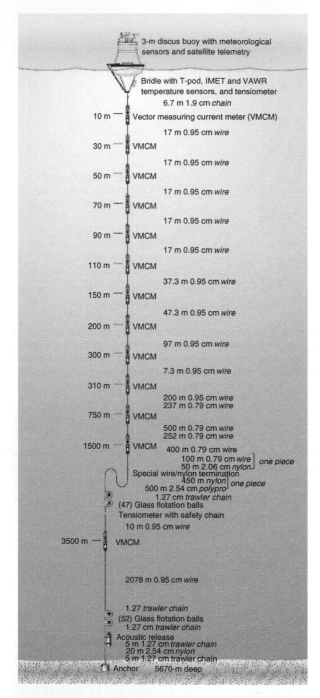

Figure 6 A mooring design that combines the features of an inverse catenary mooring with those of a semi-taut mooring in order to improve the quality of deep (3500 m) current measurements made from a surface mooring. Design by G. Tupper.

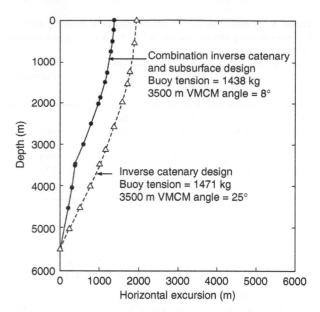

Figure 7 A comparison of the shapes of two mooring designs subjected to the same ocean current forcing, showing the differences in mooring inclination at 3500-m depth.

release near the anchor and another immediately above the 3500 m instrument. The compliance of the mooring consists of 1500 m of nylon and polypropylene inserted between the 3500 m instrument and the base of the wire at 2000 m. The combination of nylon and polypropylene gives the mooring enough stretch (from the nylon) and built-in buoyancy (from the polypropylene) to handle the range of expected current conditions. The polypropylene actually performs a double duty in that during low current periods the buoyant polypropylene keeps the excess nylon from tangling with the lower part of the mooring; when the currents increase, that buoyant member becomes available in the form of extra scope. The shape of the combination mooring design is compared with the shape of an inverse catenary design in **Figure 7**.

Surface moorings can also be used as a communications link between the ocean surface and points along the mooring all the way to the seafloor. With the appropriate mooring components, information can be passed in both directions. Data collected by instrumentation deployed on the mooring line or in close proximity to the mooring can be sent to the buoy, where it is transmitted via satellite to a receiving station. Two-way communications between a shore station and the buoy via satellite permit buoy systems to be reprogrammed so as to modify instrument sampling schemes, as well as to repair system malfunctions. Being able to diagnose and repair a buoy data collection system without having to dispatch a vessel to the site is of great value.

catenary-type moorings. The upper 2000 m of the mooring is similar to any surface mooring, with the instrumentation at the appropriate depths and wire rope in between. The lower part of the mooring from the bottom up to the 3500 m instrument is all wire with a cluster of glass ball flotation just above the

Telemetry of data from subsurface instruments on surface moorings is possible through various techniques. One approach is to utilize electromechanical (EM) cable for the transmission of an electrical signal. EM cables typically have two elements, a strength member and conductors. One type of EM cable has electrical conductors in the center with an outer armor of steel wire that provides strength, as well as fishbite protection of the conductors. Another EM cable design that has been used successfully utilizes 3×19 oceanographic cable with three conductors laid in the valleys that are formed by the three strands. The plastic jacket is then extruded over the wire and conductors. A disadvantage of this design is that the conductors are on the outside of the strength member and are more susceptible to fishbite damage than in a cable with the outer armor. Mooring cables are also being designed with optical fibers to take advantage of their capability to transmit over long distances with minimal losses, and its inherently high data-carrying capacity.

Not all communications rely solely on special purpose cables. Through the use of high-speed acoustic modems, signals can be sent from instrumentation deployed on the ocean floor to an acoustic transducer located on the mooring. Those signals are then transmitted up the mooring line to the surface buoy and then to a receiving station via satellite. Data from ocean-bottom seismometers that record undersea earthquakes have been monitored in this manner as part of a prototype tsunami-warning network. **Figure 8** depicts ocean-bottom seismographs communicating acoustically with a surface mooring that is linked with a satellite network.

The interface where a mooring transitions from just below the ocean surface to a buoy on the surface is a challenging area in the design of a mooring because the components used in that section are subject to excessive wear and fatigue failure from the constant motion of the surface buoy. To reduce the wear at this juncture, a universal joint is employed between the buoy and the mooring. It is extremely challenging to pass cables with electrical conductors through this interface. The universal joint is configured with a central hole, which provides an unbending pathway for conductors that must pass through the universal. Getting the vulnerable conductors through this dynamic near-surface region and into the buoy is not a trivial part of the design. Several approaches have been taken with reasonable success.

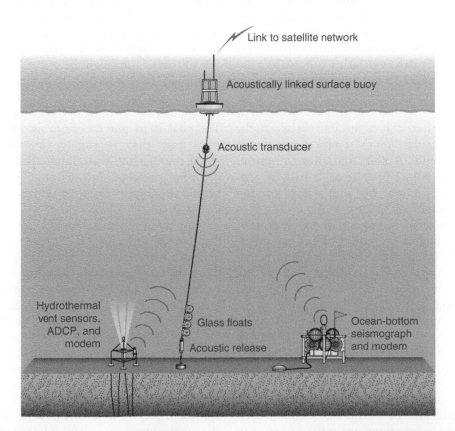

Link to satellite network

Acoustically linked surface buoy

Acoustic transducer

Hydrothermal vent sensors, ADCP, and modem

Glass floats

Acoustic release

Ocean-bottom seismograph and modem

Figure 8 A moored buoy system configured to receive acoustically transmitted data from seafloor instruments in near-real time and able to communicate data and commands between shore-based labs and the observatory via a satellite network. Illustration by J. Doucette/WHOI Graphics.

One technique is to utilize a special chain assembly directly below the buoy consisting of chain that is wrapped with a spiral shaped multiconductor cable and completely encapsulated with urethane. This component has the required strength and flexibility from the chain while the spiraled cable configuration and urethane protects the conductors from bending strain produced by the buoy motion.

Another approach that has been used, particularly in shallow-water applications, is an ultrastretchy rubber hose that is reinforced with nylon and is capable of stretching to twice its unstretched length. Electrical conductors are embedded in the wall of the hose. The angle that the conductors make with respect to the axis of the hose is a critical part of the hose design in order to allow the hose to stretch and not damage the conductors. The system also allows power generated by solar panels on the surface buoy to flow to the bottom instruments.

It is possible to use conventional plastic-jacketed wire rope, without copper electrical conductors, to send a signal up the mooring cable through the use of inductively coupled modems. In such a system, the signal is applied to the primary winding of a toroidal transformer. The mooring wire only has to pass through the toroid, to form a single-turn secondary that conveys the data along the mooring cable. The ends of the mooring cable are grounded to the seawater, which permits a current to flow through the mooring wire and seawater. So as not to have to thread the mooring cable through the toroid, it is split and clamped around the cable. It is not necessary to break the cable at the instrument position or to provide any electrical connection between the sensor and the cable. Another advantage of this system is the flexibility it offers for sensor placement. Since it does not require discrete cable lengths, sensors can be clamped along the wire at any location and easily repositioned if necessary.

Surface moorings have capabilities which make them good candidates for use with ocean observatories. The moorings have the flexibility to be deployed nearly anywhere in the world's oceans and offer near-real-time access due to the satellite communications capability from the surface buoy.

Moored observatories can vary in complexity depending on their intended application. If data must be collected from high-bandwidth sensors, such as those used in acoustic arrays or for continuous seismic monitoring, then high-speed satellite links may be required in conjunction with fiber optic cables connected to junction boxes on the seafloor. Such systems typically require large quantities of power to operate and could require active power generation. The buoy size increases to accommodate such capabilities as does the mooring hardware necessary to keep the structure on station. Other systems, which do not require high-speed satellite communications or have only a moderate number of sensors, do not have the same power requirements as a larger system and can make use of smaller moored platforms with smaller hardware.

Mooring Deployments

Deep-ocean surface and subsurface moorings are typically deployed using an anchor-last technique. As the name implies, the anchor is the last component to be deployed. The entire mooring, starting at the top, is put over the side and strung out behind the deployment vessel and towed into position. At the appropriate location, the anchor is dropped.

If the current and the wind are from the same direction, the deployment begins by positioning the ship down-current of the desired anchor-drop position. By doing this, the ship can maintain steerage as it slowly steams against the current while the mooring components are deployed and are carried away from and behind the ship by the wind and current. When the wind and current are opposing each other, it becomes necessary to alter the deployment plan. In such cases, the important factor is the relative speed of the ship, which is typically $50–100\,\mathrm{cm\,s^{-1}}$ through the water. Depending on the length of the mooring, its complexity, and the wind and current conditions, the start position could be as much as 10 km from the anchor-drop position. The goal is to put the mooring line over the side at a rate that is slightly less than the ship's speed through the water and, thus, have the entire mooring stretched out without kinks and loops behind the ship by the time it arrives at the anchor-drop site.

With the ship at the position for the start of the deployment sequence, the upper buoyancy of the mooring is lowered into the water. **Figures 9(a)–9(g)** illustrate the deployment sequence of a deep-water surface mooring. The mooring components are attached in series and paid out with the assistance of a winch. Instruments are attached to the mooring at the appropriate locations between premeasured mooring line shots. The last component put in line is the anchor. The ship tows the mooring into position with the anchor still on deck and actually steams past the desired anchor position by a distance equal to approximately 7–10% of the water depth. The anchor is deployed by either sliding it off the deck of the ship and into the water by means of a steel tip plate or it is placed into the water with a crane and

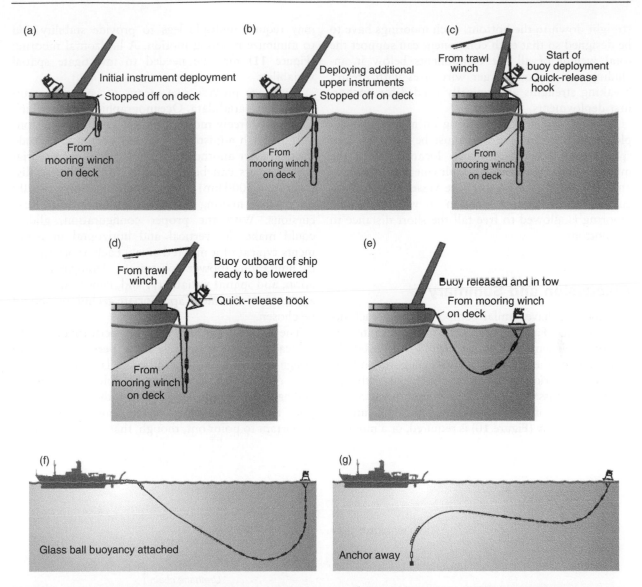

Figure 9 Surface mooring deployment sequence. (a) The first instrument is lowered into the water. (b) Instrumentation in the upper part of the mooring is lowered into the water before deploying the buoy. (c) The upper part of the mooring is attached to the surface buoy. (d) The surface buoy is placed into the water. (e) The ship steams forward slowly as additional mooring line and instrumentation are deployed. (f) The entire mooring is in tow behind the ship as the glass ball buoyancy is deployed. (g) The anchor free-falls to the ocean bottom, pulling the buoy along the surface.

mechanically released once it is just below the surface. As the anchor falls to the bottom, the mooring is pulled under with it. The mooring line takes the path of least resistance, following the anchor as it descends, resulting in the top of the mooring moving toward the anchor-drop position as the anchor falls to the bottom. The normal drag on the mooring line is greater than the tangential drag; therefore, a water-sheave effect takes place as the anchor falls to the bottom. The anchor does not, however, fall straight down but rather falls back a distance equal to a small percentage of the water depth, hence the reason for steaming past the desired anchor position before deploying the anchor. Depending on the design of the

mooring, the anchor can fall at a rate of approximately $100\,\mathrm{m\,min^{-1}}$.

Some situations make anchor-last deployments difficult or near impossible. An application requiring the deployment of moorings through the ice in high latitudes is one example where an anchor-first deployment could be used. In these cases, there is not enough open water to completely stretch the mooring out behind the vessel. Instead, the vessel breaks the ice where the mooring is to be deployed, creating a small pool. Starting with the anchor, the mooring is deployed vertically through the opening in the ice. The upper buoyancy is the last component to go into the water and the mooring is allowed to drop straight down to the bottom. Such moorings have to be designed so that each component can support the total weight of all the components below it, including the anchor. Larger wire sizes with greater breaking strengths are typically needed for anchor-first deployments.

Another application requiring an anchor-first deployment is a mooring that must be deployed in a specific depth or at a precise location. With the mooring hanging below the deployment ship and the anchor close to the bottom, the vessel can be maneuvered to the desired location, at which time the mooring is allowed to free-fall the short distance to the bottom.

Discussion and Summary

All moorings have similar components, but each design is unique. Factors such as the mooring's intended use, the environment in which it will be deployed, the water depth, the payload it must support, and the deployment period greatly affect the design. Although we have discussed vertical arrays, moorings can assume a variety of orientations. For some applications, a U-shaped array (**Figure 10**) is required, or a mooring

may require multiple legs to provide stability and to minimize mooring motion. A horizontal mooring (**Figure 11**) may be needed to investigate spatial variability.

Moorings provide one means of collecting temporal and spatial data. Oceanographic gliders, which are able to freely move both vertically and horizontally through adjustments in buoyancy and attitude control, offer another approach to collecting spatial data. Gliders can be configured to travel long distances (~ 3000 km) for extended periods (~ 200 days) while making approximately 600 vertical excursions. With the proper configuration, gliders could make the vertical and horizontal measurements typical of a moored array. Each approach has its limitations (power constraints, deployment duration, and spatial variability) and, depending on the application, the most appropriate technique should be chosen.

The ability to model mooring performance both statically and dynamically now permits extensive design studies before the mooring is taken into the field. As a result, it is possible to explore new designs and have greater confidence in how they will perform prior to cutting any wire or splicing any line. It is important to point out, though, that regardless of the

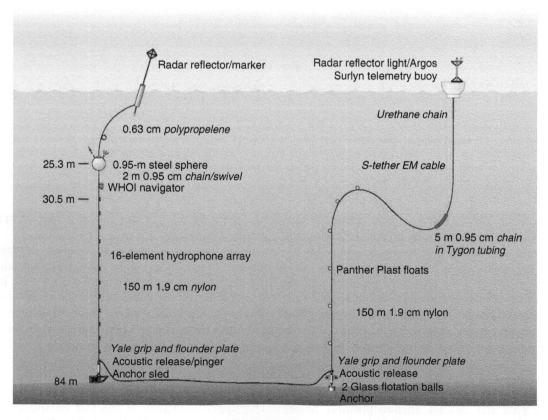

Figure 10 A U-shaped moored hydrophone array. Design by J. Kemp.

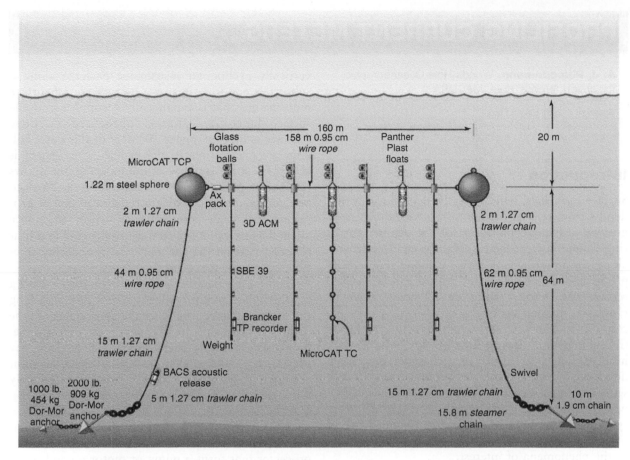

Figure 11 A two-dimensional moored array. TCP, temperature, conductivity, and pressure; BACS, binary acoustic command system; SBE, Sea-Bird Electronics; ACM, acoustic current meter; Ax pack, acceleration package. Design by R. Trask.

amount of time spent designing, modeling, and fabricating a mooring, the success of a deployment will often come down to the ability of trained personnel to pay close attention to all the details and to get the mooring safely in and out of the water while working under extremely adverse conditions at sea.

Further Reading

Berteaux HO (1991) *Coastal and Oceanic Buoy Engineering.* Woods Hole, MA: Henri Berteaux.

Berteaux HO and Prindle B (1987) *Deep sea moorings: Fishbite handbook. Woods Hole Oceanographic Institution Technical Report WHOI 87-8.* Woods Hole, MA: WHOI.

Dobson F, Hasse L, and Davis R (eds.) (1980) *Air–Sea Interaction: Instruments and Methods.* New York: Plenum.

Frye D, Ware J, Grund M, *et al.* (2005) An acoustically-linked deep-ocean observatory. *Proceedings of Oceans 2005 – Europe*, vol. 2, pp. 969–974. Woods Hole, MA: WHOI.

Morrison AT, III, Billings JD, and Doherty KW (2000) The McLane Moored Profiler: An autonomous platform for oceanographic measurements. In: *OCEANS 2000 MTS/ IEEE Conference and Exhibition*, Providence, RI, 11 Sep.–14 Sep. 2000, vol. 1, pp. 353–358. East Falmouth, MA: McLane Research Laboratories Inc.

Warren BA and Wunsch C (eds.) (1981) *Evolution of Physical Oceanography, Scientific Surveys in Honor of Henry Stommel.* Cambridge, MA: Massachusetts Institute of Technology Press.

PROFILING CURRENT METERS

A. J. Plueddemann, Woods Hole Oceanographic
Institution, Woods Hole, MA, USA

Introduction

Surface moorings, which may be deployed in water depths from tens of meters to thousands of meters, provide a suitable platform for the deployment of single-point current meters. Placing multiple sensors along the mooring line gives a discretized velocity profile and thus single-point sensors can be used to resolve the vertical structure of ocean currents. However, in deep water the number of sensors necessary to obtain sufficient vertical resolution throughout the water column quickly becomes prohibitive. Thus, a variety of techniques have been developed to obtain velocity profiles from a single sensor. Ideally, such profiles resolve the oceanic velocity fine structure (vertical scales of 1–10 m) and are synoptic in the sense that they are obtained over a time short compared to the characteristic timescale of the phenomena of interest.

There are three basic approaches to velocity profiling: vertically cycling single-point sensors, free-fall probes, and acoustic (Doppler) profilers. A fourth category involves the combination of these approaches. Single-point sensors may be cycled between the surface and some depth by use of a winch or may be cycled within the water column along a mooring line. The simplest free-fall probes are streamlined objects that are dropped through the water column and tracked acoustically. Acoustic Doppler techniques rely on a measurement of Doppler shift from a range-gated transmission that ensonifies the water column.

Vertically Cycling Sensors

It is a relatively simple matter to vertically cycle a single-point sensor from a ship using a winch, but interpretation of the resulting velocity is complicated by motion of the ship relative to the Earth and motion of the sensor relative to the ship. Cycling along a mooring line mitigates these problems, but the energetic vertical motion and potentially large inclinations of a surface mooring line are undesirable (wave-driven profilers are an exception, see below). Thus, the most common implementation involves

vertically cycling an instrument package along a subsurface mooring line or a taut section of surface mooring line that is dynamically decoupled from the surface buoy. A limitation introduced by these schemes is that the upper 20–30 m of the water column are inaccessible to the profiler. The benefit is that the relative stability of the mooring line allows the motion of the instrument relative to the Earth to be ignored, and the problem reduces to one of propelling the single-point velocity sensor or choice along the line. Solutions to this problem fall into four classes based on the nature of the propulsion: traction, buoyancy, waves, and currents.

Traction profilers use an electric motor to drive a traction wheel that propels the sensor package along the mooring line. The package is designed to be neutrally buoyant at mid-depth of the profile. Buoyancy-driven profilers achieve propulsion by varying the displacement of the package without changing its mass. This may be done by alternately filling and emptying a bladder external to the pressure housing with a fluid or gas. Alternatively, a piston may be driven in and out of a flooded chamber. Cycling is typically controlled by a microprocessor that turns a pump or motor on and off at preset times and/or pressures. Mechanical stops attached to the mooring line limit travel at the upper and lower extremes of the profiling range. Traction and buoyancy-driven systems are similar in that they carry their energy source with them in the form of batteries, and they require roughly the same amount of energy to propel a given instrument package. Traction systems are more easily configured to profile to great depths. Both techniques are capable of providing a total vertical travel distance of order 1 Mm per deployment. In a typical configuration a traction system may be configured for 100 round trips to 5000 m, while a buoyancy-driven system may be configured for a 1500 round trips to 300 m.

Wave-driven profilers make use of the vertical motion of a surface mooring to travel downward. During the down cycle the clamp attaching the sensor to the mooring also acts as a ratchet that allows the mooring line to pass through only when it is moving upward relative to the sensor. The sensor package is buoyant, but its inertia tends to keep it in place while the mooring line moves upward during the passage of a wave crest. With successive waves the sensor 'crawls' down the line until it reaches a mechanical stop. The ratchet is then released and the package rises to the starting point, where the ratchet is reset. In a typical configuration such a system may profile to 200 m depth every two hours. A current-driven profiler uses a similar technique, but the relative motion between the mooring line and the sensor is due to the generation of lift from the passage of ambient currents over a wing that is incorporated into the sensor package. Wave- and lift-driven devices are distinguished by the ability to draw on the surrounding environment for power, so that deployment durations are in principle limited only by the power and storage requirements of the sensors. However, these systems are likely to have a more limited vertical range than traction or buoyancy-driven systems and may be ineffective where waves and currents are small.

Free-fall Probes

A probe that follows the horizontal motion of the surrounding water during a descent to the ocean bottom and return to the surface can be used to determine the vertically averaged current. In a typical implementation the probe itself is buoyant, but carries a weight that is dropped upon encountering the bottom. The depth-average (vector) current $\bar{U}$ is computed from the horizontal distance between the drop and recovery points divided by the travel time. In principle, velocity precision is limited only by the precision of the position fixes, but in practice the fact that the probe may not faithfully follow the water must be considered. Vertically averaged currents over variable depth ranges can be obtained by programming weights to drop before the bottom is reached. A velocity profile is obtainable if the motion of the probe can be tracked during its descent. Tracking is done acoustically, often using inexpensive, expendable beacons. Two or more beacons are configured to 'ping' synchronously relative to a known timing reference, deployed from a ship, and then surveyed to accurately determine their relative positions. A transponder in the probe is used to detect the arrival times of signals from the beacons relative to the same time base, allowing the probe's horizontal position relative to the beacons to be determined. Vertical position is determined by a pressure sensor. The result is a profile of the total (vector) velocity (eqn [1], where $u(z)$ is the baroclinic component).

$$U(z) = \bar{U} + u(z) \qquad [1]$$

Velocity precision depends on the accuracy of the beacon positions and the precision of the timing; errors are typically near $1 \, \text{cm s}^{-1}$. Vertical resolution is limited by the product of the fall rate and the time between pings, which is typically several meters.

Temperature and salinity sensors can easily be added to the probe, providing a complete hydrographic profile. Since return to the surface is not essential when tracking is used, an alternative to the recoverable probe is to use inexpensive, expendable probes, and reverse the roles of the beacons and the transponder.

A conventional velocity sensor attached to a free-fall probe measures the relative velocity between the probe and the water (eqn [2]).

$$U_{\text{rel}} = \bar{U} + u(z) - U_{\text{probe}} \qquad [2]$$

Acoustic tracking can be used to determine U_{probe} and recover the total velocity profile. The advantages over tracking alone are higher precision and higher vertical resolution. Alternatively, the measured velocity can be used as a basis for a total velocity estimate. In this approach, U_{rel} is used as the input to a predetermined transfer function, which accounts for the probe's response to lateral forces and produces an estimate of the acceleration of the probe's center of mass. This information is used to estimate U_{probe}, which is then added to U_{rel} to produce a total velocity profile. The complication of tracking is eliminated, but uncertainties in the transfer function may result in larger errors.

Without a means of determining U_{probe}, only the velocity shear (or the baroclinic velocity relative to an unknown constant) can be determined. However, this approach has significant utility because the probe motion acts as a high-pass filter in vertical wavenumber space, i.e., for vertical wavelengths larger than the probe length (typically 2–3 m) the sensor follows the water, while for smaller wavelengths it does not. Thus, an appropriate (fast response) single-point sensor can be used to measure the velocity profile at high vertical wavenumber. This technique is routinely used to detect velocity microstructure.

The necessity of tracking a free-fall probe or using a transfer function for the probe's large-wavelength response could be eliminated if the sensor was capable of directly detecting the total water velocity during its descent. This can be accomplished by exploiting electromagnetic induction. The motion of sea water (a conductor) in the presence of the Earth's vertical magnetic field B_z induces horizontal electric currents E_h in the water column that can be measured. For a probe that perfectly follows horizontal water motion the relationship is given by eqn [3].

$$E_h(z) = B_z \mathbf{k} \times (U(z) - \bar{U}^*) \qquad [3]$$

Here $\mathbf{k}$ is the vertical unit vector and $\bar{U}^*$ is a weighted vertical integral of the horizontal velocity. In general, the depth-average contribution $\bar{U}^*$ is not known, and only the baroclinic component $u(z)$ can be determined. The electromagnetic induction technique yields baroclinic velocity profiles with vertical resolution of 5–10 m and precision of about $1\,\mathrm{cm\,s^{-1}}$. The technique has been implemented both on small, expendable probes (similar in operation to XBT, expendable bathythermographs) and on larger, more complex recoverable probes. By adding a self-contained acoustic velocity profiler to the latter, it is possible to estimate $\bar{U}^*$ and determine the total velocity profile (see Combined Approaches below).

Acoustic Doppler Sensors

An acoustic Doppler sensor estimates fluid velocity by detecting the Doppler frequency shift of the acoustic reverberation (or 'backscatter') from objects in the water column. For an acoustic Doppler current profiler (ADCP) operating in the 50–500 kHz frequency range, the primary scatterers are biological (zooplankton and micronekton). The scatterers are assumed to be drifting passively with the surrounding water, although this may not always be the case. During operation, an acoustic transducer emits a pulse of acoustic energy along a narrow beam (3–4° half-power beam width) ensonifying a volume of fluid determined by the beam width, the pulse duration, and the distance from the transducers. As the time after transmission increases, the returned signal comes from successively more distant sample volumes known as range bins. Backscattered energy from each range bin arrives at the transducer with a Doppler shift proportional to the average speed of the scatterers within the volume. For a transmission at frequency f_t through a medium with sound speed c, the velocity in the direction of the beam is related to the mean Doppler shift Δf eqn [4].

$$U_{\mathrm{beam}} = -c\Delta f/2f_t \qquad [4]$$

Estimation of U_{beam} for successive range bins results in a profile of water velocity as a function of distance along the beam.

The typical ADCP configuration consists of four downward-slanting beams separated by 90° in azimuth and inclined at 20° or 30° from vertical. The four beams form two coplanar pairs that can be combined to estimate horizontal and vertical velocity components as long as the horizontal scale of the motion is greater than the beam separation. The along-beam extent of a range bin is related to the transmitted pulse duration T_p by $\Delta r_p = cT_p/2$. The vertical resolution is set by the vertical extent of the range bin, $\Delta z_p = \Delta r_p \cos\theta$, where θ is the angle of

the beam from vertical. Instrument tilt and heading are used to compensate for vertical misalignment of range bins between coplanar beam pairs and convert the measured velocities into geographic components.

The standard deviation of horizontal velocity estimates from an ADCP with transmitter frequency f_t has the general form of eqn [5].

$$\sigma_v \sim c\left[f_t T_p \sin\theta\right]^{-1} \qquad [5]$$

The precision [5], based on a single transmission, is often too large for practical applications and is reduced by averaging over many transmissions. In the field, precision may be adversely affected by motion of the scatterers and motion of the measurement platform (the most significant difficulty in computing geographic velocity components is compass error). It is evident from [5] that velocity precision can be increased for a given vertical resolution by using a higher transmission frequency. However, profiling range R decreases with increasing frequency. It is also possible to increase precision for fixed f_t and T_p by using wideband transmission or coded pulses, and this technique is now routinely implemented. Commonly used configurations include high-precision ADCPs ($\sigma \approx 2\,\mathrm{cm\,s^{-1}}$; $R \approx 20\,\mathrm{m}$) operating near 1 MHz and long-range ADCPs ($\sigma \approx 10\,\mathrm{cm\,s^{-1}}$; $R \approx 500\,\mathrm{m}$) operating near 75 kHz. Very high-resolution systems ($\sigma \approx 0.1\,\mathrm{cm\,s^{-1}}$; $R < 10\,\mathrm{m}$), which utilize a different approach to data processing, are also available.

By selecting an operating frequency matched to the desired profiling range, it would appear that ADCPs could provide full water column velocity profiles in depths up to 500 m. However, difficulties arise in using ADCPs near surface and bottom boundaries. The first one or two range bins (nearest the transducer) may be corrupted by transient signals from the pulse transmission that saturate the system electronics. Range bins near the surface (or the bottom if the instrument is down-looking) are corrupted by reflections from the side-lobes of the acoustic beam. As a result, many applications use single-point sensors near the surface and bottom boundaries combined with ADCPs that profile the central water column.

ADCPs mounted in a frame or housing that sits directly on the seafloor generally provide the highest-quality current profiles, because platform motion is eliminated and compass error is reduced to a constant (rather than a function of direction). On subsurface moorings, ADCPs are often deployed as the uppermost element, facing upward into undisturbed water. This is an attractive option when bottom mounting is impractical owing to deep water and surface conditions prohibit use of a surface element. Platform motion is increased relative to a bottom mount, but remains significantly less than on a surface mooring. On surface moorings, ADCPs may be mounted within the buoy bridle facing downward, or attached to the mooring line facing either upward or downward. Deployment in a buoy bridle is attractive because near-surface currents can be detected, but performance may be degraded as a result of wave motion. It has been shown that small sensors mounted in-line along a mooring do not affect ADCP performance, and thus the deployment of relatively inexpensive point sensors measuring temperature (or temperature and conductivity) along with one or more ADCPs has become standard practice on both surface and subsurface moorings.

Practical issues important to shipboard ADCP operation include installation method, transducer alignment, platform motion, and navigation. The installation must account for the presence of bubbles, which are generated under the hull in rough seas and swept past the transducers, interfering with acoustic transmissions. This problem may be alleviated by extending the transducers below the bubble layer, or by using a faired housing, or both. Separating the ADCP from the ship and deploying it in a towed body eliminates interference from bubbles and reduces platform motion, but the complexity of operation is increased. After installation, the transducers must be 'calibrated' to account for imperfect mechanical alignment, which may result in both magnitude and direction errors in the observed current. The accuracy of the calibration depends principally on the quality of the shipboard navigation and compass. Of course, the accuracy of navigation also determines the accuracy of the absolute velocity obtained from a shipboard ADCP. A distinct advantage of shallow water applications is that the speed of the ship over the earth can be estimated directly from the ADCP in 'bottom track' mode, alleviating the dependence on independent navigation. However, the availability of more accurate navigation (via the Differential Global Positioning System) has allowed absolute velocity profiling from ships in the absence of bottom tracking to be done routinely.

Combined Approaches

A combined approach that has become relatively common on deep-ocean hydrographic cruises is the use of an ADCP in conjunction with a vertical cycling CTD (conductivity- temperature- depth profiler), often called a lowered ADCP or LADCP. In the most common application, a relatively long-range ADCP (e.g., $f_t = 150\,\text{kHz}$) is mounted on a CTD/rosette frame with the transducers facing downward. As the frame is lowered during occupation of a hydrographic station, the ADCP transmits rapidly ($\approx 1\,\text{Hz}$), collecting a series of overlapping profiles. Each profile is relative to an unknown (but assumed constant) velocity resulting from horizontal motion of the frame during the cast. By separating the instrument motion into a portion U_{ship} due to ship drift and a portion U_{frame} due to the motion of the frame relative to the ship, the relative velocity observed by the LADCP can be written as eqn [6].

$$U_{\text{rel}}(z) = \bar{U} + u(z) - (U_{\text{ship}} + U_{\text{frame}}) \qquad [6]$$

In post-processing these profiles are differentiated with depth to eliminate the unknowns $\bar{U}$, U_{ship}, and U_{frame} (all assumed to be constant for a given profile) and the depth bins are indexed to pressure using the pressure record from the CTD. The overlapping shear profiles are then averaged together in common pressure bins and integrated to obtain the baroclinic velocity profile $u(z)$. If the profiles are continuous during the cast (i.e., not affected by drop-outs due to low scattering strength or acoustic interference near the bottom boundary) the vertical integral of U_{frame} will be zero. The depth-average component can then be estimated from eqn [7], where $\langle U_{\text{rel}}(z) \rangle$ indicates a vertical average (taken to be equivalent to a time average over the cast) and $\langle U_{\text{ship}} \rangle$ is estimated from position fixes at the start and end of the cast.

$$\bar{U} = \langle U_{\text{rel}}(z) \rangle + \langle U_{\text{ship}} \rangle \qquad [7]$$

In practice a small correction due to $\langle u(z) \rangle \neq 0$ may be included.

Another combined approach merges a free-fall probe fitted with an electromagnetic velocity sensor with an ADCP. In this application, a relatively long-range ADCP is mounted at the bottom of the probe with transducers facing downward. Rather than producing a profile of the water column velocity, the ADCP is configured to detect the motion of the instrument relative to the bottom. Of course, this is only effective when the probe is within acoustic range of the bottom (typically a few hundred meters). Within this depth interval, the ADCP provides an independent measure of $U(z)$, the motion of the instrument relative to the Earth, and the vertically averaged flow $\bar{U}*$ can be determined. The velocity profile $U(z)$ can then be estimated over the entire water column from eqn [3].

See also

Sonar Systems.

Further Reading

Anderson SP, Terray EA, Rizoli-White JA, and Williams AJ (1999) *Proceedings of the IEEE Sixth Working Conference on Current Measurement*, 333pp. Piscataway, NJ: IEEE Press.

Baker JD (1981) Ocean instruments and experiment design. In: Warren BA and Wunsch C (eds.) *Evolution of Physical Oceanography*. Cambridge, MA: MIT Press.

Collar PG (1993) *A Review of Observational Techniques and Instruments for Current Measurement in the Open Sea*. Institute of Ocean Science, Deacon Laboratory, Report No. 304.

Emery WJ and Thomson RE (1998) *Data Analysis Methods in Physical Oceanography*. New York: Elsevier.

Pinkel R (1980) Acoustic Doppler techniques. In: Dobson EF, Hasse L, and Davis R (eds.) *Air–Sea Interaction*, pp. 171–199. New York: Plenum Press.

SINGLE POINT CURRENT METERS

P. Collar and G. Griffiths, Southampton
Oceanography Centre, Southampton, UK

Introduction

A current meter estimates the speed and direction of water moving relative to the instrument. The single point current meter is, therefore, only part of a measurement system that includes the mooring or mounting hardware or technique. This article begins with a discussion of the interaction between the current meter, the method of mounting and the characteristics of the currents within the environment being studied. This is followed by an introduction to the principles of current meter design, which are largely independent of the chosen implementation technology. Some examples of commonly used instruments follow, with an assessment of the strengths and weaknesses of the different types of sensor. The importance of direction measurement and calibration are discussed as prerequisites to making accurate observations. (For typical current meter moorings *see* Moorings.) This article concludes with a note on the future for current measurement systems.

The first self-recording current meters were ingenious mechanical devices such as the Pillsbury instrument (first used in 1884) and the Ekman current meter, available in 1904. However, the slowness of progress during the first half of the twentieth century is reflected in the view of the German hydrographer Bohnecke in 1954 that 'The subject of current measurements has kept the oceanographers busy for more than a hundred years without having found – this must honestly be admitted – an entirely satisfactory solution.' In the 1960s and 1970s the growing need for current measurements in the deep ocean provided a stimulus for the development of robust, self-contained recording instruments capable of deployment over periods of months.

Measurement of current in the open sea is usually achieved by mounting the instrument on a mooring. Movement of the mooring makes true fixed-point, or Eulerian, measurement impossible, although careful attention to mooring design can generally provide an acceptably good approximation to a fixed point. In some circumstances a fixed measurement platform can be used, for example in shallow seas, or at the deep ocean floor. Care then needs to be taken to avoid, as far as possible, disturbance to the flow by the sensor itself, and by any supporting structure. In the case of moored current meters the design of the mooring must minimize vibration, which can lead to the sensor sampling in its own turbulent wake, thereby generating significant errors. With proper attention to design of the mooring or platform, and selection of an appropriate current meter, it should be possible to make most deep-sea measurements to within about $1 \, \text{cm s}^{-1}$ in speed and $2\text{–}5°$ in direction, and with rather better precision in the case of bottom-mounted instruments. Many of the issues of current meter data quality have been reviewed in the literature.

Particular problems arise in the case of near-surface measurements. Wave orbital motion decays exponentially with depth but may be considered significant, if somewhat arbitrarily, to a depth equivalent to half the wavelength of the dominant surface waves. In the open ocean the influence of surface waves can thus easily extend to a depth of several tens of meters. Within this region the difficulty presented by the lack of a fixed Eulerian frame of reference for current measurement is compounded by the presence of three-dimensional wave orbital velocities. These can be large compared with the horizontal mean flow, making it difficult to avoid flow obstruction by the sensor itself and necessitating a linear response over a large dynamic range. If instruments are suspended some way beneath a surface buoy large errors can result from vertical motion induced by the surface buoy relative to the local water mass.

However, a more fundamental problem arises in the surface wave zone. This may be illustrated by reference to a water particle undergoing progressive wave motion in a simple small amplitude wave. Neglecting any underlying current, a particle at depth z experiences a net Lagrangian displacement, or Stokes drift, in the direction of wave travel of $O[a^2 \sigma k \exp(-2kz)]$, where a is the wave amplitude, k is the wavenumber and σ is its angular frequency. In 10 s waves of amplitude 2 m this amounts to about $4.5 \, \text{cm s}^{-1}$ at a depth of 10 m. However, a fixed instrument, even if perfect in all respects will not be able to detect the Stokes drift. Nevertheless an instrument that is moving in a closed path in response to wave action, but unable to follow drifting particles, will record some value of current speed related in a complex but generally unknown fashion to the drift. Close to the surface, where the path of a current sensor over a wave cycle can be more easily arranged to approximate the path of a water particle, the value recorded by the instrument should more closely resemble the surface value of the local Stokes drift. This has been verified in laboratory measurements involving simple waves, but is not easily tested in the open sea. The reader is referred to the bibliography for discussion of these points.

Current Meter Design

Fluid motion can be sensed in a number of ways: techniques most frequently employed nowadays include the rotation of a mechanical rotor,

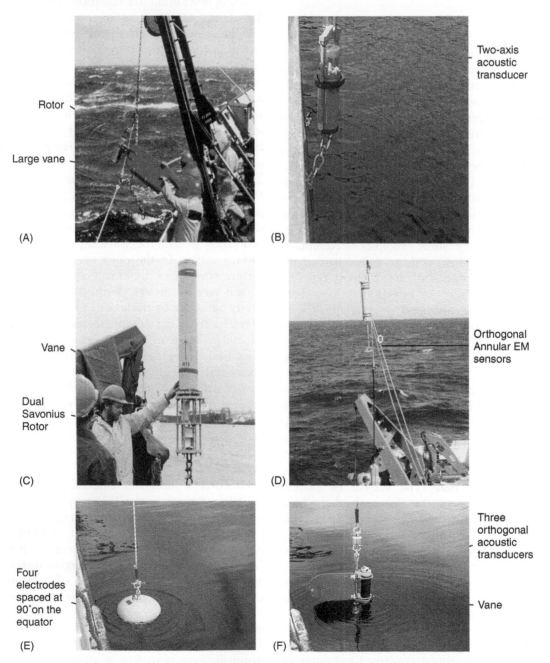

Figure 1 Current meters based on different sensors. (A) Aanderaa RCM4 deep ocean rotor-vane instrument; (B) Aanderaa RCM9 single cell Doppler current meter; (C) EGG Vector averaging current meter with dual Savonius rotor (at the base) and small vane (immediately above); (D) Vector averaging electromagnetic current meter based on an annular sensor; (E) Interocean S4 electromagnetic current meter; (F) Nortek Aquadopp high precision single cell Doppler instrument, capable of measuring horizontal and vertical currents.

Table 1 Main characteristics of some contemporary current meters and a vector averaging current meter (VACM) from the 1970s

Type	Speed accuracy	Resolution $(cm\,s^{-1})$	Range $(cm\,s^{-1})$	Direction accuracy	Depth rating (m)	Weight (kg)	Data capacity (records)
Aanderaa RCM9 MkII single-point Doppler	$\pm 0.5\,cm\,s^{-1}$	0.3	0–300	$\pm 5°$ for 0–15° tilt	2000	Air: 17 Water: 12	up to 36 100
Aanderaa RCM8 vector averaging rotor-vane	$\pm 1.0\,cm\,s^{-1}$ or $\pm 2\%$ of speed, whichever greater	Not specified	2–295	$\pm 5°$ for speeds 5–100 $cm\,s^{-1}$ $\pm 7.5°$ for 2.5–5 $cm\,s^{-1}$ and 100–200 $cm\,s^{-1}$	6000	Air: 29.3 Water: 22.7	up to 43 600
InterOcean S4 electromagnetic sensor	$\pm 1.0\,cm\,s^{-1}$ or $\pm 2\%$ of speed, whichever greater	0.03–0.43 depending on range	0–350	$\pm 2°$ for 0–5° tilt $\pm 4°$ for 15–25° tilt	S4 1000 m S4Deep 6000 m	S4 Air: 11 S4 Water: 1.5 S4Deep Air: 34.5 S4Deep Water: 10.5	S4 348 000 S4A 7 000 000
Sontek Argonaut-ADV acoustic travel time	$\pm 0.5\,cm\,s^{-1}$	0.01	0–600	optional extra, at $\pm 2°$	60 m	Air: 3.2 Water: 0.45.	> 100 000
FSI 3D ACM acoustic travel time	$\pm 1.0\,cm\,s^{-1}$ or $\pm 2\%$ of speed, whichever greater	0.01	0–300	$\pm 2.5°$ at unspecified tilt	1000 m	Not specified	200 000
EG & G VACM (1970s design)	$2.6\,cm\,s^{-1}$ threshold accuracy not stated	Not stated	2.6–309 $cm\,s^{-1}$	$\pm 2.8°$	6096 m	Air: 72.5 kg Water: 34.9 kg	50 925–76 388

Source: Company specification sheets.

electromagnetic sensing, acoustic travel time measurement, and measurement of the Doppler frequency of back-scattered acoustic energy. **Figure 1** shows examples of practical current meters based on these techniques, and **Table 1** shows the main characteristics of some commonly used instruments. The evolution in design of experimental and commercial instruments from 1970 to 2000 can be traced by comparing the descriptions of the current sensors and *in situ* processing to the 3-D current mapping discussed in contemporary literature. Acoustic Doppler and correlation back-scatter techniques can also measure current profiles, as discussed elsewhere in this volume (*see* Profiling Current Meters).

In quasi-steady flow, relatively unsophisticated instruments often produce acceptable results. However, in circumstances in which an instrument may need to cope with a broad frequency band of fluid motions, as in the wave zone or when subject to appreciable mooring motion, there are implications for the design of the sampling system.

If the sensor is to determine horizontal current it should be completely insensitive to any vertical component, while responding linearly to horizontal components across a frequency band which includes the wave spectrum.

Then if, as is usual, the sensor output is sampled in a discrete manner, the provisions of the Nyquist sampling theorem must be observed, i.e., the sampling rate must be at least twice the highest frequency component of interest, whereas negligible spectral content should exist at frequencies above the highest frequency of interest. The highest frequencies that need to be measured are encountered in velocity fluctuations in small-scale turbulence, for example in measurements of Reynolds stress from the time-averaged product of a horizontal velocity component with the vertical velocity. A frequency response to at least 50 Hz is generally required, perhaps even higher frequencies if the measurements are being made from a moving platform. Satisfying spatial sampling criteria is as important as satisfying temporal sampling requirements. Hence, the sampling path length, or sampling volume of the sensor must be less than the spatial scale corresponding to the highest frequencies of interest. In this case, specialist turbulence dissipation probes that employ miniature sensors measuring velocity shear are used (*see* Turbulence Sensors).

Experiments involving the use of laser back-scatter instruments have been carried out at sea, for example to measure fine-scale turbulence near the ocean floor, but their characteristics are generally better matched to high resolution studies in fluid dynamics in the laboratory.

Vector Averaging

Apart from the study of turbulence, the existence of significant wave energy and instrument motion down to periods of 1 s means that a sampling rate (f_s) of ≥ 2 Hz is often used. At this frequency substantial amounts of data are generated and, unless the high frequency content is specifically of interest, it is usual to average before storing data. If done correctly this involves the summation of orthogonal Cartesian components individually prior to computation of the magnitude. Any other form of averaging can produce erroneous results.

If the instrument makes a polar measurement, for example if it measures flow by determining instantaneous rotor speed V_i and the instrument is aligned with the current using a vane whose measured angle relative to north is θ_i the averages are formed:

$$\bar{E} = \frac{1}{n}\sum_{i=1}^{n} V_i \sin \theta_i$$

$$\bar{N} = \frac{1}{n}\sum_{i=1}^{n} V_i \cos \theta_i$$

If on the other hand the instrument measures orthogonal velocity components X_i, Y_i directly, as for example in electromagnetic or acoustic sensors, it forms:

$$\bar{E} = \frac{1}{n}\sum_{i=1}^{n}(X_i \cos \theta_i + Y_i \sin \theta_i)$$

$$\bar{N} = \frac{1}{n}\sum_{i=1}^{n}(-X_i \sin \theta_i + Y_i \cos \theta_i)$$

where θ_i is the instantaneous angle between the Y axis and north; n is chosen so as to reduce noisy contributions from, for example, the wave spectrum; a value of $nf_s^{-1} > 50s$ is usual.

The averaged magnitude and direction are then given by:

$$\bar{U} = \left((\bar{E})^2 + (\bar{N})^2\right)^{0.5}$$

$$\bar{\theta} = \tan^{-1}\frac{\bar{E}}{\bar{N}}$$

Mechanical Current Meters

At first, mechanical current meters were relatively simple. For example, the early, mechanically encoded Aanderaa current meter of the 1960s combined a scalar average of speed with a spot measurement of the direction (**Figure 1A**). Speed was measured by a rotor consisting of six impellers of cylindrical shape mounted between circular end plates. The rotor shaft ran in ball-race bearings at each end, and at the lower end two magnets communicated the rotation to an internal recording device. The large plastic vane, with a counterweight at the rear end, aligned the instrument with the current.

As experience in a range of deployment conditions and types of mooring widened, such sampling schemes were found to be unsuitable when the sensor experienced accelerating flow as a result of wave motion or mooring movement. The introduction of vector averaging schemes followed, initially in the vector averaging current meter (VACM) (**Figure 1C**), and provided a substantial improvement in accuracy in such conditions. Improved sampling regimes were facilitated in later instruments by low power microprocessor technology. It was also realized that it is necessary to understand fully the behavior of speed/velocity and direction sensors in unsteady flow conditions.

By the time the dual orthogonal propeller vector measuring current meter (VMCM) was developed in the late 1970s sufficient was understood about the pitfalls of near-surface current measurement to realize that rotor design required a combination of modeling and experimental testing in order to ensure a linear response. For example, the propellor in the VMCM was designed to avoid nonlinearity due to the different response times to accelerating and decelerating flows that had been found in the 'S'-shaped Savonius rotor of the VACM. Today, mechanical current meter development might be regarded as mature.

Electromagnetic Current Meters

In electromagnetic current meters an alternating current (a.c.) or switched direct current (d.c.) magnetic field is imposed on the surrounding sea water using a coil buried in the sensing head, and measurements of the potential gradients arising from the Faraday effect are made using orthogonally mounted pairs of electrodes, as illustrated in **Figure 2**. Some electromagnetic techniques make use of the Earth's field, but in self-contained instruments simple d.c. excitation is avoided. This is because unwanted potential differences arising, for example, from electrochemical effects can exceed flow-induced potential differences, which are typically between 20 and $100 \mu V m^{-1} s^{-1}$, by two orders of magnitude. Flow-field characteristics around the sensor head, including hydrodynamic boundary layer thickness

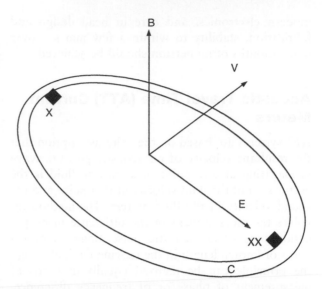

Figure 2 Sketch showing the Faraday effect, which forms the basis of the electomagnetic current meter. The effect results in a potential difference $E = BVL$ induced between two electrodes (X and XX) with a separation L when a conductor (sea water) moves at a resolved velocity V perpendicular to the line X–XX and perpendicular to a magnetic field with a flux density of B induced by coil C.

and flow separation, are of critical importance in determining the degree of sensor linearity as well as the directional response. Modeling techniques can help to evaluate specific cases.

Forms of sensor head that have been considered or used include various solids of revolution, such as spheres, cylinders, and ellipsoids. Although hydrodynamic performance weighs heavily in choice of shape, this may be balanced by consideration of ease of fabrication and robustness. One neat solution incorporates the entire instrument within a spherical housing that can be inserted directly into a mooring line (**Figure 1E**). For a smooth sphere, the resulting instrument dimensions would normally give rise to a transition from a laminar to a turbulent boundary layer over the instrument at some point within its working velocity range, at a Reynolds number of $\sim 10^5$, but this is forestalled by use of a ribbed surface so as to introduce a fully turbulent boundary layer at all measurable current speeds. Good linearity is thereby achieved.

Lower flow disturbance can be achieved using an open form of head construction (**Figure 1D**) which has been shown to provide excellent linearity and off-axis response, the only disadvantage relative to solid heads being greater complexity in construction and perhaps some reduction in robustness.

Unlike mechanical current meters electromagnetic instruments have no zero velocity threshold. In the past zero stability has presented a problem, but with modern electronics, and care in head design and fabrication, stability to within a few mm s^{-1} over many months of immersion should be achieved.

Acoustic Travel Time (ATT) Current Meters

ATT systems are based on the valid assumption that the resultant velocity of an acoustic pressure wave propagating at any point in a moving fluid is the vector sum of the fluid velocity at that point and the sound velocity in the fluid at rest. The method involves the measurement of the difference in propagation time of an acoustic pulse along reciprocal paths of known length in the moving fluid, although the principle can be realized equally in terms of measurement of phase or of frequency difference. Using reciprocal paths removes the need to know the precise speed of sound. The three techniques present differing design constraints. Typically an acoustic path length l may be of order 10 cm. For resolution of currents Δv to $1 \, \text{cm s}^{-1}$, the required time discrimination of acoustic pulse arrivals can be calculated from:

$$\Delta t = \Delta v \cdot l/c^2$$

or about $4 \times 10^{-10} \, \text{s}$ (since the sound speed, c, is about $1500 \, \text{m s}^{-1}$) requiring stable, wide-band detection in the electronic circuitry. In contrast, phase measurement, made on continuous wave signals, is effected within a narrow bandwidth, thereby relaxing the front-end design in the receiver. Phase measurement provides good zero stability and low power consumption, but the pathlength may be constrained by the need to avoid phase ambiguity.

Whichever method is chosen, hydrodynamic considerations are important in achieving accuracy: rigid mounting arrangements which do not disturb the flow significantly are required for the transducers at each end of the acoustic path. Techniques for minimizing flow obstruction have included the use of mirrors to route sound paths away from wakes and, with the development of substantial *in situ* processing, the use of redundant acoustic paths. For a given instrument orientation, the least-disturbed paths can be selected for processing.

ATT techniques have been implemented in various forms for a range of applications, including miniature probes for laboratory tanks, profiling instruments and self-recording current meters. Of the three basic methods, the measurement of frequency difference seems to have been the least exploited, although it has been successfully used in such diverse applications as a miniature profiling sensor for turbulence measurement, and a buoy-mounted instrument with 3 m path length providing surface current measurements.

ATT current meters offer well-defined spatial averaging, high resolution of currents (better than 1 mm s^{-1}), potentially good linearity and high frequency response. The main disadvantage, tackled with varying degrees of success in individual types of instrument, is associated with disturbance of flow in the acoustic path by transducers, support struts, and the instrument housing.

Remote Sensing Single-Point Current Meters

One current measurement technique that avoids flow obstruction altogether is that of acoustic back-scatter, using either Doppler shift or spatial or temporal cross-correlation. In the past, these computationally intensive techniques were restricted to use in current profilers, where the relatively expensive instrument could nevertheless substitute for an array of less-expensive single point current meters. Nowadays, the availability of low cost, low-power yet high-performance digital signal processing circuits has made it possible and economic to produce single-point acoustic back-scatter current meters (**Figure 1B & F**). Such instruments provide a combination of several desirable specifications, including: rapid data output rate, with 25 Hz being common; a dynamic range extending from 1 mm s^{-1} to several m s^{-1}; an accuracy of $\pm 1\%$ or $\pm < 5$ mm s^{-1}; a typical sampling volume of a few cubic centimeters and the capability of operating within a few millimeters of a boundary. These characteristics make this class of instrument almost ideal for current measurement within boundary layers, e.g., in the surf zone, while also enabling the collection of concurrent velocity and directional wave spectrum information through sensing the wave orbital velocity components.

Directional Measurement

The directional reference for measurement of current is invariably supplied by a magnetic compass, two main types of which are in common use. The first type is the traditional bar magnet, often mounted on an optically read encoded disk. The entire assembly is mounted on jeweled bearings, with arrangements for damping and gimballing. In the fluxgate compass, the second type of sensor, a soft magnetic core is driven into saturation by an a.c. signal. Orthogonal secondary windings detect the out-of-balance harmonic signals caused by the polarizing effect of the Earth's field and, from an appropriately summed output, the orientation of the sensor relative to the Earth's field can be determined. In current meters a gimballed two-component system may be used, but as in the case of the magnet compass, this does require that the system will respond correctly to any rotational and translational motions arising from mooring or platform motion.

Calibration, Evaluation and Intercomparison

The calibration, evaluation, and intercomparison of current measuring instruments are closely related and are central to the issue of data quality assurance. Basic velocity calibration can be carried out in a tank of nominally still water by moving the instrument, usually suspended from a moving carriage, at a constant, independently measured velocity. Compass calibration is done, typically to a precision of $\sim 1°$, in an area free from stray magnetic fields either using a precisely orientated compass table equipped with a vernier scale or by invoking a self-calibration program built into the instrument that obviates the need

for an accurate heading reference. Modern instruments can correct for heading-dependent errors in real time as well as correcting for a user-supplied magnetic variation. However, older instruments usually require the corrections to be applied at the post-processing stage.

There is a variety of practices relating to routine calibration, ranging from checks before and after every deployment to almost complete lack of checks. It has been argued that sensitivities of acoustic and electromagnetic sensors are determined by invariant physical dimensions and stable electronic gains, whereas mechanical instruments require only a simple in-air test to ensure free revolution of the rotor. However, good practice is represented by regular calibration checks in water.

Current meters generally behave well in steady flows but, as remarked above, in the near surface zone, or in the presence of appreciable mooring or platform motion, substantial differences can occur in data recorded by different instruments at the same nominal place and time. The fact is that no amount of simple rectilinear calibration in steady flow conditions can reveal the instrument response to the complex broadband fluid motions experienced in the

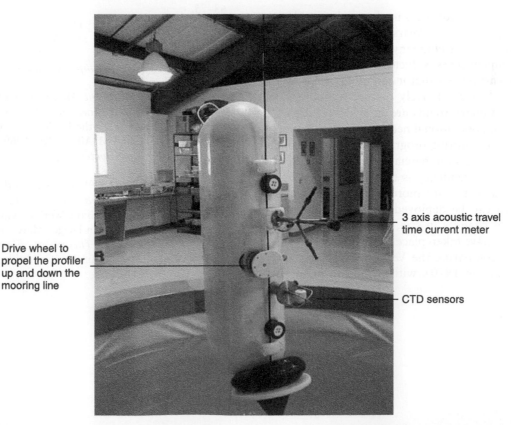

Drive wheel to propel the profiler up and down the mooring line

3 axis acoustic travel time current meter

CTD sensors

Figure 3 Acoustic travel time current meter as one instrument among many on a package capable of crawling up and down a wire mooring to obtain profiles of properties in water depths of up to 5000 m. (Illustration courtesy of McLane Research Inc.)

sea and as yet there are no standard instruments or procedures for more comprehensive calibration. Some efforts have, however, been made to model the errors incurred in some specific instruments, with a view to the prediction of performance at sea from dynamic simulation data acquired in the laboratory test tank.

Laboratory tests in controlled conditions thus provide a necessary, though insufficient basis for judging performance, and when a new instrument, or technique, is first used at sea considerable effort is put into intercomparisons with other, longer established instruments or techniques. Not surprisingly, most of the impetus for testing and intercomparison has come from the scientific community; the costs of providing anything other than basic performance data in controlled flow conditions is, with some justification, considered prohibitive by manufacturers. Extensive information on the performance at sea of instruments of many types is, therefore, to be found in the scientific literature, although cheaper instruments are generally less well represented.

Evolutionary Trends

As a result of the advances in electronics and battery technology in recent years, and the painstaking evaluation work accompanying the introduction of new instrument types, sufficient is now known about current measurement that it can in this sense at least be regarded as a relatively mature technology. Yet clear evolutionary trends are in evidence, driven by an increasing operational need for data in support of large-scale monitoring programmes. A further factor is the growing commercial involvement in data gathering. The tendency is towards cheaper, lighter instruments which are more easily handled at sea, and which can be deployed in larger numbers. An example of changes in size, recording capacity and weight that have taken place over the past 25 years is shown by comparing the Vector Averaging Current Meter from the 1970 s with a modern acoustic or electromagnetic current meter of similar performance (**Figure 1** and **Table 1**).

Another trend brought about by the growth of processing capability *in situ* is towards the incorporation of current measurement within a complete measurement system embracing a range of physical, chemical, and biological parameters (**Figure 3**). Operational requirements for current data may also in time result in the routine deployment of telemetering systems. At present, satellite telemetry of surface and near-surface measurements is well established, but telemetry of midwater measurements is not yet common practice.

See also

Moorings. Profiling Current Meters. Sonar Systems. Turbulence Sensors.

Further Reading

Appell GF and Curtin TB (1991) Special issue on current measurement. *IEEE Journal of Oceanic Engineering* 16(4): 305–414.

Collar PG, Carson RM, and Griffiths G (1983) Measurement of near-surface current from a moored wave-slope follower. *Deep-Sea Research* 30A(1): 63–75.

Dobson F, Hasse L, and Davis R (1980) *Air–Sea Interaction: Instruments and Methods*. New York: Plenum Press.

Hine A (1968) *Magnetic Compasses and Magnetometers*. London: Adam Hilger.

Howarth MJ (1989) *Current Meter Data Quality*. Cooperative Research Report No. 165. Copenhagen: International Council for the Exploration of the Sea.

Myers JJ, Holm CH, and McAllister RF (1969) *Handbook of Ocean and Underwater Engineering*. New York: McGraw-Hill.

Pinkel R and Smith JA (1999) *Into the Third Dimension: Development of Phased Array Doppler Sonar*. Proceedings of the IEEE Sixth Working Conference on Current Measurement. San Diego, March 1999.

Shercliff JA (1962) *The Theory of Electromagnetic Flow Measurement*. Cambridge: Cambridge University Press.

UNDERWATER VEHICLES

DEEP SUBMERGENCE, SCIENCE OF

D. J. Fornari, Woods Hole Oceanographic Institution, Woods Hole, USA

Introduction

The past half-century of oceanographic research has demonstrated that the oceans and seafloor hold the keys to understanding many of the processes responsible for shaping our planet. The Earth's ocean floor contains the most accurate (and complete) record of geologic and tectonic history for the past 200 million years that is available for a planet in our solar system. For the past 30 years, the exploration and study of seafloor terrain throughout the world's oceans using ship-based survey systems and deep submergence platforms has resulted in unraveling plate boundary processes within the paradigm of sea floor spreading; this research has revolutionized the Earth and Oceanographic sciences. This new view of how the Earth works has provided a quantitative context for mineral exploration, land utilization, and earthquake hazard assessment, and provided conceptual models which planetary scientists have used to understand the structure and morphology of other planets in our solar system.

Much of this new knowledge stems from studying the seafloor – its morphology, geophysical structure, and characteristics, and the chemical composition of rocks collected from the ocean floor. Similarly, the discoveries in the late 1970s of deep sea 'black smoker' hydrothermal vents at the midocean ridge (MOR) crest (**Figure 1**) and the chemosynthetic-based animal communities that inhabit the vents have changed the biological sciences, provided a quantitative context for understanding global ocean chemical balances, and suggest modern analogs for the origin of life on Earth and extraterrestrial life processes. Intimately tied to these research themes is the study of the physical oceanography of the global ocean water masses and their chemistry and dynamics, which has resulted in unprecedented perspectives on the processes which drive climate and climate change on our planet. These are but a few of the many examples of how deep submergence research has revolutionized our understanding of our Earth and ocean history, and provide a glimpse at the diversity of scientific frontiers that await exploration in the years to come.

Enabling Deep Submergence Technologies

The events that enabled these breakthroughs was the intensive exploration that typified oceanographic expeditions in the 1950s to 1970s, and focused development of oceanographic technology and instrumentation that facilitated discoveries on many disciplinary levels. Significant among the enabling technologies were satellite communication and global positioning, microchip technology and the widespread development of computers that could be taken into the field, and increasingly sophisticated geophysical and acoustic modeling and imaging techniques. The other key enabling technologies which supplanted traditional mid-twentieth century methods for imaging and sampling the seafloor from the beach to the abyss were submersible vehicles of various types, remote-sensing instruments, and sophisticated acoustic systems designed to resolve a wide spatial and temporal range of ocean floor and oceanographic processes.

Oceanographic science is by nature multidisciplinary. Science carried out using deep submergence vehicles of all types has traditionally involved a wide range of research components because of the time and expense involved with conducting field research on the seafloor using human occupied submersibles or remotely operated vehicles (ROVs), and most recently autonomous underwater vehicles (AUVs) (**Figures 2–6**). Through the use of all these vehicle systems, and most recently with the advent of seafloor observatories, deep submergence science is poised to enter a new millennium where scientists will gain a more detailed understanding of the complex linkages between physical, chemical, biological, and geological processes occurring at and beneath the seafloor in various tectonic settings.

Understanding the temporal dimension of seafloor and sub-seafloor processes will require continued use of deep ocean submersibles and utilization of newly developed ROVs and AUVs for conducting time-series and observatory-based research in the deep ocean and at the seafloor. These approaches will provide new insights into intriguing problems concerning the interrelated processes of crustal generation, evolution, and transport of geochemical fluids in the crust and into the oceans, and origins and proliferation of life both on Earth and beyond.

Since the early twentieth century, people have been venturing into the ocean in a wide range of diving vehicles from bathyscapes to deep diving submersibles.

Figure 1 Hydrothermal vents on the southern East Pacific Rise axis at depths of 2500–2800 m. (A) Titanium fluid sampling bottles aligned along the front rail of *Alvin*'s basket in preparation for fluid sampling. (B) *Alvin*'s manipulator claw preparing to sample the hydrothermal sulfide chimney. (C) View from inside Alvin's forward-looking view port of the temperature probe being inserted into a vent orifice to measure the fluid temperature. (D) Hydrothermal vent after a small chimney was sampled which opened up the orifice through which the fluids are exiting the seafloor. Photos courtesy of Woods Hole Oceanographic Institution – Alvin Group, D. J. Fornari and K. Von Damm and M. Lilley.

Even in ancient times, there was written and graphic evidence of the human spirit seeking the mysteries of the ocean and seafloor. There is unquestionably the continuing need to take the unique human visual and cognitive abilities into the ocean and to the seafloor to make observations and facilitate measurements. For about the past 40 years submersible vehicles of various types have been developed largely to support strategic naval operations of various countries. As a result of that effort, the US deep-diving submersible *Alvin* was constructed. *Alvin* is part of the National Deep Submergence Facility (NDSF) of the University National Oceanographic Laboratories System (UNOLS) operated by the Woods Hole Oceanographic Institution (**Figure 7**). Alvin provides routine scientific and engineering access to depths as great as 4500 m. The US academic research community also has routine,

observational access to the deep ocean and seafloor down to 6000 m depth using the ROV and tethered vehicles of the NDSF (**Figure 7**). These vehicle systems of the NDSF include the ROV *Jason*, and the tethered optical/acoustic mapping systems *Argo II* and DSL-120 sonar (a 120 kHz split-beam sonar system capable of providing 1–2 m pixel resolution back-scatter imagery of the seafloor and phase-bathymetric maps with ~4 m pixel resolution (**Figure 7**). These fiberoptic-based ROV and mapping systems can work at depths as great as 6000 m.

Alvin has completed over 3600 dives (more than any other submersible of its type), and has participated in making key discoveries such as: imaging, mapping, and sampling the volcanic seafloor on the MOR crest; structural, petrological, and geochemical studies of transforms faults; structural

(A)

(A)

(B)

(B)

Figure 3 (A) The ROV *Jason* being lifted off the stern of a research vessel at the start of a dive. (B) ROV *Jason* recovering amphora on the floor of the Mediterranean Sea. Photos courtesy of Woods Hole Oceanographic Institution – Alvin Group and R. Ballard.

Figure 2 (A) The submersible *Alvin* being lifted onto the sternof R/V *Atlantis*, its support ship. (B) *Alvin* descending to the seafloor. Photos courtesy of Woods Hole Oceanographic Institution– Alvin Group and R. Catanach.

studies of portions of deep-sea trenches off Central America; petrological and geochemical studies of volcanoes in back-arc basins in the western Pacific Ocean; sedimentary and structural studies of submarine canyons, discovering MOR hydrothermal vents; and collecting samples and making time series measurements of biological communities at hydrothermal vents in many MOR settings in the Atlantic and Pacific Oceans. In 1991, scientists in *Alvin* were also the first to witness the vast biological repercussions of submarine eruptions at the MOR axis,

which provided the first hint that a vast subsurface biosphere exists in the crust of the Earth on the ocean floor (**Figure 7**).

Deep Submergence Science Topics

Some of the recent achievements in various fields of deep submergence science include the following.

1. Discoveries of deep ocean hydrothermal communities and hot ($>350°C$) metal-rich vents on many segments of the global mid-ocean ridge (MOR);
2. Documentation of the immediate after-effects of submarine eruptions on the northern East Pacific

Figure 4 Photographs taken from the ROV *Jason* at hydrothermal vents on the Mid-Atlantic Ridge near 37°N on the summit ofLucky Strike Seamount at a depth of 1700 m. All photographs are of a vent named 'Marker d4.' (A) Overall view of vent lookingNorth. White areas are anhydrite and barite deposits, yellowish areas are covered with clumps of vent mussels. (B) Close-up of theside of the vent, *Jason*'s sampling basket is in the foreground. (C) Close-up view of mussels on the side of the hydrothermalchimney; the hydrothermal vent where hot fluids are exiting the mound is at the upper right, where the image is blurry because ofthe shimmering effect of the hot water. Nozzles of a titanium sampling bottle are at the middle-right edge. (D) Insertinga self-recording temperature probe into a beehive chimney. (E) Titanium fluid sampling bottles being held by *Jason*'s manipulatorduring sampling. (F) Close-up of nozzles of sampling bottle inserted into vent orifice during sampling. Photos (D)}(F) are framegrabs of *Jason* video data. Photos courtesy of Woods Hole Oceanographic Institution – ROV Group, D. J. Fornari, S. Humphris, andT. Shank.

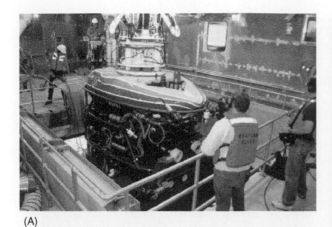

(A)

(B)

Figure 5 (A) ROV *Tiburon* of the Monterey Bay Aquarium Research Institute (MBARI) in the hanger of its support ship R/V *Western Flyer*. This electric-powered ROV can dive to 4000 m. A steel-armored, electro-optical cable connects *Tiburon* to the R/V *Western Flyer* and delivers power to the vehicle. Electric thrusters allow fine maneuvering while minimizing underwater noise and vehicle disturbance. A variable buoyancy control system, together with the syntactic foam pack, enables *Tiburon* to hover inches above the seafloor without creating turbulence, to pick up a rock sample, or maneuver quickly to follow an animal. (B) ROV *Ventana* of the MBARI being launched from its support ship R/V *Pt. Lobos*. This ROV gives researchers the opportunity to make remote observations of the seafloor to depths of 1850 m. The vehicle has two manipulator arms } a seven-function arm with five spatially correspondent joints and another seven-function robot arm with six spatially correspondent joints. Both arms can use a variety of end effectors to suit the type of work being done. *Ventana* is also equipped with a conductivity, temperature and density (CTD) package including a dissolvedoxygen sensor and a transmissometer. Photos courtesy of MBARI.

Rise, Axial Seamount, Gorda Ridge and CoAxial Segment of the Juan de Fuca Ridge;
3. Utilization of Ocean Drilling Program bore holes and specialized vehicle systems (e.g. Scripps Institution's Re-Entry Vehicle) (**Figure 8**) and instrument suites (e.g. CORKs) (**Figure 9**) for a wide range of physical properties, fluid flow and seismological experiments;

Figure 6 The autonomous underwater vehicle (AUV) ABE (Autonomous Benthic Explorer) of the Woods Hole Oceanographic Institution's Deep Submergence Laboratory, which can survey the seafloor completely autonomously to depths up to 4000 m and is especially well suited to working in rugged terrain such as is found on the Mid Ocean Ridge. Photo courtesy of Woods Hole Oceanographic Institution.

4. Discoveries of extensive fluid flow and vent-based biological communities along continental margins and subduction zones;
5. Initial deployment of ocean floor observatories of various types which enable the monitoring and sampling of geological, physical, biological and chemical processes at and beneath the seafloor (**Figures 10** and **11**).

These studies have revolutionized our concepts of deep ocean processes and highlighted the need for more detailed, time-series, multidisciplinary research.

Within the field of biological oceanography, major recent advances have come from the study of the new life forms and chemoautotrophic processes discovered at hydrothermal vents (**Figure 10**). These advances have fundamentally altered

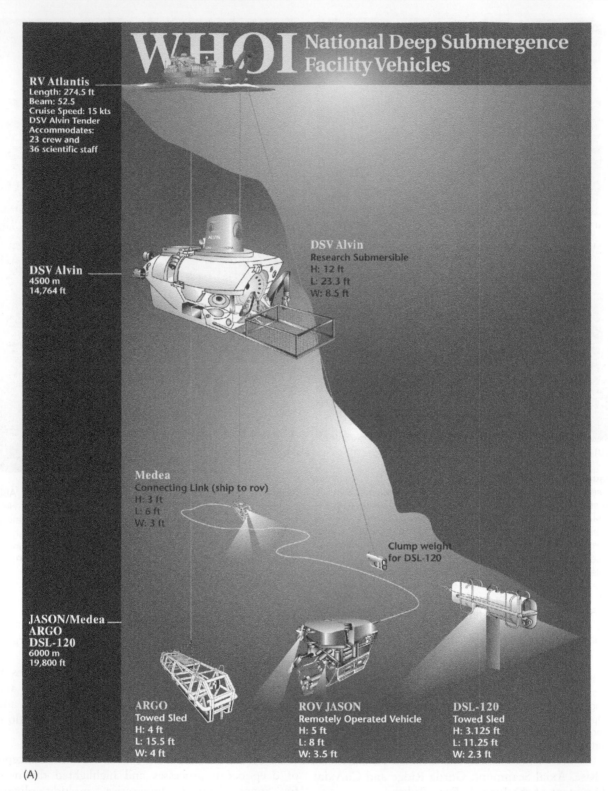

WHOI National Deep Submergence Facility Vehicles

RV Atlantis
Length: 274.5 ft
Beam: 52.5
Cruise Speed: 15 kts
DSV Alvin Tender
Accommodates:
23 crew and
36 scientific staff

DSV Alvin
4500 m
14,764 ft

DSV Alvin
Research Submersible
H: 12 ft
L: 23.3 ft
W: 8.5 ft

Medea
Connecting Link (ship to rov)
H: 3 ft
L: 6 ft
W: 3 ft

JASON/Medea
ARGO
DSL-120
6000 m
19,800 ft

Clump weight
for DSL-120

ARGO
Towed Sled
H: 4 ft
L: 15.5 ft
W: 4 ft

ROV JASON
Remotely Operated Vehicle
H: 5 ft
L: 8 ft
W: 3.5 ft

DSL-120
Towed Sled
H: 3.125 ft
L: 11.25 ft
W: 2.3 ft

(A)

Figure 7 (A) Summary of the US National Deep Submergence Facility (NDSF) vehicles operated for the University NationalOceanographic Laboratories System (UNOLS) by the Woods Hole Oceanographic Institution (WHOI). (B) Montage of the NDSFvehicles showing examples of the various types of data they collect. The figure also shows the nested quality of the surveysconducted by the various vehicles which allows scientists to explore and map features with dimensions of tens of kilometers (topright multibeam sonar map), to detailed sonar back scatter and bathymetry swaths which have pixel resolution of 1–2 m (DSL-120sonar), which are then further explored with the Argo II imaging system, or sampled using *Alvin* or ROV Jason. Graphics by P.Oberlander, WHOI; photos courtesy of WHOI – Alvin and ROV Group.

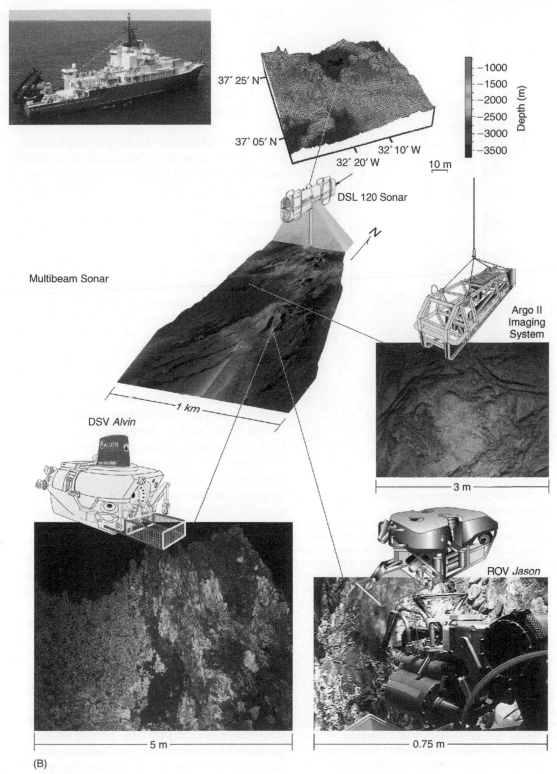

Figure 7   (Continued)

biological classification schemes, extended the known thermal and chemical limits of life, and have pushed the search for origins of life on Earth as well as for new life forms on other planetary bodies.

Recent marine biological studies show that: (1) the biodiversity of every marine community is vastly greater than previously recognized; (2) both sampling statistics and molecular tools indicate that the large majority of marine species have not been

(A)

Figure 8 (A) The Scripps Institution of Oceanography's Control Vehicle is a specialized ROV that can place instrument strings inside deep-sea boreholes, using a conventional oceanographic vessel capable of dynamic positioning and equipped with a winch carrying 17.3 mm (0.68 in) electromechanical cable with a single coax (RG8-type). The control vehicle is 3.5 m tall and weighs about 500 kg in water (1000 kg in air). It consists of a stainless-steel frame that contains two orthogonal horizontal hydraulic thrusters, a compass, a Paroscientific pressure gauge, four 250 W lights, a video camera, sonar systems, electronic interfaces to electrical releases and to a logging probe, and electronics to control all these sensors and handle data telemetry to and from the ship. Telemetry on the tow cable's single coax is achieved by analog frequency division multiplexing over a frequency band extending from 20 kHz to about 800 kHz. The sonars include a 325 kHz sector-scanning sonar, a 23.5 kHz narrow beam acoustic altimeter, and a 12 kHz sonar for long baseline acoustic navigation. This system was used successfully in the 1998 Ocean Seismic Network Pilot Experiment at ODP Hole 843B, 225 km south-west of the island of Oahu, Hawaii. In 1999, the control vehicle's analog telemetry module was converted into a digital system using fiberoptic technology thus providing bandwidth capabilities in excess of 100 Mbaud. (B) Cartoon showing the configuration of the control vehicle deployed from a research ship as it enters an Ocean Drilling Program bore hole to insert an instrument string. Photo and drawing courtesy of Scripps Institution of Oceanography – Marine Physical Laboratory, F. Spiess, and C. de Moustier.

described; (3) the complexity of biological communities is far greater than previously realized; and (4) the response of various communities to both natural and anthropogenic forcing is far more complex than had been understood even a decade ago. Focused studies over long time periods will be required to characterize these communities fully.

The field of marine biology is heading toward a more global time-series approach as a function of recent discoveries largely in the photic zone of the oceans and in the deep ocean at MOR hydrothermal vents. The marine biological, chemical, and physical oceanographic research which will be carried out in the next decade and beyond will certainly have a profound impact on our understanding of the complex food webs in the ocean which control productivity at every level and have direct implications for commercial harvesting of a wide range of resources from the ocean. Meeting the challenge of deciphering the various chemical, biological, and physical influences on these phenomena will require a better understanding and resolution of the causes and consequences of change on scales from hours to millennia. Understanding ocean ecosystems and their constituents will improve dramatically in response to emerging molecular, chemical, optical, and acoustical technologies. Given the relative paucity of information on deep-sea fauna in general, and especially the relatively recent discovery of chemosynthetic ecosystems at MOR hydrothermal vents, this will continue to be a focus for deep ocean biological research in the coming decade and beyond. Time-series and observatory-based research and sampling techniques will be required to answer the myriad of questions regarding the evolution and physiology of these unique biological systems (**Figures 10** and **11**).

Present and future foci for deep submergence science is MOR crests, hydrothermal systems, and the volcano–tectonic processes that create the architecture of the Earth's crust. Geochemists from the marine geology and geophysics community have emphasized the need for studies of: (1) the flux-frequency distribution for ridge-crest hydrothermal activity (heat, fluid, chemistry); (2) the role played by fluid flow in gas hydrate accumulation and determination of how important hydrates are to climate change; (3) slope stability; and (4) determination of how much of a role the microbial community plays in subsurface chemical and physical transformations. Another focus involves subduction zone processes, including: an assessment of the fluxes of fluids and solids through the seismogenic zone; long-term monitoring of changes in seismicity, strain, and fluid flux in the seismogenic zone;

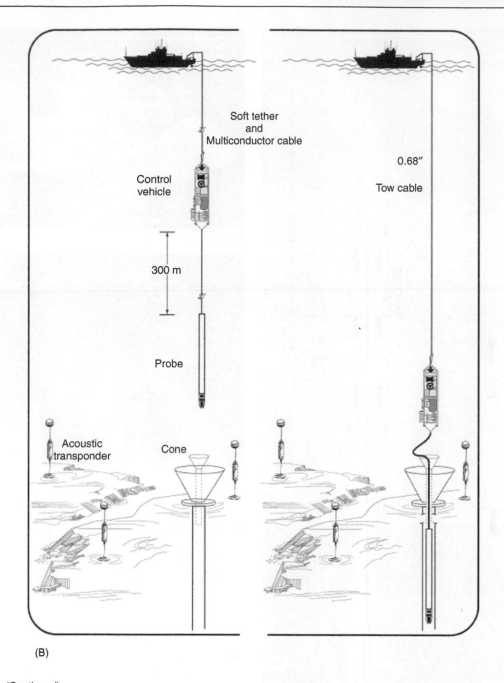

Soft tether
and
Multiconductor cable

Control
vehicle

300 m

Probe

Acoustic
transponder

Cone

0.68″

Tow cable

(B)

Figure 8b   (Continued)

and determining the nature of materials in the seis-mogenic zone.

To answer these types of questions, the marine geology and geophysics community requires systematic studies of temporal evolution of diverse areas on the ocean floor through research that includes mapping, dating, sampling, geophysical investigations, and drilling arrays of crustal holes (**Figures 8, 9** and **11**). The researchers in this field stress that the creation of true seafloor observatories at sites with different tectonic variables, with continuous monitoring of geological, hydrothermal, chemical,

and biological activity will be necessary. Whereas traditional geological and geophysical tools will continue to provide some means to address aspects of these problems, it is clear that an array of deep submergence vehicles, *in situ* sensors, and ocean floor observatory systems will be required to address these topics and unravel the variations in the processes that occur over short (seconds/minutes) to decadal time-scales. The infrastructural requirements, facility, and development needs required to support the research questions to be asked include: a capability for long-term seafloor monitoring; effective detection and

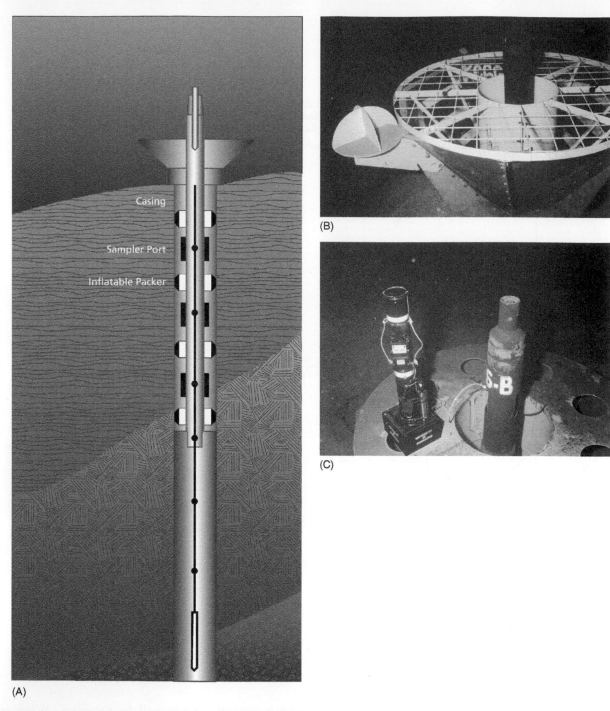

(A)

(B)

(C)

Figure 9 (A) Diagram of the upper portion of an OceanDrilling Program (ODP) borehole with a Circulation ObviationRetrofit Kit (CORK) assembly. These units serve the samepurpose as a 'cork' which seals a bottle; in the deep-seacase, the bottle is the seafloor which contains fluids that arecirculating in the ocean crust. The CORK allows scientists toaccess the circulating fluids and make controlled hydrologicmeasurements of the pressures and physical properties of thefluids. (B) A CORK observatory on the ocean floor in ODP hole858G off the Pacific north-west coast. (C) A CORK with instrumentsinstalled to measure sub-seafloor fluid circulation processes.Diagram courtesy of Woods Hole Oceanographic Institutionand J. Doucette; photo courtesy of K. Becker andE. Davis.

bathyurid crab (center). (F) A time-series temperature probe (with black and yellow tape), deployed at a hydrothermal vent (Tube worm pillar) on the East Pacific Rise crest near 9°49.6′N at a depth of 2495 m. The vent is surrounded by a large community of tube worms. (G) Seafloor markers along the Bio-Geo Transect, a series of 210 markers placed on the seafloor in 1992 to monitor the changes in hydrothermal vent biology and seafloor geology over a 1.4 km long section of the East Pacific Rise axis that have occurred after the 1991 volcanic eruption at this site. Photographs courtesy of T. Shank, Woods Hole Oceanographic Institution and R. Lutz, Rutgers University, D. J. Fornari and Woods Hole Oceanographic Institution – Alvin Group.

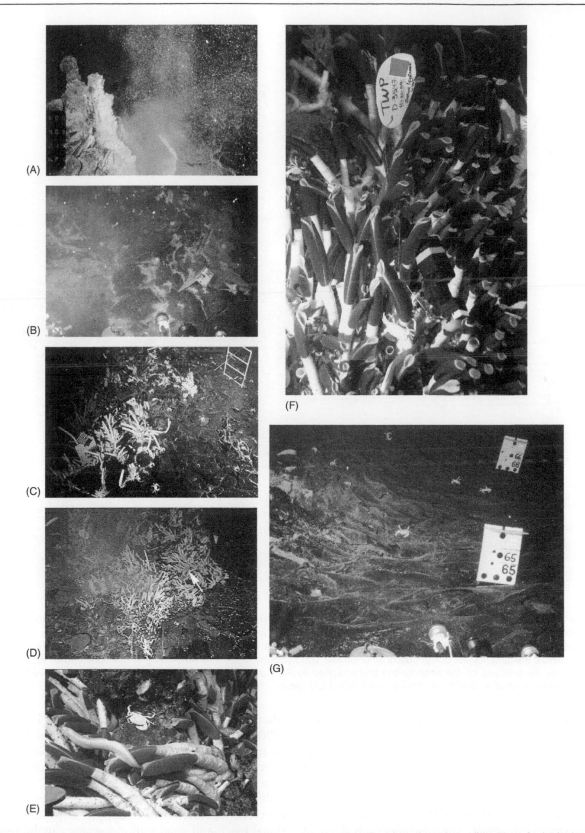

Figure 10   Time-series sequence of photographs taken of the same area of seafloor from the submersible *Alvin* of a hydrothermal vent site on the East Pacific Rise axis near 9°49.8'N at a depth of 2500 m. (A) 'Snow blower' vent spewing white bacterial by-product during the 1991 eruption. (B) Same field of view as (A) about 9 months later. Diffuse venting is still occurring as is bacterial production. White areas in the crevices of the lava flow are juvenile tube worms. (C) Patches of Riftia tube worms colonizing the vent area ~18 months after the March 1991 eruption. (D) Tube worm community has continued to develop, and the venting continues over 5 years after the eruption. (E) Close-up photograph of zoarcid vent fish (middle-left), tube worms, mussels (yellowish oblong individuals) and

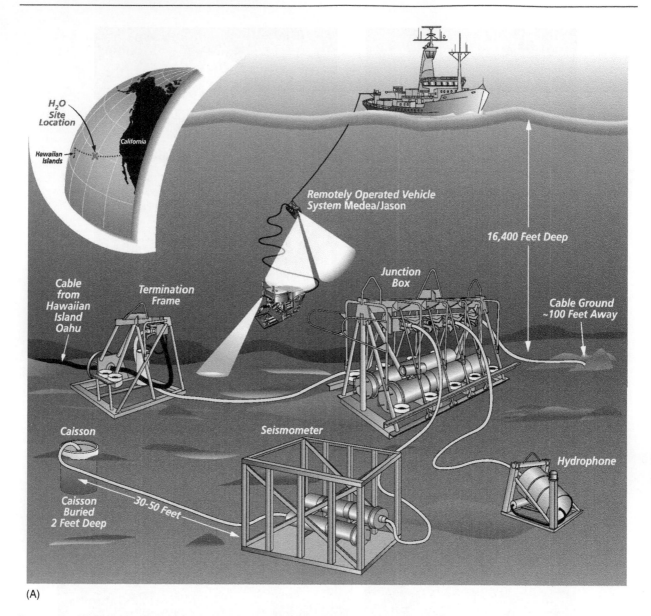

(A)

Figure 11 (A) Diagram of the deployment of the Hawaii-2 Observatory (H2O) one of the first long-term, deep seafloor observatories deployed in the past few years. Scientists used the ROV *Jason* to splice an abandoned submarine telephone cable into a termination frame which acts as an undersea telephone jack. Attached by an umbilical is a junction box, which serves as an electrical outlet for up to six scientific instruments. An ocean bottom seismometer and hydrophone are now functioning at this observatory. (B) The H2O junction box as deployed on the seafloor and photographed by ROV *Jason*. Drawing courtesy of Jayne Doucette, Woods Hole Oceanographic Institution (WHOI), A. Chave, and WHOI–ROV group.

response capability for a variety of seafloor events (volcanic, seismic, chemical); adequately supported, state-of-the-art seafloor sampling and observational facilities (e.g. submersible, ROVs, and AUVs), and accurate navigation systems, software, and support for shipboard integration of data from mutiscalar and nested surveys (**Figures 7B, 11** and **12**).

As discussed above, the disciplines involved in deep-submergence science are varied and the scales of investigation range many orders of magnitude from molecules and micrometer-sized bacteria to segment-scales of the MOR system (10s to 100s of

kilometers long) at depths that range from 1500 m to 6000 m and greater in the deepest trenches. Clearly, the spectrum of scientific problems and environments where they must be investigated require access to the deep ocean floor with a range of safe, reliable, multifaceted, high-resolution vehicles, sensors, and samplers, operated from support ships that have global reach and good station-keeping capabilities in rough weather. Providing the right complement of deep-submergence vehicles and versatile support ships from which they can operate, and the funding to operate those facilities cost-effectively, is both a

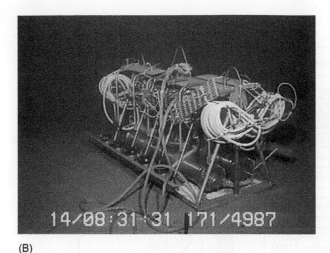

(B)

Figure 11   (Continued).

requirement and a challenge for satisfying the objectives of deep-sea research in the coming decade and into the twenty-first century.

To meet present and future research and engineering objectives, particularly with a multidisciplinary approach, deep submergence science will require a mix of vehicle systems and infrastructures. As deep submergence science investigations extend into previously unexplored portions of the global seafloor, it is critical that scientists have access to sufficient vehicles with the capability to sample, observe and make time-series measurements in these environments. Submersibles, which provide the cognitive presence of humans and heavy payload capabilities will be critical to future observational, time-series and observatory-based research in the coming decades. Fiberoptic-based ROVs and tethered systems, especially when used in closely timed, nested investigations offer unparalleled maneuverability, mapping and sampling capabilities with long bottom times and without the limitation of human/vehicle endurance. AUVs of various designs will provide unprecedented access to the global ocean, deep ocean and seafloor without dedicated support from a surface ship.

AUVs represent vanguard technology that will revolutionize seafloor and oceanographic measurements and observations in the decades to come. Over approximately the past 5 years scientists at several universities and private laboratories have made enormous advances in the capabilities and field-readiness of AUV systems. One such system is the Autonomous Benthic Explorer (ABE) developed by engineers and scientists at the Woods Hole Oceanographic Institution (**Figure 6**). ABE can survey the seafloor completely autonomously and is especially well suited to working in rugged terrain such as is found on the MOR. ABE maps the seafloor and the

water column near the bottom without any guidance from human operators. It follows programmed track-lines precisely and follows the bottom at heights from 5 to 30 m, depending on the type of survey conducted. ABE's unusual shape allows it to maintain control over a wide range of speeds. Although ABE spends most of its survey time driving forward at constant speed, ABE can slow down or even stop to avoid hitting the seafloor. In practice, ABE has surveyed areas in and around steep scarps and cliffs, and not only survived encounters with the extreme terrain, but also obtained good sensor data throughout the mission.

Recently ABE has been used for several geological and geophysical research programs on the MOR in the north-east and south Pacific which have further proved its reliability as a seafloor survey vehicle, and pointed to its unique characteristics to collect detailed, near-bottom geological and geophysical data, and to ground-truth a wide range of seafloor terrains (**Figure 12**). These new perspectives on seafloor geology and insights into the geophysical properties of the ocean crust have greatly improved our ability to image the deep ocean and seafloor and have already fostered a paradigm shift in field techniques and measurements which will surely result in new perspectives for Earth and oceanographic processes in the coming decades.

Conclusions

One of the most outstanding scientific revelations of the twentieth century is the realization that ocean processes and the creation of the Earth's crust within the oceans may determine the livability of our planet in terms of climate, resources, and hazards. Our discoveries may even enable us to determine how life itself began on Earth and whether it exists on other worlds. The next step is toward discovering the linkages between various phenomena and processes in the oceans and in exploring the interdependencies of these through time. Marine scientists recognize that technological advances in oceanographic sensors and vehicle capabilities are escalating at a increasingly rapid pace, and have created enormous opportunities to achieve a scope of understanding unprecedented even a decade ago. This new knowledge will build on the discoveries in marine sciences over the last several decades, many of which have been made possible only through advances in vehicle and sensor technology. With the rapidly escalating advances in technology, marine scientists agree that the time is ripe to focus efforts on understanding the connections in terms of

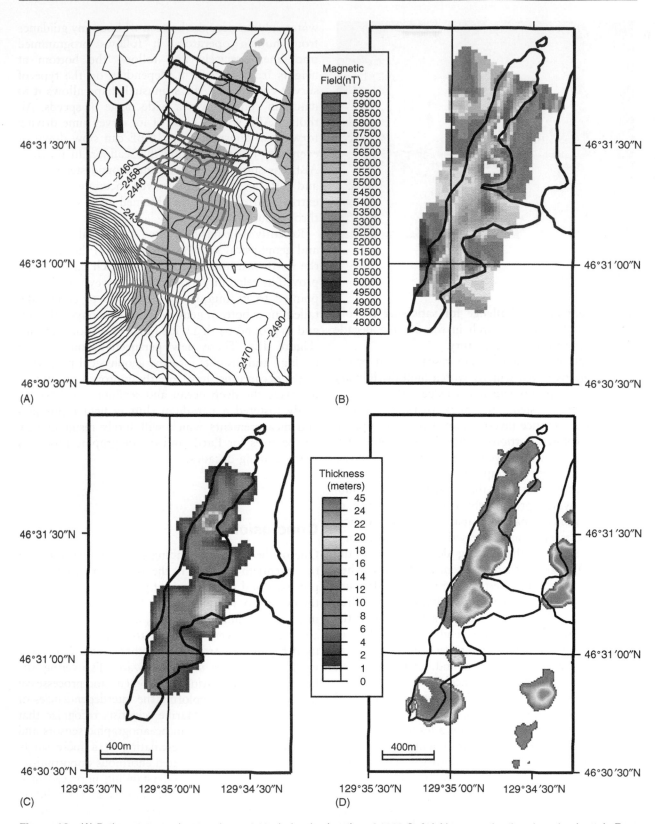

Figure 12   (A) Bathymetry map (contour interval 10 m) showing location of 1993 CoAxial lava eruption (gray) on the Juan de Fuca Ridge off the coast of Washington, and ABE tracklines (each color is a separate dive). (B) Magnetic field map based on ABE tracklines showing strong magnetic field over new lava flow. (C) Computed lava flow thickness assuming an average lava magnetization of 60 A/m compared with (D) lava flow thickness determined from differential swath bathymetry. Figure courtesy of M. Tivey, Woods Hole Oceanographic Institution.

interdependency of phenomena at work in the world oceans and their variability through time.

See also

Manned Submersibles, Deep Water. Remotely Operated Vehicles (ROVs).

Further Reading

Becker K and Davis EE (2000) Plugging the seafloor with CORKs. *Oceanus* 42(1): 14–16.

Chadwick WW, Embley RW, and Fox C (1995) Seabeam depth changes associated with recent lava flows, coaxial segment, Juan de Fuca Ridge: evidence for multiple eruptions between 1981–1993. *Geophysical Research Letters* 22: 167–170.

Chave AD, Duennebier F, and Butler R (2000) Putting H2O in the ocean. *Oceanus* 42(1): 6–9.

de Moustier C, Spiess FN, Jabson D, *et al.* (2000) Deep-sea borehole re-entry with fiber optic wire line technology. *Proceedings 2000 of the International Symposium on Underwater Technology.* pp. 23–26.

Embley RW and Baker E (1999) Interdisciplinary group explores seafloor eruption with remotely operated vehicle. *Eos Transactions of the American Geophysical Union* 80(19): 213–219 222.

Fornari DJ, Shank T, Von Damm KL, *et al.* (1998) Time-series temperature measurements at high-temperature hydrothermal vents, East Pacific Rise 9°49'–51'N: monitoring a crustal cracking event. *Earth and Planetary Science Letters* 160: 419–431.

Haymon RH, Fornari DJ, Von Damm KL, *et al.* (1993) Direct submersible observation of a volcanic eruption on the Mid-Ocean Ridge: 1991 eruption of the East Pacific Rise crest at 9°45'–52'N. *Earth and Planetary Science Letters* 119: 85–101.

Humphris SE, Zierenberg RA, Mullineaux L, and Thomson R (1995) *Seafloor Hydrothermal systems: Physical, Chemical, Biological, and Geological Interactions.* American Geophysical Union Monograph, vol. 91, 466 pp.

Ryan WBF (chair) *et al.*, (Committee on seafloor observatories: challenges and opportunities) (2000) *Illuminating the Hidden Planet, the Future of Seafloor Observatory Science.* Washington, DC: Ocean Studies Board, National Research Council, National Academy Press.

Shank TM, Fornari DJ, Von Damm KL, *et al.* (1998) Temporal and spatial patterns of biological community development at nascent deep-sea hydrothermal vents along the East Pacific Rise, 9°49.6'N–9°50.4'N. *Deep Sea Research, II* 45: 465–515.

Tivey MA, Johnson HP, Bradley A, and Yoerger D (1998) Thickness measurements of submarine lava flows determined from near-bottom magnetic field mapping by autonomous underwater vehicle. *Geophysical Research Letters* 25: 805–808.

UNOLS (University National Laboratory System) (1994) *The Global Abyss: An Assessment of Deep Submergence Science in the United States.* Narragansett, RI: UNOLS Office, University of Rhode Island.

Von Damm KL (2000) Chemistry of hydrothermal vent fluids from 9–10°N, East Pacific Rise: 'Time zero' the immediate post-eruptive period. *Journal of Geophysical Research* 105: 11203–11222.

GLIDERS

C. C. Eriksen, University of Washington, Seattle, WA, USA

Introduction

Underwater gliders are a recently developed class of autonomous underwater vehicle (AUV) driven by buoyancy changes to fly along saw-tooth trajectories through the ocean. Gliders quickly have become the AUVs with the highest endurance and longest range. They are able to sample the ocean interior at comparatively low cost because they can operate independently of ships, some for the better part of a year under global remote control.

Sampling the ocean on space and timescales as fine as those dominating its variability and over long ranges and durations has been a challenge throughout the history of ocean science. Traditionally, most knowledge of the ocean interior has been collected by ships. Starting in the late twentieth century, moored observations have complemented those from ship-based surveys, and more recently satellite remote sensing has provided images of sea surface conditions globally on ever-finer space and timescales. Were these techniques inexpensive, there would be no particular need for autonomous data collection by moving platforms. Unfortunately, ships stay at sea only for a month or two and rarely are directed to sample the same region for multiple cruises. Moorings sampling the open ocean number in the hundreds globally, typically in the dozens in the deep sea. This means variability of the ocean interior is generally undersampled. Undersampling is arguably the principal impediment to description and understanding of the ocean.

Gliders have been developed to address the shortcomings of conventional observing means so that such dominant sources of variability as mesoscale eddies, fronts, and boundary currents can be resolved simultaneously in space and time and over sufficiently long periods and wide domains to allow them to be understood. They are able to make progress in solving the sampling problem simply by the numbers their economy affords. The salient characteristic of gliders is that they cost roughly the equivalent of a few days of research vessel operation to build and can be operated for a few months for the cost of a day of ship time.

The tasks gliders do best are those for which ships are least suited: intensive, regular, and sustained observations of oceanic properties that are readily measured by electronic means. Observations that require large or power-hungry instruments, physical collection of samples, or specialized labor are unsuitable for gliders and continue the need for ship support. Gliders excel at measuring standard properties typically collected by a conductivity–temperature–depth (CTD) package. Among their many advantages is their ability to function well through the most severe seas the ocean has to offer, day and night, around the globe. They are a means to erasing the fair-weather bias of ship-based observations.

One may sample the ocean at deliberately chosen locations and times with gliders, examine the data very shortly after it is collected, and alter the sampling plan as often as a glider reaches the sea surface and communicates. The density of observations can readily be scaled to the phenomena of interest by adjusting the number and distribution of platforms to address local, regional, or global variability on diurnal, fortnightly, or seasonal scales, for example. Of course, gliders are not without limitation: long range and high endurance are achieved at the cost of traveling slowly through the ocean. When they cease to communicate, they are lost.

A Short History

John Swallow began the era of exploring the ocean interior autonomously in the 1950s with acoustically tracked floats ballasted to follow water motions at depth, famously deployed in the various North Atlantic locales. In following decades, acoustic float-tracking technology extended to ocean basin scale. Russ Davis and Doug Webb turned to satellite tracking to follow the motion of Autonomous Lagrangian Circulation Explorer (ALACE) floats that periodically adjusted their buoyancy to reach the sea surface for communication, and then returned to a 'parking' depth to drift. ALACE floats were the precursor to Argo floats, platforms used in an ambitious international program to seed the ocean with 3000 profiling floats. While Argo floats continue to track horizontal water movements, the focus of the Argo program is largely on CTD profiles collected periodically to describe large-scale features of ocean circulation.

Gliders are the descendants of profiling floats. They can be thought of as profiling floats with wings.

Their essential difference is that their sampling location is controlled, subject to a limited ability to navigate through currents. Floats randomly sample where currents carry them. While gliders do not offer a Lagrangian view of the ocean, they are commanded to describe oceanic conditions at deliberate times and locations, adding to the Eulerian description of the ocean.

Henry Stommel recognized the potential of underwater gliders (the idea was shared with him by his friend, neighbor, and colleague Doug Webb), and published a science fiction article describing them in 1989. Written as the retrospective of a scientist working at a glider network control station in 2021, the article has proved remarkably prescient in describing the gliders that were developed over the next decade and a half. Stommel envisioned a fleet of about 1000 gliders plying the oceans, continuously collecting profiles and communicating them ashore via satellite, while being controlled to variously occupy hydrographic sections or chase features within the ocean. The fundamental vision was of a distributed network of relatively inexpensive platforms. His optimism on the pace of development can understandably be blamed on his enthusiasm, for he predicted the maiden voyage at sea of a glider to have taken place in 1994 and last 198 days. In reality, the first ocean deployment longer than a week took place in 1999 and the longest mission as of this writing lasted 217 days, accomplished in 2005.

The Autonomous Oceanographic Sampling Network program set up by the US Office of Naval Research (ONR) in 1995 provided 5 years of support that resulted in gliders becoming a reality. The three operational gliders, *Slocum* (Webb Research Corp.), *Spray* (Scripps Institution of Oceanography and Woods Hole Oceanographic Institution), and *Seaglider* (University of Washington), were developed under this program and have subsequently been supported both by ONR and the US National Science Foundation. Four key technical elements enabled the successful development of underwater gliders: small reliable buoyancy engines, low-power computer microprocessors, Global Positioning System (GPS) navigation, and low-power duplex satellite communication.

Seaglider, *Slocum*, and *Spray* were designed to fulfill similar missions, hence understandably share many elements (**Figure 1**). All three are of similar size, roughly 50 kg in mass, in order that both manufacturing and operating costs would be relatively small. The vehicles can be launched and recovered from small vessels by two people without power-assisted equipment. They are designed to carry out missions from a few weeks (*Slocum*) to several months (*Seaglider* and *Spray*) duration while traveling at ~ 0.5 kt (~ 0.25 m s^{-1}). All are battery-powered, navigate by dead-reckoning underwater between GPS navigational fixes obtained at the sea surface, and send data and receive commands via a constellation of low Earth-orbit satellites on long-range missions.

Derivatives of these three gliders are currently under development, including a version powered in part by extracting thermal energy from ocean stratification, one that operates under ice, and one capable of making open ocean full-depth dives. In addition, parallel development of a vehicle considerably larger, faster, and capable of flying along a shallower glider path is underway.

Design Considerations

The three operational gliders were designed to address the sampling deficiencies of infrequent ship-based surveys and widely separated moorings in describing oceanic internal structure. Mesoscale eddies with characteristic space and timescales of $O(100 \text{ km})$ and $O(1 \text{ month})$ provide the principal noise to ocean general circulation and climate variability. Fronts, eddies, internal waves, diurnal cycles, and tidal fluctuations contribute to variability on smaller scales in problems of regional focus. Measurement platforms capable of resolving these features require sampling on scales $O(10 \text{ km})$ and $O(10 \text{ h})$ while the need to observe many realizations of random processes calls for sampling to be carried out for several weeks or months over tens or hundreds of kilometers. In order to observe at much lower cost than with ships or moorings, the three operational gliders were designed with endurance and range capability to meet these demands.

Gliding is the conversion of a buoyancy force into a body's forward as well as vertical motion. While aeronautical gliders are heavier than the air around them, hence always fall through it, underwater gliders are designed to alternately be heavier and lighter than the surrounding ocean so that they may glide both when diving and climbing. On average, underwater gliders displace an amount of water equal to their mass. They accomplish gliding by adjusting their volume displacement downward to dive and upward to climb. Gliders pitch themselves downward and are supported by wings to descend along a slanting path (conversely to ascend). Wings are glider propellers. They convert the vertical force of buoyancy into forward motion.

The desire to obtain vertical as well as horizontal structure of the ocean is conveniently and efficiently

several months (Seaglider and Spray duration while
traveling are 3.5 days and 6 km d^{-1}). All are battery-
powered, mission-life devices. Underwater
gliders carry no propeller, but obtained at the sea
surface, and send data and receive commands via a
equivalent of low-Earth-orbiting satellites on long
survey missions.

Figure 1 The operational gliders (clockwise from left): Seaglider pitched down at the sea surface with its trailing antenna raised (photo by Keith Van Thiel, University of Washington), top view of Spray at the sea surface (photo by Robert Todd, Scripps Institution of Oceanography), and rear view of Slocum climbing (photo by David Fratantoni, Woods Hole Oceanographic Institution).

provided by gliding, provided glide slopes are steep compared to slopes of oceanic property surfaces. The operational gliders variously operate along glide slopes in the range 0.25 to unity (e.g., $\sim 14-45°$), similar to the glide slope of a NASA space shuttle. By contrast, high-performance sail planes attain slopes as gentle as 0.02, not far from the higher slopes of property surfaces within the ocean.

Sinking through the ocean is gratis, while climbing through it requires energy input. The force of ambient pressure can be used to partially collapse vehicle volume, raise its mass density over that of seawater surrounding it, and make it sink. Diving requires only the energy to open a valve to allow a swim bladder to drain into a reservoir within the hull. To stop diving and begin ascent, displacement volume must be increased, requiring work against pressure to expand a swim bladder, for example. Work is required not only to provide sufficient potential energy to bring a glider to the sea surface, but also to provide kinetic energy to fly through the ocean. Once made buoyant, gliding eventually brings a vehicle to the sea surface where it can raise an antenna to navigate and communicate.

Kinetic energy provided through buoyancy forcing is ultimately dissipated by hydrodynamic drag in gliding. Since drag is generally quadratic with speed, halving speed quadruples endurance and doubles range. Underwater gliders achieve endurance and range 2 orders of magnitude higher than conventional propeller-driven AUVs by virtue of employing speeds an order of magnitude smaller. While a slow propeller-driven AUV might attain considerable endurance and range traveling horizontally, it too must work against the ocean's potential energy gradient (gravity) to reach the sea surface to navigate or communicate electromagnetically.

Slocum, *Spray*, and *Seaglider* were each designed to operate at speeds barely faster than typical ocean currents. Since currents are typically surface-intensified due to baroclinicity, deeper dives tend to reduce average current encountered by a glider, hence enhance the ability to navigate effectively. Gliders can navigate more effectively in regions where currents reverse with depth, despite being swift, than in regions of strong barotropic flow. For example, glider navigation in a swift equatorial current system or within a strait with strongly baroclinic flow is expected to be more effective than in weakly stratified coastal jets over the continental shelf. Gliders have operated across the Kuroshio and Gulf Stream by gliding with an upstream component to ferry across them, much as a canoeist paddles both upstream and across stream in order to cross a swift river. The finite spatial extent of ocean currents can be exploited to make up for downstream drift within them by gliding upstream in weak or opposing flows on their flanks. Navigating the swirling currents of a strong eddy is also possible if glider position within an eddy is recognized.

The operational gliders all carry a fixed energy supply in the form of primary batteries. Packs consisting of lithium cells provide the highest energy density, but are classified as hazardous material. Gliders using these are subject to various shipping restrictions, particularly as air freight. Nevertheless, lithium battery packs enable glider missions of half a year or more. Up to about one quarter of the total mass of these gliders can be devoted to carrying batteries, so packs of the equivalent of ~ 100 lithium D-cells can be used, providing roughly 10 MJ of energy, enough for ~ 1000 dive cycles using ~ 10 kJ per cycle. The total average power consumption may be as little as 0.5 W, resulting in mission endurance of several months. Recent improvements suggest that missions longer than a year are feasible for some battery-powered gliders. A buoyancy engine powered by ocean thermal stratification, as envisioned by Stommel and Webb, is under development for *Slocum*, but will be restricted to regions where sufficient ocean temperature stratification is found.

In principle, longer endurance and range can be attained by larger vehicles, since drag is determined by area while energy storage scales with volume. Instrument payload also scales with volume, and also expense. Practically, AUVs larger than ~ 60 kg require power assistance for handling at sea. Both the significant economy of manual handling by a pair of operators and the smaller unit manufacture and service costs associated with small platforms are consistent with the vision expressed by Stommel's 1989 article and also that guiding the Argo project: that of a network of a large number of relatively small, inexpensive platforms. By keeping gliders small, networks of them are tolerant to occasional vehicle loss.

Glider Dive Cycles

Gliders are fully autonomous from the time they leave the sea surface until they return. At the surface, they use a GPS fix to compute a desired heading to a target (or use a preset or communicated heading), then reduce volume displacement to start diving. Retracting a piston accomplishes buoyancy reduction in the 200-m-depth rated *Slocum*, while in *Spray* and *Seaglider* an external bladder is deflated to reduce buoyancy by opening a valve, since subbarometric pressure is maintained inside the pressure hull. Glider pitch attitude is adjusted by redistributing mass within the vehicles, both by changing buoyancy and moving fixed masses such as battery packs. Gliders turn to attain the desired heading through the use of an active rudder or by redistributing mass to roll the wings. Typical turn radii of the operational gliders are tens of meters.

Under water, navigation is by dead reckoning: simply following the chosen compass course. While course through the water varies due to environmental effects such as water-column shear of horizontal current as well as to vehicle control, the overall horizontal displacement due to gliding at various headings and speeds can be measured or inferred. The difference between displacement between surface GPS fixes and gliding displacement gives an estimate of depth-averaged current. The depth-averaged current for each dive cycle can be used to guide the choice of vehicle heading to approach a target through a prediction scheme, either on board or communicated from afar.

Hydrodynamics

The existing operational underwater gliders attain effectively steady flight in some tens of seconds due to relatively slow acceleration relative to hydrodynamic drag. The balance of buoyancy by lift and drag forces determines the glide slope. Glider design requires sufficiently small drag that the ratio of drag to lift, equal in steady flight to glide slope, can provide the desired performance. The wings of Seaglider, Slocum, and Spray are somewhat small relative to body vehicle size in comparison to those of typical aeronautical sail planes. Both considerations of how steeply to dive through the ocean's structure and ease of handling on launch and recovery guided wing sizing. Vehicle drag depends on both shape and surface texture, form, and skin drag, respectively. A significant component of drag is that induced by lift, typically parametrized as proportional to the square of attack angle. Considerations of drag led *Spray* to approximate a slender ellipsoidal form with a long cylinder capped by ellipsoidal ends and *Seaglider* to use a long smooth forebody (to maintain laminar boundary layer flow) and short after-section (where turbulence is tolerated). Tow tank, wind tunnel, and open-ocean experience have demonstrated that small appendages can account for disproportionately large contributions to total vehicle drag.

A glider with a given set of lift, drag, and induced drag coefficients can be flown over a range of speed and glide slope, the latter limited by stall at the maximum lift to drag ratio. As an example, buoyancy, power, and attack angle dependence on vertical and horizontal speed are shown in **Figure 2** for a typical Seaglider. Contours of power (solid curves) reflect the nearly quadratic nature of the drag law for this vehicle: operating at speeds of $\sim 0.2\,\mathrm{m\,s^{-1}}$ costs $\sim 0.1\,\mathrm{W}$, while operating at $\sim 0.4\,\mathrm{m\,s^{-1}}$ costs $\sim 0.4\,\mathrm{W}$ (of delivered power – the buoyancy engines are at most about 50% efficient). The maximum horizontal speed component attainable for a given power use is at a glide slope only slightly steeper than the stall glide slope, as can be seen from the departure of the power curve shape from near-circular about the origin to steeply curved to form a cone of exclusion at the stall slope. The extreme marks the most efficient glide slope for range, ~ 0.28 for speeds of $\sim 0.25\,\mathrm{m\,s^{-1}}$ to ~ 0.20 for speeds of $\sim 1\,\mathrm{m\,s^{-1}}$ for the example given in **Figure 2**.

A given buoyancy force leads to a wide range of speed, depending on glide slope, as illustrated by the dashed curves in **Figure 2**. For example, a buoyancy of $150\,\mathrm{g}$ can provide a horizontal speed of $0.22\,\mathrm{m\,s^{-1}}$ near the glide slope of maximum efficiency, or $0.33\,\mathrm{m\,s^{-1}}$ at a glide slope of 0.8 (using about 4 times as much power). The maximum total speed is highest for a given buoyancy if a glider were to be able to rise perfectly vertically (requiring no wing lift), but the maximum horizontal speed is attained at slopes of 0.78–0.84 (glide angles of 30–40° above horizontal) for the particular combination of lift, drag, and induced drag coefficients in the *Seaglider* example. The buoyancy range plotted in **Figure 2** exceeds what the *Seaglider* buoyancy engine is capable of providing in both positive and negative buoyancy (the end-to-end range is $\sim 850\,\mathrm{g}$), but diving through a light near surface layer may produce instantaneously buoyancy as high as is plotted in the figure.

Attack angles for gliders (the difference between pitch and glide angle, where the latter is always steeper) range from less than 1° for steep dive slopes to several degrees for glide slopes near stall. For an underwater flying wing, for example, glider pitch can be in the opposite sense to glide angle to produce large lift. For the *Seaglider* example, typical attack angles are a few degrees for the most efficient glide slope. Attack angle affects flow around the hull, amplifying it as a function of position.

Instrumentation

Size, power, hydrodynamic smoothness, and mass invariance are the principal constraints on instrumentation carried on gliders. For the operational

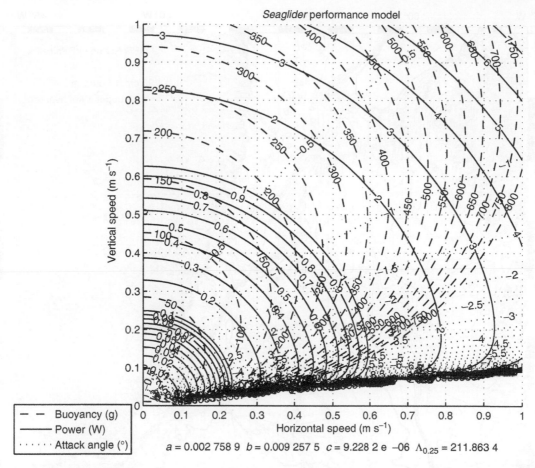

Figure 2 Model *Seaglider* performance based on steady flight with lift, drag, and induced drag coefficients a, b, and c where lift and drag forces follow $L = ql^2 a\alpha$ and $D = ql^2(bq^{-1/4} + c\alpha^2)$, where $l = 1.8$ m is the *Seaglider* body length, α is attack angle, and dynamic pressure is given by $q = \rho(u^2 + w^2)/2$ for horizontal and vertical speed components u and w for water density ρ. Solid, dashed, and dotted curves contour power, buoyancy, and attack angle, respectively, as functions of horizontal and vertical speed. Contours bunch at the stall limit. Glide slope is the ratio w/u and attack angles are labeled as the difference between vehicle pitch and glide angle, referenced to the horizon (e.g., negative attack for upward glide).

gliders, sensors and their packaging are practically limited to perhaps ~ 500-cc volume and power consumption a fraction of that used for glider propulsion. On long missions, as much as 85% of glider battery energy might be devoted to running the buoyancy engine, limiting the average power consumption available to instrumentation and operating the onboard microprocessor to less than ~ 0.1 W. Average power of this level is achieved by lowering the duty cycle of typical instruments considerably; that is, most of the time sensors are turned off and only briefly are energized for sampling. The low-power sleep current of the microprocessor is a significant constraint on mission endurance.

Oceanic fields that are readily measured electronically are well suited to gliders. Temperature, salinity, and pressure are readily measured electronically at little energy cost. In addition to scientific interest, together with measurements of vehicle pitch, volume, and heading, these can be used to estimate glider speed through the water. Other instrumentation routinely installed on gliders include dissolved oxygen sensors, bio-optical sensors for chlorophyll fluorescence, optical backscatter, and transmissivity, and acoustic sensors for both active and passive measurements. Specially adapted acoustic Doppler current profilers have also been carried on gliders. All of these instruments involve trade-offs between capability, sampling rate, and mission endurance.

Missions

The track of a *Seaglider* in the Labrador Sea (**Figure 3**) illustrates the ability of a glider to make transects in regions where depth-averaged flow is

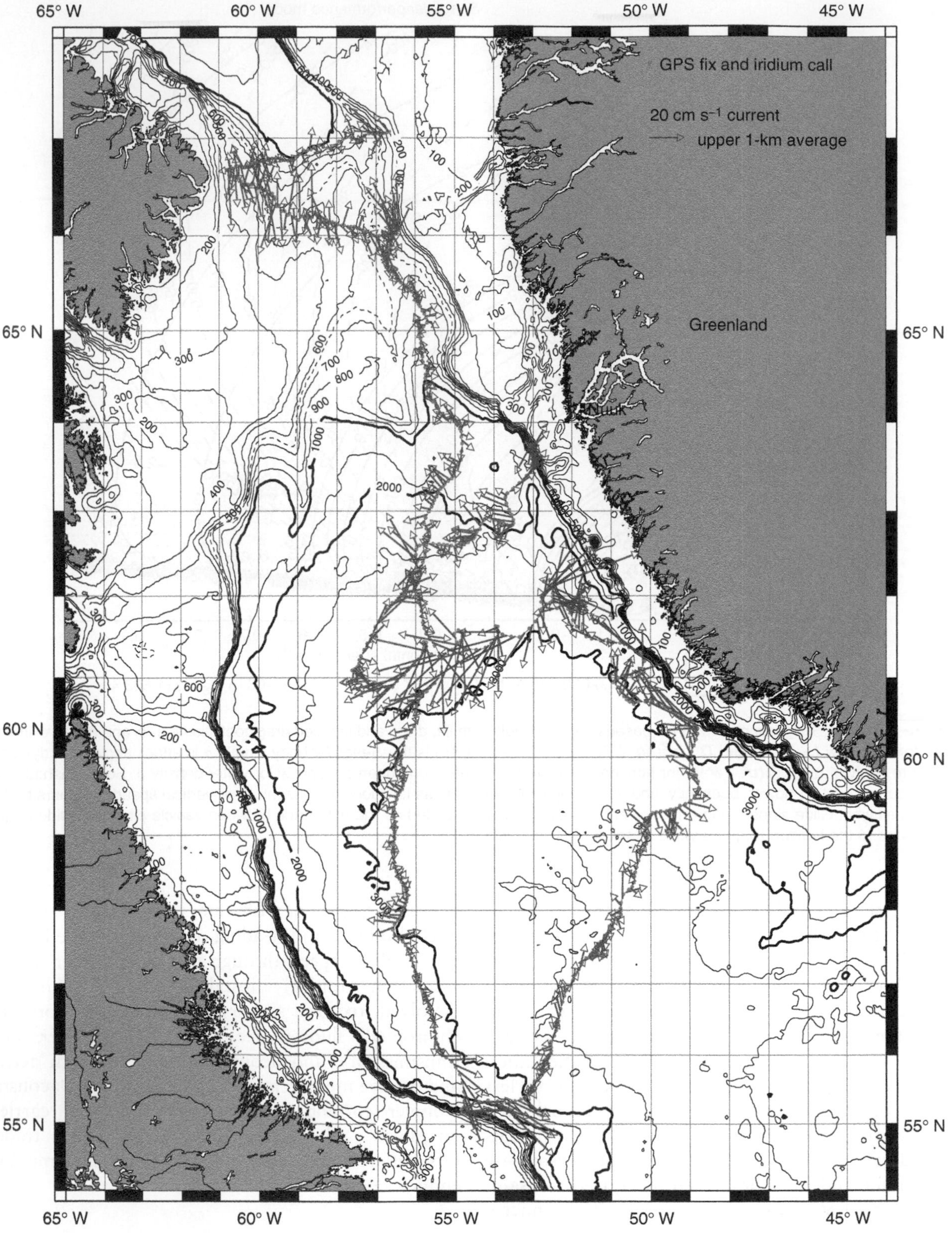

Figure 3 Track of a *Seaglider* in the Labrador Sea from 24 Sep. 2004 to 29 Apr. 2005, the longest AUV mission to date. This vehicle was launched from R/V Knorr in Davis Strait and recovered more than 7 months later offshore Nuuk, Greenland. Pink symbols mark the location of GPS fixes and Iridium calls between dive cycles. Blue arrows indicate depth-averaged currents (to the shallower of bottom depth and 1 km), where the scale length is typical glider speed through the water. Depth contour intervals are 100–1000 m, 500 m thereafter with heavy contours every 1 km.

sufficiently weak, but be hampered in its ability to navigate within strong currents. Launched from a research vessel offshore west Greenland, this vehicle made two transects across Davis Strait before being sent south to Hamilton Bank offshore southeast Labrador. It was swept across part of the continental shelf region before escaping the Labrador Current to continue northeastward on its return to Greenland near Kap Desolation. On its way north toward recovery, it was caught in an anticyclonic eddy that drifted offshore. This glider was directed mainly northward as it cycled around the eddy core four times before it finally escaped by being guided radially outward across the eddy. It continued northeastward to Fyllas Bank, offshore Nuuk, Greenland, where it was recovered from a fishing vessel, having completed the longest AUV mission to date as of this writing, lasting 217 days, traveling 3750 km through the water, and making 663 dives, the majority to 1000-m depth.

Several different types of glider missions have been undertaken, of which the example in **Figure 3** is just one: a solitary glider survey. In other applications, gliders have been used to make repeat surveys along the same track, controlled to maintain position (a 'virtual mooring'), used in numbers to survey a region intensively, used to interact with one another, to communicate with moored sensors, act as a courier for data recorded by these devices, and to control them.

Further Reading

Davis RE, Eriksen CC, and Jones CP (2003) Autonomous buoyancy-driven underwater gliders. In: Griffiths G (ed.) *Technology and Applications of Autonomous Underwater Vehicles*, pp. 37–58. New York: Taylor and Francis.

Rudnick DL, Davis RE, Eriksen CC, Fratantoni DM, and Perry MJ (2004) Underwater gliders for ocean research. *Journal of the Marine Technology Society* 38: 73–84.

Stommel H (1989) The Slocum mission. *Oceanography* 2: 22–25.

Relevant Websites

http://iop.apl.washington.edu
 – Operations Summary: Custom View, Seaglider Status, Integrative Observational Platforms, Ocean Physics Department, Applied Physics Laboratory, University of Washington.

http://spray.ucsd.edu
 – SIO IDG Spray Home.

http://www.webbresearch.com
 – Slocum Glider, Webb Research Corporation.

http://seaglider.washington.edu
 – The Seaglider Fabrication Center of the University of Washington.

MANNED SUBMERSIBLES, DEEP WATER

H. Hotta, H. Momma, and S. Takagawa,
Japan Marine Science & Technology Center, Japan

Introduction

Deep-ocean underwater investigations are much more difficult to carry out than investigations on land or in outer space. This is because electromagnetic waves, such as light and radio waves, do not penetrate deep into sea water, and they cannot be used for remote sensing and data transmission.

Moreover, deep-sea underwater environments are physically and physiologically too severe for humans to endure the high pressures and low temperatures. First of all, pressure increases by 1 atmosphere for every 10 meters depth because the density of water is 1000 times greater than that of air. Furthermore, as we have no gills we can not breathe under water. Water temperature decreases to 1°C or less in the deep sea and there is almost no ambient light at depth because sunlight can not penetrate through more than a few hundred meters of sea water. These are several of the reasons why we need either manned or unmanned submersibles to work in the deep sea.

A typical manned submersible consists of four major components: a pressure hull, propellers (thrusters), buoyant materials, and observational instruments. The pressure hull is a spherical shell made of high-strength steel or titanium. The typical internal diameter of the hull is approximately 2 m, which allows up to three people to stay at one atmosphere for 8–12 h during underwater operations. In case of emergency, a life-support system enables a stay of three to five days. Several thrusters are usually installed on the body of the submersible to give maneuverability. The buoyant material is syntactic foam, which is made of glass microballoons and an adhesive matrix. Its specific gravity is approximately $0.5 \, \mathrm{gf \, ml^{-1}}$. Observational instruments such as cameras, lights, sonar, CTD (conductivity, temperature, and depth sensors) etc., are also very important for gathering information on the deep-sea environment. It should be mentioned that the power consumption of the lights can reach as much as 15% of the total power consumption of the submersible.

History of Deep Submersibles

The first modern deep diving by humans, to a depth of 923 m, was achieved in 1934 by William Beebe, an American zoologist, and Otis Burton using the bathysphere, which means 'deep sphere'. The bathysphere was a small spherical shell made of cast iron, 135 cm in inside diameter designed for two observers. The bathysphere had an entrance hatch and a small glass view port. As the sphere was lowered by a cable and lacked thrusters, it was impossible to maneuver.

The next advance, using a free-swimming vehicle, occurred after World War II, in 1947. The bathyscaph *FNRS II* was invented by Auguste Piccard, who had been studying cosmic rays using a manned balloon in Switzerland. The principle of the bathyscaph was the same as that of a balloon. Instead of hydrogen or helium gas, gasoline was used as the buoyant material. During descent, air ballast tanks were filled with sea water, and for ascent, iron shot ballast was released. The pressure hull was made of drop-forged iron hemispheres, 2 m in inside diameter and 90 mm in thickness, allowing for two crew members. It was able to maneuver around the seafloor by thrusters driven by electric motors. Later, the second bathyscaph, *Trieste*, was sold to the US Navy, and independently at the same time, the French Navy developed the bathyscaph *FNRS III*, and later *Archimede*. In 1960, the *Trieste* made a dive into the Challenger Deep in the Mariana Trench, to a depth of 10 918 m. This historic dive was conducted by Jacques Piccard, son of Auguste Piccard, and Don Walsh from the US Navy. The bathyscaph was the first generation of deep-diving manned submersibles. It was very big and slow as it needed more than 100 kiloliter capacity gasoline tanks to provide flotation for the 2 m diameter pressure hull.

In 1964, the second generation of deep submersibles began. *Alvin* was funded by the US Navy under the guidance of the Woods Hole Oceanographic Institution (WHOI). At first, its depth capability was only 1800 m. It was small enough to be able to put on board the R/V *Lulu*, which became its support ship. Instead of gasoline flotation, syntactic foam was used. *Alvin* had horizontal and vertical thrusters to maneuver freely in three dimensions. Scientific instruments, including manipulators, cameras, sonar and a navigation system, were installed. Three observation windows were available for the three crew members. In France, the two-person 3000

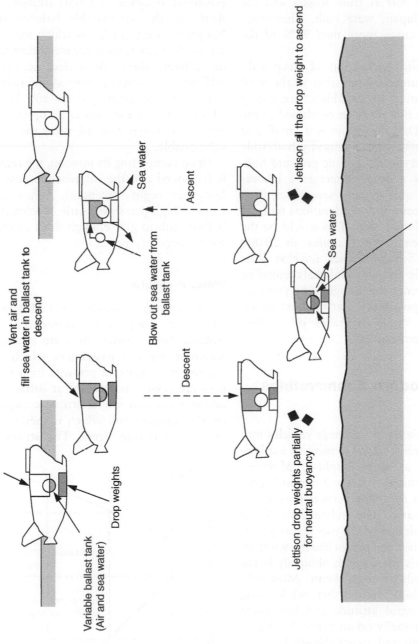

Vent air and
fill sea water in ballast tank to
descend

Sea water

Ascent

Blow out sea water from
ballast tank

Descent

Jettison all the drop weight to ascend

Sea water

Weight control by variable ballast tank
(Air and sea water)

Variable ballast tank
(Air and sea water)

Drop weights

Jettison drop weights partially
for neutral buoyancy

Figure 1 Principle of descent and ascent for a modern deep submersible.

m-class submersible *Cyana* was built. These two vehicles typified submersibles during the 1960s. At present, the depth capability of the *Alvin* has been increased to 4500 m by replacing the high-strength steel pressure hull with a titanium alloy sphere in 1973. In the 1980s, 6000 m-class submersibles, such as the *Nautile* from France, the *Sea Cliff* of the US Navy, the *Mir I* and *Mir II* from Russia and the *Shinkai 6500* from Japan, were built. They were theoretically able to cover more than 98% of the world's ocean floor.

What will the third generation of deep submersibles be like? Manned submersibles of the third generation, which would be capable of exceeding 10 000 m depth, have not yet been developed at the time of this report. One possibility is a small and highly maneuverable one- or two-person submersible with a transparent acrylic or ceramic pressure hull. Another possibility is a deep submergence laboratory, which would be able to carry several scientists and crew long distances and long durations without the assistance of a mother ship. This would be the realization of the dream like '*Nautilus*' in 20 000 Leagues Under The Sea by French novelist Jules Verne. Strong scientific and/or social goals would be needed for such a submersible design to be pursued. And there is a third possibility that the next generation will be evolutionary upgrades of existing second-generation submersibles.

Principles of Modern Submersibles

Descent and Ascent

There are several methods to submerge vehicles into the deep sea. The simplest way is to suspend a sphere by a cable, known as a bathysphere. Mobility, however, is greatly limited. A second method relies on powerful thrusters to adjust vertical position in relatively shallow water. The submersible *Deep Flight* is a high-speed design which uses thrust power coupled with fins for motion control like the wings of a jet fighter. It descends and ascends obliquely in the water column at speeds up to 10 knots. Most submersibles employ a third method that, while using weak thrusters to control attitude and horizontal movement, relies principally on an adjustable buoyancy system for descent and ascent (**Figure 1**).

When on the surface, the submersible's air ballast tank is filled with air creating positive buoyancy, hence it floats. When the dive begins, air is vented from the ballast tank and filled with sea water, thus creating negative buoyancy and sinking the vehicle. As the submersible dives deeper, buoyancy increases modestly due to the increasing water density created

by the increasing pressure. Thus the submersible slows slightly as it dives deeper (**Figure 2**).

When the submersible approaches the seafloor (50–100 m in altitude, i.e., height above the bottom), a portion of its ballast (usually lead or some other heavy material) is jettisoned to achieve neutral buoyancy. Perfect neutral buoyancy occurs when the positively buoyant materials (things which tend to float) on the submersible balance the negatively buoyant materials (things which tend to sink). This allows the vehicle to hover weightless in position and move freely about. As perfect neutral buoyancy is difficult to maintain, most submersibles have auxiliary weight-adjusting (trim and ballast) systems. This consists of a sea-water pumping system to draw in or expel water, thus adjusting the buoyancy of the submersible.

Upon completing its mission, the remaining ballast is jettisoned and the submersible now with positive buoyancy begins ascending. When resurfaced, air from a high-pressure bottle is blown into the air ballast tank to give enough draft to the submersible for the recovery operation.

Water Pressure

Water pressure increases by 0.1 MPa per 10 m depth. Thus every component sensitive to pressure must be isolated from intense pressure changes. First and foremost are the passengers which are protected against great ambient pressure by a pressure hull or pressure vessel, maintained at surface pressure. The ambient pressure exerts strong compressional force on the pressure hull which is therefore designed to avoid any tensile stress. The strongest geometric

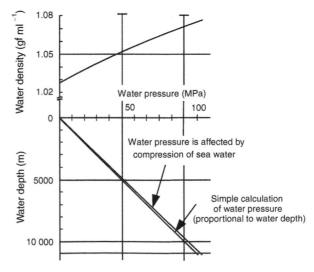

Figure 2 Relations between water depth, pressure and water density at a water temperature of 0°C and salinity of 34.5‰.

shape against outside pressure relative to volume and hence weight is a sphere, followed by a cylinder (capped at both ends). However, it is not easy to arrange instruments inside a sphere effectively.

In order to increase mobility, it is important to make submersibles small and light. The pressure hull is one of the largest and heaviest components of the submersible. The hull must be as small (and light) as possible, while affording appropriate strength against external pressure. Thus for deep-diving submersibles, a spherical pressure hull is employed whereas shallower vehicles can use a cylindrical shape if so desired.

The material used for the pressure hull is critical. In earlier vehicles, steel was used. Later, titanium alloy was the material of choice. Titanium alloy has very high tensile strength, and is resistant to corrosion and relatively light (specific gravity ~60% that of steel). Recently, the trend in submersible construction is to use nonmetallic materials, such as fiber- or graphite-reinforced plastics (FRP or GRP), or ceramics.

Components not sensitive to pressure or saline conditions need no special consideration. Though those devices which require electrical insulation need to be housed in oil-filled compartments called oil-filled pressure compensation systems (**Figure 3**). These systems do not require heavy pressure hulls and thus reduce the weight of the submersible overall. Electric motors, hydraulic systems, batteries, wiring, and power transistors are all housed in

pressure compensation systems. Technology is being developed to apply ambient pressure to electronic devices such as integrated circuits (ICs) and large scale ICs (LSIs).

Buoyancy

With the exception of some shallow-water submersibles, the total weight of the essential systems is larger than the total buoyancy. This means that extra buoyancy is needed to balance the excess weight. Wood or foam-rubber cannot be used for this purpose because they shrink under increasing water pressure. The material providing buoyancy must have a relatively small specific gravity while remaining strong under high-pressure conditions.

Historically, gasoline was used to provide buoyancy in bathyspheres as it did not lose buoyancy under pressure. However, its specific gravity was too large for practical use – huge volumes are needed to offset the weight. With the invention of syntactic foam, a superior material for deep-diving submersibles became available. Syntactic foam consists of tiny microscopic spheres of glass embedded in resin. These microballoons are 40–200 μm in diameter, and are closely packed with resin filling in the surrounding spaces. Proper selection of the balloons and resin allows the proper pressure tolerance and specific gravity to be created. For example, the syntactic foam used by the *Shinkai 6500* is tolerant up to 130 MPa with a specific gravity of 0.54 gf ml^{-1} and the foam used by the ROV *Kaiko* is tolerant up to 160 MPa with a specific gravity of 0.63 gf ml^{-1}.

Life Support

The pressure hull is a very small space where crew members must stay for up to 20 h, depending on their mission. Since the pressure hull is maintained at ambient pressure, no decompression of the occupants is needed. High-pressure oxygen bottles provide oxygen within the pressure hull, while carbon scrubbers absorb carbon-dioxide (**Figure 4**). Extra life-support is required with varying standards depending on the country (from 32 h to 120 h)

Energy

Energy for deep-diving submersibles is supplied by rechargeable (secondary) batteries. There are several types including: lead acid, nickel cadmium, nickel hydrogen, oxidized silver zinc. Batteries which contain a higher density of energy are preferable to reduce the weight and volume of the submersible. However, such batteries are very expensive. These batteries are housed in oil-filled pressure-compensated systems to

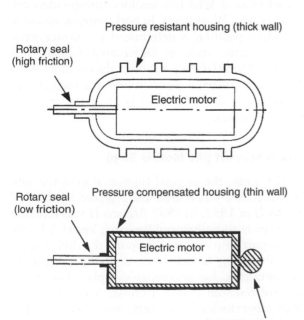

Figure 3 Pressure resistant and pressure compensated housings for electric motors.

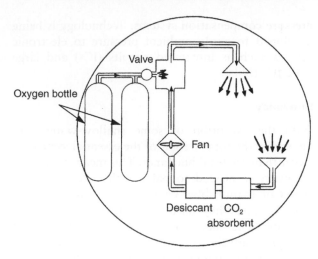

Figure 4 Life support system for a deep submersible.

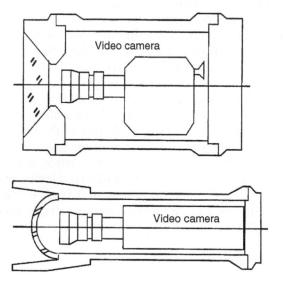

Figure 5 Examples of inside alignments of the pressure case.

reduce weight. Recent developments in fuel cell technology offer the promise of higher density energy coupled with higher efficiency. This has great potential for submersible applications.

Instrumentation

There are many scientific and observational instruments employed on research submersibles. Due to the limited payload, these instruments must be as light and small as possible. For example, bulky camera bodies must be streamlined and aligned with lenses in a compact manner, thus reducing the size and weight of their pressure housing (see **Figure 5**). Furthermore, physical conditions such as extremely cold temperatures or the differential absorption of

white light must be considered. Thus, for example, engineers must consider both the compactness of video cameras and their color sensitivity.

Manipulator

Pressure hulls, thrusters, batteries, and buoyancy materials are all essential parts of the modern submersible. One of the most important tools is the manipulator arm. Manipulators extend the arms and hands of the pilot, allowing sample collection and deployment of experimental equipment. Most manipulators are driven by hydraulic pressure. The most advanced manipulators operate in a master-slave system. The operator handles a master arm (controller) which imitates human arm and hand movements, and the motions are translated to the slave-unit (manipulator) which follows precisely the motions of the master arm. There are usually one or two manipulators on a research submersible.

Navigation

Underwater navigation is one of the most crucial elements of deep-sea submersible researches. Usually, long base line (LBL) or super short base line (SSBL) acoustic navigation systems are used depending on the accuracy required. The positioning error of LBL systems is 5–15 m, and it is approximately 2–5% of slant range for SSBL systems. An advantage of SSBL systems is that seafloor transponders are not necessary, whereas at least two seafloor transponders are necessary for LBL systems. In both systems, absolute or geodetic position is determined by surface navigation systems, such as Differential Global Positioning System (DGPS). Although the position of the submersible is usually reported from the mother ship by voice, the *Shinkai 6500* has an automatic LBL navigation system.

Surface Support (the Mother Ship)

The R/V *Lulu*, the original support ship of the submersible *Alvin*, was retired and replaced by the R/V *Atlantis II* in 1983. In 1997 *Atlantis II* was replaced by the newly commissioned Research Vessel *Atlantis*. The support ship not only supports the diving operation, including launch, recovery, communication, and positioning, but also provides a place to conduct on-board research during a cruise. Accordingly, research laboratories, computers, instruments for on-board data analysis, multi-narrow beam echo sounders, etc., are necessary. Also, remotely operated vehicles (ROVs) or autonomous underwater vehicles (AUVs), can be operated during the nighttime, or in case the manned submersible cannot be operated because of bad weather.

Deep Submersibles in the World

Alvin (USA)

Alvin (**Figure 6**) was built in 1964 by Litton Industries with funds of the US Navy and operated by WHOI. Its original depth capability was 1800 m with a steel pressure hull. Later, the pressure hull was replaced by titanium alloy to increase depth capability up to 4500 m. *Alvin*'s size and weight are: length 7.1 m; width 2.6 m; height 3.7 m and weight 17 tf in air. The outside diameter of the pressure hull is 2.08 m with a wall thickness of 49 mm. It is equipped with three view ports, 120 mm in inside diameter, two manipulators with seven degrees of freedom. The original catamaran support ship, R/V *Lulu*, was replaced by the R/V *Atlantis II*. Launch and recovery take place at the stern A-frame. It has been the leading deep submersible in the world.

Nautile (France)

Nautile (**Figure 7**) was built in 1985 by IFREMER (French Institution for Marine Research and Development) in France. Depth capability of 6000 m was aimed to cover 98% of the world's ocean floor. It is 8 m long, 2.7 m wide, 3.81 m in height and weighs 19.3 tf in air. It is equipped with three view ports, a manipulator, a grabber and a small companion ROV *Robin*. The position of the submersible is directly calculated by interrogating the seafloor transponders. Still video images are transmitted to the support ship through the acoustic link.

Sea Cliff (USA)

Sea Cliff was originally built in 1968 as a 3000 m-class submersible by General Dynamics Corp. for the US Navy as a sister submersible for the *Turtle*. In 1985, the *Sea Cliff* was converted into a 6000 m-class deep submersible. It is 7.9 m long, 3.7 m wide, 3.7 m high and weighs 23 tf in air. (*Sea Cliff* and *Turtle* are currently out of commission.)

Mir I and Mir II (Russia)

The 6000 m-class submersibles *Mir I* and *Mir II* (**Figure 8**) were built in 1987 by Rauma Repola in Finland for the P.P. Shirshov Institute of Oceanology in Russia. They are 7.8 m long, 3.8 m wide, 3.65 m high and weigh 18.7 tf in air. Inside diameter of the pressure hull, which is made of high-strength steel, is 2.1 m with a wall thickness of 40 mm. Launch and recovery take place by an articulated crane over the side of the support ship, the R/V *Academik Mistilav Keldysh*. If necessary, both *Mir I* and *Mir II* are launched simultaneously to carry out cooperative or independent research. Another characteristic feature of the *Mir* is a powerful secondary battery, 100 kWh of total energy, which allows it to stay more than 20 hours underwater, or to carry out more than 14 h of continuous operation on the bottom.

Shinkai 6500 (Japan)

The *Shinkai 6500* (**Figure 9**) was built in 1989 by Mitsubishi Heavy Industries and operated by the

Figure 6 US submersible *Alvin*.

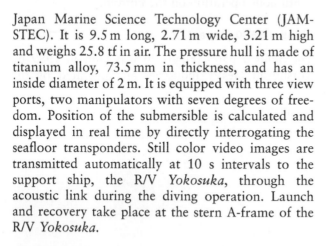

Figure 7 French 6000 m-class submersible *Nautile*.

Figure 8 Russian 6000 m-class submersible *Mir I* or *Mir II*.

Japan Marine Science Technology Center (JAM-STEC). It is 9.5 m long, 2.71 m wide, 3.21 m high and weighs 25.8 tf in air. The pressure hull is made of titanium alloy, 73.5 mm in thickness, and has an inside diameter of 2 m. It is equipped with three view ports, two manipulators with seven degrees of freedom. Position of the submersible is calculated and displayed in real time by directly interrogating the seafloor transponders. Still color video images are transmitted automatically at 10 s intervals to the support ship, the R/V *Yokosuka*, through the acoustic link during the diving operation. Launch and recovery take place at the stern A-frame of the R/V *Yokosuka*.

Major Contributions of Deep Submersibles

The dive to the Challenger Deep in the Mariana Trench by the bathyscaph *Trieste* in 1960 was one of the most spectacular achievements of the twentieth century. However, the dive was mainly for adventure rather than for science. In 1963, the US nuclear submarine *Thresher* sank in 2500 m of water off

Cape Cod in New England. After an extensive search for the submarine, the bathyscaph *Trieste* made dives to inspect the wreck in detail and recover small objects. The operation demonstrated the importance of using deep submersibles and advanced deep ocean technology to increase knowledge of the deep ocean. In 1966, hydrogen bombs were lost with a downed US B-52 bomber off Palomares, Spain. The *Alvin* showed the great utility of deep submersibles by locating and assisting in the recovery of lost objects from the sea.

Between 1973 and 1974, project FAMOUS (French–American Mid-Ocean Undersea Study) was conducted in the Mid-Atlantic Ridge off the Azores using the French bathyscaph *Archimede* and the US submersible *Alvin*. The project was the first systematic and successful use of deep submersibles for science. They discovered and sampled fresh pillow lavas and lava flows at 3000 m deep in the rift valley, where the oceanic crusts were being created, providing visual evidence of Plate Tectonics. In 1977, *Alvin* discovered a hydrothermal vent and vent animals in the East Pacific Rise off the Galapagos Islands at a depth of 2450 m. Discovery of these chemosynthetic animals, which were not dependent

Figure 9 Japanese 6000 m-class submersible *Shinkai 6500*.

on photosynthesis, had a profound impact on biology in the twentieth century.

Manned submersibles now compete with unmanned submersibles, such as ROVs and AUVs. Because of the expense of operation and maintenance, national funding is necessary for manned submersibles. However, ROVs and AUVs can be operated by private companies or institutions. In spite of the costs, the ability of the human observer to rapidly process information to make decisions provides an advantage and justifies continued use of manned submersibles.

See also

Deep Submergence: Science of. Manned Submersibles, Shallow Water. Moorings. Remotely Operated Vehicles (ROVs).

Further Reading

Beebe W (1934) *Half Miles Down*. New York: Harcourt Brace.

Busby RF (1990) *Undersea Vehicles Directory – 1990–91*, 4th edn. Arlington, VA: Busby Associates.

Funnel C (ed.) (1999) *Jane's Underwater Technology*, 2nd edn. UK: Jane's Information Group Limited.

Kaharl VA (1990) *Water Baby – The Story of Alvin*. New York: Oxford University Press.

Piccard A (1956) *Earth Sky and Sea*. New York: Oxford University Press.

Piccard J and Dietz RS (1961) *Seven Miles Down*. New York: G.P. Putnam.

MANNED SUBMERSIBLES, SHALLOW WATER

T. Askew, Harbor Branch Oceanographic
Institute, Ft Pierce, FL, USA

Introduction

Early man's insatiable curiosity to look beneath the surface of the sea in search of natural treasures that were useful in a primitive, comfortless mode of living were a true test of his limits of endurance. The fragile vehicle of the human body quickly discovered that most of the sea's depths were unapproachable without some form of protection against the destructive hostilities of the ocean.

Modern technology has paved the way for man to conquer the hostile marine environment by creating a host of manned undersea vehicles. Called submersibles, these small engineering marvels carry out missions of science, exploration, and engineering. The ability to conduct science and other operations under the sea rather than from the surface has stimulated the submersible builder/operator to further develop the specialized tools and instruments which provide humans with the opportunity to be present and perform tasks in relative comfort in ocean bottom locations that would otherwise be destructive to human life.

Over the past fifty years a depth of 1000 m has surfaced as the transition point for shallow vs deep water manned submersibles. During the prior 100 years any device that enabled man to explore the ocean depths beyond breath-holding capabilities would have been considered deep.

History

While it is difficult to pinpoint the advent of the first submersible it is thought that in 1620 Cornelius van Drebel constructed a vehicle under contract to King James I of England. It was operated by 12 rowers with leather sleeves, waterproofing the oar ports. It is said that the craft navigated the Thames River for several hours at a depth of 4 m and carried a secret substance that purified the air, perhaps soda lime?

In 1707, Dr Edmund Halley built a diving bell with a 'lock-out' capability. It had glass ports above to provide light, provisions for replenishing its air, and crude umbilical-supplied diving helmets which permitted divers to walk around outside. In 1776, Dr David Bushnell built and navigated the first submarine employed in war-like operations. Bushnell's Turtle was built of wood, egg-shaped with a conning tower on top and propelled horizontally and vertically by a primitive form of screw propeller after flooding a tank which allowed it to submerge.

In the early 1800s, Robert Fulton, inventor of the steamship, built two iron-formed copper-clad submarines, *Nautilus* and *Mute*. Both vehicles carried out successful tests, but were never used operationally.

The first 'modern submersible' was Simon Lake's *Argonaut I*, a small vehicle with wheels and a bottom hatch that could be opened after the interior was pressurized to ambient. While there are numerous other early submarines, the manned submersible did not emerge as a useful and functional means of accomplishing underwater work until the early 1960s.

It was during this same period of time that the French-built *Soucoupe*, sometimes referred to as the 'diving saucer', came into being. Made famous by Jacques Cousteau on his weekly television series, the *Soucoupe* is credited with introducing the general population to underwater science. Launched in 1959, the diving saucer was able to dive to 350 m.

The *USS Thresher* tragedy in 1963 appears to have spurred a movement among several large corporations such as General Motors (Deep Ocean Work Boat, DOWB), General Dynamics (*Star I, II, III, Sea Cliff,* and *Turtle*), Westinghouse (*Deepstar 2000, 4000*), and General Mills (*Alvin*), along with numerous other start-up companies formed solely to manufacture submersibles. Perry Submarine Builders, a Florida-based company, started manufacturing small shallow-water, three-person submersibles in 1962, and continued until 1980 (**Figures 1** and **2**). International Hydrodynamics Ltd (based in Vancouver, BC, Canada) commenced building the Pisces series of submersibles in 1962. The pressure hull material in the 1960s was for the most part steel with one or more view ports. The operating depth ranged from 30 m to 600 m, which was considered very deep for a free-swimming, untethered vehicle.

The US Navy began design work on *Deep Jeep* in 1960 and after 4 years of trials and tribulations it was commissioned with a design depth of 609 m. A two-person vehicle, *Deep Jeep* included many features incorporated in today's submersibles, such as a dropable battery pod, electric propulsion motors that operate in silicone oil-filled housings, and shaped resin blocks filled with glass micro-balloons

Figure 1 Perry-Link *Deep Diver*, 1967. Owned by Harbor Branch Oceanographic Institution. Length 6.7 m, beam 1.5 m, height 2.6 m, weight 7485 kg, crew 1, observers 2, duration 3–5 hours.

used to create buoyancy. *Deep Jeep* was eventually transferred to the Scripps Institution of Oceanography in 1966 after a stint searching for a lost 'H' bomb off Palomares, Spain. Unfortunately, *Deep Jeep* was never placed into service as a scientific submersible due to a lack of funding. The missing bomb was actually found by another vehicle, *Alvin*. *Alvin* did get funding and proved to be useful as a scientific tool.

The Nekton series of small two-person submersibles appeared in 1968, 1970, and 1971. The *Alpha*, *Beta*, and *Gamma* were the brainchildren of Doug Privitt, who started building small submersibles for recreation in the 1950s. The Nektons had a depth capability of 304 m. The tiny submersibles conducted hundreds of dives for scientific purposes as well as for military and oilfield customers. In 1982, the *Nekton Delta*, a slightly larger submersible with a depth rating of 365 m was

unveiled and is still operating today with well over 3000 dives logged.

A few of the submersibles were designed with a diver 'lock-out' capability. The first modern vehicle was the Perry-Link 'Deep Diver' built in 1967 and able to dive to 366 m. This feature enabled a separate compartment carrying divers to be pressurized internally to the same depth as outside, thus allowing the occupants to open a hatch and exit where they could perform various tasks while under the supervision of the pilots. Once the work was completed, the divers would re-enter the diving compartment, closing the outer and inner hatches; thereby maintaining the bottom depth until reaching the surface, where they could decompress either by remaining in the compartment or transferring into a larger, more comfortable decompression chamber via a transfer trunk.

Acrylic plastic was tested for the first time as a new material for pressure hulls in 1966 by the US Navy.

Figure 2   Perry PC 1204 *Clelia*, 1976. Owned and operated by Harbor Branch Oceanographic Institution. Length 6.7 m, beam 2.4 m, height 2.4 m, weight 8160 kg, crew 1, observers 2, duration 3–5 hours.

The *Hikino*, a unique submersible that incorporated a 142 cm diameter and a 0.635 cm thick hull, was only able to dive to 6 m. This experimental vehicle was used to gain experience with plastic hulls, which eventually led to the development of *Kumukahi*, *Nemo* (Naval Experimental Manned Observatory), *Makakai*, and *Johnson-Sea-Link*.

The *Kumukahi*, launched in 1969, incorporated a unique 135 cm acrylic plastic sphere formed in four sections. It was 3.175 cm thick and could dive to 92 m.

The *Nemo*, launched in 1970, and *Makakai*, launched in 1971, both utilized spheres made of 12 curved pentagons formed from a 6.35 cm flat sheet of Plexiglas™. The pentagons were bonded together to make one large sphere capable of diving to 183 m.

The *Johnson-Sea-Link*, designed by Edwin Link and built by Aluminum Company of America (ALCOA), utilized a *Nemo*-style Plexiglas™ sphere, 167.64 cm in diameter, 10.16 cm thick, and made of 12 curved pentagons formed from flat sheet and bonded together. This new thicker hull had an operational depth of 304 m.

Present Day Submersibles

The submersibles currently in use today are for the most part classified as either shallow-water or deep-water vehicles, the discriminating depth being approximately 1000 m. This is where the practicality of using compressed gases for ballasting becomes impractical. The deeper diving vehicles utilize various drop weight methods; most use two sets of weights (usually scrap steel cut into uniform blocks). One set of weights is released upon reaching the bottom, allowing the vehicle to maneuver, travel, and perform tasks in a neutral condition. The other set of weights is dropped to make the vehicle buoyant, which carries it back to the surface once the dive is complete.

The shallow vehicles use their thrusters and/or water ballast to descend to the bottom and some of the more sophisticated submersibles have variable ballast systems which allow the pilot to achieve a neutral condition by varying the water level in a pressure tank.

One advantage of the shallow vehicles is that the view ports (commonly called windows) can be much larger both in size and numbers, and where an acrylic plastic sphere is used the entire hull becomes a window.

Since the late 1960s and early 1970s acrylic plastic pressure hulls have emerged as an ideal engineering solution to create a strong, transparent, corrosion-resistant, nonmagnetic pressure hull. The limiting factor of the acrylic sphere is its ability to resist implosion from external pressure at great depths. Its strength comes from the shape and wall thickness. Therefore, the greater the depth the operator aims to reach, the thicker the sphere must be, which results in a hull that is much too heavy to be practical for use on a small submersible designed to go deeper than 1000 m.

These shallow-water submersibles, once quite numerous because of their usefulness in the offshore oilfield industry, are now limited to a few operators and mostly used for scientific investigations.

The *Johnson-Sea-Links* (J-S-Ls) stand out as two of the most advanced manned submersibles (**Figure 3**). *J-S-L I*, commissioned by the Smithsonian Institution in January 1971, was named for designer and donors Edwin A. Link and J. Seward Johnson.

Edwin Link, responsible for the submersible's unique design and noted for his inventions in the aviation field, turned his energies to solving the problems of undersea diving, a technology then still in its infancy. One of his objectives was to carry out scientific work under water for lengthy periods.

The *Johnson-Sea-Link* was the most sophisticated diving craft he had created for this purpose, and it promised to be one of the most effective of the new generation of small submersible vehicles that were being built to penetrate the shallow depths of the continental shelf (183 m or 100 fathoms).

Originally designed for a depth of 304 m, the vessel's unique features include a two-person transparent acrylic sphere, 1.82 m in diameter and 10.16 cm thick, that provides panoramic underwater visibility to a pilot and a scientist/observer. Behind the sphere, there is a separate 2.4 m long cylindrical, welded, aluminum alloy lock-out/lock-in compartment that will enable scientists to exit from its bottom and collect specimens of undersea flora and fauna. The acrylic sphere and the aluminum cylinder are enclosed within a simple jointed aluminum tubular frame, a configuration that makes the vessel resemble a helicopter rather than a conventionally shaped submarine. Attached to the frame are the vessel's ballast tanks, thrusters, compressed air, mixed gas flasks, and battery pod.

The aluminum alloy parts of the submersible, lightweight and strong, along with the acrylic capsule which was patterned after the prototype used by the US Navy on the *Nemo*, had extraordinary advantages over traditional materials like steel. They were most of all immune to the corrosive effects of sea water.

The emphasis in engineering of the submersible was on safety. Switches, connectors, and all operating gear were especially designed to avoid possible safety hazards. The rear diving compartment allows one diver to exit for scientific collections while tethered for communications and breathing air supply, while the other diver/tender remains inside as a safety backup. Once the dive is completed and the submersible is recovered by a special deck-mounted crane on the support ship, the divers can transfer into a larger decompression chamber via a transfer trunk which is bolted to the lock-out/lock-in compartment.

Now, 30 years later, the *Johnson-Sea-Links* with a 904 m depth rating, remain state of the art underwater vehicles. Sophisticated hydraulic manipulators work in conjunction with a rotating bin collection platform which allow 12 separate locations to be sampled and simultaneously documented by digital color video cameras mounted on electric pan and tilt mechanisms and aimed with lasers. Illumination is provided by a variety of underwater lighting systems utilizing zenon arc lamps, metal halide, and halogen bulbs. Acoustic beacons provide real time position

Figure 3 *Johnson-Sea-Link I* and *II*, 1971 and 1976. Owned and operated by Harbor Branch Oceanographic Institution. Length 7.2 m, beam 2.5 m, height 3.1 m, weight 10400 kg, crew 2, observers 2, duration 3–5 hours.

and depth information to shipboard computer tracking systems that not only show the submersible's position on the bottom, but also its relationship to the ship in latitude and longitude via the satellite-based global positioning system (GPS). The lock-out/lock-in compartment is now utilized as an observation and instrumentation compartment, which remains at one atmosphere.

Today's shallow-water submersibles (average dive 3–5 h) require a support vessel to provide the necessities that are not available due to their relatively small size. The batteries must be charged, compressed air and oxygen flasks must be replenished. Carbon dioxide removal material, usually soda lime or lithium hydroxide, is also replenished so as to provide maximum life support in case of trouble. Most submersibles today carry 5 days of life support, which allows time to effect a rescue should it become necessary.

The support vessel also must have a launch/recovery system capable of safely handling the submersible in all sea conditions. Over the last 30 years, the highly trained crews that operate the ship's handling systems and the submersibles, have virtually made the shallow-water submersibles an everyday scientific tool where the laboratory becomes the ocean bottom.

Operations

The two *Johnson-Sea-Links* have accumulated over 8000 dives for science, engineering, archaeology, and training purposes since 1971. They have developed into highly sophisticated science tools. Literally thousands of new species of marine life have been photographed, documented by video camera and collected without disturbing the surrounding habitat. Behavioral studies of fish, marine mammals and invertebrates as well as sampling of the water column and bottom areas for chemical analysis and geological studies are everyday tasks for the submersibles. In addition, numerous historical shipwrecks from galleons to warships like the *USS Monitor* have been explored and documented, preserving their legacy for future generations.

Johnson-Sea-Links I and II (J-S-Ls) were pressed into service to assist in locating, identifying, and ultimately recovering many key pieces of the ill-fated Space Shuttle Challenger. This disaster, viewed by the world via television, added a new dimension to the J-S-Ls' capabilities. Previously only known for their pioneering efforts in marine science, they proved to be valuable assets in the search and recovery operation. The J-S-Ls completed a total of 109 dives,

including mapping a large area of the right solid rocket booster debris at a depth of 365 m. The vehicles proved their worth throughout the operation by consistently performing beyond expectations. They were launched and recovered easily and quickly. They could work on several contacts per day, taking NASA engineers to the wreckage for first-hand detailed examination of debris while video cameras recorded what was being seen and said. Significant pieces were rigged with lifting bridles for recovery. The autonomous operation of the J-S-Ls, a dedicated support vessel, and highly trained operations personnel made for a successful conclusion to an operation that had a significant impact on the future of the US Space Program.

Summary

There is no question that the manned submersible has earned its place in history. Much of what are now cataloged as new species were discovered in the last 30 years with the aid of submersibles. The ability to conduct marine science experiments *in situ* led to the development of intricate precision instruments, sampling devices for delicate invertebrates and gelatinous organisms that previously were only seen in blobs or pieces due to the primitive methods used to collect them.

While some suggest that remotely operated vehicles (ROVs) could, and have replaced the manned submersible, in reality they are complementary. There is no substitute for the autonomous, highly maneuverable submersible that can approach and collect without contact delicate zooplankton, while observing behavior and measuring the levels of bioluminescence, or probing brine pools and cold seep regions in the Gulf of Mexico for specialized collections of biological, geological, and geochemical samples. Tubeworms are routinely marked for growth rate studies and collected individually, along with other biological species that thrive in these chemosynthetic communities. Sediments and methane ice (gas hydrates) are also selectively retrieved for later analysis.

Some new vehicles are still being produced, but have limited payloads, which restricts them to specific tasks such as underwater camera platforms or observation. Some are easily transportable but are small and restricted to one occupant; they can be carried by smaller support vessels and are more economical to operate. Man's desire to explore the lakes, oceans, and seas has not diminished. New technology will only enable, not reduce, the need for man's presence in these hostile environments.

See also

Manned Submersibles, Deep Water. Remotely Operated Vehicles (ROVs).

Further Reading

Askew TM (1980) *JOHNSON-SEA-LINK Operations Manual*. Fort Pierce: Harbor Branch Foundation.

Busby F (1976, 1981) *Undersea Vechicles Directory*. Arlington: Busby & Associates.

Forman WR (1968) *KUMUKAHI Design and Operations Manual*. Makapuu, HI: Oceanic Institute.

Forman W (1999) *The History of American Deep Submersible Operations*. Flagstaff, AZ: Best Publishing Co.

Link MC (1973) *Windows in the Sea*. Washington, DC: Smithsonian Institute Press.

Stachiw JD (1986) *The Origins of Acrylic Plastic Submersibles*. American Society of Mechanical Engineers, Asme Paper 86-WA/HH-5.

Van Hoek S and Link MC (1993) *From Sky to Sea; A Story of Ed Link*. Flagstaff, AZ: Best Publishing Co.

PLATFORMS: AUTONOMOUS UNDERWATER VEHICLES

J. G. Bellingham, Monterey Bay Aquarium Research Institute, Moss Landing, CA, USA

Introduction

Autonomous underwater vehicles (AUVs) are un-tethered mobile platforms used for survey operations by ocean scientists, marine industry, and the military. AUVs are computer-controlled, and may have little or no interaction with a human operator while carrying out a mission. Being untethered, they must also store energy onboard, typically relying on batteries. Motivations for using AUVs include such factors as ability to access otherwise-inaccessible regions, lower cost of operations, improved data quality, and the ability to acquire nearly synoptic observations of processes in the water column. An example of the first is operations under Arctic and Antarctic ice, an environment in which operations of human-occupied vehicles and tethered platforms are either difficult or impossible. Illustrating the next two points, AUVs are becoming the platform of choice for deep-water bathymetric surveying in the offshore oil industry because they are less expensive than towed platforms as well as produce higher-quality data (because they are decoupled from motion of the sea surface). Finally, the use of fleets of AUVs enables the rapid acquisition of distributed data sets over regions as large as $10\,000\,km^2$.

AUVs are a new class of platform for the ocean sciences, and consequently are evolving rapidly. The Self Propelled Underwater Research Vehicle (SPURV) AUV, built at the University of Washington Applied Physics Laboratory, was first operated in 1967. However, adoption by the ocean sciences community lagged until the late 1990s. Adoption was spurred on by two developments: AUV development teams started supporting science field programs with AUV capabilities, and AUVs that nondevelopers could purchase and operate became available. The first served the purpose of building a user base and demonstrating AUV capabilities. The second enabled scientists to obtain and operate their own vehicles. Today, a wide variety of AUVs are available from commercial manufacturers. An even larger number of companies develop subsystems and sensors for AUVs.

The most common class of AUV in use today is a torpedo-like vehicle with a propeller at its stern, and steerable control surfaces to control turns and vertical motion (see **Figure 1**). These vehicles are used when speed or efficiency of motion is an important consideration. Such torpedo-like vehicles range in weight from a few tens of kilograms to thousands of kilograms. Most typical are vehicles weighing a few hundred kilograms, with an endurance of about a day at a speed of $c.\ 1.5\,m\,s^{-1}$. Often they have parallel mid-bodies, which allow the vehicle length to be extended without large hydrodynamic consequences. This is useful when it is necessary to add new sensors or batteries to a vehicle. A disadvantage of torpedo-like vehicles is that, like an aircraft, they must maintain forward motion to generate lift over its control surfaces, and thus are not controllable at very low speed through the water.

Gliders are a class of vehicles that use changes in buoyancy rather than a propeller for propulsion. Gliders use their ability to control buoyancy to generate vertical motion. Vertical motion is translated into horizontal motion with lifting surfaces, usually wings mounted in about the middle of the vehicle (see **Figure 1**). Several types of gliders weighing about $50\,kg$ are in use today. These comparatively small vehicles are designed to move slowly, about $0.25\,m\,s^{-1}$, and operate sensors consuming a watt or less. By minimizing power consumption, these gliders can operate for periods of months using high-energy-density primary batteries. Disadvantages of this class of system are that they are limited to vertical profiling flight tracks, and can be overwhelmed by ocean currents, especially in the coastal environment or within boundary currents. However, larger gliders in development and testing will operate at higher speeds, and thus not suffer from this limitation.

A final class of AUVs uses multiple thrusters to provide capabilities similar to that of a helicopter or a ship with dynamic positioning (see **Figure 1**). The additional thrusters enable maneuvers such as hovering, translating sideways, and moving vertically. These vehicles are used when maneuverability is needed, for example, when operation near a very rough bottom is a necessity. The disadvantage is that the additional thrusters reduce efficiency for moving large distances or at high speeds.

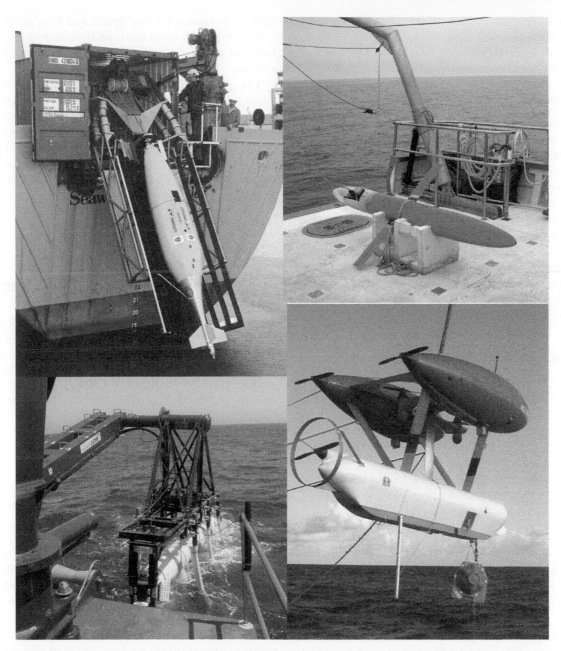

Figure 1 From top-left corner clockwise: the *Hugin* AUV, the *Spray* glider (courtesy, MBARI), the *ABE* AUV (courtesy, Dana Yoerger, WHOI), and the *Dorado* AUV (courtesy, MBARI). *Hugin* and *Dorado* are examples of propeller-driven vehicles optimized for moving through the water efficiently, and have a torpedo-like configuration. *Spray* is an example of a glider, which is a buoyancy-driven vehicle that has no propeller. *ABE* is a highly maneuverable vehicle capable of hovering or pivoting in place, and of moving straight up or down. Also illustrated are different handling strategies. The large Hugin vehicle is launched and recovered from a ship using a stern ramp. *Spray* is hand-launched and recovered. *ABE* is launched and recovered with a crane. *Dorado* is shown being launched and recovered with a capture mechanism suspended from a J-frame.

While the vehicles described above are representative of the most commonly used systems, a wide range of other vehicles are in development or are in limited use. AUVs that come to the surface and use solar panels to recharge batteries have been demonstrated in seagoing operations. Gliders which extract energy from thermal differences in the ocean for propulsion have also been tested. A hybrid vehicle is being developed for reaching the deepest portion of the ocean. The hybrid vehicle operates as a tethered platform via a disposable fiber optic link for tasks requiring human perception, in other words as a remotely operated vehicle (ROV), but operates as an AUV when that link is severed. These are just a few of the diverse AUVs being developed to answer the needs of ocean science.

Terminology

The military term for an AUV is unmanned underwater vehicle, or UUV. This phrase is ambiguous in that it can also refer to ROVs. While ROVs are unmanned, they are tethered and designed to be operated by human, and thus are not considered autonomous. However, common usage is to employ UUV as a synonym for AUV. Military terminology is relevant as many AUVs in use by the scientific community were developed under navy funding, and the military continues to be the largest single investor in AUV technology. Consequently, the technical literature on AUVs also employs military terminology.

Basics of AUV Performance

Energy is a fundamental limitation for underwater vehicles. Thus, energy efficiency is a fundamental driver for vehicle design and operations. This section outlines the relationship between vehicle speed and endurance, and its dependence on factors such as power consumed by onboard systems.

A simple but useful model for power consumption P of an AUV is as follows:

$$P = P_{prop} + H \quad \text{where} \quad P_{prop} = \frac{1}{2}\frac{C_D A \rho v^3}{\eta} \quad [1]$$

Here the total electrical power consumed by the vehicle, P, is equal to the sum of propulsion power, P_{prop} and hotel load, H. Hotel load is simply the power consumed by all subsystems other than propulsion. Propulsion power is a function of the drag coefficient of the vehicle, C_D, the area of the vehicle, A, the density of water, ρ, the speed of the vehicle, v, and the efficiency of the propulsion system, η.

What are the typical values for the coefficients in eqn [1]? Consider an 'example vehicle' which is a 12 3/4″ (0.32 m) diameter torpedo-like AUV. Note that this is a standard for a mid-size class AUV. For such a vehicle, parameter values might be $C_D = 0.2$ (based on frontal area), $A = 0.082\,\text{m}^2$, and $\eta = 0.5$. Hotel load would depend on sensors, but an overall value of 30 W might be representative, although mapping sonars would consume much more power. We use $\rho = 1027\,\text{kg m}^{-3}$. Note that these numbers, except seawater density of course, can differ greatly from vehicle to vehicle. For example,

gliders optimized for low speed and long endurance can operate with hotel loads on the order of a watt or less, but at the cost of operating a few very simple sensors.

The power relationship provides insight into a variety of AUV design considerations. For example, the speed at which the vehicle will consume the least energy per unit distance traveled (the energetically optimum speed) can be computed from observing that power divided by vehicle speed equals energy per unit distance. Finding the minimum of P/v with respect to v yields the optimum speed from an energy conservation perspective:

$$v_{opt} = \left(\frac{\eta H}{C_D A \rho}\right)^{1/3} \quad [2]$$

For the example vehicle values given above, the optimum vehicle speed is approximately $1\,\text{m s}^{-1}$.

What can we say about vehicle performance at the energetically optimum speed? Substituting eqn [2] into eqn [1] we find that the power consumed at the optimum speed is $(3/2)H$ which for our example vehicle is 45 W. If the total energy capacity of the battery system is E_{cap}, then the maximum range of the vehicle will be:

$$d_{max} = \frac{2E_{cap}}{3}\left(\frac{\eta}{C_D A \rho H^2}\right)^{1/3} \quad [3]$$

If our example vehicle carries 10 kg of high-energy-density primary batteries providing a total of $1.3 \times 10^7\,\text{J}$, it will have an endurance of 80 h, and a range of 280 km. The same vehicle with 10 kg of rechargeable batteries, with one-third the energy capacity of the high-energy-density batteries, will have its endurance and range reduced proportionately.

There are caveats to the above discussion. For example, propulsion efficiency, η, is typically a strong function of speed as the electrical motors used tend to have comparatively narrow ranges of efficiency. In practice, a vehicle's propulsion system is optimized for a particular speed and power. Also, vehicles are often operated at higher speeds than the energetically optimum speed given by eqn [2]. For example, operators of an AUV attended by a ship will be more sensitive to minimizing ship costs than to optimizing energy efficiency of the AUV.

AUV Systems and Technology

AUVs are highly integrated devices, containing a variety of mechanical, electrical, and software subsystems. **Figure 2** shows internal and external views of a deep-diving vehicle equipped with mapping

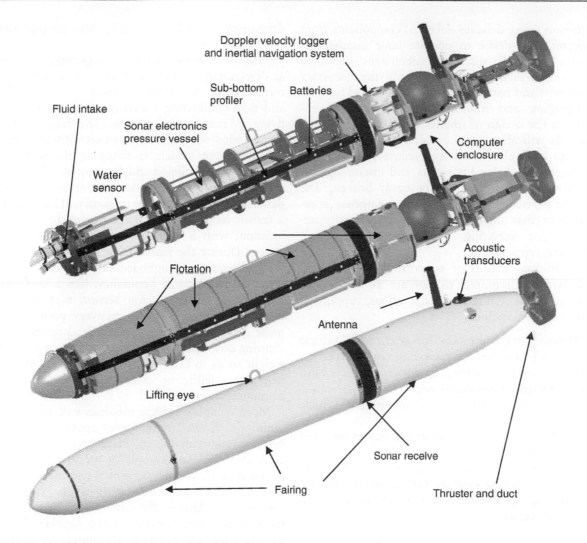

Figure 2 A propeller-driven, modular AUV with labeled subsystems. The top view shows the interior in which an internal mechanical frame supports pressure vessels, internal components, and the propulsion system. This AUV, called a *Dorado*, is a 'flooded' vehicle because the fairings are not watertight, and thus the interior spaces fill with water. Consequently Internal components must all be capable of withstanding ambient pressures. Joining rings are visible between the yellow fairing segments on the lower figure. These allow the vehicle to be separated along its axis, allowing replacement or addition of hull sections. This allows reconfiguration of the vehicle with new payloads, and if desired, with new batteries. The propulsion system on this vehicle is a ducted thruster capable of being tilted both vertically and horizontally, to steer the vehicle in the vertical and horizontal planes. Courtesy of Farley Shane, MBARI.

sonars. The anatomy of an AUV typically includes the following subsystems:

- software and computers capable of managing vehicle subsystems to accomplish specific tasks and even complete missions in the absence of human control;
- energy storage to provide power;
- propulsion system;
- a system for controlling vehicle orientation and velocity;
- sensors for measuring vehicle attitude, heading, and depth;
- pressure vessels for housing key electrical components;

- navigation sensors to determine the vehicle position;
- communication devices to allow communication of human operators with the AUV;
- locating devices to allow operators to track the vehicle and locate it for recovery or in the case of emergencies;
- devices for monitoring vehicle health (e.g., leaks or battery failure);
- emergency systems for ensuring vehicle recovery in the event of failure of primary systems.

The mechanical design of an AUV must address issues such as drag, neutral buoyancy, the highly dynamic nature of launch and recovery, and the need to protect many delicate electrical components from seawater. The desire to operate large numbers of sensors for long distances encourages the construction of larger vehicles to hold the necessary equipment and batteries. However, the need for ease of handling and minimizing logistical costs encourages the design of smaller vehicles. Operational demands will also create constraints on vehicle design; for example, launch and recovery factors will impose the need for lift points and discourage external appendages that will be easily broken. The need to service vehicle components imposes a requirement that internal components be easily accessible for servicing and testing, and when necessary, replacement.

In addition, a host of supporting software and hardware are required to operate an AUV. Depending on the nature of the operations, supporting equipment will include:

- software and computers for configuring vehicle mission plans, and for reviewing vehicle data;
- systems for communicating with the vehicle both on deck and when deployed;
- systems for recharging and monitoring vehicle batteries;
- handling gear for transporting, deploying, and recovering AUVs;
- devices for detecting locating devices on the AUV;
- acoustic tracking systems for monitoring the location of the vehicle when in the vicinity of a support vessel.

AUV Mission Software

Functionally, AUV software must address a variety of needs, including: allowing human operators to specify objectives, managing vehicle subsystems to achieve mission objectives, logging data for subsequent review, and ensuring safety of the vehicle in the event of failures or unexpected circumstances. The software must be capable of managing vehicle sensors and control systems to maintain a set heading, speed, and depth. The software might also need to support interacting with a human operator during a mission. In addition to software on the vehicle itself, AUV operators rely on a suite of software applications to configure and validate missions, to maintain vehicle subsystems such as batteries, to review data generated by the vehicle, to prepare mission summaries, and when possible, to track and manage the vehicle while underway. The exponential growth of computational power available for both onboard and off-board computers, as well as the increasingly pervasive nature of the Internet, are supporting a steady increase in software capabilities for AUVs.

Most AUV missions involve sequential tasks such as descending from the surface to a set depth, then transiting to a survey location at a set speed, and then conducting a survey which might involve flying a lawn-mower pattern. The vehicle may be commanded to maintain constant altitude over the bottom if the vehicle is mapping the seafloor. A water-column mission might require the vehicle profile in the vertical plane, moving in a saw-tooth pattern called a yo-yo. The mission will likely include a transit from the end of the survey to a recovery location, with a final ascent to the surface for recovery. During the mission, the vehicle will monitor the performance of onboard subsystems, and in the event of detection of anomalies, like a low battery level, or a failed mission sensor, may abort the mission and return to the recovery point early. A more catastrophic failure might lead to the vehicle shutting down primary systems, and dropping a drop weight so as to float to the surface, and calling for help via satellite or direct radio frequency (RF) communications.

More complex vehicle missions can involve capabilities such as adapting survey operations to obtain better measurements, or managing tasks such as AUV docking. An example of the first might be as simple as a yo-yo mission which cues its vertical inflections from water temperature in order to follow a thermocline. Also in the category of adaptive, but more demanding, surveys, is the capability of following a thermal plume to its source, for example, when an AUV is used to search for hydrothermal vents. The docking of an AUV with an underwater structure encompasses yet a different type of complexity, created by the large number of steps in the process, and the high likelihood that individual steps will fail. For example, docking involves homing on a docking structure, orienting for final approach, engaging the dock, and making physical connections to establish power and communication links. Any one of these steps might fail due to external perturbations; for example, currents or turbulence in the marine environment might cause the vehicle to miss the dock. The vehicle must be able to detect failures and execute a process to recover and try again. Docking is representative of the increasingly complex capabilities AUVs are expected to master with high reliability.

Navigation

The ability for an AUV to determine its location on the Earth is essential for most scientific applications.

However, navigation in the subsea environment is complicated by the opacity of seawater to all but very low frequency electromagnetic radiation, rendering ineffective the use of commonly used technologies such as the Global Positioning System (GPS) and other radio-based navigation techniques. Consequently, navigation underwater relies primarily on various acoustic and dead-reckoning techniques and the occasional excursion to the surface where radio-based methods can be used. There is no single method of underwater navigation that satisfies all operational needs, rather a variety of methods are employed depending on the circumstances.

Dead-reckoning methods integrate a vehicle's velocity in time to obtain an updated location. In order to dead-reckon, the vehicle must know both the direction and speed of its travel. The simplest methods use a magnetic compass to determine direction, and use speed through the water as a proxy for Earth-referenced speed. However, the large number of error sources for magnetic compasses make measurement of heading to better than a degree accuracy technically challenging. Currents pose even more of a problem, as they may be comparable to the vehicle speed in amplitude, yet are not sensed by a water-relative measurement. Dead-reckoning is improved by measuring velocity relative to the seafloor, for example, using a Doppler velocity log (DVL) or a correlation velocity log. A DVL is commonly used by AUVs to measure velocity by measuring the Doppler shift of sound reflected off the seafloor. Correlation velocity logs are more complex in concept, involving measurement of the correlation of two pulses of sounds transmitted by the vehicle, reflected off the seafloor, and received by a hydrophone array. In practice, DVLs are used when a vehicle operates close to the seafloor, perhaps within 200 m, while correlation velocity logs are used when the vehicle is operating in mid-water columns or near the surface in deep water.

Inertial navigation system (INS) technology is well developed, as it is widely used for platforms like aircraft and missiles. However, INS units appropriate for underwater use are expensive enough that they are used only when navigation requirements are stringent, for example, for producing high-accuracy maps. A modern INS includes an array of accelerometers for measuring acceleration on three axes and a laser or fiber optic gyroscope for measuring changes in orientation. Additionally, an INS will include a GPS for initializing the unit's location and orientation, and a computer for acquiring and processing data from INS component sensors. The position reported by an INS will have an error which will grow in time, and thus it is important to constrain INS

error with ancillary measurements of velocity and position. For example, combining an INS with a DVL for constraining velocity can result in a system which provides navigation accuracies better than 0.05% of distance traveled.

The two acoustic navigation methodologies most frequently used in AUV operations are ultrashort baseline navigation (USBL) and long baseline (LBL) navigation. A USBL system uses an array of hydrophone separated by a distance comparable to the wavelength of sound to measure the direction of propagation of an acoustic signal. Most often, a USBL system is mounted on a ship, and used to track a vehicle relative to the ship. With knowledge of the ship's location and orientation, the location of the AUV can also be determined. In contrast, LBL navigation acoustically measures the range between the vehicle and an array of widely separated devices of known location. A common LBL approach is to place transponders on the seafloor, and let the vehicle range off the transponders. The process of determining location using ranges from known locations is called spherical navigation, as the vehicle should be located at the intersection of spheres with the measured radius, centered on the respective transponders. An alternative LBL navigation method is to track a vehicle which pings at a preset time to an array of hydrophones at known locations. If the time of the ping is not known, the problem of solving for the vehicle location is called hyperbolic navigation, as only the difference in time of arrival of the ping at the various hydrophones can be determined, and this knowledge constrains the vehicle to be on a hyperbola between the respective receivers. If the time of the ping is known, perhaps triggered at a preselected time by a carefully calibrated clock, then the problem reduces to spherical navigation. In practice, a wide variety of USBL and LBL systems have been implemented for underwater navigation. They must all address the challenges of acoustic propagation in the ocean, which include the absorption of sound by seawater, diffraction by speed of sound variations in the underwater environment, scattering by reflecting surfaces, and acoustic noise generated by physical, geological, biological, and anthropogenic processes.

Other methods of navigation include using geophysical parameters, for example, water depth, to constrain the vehicle location in the context of known maps. These geophysically based navigation methods, similar to terrain contour mapping (TERCOM) navigation used by cruise missiles, depend on having good maps ahead of time. There are software approaches in development that simultaneously build maps and use those same maps for navigation.

These methods are called SLAM for simultaneous localization and mapping.

Using AUVs for Ocean Science

Mapping the Seafloor

AUVs are becoming the platform of choice for high-resolution seafloor maps. Obtaining high-resolution maps requires operating mapping sonars near the seafloor. Alternatives to an AUV include crewed submersibles and tethered platforms. Crewed submersibles are too valuable for routine mapping, and are reserved for other uses which require the presence of humans. Towed vehicles are used for sonar mapping, but have disadvantages as compared with AUVs, especially in deeper water. The principal problem is the high drag of the cable used for a tow sled, which in water depths of several thousand meters will limit speeds to approximately half a meter per second. Even at these slow speeds, a towed platform will stream behind the towing ship, creating several problems. Controlling the position of the towed vehicle over the bottom is very difficult, even when running on a constant heading. When surveying a defined area on the seafloor in a series of passes, the turns between passes may take longer than the actual survey passes themselves, as it is necessary to turn slowly to maintain control of the towed body. If ultrashort baseline acoustic navigation techniques are used to determine the vehicle position, then layback of the towed body behind the ship introduces significant errors as compared with having the ship directly over the sonar platform. For this reason, some commercial use of towed sonar platforms use two ships, one to tow the sonar platform, and one positioned directly over the platform to determine its precise location. Finally, surface motion of the ship will be efficiently coupled to the tow body by the tow cable. Thus, even near the seafloor, the tow body will be subject to sea state experienced by the ship. Consequently, attraction of the use of AUVs includes more economical operations and high data quality. **Figure 3** shows a cost comparison of a commercial deep-water towed survey and an equivalent AUV survey.

Sonar systems used on AUVs for mapping include multibeam sonar, side scan sonar, and sub-bottom profilers. Multibeam sonars, operating at frequencies of hundreds of kilohertz in the case of AUV-mounted systems, allow measurement of range to the seafloor in multiple sonar beams and are used to build up three-dimensional maps such as that in **Figure 4**. Side scan sonars used by AUVs also typically operate at frequencies of hundreds of kilohertz, and are used to image seafloor features. Side scan sonars are particularly useful for finding objects, for example, looking for a shipwreck resting

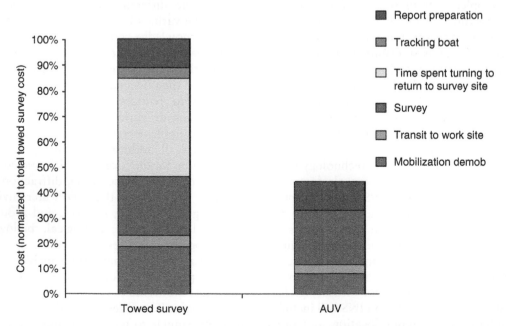

Figure 3 A comparison of the economics of deep survey taken from costs of a survey with a towed vehicle, and projected costs of the same survey with an AUV. The principal cost saving derives from the ability of the AUV to turn much faster than a deep-towed vehicle, reducing the total survey time. Also, the AUV can be acoustically tracked by its mother ship, while a towed vehicle requires a second ship for tracking because the towed vehicle will trail far behind the tow ship. Finally, mobilization and demobilization costs for the AUV can also be lower, although this depends on the size of the AUV employed.

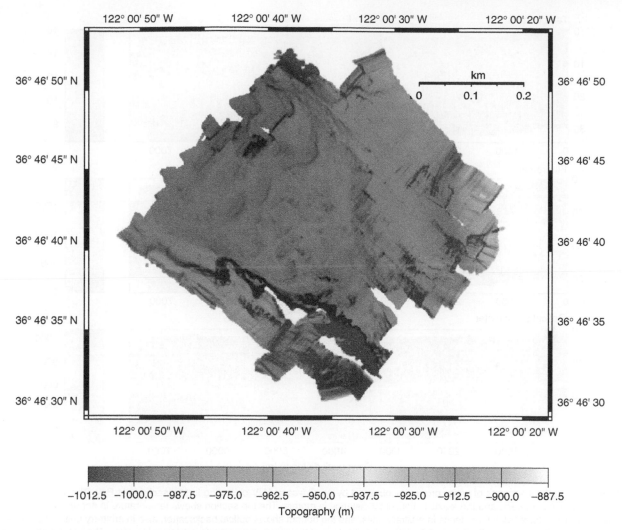

Figure 4 A bathymetric survey produced by an AUV at a depth of about 1000 m. Note the very small size of the survey area and high resolution of the bathymetry. Courtesy of Dave Caress, MBARI.

on the seafloor. Sub-bottom profilers use lower-frequency sound, ranging from 1 kHz to tens of kilohertz in the case of an AUV-mounted system, to penetrate into the seafloor. Depending on the bottom type (e.g., sandy, muddy, or rock), a sub-bottom system might penetrate tens of meters. Cumulatively sonar payloads will consume comparatively large amounts of energy, perhaps hundreds of watts. Mapping also requires high-fidelity navigation, and thus sonar-equipped AUVs will often also use more sophisticated navigation approaches, like inertial navigation. Consequently, mapping AUVs of today are larger, more sophisticated AUVs.

Observing the Water Column

AUVs provide a relatively new tool for observing the physical, chemical, and biological properties of the ocean. The smaller, buoyancy-driven gliders are unique in their combination of mobility and endurance, moving at about a quarter of a meter per second for periods of months. Larger vehicles carry more comprehensive payloads at higher speeds, but for shorter periods. Such vehicles might operate at $1.5\,\mathrm{m\,s^{-1}}$ for a day. A common flight profile is to fly the vehicle on a constant heading, while moving between two depth extremes in a saw-tooth pattern. Often the upper depth extreme will be close to the surface. This strategy allows the production of vertical sections of ocean properties, such as those in **Figure 5**. Variations of this strategy might have the vehicle moving in a lawn-mower or zigzag pattern in the horizontal plane, to develop a full three-dimensional map of ocean properties. **Figure 6** shows a visualization of an internal wave interacting with a phytoplankton layer using such a three-dimensional mapping strategy.

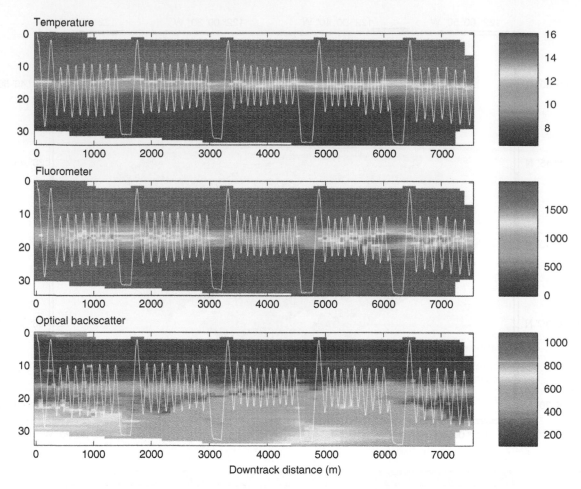

Figure 5 Vertical sections of water properties obtained by an *Odyssey* AUV operating in Massachusetts Bay. The *y*-axis of each figure is depth, in meters, and the *x*-axis is horizontal distance in meters. The top section shows temperature in degrees Celsius, the middle shows chlorophyll fluorescence in arbitrary units, and the bottom shows optical backscatter, also in arbitrary units. The path of the vehicle is shown as a white line, and the interpolated values of the measured property are plotted in color. The vehicle alternated between obtaining high-resolution observations of the thin layer of organisms at the thermocline with full water column profiles.

All AUVs are limited by the availability of sensors. Temperature, salinity, currents, dissolved oxygen, nitrate, optical backscatter properties, and chlorophyll fluorescence are examples of the growing *in situ* sensing capabilities available for AUVs. However, many important properties, for example, pH, dissolved carbon dioxide, and dissolved iron, cannot be measured reliably from a small moving platform. Furthermore, detection of marine organisms is usually accomplished by proxy; for example, chlorophyll fluorescence provides an indicator for phytoplankton abundance. *In situ* methods which directly detect, classify, and quantify marine organism abundance are not available, yet are increasingly important for understanding the structure and dynamics of ocean ecosystems.

Operations in Ice-Covered Oceans

AUVs offer unique operational capabilities for science in ice-covered oceans. Successful under-ice operation has been carried out with AUVs in both the Arctic and Antarctic. Sea ice poses special operational challenges for seagoing ocean scientists. For example, ships with ice-breaking capability can operate in the ice pack, but will typically not be able to hold station, or even assure that tethers and cables deployed over the side will not be severed. AUVs are attractive in that they provide horizontal mobility under ice, and the ability to conduct operations near the seafloor without the complications intrinsic in tether management. Challenges of operating AUVs under ice revolve around the need to assure return of the AUV to the ship for recovery, the process of recovering the AUV through the ice onto the ship, the potential for having an AUV fail and become trapped under ice, and the difficulty of carrying out tasks that would normally be accomplished having an AUV surface (e.g., obtaining a GPS update). Most safety strategies for AUVs in ice-free oceans default to bring

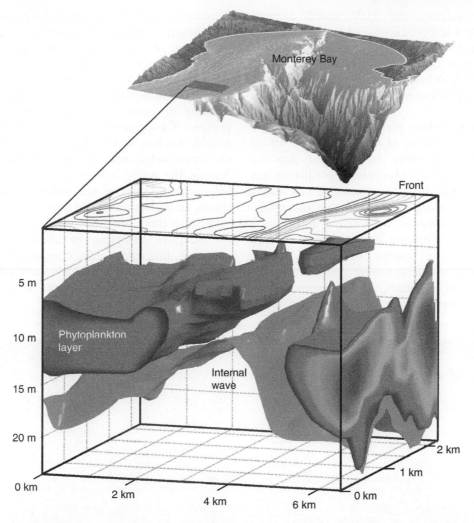

Figure 6 Interaction of layer of phytoplankton with an internal wave in Monterey Bay. Both physical and biological properties were measured by an *Odyssey* AUV, which moved in a horizontal zigzag pattern across the survey volume, while profiling constantly in the vertical plane. The phytoplankton layer, shown in green, was detected by a chlorophyll fluorescence sensor on the AUV. The cyan surface shows deflection of a level of constant density of seawater by a passing internal wave. Courtesy of John Ryan, MBARI.

the vehicle directly to the surface, for example, by dropping a weight. In the Arctic or Antarctic, this strategy could result in the vehicle becoming trapped under very thick ice, making the vehicle harder to find and potentially impossible to recover. The usual surface location devices such as RF beacons, strobes, and combinations of RF communication and satellite navigation will not work. Clearly AUV operations within the ice pack entail higher risk and a more sophisticated vehicle.

Observation Systems, Observatories, and AUVs

An understanding of power consumption of AUVs provides insight to the attractiveness of employing multiple vehicles for certain ocean observation problems. In some circumstances a survey must be accomplished within a set period. In oceanography,

time-constrained surveys are most often encountered when surveying a dynamic process. For example, if the temporal decorrelation of ocean fields associated with upwelling off Monterey Bay is about 48 h, attempts to map the ocean fields need to be accomplished within that time frame. Scales of spatial variability will also determine acceptable separation of observations: for example, decorrelation lengths in Monterey Bay are on the order of 20 km, so observations need to be spaced significantly closer to minimize errors in reconstructing the ocean field. How does this relate to the number of vehicles required to accomplish such a survey? Consider a grid survey of a 100 km × 100 km area with a resolution of 10 km. A single vehicle would have to travel c. 1000 km at a speed of nearly 6 m s^{-1}, traveling in a lawn-mower pattern. Using the example vehicle values from the 'Basics of AUV performance' section, the AUV would consume about 3500 W if it were capable of operating at such a high speed. In contrast, six of the same vehicles operating at their optimum speed would consume a total of 270 W. In other words, the six vehicles would consume 12 times less energy for the complete 48-h survey.

Autonomous mobile platforms are making observation of the interior of the ocean more affordable and more flexible, enabling the practical realization of coupled observation–prediction systems. For example, in late summer 2003, a diverse fleet of AUVs was deployed to observe and predict the evolution of episodic wind-driven upwelling in the environs of Monterey Bay. Over 21 different autonomous robotic systems, three ships, an aircraft, a coastal ocean dynamics application radar (CODAR), drifters, floats, and numerous fixed (moored) observation assets were deployed in the

Autonomous Ocean Sampling Network (AOSN) II field program (**Figure 7**). Gliding vehicles, with an endurance of weeks to months, provided a continuous presence with a minimal sensor suite. A few propeller-driven vehicles provided observations of chemical and biological ocean parameters, allowing tracking of ecosystem response to the upwelling process. Observations were fed to two oceanographic models, which provided synoptic realization of ocean fields and predicted future conditions. Among the many lessons are an improved knowledge of the scales of variability of upwelling processes, an understanding of how to scale observation systems to these processes, and insights to strategies for adaptive sampling of comparatively rapidly changing processes with comparatively slow vehicles. These lessons are particularly relevant today, given the present emphasis on developing ocean-observing systems.

Figure 7 Example of a distributed observing system using AUVs. This diagram depicts an AOSN deployment in Monterey Bay.

See also

Gliders. Remotely Operated Vehicles (ROVs).

Further Reading

Allmendinger EE (1990) *Submersible Vehicle Systems Design*. New York: SNAME.

Bradley AM (1992) Low power navigation and control for long range autonomous underwater vehicles. *Proceedings of the Second International Offshore and Polar Conference*, pp. 473–478.

Fossen T (1995) *Guidance and Control of Ocean Vehicles*. New York: Wiley.

Griffiths G (ed.) (2003) Technology and Applications of Autonomous Underwater Vehicles. London: Taylor and Francis.

IEEE (2001) Special Issue: Autonomous Ocean Sampling Networks. *IEEE Journal of Oceanic Engineering* 26(4): 437–446.

Jenkins SA, Humphreys DE, Sherman J, *et al.* (2003) *Underwater glider system study. Scripps Institution of Oceanography Technical Report No. 53*. Arlington, VA: Office of Naval Research.

Rudnick DL and Perry MJ (eds.) (2003) ALPS: Autonomous and Lagrangian Platforms and Sensors, Workshop Report, 64pp. http://www.geo-prose.com/ALPS (accessed Mar. 2008).

Relevant Website

http://www.mbari.org
    – Monterey Bay 2003 Experiment, Autonomous Ocean Sampling Network, MBARI.

PLATFORMS: BENTHIC FLUX LANDERS

R. A. Jahnke, Skidaway Institute of Oceanography, Savannah, GA, USA

Introduction

The need to better understand chemical and biological processes and solute transport mechanisms across the deep-sea floor has fueled major engineering advances in seafloor instrumentation in the last few decades. In principle, the kinds of experiments that can be conducted on the seafloor are limited only by researchers' imaginations. In reality, the majority of the instruments that have been developed seek to accomplish two basic types of operations: deploy, sample, and retrieve benthic flux chamber systems to directly measure seafloor exchange rates and deploy *in situ* sensor arrays capable of measuring the vertical distribution of solutes in near-surface pore waters from which exchange rates can be estimated based on pore water transport models. Recent advances have expanded the types of experiments and measurements that can be performed by these instruments. Generically, the instrument frames used to deploy these types of instruments have been called 'bottom landers'. This term was coined in the early 1970s in recognition of the similarity between the approximate shape and function of these devices and the more famous 'lunar lander' that carried the first men to the moon. In the following, the design strategies and basic instrumentation for conducting benthic flux chamber incubations and sensor measurements at the deep-sea floor are discussed.

Benthic Flux Measurement Strategies

In the 1960s and 1970s, it was increasingly recognized that the temperature and pressure changes that occur in bringing a sediment sample from the deep-sea floor to the sea surface may alter the sample. Examples include changes in metabolic rates of benthic populations, changes in pore water concentration gradients from which diffusive fluxes are calculated, and changes in chemical concentrations in pore waters due to pressure- or temperature-driven reactions. Additional artifacts continue to be identified to this day. Thus, analyzing deep-sea samples brought to the deck of a ship has been increasingly recognized as being not accurate enough

to examine important questions such as: What are the respiration rates of deep-sea-floor populations? What is the seafloor dissolution rate of calcium carbonate? Estimating benthic fluxes (i.e., the net exchange rate of solutes across the sediment surface) on the seafloor, at *in situ* temperature and pressures, is one strategy for avoiding sampling artifacts and improving the accuracy of deep-ocean measurements.

There are two common techniques for estimating benthic fluxes. Benthic flux chamber incubations can be performed in which a known volume of bottom water is trapped above a known area of seafloor. Any solute transported out of the sediments covered by the chamber will be trapped within the chamber waters. Hence, the concentration of this constituent in the chamber waters will increase with incubation time. Conversely, the chamber water concentration of any chemical constituent that is being transported into the sediments will decrease with incubation time. The benthic flux is directly proportional to the rate of concentration increase or decrease, the volume of bottom water enclosed within the chamber, and area of the seafloor covered by the chamber. The main advantage of this method is that there is minimal disturbance to the sediment system, maximizing the probability of accurate estimates. In addition, the fluxes obtained will reflect the net exchange due to all transport processes occurring within the spatial dimensions of the chamber incubations. Thus, this technique will include nondiffusive exchange such as that driven by the active irrigation of organism burrows. A weakness of this approach is that other than inferences about total integrated reaction rates within the sediment column supporting the observed benthic flux, little information is gained about the reaction processes and distributions themselves. Additionally, if there is significant bottom current-driven advective flow through the surface sediments, differences between hydrodynamic conditions within the chamber and the natural setting may result in significant inaccuracies in flux estimates.

The other common strategy for estimating seafloor fluxes is to measure the concentration gradient of chemical constituents very near (preferably across) the sediment–water interface. Knowing the transport processes, the flux can be calculated based on the transport rates and measured concentration gradients. In principle, this method can be applied to all types of transport processes. In practice, however, the exact nature of nondiffusive transport processes is unknown and this calculation strategy is used almost exclusively to estimate exchange due to molecular diffusion only. Thus, a limitation of this approach is that accurate benthic fluxes cannot be estimated in a location where pore water exchange due to processes other than molecular diffusion is significant. On the other hand, a strength of using concentration profiles is that the distribution of reactions can be assessed and inferences of reaction mechanisms can be made from variations in the fluxes estimated at different depths below the sediment surface.

Benthic Chamber Landers

Design and Operations

The development of benthic chamber landers followed the experiences gained from conducting chamber or 'bell jar' incubations in shallow-water areas where they could be tended by scuba divers. The major challenge in developing this instrumentation was simply to automate the required operations to function reliably in the harsh conditions of the deep-sea floor. The earliest instrument that was routinely deployed in the deep sea and has provided a large data set is the free vehicle grab respirometer (FVGR) developed in the late 1970s by Dr. K.L. Smith, Jr., at Scripps Institution of Oceanography in San Diego, California, USA (**Figure 1**).

The basic instrument consists of a structural frame upon which the critical components are mounted. To minimize weight, the frame is most often constructed of tubular aluminum. Instrument flotation is mounted on the upper portion of the instrument frame. The most common type of flotation is glass spheres (as shown in **Figure 1**), although syntactic foam flotation has also been used. The latter is much more expensive than glass spheres but is not susceptible to implosion, which is critical if manned submersibles are ever required to work in close proximity. Both types provide relatively constant buoyancy. At the top of the flotation section are mounted devices such as a flag, strobe light, radio transmitter, or satellite transmitter to help locate the instrument when it is at the sea surface.

Expendable weights are mounted at the lower portion of the tubular frame, usually adjacent to the 'feet'. As these instruments are free vehicles, these weights provide the negative buoyancy necessary to drive the instrument to the seafloor. A latching mechanism permits the weights to be released at the end of the experiment. Upon release, the flotation provides sufficient positive buoyancy to pull the instrument back to the sea surface where it may be recovered by a surface research vessel.

The specific instrument packages, controlling electronics and other operational components, are

mounted in the central part of the frame. It is in these components where the greatest differences between individual designs occur. Chamber instruments have been designed to accommodate one to four chambers simultaneously. While increasing desirable replication, multiple chambers on a single instrument tend to increase variability in the data by decreasing chamber area and increase the complexity of the control, data storage, and sampling systems.

Early chamber systems like the FVGR relied on oxygen electrodes to quantify the changes in oxygen within the chambers. Thus, the early instruments were only capable of estimating benthic oxygen demand. While electrodes continue to be widely used, many modern instruments also employ an electronically controlled sampling system so that benthic exchange of many types of solutes, such as nutrients, trace elements, and organic and inorganic carbon, can be assessed.

The utility of these instruments is demonstrated by their proliferation. Today, there are more than 20 different research groups that have developed *in situ* benthic flux chamber instruments. These groups have designed and implemented numerous modifications and alterations to Smith's initial design, yet the basic characteristics and capabilities of the chamber lander remain the same. Examples of the numerous important modifications that have been made to the early designs include several types of stirring mechanisms and time-series sampling systems so that the exchange of solutes other than oxygen can be addressed. In addition to the basic benthic flux chamber operations, complementary sampling devices have also been added. For example, one chamber instrument design has incorporated an *in situ* whole core squeezer. This device recovers a sediment core and then sequentially squeezes the surface pore waters from the sediments to provide high-resolution pore water samples. While not appropriate for all solutes due to surface exchange during squeezing, these samples provide important information concerning the pore water gradients of selected metabolites, such as oxygen and nitrate, which greatly enhances the interpretation of the flux chamber results.

Examples of results from a benthic flux chamber deployment at c. 3000 m on the continental rise of the eastern US seaboard is displayed in **Figure 2**. On each plot, the concentration is on the vertical axis and the horizontal axis represents incubation time. These results demonstrate the range of possible responses. Solutes that are taken up by the sediments tend to decrease with incubation time. An example of this is oxygen which is consumed in the sediments through benthic respiration. Most other components are produced in the sediments through the

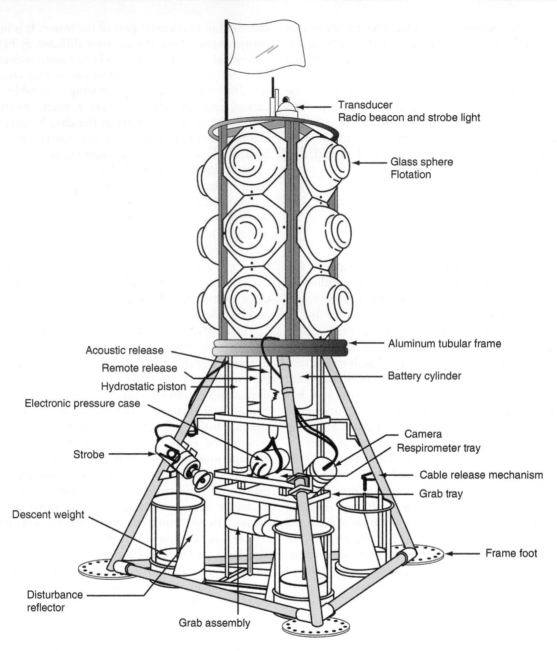

Figure 1 Schematic of the free vehicle grab respirometer developed by Dr. Kenneth L. Smith, Jr. (Scripps Institution of Oceanography).

decomposition or dissolution of biogenic debris. This production supports a flux out of the sediments and the concentrations of these constituents increase with incubation time. Specific examples of these include phosphate, released from degrading organic tissue, total inorganic carbon (TIC) released through respiration and the dissolution of $CaCO_3$, titration alkalinity (TA) produced by the dissolution of $CaCO_3$, and silicate which is produced by the dissolution of opal. Some species, such as nitrate, may increase or decrease with incubation time depending on the relative rates of the competing sedimentary processes

that produce or consume them. For example, nitrate is produced by the oxidation of ammonium (nitrification) and is consumed by denitrifying bacteria below the oxic zone. Whether there is a flux out of or into the sediments depends on the ratio of these rates.

Direct estimates of benthic fluxes can be made from these results from the relationship

$$\text{Benthic flux} = (S \times V)/A$$

where S is the slope of the concentration versus time results, V is the volume of the bottom water trapped

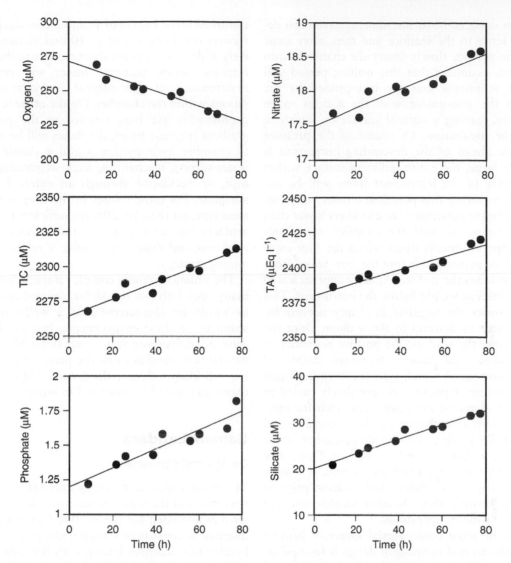

Figure 2 Example benthic flux chamber results from 2927 m on the continental rise of the US eastern seaboard using the chamber system of Dr. Richard A. Jahnke (Skidaway Institute of Oceanography).

in the chamber, and A is the sediment surface enclosed by the chamber.

Note that if the chamber has vertical sides, V/A = height of the water column trapped within the chamber.

Major Design Controversies and Differences

Despite the numerous advantages of benthic flux incubations for estimating seafloor exchange, there are a variety of limitations and concerns that need to be addressed in evaluating results from chamber incubations and in assessing the relative merits of the different instrument designs.

As shown in the example results, concentration changes are required to evaluate the solute exchange rate. However, the concentration changes will also alter the near-surface gradients, altering the flux. In the extreme case, where concentration changes are large, such as complete oxygen depletion within the chamber, changes in chamber water chemistry may alter near-surface chemical reactions and/or exchange processes, greatly altering the chemical flux. To minimize this source of uncertainty, chamber deployments must be designed to minimize the concentration changes and maintain the chamber sediments as near to their natural state as possible. Thus, it is important that the most precise analytical procedures be employed, so that accurate fluxes can be quantified without large concentration changes.

Maintaining natural benthic fluxes requires that the sediment surface not be disturbed during chamber deployment. This need has resulted in several different types of deployment strategies and instrument designs. Most instruments have been designed to settle to the seafloor and then, after some preset time period, slowly insert the chambers into the bottom, assuming that this waiting period will allow the sediments that are resuspended by the impact of the instrument with the bottom to be swept away, leaving a natural surface on which to conduct the incubation. Of course, if the pressure disturbance ahead of the descending instrument is sufficiently large, the entire sediment surface within the perimeter of the instrument frame will be impacted. Recognizing this potential artifact, some instrument designs positioned the chambers lower than the main frame, so that the chamber would encapsulate the sediments upon which the flux incubation will be performed before the 'bow wave' from the frame reaches the surface. Another approach is to suspend a descent weight below the instrument. This weight provides the negative buoyancy needed for the instrument to descend to the seafloor. Once the weight reaches the seafloor, the positive buoyancy of the instrument itself causes it to remain suspended above the bottom. A winch system is then activated and the instrument package is very slowly pulled to the seafloor. A recent innovative approach for minimizing bottom disturbance has been implemented on the ROVER lander, a benthic flux chamber device capable of 'crawling' around the seafloor. Thus, once the instrument has impacted the bottom, it simply crawls laterally to an undisturbed location prior to conducting benthic flux chamber incubations and pore water profiling operations.

Perhaps the most controversial aspect of benthic flux incubations and instrument design is focused on mechanisms and requirements for simulating natural flow conditions within the chamber. Exchange across a homogeneous muddy sediment surface requires transport across the hydrodynamic boundary layer including the molecular diffusive sublayer. Hydrodynamic conditions within the chamber control the thickness of this layer. Changes in the hydrodynamic regime will alter the thickness of the layer and, at least temporarily, will alter the exchange rate. However, since the benthic flux is supported by metabolic and chemical reactions in the sediments that are not influenced by the diffusive sublayer thickness, the seafloor exchange rate will eventually revert to its natural rate. Thus, the need to accurately reproduce the natural hydrodynamic conditions with the benthic flux chamber depends critically upon the response time of the surface pore water gradients and the length of the chamber incubations. If controlled by molecular diffusion, the time required for pore water concentration gradients to recover after disturbance scales as the square of the thickness of the

disturbed layer. Layers of 1- and 3-mm thickness will recover on timescales of c. 10 and 80 min, respectively. If deployments are short relative to the gradient response times, accurate fluxes will require the maintenance of near-natural hydrodynamic conditions within the chamber. On the other hand, if the deployments are long relative to the pore water gradient response times, the fluxes will be insensitive to chamber hydrodynamics and a simple chamber water-stirring mechanism, such as rotating rods, a disk, or circulation through an external pump, is adequate. For most solutes in the deep sea, deployments greater than 10–20 h are sufficient to minimize artifacts due to changes in the diffusive sublayer thickness and thus only require a relatively simple mixing strategy.

The situation is more complex if the sediments have many open burrows or exhibit elevated permeability, as would be characteristic of a well-sorted sandy sediment. In these environments, bottom flows may drive advective pore water exchange. Alternation of natural flow conditions by the presence of the chamber will likely influence the accuracy of the flux estimates and must be considered in interpreting results.

Sensor Landers

Design and Operations

The basic instrument design and field deployment operations of the sensor landers are the same as that already discussed for the benthic flux chamber. The instrument consists of a frame, identical to that of a benthic flux chamber lander, with flotation mounted at the top and expendable descent weights attached near the feet. The major difference is that the benthic chamber is replaced with an instrument package capable of inserting microelectrodes or optode with high vertical resolution into the surface sediments. An example of the basic instrument package is presented in **Figure 3**.

Each of these types of packages consists of three basic components: the sensing electrodes themselves; a mechanism for moving the electrodes vertically across the sediment–water interface; and the data storage and controlling electronics. In the design pictured in **Figure 3**, the sensing electrodes are attached directly to the main pressure case. Located within the pressure case are the power and electronics necessary to operate the motor and electrodes and to provide for data storage and retrieval. The motor drive system is positioned outside the case to move the case vertically, so that profiles of the measured components are obtained. The vertical resolution of the profiles is controlled by the size of

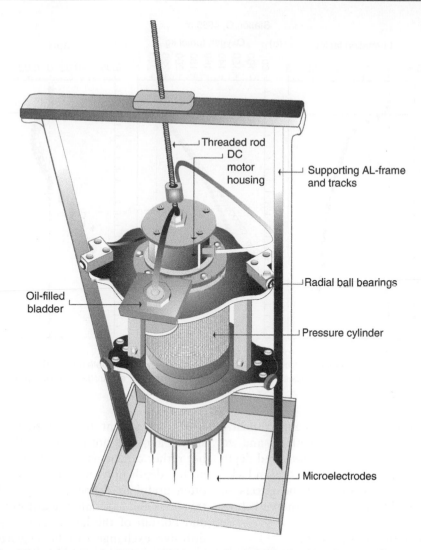

Threaded rod

DC motor housing

Supporting AL-frame and tracks

Oil-filled bladder

Radial ball bearings

Pressure cylinder

Microelectrodes

Figure 3 Schematic of the microelectrode deployment apparatus developed by Dr. Clare Reimers (Rutgers University).

the electrode tip and the precision of the motor drive assembly.

Early designs primarily employed oxygen and electrical resistivity electrodes. The latter is required to assess the porosity and tortuosity that influence the rate of diffusion in sediments. Porosity is a measure of the proportion of the sediment comprised of void space between sediment particles and is generally 70–90% in muds and 30–60% in sands. Tortuosity is the actual distance a solute would have to travel around sediment particles relative to the straight-line distance between two points. In recent years, numerous other sensors have been developed and implemented for deep-sea lander use. These include pH, pCO$_2$, total CO$_2$, calcium, ammonium, and nitrate. It is anticipated that the development of other sensors will continue and the types of measurements possible in the future will continue to expand.

An example of oxygen, resistivity (formation factor), and pH results obtained from a sensor lander deployment is provided in **Figure 4**. Because oxygen is consumed within the sediments, primarily due to the respiration by benthic organisms but possibly also due to chemical consumption, oxygen concentrations decrease with increasing depth in the sediments. This downward concentration gradient implies a benthic flux into the sediment. Contrastingly, in the example provided in **Figure 4**, pH first increases and then decreases with sediment depth. This more complicated profile shape is due to competing reactions downcore. In this example, pH first increases with depth due to the dissolution of calcium carbonate and then decreases with depth due to continued production of carbon dioxide.

Formation factor values increase with sediment depth due to the compaction of sediment particles. This decrease in resistivity implies a decrease in sediment porosity and increase in sediment tortuosity, both of which tend to decrease the effective diffusion coefficient with depth in the sediments.

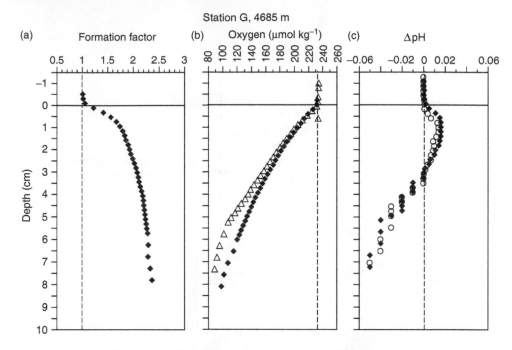

Figure 4 Examples of deep-sea-floor microlectrode oxygen, pH, and resistivity results obtained by Dr. Burke Hales (Oregon State University). Different symbols display simultaneous results from separate, individual electrodes installed a few centimeters apart on the instrument.

These results can be used to estimate the diffusive flux across the sediment–water interface and to evaluate the reaction rate within the measured depth interval. The benthic flux can be estimated from Fick's first law corrected for porosity and the effects of sediment particles on diffusion rates as shown below:

$$\text{Benthic flux} = \phi D_s \delta C/\delta z$$

where ϕ = surface porosity, D_s = effective sediment diffusion coefficient, and $\delta C/\delta z$ = vertical concentration gradient at the sediment–water interface.

Porosity near the sediment surface is generally estimated from a regression of the measured resistivity and directly measured porosities at wider-spaced depth intervals. The latter are generally determined from the weight loss upon drying bulk sediments. The effective diffusion coefficient can be estimated by dividing the molecular diffusion coefficient by the porosity and tortuosity (i.e., $D_s = D/(\phi\theta)$, where D = molecular diffusion in free solution, and θ = tortuosity). This expression requires independent measurements of porosity and tortuosity at the same vertical scale as the concentration profile. Such measurements are often not available. A common alternate but less-accurate strategy in fine-grained sediments is to approximate the effective diffusion coefficient as the molecular diffusion coefficient times the square of the porosity. This relationship has been empirically derived from numerous individual studies.

It is important to note that the flux equation shown above is not limited to the sediment surface but rather can be used to evaluate the diffusive flux at any depth horizon within the profile. Thus, it is often useful to define a sediment layer of a small thickness and calculate the diffusive fluxes at the top and bottom of the layer. Assuming that horizontal diffusive exchange can be neglected and that the profile is in steady state, the difference in these fluxes is a measure of the net production or consumption rate of the measured solute within the layer. Thus, by interpreting the vertical variations in the flux, one can evaluate the distributions of reactions in the sediments and potentially make inferences about the processes and mechanisms controlling solute diagenesis and benthic flux.

Advantages, Limitations, and Design Concerns

Unlike benthic flux chambers that require a significant incubation time, sensor landers can obtain a profile relatively quickly, usually within 1–2 h. The exact required time is determined by the number of sampling depths and the response time of the sensors employed. Thus, sensor landers can be used to obtain *in situ* profiles relatively rapidly and estimate diffusive fluxes at the deep-sea floor.

There are also several limitations to this approach. Because the measurements are generally made within several hours of the instrument reaching the seafloor, the accuracy of the flux estimate can be severely compromised by physical disturbance of the sediment surface. Potential disturbance can be caused by the instrument itself or by the bow wave that precedes the instrument as it settles. Video recordings of these instruments settling onto the bottom reveal significant resuspension of surface sediments. While the resuspended sediments are generally allowed to settle back to the sea floor or be advected away prior to making measurements, the effect of this disturbance on the profiles is still a concern. The only instrument to completely avoid this potential problem is the ROVER lander (discussed in the next section).

Maintaining natural hydrodynamic conditions within the benthic boundary layer is also critical to the accuracy of microelectrode flux estimates. Unlike benthic flux chamber incubations that extend over time intervals sufficient to return to initial conditions if diffusive boundary layer thicknesses are altered, electrode measurements are rapid and would record transient conditions if made directly after altering bottom hydrodynamic conditions. Because the instrument frame and electrodes themselves may alter bottom flow, such changes are a concern.

Since the electrode sensing tips are very small (generally 5–20 μm in diameter) as required to achieve fine vertical resolution, they also respond to horizontal variations that may be caused by burrowing organisms or physical inhomogeneities. Because the geochemical questions being asked often require knowledge of the mean benthic flux for a known area or region, numerous replicate profiles are often required to estimate the average profile and benthic flux.

Special Landers and Lander-Based Research Strategies

In addition to the basic landers discussed above, a variety of instruments have been developed in the last decade for special purposes and it is likely that new types will continue to be developed in the future. For example, simple benthic flux chamber landers have been developed to measure advective pore water flows around hydrothermal vent and mid-ocean ridge systems. In these types of chambers, osmotic pumps are used to continuously add a tracer and remove a sample. Pore water advection rates as low as $0.1 \, \mathrm{mm \, yr^{-1}}$ can be measured with this system. For seafloor microbial studies, a lander has been developed capable of injecting radiolabeled tracers continuously throughout the upper 70 cm of the sediment column. The sediments surrounding the line of injection are cored and recovered at the end of

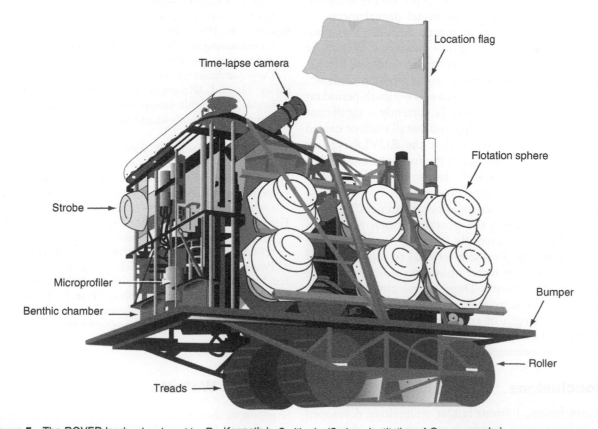

Figure 5 The ROVER lander developed by Dr. Kenneth L. Smith, Jr. (Scripps Institution of Oceanography).

a preset incubation period and returned to the ship for analysis. Other lander instruments are being developed to address specific research questions and to operate in particularly demanding environments. For example, the benthic exchange of certain trace metals is sensitive to the redox conditions of the sediment–water interface. To accurately measure benthic fluxes of these metals, the 'oxystat' lander has been developed in which the chamber waters are circulated through gas-permeable tubing in contact with bottom waters. Exchange of oxygen across the tubing wall keeps the oxygen concentrations in the chambers at near-natural levels, minimizing potential changes to the metal fluxes. Another lander device is the 'integrated sediment disturbing lander', in which a set of small plow blades is rotated through the surface sediments within the chamber to evaluate the impact of sediment disturbance, such as would occur following a bottom trawl or large storm, on benthic solute exchange.

Perhaps the most innovative lander recently developed is the ROVER (**Figure 5**). ROVER is capable of performing repeated benthic flux chamber incubations and microelectrode profiling measurements. Most importantly, after a measurement cycle is complete, the instrument uses a tractor-tread propulsion system to move c. 5 m, so that the next measurement is performed on a natural, undisturbed surface. Since this instrument is crawling laterally on the sediment surface, it eliminates the potential disturbances discussed for free-fall instruments. ROVER was constructed to examine temporal variations in seafloor fluxes and is capable of performing duplicate benthic flux chamber and microelectrode profiling measurements at 30 individual sites over a 6-month period on a single deployment. There is currently a significant effort underway to develop and install seafloor cabled observatories. It is likely that the ROVER-type of instrumented systems will in the future be attached to seafloor cable nodes and observatories and provide long-term, near-real-time observations.

Finally, an eddy correlation instrument is being developed in which the benthic oxygen flux is assessed by simultaneously directly measuring the concentration and velocity normal to the seafloor of a small volume of water above the sediment surface by using an acoustic Doppler velocimeter and an oxygen microelectrode in concert. By integrating, for a sufficient length of time, the instantaneous fluxes toward and away from the sediments, the net benthic flux can be estimated.

Conclusions

In conclusion, bottom lander technology developed in the last several decades now permits a wide variety of remote sampling procedures and incubation experiments on the seafloor. These capabilities have greatly improved our understanding of the benthic processes, especially at abyssal depths where pressure- and temperature-driven artifacts have hindered earlier studies.

See also

Grabs for Shelf Benthic Sampling. Sensors for Micrometeorological and Flux Measurements.

Further Reading

Berelson WM, Hammond DE, Smith KL, Jr., et al. (1987) In situ benthic flux measurement devices: Bottom lander technology. MTS Journal 21: 26–32.

Berg P, Roy H, Janssen F, et al. (2003) Oxygen uptake by aquatic sediments with a novel non-invasive eddy-correlation technique. Marine Ecology Progress Series 261: 75–83.

Greeff O, Glud RN, Gundersen J, Holby O, and Jorgensen BB (1998) A benthic lander for tracer studies in the sea bed: In situ measurements of sulfate reduction. Continental Shelf Research 18: 1581–1594.

Jahnke RA and Christiansen MB (1989) A free-vehicle benthic chamber instrument for sea floor studies. Deep-Sea Research 36: 625–637.

Reimers CE (1987) An in situ microprofiling instrument for measuring interfacial pore water gradients: Methods and oxygen profiles from the North Pacific Ocean. Deep-Sea Research 34: 2019–2035.

Sayles FL and Dickinson WH (1991) The ROLAI^{2}D lander: A benthic lander for the study of exchange across the sediment–water interface. Deep-Sea Research 38: 505–529.

Smith KL, Jr., Glatts RC, Baldwin RJ, et al. (1997) An autonomous bottom-transecting vehicle for making long time-series measurements of sediment community oxygen consumption to abyssal depths. Limnology and Oceanography 42: 1601–1612.

Tengberg A, De Bovee F, Hall P, et al. (1995) Benthic chamber and profiling landers in oceanography – a review of design, technical solutions and functioning. Progress in Oceanography 35: 253–294.

Tengberg A, Stahl H, Gust G, et al. (2004) Intercalibration of benthic flux chambers. Part I: Accuracy of flux measurements and influence of chamber hydrodynamics. Progress in Oceanography 60: 1–28.

Relevant Website

http://www.cobo.org.uk
　– Coastal Ocean Benthic Observatory.

REMOTELY OPERATED VEHICLES (ROVS)

K. Shepherd, Institute of Ocean Sciences, British Colombia, Sidney, Canada

Introduction

Remotely operated vehicles (ROVs) are vehicles that are operated underwater and remotely controlled from the surface. All types of this vehicle are connected to the surface platform by a cable that provides power and control communication to the vehicle. There are three basic types of vehicle: free-swimming tethered vehicles; bottom-crawling tethered vehicles; and towed vehicles. The free-swimming vehicle is the most common. It has thrusters that allow maneuvering in three axes, and provides visual feedback through onboard video cameras. It is often used for mid-water or bottom observation or intervention. Bottom-crawling vehicles move with wheels or tracks and can only maneuver on the bottom. Visual feedback is provided by onboard video cameras. Bottom crawlers are usually used for cable or pipeline work, such as inspection and burial. Towed vehicles are carried forward by the surface ship's motion, and are maneuvered up and down by the surface-mounted winch. Towed vehicles usually carry sonar, cameras, and sometimes sample equipment.

Remotely operated vehicles were first introduced to the offshore community in 1953. Over the next 22 years, several more vehicles were built to fulfill military and other government research requirements. In 1975, the first commercial vehicle was built for the offshore oil industry. Since 1975, over 90% of the ROVs produced have been developed for commercial offshore work that includes oil and gas drilling support, as well as pipeline and telecommunications cable inspection, burial, and repair. As the depths for oil exploration and production have increased, the commercial ROV industry has been pressed to keep pace. Current exploration depths are now reaching 3000 m. The remaining vehicles are used to support military and scientific research and intervention. Military applications include submarine rescue, mapping, reconnaissance, recovery, and mine countermeasures. Scientific applications are far-ranging and cover many different fields including biology, physics, geology, and chemistry. Depths for this work range from a few metres to 10 000 m.

Basic Design Characteristics

ROV systems are built in many different configurations and sizes. However, there are many common design characteristics that consist of some or all of the components described in the sections below.

Vehicle

Vehicles range in size from 20 cm in length and a mass of a few kilograms, to several metres in length and masses of thousands of kilograms. The vehicle itself can be broken down into several subsystems.

Frame The vehicle frame is typically an open frame constructed of aluminum. Components are bolted to the frame. The frame provides structural support and protection, and provides a method of connecting the buoyancy, propulsion, and other vehicle systems.

Buoyancy Buoyancy control is critical to the proper performance of the vehicle. ROVs typically have fixed buoyancy provided by syntactic foam, or some other type of noncompressible foam. This flotation counteracts the weight of the vehicle frame and mechanical components. Smaller variations in buoyancy are provided by vertical thrusters. This type of vehicle is usually ballasted so that it will float to the surface if the tether is accidentally severed. This also improves operations, as the vertical thrusters are usually forcing the vehicle down, with the thruster wash moving upward away from the bottom. If the thrust is directed toward the seabed, the silt is easily stirred up, destroying visibility.

Propulsion The propulsion system consists of thrusters that control the vehicle motion in three axes. A minimum of two fore/aft thrusters control forward and reverse motion and speed and, by the direction of thrust, the vehicle heading. Vertical thrusters control the vertical motion of the vehicle. Lateral thrusters may be used to allow the vehicle to maneuver sideways while maintaining a constant heading.

Vision   Video cameras contained in pressure-proof housings with acrylic or glass faceplates are the primary source of vision. Multiple cameras are used on larger vehicles to give a wider field of view, or a different perspective. High-resolution, state-of-the-art television broadcast-quality cameras are now being integrated to provide high-quality images that some clients require. In some cases, stereo vision is implemented to help improve spatial awareness and operator efficiency.

Control   Control of the vehicle is most often implemented with computer control. A computer on the surface communicates with a computer mounted on the vehicle. Control input, from human or computer, is fed into the surface control computer. The vehicle computer then issues the control commands and provides feedback to the operator. This system is referred to as a telemetry system. A second type of control, most often found on small, less sophisticated vehicles, is hardwire control. In this case the vehicle thrusters, lights, etc., are wired directly to surface controls. This eliminates the requirement for control computers but restricts the amount of control that can be implemented, and can also limit the tether length.

Manipulators   ROVs are usually fitted with some type of manipulator. Smaller vehicles, if fitted with a manipulator, will often carry a small arm with one or two functions. Large vehicles will be fitted with two powerful manipulators. These will range from simple five-function, rate (the function direction is either on or off) control arms to complex seven- or eight-function arms with force feedback and spatially correspondent control. Manipulator technology has evolved steadily during the history of the ROV. Reliability and efficiency have improved as a result.

Other sensors   ROVs are usually fitted with additional sensors. Scanning sonars are common and give an acoustic image of the area surrounding the vehicle. The range of the sonar will vary depending upon the system used, but generally it will reach past 100 m – well beyond the visual range of the cameras. Altimeters are similar to echo sounders and give vehicle height above the bottom. Depth sensors are implemented on nearly every vehicle. They range from precision sensors to hand-held units strapped to the vehicle frame in front of the camera.

Tether Management System (TMS)

Some ROVs operate with a neutrally buoyant tether cable connecting them directly to the ship or work platform. The buoyancy of this cable can be modified by adding floats and weights. A common alternative approach for larger vehicles is to use a tether management system (see **Figure 1**). The TMS can be designed in several different configurations.

- One approach is to use a 'cage', which houses the vehicle for launch and recovery, and has a winch that pays out or retracts tether as needed. The vehicle is clamped into the cage and is launched from the support vessel. The main winch, mounted on the support vessel, lowers the complete package to the working depth and then suspends it several meters above the worksite. The vehicle is released, tether is paid out by the TMS, and the ROV flies out to perform its work. Upon completion of the work, the vehicle returns to the cage and is clamped in place, and the complete package is recovered.
- The 'top hat' configuration has a smaller TMS with an integral winch that sits on top of the vehicle. Once at operating depth the ROV unlatches from the TMS and descends to the worksite. Upon completion of the work, the ROV latches to the bottom of the TMS and the complete package is recovered.

Tether Cable

The term tether is usually used to refer to the cable directly connected to the vehicle. The vehicle tether is the greatest advantage that an ROV has over other types of systems, such as autonomous underwater vehicles (AUVs) and manned submersibles. It delivers power continuously to the vehicle as well as delivering control data. It also allows a tremendous volume of data to be transmitted in real time from the vehicle to the surface. This includes many channels of high-resolution video, acoustic sonar data, vehicle feedback information, and other data. The tether is also the greatest liability of the ROV: it is adversely affected by currents, it has high drag, and it can easily be damaged during operations. It is most often neutrally buoyant by design, or is made neutral by adding floats.

Umbilical Cable

The term umbilical usually refers to the cable, commonly steel armored, that connects the support vessel to the TMS. This cable will have a fiberoptic bundle, or coaxial cable for command, control, and data

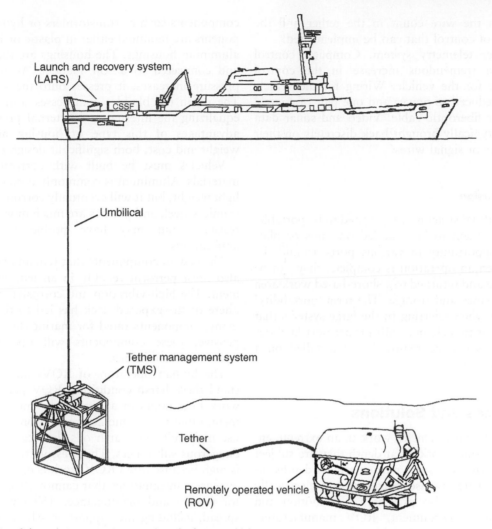

Figure 1 Use of the tether management system for larger vehicles.

transmission. This core will be surrounded by power conductors, used to provide the vehicle with power. Finally, it will have a protective jacket and steel or synthetic strength member. This cable will be paid in and out from a deck mounted winch, to control the depth of the TMS or vehicle.

Launch and Recovery System (LARS) and Winch

Most ROV systems come complete with an integrated LARS. Small vehicles can be deployed and recovered by hand, while medium to large vehicles employ either a crane or an A-frame. Large systems typically have a purpose-built LARS that is integrated with the umbilical winch. With a self-contained system the vehicle can be installed upon many different platforms that are not equipped with launch and recovery gear.

Surface Control Station(s)

Surface control stations usually contain at a minimum a video monitor, videocassette recorder, and joystick for vehicle control. As systems become larger and more complex, the amount of surface equipment grows to include electrical distribution systems, surface control computers, and consoles for copilots and navigators.

Control System

Control systems cover as wide a range of design as there are vehicles. The control systems can be broken down into two basic types.

● Hardwired control. In this configuration each individual ROV component is connected directly to the surface, through the tether, with its own set of dedicated wires. This approach is simple, robust, and inexpensive. It does limit tether length and increases the wire count in the tether and the amount of control that can be implemented.

- Computer telemetry system. Computer control allows a tremendous increase in the control available for the vehicle. Wiring for the vehicle can be reduced to power and one pair of control wires, or fiberoptic cable. Video and sonar data are still typically brought back discretely on their own fiber or signal wires.

Portable Design

Almost all ROV systems are designed to be portable. This allows them to be installed on ships or platforms of opportunity in various ports around the world. When an operation is complete, they can be demobilized and returned to a shore-based work area for maintenance and storage. The term 'portability' is stretched when referring to the large systems that weigh tens of tonnes, but with proper port facilities these systems can be removed and installed on a variety of vessels.

Challenges and Solutions

Remotely operated vehicles work in an extreme environment. While working at depth they are subject to high external pressure, particularly as depths increase. Sea water is also corrosive and electrically conductive. Ships also present a high-motion and high-vibration environment. ROV manufacturers and operators have dealt with these challenges in several ways, as described below.

Some components must be protected from the pressure and water by being mounted in a pressure-proof housing. Pressure-proof housings are typically made of a corrosion-resistant material such as stainless steel or anodized aluminum. As greater pressures are encountered, the strength of these two materials is no longer adequate and housings are made from more exotic materials, such as titanium, composites, or ceramics.

Electrical components such as cameras, lights and sonars are mounted outside the main pressure housings. They must be connected to the main telemetry pressure housing by an electrical cable. The cable penetrations, where the wires enter the pressure proof housings, must be carefully designed. Improper design can result in cables being extruded into the housing or, worse, failure of the seal and flooding of the housing.

Pressure-compensated housings are often used for components that can withstand the pressure but require protection from the water. In this case, components such as transformers or hydraulic components are mounted either in plastic or in thin-wall aluminum housings. The housings are then oil filled and connected to a soft bladder. As the external pressure increases, it presses onto the soft bladder. The oil in the bladder compresses somewhat, thus equalizing the internal and external pressures. The advantages of this type of housing are reduced weight and cost, both significant design constraints.

Vehicles must be built with corrosion-resistant materials. Aluminum is commonly used owing to its light weight, but it will eventually corrode. Titanium, stainless steel, and plastics are much more corrosion-resistant, but may have problems in specific applications.

The system components that remain at the surface also must perform reliably in an extreme environment. The high-vibration and corrosive, wet atmosphere of the exposed deck has led to the design of many components rated for marine duty. While expensive, these components will operate reliably under such conditions.

The human operators of ROVs must also withstand these harsh conditions. ROV personnel must work long hours in a continually moving environment, often in wet and cold conditions. The systems use high voltages, harsh oils, lubricants, and other dangerous substances. The pressure to perform well is high because often ROV work is carried out upon expensive installations that cannot afford downtime for repairs and maintenance. The complete ROV spread, including the support vessel, is expensive to hire and there is no tolerance for unreliable people or vehicles.

Scientific Research Vehicles

Remotely operated vehicles have been supporting scientific operations since the mid-1980s. Some ROVs were originally funded to complement manned submersible work, but a few were developed as replacements for existing submersibles, or as stand-alone vehicles for smaller institutions. The strengths and weaknesses of ROVs do not allow them to be direct replacements for manned submersibles.

Manned submersibles (See Manned Submersibles, Deep Water, Manned Submersibles, Shallow Water). refer to manned vehicle article) were the dominant technology for ocean floor scientific research for decades. ROVs have entered the field, and have gained acceptance because of their distinct advantages in many areas. They have unlimited power and can therefore remain on the bottom for extended periods, efficiently performing large surveys, extended time series experiments, and multidisciplinary operations. A tremendous volume of data is transmitted to the surface, with many channels of real time video, sonar, CTD (conductivity–temperature–depth) data, and other information. In fact, properly managing the data can be a challenge. Many scientists can participate in the operations, which is an advantage. Operations often cover many disciplines, often with unexpected results. Key people can always be on hand to discuss and decide upon modifications to the operational plan as the operation unfolds. Some of the advantages that manned submersibles have will be difficult to replace with the ROV. It is difficult to replace the human eye with remote telepresence. The surface ship motion and control will always influence the ROV operations.

The current (as at 2000) high-profile science ROVs are briefly described in **Table 1**. Each of these vehicles is unique. Some have been developed with a specific focus, and are therefore better at some tasks than others. Every vehicle design is a compromise

between the many elements that can be incorporated into an ROV.

Conclusion

The efficiency of ROVs will continue to improve in two major ways. (i) The efficiency of the work will improve with better integration of ROV capabilities into offshore component design. (ii) The efficiency of the ROV itself will also improve with advances in hydraulic components and design, control system components and design, and higher-voltage cables and motors. Electric vehicles that do not have large hydraulic systems are beginning to enter the market. Altogether, this will result in smaller, lighter cables, which will reduce systems size and cost and will have less effect upon the vehicle as it is operating in currents, or traveling at speed. The multidisciplinary vehicle will remain as the dominant vehicle type, but specialized vehicles will also become more widespread as more and more tasks are assigned to ROVs

Table 1   Summary of some scientific ROVs and their characteristics

Propulsion	TMS	Operator[a]	Power	Manipulators	Depth
Jason/Medea http://www.marine.whoi.edu/ships/rovs/jason_med.htm					
Electric	Depressor weight; 50 m fixed tether length	WHOI	9 kW	Single electric manipulator	6000 m; significant vehicle upgrades planned for 2002
ROPOS http://www.ropos.com					
Hydraulic	Cage; 250 m tether	CSSF	22 kW	Two hydraulic manipulators	5000 m
Tiburon http://www.mbari.org/dmo/vessels/tiburon.html)					
Electric	None	MBARI	15 kW	Two hydraulic manipulators	4000 m
Ventanna http://www.mbari.org/dmo/ventanna/ventanna.html					
Hydraulic	None	MBARI	30 kW	Two hydraulic manipulators	2000 m
Victor http://www.ifremer.fr/victor/victor_uk.html					
Electric	Depressor weight	IFREMER	20 kW	Two	6000 m
Dolphin 3K http://www.jamstec.go.jp/jamstec-e/rov/3k.html					
Hydraulic	None	JAMSTEC	?	Two	3300 m
HYPER-DOLPHIN http://www.jamstec.go.jp/jamstec-e/rov/hyper.html					
Hydraulic	None	JAMSTEC	56 kW	Two	3000 m
KAIKO http://www.jamstec.go.jp/jamstec-e/rov/kaiko.html					
Hydraulic	None	JAMSTEC	?	Two	11 000 m

[a]WHOI, Woods Hole Oceanographic Institution; CSSF, Canadian Scientific Submersible Facility; MBARI, Monterey Bay Aquarium Research Institute; IFREMER, l'Institut Francais de Recherche pour l'Exploitation de la Mer; JAMSTEC, Japan Marine Science and Technology Center

and the scope of work increases. Designers of offshore equipment are more commonly incorporating ROV intervention technology into the original equipment. This has great benefits in improving ROV efficiency. For many years, ROVs have been challenged with attempting to work with components designed for human hands or for dry land manipulation. Once thought and design are applied to ROV intervention techniques, all parties benefit from the increased efficiency. ROVs have evolved, and are still evolving, to fill a requirement for reliable, efficient vehicles in an environment that is inaccessible to humans. As these vehicles develop, and the engineering progresses on the vehicles as well as on their worksites, they will continue to fulfill a unique and expanding role in the underwater world.

See also

Manned Submersibles, Deep Water. Manned Submersibles, Shallow Water. Towed Vehicles.

Further Reading

Marine Technology Society (1984) *ROV'84 Technology Update – An International Perspective. Proceedings of ROV'84 Conference and Exposition, San Diego.* San Diego: Marine Technology Society.

Vadus JR and Busby RF (1979) *Remotely Operated Vehicles: An Overview.* Washington, DC: US Department of Commerce, National Oceanographic and Atmospheric Administration, Office of Ocean Engineering.

SEISMOLOGY SENSORS

L. M. Dorman, University of California, San Diego, La Jolla, CA, USA

Introduction

A glance at the globe shows that the Earth's surface is largely water-covered. The logical consequence of this is that seismic studies based on land seismic stations alone will be severely biased because of two factors. The existence of large expanses of ocean distant from land means that many small earthquakes underneath the ocean will remain unobserved. The difference in seismic velocity structure between continent and ocean intruduces a bias in locations, with oceanic earthquakes which are located using only stations on one side of the event being pulled tens of kilometers landward. Additionally, the depths of shallow subduction zone events, which are covered by water, will be very poorly determined. Thus seafloor seismic stations are necessary both for completeness of coverage as well as for precise location of events which are tectonically important. This paper summarizes the status of seafloor seismic instrumentation.

The alternative methods for providing coverage are temporary (pop-up) instruments and permanently connected systems. The high costs of seafloor cabling has thus far precluded dedicated cables of significant length for seismic purposes, although efforts have been made to use existing, disused wires. Accordingly, the main emphasis of this report will be temporary instruments.

Large ongoing programs to investigate oceanic spreading centers (RIDGE) and subductions (MARGINS) have provided impetus for the upgrading of seismic capabilities in oceanic areas.

The past few years has seen a blossoming of ocean bottom seismograph (OBS) instrumentation, both in number and in their capabilities. Active experimental programs are in place in the USA, Europe, and Japan. Increases in the reliability of electronics and in the capacity of storage devices has allowed the development of instruments which are much more reliable and useful. Major construction programs in Japan and the USA are producing hundreds of instruments, a number which allows

imaging experiments which have been heretofore associated with the petroleum exploration industry. This contrasts sharply with the severely underdetermined experiments which have characterized earthquake seismology. One change over the past decade has been the disappearance of analog recording systems.

A side effect of this rapid change is that a review such as this provides a snapshot of the technology, rather than a long-lasting reference. The technical details reported below are for instruments at two stages of development: existing instruments (UTIG, SIO/ONR, SIO/IGPP-SP, GEOMAR, LDEO-BB) and instruments still in design and construction (WHOI-SP, WHOI-BB SIO/IGPP-BB) (see **Tables 1** and **2**; **Figures 1–6**). The latter construction project has the acronym 'OBSIP' for OBS instrument pool, and sports a polished, professionally designed web site at http://www.obsip.org.

OBS designs are roughly divided into two categories which for brevity will be called 'short-period' (SP) and 'broadband' (BB). The distinction blurs at times because some instruments of both classes use a common recording system, a possibility which emerges when a high data-rate digitizer has the

Figure 1 The UTIG OBS, a particularly 'clean' mechanical design, which has been in use for many years, with evolving electronics. The anchor is 1.2 m on each side. (Photograph by Gail Christeson, UTIG.)

Table 1 Characteristics of short period ocean bottom seismometers

Parameter	UTIG	WHOI-SP	SIO-IGPP-SP	GEOMAR-SP	JAMSTEC-ORI[a]
Contact/website	Nakamura/http://www.ig.utexas.edu/research/projects/obs/obs.html	Detrick/Collins http://www.obsip.org	Orcutt/Babcock http://www.obsip.org	Flüh/Bialas http://www.geomar.de	–
Seismic sensor(s)	3-component 4.5 Hz Mark Products L-15B or Oyo GS-11D	2 Hz Mark Products L-22 4.5 Hz or L-28 4.5 Hz, vertical component	2 Hz Mark Products L-22, vertical component	optionally uses Webb BB sensor	4.5 Hz vertical
Frequency response	4.5–100 Hz	2–X Hz	2–X Hz	0.05–30 Hz/0.01–X Hz	4.5–100 Hz
Nominal sensitivity	2.5 nm s^{-1}	–	–	–	–
Hydrophone	OAS E-2PD crystal	Hightech crystal	Hightech HYI-90-U	OAS E4SD or Cox-Webb DPG	–
Frequency Response	3–100 Hz	5–X Hz	50 mHz-X	0.05–30 Hz	–
Digitizer type, dynamic range, sample rates	14 bits + gain-ranging, 126 dB,112 dB re electronic noise	Quanterra Q330 24-bit,126 dB, 1–200 Hz	Cirrus/Crystal CS5321-CS5322 24-bit, 124–130 dB	oversampling, 120 dB, 25–200 Hz	–
Recording medium and capacity[b]	Disk, semi-continuous	Disk, data download through pressure case	9 Gbyte disk, data download through pressure case	1 Gbyte DAT or 2 Gbyte semiconductor memory	–
Clock type and drift[c]	10 ms	Seascan Precision Timebase < 0.5 ms d^{-1}	Seascan Precision Timebase < 0.5 ms/d^{-1}	Seascan Precision Timebase < 0.5 ms d^{-1}	–
Endurance	8 weeks–6 months	90 days, alkaline/1 year, lithium	80–180 days at 250/31.25 Hz sampling 5 days on NiCad rechargeable cells	300 days	–
Power consumption	550 mW	–	420 mW at 31.25 Hz, 1.6 W at 250 Hz for 2 channels	230–250 mW	–
Release type	Burnwire release, acoustically controlled, acoustic release, two backup timers	Burnwire, Edgetech acoustics	Double burnwire, Edgetech acoustics	Acoustic release with back-up timer	–
Mechanical configuration, launch/recovery weights	Single 43 cm diameter glass sphere, 85 kg/35 kg	63 kg/43 kg	110 kg/80 kg	Vertical cylinders, 175 kg/125 kg	Single 43 cm diameter glass sphere
Number available	37–39[d]	15 now, 40 more under construction	14 now, 74 more under construction	27 OBH + 11 OBS	100?
Total			51		

[a]Information on these instruments is incomplete.
[b]2 gigabytes is about 22 days of data sampling four channels at 128 Hz or 176 days sampling four channels at 16 Hz.
[c]1 ms d^{-1} is about 1×10 E-8.
[d]Includes instruments of the same design operated by IRD (formerly ORSTOM) and National Taiwan Ocean University.

Table 2 Characteristics of broadband ocean bottom seismometerrs

Parameter	SIO/ONR[a]	WHOI-BB	LDEO-BB[e]	SIO/IGPP-BB	ORI-BB[f]
Manager	Dorman/ Sauter http:// www-mpl. ucsd.edu/obs	Detrick/ Collins http:// www.obsip.org	Webb/ http:// www.obsip.org	Orcutt/Babcock http://www. obsip.org	–
Seismic sensor	1 Hz Mark Products L4C-3D or PMD 2123	Guralp CMG-3ESP	1 Hz Mark Products L4C-3D	Kinemetrics	PMD 2023
Frequency response	0.033–32 Hz[b]	0.033–50 Hz[b]	0.005–30 Hz[b]	0.02–50 Hz	0.033–50 Hz
Hydrophone	Cox-Webb DPG	HighTech	Cox-Webb DPG for low frequency hydrophone for high frequency	High Tech HYI-90-U or Cox-Webb DPG	–
Frequency response	0.001–5 Hz	0.033–X Hz	0.001–60 Hz	0.05–15 kHz or 0.01–32 Hz	–
Digitizer type, Dynamic range, sample rates	16-bit + gain-ranging, 126 dB, X-128 sps	Quanterra QA330 24-bit, 135 dB	nominal 24-bit ~ 135 dB, 1, 20, 40, 100, 200 sps	24-bit Cirrus/ Crystal CS 5321–CS-5322, 130 dB	20-bit
Recording medium and capacity[c]	Disk, 9–27 Gbyte	Disk, 2 Gbyte	Disk, 18 Gb, 72 Gb planned	Disk, 9 Gbyte	4 × 6.4 Gbyte disks
Clock drift[d]	< 1 ms d^{-1}	< 1 ms d^{-1}	0.5 ms d^{-1}	< 1 ms d^{-1}	–
Endurance	6–12 months	6–12 + months	Up to 15 months	6–12 months	9 months?
Power consumption	400 mW	1.5 W at 20 sps	–	–	~ 600 mW
Release type	Two EG&G 8242 acoustic releases	One EG&G 8242 acoustic relase with back-up acoustic burnwire release	Acoustically controlled burnwire?	EG&G acoustically controlled burnwire	Acoustically controlled burn-plate
Mechanical configuration, launch/ recovery weights	Fiberglass frame, aluminum pressure cases, glass flotation (ONR OBS-style)	ONR OBS-style, floating above anchor, 570 kg/ 472 kg	Aluminum pressure cases, plastic plate frame, 215/145 kg	178/138 kg	50 cm diameter, titanium sphere
Number available	14	25 under construction	64 under construction	15–20 (using recording package from SP instrument)	15
Total			~ 133		

[a]These instruments incorporate a fluid flowmeter/sampler in the instrument frame (see Tryon *et al.* 2001).
[b]Seismometers are free from spurious resonances below 20 Hz.
[c]2 gigabytes is about 22 days of data sampling four channels at 128 Hz or 176 days sampling four channels at 16 Hz.
[d]1 ms is about 1×10^{-8}.
[e]1 See Webb *et al.*, 2001.
[f]1 See Shiobara *et al.*, 2000.

capability of operation in a low-power, high endurance mode.

Short Period (SP) Instruments

The SP instruments (**Table 1**) are light in weight and easy to deploy, typically use 4.5 Hz geophones, commonly only the vertical component, and/or hydrophones, may have somewhat limited recording capacity and endurance, and are typically used in active-source seismic experiments and for micro-earthquake studies with durations of a week to a few months. Two types (UTIG and JAMSTEC) are single-sphere instruments.

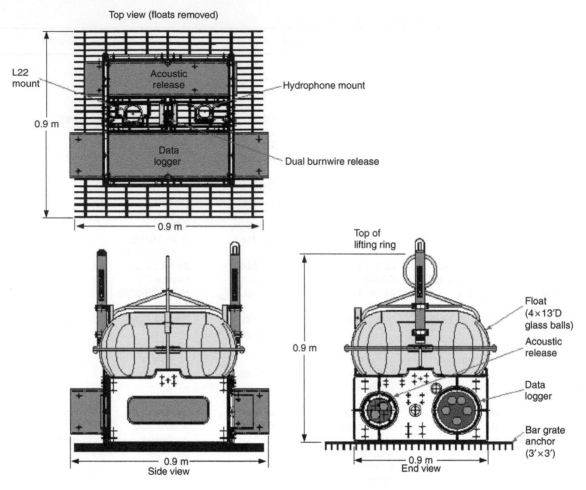

Figure 2   The IGPP-SP instrument. (Figure from Babcock, Harding, Kent, and Orcutt.)

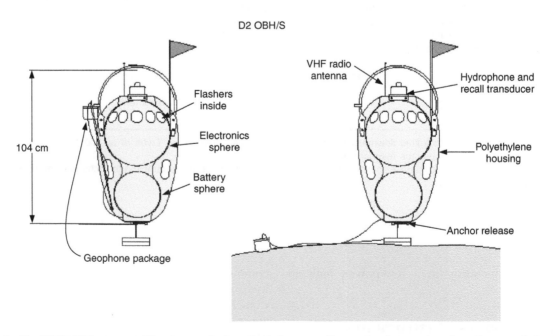

Figure 3   The WHOI-SP instrument. The change of orientation between seafloor and surface modes allows the acoustic transducer an unobstructed view of the surface while on the seafloor and permits acoustic ranging while the instrument is on the surface. (Figure by Beecher Wooding and John Collins.)

Figure 4 The GEOMAR OBH/S. The shipping/storage container is equipped with an overhead rail so that it serves as an instrument dispenser. The OBS version is shown here. (Photograph by Michael Tryon, UCSD.)

Broadband (BB) Instruments

The BB instruments (**Table 2**) provide many features of land seismic observatories, relatively high dynamic range, excellent clock stability $-<1\,\mathrm{ms\,d}^{-1}$ drift. This class of instruments can be equipped with hydrophones useful down to a millihertz. The BB instruments are designed in two parts, the main section contains the recording package, and release and recovery aids, while the sensor package is physically separated from the main section. This configuration allows isolation from mechanical noise and and permits tuning of the mechanical resonance of the sensor–seafloor system.

Figure 5 The SIO-ONR OBSs (Jacobson *et al*. 1991, Sauter *et al*, 1990) being launched in Antarctica. The anchor serves as a collector for the CAT fluid fluxmeter (Tryon *et al.* 2001). The plumbing for the flowmeter is in the light-colored box at the right-hand end of the instrument. (Photograph by Michael Tryon, UCSD.)

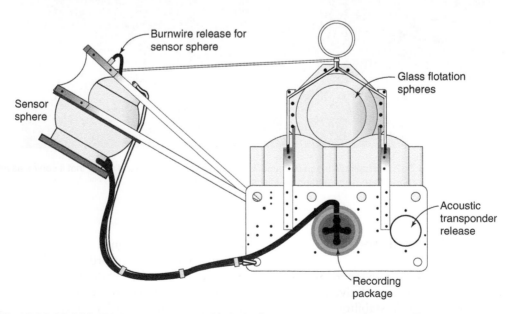

Figure 6 The LDGO-BB OBS. This is based on the Webb design in use during the past few years. The earlier version established a reputation for high reliability and was the lowest noise OBS and lightest of its time period. The main drawback of the earlier version was its limited (16-bit) dynamic range. (Figure from S. Webb, LDEO.)

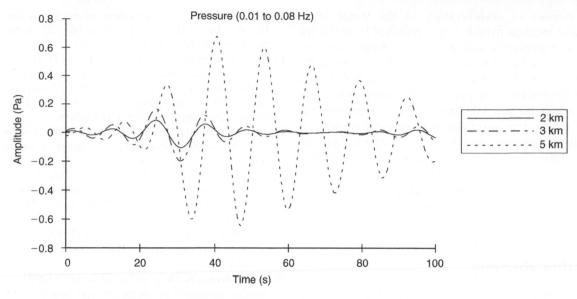

Figure 7 Synthetic seismograms of pressure at three ocean depths.

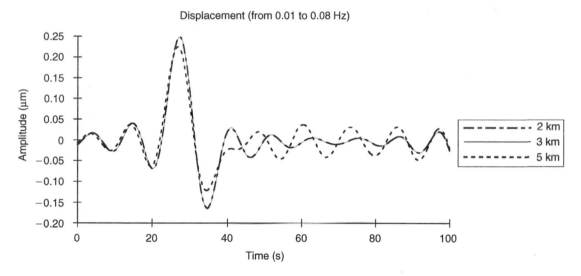

Figure 8 Synthetic seismograms of vertical motion at the same depths as **Figure 7**; note the reduction in the effects of the water column reverberations. (From Lewis and Dorwan, 1998).

Sensor Considerations

Emplacement of a sensor on the seafloor is almost always suboptimal in comparison with land stations. Instruments dropped from the sea surface can land tens of meters from the desired location. The seafloor material is almost always softer than the surficial sediments (these materials have shear velocities as low as a few tens of meters per second. The sensor is thus almost always poorly coupled to the seafloor and sensor resonances can occur within the frequency range of interest for short period sensors. Fortunately, these resonances have little effect on lower frequencies. However, lower frequency sensors are affected, since a soft foundation permits tilt either in response to sediment deformation by the weight of the sensor or in response to water currents. The existing Webb instruments combat this problem by periodic releveling of the sensor gimbals. The PMD sensors have an advantage here in that the mass element is a fluid and the horizontal components self-level to within 5°.

Why not use hydrophones then? These make leveling unnecessary and are more robust mechanically. In terms of sensitivity, they are comparable to seismometers. The disadvantage of hydrophones lies in the physics of reverberation in the water layer. A pulse incident from below is reflected from the sea surface completely, and when it encounters the seafloor it is reflected to a significant degree. Since the seafloor has a higher acoustic impedance than water, the reflected pressure pulse has the same sign as the incident pulse and the signal is large. However, the seafloor motion associated with pressure pulses traveling in opposite directions is opposite in sign, so cancellation occurs. Unfortunately, the frequency range in which these reverberations are troublesome is in the low noise region. **Figures 7** and **8** show synthetic seismograms of pressure and vertical motion illustrating this effect.

Further Reading

Barash TW, Doll CG, Collins JA, Sutton GH, and Solomon SC (1994) Quantitative evaluation of a passively leveled ocean bottom seismometer. *Marine Geophysical Researches* 16: 347–363.

Dorman LM (1997) Propagation in marine sediments. In: Crocker MJ (ed.) *Encyclopedia of Acoustics*, pp. 409–416. New York: John Wiley.

Dorman LM, Schreiner AE, and Bibee LD (1991) The effects of sediment structure on sea floor noise. In: Hovem J, *et al.* (eds.) *Proceedings of Conference on Shear Waves in Marine Sediments*, pp. 239–245. Dordrecht: Kluwer Academic Publishers.

Duennebier FK and Sutton GH (1995) Fidelity of ocean bottom seismic observations. *Marine Geophysical Researches* 17: 535–555.

Jacobson RS, Dorman LM, Purdy GM, Schultz A, and Solomon S (1991) *Ocean Bottom Seismometer Facilities Available*, EOS, Transactions AGU, 72, pp. 506, 515.

Sauter AW, Hallinan J, Currier R *et al.* (1990) *Proceedings of the MTS Conference on Marine Instrumentation*, pp. 99–104.

Tryon M, Brown K, Dorman L, and Sauter A (2001) A new benthic aqueous flux meter for very low to moderate discharge rates. *Deep Sea Research* (in press).

Webb SC (1998) Broadband seismology and noise under the ocean. *Reviews in Geophysics* 36: 105–142.

VEHICLES FOR DEEP SEA EXPLORATION

S. E. Humphris, Woods Hole Oceanographic Institution, Woods Hole, MA, USA

Introduction

Exploring the deep sea has captured the imagination of humankind ever since Leonardo da Vinci made drawings of a submarine more than 500 years ago, and Jules Verne published *20 000 Leagues under the Sea* in 1875. Since the early twentieth century, people have been venturing into the ocean in bathyspheres and bathyscaphs. However, it was not until 1960 that the dream to go to the bottom of the deepest part of the ocean was realized, when Jacques Piccard and a US Navy lieutenant, Don Walsh, descended to the bottom of the Mariana Trench (10 915 m or 6.8 mil) in *Trieste* (**Figure 1**). This vehicle consisted of a float chamber filled with gasoline for buoyancy, and a separate pressure sphere for the personnel, allowing for a free dive rather than a tethered one. Containers filled with iron shot served as ballast to make the submersible sink. After a 5-h trip to the bottom, and barely 20 min of observations there, the iron shot was released and *Trieste* floated back to the surface.

Since that courageous feat almost 50 years ago, dramatic advances in deep submergence vehicles and technologies have enabled scientists to routinely explore the ocean depths. For many years, researchers have towed instruments near the seafloor to collect various kinds of data (e.g., acoustic, magnetic, and photographic) remotely. With the development of sophisticated acoustic and imaging systems designed to resolve a wide range of ocean floor features, towed vehicle systems have become increasingly complex. Some now use fiber-optic, rather than coaxial, cable as tethers and hence are able to transmit imagery as well as data in real time. Examples of deep-towed vehicle systems are included in **Table 1** and **Figure 2,** and they tend to fall into two categories. Geophysical systems, such as *SAR* (IFREMER, France), *TOBI* (National Oceanography Centre, Southampton, UK), and *Deep Tow 4KS* (JAMSTEC, Japan), collect sonar imagery, bathymetry, sub-bottom profiles, and magnetics data, as they are towed tens to hundreds of meters off the bottom. Imaging systems, such as *TowCam* (WHOI, USA), *Scampi* (IFREMER, France), and *Deep Tow 6KC* (JAMSTEC, Japan), are towed a few meters off the bottom and provide both video and digital imagery of the seafloor.

However, since the 1960s, scientists have been transported to the deep ocean and seafloor in submersibles, or human-occupied vehicles (HOVs), to make direct observations, collect samples, and deploy instruments. More recently, two other types of deep submergence vehicles – remotely operated vehicles (ROVs) and autonomous underwater vehicles (AUVs) – have been developed that promise to greatly expand our capabilities to map, measure, and sample in remote and inhospitable parts of the ocean, and to provide the continual presence necessary to study processes that change over time.

Human-Occupied Vehicles

The deep-sea exploration vehicles most familiar to the general public are submersibles, or HOVs. This technology allows a human presence in much of the world's oceans, with the deepest diving vehicles capable of reaching 99% of the seafloor.

There exist about 10 submersibles available worldwide for scientific research and exploration that can dive to depths greater than 1000 m (**Table 2** and **Figure 3**). All require a dedicated support ship. These battery-operated vehicles allow two to four individuals (pilot(s) and scientist(s)) to descend into the ocean to make observations and gather data and samples. The duration of a dive is limited by battery life, human

Figure 1 The bathyscaph *Trieste* hoisted out of the water in a tropical port, around 1959. Photo was released by the US Navy Electronics Laboratory, San Diego, California (US Naval Historical Center Photograph). Photo #NH 96801: US Navy Bathyscaphe *Trieste* (1958–63).

Table 1 Examples of deep-towed vehicle systems for deep-sea research and exploration (systems that can operate at depths ≥1000 m)

Vehicle	Operating organization	Maximum operating depth (m)	Purpose
TowCam	WHOI, USA	6500	Photo imagery; CTD; volcanic glass samples; water samples
Deep-Tow Survey System	COMRA, China	6000	Sidescan, bathymetry; sub-bottom profiling
DSL-120A	HMRG, USA	6000	Sidescan; bathymetry
IMI-30	HMRG, USA	6000	Sidescan; bathymetry; sub-bottom profiling
Scampi	IFREMER, France	6000	Photo and video imagery
Système Acoustique Remorqué (SAR)	IFREMER, France	6000	Sidescan; sub-bottom profiling; magnetics; bathymetry
SHRIMP	NOC, UK	6000	Photo and video imagery
TOBI	NOC, UK	6000	Sidescan; bathymetry; magnetics
BRIDGET	NOC, UK	6000	Geochemistry
Deep Tow 6KC	JAMSTEC, Japan	6000	Photo and video imagery
Deep Tow 4KC	JAMSTEC, Japan	4000	Photo and video imagery
Deep Tow 4KS	JAMSTEC, Japan	4000	Sidescan; sub-bottom profiling

WHOI, Woods Hole Oceanographic Institution, USA; COMRA, China Ocean Mineral Resources R&D Association; HMRG, Hawai'i Mapping Research Group, USA; IFREMER, French Research Institute for Exploration of the Sea; NOC, National Oceanographic Centre, Southampton, UK; JAMSTEC, Japan Marine Science & Technology Center.

Figure 2 Examples of deep-towed vehicle systems. (a) SHRIMP, (b) Deep Tow, (c) Tow Cam, and (d) DSL-120A. (a) Courtesy of David Edge, National Oceanography Centre, UK. (b) © JAMSTEC, Japan, with permission. (c) Photo by Dan Fornari, WHOI, USA. (d) Courtesy of WHOI, USA.

Table 2 HOVs for deep-sea research and exploration (vehicles that can operate at depths ≥ 1000 m)

Vehicle	Operating organization	Maximum operating depth (m)
HOV (under construction)	COMRA, China	7000
Shinkai 6500	JAMSTEC, Japan	6500
Replacement HOV (in planning stages)	NDSF, WHOI, USA	6500
MIR I and II	P.P. Shirshov Institute of Oceanology, Russia	6000
Nautile	IFREMER, France	6000
Alvin	NDSF, WHOI, USA	4500
Pisces IV	HURL, USA	2170
Pisces V	HURL, USA	2090
Johnson-Sea-Link I and II	HBOI, USA	1000

Abbreviations as in **Table 1**; NDSF, National Deep Submergence Facility; HURL, Hawaii Undersea Research Laboratory; HBOI, Harbor Branch Oceanographic Institution, USA.

endurance, and safety protocols, and typically does not exceed 8–10 h, including transit time to and from the working depth (about 4 h for a seafloor depth of 4000 m). (The Russian MIR submersibles are an exception; they operate on a 100-kWh battery that can accommodate dive times in excess of 12 h.) Housed in a personnel sphere (**Figure 4**), the divers are maintained at atmospheric pressure despite the ever-increasing external pressure with depth (1 atm every 10 m). Cameras on pan and tilt mounts with zoom and focus controls are located on the exterior of the vehicles, as well as quartz iodide and/or metal halide lights to illuminate the area. Submersibles are also equipped with robotic arms that can be used to manipulate equipment or pick up samples, and a basket, usually mounted on the front of the vehicle, to transport instruments, equipment, or samples. These vehicles can handle heavy payloads, maintaining neutral buoyancy as their weight changes through a variable ballast control system. All these capabilities, together with their slow speeds (1–2 knots), make submersibles best suited to detailed observations, imaging, and sampling in localized areas, rather than operating in a survey mode.

Many significant discoveries during the past four decades of marine research have resulted from observations and samples taken from submersibles. Through direct observations from submersibles, biologists have discovered many previously unknown animals, and have documented that gelatinous animals (cnidarians, ctenophores, etc.) form a dominant ecological component of mid-water communities. These soft-bodied, fragile animals would have been destroyed by the trawl nets used in earlier days to sample these depths. Submersibles have enabled geologists to explore the global mid-ocean ridge system, and have provided them with a detailed view of the nature of volcanic and tectonic activity during the formation of oceanic crust. Submersibles played an important role in the discovery of hydrothermal vents

and their exotic communities of organisms, and continue to be used extensively for investigation of these extreme deep-sea environments.

HOVs will continue to provide important capabilities for deep-sea research at least for the foreseeable future. Although rapid progress is being made in videography and photography to develop capabilities that match those of the human eye, there is still no substitute for the direct, three-dimensional view that allows divers to make contextual observations and integrate them with the cognitive ability of the human brain. In recognition of this continuing need, there are two submersibles that are under construction or in the planning stages. The China Ocean Mineral Resources R&D Association (COMRA) is constructing their first submersible that will have a maximum operating depth of 7000 m. It is expected to be operational in 2007. In the United States, over 40 years after the submersible Alvin was delivered in 1964, the National Deep Submergence Facility at Woods Hole Oceanographic Institution is in the planning stages for a new and improved replacement HOV with an increased operating depth of 6500 m.

Remotely Operated Vehicles

Over the past 20 years, marine scientists have begun to routinely use ROVs to collect deep-sea data and samples. ROVs were originally developed for use in the ocean by the military for remote observations, but were adapted in the mid-1970s by the offshore energy industry to support deep-water operations. There are many ROVs commercially operated today, ranging from small, portable vehicles used for shallow-water inspections to heavy, work-class, deepwater ROVs used by the offshore oil and gas industry in support of subsea cable laying, retrieval, and repair.

Figure 3 Examples of HOVs used to conduct scientific research. (a) *Shinkai 6500*, (b) *Sea Link*, and (c) *Nautile*. (a) © JAMSTEC, Japan, with permission. (b) Courtesy of Harbor Branch Oceanographic Institution, USA. (c) © IFREMER, France, with permission; O. Dugornay.

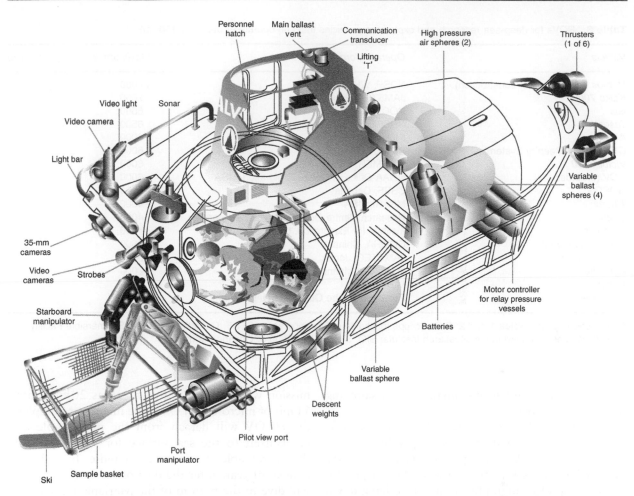

Figure 4 Cutaway illustration of the submersible *Alvin* showing the major components of an HOV. Illustration by E. Paul Oberlander, WHOI, USA.

There are about a dozen ROVs that are available to the international scientific community (**Table 3** and **Figure 5**). While some of these have dedicated support ships, many can operate in the 'flyaway' mode; that is, they can be shipped to, and operated on, a number of different ships. Unlike the HOVs, ROVs are unoccupied, and are tethered to a support ship usually by a fiber-optic cable that has sufficient bandwidth to accommodate a wide variety of oceanographic sensors and imaging tools. The cable provides power and communications from the ship to the ROV, allowing control of the vehicle by a pilot on board the ship. The pilot can also use the manipulator arm(s) to collect samples and perform experiments. The cable transmits images and data from the ROV to the control room on board the ship where monitors display the images of the seafloor or water column in real time. These capabilities, together with their excellent power and lift, allow ROVs to perform many of the same operations as HOVs.

Obvious advantages of using ROVs are that they remove the human risk factor from deep-sea research and exploration and, through the shipboard control room (**Figure 6**), allow a number of scientists and engineers to discuss the incoming data and make collective decisions about the operations. Another distinct advantage is their ability to remain underwater for extended periods of time because power is provided continuously from the ship. This endurance means that scientists can make observations over periods of many days, instead of a few hours a day, and gives them the flexibility to react to unexpected events. The disadvantage of an ROV is that its tether constrains operations because the range of the vehicle with respect to the ship cannot exceed a few hundred meters. Movement of the ship must therefore be carefully coordinated with the movements of the vehicle – this requires a ship equipped with a dynamic positioning system. In addition, the tether is heavy and produces drag on the vehicle, making it less maneuverable and vulnerable to entanglement in rugged terrain. However, with careful tether management, ROVs are well suited to mapping and surveying small areas, as well as to making

Table 3 ROVs for deep-sea research and exploration (vehicles that can operate at depths ≥ 1000 m)

Vehicle	Operating organization	Maximum operating depth (m)
Nereus (hybrid) (under construction)	NDSF, WHOI, USA	11 000
Kaiko 7000	JAMSTEC, Japan	7000
Isis	NOC, UK	6500
Jason II	NDSF, WHOI, USA	6500
ATV	SIO, USA	6000
CV (Wireline Reentry System)	SIO, USA	6000
Victor 6000	IFREMER, France	6000
ROV (on order)	NOAA Office of Ocean Exploration, USA	6000
ROPOS	CSSF, Canada	5000
Tiburon	MBARI, USA	4000
Quest	Research Centre Ocean Margins, Germany	4000
Hercules	Institute for Exploration, USA	4000
Sea Dragon 3500	COMRA, China	3500
Hyper Dolphin	JAMSTEC, Japan	3000
Aglantha	Institute of Marine Research, Norway	2000
Ventana	MBARI, USA	1500
Cherokee	Research Centre Ocean Margins, Germany	1000

Abbreviations as in **Tables 1** and **2**; SIO, Scripps Institution of Oceanography; CSSF, Canadian Scientific Submersible Facility; MBARI, Monterey Bay Aquarium Research Institute; SIO, Scripps Institution of Oceanography.

more detailed observations, imaging, and sampling of specific features.

While many of the ROVs available to the scientific community have a wide range of capabilities, a few are purpose-built. For example, the Wireline Reentry System known as *CV*, and operated by Scripps Institution of Oceanography, is a direct hang-down vehicle designed specifically for precision placement of heavy payloads on the seafloor or in drill holes (**Figure 7**). Unlike conventional, near-neutrally buoyant ROVs, the Wireline Reentry System can handle payloads of a few thousand kilograms, depending on the water depth. It has been used, for example, to install seismometer packages in, and recover instruments packages from, seafloor drill holes in water depths up to 5500 m, as well as to deploy precision acoustic ranging units on the axis of the mid-ocean ridge.

Another ROV being built at Woods Hole Oceanographic Institution for a specific purpose is *Nereus* (**Figure 8**). More correctly referred to as a hybrid remotely operated vehicle, or HROV, because it will be able to switch back and forth to operate as either an AUV or an ROV on the same cruise, *Nereus* will be capable of exploring the deepest parts of the world's oceans, as well as bringing ROV capabilities to ice-covered oceans, such as the Arctic. The HROV will use a lightweight fiber-optic micro-cable, only 1/32 of an inch in diameter, allowing it to operate at great depth without the high-drag and expensive cables typically used with ROV systems. Once the HROV reaches the bottom, it will conduct its mission while paying out as much as 20 km (about 11 mi) of micro-cable. Once the mission is complete, the HROV will detach from the micro-cable and guide itself to the sea surface for recovery, while the micro-cable is recovered for reuse. In 2008–09, almost 50 years after the dive of the *Trieste*, *Nereus* will dive to the bottom of the Mariana Trench.

Autonomous Underwater Vehicles

Although the concept of AUVs has been around for more than a century, it is only in the last decade or two that AUVs have been applied to deep-sea research and exploration. AUV technology is in a phase of rapid growth and expanding diversity. There are now more than 50 companies or institutions around the world operating AUVs for a variety of purposes. For example, the offshore gas and oil industry uses them for geologic hazards surveys and pipeline inspections, the military uses them for locating mines in harbors among other applications, and AUVs have been used to search for cracks in the aqueducts that supply water to New York City.

There are currently about a dozen AUVs being used specifically for deep-sea exploration (**Table 4** and **Figure 9**), although the numbers continue to increase. These unoccupied, untethered vehicles are preprogrammed and deployed to drift, drive, or glide through the ocean without real-time intervention from human operators. All power is supplied by energy systems carried within the AUV. Data are

Figure 5 Examples of ROVs used for deep-sea research. (a) ROV *Kaiko*, (b) ROV *Jason II*, (c) ROV *Tiburon*, and (d) ROV *Victor 6000*. (a) © JAMSTEC, Japan, with permission. (b) Photo by Tom Bolmer, WHOI, USA. (c) Photo by Todd Walsh © 2006, MBARI, USA, with permission. (d) © IFREMER, France, with permission; M. Bonnefoy.

recorded and are then either transmitted via satellite when the AUV comes to the surface, or are downloaded when the vehicle is recovered. They are generally more portable than HOVs and ROVs and can be deployed off a wide variety of ships. By virtue of their relatively small size, limited capacity for scientific payloads, and autonomous nature, AUVs do not have the range of capabilities of HOVs and ROVs. They are, however, much better suited than HOVs and ROVs to surveying large areas of the ocean that would take years to cover by any other means. They can run missions of many hours or days on their battery power and, with their streamlined shape, can travel many kilometers collecting data of various types depending on which sensors they are carrying.

Hence, AUVs are frequently used to identify regions of interest for further exploration by HOVs and ROVs.

Unlike HOVs and ROVs that are designed with the flexibility to carry different sensors and equipment for different purposes, AUV system design and attributes are driven by the specific research application. Some, such as the autonomous drifters and gliders (essentially drifters with wings and a buoyancy change mechanism that allow the vehicle to change heading, pitch, and roll, and to move horizontally while ascending and descending in the water column), are designed for research in the water column to better understand the circulation of the ocean and its influence on climate. While satellites provide

Figure 6 Portable control van for the ROV *Jason II* constructed from two shipping containers assembled on board the R/V *Knorr*. © Dive and Discover, WHOI, USA.

global coverage of conditions at the sea surface, AUVs are likely to be the only way to continuously access data from the ocean depths. Equipped with oceanographic sensors that measure temperature, salinity, current speed, and phytoplankton abundance, drifters and gliders profile the water column by sinking to a preprogrammed depth, and then rising to the surface where they transmit their data via satellite back to the scientist on shore. By deploying hundreds to thousands of these vehicles, scientists will achieve a long-term presence in the ocean, and will be able to make comprehensive studies of vast oceanic regions.

Other, more sophisticated AUVs are also used to investigate water column characteristics, and ephemeral or localized phenomena, such as algal blooms. The first of the *Dorado Class* of AUVs, operated by Monterey Bay Aquarium Research Institute, was deployed in late 2001 to measure the inflow of water into the Arctic basin through the Fram Strait. *Autosub*, operated by the National Oceanography Centre, Southampton, UK, was deployed to measure flow over the sills in the Strait of Sicily. The *REMUS* (Remote Environmental Monitoring UnitS) class of AUVs is extremely versatile and they have been used on many types of missions. The standard configuration includes an up- and down-looking acoustic Doppler current profiler (ADCP), sidescan sonar, a conductivity–temperature (CT) profiler, and a light scattering sensor. However, many other instruments have been integrated into it for specific missions, including fluorometers, bioluminescence sensors, radiometers, acoustic modems, forward-looking sonar, altimeters, and acoustic Doppler velocimeters. *REMUS* can also carry a video plankton recorder, a plankton pump, video cameras,

Figure 7 The Wireline Reentry System, known as the *CV*, operated by Scripps Institution of Oceanography. This specialized ROV can precisely place heavy payloads on the seafloor and in drill holes. The control vehicle, which weighs about 500 kg in water, is deployed at the end of a 17.3-mm (0.68") electromechanical (coax) or electro-optico-mechanical (three copper conductors, three optical fibers) oceanographic cable. The vehicle consists of a steel frame equipped with two horizontal thrusters mounted orthogonal to each other to control lateral position. The vertical position is controlled by winch operation. Instrumentation includes a compass, pressure gauge, lights, video camera, sonar systems, and electronic interfaces to electrical releases and to a logging probe. Courtesy of Scripps Institution of Oceanography – Marine Physical Laboratory, USA.

electronic still cameras, and, most recently, a towed acoustic array.

Still other AUVs are designed specifically for near-bottom work. They have proved particularly useful for near-bottom surveying and mapping, which can be accomplished autonomously while the support ship simultaneously conducts other, more traditional, operations. One of the earliest vehicles to provide this capability was the *Autonomous Benthic Explorer* (*ABE*) developed at Woods Hole Oceanographic Institution. *ABE* was designed to be extremely stable in pitch and roll and to be reasonably efficient in forward travel. All the buoyancy is built into the two upper pods, while the majority of the weight (the batteries and the main pressure housing) is in the central lower section. The three-hull structure also allows the seven vertical and lateral thrusters to be

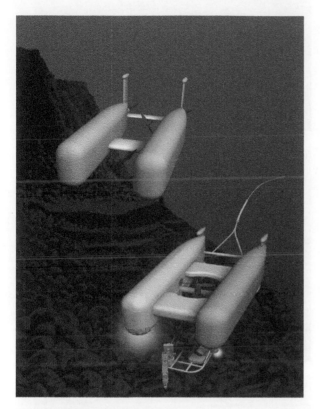

Figure 8 Schematic illustration of the HROV, *Nereus*, currently under construction at Woods Hole Oceanographic Institution, USA, in its autonomous mode (upper) and its ROV mode (lower). Illustration by E. Paul Oberlander, WHOI, USA.

placed between the hulls where they are protected. *ABE* is most efficient traveling forward, but it can also move backward, up or down, left or right, and can hover and turn in place. Equipment that it usually carries includes temperature and salinity sensors, an optical backscatter sensor, a magnetometer to measure near-bottom magnetic fields, and an acoustic altimeter to make bathymetric measurements and for its automated bottom following. *ABE* can dive to depths of 5500 m for 16–34 h, and it uses acoustic transponder navigation to follow preprogrammed track lines automatically. Its capability to maintain a precise course over rugged seafloor terrain gives it the ability to make high-precision seafloor bathymetric maps with features a few tens of centimeters tall and less than a meter long being identifiable.

Other AUVs have been specifically developed for high-resolution optical and acoustic imaging of the seafloor. For example, *SeaBED*, also developed at Woods Hole Oceanographic Institution, was designed specifically to further the growing interests in seafloor optical imaging – specifically, high-resolution color imaging and the processes of photo-mosaicking and three-dimensional image reconstruction. In addition to requiring high-quality sensors, this imposes additional constraints on the ability of the AUV to carry out structured surveys, while closely following the seafloor. The distribution of the four thrusters, coupled with the passive stability inherent in a two-hulled vehicle with a large metacentric height, allows *SeaBED* to survey close to the seafloor, even in very rugged terrain.

In the future, AUVs will play an important role in the development of long-term seafloor observatories.

Table 4 Examples of AUVs for deep-sea research and exploration (vehicles that can operate at depths ≥ 1000 m)

Vehicle	Operating organization	Maximum operating depth (m)
Dorado Class	MBARI, USA	6000
CR-01, CR-02	COMRA, China	6000
Sentry	WHOI, USA	6000
REMUS Class	WHOI, USA	6000
Autosub 6000	NOC, UK	6000
Autonomous Benthic Explorer	NDSF, WHOI, USA	5500
Explorer 5000	Research Centre Ocean Margins, Germany	5000
Jaguar/Puma	WHOI, USA	5000
Urashima (hybrid)	JAMSTEC, Japan	3500
Aster x	IFREMER, France	3000
Bluefin AUV	Alfred Wegener Institute, Germany	3000
Bluefin 21 AUV	SIO, USA	3000
Odyssey Class	MIT, USA	3000
SeaBED	WHOI, USA	2000
Autosub 3	National Oceanography Centre, UK	1600
Spray Gliders	WHOI, USA	1500
Seaglider	Univ. of Washington, USA	1000

Abbreviations as in **Tables 1–3**.

Figure 9 Examples of AUVs used in oceanographic research. (a) The *Spray Glider*, (b) *Urashima*, (c) *Autosub*, (d) *SeaBED*, (e) *Dorado Class*, and (f) *ABE*. (a) Photo by Jane Dunworth-Baker, WHOI, USA. (b) © JAMSTEC, Japan, with permission. (c) Courtesy of Gwyn Griffiths, National Oceanography Centre, Southampton, UK. (d) Photo by Tom Kleindinst, WHOI, USA. (e) Photo by Todd Walsh © 2004, MBARI, USA, with permission. (f) Photo by Dan Fornari, WHOI, USA.

Apart from providing the high-resolution maps needed to optimally place geological, chemical, and biological sensors as part of an observatory, AUVs will also operate in a rapid response mode. It is envisaged that deep-sea observatories will include

docking stations for AUVs, and there are a number of research groups currently working on developing this technology. When an event – most likely a seismic event – is detected, scientists on shore will be able to program the AUV, via satellite and a cable to a surface buoy, to leave its dock and conduct surveys in the vicinity of the event. The AUV will then return to its dock and return the data to shore for assessment by scientists as to whether further investigation with ships is warranted.

Navigating Deep-Sea Vehicles

Unlike glider and drifter AUVs that can come to the sea surface and determine their positions using a Global Positioning System (GPS), deep-sea vehicles working at the bottom of the ocean have no such reference system because the GPS system's radio frequency signals are blocked by seawater. The technique that has been the standard for three-dimensional acoustic navigation of deep-sea vehicles is long-baseline (LBL) navigation – a technique developed more than 30 years ago. LBL operates on the principle that the distance between an underwater vehicle and a fixed acoustic transponder can be related precisely to the time of flight of an acoustic signal propagating between the vehicle and transponder. Two or more acoustic transponders are dropped over the side of the surface ship and anchored at locations selected to optimize the acoustic range and geometry of planned seafloor operations. Each transponder is a complete subsurface mooring comprised of an anchor, a tether, and a buoyant battery-powered acoustic transponder. The positions of the transponders on the seafloor are determined by using the GPS on board the ship and ranging to them acoustically while the ship circles the point where each transponder was dropped. The positions of the transponders on the seafloor can be determined this way with an accuracy of about 10 m.

Transponders have accurate clocks to measure time very precisely, and they are synchronized with the clocks on the vehicle and on the ship. Each transponder is set to listen for acoustic signals (or pings) transmitted either from the deep-sea vehicle or the ship at a specific frequency. When each transponder hears these acoustic signals, it is programmed to transmit an acoustic signal back to the vehicle and the ship. Each transponder pings at a different frequency, so the ship and the vehicle can discern which transponder sent it. The time of flight of the acoustic signals gives a measure of distance to each transponder, and using simple triangulation, the unique point in three-dimensional space where all distances measured from all the transponders and the ship intersect can be calculated. More recently, conventional LBL navigation has been combined with Doppler navigation data, which measures apparent bottom velocity of the vehicle, for better short-term accuracy.

The Future

The technological breakthroughs in deep-sea vehicle design over the last 40 years have resulted in unprecedented access to the deep ocean. While each type of vehicle has its own advantages and disadvantages, the complementary capacities of all types of deep-submergence vehicles provide synergies that are revolutionizing how scientists conduct research in the deep ocean. They are learning how to exploit those synergies by using a nested survey strategy that employs a combination of tools in sequence for investigations at increasingly finer scales: ship-based swath-mapping systems and towed vehicle systems for reconnaissance over large areas to identify features of interest, followed by more detailed, high-resolution mapping, imagery, and chemical sensing with AUVs, and finally, seafloor observations and experimentation using HOVs and ROVs. A demonstration of the power of such an approach occurred on a cruise to the Galápagos Rift in 2002. The investigative strategy was directed toward ensuring that all potential sites of hydrothermal venting in the rift valley were identified and investigated visually with the HOV *Alvin*. The AUV *ABE* was deployed at night to conduct high-resolution mapping of the seafloor and collect conductivity–temperature–depth (CTD) data in the lower water column to detect sites of venting. Upon its recovery in the morning, micro-bathymetry maps and temperature anomaly maps were quickly generated, compiled with previous data, and then given to the scientists diving in *Alvin* that day for their use in directing the dive. Today, the vehicles are being deployed in various combinations to attack a range of multidisciplinary problems.

Deep-sea vehicles will also play indispensable roles in establishing and servicing long-term seafloor observatories that will be critical for time-series investigations to understand the dynamic processes going on beneath the ocean. AUVs will undertake a variety of mapping and sampling missions while using fixed observatory installations to recharge batteries, offload data, and receive new instructions. They will be used to extend the spatial observational capability of seafloor observatories through surveying activities, and will document horizontal variability in seafloor and water column properties – necessary for establishing the context of point measurements made by fixed instrumentation. HOVs and ROVs will be required to install, service, and repair equipment and instrumentation on the seafloor and in drill holes, as well as collect samples as part of time-series measurements. The additional capabilities that these vehicles will need for service and repair activities will likely build on ROV tools that are currently being used in the commercial undersea cable industry.

Deep-sea vehicles will clearly have a role to play in deep-sea research for the foreseeable future, and they will be at the vanguard of a new era of ocean exploration.

See also

Gliders. Platforms: Autonomous Underwater Vehicles.

Further Reading

Bachmayer R, Humphris S, Fornari D, et al. (1998) Oceanographic exploration of hydrothermal vent sites on the Mid-Atlantic Ridge at 37°N 32°W using remotely operated vehicles. *Marine Technology Society Journal* 32: 37–47.

Davis RE, Eriksen CE, and Jones CP (2002) Autonomous buoyancy-driven underwater gliders. In: Griffiths G (ed.) *The Technology and Applications of Autonomous Underwater Vehicles*, pp. 37–58. London: Taylor and Francis.

De Moustier C, Spiess FN, Jabson D, et al. (2000) Deep-sea borehole re-entry with fiber optic wireline technology. *Proceedings of the 2000 International Symposium on Underwater Technology*, Tokyo, 23–26 May 2000, pp. 379–384.

Fornari D (2004) Realizing the dreams of da Vinci and Verne. *Oceanus* 42: 20–24.

Fornari DJ, Humphris SE, and Perfit MR (1997) Deep submergence science takes a new approach. *EOS, Transactions of the American Geophysical Union* 78: 402–408.

Fryer P, Fornari DJ, Perfit M, et al. (2002) Being there: The continuing need for human presence in the deep ocean for scientific research and discovery. *EOS, Transactions of the American Geophysical Union* 83(526): 532–533.

Funnell C (2004) *Jane's Underwater Technology 2004–2005*, 800pp, 23rd edn. Alexandria, VA: Jane's Information Group.

National Research Council (2004) *Exploration of the Seas: Voyage into the Unknown*. Washington, DC: National Academies Press.

National Research Council (2004) *Future Needs of Deep Submergence Science*. Washington, DC: National Academies Press.

Reves-Sohn R (2004) Unique vehicles for a unique environment. *Oceanus* 42: 25–27.

Rona P (2001) Deep-diving manned research submersibles. *Marine Technology Society Journal* 33: 13–25.

Rudnick DL, Davis RE, Eriksen CC, Fratantoni DM, and Perry MJ (2004) Underwater gliders for ocean research. *Marine Technology Society Journal* 38: 48–59.

Shank T, Fornari D, Yoerger D, et al. (2003) Deep submergence synergy: *Alvin* and *ABE* explore the Galápagos Rift at 86°W. *EOS, Transactions of the American Geophysical Union* 84(425): 432–433.

Yoerger D, Bradley AM, Walden BB, Singh H, and Bachmayer R (1998) Surveying a subsea lava flow using the *Autonomous Benthic Explorer* (*ABE*). *International Journal of Systems Science* 29: 1031–1044.

Relevant Websites

http://auvlab.mit.edu
– AUV Lab Vehicles, AUV Lab at MIT Sea Grant.

http://www.ropos.com
– Canadian Scientific Submersible Facility.

http://www.comra.org
– China Ocean Mineral Resources R&D Association.

http://divediscover.whoi.edu
– Dive and Discover: Expeditions to the Seafloor.

http://www.soest.hawaii.edu
– Hawai'i Undersea Research Laboratory (HURL), School of Ocean and Earth Science and Technology.

http://www.ifremer.fr
– IFREMER Fleet.

http://www.mbari.org
– Marine Operations: Vessels and Vehicles, Monterey Bay Aquarium Research Institute.

http://www.mpl.ucsd.edu
– Marine Physical Laboratory, Scripps Institution of Oceanography.

http://www.noc.soton.ac.uk
– National Oceanography Centre, Southampton.

http://www.jamstec.go.jp
– Research Vessels, Facilities, and Equipment, JAMSTEC.

http://www.apl.washington.edu
– *Seaglider*, Applied Physics Laboratory, University of Washington.

http://www.whoi.edu
– Ships and Technology: National Deep Submergence Facility, Woods Hole Oceanographic Institution.

http://www.rcom.marum.de
– Technology page, MARUM.

SENSORS: METEOROLOGY

SENSORS FOR MEAN METEOROLOGY

K. B. Katsaros, Atlantic Oceanographic and
Meteorological Laboratory, NOAA, Miami, FL, USA

Introduction

Basic mean meteorological variables include the following: pressure, wind speed and direction, temperature, and humidity. These are measured at all surface stations over land and from ships and buoys at sea. Radiation (broadband solar and infrared) is also often measured, and sea state, swell, wind sea, cloud cover and type, and precipitation and its intensity and type are evaluated by an observer over the ocean. Sea surface temperature and wave height (possibly also frequency and direction of wave trains) may be measured from a buoy at sea; they are part of the set of parameters required for evaluating net surface energy flux and momentum transfer. Instruments for measuring the quantities described here have been limited to the most common and basic. Precipitation is an important meteorological variable that is measured routinely over land with rain gauges, but its direct measurement at sea is difficult because of ship motion and wind deflection by ships' superstructure and consequently it has been measured routinely over the ocean only from ferry boats. However, it can be estimated at sea by satellite techniques, as can surface wind and sea surface temperature. Satellite methods are included in this article, since they are increasing in importance and provide the only means for obtaining complete global coverage.

Pressure

Several types of aneroid barometers are in use. They depend on the compression or expansion of an evacuated metal chamber for the relative change in atmospheric pressure. Such devices must be compensated for the change in expansion coefficient of the metal material of the chamber with temperature, and the device has to be calibrated for absolute values against a classical mercury in glass barometer, whose vertical mercury column balances the weight of the atmospheric column acting on a reservoir of mercury. The principle of the mercury barometer was developed by Evangilista Toricelli in the 17th century, and numerous sophisticated details were worked out over a period of two centuries. With modern manufacturing techniques, the aneroid has become standardized and is the commonly used device, calibrated with transfer standards back to the classical method. The fact that it takes a column of about 760 mm of the heavy liquid metal mercury (13.6 times as dense as water) illustrates the substantial weight of the atmosphere. Corrections for the thermal expansion or contraction of the mercury column must be made, so a thermometer is always attached to the device. Note that the word 'weight' is used, which implies that the value of the earth's gravitational force enters the formula for converting the mercury column's height to a pressure (force/unit area). Since gravity varies with latitude and altitude, mercury barometers must be corrected for the local value of the acceleration due to gravity.

Atmospheric pressure decreases with altitude. The balancing column of mercury decreases or the expansion of the aneroid chamber increases as the column of air above the barometer has less weight at higher elevations. Conversely, pressure sensors can therefore be used to measure or infer altitude, but must be corrected for the variation in the atmospheric surface pressure, which varies by as much as 10% of the mean (even more in case of the central pressure in a hurricane). An aneroid barometer is the transducer in aircraft altimeters.

Wind Speed and Direction

Wind speed is obtained by two basic means, both depending on the force of the wind to make an object rotate. This object comprises either a three- or four-cup anemometer, half-spheres mounted to horizontal axes attached to a vertical shaft (**Figure 1A**). The cups catch the wind and make the shaft rotate. In today's instruments rotations are counted by the frequency of the interception of a light source to produce a digital signal.

Propeller anemometers have three or four blades that are turned by horizontal wind (**Figure 1B**). The propeller anemometer must be mounted on a wind vane that keeps the propeller facing into the wind. For propeller anemometers, the rotating horizontal shaft is inserted into a coil. The motion of the shaft generates an electrical current or a voltage difference that can be measured directly. The signal is large enough that no amplifiers are needed.

Both cup and propeller anemometers, as well as vanes, have a threshold velocity below which they do

Figure 1 (A) Cup anemometer and vane; (B) propeller vane assembly; (C) three-dimensional sonic anemometer; (D) radiation shield for temperature and humidity sensors. (Photographs of these examples of common instruments were provided courtesy of R.M. Young Company.)

not turn and measure the wind. For the propeller anemometer, the response of the vane is also crucial, for the propeller does not measure wind speed off-axis very well. These devices are calibrated in wind tunnels, where a standard sensor evaluates the speed

in the tunnel. Calibration sensors can be fine cup anemometers or pitot tubes.

The wind direction is obtained from the position of a wind vane (a vertical square, triangle, or otherwise shaped wind-catcher attached to a horizontal shaft, **Figure 1A** and **B**). The position of a sliding contact along an electrical resistance coil moved by the motion of the shaft gives the wind direction relative to the zero position of the coil. The position is typically a fraction of the full circle (minus a small gap) and must be calibrated with a compass for absolute direction with respect to the Earth's north.

Other devices such as sonic anemometers can determine both speed and direction by measuring the modification of the travel time of short sound pulses between an emitter and a receiver caused by the three-dimensional wind. They often have three sound paths to allow evaluation of the three components of the wind (**Figure 1C**). These devices have recently become rugged enough to be used to measure mean winds routinely, and have a high enough frequency response to also determine the turbulent fluctuations. The obvious advantage is that the instrument has no moving parts. Water on the sound transmitter or receiver causes temporary difficulties, so a sonic anemometer is not an all-weather instrument. The sound paths can be at arbitrary angles to each other and to the natural vertical. Processing of the data transforms the measurements into an Earth-based coordinate system. The assumptions of zero mean vertical velocity and zero mean cross-wind velocity allow the relative orientation between the instrument axes and the Earth-based coordinate system to be found. Difficulties arise if the instrument is experiencing a steady vertical velocity at its location due to flow distortion around the measuring platform, for instance.

Cup and propeller anemometers are relatively insensitive to rain. However, snow and frost are problematic to all wind sensors, particularly the ones described above with moving parts. Salt contamination over the ocean also causes deterioration of the bearings in cup and propeller anemometers. Proper exposure of wind sensors on ships is problematic because of severe flow distortion by increasingly large ships. One solution has been to have duplicate sensors on port and starboard sides of the ship and selecting the valid one on the basis of the recording of the ship's heading and the relative wind direction.

Temperature

The measurements of both air and water temperature will be considered here, since both are important in air–sea interaction. Two important considerations for measuring temperature are the exposure of the sensor and shielding from solar radiation. The axiom that a 'thermometer measures its own temperature' is a good reminder. For the thermometer to represent the temperature of the air, it must be well ventilated, which is sometimes assured by a protective housing and a fan pulling air past the sensor. Shielding from direct sunlight has been done traditionally over land and island stations by the use of a 'Stephenson screen,' a wooden-roofed box with slats used for the sides, providing ample room for air to enter. Modern devices have individual housings based on the same principles (**Figure 1D**).

The classic measurements of temperature were done with mercury in glass or alcohol in glass thermometers. For sea temperature, such a thermometer was placed in a canvas bucket of water hauled up on deck. Today, electronic systems have replaced most of the glass thermometers. **Table 1** lists some of these sensors (for details see the Further Reading section).

The sea surface temperature (SST) is an important aspect of air–sea interaction. It enters into bulk formulas for estimating sensible heat flux and evaporation. The temperature differences between the air at one height and the SST is also important for determining the atmospheric stratification, which can modify the turbulent fluxes substantially compared with neutral stratification.

The common measure of SST is the temperature within the top 1 or 2 m of the interface, obtained with any of the contact temperature sensors described in **Table 1**. On ships, the sensor is typically placed in the ship's water intake, and on buoys it may even be placed just inside the hull on the bottom, shaded side of the buoy. Because the heat losses to the air occur at the air–sea interface, while solar heating penetrates of the order of tens of meters (depth depending on sun angle), a cool skin, 1–2 mm in depth and 0.1–0.5 °C cooler than the lower layers, is often present just below the interface. Radiation thermometers are sometimes used from ships or piers to measure the skin temperature directly (*see* Radiative Transfer in the Ocean).

Humidity

The Classical Sling Psychrometer

An ingenious method for evaluating the air's ability to take up water (its deficit in humidity with respect to the saturation value, see 68 on Evaporation and Humidity) is the psychrometric method. Two thermometers (of any kind) are mounted side by side, and one is provided with a cotton covering (a wick) that is wetted with distilled water. The sling psychrometer (**Figure 2**) is vigorously ventilated by swinging it in the air. The air passing over the sensors changes their temperatures to be in equilibrium with the air; the dry bulb measures the actual air

Table 1 Electronic devices for measuring temperature in air or water

Name	Principle	Typical use
Thermocouple	Thermoelectric junctions between two wires (e.g. Copper-Constantan) set up a voltage in the circuit, if the junctions are at different temperatures. The reference junction temperature must be measured as well	Good for measuring differences of temperature
Resistance thermometer	$R = R_{Ref}(1 + \alpha^T)$ Where R is the electrical resistance, R_{Ref} is resistance at a reference temperature, and α is the temperature coefficient of resistance	Platinum resistance thermometers are used for calibration and as reference thermometers
Thermistor	$R = a \exp (b/T)$ Where R is resistance, T is absolute temperature, and a and b are constants	Commonly used in routine sensor systems
Radiation thermometer	Infrared radiance in the atmospheric window, 8–12 μm, is a measure of the equivalent black body temperature	Usually used for measuring water's skin temperature

temperature, the wet bulb adjusts to a temperature that is intermediate between the dew point and air temperature. As water from the wick is evaporated, it takes heat out of the air passing over the wick until an equilibrium is reached between the heat supplied to the wet bulb by the air and the heat lost due to evaporation of water from the wick. This is the wet bulb temperature. The *Smithsonian Tables* provide the dew point temperature (and equivalent saturation humidity) corresponding to the measured 'wet bulb temperature depression,' i.e. the temperature difference between the dry bulb and the wet bulb thermometers at the existing air temperature.

Resistance Thermometer Psychrometer

A resistance thermometer psychrometer consists of stainless steel-encased platinum resistance thermometers housed in ventilated cylindrical shields. Ventilation can be simply due to the natural wind (in which case errors at low wind speeds may develop), or be provided by a motor and a fan (typically an air speed of 3 m s^{-1} is required). A water reservoir must be provided to ensure continuous wetting of the wet bulb. The reservoir should be mounted below the psychrometer so that water is drawn onto the wet bulb with a long wick. (This arrangement assures that the water has had time to equilibrate to the wet bulb temperature of the air.)

If these large wet bulbs collect salt on them over time, the relative humidity may be in error. This is not a concern for short-term measurements. A salt solution of 3.6% on the wet bulb would result in an overestimate of the relative humidity of approximately 2%.

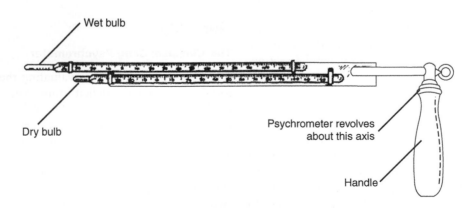

Figure 2 Sling pyschrometer. (Reproduced with permission from Parker, 1977.)

Capacitance Sensors of Humidity

The synoptic weather stations often use hygrometers based on the principle of capacitance change as the small transducer absorbs and desorbs water vapor. To avoid contamination of the detector, special filters cover the sensor. Dirty filters (salt or other contaminants) may completely mask the atmospheric effects. Even the oil from the touch of a human hand is detrimental. Two well-known sensors go under the names of Rotronic and Humicap. Calibration with mercury in glass psychrometers is useful.

Exposure to Salt

As for wind and temperature devices, the humidity sensors are sensitive to flow distortion around ships and buoys. Humidity sensors have an additional problem in that salt crystals left behind by evaporating spray droplets, being hygroscopic, can modify the measurements by increasing the local humidity around them. One sophisticated, elegant, and expensive device that has been used at sea without success is the dew point hygrometer. It depends on the cyclical cooling and heating of a mirror. The cooling continues until dew forms, which is detected by changes in reflection of a light source off the mirror, and the temperature at that point is by definition the dew point temperature. The problem with this device is that during the heating cycle sea salt is baked onto the mirror and cannot be removed by cleaning.

Several attempts to build devices that remove the spray have been tried. Regular Stephenson screen-type shields provide protection for some time, the length of which depends both on the generation of spray in the area, the height of the measurement, and the size of the transducer (i.e. the fraction of the surface area that may be contaminated). One of the protective devices that was successfully used in the Humidity Exchange over the Sea (HEXOS) experiment is the so-called 'spray flinger.'

Spray-Removal Device

The University of Washington 'spray flinger' (**Figure 3**) was designed to minimize flow modification on scales important to the eddy correlation calculations of evaporation and sensible heat flux employing data from temperature and humidity sensors inside the housing. The design aims to ensure that the droplets removed from the airstream do not remain on the walls of the housing or filter where they could evaporate and affect the measurements. The device has been tested to ensure that there are no thermal effects due to heating of the enclosure, but this would be dependent on the meteorological conditions encountered, principally insolation. The housing should be directed upwind.

Although there is a slow draw of air through the unit by the upwind and exit fans (1–2 m s^{-1}), it is mainly a passive device with respect to the airflow. Inside the tube, wet and dry thermocouples or other temperature and humidity sensors sample the air for mean and fluctuating temperature and humidity. Wind tunnel and field tests showed the airflow inside the unit to be steady and about one-half the ambient wind speed for wind directions <40° off the axis. Even in low wind speeds there is adequate ventilation

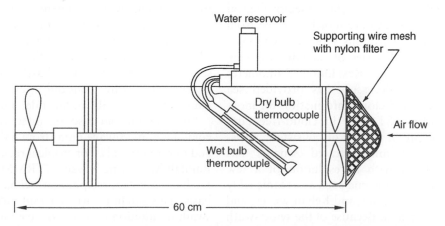

Figure 3 Sketch of aspirated protective housing, the 'spray flinger', used for the protection of a thermocouple psychrometer by the University of Washington group. The system is manually directed upwind. The spray flinger is a 60 cm long tube, 10 cm in diameter, with a rotating filter screen and fan on the upwind end, and an exit fan and the motor at the downwind end. The filter is a single layer of nylon stocking, which is highly nonabsorbent, supported by a wire mesh. Particles and droplets are intercepted by the rotating filter and flung aside, out of the airstream entering the tube. The rotation rate of the filter is about 625 rpm. Inspection of the filter revealed that this rate of rotation prevented build-up of water or salt. The nylon filter needs to be replaced at least at weekly intervals. (Reproduced with permission from Katsaros et al., 1994.)

for the wet bulb sensor. Comparison between data from shielded and unshielded thermocouples (respectively, inside and outside the spray flinger) show that the measurements inside are not noticeably affected by the housing.

A quantitative test of the effectiveness of the spray flinger in removing aerosols from the sample airstream was performed during HEXMAX (the HEXOS Main Experiment) using an optical particle counter to measure the aerosol content with diameters between 0.5 and 32 μm in the environmental air and at the rear of the spray flinger. Other devices have been constructed, but have had various difficulties, and the 'spray flinger' is not the final answer. Intake tubes that protect sensors have also been designed for use on aircraft.

Satellite Measurements

With the global ocean covering 70% of the earth's surface, large oceanic areas cannot be sampled by *in situ* sensors. Most of the meteorological measurements are taken by Voluntary Observing Ships (VOS) of the merchant marine and are, therefore, confined to shipping lanes. Research vessels and military ships may be found in other areas and have contributed substantially to our knowledge of conditions in areas not visited by VOS. The VOS report their observations on a 3 hour or 6 hour schedule. Some mean meteorological quantities such as SST and wind speed and direction are observable by satellites directly, while others can be inferred from less directly related measurements. Surface insolation and precipitation depend on more complex algorithms for evaluation. Satellite-derived surface meteorological information over the ocean are mostly derived from polar-orbiting, sun-synchronous satellites. The famous TIROS and NOAA series of satellites carrying the Advanced Very High Resolution Radiometer (AVHRR) and its predecessors has provided sea surface temperature and cloud information for more than three decades (SST only in cloud-free conditions). This long-term record of consistent measurements by visible and infrared sensors has provided great detail with a resolution of a few kilometers of many phenomena such as oceanic eddy formation, equatorial Rossby and Kelvin waves, and the El Niño phenomenon. Because of the wide swath of these short wavelength devices, of the order of 2000 km, the whole earth is viewed daily by either the ascending or descending pass of the satellite overhead, once in daytime and once at night.

Another mean meteorological variable observable from space is surface wind speed, with microwave radiometers and the wind vector from scatterometers, active microwave instruments. Both passive and active sensors depend on the changing roughness of the sea as a function of wind speed for their ability to 'sense' the wind. The first scatterometer was launched on the Seasat satellite in 1978, operating for 3 months only. The longest record is from the European Remote Sensing (ERS) satellites 1 and 2 beginning in 1991 and continuing to function well in 2000. Development of interpretation of the radar returns in terms of both speed and direction depends on the antennae viewing the same ocean area several times at different incidence angles relative to the wind direction. A recently launched satellite (QuikSCAT in 1999) carries a new design with a wider swath, the SeaWinds instrument. Scatterometers are providing surface wind measurements with accuracy of $\pm 1.6 \, \text{m s}^{-1}$ approximately in speed and $\pm 20°$ in direction at 50 km resolution for ERS and 25 km for SeaWinds. They view all of the global ocean once in 3 days for ERS and in approximately 2 days for QuikSCAT. Microwave radiometers such as the Special Sensor Microwave/Imager (SSM/I), operational since 1987 on satellites in the US Defense Meteorological Satellite Program, have wider swaths covering the globe daily, but they are not able to sense the ocean surface in heavy cloud or rainfall areas and do not give direction. They can be assimilated into numerical models where the models provide an initial guess of the wind fields, which are modified to be consistent with the details of the radiometer-derived wind speeds.

Surface pressure and atmospheric surface air temperature are not yet amenable to satellite observations, but surface humidity can be inferred from total column water content. From the satellite-observed cloudiness, solar radiation at the surface can be inferred by use of radiative transfer models. This is best done from geostationary satellites whose sensors sweep across the Earth's surface every 3 hours or more often, but only view a circle of useful data extending $\pm 50°$ in latitude, approximately.

Precipitation can also be inferred from satellites combining microwave data (from SSM/I) with visible and infrared signals. For tropical regions, the Tropical Rainfall Measuring Mission (TRMM) on a low-orbit satellite provides precipitation estimates on a monthly basis. This satellite carries a rain radar with 500 km swath in addition to a microwave radiometer.

Developments of multispectral sensors and continued work on algorithms promises to improve the accuracy of the satellite information on air–sea interaction variables. Most satellite programs depend on the simple *in situ* mean meteorological measurements described above for calibration and validation.

A good example is the important SST record provided by the US National Weather Service and used by all weather services. The analysis procedure employs surface data on SST from buoys, particularly small, inexpensive, free-drifting buoys that are spread over the global oceans to 'tie-down' the correction for atmospheric interference for the satellite estimates of SST. The satellite-observed infrared radiances are modified by the transmission path from the sea to the satellite, where the unknown is the aerosol that can severely affect the interpretation. The aerosol signal is not directly observable yet by satellite, so the surface-measured SST data serve an important calibration function.

Future Developments

New measurement programs are being developed by international groups to support synoptic definition of the ocean's state similarly to meteorological measurements and to provide forecasts. The program goes under the name of the Global Ocean Observing System (GOOS). It includes new autonomous buoys cycling in the vertical to provide details below the interface, a large surface drifter component, and the VOS program, as well as certain satellite sensors. The GOOS is being developed to support a modeling effort, the Global Ocean Data Assimilation Experiment (GODAE), which is an experiment in forecasting the oceanic circulation using numerical models with assimilation of the GOOS data.

See also

Sensors for Micrometeorological and Flux Measurements.

Further Reading

Atlas RS, Hoffman RN, Bloom SC, Jusem JC, and Ardizzone J (1996) A multiyear global surface wind velocity data set using SSM/I wind observations. *Bulletin of the American Meteorological Society* 77: 869–882.

Bentamy A, Queffeulou P, Quilfen Y, and Katsaros KB (1999) Ocean surface wind fields estimated from satellite active and passive microwave instruments. *IEEE Transactions Geosci Remote Sens* 37: 2469–2486.

de Leeuw G (1990) Profiling of aerosol concentrations, particle size distributions, and relative humidity in the atmospheric surface layer over the North Sea. *Tellus* 42B: 342–354.

Dobson F, Hasse L, and Davies R (eds.) (1980) *Instruments and Methods in Air–Sea Interaction*, pp. 293–317. New York: Plenum.

Geernaert GL and Plant WJ (eds.) (1990) *Surface Waves and Fluxes*, 2, pp. 339–368. Dordrecht: Kluwer Academic Publishers.

Graf J, Sasaki C, *et al.* (1998) NASA Scatterometer Experiment. *Asta Astronautica* 43: 397–407.

Gruber A, Su X, Kanamitsu M, and Schemm J (2000) The comparison of two merged rain gauge-satellite precipitation datasets. *Bulletin of the American Meteorological Society* 81: 2631–2644.

Katsaros KB (1980) Radiative sensing of sea surface temperatures. In: Dobson F, Hasse L, and Davies R (eds.) *Instruments and methods in Air–Sea Interaction*, pp. 293–317. New York: Plenum Publishing Corp.

Katsaros KB, DeCosmo J, Lind RJ, *et al.* (1994) Measurements of humidity and temperature in the marine environment. *Journal of Atmospheric and Oceanic Technology* 11: 964–981.

Kummerow C, Barnes W, Kozu T, Shiue J, and Simpson J (1998) The tropical rainfall measuring mission (TRMM) sensor package. *Journal of Atmospheric and Oceanic Technology* 15: 809–817.

Liu WT (1990) Remote sensing of surface turbulence flux. In: Geenaert GL and Plant WJ (eds.) *Surface Waves and Fluxesyyy*, 2, pp. 293–309. Dordrecht: Kluwer Academic Publishers.

Parker SP (ed.) (1977) *Encyclopedia of Ocean and Atmospheric Science*. New York: McGraw-Hill.

Pinker RT and Laszlo I (1992) Modeling surface solar irradiance for satellite applications on a global scale. *Journal of Applied Meteorology* 31: 194–211.

Reynolds RR and Smith TM (1994) Improved global sea surface temperature analyses using optimum interpolation. *Journal of Climate* 7: 929–948.

List RJ (1958) *Smithsonian Meteorological Tables* 6th edn. City of Washington: Smithsonian Institution Press.

van der Meulen JP (1988) On the need of appropriate filter techniques to be considered using electrical humidity sensors. In: *Proceedings of the WMO Technical Conference on Instruments and Methods of Observation (TECO-1988)*, pp. 55–60. Leipzig, Germany: WMO.

Wentz FJ and Smith DK (1999) A model function for the ocean-normalized radar cross-section at 14 GHz derived from NSCAT observations. *Journal of Geophysical Research* 104: 11 499–11 514.

SENSORS FOR MICROMETEOROLOGICAL & FLUX MEASUREMENTS

J. B. Edson, Woods Hole Oceanographic Institution, Woods Hole, MA, USA

Introduction

The exchange of momentum, heat, and mass between the atmosphere and ocean is the fundamental physical process that defines air–sea interactions. This exchange drives ocean and atmospheric circulations, and generates surface waves and currents. Marine micrometeorologists are primarily concerned with the vertical exchange of these quantities, particularly the vertical transfer of momentum, heat, moisture, and trace gases associated with the momentum, sensible heat, latent heat, and gas fluxes, respectively. The term flux is defined as the amount of heat (i.e., thermal energy) or momentum transferred per unit area per unit time.

Air–sea interaction studies often investigate the dependence of the interfacial fluxes on the mean meteorological (e.g., wind speed, degree of stratification or convection) and surface conditions (e.g., surface currents, wave roughness, wave breaking, and sea surface temperature). Therefore, one of the goals of these investigations is to parametrize the fluxes in terms of these variables so that they can be incorporated in numerical models. Additionally, these parametrizations allow the fluxes to be indirectly estimated from observations that are easier to collect and/or offer wider spatial coverage. Examples include the use of mean meteorological measurements from buoys or surface roughness measurements from satellite-based scatterometers to estimate the fluxes.

Direct measurements of the momentum, heat, and moisture fluxes across the air–sea interface are crucial to improving our understanding of the coupled atmosphere–ocean system. However, the operating requirements of the sensors, combined with the often harsh conditions experienced over the ocean, make this a challenging task. This article begins with a description of desired measurements and the operating requirements of the sensors. These requirements involve adequate response time, reliability, and survivability. This is followed by a description of the sensors used to meet these requirements, which includes examples of some of the obstacles that marine researchers have had to overcome. These obstacles include impediments caused by environmental conditions and engineering challenges that are unique to the marine environment. The discussion is limited to the measurement of velocity, temperature, and humidity. The article concludes with a description of the state-of-the-art sensors currently used to measure the desired fluxes.

Flux Measurements

The exchange of momentum and energy a few meters above the ocean surface is dominated by turbulent processes. The turbulence is caused by the drag (i.e., friction) of the ocean on the overlying air, which slows down the wind as it nears the surface and generates wind shear. Over time, this causes faster-moving air aloft to be mixed down and slower-moving air to be mixed up; the net result is a downward flux of momentum. This type of turbulence is felt as intermittent gusts of wind that buffet an observer looking out over the ocean surface on a windy day.

Micrometeorologists typically think of these gusts as turbulent eddies in the airstream that are being advected past the observer by the mean wind. Using this concept, the turbulent fluctuations associated with these eddies can be defined as any departure from the mean wind speed over some averaging period (eqn [1]).

$$u(t) = U(t) - \bar{U} \qquad [1]$$

In eqn [1], $u(t)$ is the fluctuating (turbulent) component, $U(t)$ is the observed wind, and the overbar denotes the mean value over some averaging period. The fact that an observer can be buffeted by the wind indicates that these eddies have some momentum. Since the eddies can be thought to have a finite size, it is convenient to consider their momentum per unit volume, given by $\rho_a U(t)$, where ρ_a is the density of air. In order for there to be an exchange of momentum between the atmosphere and ocean, this horizontal momentum must be transferred downward by some vertical velocity. The mean vertical velocity associated with the turbulent flux is normally assumed to be zero. Therefore, the turbulent transfer of this momentum is almost exclusively via the turbulent vertical velocity, $w(t)$, which we associate with overturning air.

The correlation or covariance between the fluctuating vertical and horizontal wind components is the most direct estimate of the momentum flux. This approach is known as the eddy correlation or direct covariance method. Computation of the covariance involves multiplying the instantaneous vertical velocity fluctuations with one of the horizontal components. The average of this product is then computed over the averaging period.

Because of its dependence on the wind shear, the flux of momentum at the surface is also known as the shear stress defined by eqn [2], where $\hat{i}$ and $\hat{j}$ are unit vectors, and v is the fluctuating horizontal component that is orthogonal to u.

$$\tau_0 = \hat{i}\overline{uw} - \hat{j}\overline{vw} \qquad [2]$$

Typically, the coordinate system is rotated into the mean wind such that u, v, and w denote the longitudinal, lateral, and vertical velocity fluctuations, respectively. Representative time series of longitudinal and vertical velocity measurements taken in the marine boundary layer are shown in **Figure 1**. The velocities in this figure exhibit the general trend that downward-moving air (i.e., $w < 0$) is transporting eddies with higher momentum per unit mass (i.e., $\rho_a u > 0$) and vice versa. The overall correlation is therefore negative, which is indicative of a downward flux of momentum.

Close to the surface, the wave-induced momentum flux also becomes important. At the interface, the turbulent flux actually becomes negligible and the momentum is transferred via wave drag and viscous shear stress caused by molecular viscosity. The ocean is surprisingly smooth compared to most land surfaces. This is because the dominant roughness elements that cause the drag on the atmosphere are the wind waves shorter than 1 m in length. Although the longer waves and swell give the appearance of a very rough surface, the airflow tends to follow these waves and principally act to modulate the momentum flux supported by the small-scale roughness. Therefore, the wave drag is mainly a result of these small-scale roughness elements.

Turbulence can also be generated by heating and moistening the air in contact with the surface. This increases the buoyancy of the near-surface air, and causes it to rise, mix upward, and be replaced by less buoyant air from above. The motion generated by this convective process is driven by the surface buoyancy flux (eqn [3]).

$$B_0 = \rho_a c_p \overline{w\theta_v} \qquad [3]$$

In eqn [3] c_p is the specific heat of air at constant pressure and θ_v is the fluctuating component of the virtual potential temperature defined in eqn [4], where $\bar{\Theta}$ and θ are the mean and fluctuating components of the potential temperature, respectively, and q is the specific humidity (i.e., the mass of water vapor per unit mass of moist air).

$$\theta_v = \theta + 0.61\bar{\Theta}q \qquad [4]$$

These quantities also define the sensible heat (eqn [5]) and latent heat (eqn [6]) fluxes.

$$H_0 = \rho_a c_p \overline{w\theta} \qquad [5]$$

$$E_0 = L_e \overline{w\rho_a q} \qquad [6]$$

where L_e is the latent heat of evaporation. The parcels of air that are heated and moistened via the buoyancy flux can grow into eddies that span the entire atmospheric boundary layer. Therefore, even in light wind conditions with little mean wind shear, these turbulent eddies can effectively mix the marine boundary layer.

Conversely, when the air is warmer than the ocean, the flow of heat from the air to water (i.e., a downward buoyancy flux) results in a stably stratified boundary layer. The downward buoyancy flux is normally driven by a negative sensible heat flux. However, there have been observations of a downward latent heat (i.e., moisture) flux associated with the formation of fog and possibly condensation at the ocean surface. Vertical velocity fluctuations have

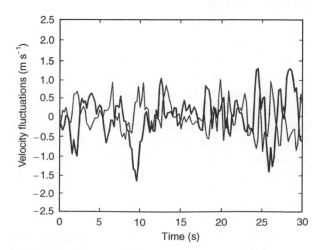

Figure 1 Time-series of the longitudinal (thick line) and vertical velocity (thin line) fluctuations measured from a stable platform. The mean wind speed during the sampling period was $10.8\,\mathrm{m\,s^{-1}}$.

to work to overcome the stratification since upward-moving eddies are trying to bring up denser air and vice versa. Therefore, stratified boundary layers tend to dampen the turbulent fluctuations and reduce the flux compared to their unstable counterpart under similar mean wind conditions. Over the ocean, the most highly stratified stable boundary layers are usually a result of warm air advection over cooler water. Slightly stable boundary conditions can also be driven by the diurnal cycle if there is sufficient radiative cooling of the sea surface at night.

Sensors

Measurement of the momentum, sensible heat, and latent heat fluxes requires a suite of sensors capable of measuring the velocity, temperature, and moisture fluctuations. Successful measurement of these fluxes requires instrumentation that is rugged enough to withstand the harsh marine environment and fast enough to measure the entire range of eddies that transport these quantities. Near the ocean surface, the size of the smallest eddies that can transport these quantities is roughly half the distance to the surface; i.e., the closer the sensors are deployed to the surface, the faster the required response. In addition, micrometeorologists generally rely on the wind to advect the eddies past their sensors. Therefore, the velocity of the wind relative to a fixed or moving sensor also determines the required response; i.e., the faster the relative wind, the faster the required response. For example, planes require faster response sensors than ships but require less averaging time to compute the fluxes because they sample the eddies more quickly.

The combination of these two requirements results in an upper bound for the required frequency response (eqn [7]).

$$\frac{fz}{U_r} \approx 2 \qquad [7]$$

Here f is the required frequency response, z is the height above the surface, and U_r is the relative velocity. As a result, sensors used on ships, buoys, and fixed platforms require a frequency response of approximately 10–20 Hz, otherwise some empirical correction must be applied. Sensors mounted on aircraft require roughly an order of magnitude faster response depending on the sampling speed of the aircraft.

The factors that degrade sensor performance in the marine atmosphere include contamination, corrosion, and destruction of sensors due to sea spray and salt water; and fatigue and failure caused by long-term operation that is accelerated on moving

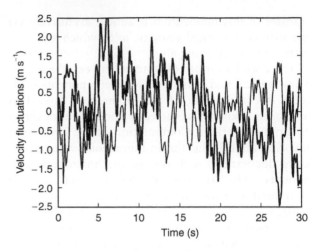

Figure 2   Time-series of the longitudinal (thick line) and vertical velocity (thin line) fluctuations measured from a 3 m discus buoy. The mean wind speed during the sampling period was 10.9 m s^{-1}. The measured fluctuations are a combination of turbulence and wave-induced motion of the buoy.

platforms. Additionally, if the platform is moving, the motion of the platform will be sensed by the instrument as an additional velocity and will contaminate the desired signal (**Figure 2**). Therefore, the platform motion must be removed to accurately measure the flux. This requires measurements of the linear and angular velocity of the platform. The alternative is to deploy the sensors on fixed platforms or to reduce the required motion correction by mounting the sensors on spar buoys, SWATH vessels, or other platforms that are engineered to reduced the wave-induced motion.

Shear Stress

The measurement of momentum flux or shear stress has a long history. The earliest efforts attempted to adapt many of the techniques commonly used in the laboratory to the marine boundary layer. A good example of this is the use of hot-wire anemometers that are well-suited to wind tunnel studies of turbulent flow. Hot-wire anemometry relies on very fine platinum wires that provide excellent frequency response and satisfy eqn [7] even close to the surface. The technique relies on the assumption that the cooling of heated wires is proportional to the flow past the wire. Hot-wire anemometers are most commonly used in constant-temperature mode. In this mode of operation, the current heating the wire is varied to maintain a constant temperature using a servo loop. The amount of current or power required to maintain the temperature is a measure of the cooling of the wires by the wind.

Unfortunately, there are a number of problems associated with the use of these sensors in the marine environment. The delicate nature of the wires (they are typically 10 μm in diameter) makes them very susceptible to breakage. Hot-film anemometers provide a more rugged instrument with somewhat slower, but still excellent, frequency response. Rather than strands of wires, a hot-film anemometer uses a thin film of nickel or platinum spread over a small cylindrical quartz or glass core. Even when these sensors are closely monitored for breakage, aging and corrosion of the wires and films due to sea spray and other contaminants cause the calibration to change over time. Dynamic calibration in the field has been used but this requires additional sensors. Therefore, substantially more rugged anemometers with absolute or more stable calibrations have generally replaced these sensors in field studies.

Another laboratory instrument that meets these requirements is the pitot tube; this uses two concentric tubes to measure the difference between the static pressure of the inner tube, which acts as a stagnation point, and the static pressure of the air flowing past the sensor. The free stream air also has a dynamic pressure component. Therefore, the difference between the two pressure measurements can be used to compute the dynamic pressure of the air flow moving past the sensor using Bernoulli's equation (eqn [8]).

$$\Delta p = \frac{1}{2}\rho_a \alpha U^2 \qquad [8]$$

Here α is a calibration coefficient that corrects for departures from Bernoulli's equations due to sensor geometry. A calibrated pitot tube can then be used to measure the velocity.

The traditional design is most commonly used to measure the streamwise velocity. However, three-axis pressure sphere (or cone) anemometers have been used to measure fluxes in the field. These devices use a number of pressure ports that are referenced against the stagnation pressure to measure all three components of the velocity. This type of anemometer has to be roughly aligned with the relative wind and its ports must remain clear of debris (e.g., sea spray and other particulates) to operate properly. Consequently, it has been most commonly used on research aircraft where the relative wind is large and particulate concentrations are generally lower outside of clouds and fog.

The thrust anemometer has also been used to directly measure the momentum flux in the marine atmosphere. This device measures the frictional drag of the air on a sphere or other objects. The most

successful design uses springs to attach a sphere and its supporting structure to a rigid mount. The springs allow the sphere to be deflected in both the horizontal and vertical directions. The deflection due to wind drag on the sphere is sensed by proximity sensors that measure the displacement relative to the rigid mount. Carefully calibrated thrust anemometers have been used to measure turbulence from fixed platforms for extended periods. They are fairly rugged and low-power, and have adequate response for estimation of the flux. The main disadvantages of these devices are the need to accurately calibrate the direction response of each sensor and sensor drift due to aging of the springs.

A very robust sensor for flux measurements, particularly for use on fixed platforms, relies on a modification of the standard propellor vane anemometer used to measure the mean wind. The modification involves the use of two propellers on supporting arms set at 90° to each other. The entire assembly is attached to a vane that keeps the propeller pointed into the wind. The device is known as a K-Gill anemometer from the appearance of the twin propeller-vane configuration (**Figure 3**). The twin propellers are capable of measuring the instantaneous vertical and streamwise velocity, and the vane reading allows the streamwise velocity to be broken down into its u and v components. This device is also very robust and low power. However, it has a complicated inertial response on moving

Figure 3 The instrument at the far right is K-Gill anemometer shown during a deployment on a research vessel. The instrument on the left is a sonic anemometer that is shown in more detail in **Figure 4**. Photograph provided by Olc Persson (CIRES/NOAA/ETL).

platforms and is therefore most appropriate for use on fixed platforms. Additionally, the separation between the propellers (typically 0.6) acts as a spatial filter (i.e., it cannot detect eddies smaller than the separation), so it cannot be used too close to the surface and still satisfy eqn [7]. This is generally not a problem at the measurement heights used in most field deployments.

Over the past decade, sonic anemometers have become the instrument of choice for most investigations of air–sea interaction. These anemometers use acoustic signals that are emitted in either a continuous or pulsed mode. At present, the pulse type sonic anemometers are most commonly used in marine research. Most commercially available devices use paired transducers that emit and detect acoustic pulses (**Figure 4**). One transducer emits the pulse and the other detects it to measure the time of flight between them. The functions are then reversed to measure the time of flight in the other direction. The basic concept is that in the absence of any wind the time of flight in either direction is the same.

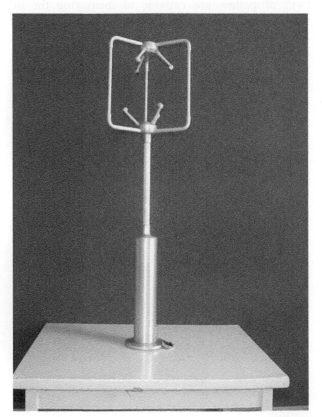

Figure 4 A commercially available pulse-type sonic anemometer. The three sets of paired transducers are cabable of measuring the three components of the velocity vector. This type of device produced the time-series in **Figure 1** and **Figure 2**. The sonic anemometer measures 0.75 m for top to bottom.

However, the times of flight differ if there is a component of the wind velocity along the path between the transducers. The velocity is directly computed from the two time of flight measurements, t_1 and t_2, using eqn [9], where L is the distance between the transducers.

$$U = \frac{L}{2}\left(\frac{1}{t_1} - \frac{1}{t_2}\right) \qquad [9]$$

Three pairs of transducers are typically used to measure all three components of the velocity vector. These devices have no moving parts and are therefore far less susceptible to mechanical failure. They can experience difficulties when rain or ice covers the transducer faces or when there is a sufficient volume of precipitation in the sampling volume. However, the current generation of sonic anemometers have proven themselves to be remarkably reliable in long-term deployments over the ocean; so much so that two-axis versions of sonic anemometers are also beginning to replace cup and propellor/vane anemometers for mean wind measurements over the ocean.

Motion Correction

The measurement of the fluctuating velocity components necessary to compute the fluxes is complicated by the platform motion on any aircraft, seagoing research vessel, or surface mooring. This motion contamination must be removed before the fluxes can be estimated. The contamination of the signal arises from three sources: instantaneous tilt of the anemometer due to the pitch, roll, and yaw (i.e., heading) variations; angular velocities at the anemometer due to rotation of the platform about its local coordinate system axes; and translational velocities of the platform with respect to a fixed frame of reference. Therefore, motion sensors capable of measuring these quantities are required to correct the measured velocities. Once measured, these variables are used to compute the true wind vector from eqn [10].

$$\mathbf{U} = T(\mathbf{U}_m + \mathbf{\Omega}_m \times \mathbf{R}) + \mathbf{U}_p \qquad [10]$$

Here $\mathbf{U}$ is the desired wind velocity vector in the desired reference coordinate system (e.g., relative to water or relative to earth); $\mathbf{U}_m$ and $\mathbf{\Omega}_m$ are the measured wind and platform angular velocity vectors respectively, in the platform frame of reference; T is the coordinate transformation matrix from the platform coordinate system to the reference coordinates; $\mathbf{R}$ is the position vector of the wind sensor with respect to the motion sensors; and U_p is the translational velocity vector of the platform measured at the location of the motion sensors.

A variety of approaches have been used to correct wind sensors for platform motion. True inertial navigation systems are standard for research aircraft. These systems are expensive, so simpler techniques have been sought for ships and buoys, where the mean vertical velocity of the platform is unambiguously zero. These techniques generally use the motion measurements from either strapped-down or gyro-stabilized systems.

The strapped-down systems typically rely on a system of three orthogonal angular rate sensors and accelerometers, which are combined with a compass to get absolute direction. The high-frequency component of the pitch, roll, and yaw angles required for the transformation matrix are computed by integrating and highpass filtering the angular rates. The low-frequency component is obtained from the lowpass accelerometer signals or, more recently, the angles computed from differential GPS. The transformed accelerometers are integrated and highpass filtered before they are added to lowpass filtered GPS or current meter velocities for computation of U_p relative to earth or the sea surface, respectively. The gyro-stabilized system directly computes the orientation angles of the platform. The angular rates are then computed from the time-derivative of the orientation angles.

Heat Fluxes

The measurement of temperature fluctuations over the ocean surface has a similar history to that of the velocity measurements. Laboratory sensors such as thermocouples, thermistors, and resistance wires are used to measure temperature fluctuations in the marine environment.

Thermocouples rely on the Seebeck effect that arises when two dissimilar materials are joined to form two junctions: a measuring junction and a reference junction. If the temperature of the two junctions is different, then a voltage potential difference exists that is proportional to the temperature difference. Therefore, if the temperature of the reference junction is known, then the absolute temperature at the junction can be determined. Certain combinations of materials exhibit a larger effect (e.g., copper and constantan) and are thus commonly used in thermocouple design. However, in all cases the voltage generated by the thermoelectric effect is small and amplifiers are often used along with the probes.

Thermistors and resistance wires are devices whose resistance changes with temperature. Thermistors are semiconductors that generally exhibit a large negative change of resistance with temperature (i.e., they have a large negative temperature coefficient of resistivity). They come in a variety of different forms including beads, rods, or disks. Microbead thermistors are most commonly used in turbulence studies; in these the semiconductor is situated in a very fine bead of glass. Resistance wires are typically made of platinum, which has a very stable and well-known temperature–resistance relationship. The trade-off is that they are less sensitive to temperature change than thermistors. The probe supports for these wires are often similar in design to hot-wire anemometers, and they are often referred to as cold-wires.

All of these sensors can be deployed on very fine mounts (**Figure 5**), which greatly reduces the adverse effects of solar heating but also exposes them to harsh environments and frequent breaking. Additionally, the exposure invariably causes them to become covered with salt from sea spray. The coating of salt causes spurious temperature fluctuations due to condensation and evaporation of water vapor on these hygroscopic particles. These considerations generally require more substantial mounts and some

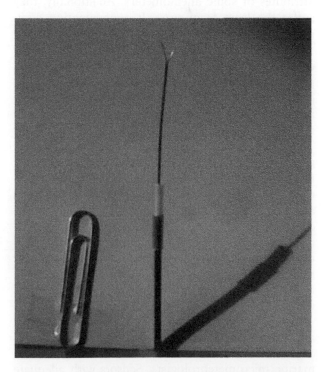

Figure 5 A thermocouple showing the very fine mounts used for turbulence applications. The actual thermocouple is situated on fine wires between the probe supports and is too small to be seen in this photograph.

sort of shielding from the radiation and spray. While this is acceptable for mean temperature measurements, the reduction in frequency response caused by the shields often precludes their use for turbulence measurements.

To combat these problems, marine micrometeorologists have increasingly turned to sonic thermometry. The time of flight measurements from sonic anemometers can be used to measure the speed of sound c along the acoustic path (eqn [11]).

$$c = \frac{L}{2}\left(\frac{1}{t_1} + \frac{1}{t_2}\right)$$ [11]

The speed of sound is a function of temperature and humidity and can be used to compute the sonic temperature T_s defined by eqn [12], where U_N is velocity component normal to the transducer path.

$$T_s = T(1 + 0.51q) = \frac{c^2 + U_N^2}{403}$$ [12]

The normal wind term corrects for lengthening of the acoustic path by this component of the wind. This form of velocity crosstalk has a negligible effect on the actual velocity measurements, but has a measurable effect on the sonic temperature.

Sonic thermometers share many of the positive attributes of sonic anemometers. Additionally, they suffer the least from sea salt contamination compared to other fast-response temperature sensors. The disadvantage of these devices is that velocity crosstalk must be corrected for and they do not provide the true temperature signal as shown by eqn [12]. Fortunately, this can be advantageous in many investigations because the sonic temperature closely approximates the virtual temperature in moist air $T_v = T(1 + 0.61q)$. For example, in many investigations over the ocean an estimate of the buoyancy flux is sufficient to account for stability effects. In these investigations the difference between the sonic and virtual temperature is often neglected (or a small correction is applied), and the sonic anemometer/thermometer is all that is required. However, due to the importance of the latent heat flux in the total heat budget over the ocean, accurate measurement of the moisture flux is often a crucial component of air–sea interaction investigations.

The accurate measurement of moisture fluctuations required to compute the latent heat flux is arguably the main instrumental challenge facing marine micrometeorologists. Sensors with adequate frequency response generally rely on the ability of water vapor in air to strongly absorb certain wavelengths of radiation. Therefore, these devices require a narrowband source for the radiation and a detector to measure the reduced transmission of that radiation over a known distance.

Early hygrometers of this type generated and detected ultraviolet radiation. The Lyman α hygrometer uses a source tube that generates radiation at the Lyman α line of atomic hydrogen which is strongly absorbed by water vapor. This device has excellent response characteristics when operating properly. Unfortunately, it has proven to be difficult to operate in the field due to sensor drift and contamination of the special optical windows used with the source and detector tubes. A similar hygrometer that uses krypton as its source has also been used in the field. Although the light emitted by the krypton source is not as sensitive to water vapor, the device still has more than adequate response characteristics and generally requires less maintenance than the Lyman α. However, it still requires frequent calibration and cleaning of the optics. Therefore, neither device is particularly well suited for long-term operation without frequent attention.

Commercially available infrared hygrometers are being used more and more in marine micrometeorological investigations (**Figure 6**). Beer's law

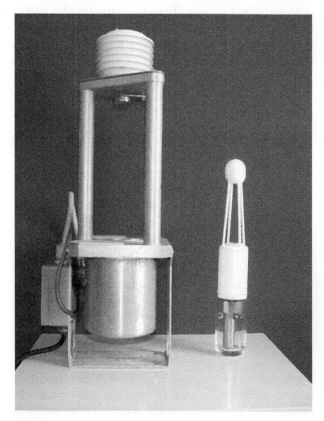

Figure 6 Two examples of commercially available infrared hygrometers. The larger hygrometer is roughly the height of the sonic anemometer shown in **Figure 4**.

(eqn [13]) provides the theoretical basis for the transmission of radiation over a known distance.

$$T = e^{-(\gamma+\delta)D} \qquad [13]$$

Here T is the transmittance of the medium, D is the fixed distance, and γ and δ are the extinction coefficients for scattering and absorption, respectively. This law applies to all of the radiation source described above; however, the use of filters with infrared devices allows eqn [13] to be used more directly. For example, the scattering coefficient has a weak wavelength dependence in the spectral region where infrared absorption is strongly wavelength dependent. Filters can be designed to separate the infrared radiation into wavelengths that exhibit strong and weak absorption. The ratio of transmittance of these two wavelengths is therefore a function of the absorption (eqn [14]), where the subscripts s and w identify the variables associated with the strongly and weakly absorbed wavelengths.

$$\frac{T_s}{T_w} \approx e^{-\delta_s D} \qquad [14]$$

Calibration of this signal then provides a reliable measure of water vapor due to the stability of current generation of infrared sources.

Infrared hygrometers are still optical devices and can become contaminated by sea spray and other airborne contaminants. To some extent the use of the transmission ratio negates this problem if the contamination affects the two wavelengths equally. Obviously, this is not the case when the optics become wet from rain, fog, or spray. Fortunately, the devices recover well once they have dried off and are easily cleaned by the rain itself or by manual flushing with water. Condensation on the optics can also be reduced by heating their surfaces. These devices require longer path lengths (0.2–0.6 m) than Lyman α or krypton hygrometers to obtain measurable absorption (**Figure 6**). This is not a problem as long as they are deployed at heights $\gg D$.

Conclusions

The state of the art in sensor technology for use in the marine surface layer includes the sonic anemometer/thermometer and the latest generation of infrared hygrometers (**Figure 7**). However, the frequency response of these devices, mainly due to spatial averaging, precludes their use from aircraft. Instead, aircraft typically rely on gust probes for measurement of the required velocity fluctuations, thermistors for temperature fluctuations, and Lyman α hygrometers

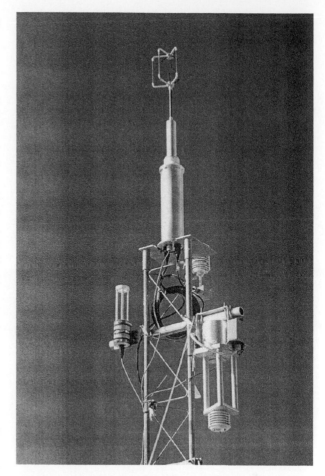

Figure 7 A sensor package used to measure the momentum, sensible heat, and latent heat fluxes from a moving platform. The cylinder beneath the sonic anemometer/thermometer holds 3-axis angular rate sensors and linear accelerometers, as well as a magnetic compass. Two infrared hygrometers are deployed beneath the sonic anemometer. The radiation shield protects sensors that measure the mean temperature and humidity. Photograph provided by Wade McGillis (WHOI).

for moisture fluctuations. Hot-wire and hot-film anemometers along with the finer temperature and humidity devices are also required to measure directly the viscous dissipation of the turbulent eddies that occurs at very small spatial scales.

Instruments for measuring turbulence are generally not considered low-power when compared to the mean sensors normally deployed on surface moorings, so past deployments of these sensors were mainly limited to fixed platforms or research vessels with ample power. Recently, however, sensor packages mounted on spar and discus buoys have successfully measured motion-corrected momentum and buoyancy fluxes on month- to year-long deployments with careful power management. The use of these sensor packages is expected to continue owing to the desirability of these measurements and technological advances leading to improved power sources and reduced power consumption by the sensors.

See also

Moorings. Satellite Remote Sensing Microwave Scatterometers. Sensors for Mean Meteorology. Ships. Turbulence Sensors.

Further Reading

Ataktürk SS and Katsaros KB (1989) The K-Gill, a twin propeller-vane anemometer for measurements of atmospheric turbulence. *Journal of Atmospheric and Oceanic Technology* 6: 509–515.

Buck AL (1976) The variable path Lyman-alpha hygrometer and its operating characteristics. *Bulletin of the American Meteorological Society* 57: 1113–1118.

Crawford TL and Dobosy RJ (1992) A sensitive fast-response probe to measure turbulence and heat flux from any airplane. *Boundary-Layer Meteorology* 59: 257–278.

Dobson FW, Hasse L, and Davis RE (1980) *Air–Sea Interaction; Instruments and Methods*. New York: Plenum Press.

Edson JB, Hinton AA, Prada KE, Hare JE, and Fairall CW (1998) Direct covariance flux estimates from mobile platforms at sea. *Journal of Atmospheric and Oceanic Technology* 15: 547–562.

Fritschen LJ and Gay LW (1979) *Environmental Instrumentation*. New York: Springer-Verlag.

Kaimal JC and Gaynor JE (1991) Another look at sonic thermometry. *Boundary-Layer Meteorology* 56: 401–410.

Larsen SE, Højstrup J, and Fairall CW (1986) Mixed and dynamic response of hot wires and measurements of turbulence statistics. *Journal of Atmospheric and Oceanic Technology* 3: 236–247.

Schmitt KF, Friehe CA, and Gibson CH (1978) Humidity sensitivity of atmospheric temperature sensors by salt contamination. *Journal of Physical Oceanography* 8: 141–161.

Schotanus P, Nieuwstadt FTM, and de Bruin HAR (1983) Temperature measurement with a sonic anemometer and its application to heat and moisture fluxes. *Boundary-Layer Meteorology* 26: 81–93.

SENSORS: TURBULENCE

TURBULENCE SENSORS

N. S. Oakey, Bedford Institute of Oceanography,
Dartmouth, Nova Scotia, Canada

Introduction

This article describes sensors and techniques used to
measure turbulent kinetic energy dissipation in the
ocean. Dissipation may be thought of simply as the
rate at which turbulent mechanical energy is con
verted into heat by viscous friction at small scales.
This is a complicated indirect measurement requiring
mathematical models to allow us to envisage and
understand turbulent fields. It will require using this
theory to understand how sensors might be developed using basic principles of physics to measure
properties of a turbulent field to centimeter scales.
Instruments must be used to carry these sensors into
the ocean so that the researcher can measure its
turbulent characteristics in space and time. It is also
this sensor–instrument combination that converts the
sensor output into a quantity, normally a voltage
varying in time, that is used by the experimenter to
calculate turbulent intensity. Thus, both the characteristics of sensors and the way in which the sensor–
instrument combination samples the environment
must be understood and will be discussed below.

Understanding Turbulence in the Ocean

There is no universally accepted definition of turbulence. Suppose that one stirs a bowl of clear water
and injects some colored dye into it. One sees that
filaments of dye become stretched, twisted and con-
torted into smaller and smaller eddies and eventually
the bowl becomes a uniform color. This experiment
leads to one definition of turbulence. It includes the
concept that eddies in the water are distributed
randomly everywhere in space and time, that energy
is transferred from larger to smaller eddies, and that
over time the mean separation of the dyed particles
increases. In contrast, the ocean is typically stratified
through a density that is determined by the temperature and salt in the water as well as the pressure.
In this environment, a vertical shear in the velocity in
the water column can be large enough to overcome
the stability. Energy from the mean flow is converted

into large-scale eddies determined by flow boundary
conditions that characterize turbulent kinetic energy
at its maximum scales. Further vortex stretching
creates smaller and smaller eddies resulting in a turbulent cascade of energy (velocity fluctuations) to
smaller scales until viscous forces begin to dominate
where the energy is eventually dissipated as heat.
This article focuses on sensors to measure this dissipation process directly by measuring the effect of
viscosity on the turbulent cascade.

The irregular and aperiodic velocity fluctuations in
space and time characteristic of turbulence, accompanied by energy transfer between scales and associated fluid mixing, may be described mathematically
through nonlinear terms in the Navier–Stokes equation. Nevertheless, it is difficult to solve numerically
in oceanographic applications. At the dissipation
scales, typically a few meters and smaller, we normally assume that the turbulent field is homogenous
and that it has definable statistical averages in all
parts of the field. We further assume that direction is
unimportant (isotropy) and statistical distributions
depend only on separation distances between points.
With the turbulence controlled only by internal
parameters, we assume the nature of the nonlinear
cascade of energy from large to small scales generates
a universal velocity spectrum. An example of this
spectrum is shown schematically in **Figure 1A**. At
low wavenumbers, k, no energy is taken out by viscous dissipation, so the energy flux, ε, across each
wave number, or down the cascade, is constant.
Through dimensional arguments, the three-dimensional turbulent energy spectrum, $E(k)$ in this region
(called the inertial subrange) as a function of wavenumber k is given by

$$E(k) = \alpha \varepsilon^{2/3} k^{-5/3} \qquad [1]$$

where α is a constant determined experimentally to
be approximately 1.5. In practice the three-dimensional spectrum given in eqn [1] cannot easily be
measured and one must use the one-dimensional
analogy where k is replaced by a component k_i.

At higher wavenumbers or smaller scales the velocity gradient spectrum (obtained by multiplying the
spectrum in eqn [1] by the square of the wave
number k^2) shows more clearly where dissipation
occurs. **Figure 1B** shows the spectra of velocity shear
for velocity fluctuations for one component of k for
values of ε most typically found in the ocean. In this
case, the spectra of fluctuations transverse to the

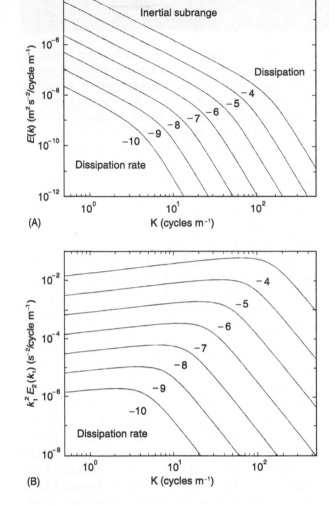

Figure 1 (A) The universal, velocity spectra for dissipation rates that typically occur in the ocean. Power density in velocity is plotted as a function of wavenumber. The shape of the spectrum remains the same but, as the energy in the turbulent field increases, the spectrum moves to higher wavenumbers and to higher intensities. (B) The equivalent universal, velocity shear spectra. (A) and (B) both show the inertial subrange and dissipation region, but in (B), the dissipation portion is more strongly emphasized.

measurement direction are shown but the picture for along-axis fluctuations would look almost identical. At the highest wavenumbers (smallest scales), viscous dissipation reduces the energy per unit wavenumber to zero. At small scale, it is assumed that turbulent motion is determined only by kinematic viscosity, $\nu (= 1.3 \times 10^{-6}$ m^2 s^{-1} at 10°C), and the rate, ε, at which energy passed down from larger eddies, must be dissipated. By dimensional arguments the length scale at which viscous forces equal inertial forces, and viscosity dissipates the turbulent energy as heat, is given by viscous cutoff scale

$$L_\nu = 2\pi \left(\nu^3 / \varepsilon \right)^{1/4} \qquad [2]$$

The factor 2π gives a length scale from the radian wave number. This is an important scale for the design of instruments and sensors because it defines the smallest diameter eddies that must be measured.

The dissipation ε is given by integrating the spectrum shown in **Figure 1B**.

$$\varepsilon = 15\nu \int_0^\infty k_1^2 E_1(k_1) \mathrm{d}k_1 = 7.5\nu \int_0^\infty k_1^2 E_2(k_1) \mathrm{d}k_1 \qquad [3]$$

$E_1(k_1)$ is the one-dimensional wavenumber spectrum of longitudinal velocity, and $E_2(k_1)$ is the one-dimensional spectrum of transverse velocity and one assumes isotropy to estimate the factors 15 and 7.5, respectively. In practice, the upper integration limit may be replaced with the viscous cutoff scale. For the transverse turbulent velocity u, the shear variance in the z direction, $\overline{(\mathrm{d}u/\mathrm{d}z)^2}$ is equivalent to the integral of equation [3] and ε is given by

$$\varepsilon = \frac{15}{2}\nu \overline{\left(\frac{\mathrm{d}u}{\mathrm{d}z}\right)^2} \qquad [4]$$

These assumptions are important to the way in which sensors are designed. A common way to observe turbulent fields is by making measurements of velocity and other mixing quantities along a trajectory through a turbulent field assuming that it is frozen in space and time. Measurement along a line, recorded as a time-series (**Figure 2**), is interpreted as spatial variability by assuming stationarity and using the known sensor velocity to convert into distance. Standard Fourier transform techniques allow one to generate spectra similar to those in **Figure 1** from which dissipation, ε, may be estimated.

If there is a temperature gradient in the water column when turbulence is generated, the velocity field strains the temperature field, creating strongly interleaved temperature filaments over the vertical range of the overturn. The temperature microstructure intensity depends not only on the mean gradient but also on the energy in the turbulent field, in particular dissipation, ε. Temperature fluctuations recorded as a time-series (**Figure 2**) can be represented by spectra similar to those shown in **Figure 1**. As with the velocity fluctuations, there is a subrange where diffusive and viscous effects are unimportant where temperature fluctuations are transferred towards higher wave numbers. Temperature spectra persist to length scales smaller than the viscous cutoff scale. In this range, not only kinematic viscosity, ν, and dissipation, ε, are important but also, thermal diffusion, $\kappa_T (\approx 1.4 \times 10^{-7}$ m^s s^{-1}). The cutoff

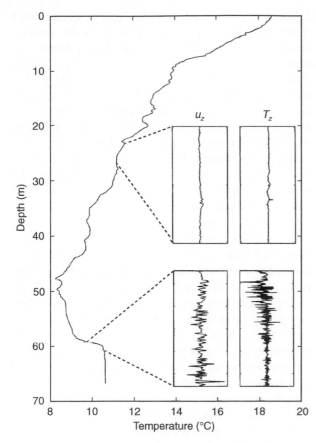

Figure 2 A representative vertical profile of temperature is shown from the surface to bottom obtained with a vertical falling instrument. In panels at the right are shown expanded portions of the velocity shear (u_z) and the gradients in temperature (T_z). The panels represent small sections of the vertical record that are treated as time-series to calculate spectra similar to those in **Figure 1**. The upper panel of U_z at mid-depth is a region of low dissipation and the one below represents higher dissipation.

wavelength for temperature fluctuations is given by

$$L_T = 2\pi\left(\nu k_T^2/\varepsilon\right)^{1/4} \qquad [5]$$

Under restricted circumstances, the temperature gradients or temperature microstructure can be measured in the ocean to this scale. Under these circumstances one can determine L_T and hence estimate dissipation ε.

The sensors used most commonly in oceanography to measure dissipation make use of the above ideas. Velocity fluctuations may be used to determine dissipation ε directly using eqns [3] or [4]. Measuring temperature fluctuations allows dissipation to be calculated indirectly from eqn [5]. The units used to express dissipation in the ocean are $W\,kg^{-1}$ (watts of mechanical energy converted into heat per kilogram of sea water). Typical values range from $10^{-10}\,W\,kg^{-1}$ in the deep ocean to $10^{-4}\,W\,kg^{-1}$ in active boundary

layers on continental shelves. (To put these numbers into a simple perspective, energy dissipated in the ocean may range from the almost insignificant rate of $100\,W\,km^{-3}$ to the very large rate of $100\,MW\,km^{-3}$.) Present sensors and instruments are capable of measuring over this range of dissipation. Regions of higher dissipation such as river outflows and tidal channels are not normally measurable with sensors and instruments described here.

Measuring Dissipation in the Ocean

The most common technique of estimating dissipation in the ocean involves measuring small-scale velocity and temperature fluctuations. This may be accomplished by dropping a profiler vertically, towing one horizontally or setting it at a fixed position and measuring the fluctuations in velocity and temperature as the water moves past the sensors. This allows a time-series of turbulent velocity fluctuations to be recorded. A typical platform used to measure dissipation in the ocean is a vertical profiler that falls typically at a speed of $0.5–1.0\,ms^{-1}$. There have been many such instruments built and each one typically carries a number of sensors to measure some components of the turbulent velocity as well as temperature microstructure. A sample time-series for a vertical profiler is shown in **Figure 2**. Assuming that the turbulent field is isotropic, homogeneous and stationary one can use the mean flow velocity to determine the wavenumber scale and calculate the one-dimensional turbulence spectrum, $E_1(k_1)$ or $E_2(k_1)$, as defined above and from this determine the dissipation, ε, using eqns [3] and [4].

As the turbulent dissipation gets larger, the wavenumber at cut-off gets larger. Alternatively, L_ν and L_T become smaller as shown in **Figure 3(A)**. As one tries to measure higher dissipation one must have a sensor with better spatial resolution and higher frequency response. We convert from wavenumber, k (cycles m^{-1}), to frequency using the relationship $f = kV$ (Hz) where V (m s^{-1}) is the flow speed past the sensor. The cutoff frequencies corresponding to L_ν and L_T are given by $f_{ci} = V/L_i$. In practice, one does not have to measure the microstructure variance to the cutoff frequency because of the universal characteristic of the dissipation curves. A usual compromise is to consider that if 90% of the dissipation curve is measured then a satisfactory measure of dissipation can be achieved. This is summarized in **Figure 3A** which shows the sampling frequency that must be achieved to resolve a particular dissipation. (It must be remembered that to resolve the energy at any frequency one must sample at least twice that frequency.)

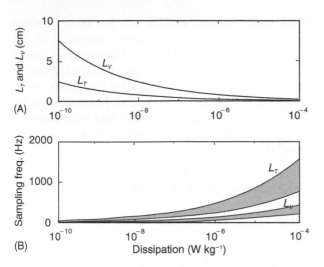

Figure 3 (A) The decrease in the cutoff scale with increasing dissipation for viscous dissipation, L_v (eqn [2]) and thermal dissipation, L_T (eqn [5]). (B) The sampling frequency that is required for a particular ε for viscous dissipation, L_v and for thermal cutoff, L_T. The upper and lower boundaries of the shaded bands correspond to measurement at flow speeds of 1.0 and 0.5 m s^{-1}, respectively.

Turbulence Dissipation Sensors

Airfoil Probes

One of the most commonly used turbulence sensors to measure turbulent velocity fluctuations is called an airfoil probe. This sensor is an axially symmetrical airfoil made of flexible rubber surrounding a sensitive piezoelectric crystal. The sensitive tip of the probe approximates a parabola of revolution, several millimeters in diameter and about 1 cm long. The crystal generates a voltage proportional to the magnitude of a force applied perpendicular to its axis. The crystal is rigid in one transverse direction so responds to a cross force only in one direction. Thus, two sensors are required to measure the two transverse components of turbulent velocity fluctuations. The sensor is placed on the leading end of an instrument that is moving relative to the water at a mean speed V. In a mean flow along the axis of the shear probe, no lift will be generated and no force applied to the crystal. If there is an off-axis turbulent velocity, a lift will be generated which will apply a force to the piezoelectric crystal through the flexible rubber tip. Thus, the sensor will provide a voltage that is linearly proportional to the turbulent velocity. The effective resolution of the sensor is of order 1 cm, the smallest scale of turbulence that can be effectively measured by this sensor. From **Figure 3**, it can be seen that for values above 10^{-5} W kg^{-1} this type of sensor will begin to underestimate dissipation. Normally the signal from the sensor is

differentiated to emphasize the high frequency part of the turbulence spectrum. This gives the velocity shear, and analysis of this signal allows direct generation of spectra similar to the theoretical ones shown in **Figure 1B**. For this reason, airfoil probes are often called shear probes. These sensors measure the component of turbulence perpendicular to the drop direction of the instrument. As such, it is eqn [4]. Which is most relevant to calculating dissipation, ε. **Figure 3B**, shows that to measure dissipation to 10^{-5} W kg^{-1} in a flow speed of 1 m s^{-1} (along the axis of the sensor) the output must be sampled to at least as rapidly as 200 Hz.

Of the many instruments that use this sensor to measure dissipation, the most common are vertical profilers. Those used near the surface are often called tethered free-fall profilers because they have a light, loose line attached to the instrument for quick recovery and redeployment. The line is usually a data link to the ship where data are recorded on computers for analysis. Because of the intermittent nature of turbulence, it is important to have many profiles (or independent samples) in measuring dissipation to be able to obtain a statistically robust average value. For deeper measurements of dissipation, free-fall profilers are used that have no tether line. They are deployed to a predetermined depth in the ocean where their buoyancy is changed to allow them to return to the surface. These instruments record internally and can be inherently quieter than tethered free-fall instruments but are slower to recover and redeploy. In practice, both types of profilers can measure dissipation as low as 10^{-10} W kg^{-1}. In shallow regions of high dissipation such as the bottom boundary layer of tidally generated flow over banks, in bottom river channels or in active regions such as the Mediterranean outflow tethered free-fall instruments have been most successful. Where the dissipation exceeds 10^{-5} W kg^{-1}, these profilers and the shear probe sensor give limited results.

The airfoil probe has also been used successfully to obtain dissipation measurements horizontally. It has been used as a sensor on a towed fish pulled horizontally at speeds of order 1 m s^{-1}. The results look similar to those in **Figure 2** where the depth axis is replaced by a horizontal axis and similar techniques to those described above are used to extract dissipation. Because of towline vibration, a towed instrument is generally noisier than a free-fall profiler. If the vibration noise of the platform is transferred to the airfoil sensor, it will generate velocity signals relative to the sensor indistinguishable from turbulence in the water with the sensor not vibrating. Generally, a measurable dissipation lower limit for these instruments of 10^{-9} W kg^{-1} would be considered good. These shear probes have also been mounted on submarines for horizontal measurements. Nevertheless, this platform has had only limited use because of vehicle noise and expensive operating costs. More recently, unmanned submarines called autonomous underwater vehicles have been used as suitable platforms for turbulent kinetic energy measurements. They are expected to have similar noise characteristics to towed instruments. Another interesting way of obtaining horizontal measurements is to place shear probes on a moored instrument. The turbulence in the water is measured as it flows past the sensor at a speed V m s^{-1}. In this case, the water velocity must typically be faster than 0.1 m s^{-1} for the measurements to be within the sensor capabilities and mooring vibrations generate similar problems to towed instruments.

Thin Film Sensors

One of the original sensors used to measure turbulence and dissipation is called a hot film sensor. In these sensors, a platinum or nickel film is deposited on the surface near the conical tip of a glass rod of order 1 mm diameter and covered with a thin film of quartz to insulate it from the water. The film is heated to several degrees centigrade above the ambient temperature and special electronics are used to maintain a constant thin film temperature. Water flowing across the probe cools the platinum. Fluctuations in the current, required to keep the sensor at a constant temperature, are a measure of the turbulent velocity fluctuations along the axis of the sensor. This sensor measures the $E_1(k_1)$ component of the turbulent field as opposed to the $E_2(k_1)$ component measured by the shear probe. Therefore, the first part of the eqn [3] is relevant to estimating dissipation. The primary advantage of this sensor over the shear probe is that it has much smaller spatial resolution and a much higher frequency response. As one can see from **Figure 3**, this allows one to measure to higher dissipation rates. The disadvantages of this probe are that the electronics to run it are much more complicated than for shear probes and the sensors are more difficult to fabricate and quite expensive. They also require a lot of power to heat since they are very low in resistance (of order 5–10 Ω). Because the quartz insulation must be extremely thin to provide good heat transfer, thin films are also very fragile and easily damaged by impact with particles in the water. These probes do not provide an output voltage that is linear with turbulent velocity fluctuations. They also tend to be noisy and subject to fouling. They are seldom used today in ocean measurements.

Pitot Tubes

Another recently developed sensor used to measure dissipation makes use of a Pitot tube. If a Pitot tube is placed in water flowing at a speed W along its axis, the pressure generated by the flow is proportional to W^2. This technique has been applied to turbulence measurements by carefully designing an axisymmetric port a few millimeters in diameter on the tip of a sensor of order 1 cm in diameter. By connecting the port to a very sensitive differential pressure sensor, fluctuations in pressure along the axis of the probe can be measured. Using suitable electronic circuits, a signal is produced that is linearly proportional to along-axis fluctuations in turbulent velocity. In this sense, it is similar to the heated-film sensor and different from the shear probe which measures fluctuations perpendicular to the mean flow. This sensor has been used in conjunction with a pair of shear probes to simultaneously measure all three components of turbulent velocity fluctuations.

Temperature Microstructure Sensors

As outlined above, if there is turbulent mixing occurring in a region where there is a temperature gradient, the turbulent velocity will cause the temperature to be mixed. If, for example, warmer fluid overlays colder fluid, turbulence will move parcels of warm fluid down and cold fluid up. A temperature sensor that traverses a patch of fluid such as this will measure fluctuations in temperature as shown in **Figure 2**. The spectrum of these fluctuations can be used to determine the dissipation using eqn [5]. Because the molecular diffusivity of heat for water is much smaller than the molecular viscosity, the scale at which temperature fluctuations cease is about a factor of three smaller than the scale at which velocity fluctuations cease. This is shown clearly in **Figure 3A** that compares L_T and L_v. These facts place a severe restriction on the speed and size of a temperature sensor compared to a shear probe, or alternatively limits the speed that an instrument may fall. For the same fall speed, a temperature sensor must be sampled at a much higher rate than a shear probe. The simplest temperature sensor with the precision and noise level to measure temperature microstructure in the ocean is the thermistor. The smallest thermistors that are used in sea water are a fraction of millimeter in diameter and have a frequency response of order 10 ms. At a flow speed of 1 m s^{-1}, one is able to delimit the spectrum of temperature for dissipations up to about 10^{-7} W kg^{-1}. Some success has been obtained by using very slow moving profilers that fall or rise at about 0.1 m s^{-1}. An alternative to the thermistor is a thin film thermometer. It is similar to the hot film velocity sensor described above and is constructed identically. Used as a thermometer, the change in the resistance of this sensor is a measure of change of temperature. Thin film sensors are faster than thermistors, typically with a time-constant of 2 ms which means that for any sensor velocity the temperature fluctuations may be measured to a higher wave number. These sensors are nevertheless at least an order of magnitude noisier than thermistors, which means that they are suitable for measuring microstructure only in regions where there are strong mean gradients. Using thermometry to measure dissipation is subject to large errors because, as indicated in eqn [5], dissipation is proportional to $(L_T)^4$, and this requires accuracy in determining L_T that is seldom achieved.

Some success has been made using sensors that measure conductivity as a proxy for temperature. These sensors make use of the fact that the conductivity of sea water is determined by both salt and temperature and in most cases, the temperature causes most of the fluctuations. The techniques used are similar to those described above for temperature. Some of the sensors are smaller and faster than thermistors and less noisy than thin film thermometers. They are still limited to the same constraints as thermometers in that they must fully resolve the spectrum in order to estimate L_T and utilize eqn [5].

Acoustic Current Meters

Acoustic techniques have also been used to measure water velocity in the ocean and indirectly to infer dissipation rates. One such technique utilizes an acoustic Doppler current meter optimized to measure vertical velocity fluctuations in the water column. In these instruments, a sound pulse is transmitted into the water and the sound scattered back to a sound receiver. The back-scattered pulse contains information about the water velocity because of the Doppler shift in the sound frequency. This technique is unable to measure to dissipation scales but instead, measures vertical velocities in the $k^{-5/3}$ wavenumber range defined by eqn [1]. By suitably defining a turbulent timescale, dissipation is estimated from the intensity in the fluctuations in the vertical velocity. This technique is very useful in studying turbulence in regions of intense mixing such as tidally driven flows.

In another technique, an array of small acoustic transmitters and receivers is configured such that the transit time of a pulse of sound can be measured over a short distance of around 10–20 cm. Velocity fluctuations in the water change the transit time and allow water velocity fluctuations to be inferred. This configuration of sensors is generally mounted as a fixed array on a platform on the bottom and has been used to measure turbulent mixing in many places on continental shelves. This technique has the advantage over profiling dissipation sensors of measuring three components of velocity fluctuations over long periods of time at a single place. Dissipation is estimated from the $k^{-5/3}$ wavenumber range.

Conclusions

The measurement of mixing rates in the ocean is important to our understanding of the distributions of temperature, salinity, and nutrients in the ocean. We need to understand this to include them correctly in climate and biological ocean models. The way in which energy is converted from sources at large scale and dissipated at small scales has required the development of a variety of ocean sensors. Some of these are described briefly above. It is hoped that enough of the key words and ideas have been put forward for the reader to understand some of the principles involved in turbulence measurement and at least some of the sensors and techniques used.

Nomenclature

α	An experimentally determined spectral constant
$E(k)$	Energy spectral density
$E_1(k_1)$	One-dimensional energy wavenumber spectrum – fluctuations along the axis of measurement
$E_2(k_1)$	One-dimensional energy wavenumber spectrum – fluctuations perpendicular to the axis of measurement
e	Dissipation of turbulent kinetic energy
f	measurement or sampling frequency
K_T	molecular diffusivity of heat
L_v	viscous cutoff scale
L_T	temperature cutoff scale
u	horizontal velocity fluctuation
v	kinematic viscosity
V	flow velocity along axis of sensor
W	drop velocity
z	distance coordinate (normally vertical)

See also

Profiling Current Meters.

Further Reading

Bradshaw P (1971) *An Introduction to Turbulence and Its Measurement*. Oxford, New York, Toronto, Sydney, Paris, Braunschweig: Pergamon Press.

Dobson F, Hasse L, and Davis R (1980) *Air–Sea Interaction Instruments and Methods*. New York and London: Plenum Press.

Frost W and Moulden TH (1977) *Handbook of Turbulence*, vol. 1: Fundamentals and Applications. New York: Plenum Press

Hinze JO (1959) *Turbulence*. New York, Toronto, London: McGraw-Hill.

Journal of Atmospheric and Oceanic Technology (1999) 16(11), Special Issue on Microstructure Sensors.

Neumann G and Pierson WJ (1966) *Principles of Physical Oceanography*. Englewood Cliffs, NJ: Prentice Hall.

Patterson GK and Zakin JL (1973) Turbulence in liquids. *Proceedings of the Third Symposium*, 414pp., Department of Chemical Engineering, University of Missouri-Rolla.

Summerhayes CP and Thorpe SA (1996) *Oceanography*, pp. 280–299. New York: John Wiley.

SENSORS: OPTICAL

ABSORBANCE SPECTROSCOPY FOR CHEMICAL SENSORS

R. Narayanaswamy, The University of Manchester, Manchester, UK

F. Sevilla, III, University of Santo Tomas, Manila, The Philippines

Introduction

Chemical sensors have introduced an alternative technology for chemical measurements. These analytical devices generate an electrical signal in response to the presence of a specific substance, the signal being related to the concentration of the analyte. The application of these sensors has simplified chemical analysis, since the need for obtaining a laboratory sample is eliminated and a real-time and on-site measurement can be carried out. Furthermore, the configuration of these devices enables the detection with great sensitivity of low concentrations of chemical species.

A number of analytical principles have been exploited in the development of chemical sensors. Among them are the optical methods of chemical analysis, which rely on the interaction of electromagnetic radiation with matter for the quantitation of a large number of substances. These methods provide a rapid and nondestructive tool for the measurement of chemical species. Optical methods have indeed played an important role, and continue to do so, in various field of chemical analysis.

One group of chemical sensors that are based on the optical methods of chemical analysis is called optodes (or optrodes). Other types of optical sensing techniques that utilize waveguides, for example, surface plasmon resonance, also exist but are not reviewed here. The basic concept of optodes is depicted diagrammatically in **Figure 1**. An optode consists of an optical fiber, an optoelectronic instrumentation that incorporates a light source and a phototransducer, and a solid-phase molecular recognition element. Light from a suitable source is launched into the optical fiber and directed to the sensing zone which contains the molecular recognition element. The molecular recognition element reacts with the analyte, resulting in a modification of its optical property. This change is probed by the supplied radiation which is subsequently guided via the optical fiber to the phototransducer. The phototransducer generates an electrical signal that is related to the concentration of the analyte.

This article focuses on chemical sensors that are based on absorbance spectroscopy. The various principles involved are reviewed, and some applications are presented.

Absorbance and Reflectance Spectroscopy

Ultraviolet (UV)/visible absorbance spectroscopy has been employed extensively in analytical chemistry. The basis for analysis here is the ability of the analyte or its derivative to absorb radiation that impinges into the measuring system. The radiation can have a wavelength occurring in the UV (200–400 nm), visible (400–780 nm), near-infrared (780–3000 nm), or infrared (3–50 μm) region. The absorption of radiation causes a reduction in the intensity of the radiation after it has passed through the system (**Figure 2**). This phenomenon is mathematically described by the Beer–Lambert law, as expressed by the following equation:

$$A = \log\frac{I_0}{I_t} = \varepsilon c l \qquad [1]$$

where A is the absorbance, c is the concentration of the absorbing species, I_0 and I_t are the intensity of incident and transmitted light, respectively, ε is the

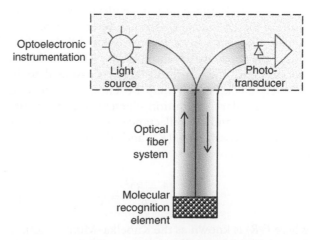

Figure 1 Basic design of optical fiber chemical sensor.

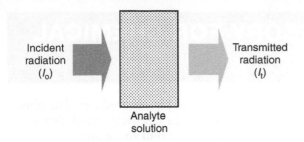

Figure 2   Absorption of radiation.

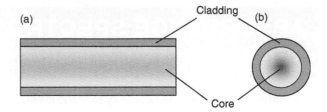

Figure 3   Structure of an optical fiber. (a) Longitudinal cross-section view, and (b) latitudinal cross-section view.

molar absorptivity of the species, and l is the optical path length of the absorbing species.

Absorbance spectroscopy is applicable only when the measurand system containing the analyte is transparent. However, if the medium is optically dense or even opaque, in which case the absorbance measurements would produce a high background, a technique that is complementary to absorptiometry (viz., reflectometry) can be employed for analytical measurements. In this case, the radiation infringes on the boundary interface of two media having different dielectric constants and reflection occurs.

Two distinct types of reflection are possible, namely (1) specular (or mirror-type) reflection and (2) diffuse reflection. Specular reflection occurs at the interface of a medium with no transmission through it, and reflection is at the same angle as the incident light; whereas, in diffuse reflection, the light penetrates the medium and subsequently reappears at the surface after partial absorption and multiple scattering within the medium. Of these two processes, diffuse reflection has found to be useful in chemical measurements. Specular reflection is minimized or eliminated through appropriate sample preparation and optical engineering.

The distribution of diffusely reflected light is rather homogeneous and largely independent of size and shape of the particles. The optical characteristics of diffuse reflectance are dependent on the composition of the system. Among several theoretical models that have been proposed for diffuse reflectance, the most widely used is the Kubelka–Munk theory. Here, the scattering layer is assumed to be infinitely thick, which may be effectively the case with molecular recognition element utilized in optical sensors, and the reflectance (R) is related to the absorption coefficient (K) and the scattering coefficient (S), as follows:

$$F(R) = \frac{(1-R)^2}{2R} = \frac{K}{S} \qquad [2]$$

where $F(R)$ is known as the Kubelka–Munk function. The absorption coefficient K can be expressed in

terms of the molar absorptivity (ε) and the concentration (c) of the absorbing species. Thus, eqn [2] can be rewritten as

$$F(R) = \frac{(1-R)^2}{2R} = \frac{\varepsilon c}{S} = kc \qquad [3]$$

where $k = \varepsilon/S$, and S is assumed to be independent of concentration. Equation [3] is analogous to the Beer–Lambert relationship (eqn [1]) and holds true within a range of concentrations for solid solutions in which the absorber is adsorbed onto the surface of a scattering particle. The reflectance values (R) are generally evaluated relative to the reflectance of standard reference materials such as barium sulfate.

Optical Fibers

Almost all optical sensors employ optical fibers to transmit light to and from the molecular recognition element. The fiber couples the optoelectronic instrumentation to the molecular recognition element, resulting in an integrated analytical system. This integration has contributed to the simplification of the chemical measurement process.

Optical fibers consist of a cylinder (known as the 'core') of transparent dielectric with a certain refractive index (n_1), surrounded by a thin film of another dielectric (called the 'cladding') of a lower refractive index (n_2). Most optical fibers are then covered with a protective jacket that has no influence on the wave-guiding properties of the optical fiber. The basic structure of an optical fiber is shown in **Figure 3**. The common materials used in optical fibers include plastic (poly(methyl methacrylate)), glass, and quartz.

Incident light is transmitted through the fiber when it impinges the core–cladding interface at an angle greater than the critical angle, so that there is total internal reflection at the core–cladding interface. A series of total internal reflection takes place until the light reaches the other end of the fiber (**Figure 4**). The optical fiber light transmission characteristics are described by its numerical aperture

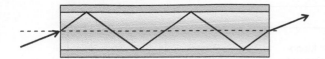

Figure 4 Transmission of a light beam inside an optical fiber through total internal reflection.

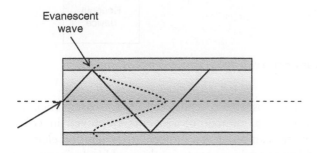

Figure 5 Evanescent wave due to a light beam being transmitted inside an optical fiber.

(NA), which is directly proportional to the sine of the half angle (α) of the acceptance cone of light entering it which, in turn, is related to n_1 and n_2, as in eqn [4]:

$$\text{NA} = \sin \alpha = (n_1^2 - n_2^2)^{1/2} \quad [4]$$

where n_0 is the refractive index of the surrounding medium, for example, air.

Basically there are three kinds of optical fibers in use for sensing purposes. These are the multimode step index, the multimode graded index, and the single-mode step index fibers. These fibers differ in the number of light beams that travel through the length of the waveguide. The light rays accepted into the fiber interact, and only those which undergo constructive propagation traverse through the fiber.

A small portion of light that is transmitted through an optical fiber or a waveguide by total internal reflection extends outside core and is referred to as the 'evanescent wave'. In an optical fiber, the evanescent wave penetrates the cladding material (**Figure 5**). The intensity (I) of the evanescent wave decreases exponentially with increasing distance (d) from the surface according to eqn [5]:

$$I = I_0 e^{-d/d_p} \quad [5]$$

where I_0 is the electric field intensity at the core–cladding interface and d_p, the depth of penetration, is the distance from the interface of the point at which the electric field intensity has reduced to 1/e of its value at the interface. This characteristic depth of penetration is dependent on wavelength of the propagating light, refractive indices of the core and the cladding, and the acceptance angle of the light

beam entering the fiber core. The distance d_p is typically of the order of a fraction of a wavelength of light.

The incorporation of optical fibers in chemical sensors imparts a number of advantages to optical sensors over the conventional devices in many areas of application. Optical sensors are electrically passive and immune to electromagnetic disturbances. They are geometrically flexible, corrosion-resistant, and capable of being miniaturized. They are compatible with telemetry and capable of operation in remote and hostile environments. They can be of low-cost, of rugged construction, and intrinsically safe. The optical fibers used in these sensors are capable of transmission of optical signals over great distances with low attenuation of optical power. Thus optical sensors are capable of measurements of samples in their dynamic environment, no matter how distant, difficult to reach, or harsh that environment is. Intrinsic safety aspects are imparted to these sensing devices by the low optical power utilized in them. Furthermore, the chemical sensing process itself is nonelectrical. With these sensors only very small sample volumes are needed for analysis, which has the advantages of nonperturbation of samples in real-time monitoring applications.

However, optical sensing devices possess certain limitations, such as interference from ambient light, limited dynamic range, long response times, limited specificity, and nonreversibility. Many of these limitations can be eliminated or reduced by the use of appropriate instrumentation and sensing phases, and thus the sensor devices can be used advantageously in specific applications.

Optoelectronic Instrumentation

Optical fiber sensors involve similar instrumentation as those involved in spectrophotometers. These sensors require both optical and electrical components, including a light source; a wavelength selector; a photodetector; and a readout device. A block diagram of the basic instrumentation associated with optical sensors is presented in **Figure 6**. The radiation supplied by the light source is first made monochromatic by a wavelength selector before it enters into optical fiber sensor system. Then, the radiation emanating from the sensor is directed to the photodetector which subsequently generates an electrical signal. This magnitude of the final signal is displayed in the readout device.

Several types of sources, including incandescent lamps (tungsten and quartz–halogen), lasers, light-emitting diodes (LEDs), and laser diodes, have been

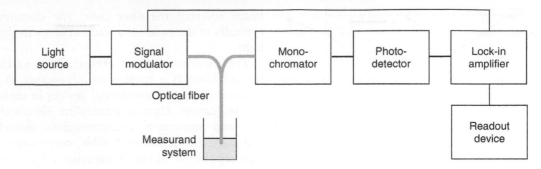

Figure 6 Schematic diagram of a typical instrumentation system associated with optical fiber chemical sensors.

used. Each of these types of source has its own advantages and disadvantages. The detection of light is carried out using a photocounting device, which converts optical signals into electrical signals that can be amplified electronically. The photodetectors used in optical sensors include photomultiplier tubes, p–i–n photodiodes, avalanche photodiodes, and photodiode arrays.

A wavelength selector, such as a filter or a monochromator, isolates the desired wavelength for the measurement. Optical couplers and lenses are used to focus the beam of light to the optical fiber and to direct light to the detector. It should be emphasized that efficient coupling of light ensures the attainment of high sensitivity of sensor signals. It is also common practice to exclude extraneous light reaching the detector by suitable modulation of the light source and by synchronizing the detector to this modulation frequency so as to detect only those source signals. Instrumental drift may be eliminated or reduced by the use of a suitable referencing system. Drifts in sensor response due to aging of the sensing phase and variable light source may be encountered. But these contributions may not be significant due to short periods of measurement periods used with the sensing devices.

Molecular Recognition Element

The molecular recognition element constitutes the primary sensing unit of a chemical sensor. It interfaces the chemical sensor with the measurand system, generating an optical signal in response to the presence of the analyte. It transforms the analyte into another substance with distinct optical properties, so that it can be considered as a chemical transduction unit. The transformation can be detected and quantified by the instrumentation system via the optical fiber. However, if the analyte itself possesses a detectable optical characteristic, then that property can be directly measured using optical fibers and quantified to the concentrations of the analyte.

For absorbance- and reflectance-based chemical sensors, a number of reagents employed in colorimetry can serve as a molecular recognition element. Chromogenic reagents, such as pH colorimetric indicators and chelating reagents, have been used as chemical transduction elements in optical sensors. Novel materials, such as conducting polymers, molecularly imprinted polymers, and nanoparticles, have also been exploited for molecular recognition in optical fiber chemical sensors.

The molecular recognition element is often employed in the solid state. The reagent(s) is usually immobilized onto an inert and stable solid material by physical methods (adsorption, entrapment, and electrostatic attraction) or by chemical means (covalent bond formation). Among the solid supports that have been used to immobilize the molecular recognition element are glass, silica gel, or organic polymers. The physical methods are simple and economical to carry out, but they do not necessarily produce stable reagent matrices. On the other hand, the chemical means of immobilization produce a strongly bound reagent, but this is often achieved after several reaction steps in the synthesis or modification of the reagent and/or the support material in order to realize stable chemical bond(s) between them.

The sensor response characteristics in chemical transducer-based sensors depend on the manner in which the analyte–reagent interaction takes place. In a simple system, where direct indicators are employed, the analyte concentrations can be correlated to the optical changes that occur in the reagent phase or in the product or both. The correlated sensor signal may also be dependent on the equilibrium constant of the analyte–reagent reaction. For example, pH can be measured by monitoring the changes in optical property of acid-base indicators. In many sensors, reversible reactions are preferred in the chemical transduction process because they can be used in continuous monitoring applications. In this case, the response time of these sensors (i.e., the time to reach the reaction equilibrium) is dependent on mass transfer processes. Irreversible reactions may also be employed in sensors, which can result in 'one-shot' devices. Although such sensors would be of limited merit, measurements with high sensitivity can be attained here. In certain cases, the reagent phases can be regenerated by the use of another chemical reaction and the sensor reused. Indirect chemical reactions involving two or more reagents and/or reactions can also be adapted as optical chemical transducers. Many enzyme-based reactions fall into this category.

Sensor Design

A variety of configurations have been utilized in sensors based on absorbance and other spectroscopic measurements. The sensor designs can be classified into two categories: extrinsic and intrinsic sensors. In extrinsic sensors, the optical fiber acts only as a light guide between the light source and chemical sensor and between the sensor and the detector. In intrinsic sensors, the optical fiber becomes a part of the transducer.

In absorbance-based sensors, light is usually fed at one end of an optical fiber, guided to a sensing cell, collected through a second optical fiber, and detected at the other end of this fiber. If no chemical transduction is employed (**Figure 7(a)**), the light interacts directly with the measurand system. This type of sensors has been described as 'spectroscopic sensors', since they are based on the spectral properties of the analyte. In sensors involving a molecular recognition element, the reagent phase is often translucent and is set at one end of the feed fiber and of the collector fiber (**Figure 7(b)**). The analyte reacts with the immobilized reagent, and the product absorbs the light that passes through the reagent phase.

In reflectance-based sensors, chemical transduction is often employed and the molecular recognition element is placed at one end of an optical fiber system. A single optical fiber or optical fiber bundles can be used in this type of optical sensors. In single fiber sensors (**Figure 8(a)**), the source light and detected light travel through the same optical fiber, and are discriminated either temporally or by wavelength with the aid of a beam splitter. Optical fiber bundles are often configured as a bifurcated system (**Figure 8(b)**), wherein the incident radiation travels through one branch and the reflected light is directed to the photodetector through the other branch.

Evanescent wave interactions have been exploited in optical fiber chemical sensors. In these devices, the cladding material of the optical fiber is removed and replaced with the analyte system itself (**Figure 9(a)**)

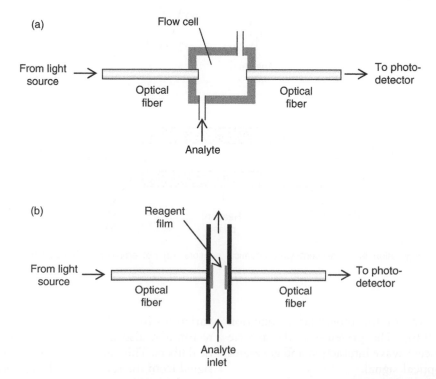

Figure 7 Typical configuration of absorbance-based chemical sensors: (a) employing no chemical transduction, and (b) with chemical transduction.

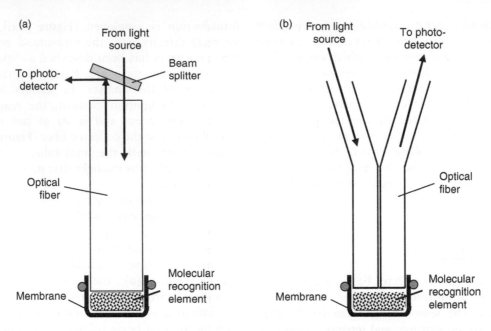

Figure 8 Typical configuration of reflectance-based chemical sensors: (a) employing a single fiber, and (b) employing a bifurcated optical fiber bundle.

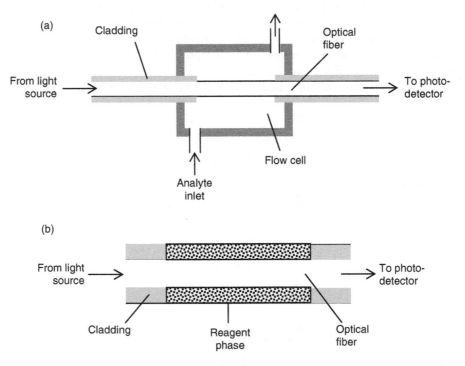

Figure 9 Typical configuration of evanescent-wave chemical sensors: (a) not employing chemical transduction, and (b) with chemical transduction.

or with a thin layer of the molecular recognition element (**Figure 9(b)**). The presence of the analyte affects the evanescent wave interaction and generates changes in the optical signal.

In optical fiber absorptiometry or reflectometry, reference signals are employed for correcting variations in the background caused by source, detector, and also intrinsic absorption of light by optical fibers. This can be done by subtracting the blank signal from the sensor signal electronically, or by the use of two identical optical fibers, one of which is used as a reference.

Sensor Applications

A number of optical fiber pH sensors have been developed based on the absorption characteristics of a colorimetric acid-base indicator. The reagent is immobilized on a finely powdered solid support or on a membrane that is deposited on the tip of the optical fiber and held in place by an enclosing membrane. One of the earliest reported optical fiber sensor for pH was based on the measurement of the absorbance of phenol red covalently bound to polyacrylamide mixed with polystyrene microspheres to scatter the light within the reagent phase. The sensor was configured as a probe employing a bifurcated optical fiber, with one arm for the probing radiation and the other arm for the scattered radiation from the reagent phase. The sensor measures pH in the range 7.0–7.4 to 0.01 pH units. This example illustrates the feasibility of the approach though the sensors may not be directly applicable to pH measurements in seawater. Different indicators may be employed for such applications.

Absorbance-based optical sensing has been applied for the measurement of metal ions. These ion sensors involve the use of immobilized metal-ion-selective reagents interfaced to optical fibers. These sensors rely on the fact that the metal ion (M) reacts with the immobilized reagent (R) to form a metal complex (MR), accompanied by either an enhancement or change of color of the immobilized reagent. This change can be correlated to metal ion concentrations. Thus, an optical sensor for micromolar amounts of cobalt was developed employing a reagent phase consisting of pyrogallol red held on a cellulose acetate film. Likewise, a fiber optic sensor for copper ions occurring in ppm levels was devised based on α-benzoinoxime, a highly selective colorimetric reagent for copper ion, immobilized on hydrophobic reagent Amberlite XAD-2 microspheres.

Optical sensors based on absorbance measurement have been developed for organic compounds, such as pesticides, occurring in environmental water systems. In the case of pesticide sensors, more than one type of transduction system is involved – an enzyme reagent and a pH indicator co-immobilized on suitable polymeric material. In the absence of pesticides, the reaction of the immobilized enzyme, such as acetylcholine esterase, with its specific substrate will be accompanied by a change in pH that can then be measured by the pH sensor using absorbance or reflectance. However, in the presence of the pesticides, the substrate/immobilized enzyme reaction is inhibited and the degree of inhibition, transduced in the pH sensor, can be correlated to the pesticide concentration. Organophosphate and carbamate pesticides, and also many toxic metal ions, have been quantified using this principle.

Absorbance measurement has also been used in gas-phase sensing. A nitrogen dioxide sensor has been constructed using optical fibers and measurement of absorbance of the gas at $0.5 \, \mu m$ using an argon-ion laser source. Real-time measurement in the lower ppm concentration range has been conducted at remote locations that are 20 km away. A similar device has been used for the measurement of methane at low concentrations by recording absorbance at $1.33 \, \mu m$.

Many sensors for gases and vapors such as ammonia, carbon dioxide, humidity, hydrogen cyanide, hydrogen sulfide, etc., have developed based on the use of analyte-specific immobilized reagents and measurements of absorbance or reflectance. For instance, the optical sensing of ammonia can be carried out through the change in color of polyaniline deposited on a polystyrene substrate. An optical sensor for gaseous hydrogen sulfide was based on the development of a grayish color on a cellulose membrane impregnated with lead acetate.

Conclusions

As described above, many types of optical sensor designs have been studied for a variety of analytes using different types of transduction reactions in the development of absorbance- and reflectance-based sensors. A current trend in this development is to construct multianalyte sensing systems based on the use of single or a few reagent phases, together with the employment of appropriate signal-processing techniques such as pattern recognition and artificial neural networks. Most of the optical sensors described above can be designed for use in oceanographic measurements in the analysis of heavy metal ions, dissolved gases, and other species.

Applications of absorbance-based sensors to ocean sciences have yet to demonstrate their potential. Most of the applications published in the literature describe 'proof-of-concept' studies with the sensors and with very little or no practical demonstration in such areas. Some of the recent studies in this area have been focused on the monitoring of dissolved CO_2 in seawater using absorbance and fluorescence measurements with transducers that incorporate pH-sensitive indicators. The use of infrared absorption measured through the evanescent waves in optical fibers for subsea monitoring of organic compounds (at ppm levels) has also been demonstrated. Parameters such as sensitivity, specificity, lifetime, aging, etc., need to be investigated and addressed before the sensors can be used for seawater measurements. These studies would present great challenges in order to realize practical sensors. However, there is substantial interest in new sensors for oceanographic applications including monitoring of nutrients and pollutants, and optical sensors have great potential here. The devices are clearly attractive in concept and require expertise from several scientific disciplines including analytical chemistry, polymer chemistry, environmental chemistry, fiber optics, and opto-electronics.

See also

Fluorometry for Chemical Sensing. Inherent Optical Properties and Irradiance. Wet Chemical Analysers.

Further Reading

Andres R, Kuswandi B, and Narayanaswamy R (2001) Optical fiber biosensors based on immobilized enzymes – a review. *The Analyst* 126: 1469–1491.

Eggins BR (2002) *Chemical Sensors and Biosensors*. London: Wiley.

Hales B, Burgess L, and Emerson S (1997) An absorbance-based fiber-optic sensor for CO_2(aq) measurement in pure waters of sea floor sediments. *Marine Chemistry* 59: 51–62.

Harmer AL and Narayanaswamy R (1988) Spectroscopic and fibre-optic transducers. In: Edmunds TE (ed.) *Chemical Sensors*, ch. 13. Glasgow: Blackie.

Lieberzeit PA and Dickert FL (2007) Sensor technology and its applications in environmental analysis. *Analytical and Bioanalytical Chemistry* 387: 237–247.

Mizaikoff B (1999) Mid-infrared evanescent wave sensor – a novel approach for sub-sea monitoring. *Measurement Science and Technology* 10: 1185–1194.

Narayanaswamy R (1991) Current developments in optical biochemical sensors. *Biosensors and Bioelectronics* 6: 467–475.

Narayanaswamy R (1993) Chemical transducers based on fibre optics for environmental monitoring. *The Science of the Total Environment* 135: 103–113.

Narayanaswamy R (1993) Optical chemical sensors: Transduction and signal processing. *Analyst* 118: 317–322.

Narayanaswamy R and Sevilla FS, III (1988) Optical fibre sensors for chemical species. *Journal of Physics E: Scientific Instruments* 21: 10–17.

Narayanaswamy R and Wolfbeis OS (2004) *Springer Series on Chemical Sensors and Biosensors, Vol. 1: Optical Sensors – Industrial Environmental and Diagnostic Applications*. Berlin: Springer.

Oehme I and Wolfbeis OS (1997) Optical sensors for determination of heavy metal ions. *Microchimica Acta* 126: 177–192.

Orellana G and Moreno-Bondi MC (2005) *Springer Series on Chemical Sensors and Biosensors, Vol. 3: Frontiers in Chemical Sensors – Novel Principles and Techniques*. Berlin: Springer.

Rogers KR and Poziomek EJ (1996) Fiber optic sensors for environmental monitoring. *Chemosphere* 33: 1151–1174.

Seitz WR (1988) Chemical sensors based on immobilised indicators and fiber optics. *CRC Critical Reviews in Analytical Chemistry* 19: 135–171.

Sevilla F, III and Narayanaswamy R (2003) Optical chemical sensors and biosensors. In: Alegret S (ed.) *Integrated Analytical Systems*, ch. 9. Amsterdam: Elsevier.

Tokar JM and Dickey TDN (2000) Chemical sensor technology. Current and future applications. *Ocean Science and Technology* 1: 303–329.

Varney MS (2000) *Chemical Sensors in Oceanography*. London: Taylor and Francis.

Wise DL and Wingard LB, Jr. (1991) *Biosensors with Fiberoptics*. Clifton, NJ: Humana.

Wolfbeis OS (1991) *Fiber Optic Chemical Sensors and Biosensors*, vols. I and II. Boca Raton, FL: CRC Press.

BIO-OPTICAL MODELS

A. Morel, Université Pierre et Marie Curie,
Villefranche-sur-mer, France

Copyright © 2001 Elsevier Ltd.

Introduction

The expression 'bio-optical state of ocean waters' was coined, in 1978, to acknowledge the fact that in many oceanic environments, the optical properties of water bodies are essentially subordinated to the biological activity, and ultimately to phytoplankton and their derivatives. More recently the adjective bio-optical has been associated with nouns like model or algorithms. At least two meanings can be distinguished under the term 'bio-optical model.'

A bio-optical model can designate a tool used to analyze, and then to predict, the optical properties of biological materials, such as phytoplanktonic or heterotrophic unicellular organisms, the most abundant living organisms in the ocean. Such models are based on various fundamental theories of optics which apply to a single particle, and make use of a set of rigorous equations. The optical properties which can be 'modeled' belong to the category of the inherent optical properties (IOP, see Radiative Transfer in the Ocean). Defined at the level of a single cell, the extension of IOPs to a collection of cells (a population) or to an assemblage of populations is straightforward from conceptual and numerical viewpoints. The computation of IOPs are carried out by using some physical characteristics of the organisms, or of the population (such as cell size, size distribution, chemical composition which governs the complex index of refraction).

Bio-optical models can also refer to various ways of describing and forecasting the 'bio-optical state' of the ocean, namely the optical properties of a water body as a function of the biological activity within this water. Both the IOPs and the apparent optical properties (AOPs) of the water are aimed at in such approaches. In contrast to the first kind of theoretical models, these models are essentially empirical, descriptive, and actually derived from field measurements. They initially rest on observations of some regular variations in the oceanic optical properties along with its algal content in 'Case 1 waters' (see **Table 1**). The chlorophyll concentration, [Chl], is commonly used as an index to quantify the algal content, and more generally the bio-optical state of ocean water. Once identified, and if recognized as statistically significant, such empirical relationships (between optical properties and [Chl] can be inverted, and thereafter used as predictive tools or model.

It is worth remarking that regular trends generally vanish in so-called Case 2 waters (**Table 1**). Indeed, in these waters the optical properties are no longer influenced just by phytoplankton and related particles, as they are in Case 1 waters. They are also, and independently, determined by other substances of terrestrial origin, notably by sediments and colored dissolved (organic) matter, carried from land into coastal zones and not correlated to [Chl]. Therefore, bio-geo-optical models, that might be developed and locally useful in such areas, are not of general applicability.

The two kinds of models are not disconnected. To the extent that the IOPs at the level of particles are additive, the first models, in principle, may be utilized to reconstruct the IOPs of a water body containing any assemblage of organisms and other (living or detritus) biogenic particles. Then these bulk IOPs can be combined through the radiative transfer equation (RTE) with the appropriate boundary conditions (the illumination conditions at the surface and the reflectance properties of the bottom, in particular), with a view to computing the AOPs at various depths within the water column. In this way, the result of the second category of models, the descriptive models, can be understood or interpreted.

Because empirical bio-optical models generally refer to the trophic level, depicted by [Chl], they are in essence restricted to upper oceanic layers, where the photosynthetic activity takes place, where the vegetal biomass is confined, and [Chl] is measurable. In addition, a considerable effort in developing bio-optical models originates from the need to interpret the satellite ocean color data in terms of chlorophyll concentration, which is only detectable in the upper oceanic layer. Possible relationships between optical properties and heterotrophic activity (bacterial abundance), or particulate organic carbon and minerogenic contents in the interior of the ocean, are not examined here.

Finally, it must be added that the spectral domain encompassed by bio-optical models is that of visible (or photosynthetic) radiation, namely the 400–700 nm domain, occasionally slightly extended toward the near infrared and near ultraviolet regions.

Table 1 Concepts and quantities used in bio-optical models

Case 1/Case 2 water. Case 1 waters are those waters in which phytoplankton and their accompanying and covarying retinue of material are the principal agents responsible for the variations in optical properties of the water bodies. The accompanying material includes living heterotrophic organisms, such as bacteria or virus, various debris of biological origin, and dissolved organic matter excreted by organisms or liberated by decaying detritus. Such waters are typical of the open ocean, far from land influence. Conversely, Case 2 waters are influenced not only by unicellular algae and related particles or substances, but also by other optically significant components, from terrestrial origin, such as inorganic and organic particles in suspension, yellow substances resulting from land drainage, and sediments resuspended from bottom

Quantity	Units	Symbol
Absorption coefficient	m^{-1}	a
Scattering coefficient	m^{-1}	b
Volume scattering function	$m^{-1}sr^{-1}$	$\beta(\theta)$
Back-scattering coefficient	m^{-1}	b_b
Back-scattering efficiency (the ratio b_b/b)	–	$\tilde{b}_b$
Attenuation coefficient ($c = a + b$)	m^{-1}	c
Chlorophyll-specific (absorption or scattering) coefficients of phytoplankton	$m^2 (mg\ Chl)^{-1}$	a_ϕ^*, b_ϕ^*
Efficiency factors for absorption and scattering (subscripts a, b, respectively), defined as the ratios of energy absorbed within the particle, or scattered out from the particle, to the energy impinging onto its geometrical cross-section		Q_a, Q_b
Relative size of a spherical particle, defined as $\alpha = \pi D n_w(\lambda_0)^{-1}$ D diameter, n_w refractive index of water, and λ_0, wavelength in vacuo	–	α
Relative (complex) refractive index of the particle, defined as the ratio of the index of the substance forming the particle to the refractive index of water (n, real part, n' imaginary part)	–	$m = n - in'$
Van de Hulst parameter, defined as $\rho = 2\alpha(n - 1)$	–	ρ
Depth of the euphotic layer where PAR is reduced to 1% of its surface value	m	Z_{eu}
Photosynthetic available radiation (within the 400–700 nm range)	photons $s^{-1}m^{-2}$	PAR
Attenuation coeffcient for downward irradiance (also K)	m^{-1}	K_d

Optical Models for Individual Particle or Population of Particles

In the open ocean Case 1 waters, phytoplankton with their accompanying retinue of living and detrital particles, are the principal agents responsible for the determination of the optical properties. The size of these particles extends from less than $0.1\,\mu m$ (virus, colloids, debris), to less than $1\,\mu m$ for heterotrophic bacteria and picoplanktonic algal species, and from 1 to tens or even several hundreds of micrometers for phytoplankton, protists, and large heterotrophic organisms (and actually up to tens of meters for whales). It is well known that the size distribution function of marine particles is rather monotonic, with numbers continuously increasing toward smaller sizes (**Figure 1**). A simple function, often sufficient to approximately describe the size distribution of oceanic particulate matter, is a power law (known as Junge distribution),

$$dn(x)/dx = N(x) = kx^{-j} \qquad [1]$$

where x is the size (e.g., the diameter, if the particles can be considered as spherical), $N(x)$ is the distribution function, i.e., the number of particles per unit of volume, having a given size x, and within a dx interval (around x), k is a scaling factor, and j is an exponent, with typical values around 4 for oceanic particles. Such a distribution means that an increase by a factor of 10 in size corresponds to a reduction in number (in frequency of occurrence) by a factor of 10 000.

As a consequence of this abruptly decreasing number of particles with size (combined with optical theories, see below), the particles which are the most optically significant are in the size range of 1–10 μm, and thus include most of the small heterotrophic organisms, phytoplanktonic cells, and various small debris. This is true for the scattering properties, but not for the back-scattering coefficient predominantly due to smaller particles (**Figure 1**). It is also true for absorption, even if in this process heterotrophs play a minor role (they are rather colorless); in contrast, algal cells containing a variety of pigments (chlorophylls, carotenoids, and occasionally phycobilins) are strongly absorbing bodies.

The interaction between radiation and an optical object like a particle is conveniently described by dimensionless numbers, called efficiency factors for absorption and for scattering, and denoted Q_a and Q_b, respectively (**Table 1**). One advantage of these

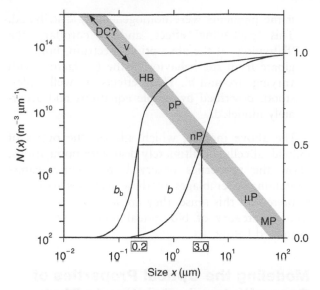

Figure 1 Schematic plot (shaded stripe, left-hand side ordinate log-scale) of the approximate numerical concentration of particles according to their size in oceanic waters (eqn [1] with $j = 4$). Approximate abundance versus mean size of major groups of microorganisms or particles are also indicated, with notations as follows: DC, debris and colloids; V, viruses; HB, heterotrophic bacteria; pP, picophytoplankton; nP nanoplankton; μP microplankton; MP, large macroplankton. (Adapted from Stramski and Kiefer (1991) Light scattering by microorganisms in the open ocean. *Progress in Oceanography* 28: 343–383.) Linear plot (right-hand ordinate scale, from 0 to 1) of the progressive value of the scattering coefficient, b, when the upper limit of the integral (eqn [3]) is increasing; the progressive value is relative and normalized by its final value (unity). A similar curve is drawn for the backscattering coefficient, b_b. For these computations, the relative refractive index of the spherical particles is 1.05, and the wavelength is 550 nm. (Adapted from Morel A and Ahn Y-H (1991) Optics of heterotrophic nanoflagellates and ciliates: atentative assessment of their scattering role in oceanic waters compared to those of bacterial and algal cells. *Journal of Marine Research* 49: 177–202.)

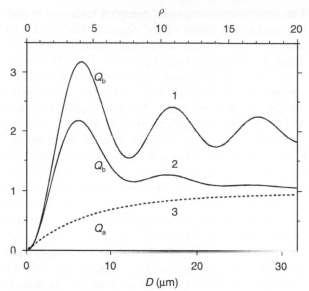

Figure 2 Efficiency factor for scattering, Q_b, as a function of the parameter ρ (defined in **Table 1**), or of the diameter, D, when the relative index of the particle is set equal to 1.05, and the wavelength λ is 675 nm. Curve represents Q_b for a nonabsorbing particle. Curve 2 represents Q_b for an absorbing particle, with $n' = 0.0075$, $n = 1.05$, and $\lambda = 675 nm$. Curve 3 represents the efficiency factor for absorption Q_a for the same n, n', and λ values as for curve 2. Note that when the size increases, Q_b oscillates around, and tends asymptotically, toward 2, if the particle is not absorbing, or toward 1 when it is absorbing; Q_a tends also toward 1.

factors lies in the fact that theories are available by which their values can be predicted as a function of the relative size (α, defined in **Table 1**), and the relative complex index of refraction of the particle (m, see **Table 1**). The Mie-Lorenz theory provide for spherical particles accurate Q values, and the angular values of the volume scattering function (*see* Radiative Transfer in the Ocean). When the refractive index of the particle is close to that of the surrounding medium (as is the case for most of the watery oceanic particles in suspension in water), the so-called van de Hulst approximation can apply and provide Q-factors more rapidly than via Mie computations.

If the particles are made of the same substance (same refractive index, m), and are assumed to be spherical, with the same diameter, D, the absorption and scattering coefficients, a and b, of the medium which contains these particles, are simply expressed as

$$a \text{ or } b = N\pi(D^2/4)\,Q_{a \text{ or } b} \qquad [2]$$

where N is the number of particles per unit volume, $\pi(D^2/4)$ represents the geometrical cross-section of a single (spherical) particle. The Q factors are simultaneously functions of D and m, through the parameter ρ (*see* **Table 1**).

For perfectly transparent particles ($n' = 0$), a and Q_a are obviously 0. In this case, Q_b after oscillations tends asymptotically toward 2 for increasing size (**Figure 2**); such a particle is able to remove from the radiative field by scattering twice the amount of radiation intercepted by its geometrical cross-section (this is often called the 'extinction paradox'). For absorbing particles, Q_a increases with increasing size, and Q_b, after some oscillations tends toward 1, like Q_a (**Figure 2**).

If the particles are not uniform in size, and the population is characterized by a size distribution function, $N(D)$ the eqn [2] must be integrated over the appropriate size interval, according to

$$a, \text{ or } b = (\pi/4) \int N(D)D^2 Q_{a \text{ or } b}(D,m)\mathrm{d}(D) \qquad [3]$$

The same parameters (size, complex refractive index) are used in the frame of the rigorous Mie theory, to compute the volume-scattering function (VSF, see **Table 1**) of any individual particle. The VSF for the entire particle population is simply obtained by adding the individual VSF with the appropriate weight, as derived from the distribution function $N(D)$. All the bio-optical models of the scattering properties of marine particles rest on such an approach. Their limitations do not originate from theory, but from the present lack of accurate information about the sizes and composition of the suspended material (in Case 1, and even more in Case 2 waters). If the actual pattern of the particle VSF is globally understood and can be reconstructed, the predictive skill of the models (actually the available information used as inputs) are still insufficient to allow the evolution of the back-scattering coefficient to be safely parameterized as a function of the bio-optical state (see reflectance modeling). What can be predicted, however, is the extremely low back-scattering efficiency exhibited by most of the unicellular organisms with low refractive index (such as algal cells, for instance). Also that this efficiency increases with the decreasing particle size is a theoretical evidence; as a consequence, the particles responsible for the formation of the scattering and back-scattering coefficients do not belong to the same size range (**Figure 1**).

In summary, theoretical models are available which account very well for most of the observed optical properties. They have been validated in particular through *in vitro* experiments and by using various cells grown in culture (heterotrophic or photo-autotrophic organisms). Several phenomena, predicted through theories and subsequent bio-optical models, are worth mentioning.

- Scattering by small (0.4–2 μm) organisms depends on the wavelength according to a λ^{-2} law. This spectral dependency is perfectly verified for heterotrophic bacteria (almost nonabsorbing bodies); for small phytoplanktonic cells (such as *Prochlorococcus* and *Synechococcus*), the presence of pigments results in localized features (minima) superimposed onto the general λ^{-2} spectral pattern.
- For larger organisms, the scattering spectrum may exhibit various shapes, including a 'flat' (λ^0) shape when the size exceeds approximately 10 μm. For algal cells, the various absorbing pigments always influence the scattering spectrum, by introducing minima and maxima in scattering throughout the absorption bands of these pigments.
- Absorption by pigmented cells in suspension differs from that of a 'solution' of the same material,

if the pigments were homogeneously distributed. This 'packaging' effect, and its corollary, the 'flattening' of the absorption spectrum, both originate from the behavior of the Q_a factors with varying size and n'. These effects are well understood, described by simple equations and accurately modeled.

The above models, which address the optics of individual cells, or ultimately deal with populations, allow the properties observed in oceanic waters containing assemblages of these particles to be interpreted. In this sense, they are able to support the second category of bio-optical models, which are examined below.

Modeling the Optical Properties of Ocean Waters in Relation to Their Biological State

In oceanic Case 1 waters, far from significant terrigenous influences, the origin of all materials present is necessarily to be found in the first link of the food chain, namely in photosynthesizing phytoplanktonic organisms. Heterotrophic organisms, as well as inanimate detritus or dissolved organic matter are related to algal biomass, and to the initial creation of organic matter (and particles) through photosynthesis. Therefore, the water optical properties are logically studied as a function of this vegetal biomass.

Because chlorophyll *a* is the single ultimate photosynthetic pigment, the most abundant in all living plants, and because it is easily determined, its concentration in the water is a convenient, albeit imperfect, index of the bio-optical state. It is worth recalling that the chlorophyll concentration in oceanic waters, used as descriptor of the bio-optical state, and to which the optical properties are to be related, varies within about 3 orders of magnitude (say 0.02 and 20 mg m^{-3}, between oligotrophic zones and eutrophic conditions in upwelling areas).

In practice, [Chl] and the optical properties within the upper layers must be measured simultaneously at sea to examine if statistically significant correlations can be found between some IOPs or AOPs and [Chl]. When such correlations are expressed mathematically (in general through nonlinear laws, and with a certain confidence interval), the corresponding expression can be used as a model or an algorithm. In contrast to what occurred for the previous category of bio-optical models (which are based on exact physical laws), the models examined below are in essence 'empirical'. Varying uncertainties are therefore attached to each model; depending on inclusion of new data, the numerical formulation or the input parameters of these models are still liable to further evolution.

Inherent Optical Properties and [Chl]

The absorption coefficient of oceanic water Beside the fixed contribution of water itself (a_w) to this coefficient, the varying biological contribution (sometimes denoted a_{bio}) can itself be partitioned into a component due to particulate material, a_p, and another due to colored dissolved organic matter, a_{cdom} (all these coefficients are spectral quantities, even if the symbol λ is omitted, when not necessary)

$$a = a_w + a_{cdom} + a_p \qquad [4]$$

In turn, a_p can be divided according to

$$a_p = a_\phi + a_{nap} \qquad [5]$$

where a_ϕ and a_{nap} are the partial absorption coefficients by phytoplanktonic cells, and 'nonalgal' particles, respectively. This nonalgal compartment includes colored debris and all kinds of heterotrophic organisms. Techniques are available to discriminate between a_p and a_{nap} and to determine their spectra; such measurements are performed with particles retained on a filter, before and after methanol extraction. The algal absorption a_ϕ spectrum is indirectly obtained by difference. Typical shapes of these spectra are shown in **Figure 3(A)** when [Chl] is $1\,mg\,m^{-3}$.

Statistical analyses of absorption measurements performed systematically in Case 1 waters with increasing [Chl] have demonstrated that the partial coefficients (a_ϕ and a_{nap}) increase with [Chl] in a nonlinear manner (**Table 2**, examples in **Figure 3B**). As a consequence, the chlorophyll-specific absorption coefficient of phytoplankton (a_ϕ, **Table 1**) is not a constant and decreases when [Chl] increases. Such a trend is, at least partly, due to the above-mentioned packaging effect. In addition, a regular change in the pigment composition is also at the origin of this decrease; recall that algal absorption originates from all accessory pigments, while normalization is made with respect to the sole chlorophyll a.

Most of the recent studies in oceanic waters have shown that algal cells are the dominant term in forming a_p. On average, a_ϕ would represent about 70% of a_p within the absorption blue maximum of algae (around 440 nm), and even more in the red peak (around 675 nm). The spectral a_ϕ and a_p patterns in **Figure 3(A)** are slightly changing with [Chl], as a consequence of the differences between the spectral

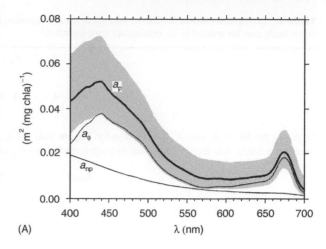

(A)

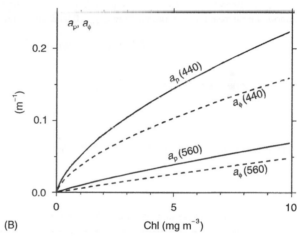

(B)

Figure 3 (A) Mean spectral absorption values, as they result from statistical analyses, when the chlorophyll concentration within the water body is $1\,mg\,m^{-3}$. The curves represents $a_p(\lambda)$, plotted inside a shaded area which represents $\pm$ 1SD, the nonalgal particle absorption spectrum, $a_{nap}(\lambda)$, and phytoplankton absorption spectrum, $a_\phi(\lambda)$ (see eqn [4]). (B) Nonlinear evolution of the mean absorption coefficients $a_p(\lambda)$ and $a_\phi(\lambda)$, with increasing chlorophyll concentration (see eqns [2.1] and [2.2] in **Table 2**); the two selected wavelengths correspond to the maximum (440 nm), and the minimum (560 nm) of algal absorption. (Adapted from Bricaud A, Morel A, Babin M *et al.* (1998) Variation of light absorption by suspended particles with chlorophyll a concentration in oceanic (case 1) waters: analysis and implications for bio-optical models. *Journal of Geophysical Research* 103: 31033–31044.)

values of the exponents (in eqns [2.1] and [2.2], **Table 2**). In comparison, nonalgal particles are less absorbing, and their absorption, regularly increasing toward the short wavelengths, is well modeled with an exponential function (**Table 2**, eqn [2.3]).

The measurements of a_{cdom} (performed on filtered water) are rather scarce in oceanic waters; they have not yet provided a clear relationship (if any) between this term and [Chl], but have shown that the spectral shape, namely a monotonous (exponential) increase of a_{cdom} toward the short wavelengths, is rather

Table 2 Bio-optical models for Case 1 waters, and corresponding equations by which the main inherent optical properties (IOP) of a water body can be related to its chlorophyll concentration[a]

$$a_p(\lambda) = A_p(\lambda)[\text{Chl}]^{Ep(\lambda)} \qquad [2.1]$$

$$a_\varphi(\lambda) = A_\varphi(\lambda)[\text{Chl}]^{E\varphi(\lambda)} \qquad [2.2]$$

Note that the terms A_p and A_φ are displayed in **Figure 3(A)**, as $a_p(\lambda)$ and $a_\varphi(\lambda)$ are shown when [Chl] is set equal to $1\,\text{mg m}^{-3}$; the exponents $Ep(\lambda)$, and $E\varphi(\lambda)$ are similar in magnitude but not equal, and they vary within the range 0.6–0.9, approximately.

$$a_{nap}(\lambda) = a_{nap}(\lambda_0)\exp[-S_{nap}(\lambda - \lambda_0)] \qquad [2.3]$$

$$a_{cdom}(\lambda) = a_{cdom}(\lambda_0)\exp[-S_{cdom}(\lambda - \lambda_0] \qquad [2.4]$$

Note that in these two last expressions, the reference wavelength (λ_0) is arbitrary (often 440 nm is adopted), and the slopes (S) of the exponential decrease are approximately $S_{nap} = 0.012\,\text{nm}^{-1}$ and $S_{cdom} = 0.015\,\text{nm}^{-1}$.

$$b(\text{Chl},\lambda_0) = b_w(\lambda_0) + b_p(\text{Chl},\lambda_0) \quad and \quad b_p(\text{Chl},\lambda_0) = B_p(\lambda_0)[\text{Chl}]^x \qquad [2.5]$$

with the exponent x in the range 0.6–0.7, approximately; with $B(\lambda_0)$, at $\lambda_0 = 550\,\text{nm}$, statistically found to vary between 0.15 and $0.45\,\text{m}^2\,(\text{mgChl})^{-1}$ (see also, **Figure 4**).

$$b_p(\text{Chl},\lambda) = b_p(\text{Chl},\lambda_0)(\lambda/\lambda_0)^y \qquad [2.6]$$

with the exponent y between -0.5 and -2 (the value -1 is commonly adopted).

$$b_p(C,\lambda_0) = B'_p(\lambda_0)[C]^{x'} \qquad [2.7]$$

with the exponent x′ close also to 1 (see also **Figure 4**).

[a]Subscripts: w, water; p, particles; ϕ, phytoplankton; nap, nonalgal particle; cdom, colored dissolved organic matter.

stable (eqn [2.4]). It is worth noting that this spectral dependency is close to that typical of a_{nap}, apart from a small difference in slope. All models presently proposed in the literature have the same exponential structure.

The scattering coefficient   In open ocean waters, the scattering coefficient, $b = b_w + b_p$, is the sum of the constant molecular scattering (b_w), and of a varying contribution, b_p, resulting from the presence of all kinds of particles, living organisms, and detritus. Unlike a_p, b_p cannot be split into algal and nonalgal contributions, as there is no experimental technique allowing such a discrimination.

Bio-optical models aim at relating this bulk coefficient b_p to [Chl]. *In situ* measurements of $b_p(\lambda)$, at fixed wavelength, λ_0 (or indirect determination via the particle attenuation coefficient $c_p(\lambda)$, from which $a_p(\lambda)$ is subtracted), have led, through least-squares analyses, to a nonlinear dependence upon [Chl] of the form

$$b_p(\lambda_0) = B_p(\lambda_0)[\text{Chl}]^x \qquad [6]$$

$B_p(\lambda_0)$ (which represents the value of b_p when Chl = $1\,\text{mg m}^{-3}$) has been found to vary within a factor 3 for Case 1 waters (see also **Figure 4**; and eqn [2.5]).

The nonlinearity in eqn [6] (a power law with $x < 1$) is such that when the algal biomass increases, b_p increases more slowly. This effect is generally attributed to a change (i.e., a decrease) in the relative contribution of detritus and of heterotrophic organisms to scattering when [Chl] increases (in eutrophic waters). This explanation is corroborated by the following observation: when b_p is studied as a function of the particulate organic carbon concentration, [C], instead of [Chl], it varies linearly with [C] (see **Table 2**, eqn [2.7]).

Spectral measurements of b_p in the open ocean are rather scarce. If the size distribution function obeys a Junge distribution (eqn [1]), the spectral dependency of the scattering coefficient can be theoretically predicted as being a power law, with an exponent (y, in **Table 2**, eqn [2.6]) which is related to j (eqn [1]), simply by: $y = 3 - j$. With typical values for j around 4, y would be around -1, as roughly observed, and

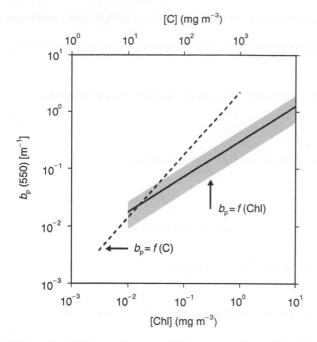

Figure 4 Particle-scattering coefficient as a function of the chlorophyll concentration in Case 1 waters, and at $\lambda = 550$ nm (eqn [2.5] in **Table 2**); the natural variability in this relationship for Case 1 waters is represented by the shaded band. The dashed line represents the empirical relationship statistically obtained between b_p and the organic carbon concentration (upper abscissa scale).

generally adopted in bio-optical modeling. Such a monotonic decrease of b_p throughout the spectrum is probably oversimplifying, particularly at high algal concentration; indeed, phytoplankton scattering spectra are, as mentioned before, featured in reponse to the pigment absorption.

Apparent Optical Properties and [Chl]

Downwelling irradiance The downwelling and upwelling irradiances (radiant flux per unit of area, *see* Radiative Transfer in the Ocean) are convenient measurements to make at sea to characterize the penetration of daylight into the water column. The depth variations of these quantities are quantified by diffuse attenuation coefficients (Radiative Transfer in the Ocean). When dealing with downward irradiance E_d, the corresponding coefficient is K_d (simply written K). It depends on the IOPs of the water and on the geometrical structure of the light field. Only by approximation, it can be seen as the sum of a term due to the water itself and a varying contribution of all materials (particulate and dissolved) originating from biological activity, so that

$$K(\lambda) \cong K_w(\lambda) + K_{bio}(\lambda) \qquad [7]$$

The first term can be (again approximately) expressed as a function of the IOPs of optically pure seawater (a_w and b_w), whereas $K_{bio}(\lambda)$, resulting from the presence of all kinds of biological materials, can be related to [Chl].

On the basis of many field measurements in oceanic waters, it has been shown that the spectral $K_{bio}(\lambda)$ values do not vary at random but are interrelated. The proposed optical classification is based on the realization that such a rather regular change affects simultaneously all the wavelengths and progressively modifies the entire spectrum. Bio-optical algorithms derived from statistical analyses of these field data allow the entire $K(\lambda)$ spectrum to be specified, as soon as $K(\lambda_0)$, at a reference wavelength λ_0 is known (**Table 3**, eqn [3.1]).

A second way of analyzing the field data consists of relating the $K_{bio}(\lambda)$ values to [Chl]. In Case 1 waters, the diffuse attenuation coefficients appear to be highly correlated to [Chl], and the statistical relationships (linear regression on log-transformed quantities) are expressed as

$$K_{bio}(\lambda) = \chi(\lambda)[Chl]^{e(\lambda)} \qquad [8]$$

with exponents $e(\lambda)$ are always <1, whatever the wavelength. The corresponding $K(\lambda)$ bio-optical model consists of a set of such nonlinear expressions based on eqns [7] and [8] (eqn [3.2], **Table 3**; **Figure 5**). More complex models have also been used (eqn [3.4]).

To the extent that K is largely determined by absorption, the nonlinear character of the correlation between K and [Chl] is not surprising and resembles that observed for a (**Figure 3B**). By integrating over the whole visible domain (the photosynthetic available radiation (PAR) domain, *see* Radiative Transfer in the Ocean), a relationship between Z_{eu}, the depth of the euphotic zone (cf. **Table 1**), and [Chl] can be obtained. This nonlinear relationship can also be derived through a direct analysis of the column-integrated chlorophyll content and of Z_{eu}, observed at sea by using a photometer able to determine the vertical PAR profile. This bio-optical algorithm is useful to predict, in Case 1 waters, the depth of the euphotic zone when the vertical chlorophyll profile has been determined (eqn [3.3], **Table 3**).

Irradiance reflectance This apparent optical property, $R(\lambda)$, is crucial in the interpretation of the remotely sensed ocean color. It is defined as the ratio of upwelling irradiance, E_u, to downwelling irradiance at the same depth (actually just beneath

Table 3 Bio-optical models, and corresponding equations by which the main apparent optical properties (AOP) in Case 1 waters can be related to the chlorophyll concentration [Chl]

$$K(\lambda) = K_w(\lambda) + M(\lambda)[K(\lambda_0) - K_w(\lambda_0)] \qquad [3.1]$$

where $K_w(\lambda)$ is the attenuation coefficient for downwelling irradiance in pure water, $M(\lambda)$ are statistically derived coefficients and λ_0 a reference wavelength.

$$K(\lambda) = K_w(\lambda) + \chi(\lambda)[Chl]^{e(\lambda)} \qquad [3.2]$$

$\chi(\lambda)$ and $e(\lambda)$ are empirical wavelength-dependent factors and exponents, derived from statistical analysis.

$$Z_{eu} = z[Chl]^\zeta \qquad [3.3]$$

z is about 38 m, and the exponent ζ is close to 1/2; this expression is computed by using eqn [3.2], combined with a standard spectrum of solar radiation at sea level.

$$K(\lambda[Chl]) = k(\lambda)\exp\left\{-k'(\lambda)\log_{10}\left([Chl]/[Chl_0]^2\right)\right\} + 0.001[Chl]^2 \qquad [3.4]$$

$[Chl_0]$ is a reference concentration (0.5 mg m^{-3}), and $k(\lambda)$ and $k'(\lambda)$ are empirical spectral parameters derived from statistical analysis.

$$\log_{10}([Chl]) = a_0 + a_1 r + a_2 r^2 + a_3 r^3 \qquad [3.5]$$

where $r = \log_{10}[R(\lambda_1)/R(\lambda_2)]$ is the decimal logarithm of a ratio of reflectances at two wavelengths; the cubic polynomial is an example (first order polynomials have also been proposed). The inverse relationships, which express various r ratios as a function of $\log_{10}$ ([Chl]), also have the same polynominal structure.

---

the surface in remote sensing applications)

$$R(\lambda) = E_u(\lambda)/E_d(\lambda) \qquad [9]$$

Other expressions involving upward radiance (instead of E_u) are also in use in ocean color remote sensing; they are geometrically related to $R(\lambda)$ as defined, and physically modeled in a similar way.

Historically, purely empirical models were the first to have been developed. As for K_d, quasi-simultaneous field measurements of $R(\lambda)$ and of [Chl] were carried out, and then statistically analyzed. The relationships obtained through such regression analyses were used as algorithms when processing ocean color data. Most of these algorithms consider the ratios of reflectance at two wavelengths, λ_1 and λ_2. Statistical analyses have demonstrated that these ratios vary regularly with [Chl]. Therefore, they can be operated as predictor of [Chl], through a certain function F:

$$[Chl] = F[R(\lambda_1)/R(\lambda_2)] \qquad [10]$$

When such ratios of reflectances (called band ratio in ocean color science) and [Chl] are plotted in log-log space, the scatterplot is sigmoid. In the so-called empirical 'chlorophyll-algorithms' in use in ocean color remote sensing, the functional dependency is

expressed through a polynomial with respect to the log-transformed quantities (**Table 3**; linear approximations have also been employed).

Thanks to radiative transfer studies, it has been shown that (to a first approximation) $R(\lambda)$ can be expressed as a function of two inherent properties, a and b_b, through

$$R(\lambda) = f.b_b/(a + b_b) \qquad [11]$$

where f is not a constant factor (because R is not an inherent property), but depends in a complex manner on the structure of the light field, and on the VSF of the particles. Its value (between about 0.3 and 0.5 for common cases in oceanic waters) can only be computed by solving the RTE with the appropriate boundary conditions. As a consequence of eqn [11], a bio-optical model for reflectance reduces to a combination of models for a and b_b, separately considered. Because a and b_b are inherent properties, the additivity principle applies (eqns [4] and [5], for a). Therefore a purely analytical model would be built if a and b_b (actually each term forming a and b_b) could be completely parameterized in terms of [Chl].

Such a model is not yet available. Modeling the back-scattering coefficient is presently an unresolved problem. This coefficient is generally expressed as

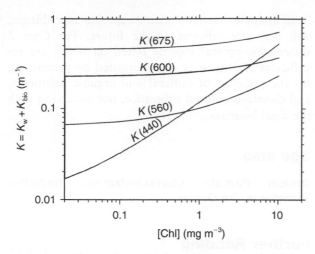

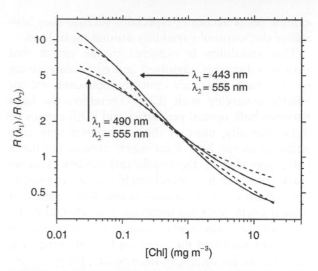

Figure 5 Diffuse attenuation coefficient for downward irradiance, at selected wavelengths, as a function of the chlorophyll concentration in Case 1 waters. The initial values (the ordinates when [Chl] = 0.02 mg m^{-3}) are almost the pure sea water values (K_w in eqn [7], see also eqn [3.1], in **Table 3**, for the term K_{bio}).

Figure 7 Example of outputs of models for Case 1 waters providing the evolution with [Chl] of ratios of reflectances at two wavelengths (as indicated) (see eqn [3.2], **Table 3**).

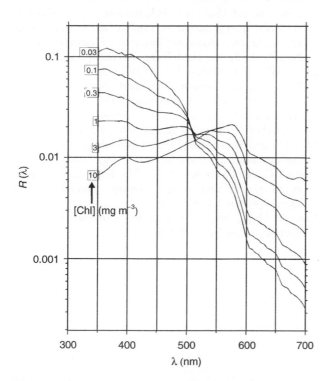

Figure 6 Example of a spectral model for reflectance as a function of [Chl] and for Case 1 waters; the various bio-optical models presently in use are not identical but very similar.

being the product $b.\tilde{b}_b$, of the scattering coefficient (which can be related to [Chl] through eqn [6]), and the dimensionless back-scattering efficiency $\tilde{b}_b$. The latter quantity is not presently well documented, and rather divergent hypotheses (including a constant

value) have been proposed to express its dependency on [Chl] and on λ.

When dealing with absorption, the role of a_{cdom} in open ocean is largely unknown, or more precisely its magnitude has not yet been successfully related to [Chl]. For this reason, in particular, a can advantageously be replaced by its proxy, K_d, which cumulates the influence of all absorbing substances, and is easily modeled (see above). Some manipulations of the RTE, however, are necessary to derive a from K_d. Such models are called 'semianalytical' (see **Figure 6**). They are also used to predict the ratios of reflectances as a function of [Chl], and reciprocally (as through eqn [10]); they result in algorithms having the same structure as those derived from the empirical approach (eqn [3.4], **Table 3**, and **Figure 7**). Other bio-optical models, involving reflectances at more than two wavelengths, have also been proposed.

Reasonable agreements have been reached between such semi-analytical approaches, empirical approaches, and independent data sets (not used when developing the models). The sigmoidal shape is well understood if not analytically reproduced with accuracy.

Conclusions

In terms of accuracy or predictive capabilities, the first category of bio-optical models, dealing with the properties of individual organisms, are robust particularly because they rest on firm theoretical bases. The second category, dealing with bio-optical properties of oceanic water bodies, in essence empirical, cannot be as accurate, especially because they also reflect the 'natural' variability around mean laws.

This variability is expected to the extent that various substances, dissolved and particulate, living or inanimate, which are optically significant, are not strictly covarying with [Chl]. Therefore, the links between bulk optical properties and [Chl] cannot be tight. Actually, most of these properties may vary within a factor 2 or 3 (or more) around the mean value provided by the (nonlinear) models. This remark holds true for coefficients like a_p, a_φ, b_p, and K. The situation is somewhat better when ratios of optical coefficients are involved, and when covariations of these coefficients reduce the amplitude of fluctuations around the mean (as for reflectance and band ratios, for instance). In summary, the predictive skill of such models is inevitably limited by the tightness of the regressions on which they are based.

It is necessary to bear in mind that further studies at sea will likely change the result of the regression analyses, and thus the numerical values of the tabulated parameters entering into the models. For this reason, only equations are given in **Tables 2** and **3**, the provisional numerical information to be found in the literature, is likely amenable to modification in the future).

Most of the bio-optical models, by which optical properties and [Chl], are related, rely on nonlinear relationships, and often on power laws of [Chl], with exponents below 1. As a result, the optical properties vary within a narrower interval than does [Chl] in Case 1 waters, and do not span more than 2 orders of magnitude (instead of 3), which is still considerable. It can be envisaged that for Case 1 waters 'regional' bio-optical models, encompassing a less wide chlorophyll concentration range, and accounting for local biological activity with its specific assemblages, will be more efficient in the future. For Case 2 waters, bio-optical models based on [Chl], are insufficient and must be supplemented by accounting for the presence of mineral and organic sediments, and dissolved colored substance, not covarying with the algal biomass.

See also

Optical Particle Characterization. Radiative Transfer in the Ocean.

Further Reading

Demers S (1991) *Particle Analysis in Oceanography.* NATO-ASI Series, vol. G 27. Berlin Heidelberg: Springer-Verlag.

Gordon HR, Brown OB, Evans RH, *et al.* (1988) A semi-analytical radiance model for ocean color. *Journal of Geophysical Research* 93: 10909–10924.

Gordon HR and Morel A (1983) Remote sensing of ocean color for interpretation of satellite visible imagery: a review. *Lecture Notes on Coastal and Estuarine Studies.* New York, Berlin, Heidelberg, Tokyo: Springer-Verlag.

Jerlov NG and Nielsen ES (1974) *Optical Aspects of Oceanography.* London, New York: Academic Press.

Morel A (1988) Optical modeling of the upper ocean in relation to its biogenous matter content (Case 1 waters). *Journal of Geophysical Research* 93: 10749–10768.

Morel A and Smith RC (1982) Terminology and units in optical oceanography. *Marine Geodesy* 5: 335–349.

Smith RC and Baker KS (1978) The bio-optical state of ocean waters and remote sensing. *Limnology and Oceanography* 23: 247–259.

Van de Hulst HC (1957) *Light Scattering by Small Particles.* New York: Wiley.

FLUOROMETRY FOR BIOLOGICAL SENSING

D. J. Suggett and C. M. Moore, University of Essex, Colchester, UK

Introduction

Oceanography has been transformed through the use of fluorescence to assay biology. Light-absorbing pigments cause many organisms to naturally fluoresce. Primarily, but not exclusively, these pigments are associated with photosynthesizing organisms, such as algae, aquatic vascular plants, and aerobic anoxygenic photosynthetic (AAP) bacteria. Fluorescence is most commonly detected as the emission that follows 'active' excitation using an actinic light source. One of the major breakthroughs for oceanography occurred in the 1960s with the detection of actively induced chlorophyll *a* fluorescence *in situ*. Chlorophyll *a* is contained by all algae and cyanobacteria and thus provides a measure of abundance. However, at typical environmental temperatures the chlorophyll *a* fluorescence emission signature largely originates from oxygen evolving photosystem II (PSII). Consequently, the chlorophyll *a* fluorescence signal contains information that can be used to characterize the photosynthetic activity of this complex. Recent technological advances have provided twenty-first-century oceanography with an array of active chlorophyll *a* induction fluorometers that are used routinely to assess photosynthetic physiology. In addition, enhanced capacities within remote sensing platforms has enabled researchers to make major steps in using 'passive' fluorescence from chlorophyll *a*, the fluorescence that is stimulated as a result of natural excitation by the sun (solar-stimulated), to assess global photosynthetic activity.

In addition to the natural (auto-)fluorescent molecules, the process of introducing compounds into cells for binding to, and thus labeling, specific molecules has further expanded the tools available to oceanographic research. Such compounds are either themselves fluorescent or become fluorescent following a specific biological reaction and have opened up many new avenues for exploring physiology and taxonomy. Here we provide a brief synthesis of fluorescence techniques used to examine the biology of marine systems.

What Is Fluorescence?

Light absorption by chromophoric molecules (pigments) raises electrons within those molecules to an excited state or higher energy level. As the molecules de-excite, most of the energy is released as heat; however, a proportion of the energy is also released as light. Fluorescence is the emission of light at a longer wavelength following light absorption at a shorter wavelength. The fluorescence yield that arises from an excitation light source is typically referred to as F. Many factors including the optical geometry of the measurement system comprising the excitation light source, sample compartment, and detector will affect F. Typically, F is measured in relative units, although, for a given optical geometry (C), F will also vary according to the absorptivity (A) of the sample and the photon flux density (PFD) of the excitation light source:

$$F = \phi_F \cdot C \cdot A \cdot PFD$$

The value of C depends on the proportion of fluoresced light that is intercepted by the detector and the efficiency of conversion of photons into an electric signal. Importantly, ϕ_F is the quantum yield of fluorescence, or the molar ratio of light emitted to light absorbed, a value that is unique to the inherent physical properties of the fluorescent molecule.

Estimating Abundance

Fluorescence is most commonly used to estimate the abundance of chlorophyll *a*. In the case of chlorophyll *a*, relaxation from excited state 1 to ground state results in dissipation of a small proportion (3–5%) of the excitation energy at wavelengths greater than *c.* 650 nm, that is, as red fluorescence. Most fluorometers deliver narrow-band blue light to correspond with wavelengths at which algae and plants exhibit maximum rates of light absorption (**Figure 1**). Subsequent detection of fluorescence is centered toward 680–700 nm to coincide with the peak emission by chlorophyll *a*. Almost all (>99%) of fluorescence at these wavelengths arises from PSII for the majority of algae and aquatic plants in nature.

When algal and plant pigments are extracted into solvents, the fluorescence that arises (F) is expected to exhibit a one-to-one relationship with the concentration of chlorophyll *a* (and their breakdown products). However, such a relationship is not

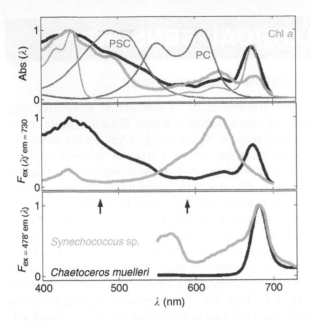

Figure 1 Absorption and fluorescence spectra for a diatom (*Chaetoceros muelleri*) and a phycocyanin-containing cyanobacteria (*Synechococcus* spp.): top panel – optical absorption spectra on intact cellular suspensions. Overlaid are absorption spectra for extracts of the predominant pigments, chlorophyll *a* (chl *a*) and photosynthetic carotenoids (PSC) for *Chaetoceros* and chl *a* and phycocyanin (PC) for *Synechococcus*; middle panel – the fluorescence emission at 730 nm following hyperspectral excitation. The two arrows are to demonstrate that excitation in the blue yields a higher F_{730} for *Chaetoceros* since excitation of chl *a* and PSC is favored while excitation in the orange yields a higher F_{730} for *Synechcoccus* since excitation of chl *a* and PC is favored. In this way, fluorometers with fluorescence yields recorded for excitation at various wavelengths can provide some taxonomic discrimination; bottom panel – hyperspectral fluorescence emission following excitation by blue light, a wavelength commonly used for excitation sources of many commercial fluorometers. Most fluorometers are designed to measure fluorescence emission at *c.* 680 nm (chl *a*, e.g., *Chaetoceros*); however, fluorescence from PC can contaminate the chl *a* signal as is the case for *Synechococcus*. Maximum values for all spectra are scaled to a value of 1 for unity.

expected for natural intact samples. Viable cells modify the relationship between F and the chlorophyll *a* concentration as a result of variability of the amount of light absorbed, the proportion of absorbed light that is transferred to PSII, referred to as the transfer efficiency (Φ_t), as well as of ϕ_F. As a result, the ratio of F to chlorophyll *a* concentration can alter according to taxonomic (genetic) as well as physiological (acclimation and stress) variability. Consequently, chlorophyll *a* fluorescence in nature does not provide an absolute measurement of the amount of chlorophyll *a* in a water sample but can be used to infer changes of abundance.

Similar arguments hold for pigmented organisms other than aquatic plants and algae that fluoresce, most notably, corals and anemones (anthazoa).

These organisms contain green fluorescence-like proteins (GFPs), a family of chromophoric molecules that contribute to the colorfulness of corals and fluoresce at wavelengths between 450 and 600 nm. Although the function of GFPs is widely debated, researchers have recently employed GFP fluorescence *in situ* to determine the abundance of polyps recruited onto coral reefs.

Phylogenetic Discrimination

In addition to chlorophyll *a*, algae and aquatic plants contain many other chromophoric molecules termed accessory pigments that act to supplement light capture. Light absorption by chlorophyll *a* is restricted to narrow wavebands centered on *c.* 440 and 670 nm; however, light spectra within aquatic systems are relatively broad (*c.* 300–750 nm). Accessory pigments absorb at wavelengths not targeted by chlorophyll *a* (**Figure 1**). Some accessory pigments have a high Φ_t with chlorophyll *a* and actively supplement the light that is harvested for photosynthesis, whereas others have a relatively low Φ_t and are termed photoprotective pigments. Importantly, a specific accessory pigment array is unique to each algal family providing a first-order taxonomic discrimination of complex algal communities. An accessory pigment that is excited with a light source tuned to the wavelength of peak absorption will induce a higher chlorophyll *a* fluorescence yield. Therefore, the chlorophyll *a* emission signature from sequential excitation of multiple wavelengths provides information on the relative abundance of specific accessory pigments and thus on certain algal families.

In a similar manner, fluorometers can be tuned to target-specific accessory pigments to assess changes in abundance of a specific algal group. Cyanophytes are a group of prokaryotic organisms that are also known as blue-green algae but are in fact bacteria that possess photosynthetic machinery, including chlorophyll *a* and accessory pigments. Blooms of cyanophytes occur frequently in coastal waters and lakes with many species producing substances that can prove highly toxic to other organisms, including humans if ingested in large quantities. The most abundant pigments in these organisms, collectively termed phycobilins, have a unique auto-fluorescence signature that can be easily detected in addition to that of chlorophyll *a* (**Figure 1**).

Prochlorophytes are oxygenic phototrophic prokaryotes and, along with the cyanophyte *Synechococcus*, represent the most abundant phytoplankton cells throughout the world's oceans. Cells of both phytoplankton groups are extremely small (0.2–2 μm) and require use of their fluorescence signatures for enumeration. Traditional microscope-based epifluorescence techniques rely on the fluorescence naturally generated by cells under actinic light; however, the fluorescence yielded by these phytoplankton groups is often too dim to make this approach viable for oceanography. Instead, flow cytometry (FCM), a technique of biomedical origin, is now commonly used. FCM employs a combination of light scattering by cells and auto-fluorescence by natural photosynthetic pigments (chlorophyll *a* and the orange fluorescing phycobilin, phycoerythrin) to enable both identification and enumeration of the different phytoplankton groups.

Modification of FCM via introduction of fluorescence tags into water samples can be used to discriminate other functional groups of aquatic organisms. For example, highly sensitive nucleic acid-specific fluorescent stains that target DNA or rRNA have also made possible the detection and enumeration of heterotrophic bacteria and most recently of viruses. Similarly, a more advanced technique, referred to as fluorescence *in situ* hybridization (FISH), can be used to selectively target regions of DNA or rRNA that consist of evolutionarily conserved and variable nucleotide sequences, thus enabling discrimination of cells at any taxonomic level ranging from kingdom to species. For example, the FISH technique has been modified to enable identification of the taxonomic and life cycle status of single coccolithophore cells collected from the ocean.

Physiological Applications

Many of the fluorescence-based approaches used to examine abundance and taxonomy have been further modified to provide insights into physiology. Applications of fluorescent tags, molecules, and dyes that can be covalently bound to sensing biomolecules are now widespread within online fiber-optic biosensors, DNA sequencing, DNA chips, protein detection, and immunoassays. Near-infrared (NIR) fluorescent dyes are often most desirable since nonspecific background fluorescence is considerably reduced in the NIR spectral range and hence sensitivity can be significantly improved. For example, NIR fluorescence dyes attached to esters can be used to examine the nature of intracellular signaling mechanisms. Recent advances in understanding the biological production of reactive oxygen species, such as singlet oxygen ($^1O_2$), superoxide (O_2^-), and hydrogen peroxide (H_2O_2), which can be destructive to proteins and lipids, has been achieved almost exclusively through use of fluorescent tags. Consequently, fluorescence-based approaches can be considered to be 'everyday' tools for examining physiology and molecular-based ecology. However, these approaches require removal of samples from nature. Perhaps the greatest advance for examining physiology *in situ* again takes advantage of the unique fluorescence characteristics of chlorophyll *a*.

Chlorophyll *a* fluorescence contains a measure of ϕ_F but represents just one of several pathways that can be used to dissipate absorbed excitation energy, the others being photochemistry (ϕ_P) and heat (ϕ_H). Conservation of energy requires that preferential use of one pathway decreases the use of the others such that $\phi_F + \phi_P + \phi_H = 1$. Therefore, an excitation protocol that changes the quantum (photon) yields of competing processes that affect ϕ_F can thus provide information on photosynthetic physiology. Such an approach is termed 'variable' fluorescence induction. Absolute quantum yields are not typically measured, nor do they need to be, provided that the fluorescence signals that are measured accurately report changes in the relative quantum yield of fluorescence.

One variable fluorescence induction approach that has become popular in oceanography is fast repetition rate (FRR) fluorometry and its derivatives. Here, the excitation that is delivered cumulatively closes the PSII reaction centers that dissipate excitons via photochemistry to raise F from an initial background (F_0) to a maximum (F_m) value (**Figure 2**). This process occurs within a single trapping event of the reaction center pool, termed a 'single turnover' (ST). A biophysical model describing the process of sequential reaction center closure is fit to this fluorescence transient to yield various parameters that describe the photochemical process (**Figure 2**). Another protocol commonly favored to assay PSII physiology is multiple turnover (MT) induction using an excitation pulse of longer duration to turn over the pool of reaction centers more than once (**Figure 2**). ST and MT protocols provide very different physiological information.

Use of variable fluorescence techniques to provide measures of photochemical efficiency, termed $F_v/F_m = (F_m - F_0)/F_m$, for algae and aquatic plants has become commonplace in oceanography (see **Figure 3**). Some of the earliest *in situ* measurements were made in the mid-1990s to demonstrate lower photosynthetic viability under limitation of nutrients, such as iron and nitrogen. Chlorophyll *a* fluorescence yields the same action spectrum as that of O_2 evolution confirming that it can provide a representative description of PSII photochemical efficiency. Several investigations have also demonstrated good agreement between F_v/F_m and

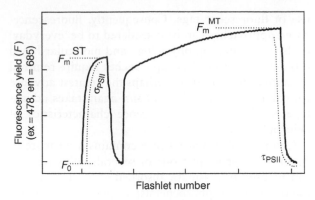

Figure 2 Fast repetition rate (FRR) style chl *a*-fluorescence induction for the microalga *Symbiodinium* spp. Each induction consists of a four-step sequence: (1) single turnover (ST) excitation from a short (*c.* 100 μs) pulse; (2) ST relaxation from a weak modulated light (*c.* 500 ms); (3) MT excitation from a longer (*c.* 600 ms) pulse; (4) MT relaxation from a weak modulated light (over 1 s). In this way the fluorescence is modulated in a controlled manner between the initial background and maximum yield, termed F_0 and F_m, respectively. A biophysical model describing the oxidation–reduction of PSII is fit to the induction profile to yield measures of various physiological parameters, notably, the effective absorption by PSII (σ_{PSII}) and the turnover time of electrons by PSII (τ). σ_{PSII} describes the target area available for light harvesting by the PSII reaction centers τ describing the rate at which acceptor molecules within PSII re-oxidize following excitation and thus the rate at which electrons can be funneled out of PSII. Both σ and τ characterize a wide range of acclimation and adaptation properties inherent to microalgae and AAPs.

the quantum yield of photosynthesis based on 'conventional' O_2 measurements. However, as has been identified from phytoplankton cultures in the laboratory, natural variability of F_v/F_m is notoriously difficult to interpret since ϕ_F can vary between taxa largely as a result of differences in photosynthetic architecture (**Figure 3**). Measurements of F_v/F_m also become problematic where a proportion of fluorescence that is induced does not originate from (active) chlorophyll *a*. One extreme example comes from cyanobacteria species containing a phycobilin called phycocyanin that often bloom in the Baltic Sea. Fluorescence emitted by phycocyanin is centered between *c.* 615 and 645 nm and has a waveband that overlaps with the fluorescence emitted by chlorophyll *a* (see **Figure 1**). Consequently, phycocyanin can lead to an elevated measure of the minimum fluorescence yield (F_0) that is detected and consequently to a lower F_v/F_m. Recently, fluorescence lifetime imaging (FLIM) has been introduced to avoid some of these issues associated with (what are essentially empirical) chlorophyll *a* induction kinetic measurments of F_v/F_m. FLIM is a technique for producing an image based on differences in the exponential decay rate of fluorescence and thus provides mechanistic information inherent to ϕ_F.

Measurements of PSII physiology as photochemical efficiency, effective absorption and electron turnover F_v/F_m, τ, and σ_{PSII} (**Figures 2** and **3**) are strictly obtained (1) in darkness either *in situ* at night and (2) in the laboratory upon discrete samples removed from the environment. Consequently, all parameters yield the characteristics inherent to the potential to process excitation energy. However, these parameters are all modified in the light as a result of changes to ϕ_P, ϕ_H, and ϕ_S, and hence the minimum and maximum fluorescence. Such measurements are important since they inform the 'functional' capacity of PSII to process electrons. Here, these parameters are denoted as F_q'/F_v', τ', and σ_{PSII}', respectively. Measurements of F_q'/F_v' represent the product of two other parameters of physiological interest, F_q'/F_m' (the PSII operating efficiency) and F_v'/F_m' (the maximum PSII yield under actinic light), which describe the contribution of photochemistry and heat to changes of F_v/F_m, respectively. Increasingly, these fluorescence parameters are being employed to address complex aspects of photosynthetic physiology that are observed in nature.

Algal and Aquatic Plant Productivity

Changes of abundance over time provide a simple index of productivity. Numerous laboratory studies have employed fluorescence to monitor phytoplankton abundance within cultures; however, such measurements rarely equate to the actual cellular growth rate since ϕ_F can vary according to the cellular growth phase. An alternative measurement of productivity can be obtained from transformation of the physiological parameters afforded from variable chlorophyll *a* fluorescence induction of PSII into photosynthesis rates.

Simple algorithms have been introduced that use FRR-style induction techniques to generate the gross photosynthetic electron transfer by PSII (PET, mol electrons (mol PSII reaction center)$^{-1}$ s^{-1}):

$$PET = \sigma_{PSII}' \cdot (F_q'/F_v') \cdot PFD$$

Measurements of PET *in situ* are relatively easy to make but surprisingly rarely reported, primarily since PET measurements are expressed in units that are difficult to interpret ecologically. Conversion of PET to a measure of productivity expressed in conventional units, such as O_2 evolution or CO_2 fixation, is not trivial for natural samples. Within PSII, four electron steps are required to evolve 1 mol of O_2; hence, PET should be proportional to 4 times the molar O_2 evolution by all functional PSII reaction centers (RCIIs). In the laboratory, the amount of O_2

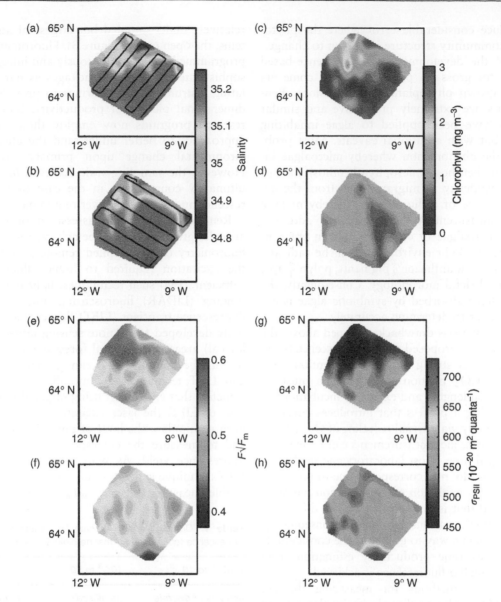

Figure 3 An example of small spatial and temporal resolution of phytoplankton physiology using active fluorescence: maps of salinity (a, b), chlorophyll (c, d), F_v/F_m (e, f), and σ_{PSII} (g, h), across a section of the Iceland-Faroes front during two surveys separated by around 2 weeks during June 2001 (Moore, unpublished). A bloom is observed along the frontal boundary and is associated with marked changes in physiological parameters (e–h) measured by active fast repetition rate fluorometry. These different 'physiological acclimation states' were associated with communities containing different phytoplankton species. Therefore, fluorescence-based physiological parameters contain information on both taxonomy (adaptation) and physiology (acclimation and/or stress).

evolved per unit chlorophyll a per ST saturating flash, termed the photosynthetic unit size of PSII (PSU_{PSII}), can be determined relatively easily. This measurement accounts both the yield of O_2 evolved per electron step and the ratio of RCIIs per unit chlorophyll a. Therefore, gross O_2 production by PSII per unit chlorophyll a ($P^{O_2}_{chl\ a}$) can be obtained as

$$P^{O_2}_{chl\ a} = PET \cdot PSU_{PSII}$$

Of major concern for $P^{O_2}_{chl\ a}$ determinations in oceanography is thus an accurate understanding of

the variability of PSU_{PSII} in nature. Direct determination of PSU_{PSII} for natural phytoplankton samples has been achieved but only when the biomass was considerably high, such as within shelf seas or during blooms, to provide an adequate O_2 signal. Alternative (indirect) methods have been proposed to estimate PSU_{PSII} from knowledge of the spectral absorption and fluorescence excitation characteristics, σ_{PSII} and the pigment array. However, such alternative methods have only received limited validation. Most oceanographic investigations to date have simply assumed values for PSU_{PSII}, an approach that can introduce considerable error where the phytoplankton community structure is known to change.

Much of the development of fluorescence-based algorithms for gross O_2 productivity has come via laboratory-grown phytoplankton suspensions. Benthic habitats are extremely productive and similar algorithms have been applied to algae inhabiting substrates but with additional caveats. Most problematic is the phenomenon whereby microalgae inhabiting the benthos (microphytobenthos) have a frustrating tendency to migrate away from the actinic light source of a fluorometer, thereby making prolonged measurements inaccurate. In addition, symbiotic microalgae of corals inhabit an environment where the light environment can be radically altered by the host anthazoa's pigments, polyp shape, and $CaCO_3$ skeletal morphology. Consequently, the amount of light absorbed by symbiotic algae is extremely difficult to determine accurately.

Despite the various drawbacks outlined above, the major promise proposed for fluorescence-based productivity is as strong as ever. Conventional O_2 evolution or CO_2 fixation measurements must be performed upon lengthy and expensive incubations of discrete samples, a process that introduces considerable error notably via internal recycling of O_2 or CO_2. However, fluorescence measurements can be made *in situ* and in real time. Most laboratory and field $P_{\text{chl } a}^{O_2}$ determinations do not correspond one-to-one with simultaneous O_2 evolution or CO_2 fixation measurements, a trend that is not surprising given difficulties associated with the various approaches. Consequently, there remains some way to go before the considerable promise of real-time productivity estimation from active chlorophyll *a* fluorescence is achieved.

New *Optode* methods for measuring the net community O_2 evolution directly *in situ* also employ fluorescence. *Optode* technology is based on the ability of selected substances to act as dynamic fluorescence quenchers in the presence of oxygen. Not only do such methods provide highly accurate and sensitive O_2 measurements but also, upon coupling to *in situ* fluorescence-based $P_{\text{chl } a}^{O_2}$ determinations, may potentially provide important new insights into ecosystem function, metabolism, and trophic coupling.

Scales of Productivity Measurements

A second advantage afforded from fluorescence over conventional O_2 evolution or CO_2 fixation productivity determinations is an increase of both spatial and temporal scale that can be investigated. Such an advantage is particularly important for the relatively under-sampled but largest of aquatic systems, the open ocean (**Figure 3**). Fluorometers can be programmed to log continuously and integrated into sophisticated bio-optical packages as part of undulating instruments or moorings to provide a multidimensional picture of productivity. Indeed, many research programs now employ these larger-scale approaches to better understand the effects of environmental change upon primary productivity. However, the sampling scale achieved *in situ* is still ultimately constrained, in the case of active fluorescence and FCM, by power requirements (**Table 1**).

Remote sensing enables assessment of productivity at scales greater than can be achieved *in situ*. Active fluorometry can be applied remotely provided that the excitation required to induce fluorescence is sufficient. Two such techniques, light detection and ranging (LIDAR) fluorosensing and laser-induced fluorescence transient (LIFT) fluorescence, were initially developed for remote sensing large-scale (ecological) areas of terrestrial forest canopies but have recently been applied to oceanography. Both LIDAR and LIFT techniques are modified induction fluorometers that excite PSII using targeted lasers. In the case of LIFT, the laser excitation signal is used to both manipulate the level of photosynthetic activity and to measure the corresponding changes in the fluorescence yield. As with FRR fluorescence, the LIFT technique generates fluorescence transients that enable measurements of F_v/F_m, τ, and σ_{PSII}. Ulti-

Table 1 Common scales of sampling (meters) with current fluorescence techniques for marine biological applications

$< 10^{-6}$ to 10^{-3}	10^{-3} to 10^3	$> 10^3$
Analysis of discrete samples:	Small-scale in situ campaigns:	Large-scale in situ campaigns*
Fluorescence imaging microscopes (epifluorescence and active chl a fluorescence induction)	Induction fluorometers	Remote sensing:
	Optodes	LIDAR and LIFT
	Flow cytometers	Satellite retrieval
Flow cytometers		
Intracellular molecular tags		

The smallest scales are dominated by techniques that perform measurements at the single-cell (or intracellular) scale upon discrete water samples removed from nature. Mesoscales are dominated by instruments that can be deployed *in situ*, for example, variable chlorophyll *a* fluorometers and modified flow cytometers (*Cytobuoy* and *FlowCytobot* technologies), while the largest scales are dominated by remote sensing techniques. The asterisk indicates that any of the techniques could be applied but that the scale afforded is limited by sampling effort and power requirements.

mately, power constraints and safety considerations also constrain measurement scales for these laser-based techniques.

Remote sensing of ocean productivity largely relies on satellite retrieval of ocean color reflectance by algal pigments. Productivity algorithms most simplistically combine chlorophyll a concentration and PFD; however, such an approach has proven difficult to retrieve quantitative information on algal physiology.

Fluorescence can be observed passively via remote sensing since algae and aquatic plants naturally fluoresce when they are excited by sunlight (solar stimulation). Chlorophyll a fluorescence is inherent to the water-leaving radiance with peak emission at 683–685 nm. By analogy to the fluorescence yield induced actively, the amount with which chlorophyll a fluorescence increases the water-leaving radiance (F) depends on several factors, including the specific absorption of chlorophyll (A^{chl}), fluorescence quantum efficiency (ϕ_F), amount of incident sunlight (PFD), in addition to various atmospheric effects (C^*):

$$F = \phi_F \cdot C^* \cdot A^{chl} \cdot PFD$$

Chlorophyll a fluorescence (F) is quantified according to the fluorescence line height (FLH) above background levels at 678 nm, several nanometers below the wavelength of peak emission to avoid atmospheric oxygen absorption at 687 nm. Estimation of ϕ_F is achieved as $F/(C^* \cdot A^{chl} \cdot PFD)$, where $(C^* \cdot A^{chl} \cdot PFD)$ is the instantaneous absorbed radiation by phytoplankton, termed ARP. A semi-analytical model that also retrieves chlorophyll a concentration from measurements of water-leaving radiance in all bands from 412 to 551 nm is used to determine the ARP. Primary productivity (PP) can then be determined as

$$PP = \phi_P \cdot ARP$$

Therefore, the relationship between ϕ_F and the quantum yield of PSII photochemistry (ϕ_P) must be known. As shown above, this relationship will be modified by ϕ_H and ϕ_S, parameters that are not easy to constrain for natural samples as a result of the huge taxonomic and physiological (both acclimation and inhibition) variability. Consequently, considerable effort is currently invested in trying to understand the relationship between ϕ_H and environment for key algal taxa. Despite these potential drawbacks, passive measurements of global chlorophyll a fluorescence show great potential for future observations of global productivity and may offer one of

our most powerful monitoring tools as the Earth is subjected to a time of extreme climate change.

Nomenclature

A	absorptivity
A^{chl}	absorption specific to chlorophyll
C	geometry of fluorometer optics
C^*	atmospheric effects
F	fluorescence yield
F_m	maximum fluorescence yield
F_0	initial fluorescence yield
F'_q/F'_m	PSII operating efficiency
F'_q/F_m	PSII photochemical efficiency (under actinic light)
F_v/F_m	PSII photochemical efficiency (dark adapted)
F'_v/F'_m	PSII maximum yield under actinic light
$P^{O_2}_{chl\ a}$	gross O_2 production by PSII per unit chlorophyll a
σ_{PSII}	PSII effective absorption cross section (dark adapted)
σ'_{PSII}	PSII effective absorption cross section (under actinic light)
τ	turnover time of electrons (dark adapted)
τ'	turnover time of electrons (under actinic light)
ϕ_F	quantum yield of fluorescence
ϕ_H	quantum yield of heat
ϕ_P	quantum yield of photochemistry
Φ_t	energy transfer efficiency between pigment molecules

See also

Aircraft Remote Sensing. Bio-Optical Models. Fluorometry for Chemical Sensing. Inherent Optical Properties and Irradiance. Optical Particle Characterization. Photochemical Processes. Satellite Oceanography, History and Introductory Concepts.

Further Reading

Andersen RA (ed.) (2005) *Algal Culturing Techniques*. New York: Elsevier Academic Press.

Barbini R, Colao F, Fantoni R, Palucci A, and Ribezzo S (2001) Differential LIDAR fluorosensor system used for phytoplankton bloom and seawater quality monitoring in Antarctica. *International Journal of Remote Sensing* 22: 369–384.

Falkowski PG and Raven JA (1997) *Aquatic Photosynthesis*. New York: Blackwell.

Geider RJ and Osborne BA (1992) *Algal Photosynthesis: The Measurement of Algal Gas Exchange*. Boca Raton, FL: Chapman and Hall.

Groben R and Medlin L (2005) *In situ* hybridization of phytoplankton using fluorescently labeled rRNA probes. *Methods in Enzymology* 395: 299–310.

Hoge FE, Lyon PE, Swift RN, *et al.* (2003) Validation of Terra-MODIS phytoplankton chlorophyll fluorescence line height. I. Initial airborne LIDAR results. *Applied Optics* 42: 2767–2771.

Huot Y, Brown CA, and Cullen JJ (2005) New algorithms for MODIS sun-induced chlorophyll fluorescence and a comparison with present data products. *Limnology and Oceanography Methods* 3: 108–130.

Jeffrey SW, Mantoura RFC, and Wright SW (1997) *Phytoplankton Pigments in Oceanography*. Paris: UNESCO Publishing.

Kolber ZS, Klimov D, Ananyev G, Rascher U, Berry J, and Osmond B (2005) Measuring photosynthetic parameters at a distance: Laser induced fluorescence transient (LIFT) method for remote measurements of photosynthesis in terrestrial vegetation. *Photosynthesis Research* 84: 121–129.

Kolber ZS, Prášil O, and Falkowski PG (1998) Measurements of variable chlorophyll fluorescence using fast repetition rate techniques: Defining methodology and experimental protocols. *Biochimica Biophysica Acta* 1367: 88–106.

Labas YA, Gurskaya NG, Yanushevich YG, *et al.* (2002) Diversity and evolution of the green fluorescent protein family. *Proceedings of the National Academy of Sciences of the United States of America* 99: 4256–4261.

Moore CM, Suggett DJ, Hickman AE, *et al.* (2006) Phytoplankton photo-acclimation and photo-adaptation in response to environmental gradients in a shelf sea. *Limnology and Oceanography* 51: 936–949.

Papageorgiou GC and Govindjee (eds.) (2004) *Chlorophylla Fluorescence: A Signature of Photosynthesis*. Amsterdam: Springer.

Suggett DJ, Maberly SC, and Geider RJ (2006) Gross photosynthesis and lake community metabolism during the spring phytoplankton bloom. *Limnology and Oceanography* 51: 2064–2076.

Veldhuis MJW and Kraay GW (2000) Application of flow cytometry in marine phytoplankton research: Current applications and future perspectives. *Scientia Marina* 64: 121–134.

Relevant Websites

http://www.cytobuoy.com
– Environmental applications of flow cytometry, CytoBuoy.

http://www.esa.int
– European Space Agency Living Planet.

http://www.chelsea.co.uk
– Fast Repetition Rate Methods Manual, Chelsea Technologies Group.

http://modis.gsfc.nasa.gov
– MODIS web, NASA.

FLUOROMETRY FOR CHEMICAL SENSING

S. Draxler and M. E. Lippitsch,
Karl-Franzens-Universität Graz, Graz, Austria

Introduction

In the context of ocean sciences, 'chemical sensing' denotes the acquisition of data on the concentration of certain chemical species (the analytes) in sea water. This acquisition process may be achieved by a 'chemical sensor' in the strict sense, i.e., a device that continuously and reversibly exhibits a change in some of its properties as a function of the concentration of a respective analyte. As an alternative, sensing may be done directly, without the detour over the properties of a sensor, by measuring some effect displayed by the analyte itself. Intermediate between these two methods is the use of an indicator that is added to a sample of sea water and changes its properties depending on the analytes' concentration. An example of a sensor is the Clark electrode, which delivers a current proportional to the concentration of oxygen. An example of direct sensing is the determination of salinity by conductivity measurements. The well-known use of litmus to monitor pH is an example for the indicator method.

'Fluorometry' refers to the measuring of fluorescence, which is the reemission of light from a compound upon exposition to light of a shorter wavelength (in this context, phosphorescence is included under the term fluorescence, although strictly speaking these are two different processes). Fluorescence occurs from a singlet excited state, while phosphorescence is originating from a triplet. This emission is characteristic for the emitting compound, but may be subject to modification by the environment. Thus, fluorometry may be applied to both sensing schemes outlined above: In a sensor device, its characteristic fluorescence properties may be modified by the concentration of the analyte. Direct sensing preferentially will be applied if the analyte itself is fluorescent.

Fluorescence

The fluorescence emission displays a number of general characteristics. Except for atoms in the vapor phase, the emission is shifted to lower wavelengths relative to the excitation. The energy loss between excitation and emission, the so-called 'Stokes shift', occurs owing to rapid decay of excitation energy to the lowest vibrational level of the excited state, as well as from subsequent decay to higher vibrational levels of the ground state. As the relaxation to the lowest vibrational level usually occurs within 10^{-12} s, the fluorescence emission spectrum does not depend on the excitation wavelength. The fluorescence can be described using the parameters spectral distribution of the emission, quantum yield, decay time, and polarization.

If the molecule is excited to its first excited electronic state, fluorescence is usually emitted directly from this state, and the fluorescence emission spectrum appears to be a 'mirror image' of the absorption spectrum (**Figure 1**). Deviations from the mirror image rule occur if the excitation goes to higher-lying states, if the molecule undergoes major geometrical rearrangements in the excite state, or if excited-state chemical reactions occur.

Fluorescence decay times usually range from some picoseconds to nanoseconds, or up to milliseconds in the case of phosphorescence.

The fluorescence decay time τ (sometimes also called 'fluorescence lifetime') describes the average time a molecule spends in the excited state before it returns to the ground state. In the simplest case, the fluorescence intensity follows an exponential law:

$$I(t) = I_0 e^{-t/\tau} \qquad [1]$$

where τ is the time it takes for the fluorescence intensity to decrease to a value of $I_0 e^{-1}$. The decay time is connected to rate constants via

$$\tau = \frac{1}{\Gamma + k} \qquad [2]$$

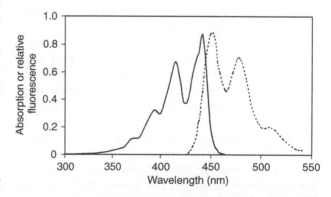

Figure 1 Typical absorption and fluorescence of an organic compound, showing 'mirror image' relationship between the absorption spectrum (----) and the fluorescence spectrum (----).

and therefore depends on all processes affecting the radiative or nonradiative rate constant.

Fluorescence decay times usually range from some picoseconds to nanoseconds. If triplet states are involved in the emission process, the measured decay time can be of milliseconds to seconds, and this emission usually is called phosphorescence.

Excitation with polarized light results in polarized emission, since fluorophores preferentially absorb photons whose electric vectors are parallel to the transition dipole moment of the fluorophores. The fluorescence polarization P after excitation with vertically polarized light is defined as

$$P = \frac{I_n - I_p}{I_n + I_p} \qquad [3]$$

where I_p and I_n are the fluorescence intensities of the vertically (n) and horizontally (p) polarized emission. In a solid environment at low fluorophore concentrations, the polarization is determined only by the relative directions between absorption and emission dipole moment in the fluorophore. In liquids, where the fluorophore molecules may move freely, a decrease in polarization can be observed owing to rotational diffusion or excitation transfer between fluorophores.

Techniques

Direct Sensing

Direct measurement of the concentration of fluorescent materials in sea water is usually done with samples taken in the field and transported to a laboratory, either on board a research ship or on land. The procedure of measuring involves three steps: excitation of the analyte, detection and identification of fluorescent light, and determination of concentration of fluorescent analyte.

To excite a given analyte a source emitting light within the analyte's absorption band is necessary. A monochromatic or at least narrowband source serves to discriminate the analyte against all compounds with different absorption. In any case, the detected fluorescence may be the superposition of the emissions from various different analytes, and to add further specificity, the fluorescence light has to be spectrally analyzed. Most analytes have a characteristic 'fingerprint' spectrum; thus, it is possible to identify the analyte from its spectrum provided a limited choice of analytes is known to be present. The main difficulty is in extracting from the intensity of the detected fluorescence the concentration of the respective analyte. This can only be done with a

certain amount of reliability if an 'internal standard' can be used for comparison. If water samples are analyzed in the laboratory, a fluorophore (e.g. quinine sulfate) added in a well-defined concentration can serve as a standard. This is not possible for *in situ* measurements. The only substance available as a standard in any case is water itself, but water is nonfluorescent. Also, scattered light cannot be used, since scattering strongly depends on the amount of small particles (dust, microorganisms, etc.) in the sample.

There is, however, a weak wavelength-shifted component, called Raman scattering, that is useful as a standard. Raman scattering does not, like fluorescence, produce a specific wavelength but rather a difference in wavelengths between excitation and emission. Thus, in shifting the excitation wavelength the scattered wavelength is also shifted. The intensity of the Raman band of water may be used as the standard against which the fluorescence intensity of the analyte is calibrated (**Figure 2**). The quinine sulfate standard is related to the Raman standard by defining the ratio of the integrated fluorescence of a $1 \mu g \ l^{-1}$ quinine sulfate solution to the integrated H_2O Raman band as the 'quinine sulfate unit' (QSU). Results in QSUs should be directly comparable among different instruments and different laboratories.

A much more refined method is three-dimensional fluorescence (or excitation–emission matrix spectroscopy, EEM). The excitation as well as the emission wavelengths are scanned over a board range and the fluorescence intensity is plotted as a function of both (**Figure 3**). Sophisticated mathematical evaluation allows identification and relative quantification of multiple analytes.

Another technique for distinguishing between multiple analytes is time-resolved fluorometry. Excitation is done with very short light pulses

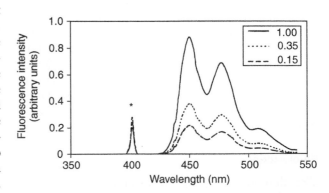

Figure 2 Fluorescence spectra of perylene at three different concentrations with the water Raman line (*) as the reference (maximum concentration 100 ng/l).

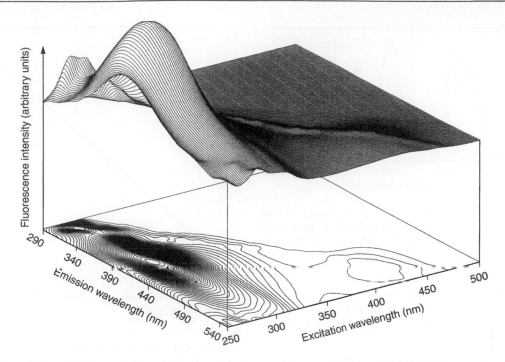

Figure 3 Three-dimensional or excitation-emission matrix spectrum. The small ridge running across the spectrum is the water Raman line. (Courtesy of M.J.P. Leiner and K. Kniely.)

(duration < 1 ns), and the fluorescence spectrum is recorded after a certain delay. Since the fluorescence decay times of various compounds are different, the spectrum will change with increasing delay. With short delays, all excited compounds contribute to the spectrum. After a short while, the fluorescence of some of the compounds has vanished while that of others persists (**Figure 4**). A careful analysis of a sequence of successively delayed spectra provides the relative concentration of the various fluorescent compounds contained in the sample.

Fluorescence polarization measurement is a rather unusual technique in ocean sciences. The only experimental difference from simple fluorometry is that excitation is done with polarized light and the emission is recorded separately for two perpendicular directions of polarization.

Indicators and Sensors

Indicators and sensors have in common that the fluorescent compound is not a constituent of the sea water under investigation but is added deliberately during the measuring process. From the standpoint of fluorometry it makes no difference whether the fluorescent indicator is dissolved in the sea water sample or is incorporated in a sensor element that is brought into contact with the water. Therefore, the following technical description applies to indicators and sensors as well.

Fluorescent chemical sensor devices consist essentially of three main building blocks: an illumination device, the sensor element proper, and a detecting apparatus. Since the fluorescence to be excited and detected is that of the sensor element alone, much less effort needs to be made to discriminate between various sources of fluorescence. This opens the possibility of using small and cheap light sources such as light-emitting diodes. Usually the sensor is used *in situ* rather than in a laboratory and the sensor element is dipped directly into the sea water. However, this implies that the excitation light has to be brought to the sensor element. This can be done by implementing a submersible sensor unit consisting of the light source, the sensor element, and even the detector. Alternatively, only the sensor element may be submersed and light can be transported via optical fiber to and from the element. In some cases even the sensor element may be part of an optical fiber.

The sensor element in most cases is a polymer membrane containing a fluorescent indicator dye. The fluorescence of the indicator can be altered by the presence of the analyte in several different ways: The analyte may interact with the excited molecule in a way that brings about changes in the fluorescence intensity. On the other hand, it binds to the indicator and thus alters the molecular properties in such a way that the absorption and/or emission are shifted spectrally. And finally, binding of the analyte may not alter the spectral properties of an indicator

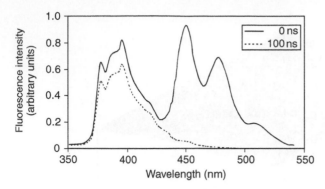

Figure 4 Time-resolved fluorometry of a mixture of two compounds with decay times 4 ns and 300 ns. Spectra taken immediately after excitation (0 ns) and with a considerable delay (100 ns).

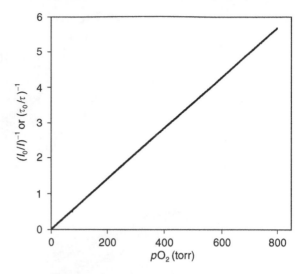

Figure 5 Stern–Volmer plot (see text) of fluorescence quenching by oxygen.

molecule but may change its mass and size. This yields a different moment of inertia and hence a different rotational diffusion and a change in fluorescence depolarization.

A reduction in fluorescence intensity is called quenching. Two types of quenching can be distinguished. Quenching is called dynamic if the interaction between the quencher (analyte) and the luminophore (indicator) occurs while the latter is in the excited state. In this case the rate constant of the nonradiative decay is increased, shortening the decay time and diminishing the quantum efficiency. Dynamic quenching is not associated with a chemical reaction between indicator and analyte. If, on the other hand, the interaction occurs while the indicator is in its electronic ground state, the quenching is called static. Static quenching may or may not include a chemical reaction.

In the simplest case, the reduction of the fluorescence intensity or the decay time is described by a linear function, the so called Stern–Volmer equation,

$$\frac{I_0}{I} = 1 - k_{SV}[A] \qquad [4]$$

where I_0 and I are the unquenched and quenched intensity, k_{SV} is the Stern–Volmer constant (a quantity characteristic of a certain combination of indicator and quencher), and $[A]$ is the concentration of the quenching analyte (**Figure 5**).

The Stern–Volmer equation shows that we have to measure two quantities, I and I_0, to extract the analyte's concentration. Because of the more reproducible conditions in a sensor membrane, instead of using an internal standard it is sufficient to recalibrate the sensor from time to time by applying a standard solution of the respective analyte. This is rather easy if the sensor is used in a laboratory environment, but

presents difficulties for *in situ* measurements, especially when the sensor equipment is intended to measure autonomously over an extended time. This difficulty can be overcome by measuring the fluorescence decay time rather than the intensity. The decay time also obeys a Stern–Volmer law but it is independent of drifts in fluorescence intensity that may be brought about by leaching or bleaching of the indicator dye or by aging effects in the light source or the detector. Thus, decay time-based sensors, despite requiring a more sophisticated optoelectronics, have much superior long-term stability and, in fact, are the only promising candidates for unattended monitoring purposes.

Static quenching implies some close association between the indicator molecule in its ground state and the analyte. Usually an equilibrium is reached between associated indicator–analyte pairs and free indicator, which is determined by the association constant characteristic for the partners. The higher the concentration of the analyte, the more the equilibrium is shifted toward the association.

The associated and the free indicator may have distinct fluorescence spectra or distinct decay times. The spectra of the two forms intersect at a certain wavelength, the so-called isosbestic point. The total fluorescence intensity at that wavelength is independent of the relative concentrations and hence independent of the analyte. Thus, this intensity can be conveniently used as an internal standard. For two distinct decay times, the decay function becomes the sum of two exponentials

$$I(t) = I_0\left(ae^{-t/\tau_a} + be^{-t/\tau_b}\right) \qquad [5]$$

where a and b represent the relative concentrations and τ_a and τ_b the respective decay times of the associated and the free indicator. The concentration of the analyte can be extracted from the decay curve by a fitting algorithm.

If the associated and free indicator have different absorption spectra, the result in fluorometry is simply a different fluorescence intensity because of a different probability of excitation. If, on the other hand, the species have different emission spectra, the result will be the superposition of the two spectra. In many cases there exists an isosbestic point, that is, a wavelength where the two emission spectra cross and where the emission intensity is independent of the relative concentrations of two species. This point may be used for internal referencing. The concentration of the analyte can be deduced from the ratio of emission intensities at two wavelengths that are outside the overlapping region of the two spectra.

Applications

Direct Measurement of Fluorescent Species in Sea Water

The most important marine fluorophore is chlorophyll. However, this article is devoted only to sensing of dissolved compounds (see Fluorometry for Biological Sensing) for fluorophores in living organisms). **Table 1** summarizes the most commonly measured substances as well as their excitation and emission wavelengths.

Dissolved organic matter (DOM) is the catch-all denotation for all sorts of organic products from metabolism and biological decomposition. The term dissolved organic carbon (DOC) is more specific and quantitative as it is related to the carbon content in the dissolved organic matter. Consequently, DOM can be quantified only in mass per unit water volume, while DOC can be specified in moles of carbon per unit volume. DOC nevertheless subsumes a multitude of compounds. They are important components of the global carbon cycle. Little is known about the detailed chemical composition of DOM, but recent investigations have shown that acyl oligosaccharides may form a significant fraction.

A large part of DOM is formed by components that can be detected via fluorescence (fluorescent organic matter, FDOM). The ratio FDOM/DOM seems to be rather constant in surface waters; accordingly, the correlation between DOC and fluorescence intensity is fairly robust. However, compounds of low molecular weight essentially lack fluorescence, at least when excited in the usual wavelength range for DOC (250–350 nm). This low-

Table 1 Dissolved chemical species in sea water measurable directly via fluorescence

Analyte	$\lambda_{ex}/\lambda_{em}$ (nm)	Decay time (ns)
Dissolved organic matter (DOM/DOC)	220–390/300–500	<2
Dissolved proteinaceous materials	220–280/300–350	<2
Humic acids	230, 300–350/420–450	<2
Polycyclic aromatic hydrocarbons	300–450/400–550	4–130
Trace plastic and epoxy compounds	UV/UV	<2
Heavy metals	X-ray	—

λ_{ex}, excitation wavelength; λ_{em}, emission (fluorescence) wavelength.

fluorescent DOC was determined at about 150 μmol in the Mid-Atlantic Bight. DOC from rivers seems to be more strongly fluorescent than DOC from marine systems. Fluorescence intensity is 2–2.5 times higher in deep waters, where photodecomposition and oxidation of high-molecular-weight matter is low. There may, in certain extreme cases, even be a negative correlation between DOC and fluorescence intensity. This shows that the use of fluorescence data to determine DOC quantitatively is not straightforward. Recalibration by comparison with chemical DOC quantification is necessary if water conditions change significantly.

Among the components of DOM, the nutrients (nitrate and phosphate) play a major role because of their importance for metabolism. They are usually nonfluorescent, but their concentrations often correlate well with overall DOC and thus also with fluorescence intensity under equivalent conditions. However, the ratio of carbon to nitrogen or phosphorus may vary in a wide range (observed C:N ratios are in the range 16–38). Specific detection of nutrients is usually impossible with direct fluorescence methods.

The situation is more favorable for proteinaceous material. The characteristic absorption and emission wavelengths (λ_{ex} 220–270 nm, λ_{em} 300–350 nm, respectively) allow differentiation from other compounds. The same is true for humic acids, which can be distinguished by their unusually long-wavelength emission (420–450 nm).

The second large bulk of fluorescent compounds in sea water is made up of pollution products. The main sources are sewage, oil spills, runoff, and atmospheric deposition. A class of special interest is that of polycyclic aromatic hydrocarbons (PAHs), which are released into the environment in large quantities. They are quite persistent, and some of them are potent carcinogens. Their fluorescence spectra are not significantly different from those of other organic material, but they are outstanding by their long fluorescence decay times. Thus they can be distinguished by time-resolved spectroscopy. The detection limit for PAHs by time-resolved fluorometry has been shown of the order of nanograms per liter in the presence of DOC at milligrams per liter.

Three-dimensional fluorometry has also been used to differentiate between various constituents of fluorescent matter in sea water. For humic acids, different excitation emission maxima were found for different types of water: coastal waters peaked at 340/440 nm, shallow transitional waters at 310/420 nm, eutrophic waters near 300/390 nm, and deep-sea samples at 340/440 nm (exitation/emission wavelengths). Although the chemical nature of the different humic species is not yet clear, these results provide means of distinguishing between water mass sources in the ocean. Three-dimensional fluorometry was found to be advantageous for identifying pollutants. It was used, for example, to detect trace plastic and epoxy compounds in the presence of other organic materials.

X-ray fluorescence is the method of choice for measuring metal contaminants. Excitation as well as emission wavelengths are in the X-ray region. The spectra obtained are rather specific for certain metals. Portable X-ray fluorescence spectrometers are available for measurement in the field.

Indicators and Sensors

Table 2 summarizes the sensor type, the range covered, and the sensitivity obtained for several analytes.

Dissolved oxygen is the analyte most readily measured using a fluorescent optical chemical sensor. Oxygen, because of its triplet ground state, is a notorious fluorescence quencher. Various organic fluorescence indicators have been proposed for measuring oxygen, but in recent years metalloorganic complexes have been found to be most suitable, especially ruthenium diimine complexes and platinum or palladium porphyrins. These indicators are outstanding because of their long fluorescence lifetimes and good quenchability by oxygen. The sensors cover the whole range of dissolved oxygen concentration encountered in aquatic environments. Compared to electrochemical sensors (Clark electrode) they have the advantage that they do not consume oxygen and hence may be used even in stagnant water. The technique of lifetime imaging has been applied to the study of spatial oxygen distribution in sediments.

Table 2 Dissolved chemical species in sea water measurable via fluorescence sensor devices

Analyte	Sensor type	Range, sensitivity
Oxygen	Decay time	0–60 ppm, 1 ppb
pH	Decay time	6–10, <0.1
Carbon dioxide	Intensity, decay time	200–1000 ppm, 1 ppm
Aluminum	Intensity	10–1000 nmol l^{-1}, 10 nmol l^{-1}

Fluorescent sensors for other analytes have been made commercially available only in the field of medical diagnostics. Fluorescent optical sensors have been developed by several laboratories for marine research, but these have not reached the market yet. A sensor system for CO_2 was developed by C. Goyet and colleagues. Using a combination of fluorescent and absorptive indicator dyes, they succeeded in measuring pCO_2 in sea water with a mean relative error of about 2% compared to results from a gas chromatograph.

Luminescent pH sensing has also been reported. In the range from pH 6 to 9, a resolution of better than 0.1 pH is readily achieved.

A fluorescent indicator rather than a true sensor has been used to determine dissolved aluminum in sea water at the nanomolar level. The indicator salicylaldehyde picolinoylhydrazone complexed with aluminum gives a fluorescence peaking at 486 nm when excited at 384 nm. In the concentration range from 38 to 930 nmol l^{-1}, a detection limit of 9.8 nmol l^{-1} and a precision of about 2% were achieved.

Numerous other sensors have been tested in the laboratory, but no proof of suitability for marine applications has been given. There are remarkable new approaches to the detection of nutrients (nitrate, phosphate), but in these cases the specifications needed for marine research have not yet been accomplished.

Conclusion

Numerous fluorescent sensing methods have been developed, but only a limited number have become standard methods in marine research. The most recent developments (decay-time sensors, three-dimensional fluorescence, lifetime imaging) offer up interesting prospects. In particular, the high long-term stability of decay time-based sensors make them prime candidates for instrumentation on unattended measuring stations needed for global ocean observing systems.

Symbols used

[A]	analyte concentration
Γ	radiative rate constant
I	fluorescence intensity
I_0	fluorescence intensity immediately after excitation
I_n	fluorescence intensity polarized perpendicular to the plane of incidence
I_p	fluorescence intensity polarized parallel to the plane of incidence
k	radiationless rate constant
k_{SV}	Stern–Volmer constant
λ_{ex}	excitation wavelength
λ_{em}	emission wavelength
Q	fluorescence quantum yield
P	fluorescence polarization
τ	fluorescence decay time
t	time

See also

Fluorometry for Biological Sensing.

Further Reading

Aluwihare LI, Repeta DJ, and Chen RF (1997) A major biopolymeric component to dissolved organic carbon in surface sea-water. *Nature* 387: 166–169.

Glud RN, Gunderden JK, and Ramsing NB (2000) Electrochemical and optical oxygen microsensors for *in situ* measurements. In: Buffle J and Horvai G (eds.) *In-situ Monitoring of Aquatic Systems*. Chichester: Wiley.

Goyet C, Walt DR, and Brewer PG (1992) Development of a fiber optic sensor for measurement of $p CO_2$ in sea water: design criteria and sea trials. *Deep-Sea Research* 39: 1015–1026.

Karabashev GS (1996) Fluorometric methods in studies and development of the ocean (a review). *Okeanologiya* 36: 165–172 (In Russian.).

Lakowicz JR (1999) *Principles of Fluorescence Spectroscopy*. Dordrecht: Kluwer Academic.

Manuelvez MP, Moreno C, Gonzalez DJ, and Garciasvargas M (1997) Direct fluorometric determination of dissolved aluminium in seawater at nanomolar level. *Analytica Chimica Acta* 355: 157–161.

Wolfbeis OS (ed.) (1991) *Fiber Optic Chemical Sensors and Biosensors*. Boca Raton, FL: CRC Press.

INHERENT OPTICAL PROPERTIES AND IRRADIANCE

T. D. Dickey, University of California, Santa Barbara, CA, USA

Introduction

Light is of great importance for the physics, chemistry, and biology of the oceans. In this article, a brief introduction is provided to the two subdisciplines focusing on light in the ocean: ocean optics and bio-optics. A few of the problems addressed by these subdisciplines are described. Several of the *in situ* sensors and systems used for observing the sub-surface light field and optical properties are introduced along with general explanations of the operating principles for measuring optical variability in the ocean. Some of the more commonly used ocean platforms and optical systems are also discussed. Finally, some examples of oceanographic optical data sets are illustrated.

Solar radiation, which includes visible radiation or light, impinges on the surface of the ocean. On average, a small fraction or percentage is reflected back into the atmosphere (roughly 6% on average) while a high fraction penetrates into the ocean. This fraction (or percentage), defined as the albedo, varies in time and space as a function of several factors including solar elevation, wave state, surface roughness, foam, and whitecaps. Radiative transfer is a branch of oceanography termed 'ocean optics,' a term that denotes studies of light and its propagation through the ocean medium. Radiative transfer processes depend on the optical properties of the components lying between the radiant source (e.g., the sun) and the radiation sink (the ocean and its constituents). Another commonly used term is 'bio-optics,' which invokes the notion of biological effects on optical properties and light propagation and vice versa. Some of the data used for examples here focus on bio-optical and physical interactions.

Solar radiation spans a broad range of the electromagnetic energy spectrum. The visible portion, roughly 400–700 nm, is of primary concern for ocean optics for many practical problems. These include: underwater visibility, photosynthesis and primary production of phytoplankton, upper ocean ecology, red tides, pollution, biogeochemistry including carbon cycling, photochemistry, light-keyed migration of organisms, heating of the oceans, global climate change, and remote sensing of ocean color. It is worth noting that optical oceanographers are also interested in the ultraviolet (UV) range and the infrared. For example, UV radiation can damage phytoplankton and can also lead to modifications in genetic material, and thus evolution, whereas the penetration of the infrared is critical for near surface radiant heating.

Solar radiation is clearly a primary driver for ocean physics, chemistry, and biology as well as their complex interactions. Thus, it is not surprising that virtually all important oceanographic problems require interdisciplinary approaches and necessarily atmospheric, physical, chemical, biological, optical, and geological data sets. These data sets should be collected concurrently (concept of synopticity) and span sufficient time and space scales to observe the relevant processes of interest. For local studies involving optics, variability at timescales of one day and one year are especially important. For global problems, variability extends well over ten orders of magnitude in space [O(millimeters) to O(10^4 kilometers)] and much longer in time for climate problems. Multiplatform observing approaches are essential. Capabilities for obtaining atmospheric and physical oceanographic data are relatively well advanced in contrast to those for chemical, biological, optical, acoustical, and geological data. This is not surprising, because of the greater complexity and nonconservative nature of the chemistry and biology of the oceans. Yet, remarkable advances are being made in these areas as well. In fact, several bio-optical, chemical, geological, and acoustical variables can now be measured on the same time and space scales as physical variables; however many more variables still need to be measured.

Fundamentals of Ocean Optics

A few operation definitions need to be introduced before discussing the various optical sensors. Light entering the ocean can be absorbed or scattered. The details of these processes comprise much of the study of ocean optics. First, it is convenient to classify bulk optical properties of the ocean as either inherent or apparent. Inherent optical properties (IOPs) depend only on the medium and are independent of the ambient light field and its geometrical distribution. Inherent optical properties include: spectral absorption coefficient, $a(\lambda)$, spectral scattering coefficient, $b(\lambda)$, and spectral beam attenuation coefficient (or, sometimes denoted as 'beam c'), $c(\lambda)$. To define these coefficients, consider a beam of monochromatic light impinging perpendicularly on a thin layer of water. The fraction of the incident radiant flux, Φ_0 (energy or quanta per unit time), which is absorbed by the medium, Φ_a/Φ_0, divided by the thickness of the layer is the absorption coefficient, a, in units of m^{-1}. Similarly, the analogous scattering coefficient, b (m^{-1}) is the fraction of the incident radiant flux scattered from its original path (primarily forward), Φ_b/Φ_0, divided by the thickness of the layer. The beam attenuation coefficient is simply $c = a + b$.

The spectral volume scattering function, $\beta(\psi, \lambda)$ represents the scattered intensity of light per unit incident irradiance per unit volume of water at some angle ψ (with respect to the exiting, nonscattered incident beam) into solid angle element $\Delta\Omega$. Units for $\beta(\psi, \lambda)$ are m^{-1} sr^{-1}. This is essentially the differential scattering cross-section per unit volume in the parlance of nuclear physics. The spectral scattering coefficient is obtained by integrating the volume scattering function over all directions (solid angles). The forward scattering coefficient, b_f, is obtained by integrating over the forward-looking hemisphere ($\psi = 0$ to $\pi/2$) and the backward scattering coefficient, b_b, is calculated by integrating the back-looking hemisphere ($\psi = \pi/2$ to π). The spectral volume scattering phase function is the ratio of $\beta(\psi, \lambda)$ to $b(\lambda)$ with units of sr^{-1}. Other important IOPs include spectral single-scattering albedo, $\omega_0(\lambda) = b(\lambda)/c(\lambda)$, and fluorescence, which can be considered a special case of scattering. (Note: fluorescence is not strictly an IOP; however it is often used as a proxy for chlorophylla.) The proportion of light which is scattered versus absorbed is characterized by $\omega_0(\lambda)$; that is, if scattering prevails then $\omega_0(\lambda)$ approaches a value of 1 and if absorption dominates, $\omega_0(\lambda)$ approaches 0.

IOPs obey simple mathematical operations. For many applications, it is often convenient to partition the total absorption coefficient, $a_t(\lambda)$, the total scattering coefficient, $b_t(\lambda)$, and the total beam attenuation coefficient, $c_t(\lambda)$, in terms of contributing constituents such that:

$$a_t(\lambda) = a_w(\lambda) + a_{ph}(\lambda) + a_d(l) + a_g(\lambda) \qquad [1]$$

$$b_t(\lambda) = b_w(\lambda) + b_{ph}(\lambda) + b_d(\lambda) + b_g(\lambda) \qquad [2]$$

$$c_t(\lambda) = c_w(\lambda) + c_{ph}(\lambda) + c_d(\lambda) + c_g(\lambda) \qquad [3]$$

The subscripts indicate contributions by pure sea water (w), phytoplankton (ph), detritus (d), and gelbstoff (g) with units of m^{-1} for all variables. Detritus is the term for particulate organic debris (e.g., fecal material, plant and animal fragments, etc.) and gelbstoff (sometimes called yellow matter or gilvin) is the term for optically active dissolved organic material. Note that eqns [1]–[3] are intended for use where bottom and coastal sediment contributions are minimal. The spectral absorption coefficient of pure sea water is well characterized and is nearly constant in space and time with greater absorption in the red than blue portions of the visible spectrum. The magnitudes of the detrital and gelbstoff absorption spectra tend to decrease monotonically with increasing wavelength and can be modeled. Phytoplankton spectral absorption varies significantly in relation to community composition and environmental changes; however characteristic peaks are typically found near wavelengths of 440 nm and 683 nm. Much bio-optics research focuses on the temporal and spatial variability of $a_{ph}(\lambda)$. Ship-based ocean water samples have often been used for studies of the IOPs, however, it is quite preferable to obtain *in situ* measurements to insure representative local values as well as to characterize temporal and spatial variability, preferably with simultaneous physical, chemical, and biological measurements.

Instrumentation

Beam transmissometers have been the most commonly used instruments for measuring IOPs with a variety of applications ranging from determinations of suspended sediment volume to phytoplankton biomass and productivity to particulate organic carbon (POC). The principle of operation involves the measurement of the proportion of an emitted beam, C, which is lost through both absorption and scattering as it passes to a detector through some pathlength ΔL. The beam attenuation coefficient, c, is then given as $-[\ln(1 - C)]/\Delta L$; or light intensity, I, is given as $I(\Delta L) = I(0) \exp(-c\Delta L)$. Similar expressions apply for absorption coefficient, a, and scattering coefficient, b. Beam transmissometers have often used a red light emitting diode (660 nm) for the collimated light source and pathlengths of 25 or 100 cm (long pathlengths are preferable for clearer waters). The wavelength of 660 nm was selected in order to minimize attenuation by gelbstoff, which attenuates strongly at shorter wavelengths but minimally in the red. One of the measurement complications for beam c is that most scattering in ocean waters is in the forward direction. The acceptance angle of the detector is thus made as small as possible ($< 1-2°$) to minimize underestimation of beam c. Corrections are done for this effect as well as for instrument temperature and pressure effects.

In situ multispectral attenuation–absorption (ac-meters) instruments have recently become commercially available. These concurrently measure spectral absorption and attenuation coefficients, which are spectral signatures of both particulate and dissolved material. These devices use dual light sources and interference filters on a single rotating wheel (nine wavelengths from 400 to 715 nm with 10 nm bandwidth). Whereas earlier beam transmissometers were open to sea water, the newer ac-meters use two tubes through which sea water is pumped. The inside of the c-tube is flat black to minimize reflections whereas the a-tube is reflective (shiny) in order to maximize internal reflection to better estimate absorption. The deployment of these meters requires minimal lateral and torsional stressing in mounting (sensitivity of beam alignment) and good flow through the tubes. Calibrations involve temperature, salinity, and pure water measurements. A correction is also needed to account for the fact that not all scattered light is collected (causing a biased overestimate of absorption).

Similar, though more capable instruments, have recently been developed to measure a and c at 100 wavelengths from 400 to 726 nm wavelengths with 3.3 nm resolution. A single white light source is used with fiberoptics to provide light to each of the two tubes (for a and b) as well as for a reference path (for correcting changes in lamp output). Each light path has its own spectrometer (using a 256 pixel photodiode array). Biofouling is an issue for all optical instruments. Closed (pumped) systems are less susceptible to fouling as they are primarily in the dark. A variety of special antifouling methods have been attempted with mixed success. These have included copper screens, copper flow tubes, bromide solutions, and other cleaning agents. Spectral scattering coefficient can be computed from ac-meter data by simply performing the difference $b = c - a$. Spectral $a - c$ meter data and relevant spectral decomposition models can be used to provide *in situ* estimates of $a_{ph}(\lambda)$ and other components (e.g., gelbstoff) for a broad range of environmental conditions as a function of time at fixed depths using moorings or as a function of depth from ships or profilers deployed from moorings.

The volume scattering function is one of the more important and challenging measurements of optical oceanography. Until recently, few instruments existed for this measurement and data collected with an instrument developed in the 1970s were commonly used for many problems. The challenge arises because dominant scattering typically occurs at small angles (roughly 50% in the forward direction between 0 and 2–6°) making it difficult to sense the weak scattering light, which is near the intense illuminating beam. Ideally, measurements should be done at several angles to better resolve the function. However, the backscatter signal, b_b, can be estimated with a few measurement angles provided the form of the volume scattering function is relatively well known. Recently developed volume scattering and backscatter instruments take advantage of this function. An important consideration is that backscattering is related to the size distribution, shapes, and composition of the particles sampled (e.g., coastal versus open ocean particles). Thus, this type of information can be obtained in principle. Radiative transfer models need both absorption and volume scattering function information to estimate the propagation of light. Also, an important remote sensing parameter, remote sensing reflectance (discussed below) is related to b_b and a. Key wavelengths are selected for the spectral measurements on the basis of the application and water types (e.g., coastal versus open ocean). Particle size distributions can also be measured using laser (Frauenhofer) diffraction instruments. These devices employ charged coupled device (CCD) array photodetectors (linear or circular ring geometries). Modified versions of some of these instruments can also measure particle settling velocities. Particles in the range of 5–500 μm can be measured with resolution dependent on the number of individual detector rings. These devices can also measure beam transmission by sensing the undeviated light.

Irradiance is the radiant flux per unit surface area (units of W m^{-2} or quanta (or photons) s^{-1} m^{-2} or mol quanta (or photons) s^{-1} m^{-2}. Note that 1 mol of photons is 6.02×10^{23} (Avogadro's number) and that one mole of photons is commonly called an einstein. It is convenient to define downwelling irradiance, E_d, as the irradiance of a downwelling light stream impinging on the top face of a horizontal surface (e.g., ideally flat light collector oriented perpendicular to the local gravity vector) and upwelling irradiance, E_u, as the irradiance of an upwelling light stream impinging on the bottom face of a horizontal surface. Irradiance reflectance, R, is E_u/E_d. Irradiance measurements are fundamentally important for quantifying the amount of light available for photosynthesis and for radiative transfer theory and computations. Another useful radiometric variable is radiance, $L = L(\theta, \Phi)$, which is defined as the radiant flux at a specified point in a given direction per unit solid angle per unit area perpendicular to the direction of light propagation (units of W (or quanta s^{-1}) m^{-2} sr^1). Zenith angle, θ, is the angle between a vertical line perpendicular to a flat plane and an incident light beam, and azimuthal angle, Φ, is the angle with respect to a reference line in the plane of the flat plate. By integrating $L(\theta, \Phi)cos\theta$ over the solid angle $(d\omega = sin\theta\, d\theta\, d\Phi)$, of the upper hemisphere (say of a flat plate from $\theta = 0$ to $\pi/2$ and Φ from 0 to 2π), one obtains E_d. Similarly, upwelling irradiance, E_u, is calculated by integrating over the bottom hemisphere. Net downward irradiance is the difference $E_d - E_u$, or the integral of $L(\theta, \Phi)cos\theta$ over the entire sphere (full solid angle 4π). Scalar irradiance, E_0, is defined as the integral of $L(\theta, \Phi)$ over the entire sphere. It should be noted that all definitions are for a specific wavelength of light. If one integrates scalar irradiance over the visible wavelengths (roughly 400–700 nm; note that the visible range is sometimes defined as 350–700 nm), then the biologically important quantity called photosynthetically available radiation (PAR) is obtained.

Apparent optical properties (AOPs) depend on both the IOPs and the angular distribution of solar radiation (i.e., the geometry of the subsurface ambient light field). To a reasonable approximation, the attenuation of spectral downwelling incident solar irradiance, $E_d(\lambda, z)$, can be described as an exponential function of depth:

$$E_d(\lambda, z) = E_d(\lambda, 0^-)exp[-K_d(\lambda, z)z] \qquad [4]$$

where z is the vertical coordinate (positive downward), $E_d(\lambda, 0^-)$ is the value of E_d just below the air-sea interface, and $K_d(\lambda, z)$ is the spectral diffuse attenuation coefficient of downwelling irradiance. $K_d(\lambda)$ (here depth dependence notation is suppressed for convenience) is one of the important AOPs. The contributions of the various constituents (e.g., water, phytoplankton, detritus, and gelbstoff) to $K_d(\lambda)$ are often represented in analogy to the absorption coefficients given in eqn [1][1]. This leads to the term 'quasi-inherent' optical property as it is suggestive that inherent optical properties, i.e., like $a_t(\lambda)$, are closely related to $K_d(\lambda)$, which is often described as a quasi-inherent optical property. However, a key point is that $K_d(\lambda)$ is in fact dependent on the ambient light field. Analogous spectral attenuation coefficients are defined for diffuse upwelling irradiance, downwelling radiance, and upwelling radiance, $K_L(\lambda)$. It should be noted that for periods of low sun angle and/or when highly reflective organisms or their products (e.g., coccolithophores and coccoliths) are present, then multiple scattering becomes increasingly more important and simple relations

between IOPs and AOPs break down. It is important to mention that spectral radiometric measurements of AOPs have been far more commonly made than measurements of IOPs.

The relationships between IOPs and AOPs are central to developing quantitative models of spectral irradiance in the ocean. Radiative transfer theory is used to provide a mathematical formalism to link IOPs and the conditions of the water environment and light forcing to the AOPs of the water column. One motivation of this research is the desire to predict AOPs given environmental forcing conditions and estimates or actual measurements of IOPs. Another related, inverse goal is to be able to estimate or determine the vertical (and ideally horizontal) structure of IOPs and their temporal variability given normalized water-leaving radiance, $L_{wn}(\lambda)$, measured remotely from satellite or airplane color imagers. The extrapolation of subsurface values of $L(\lambda)$ to the surface has been the subject of considerable research, because a major requirement of ocean color remote sensing is to match *in situ* determinations of those from satellite- or plane-based spectral radiometers which necessarily must account for the effects of clouds and aerosols. As mentioned above, the important remote sensing parameter, spectral radiance reflectance, $L_u(\lambda)/E_d(\lambda)$, has been found to be proportional $b_b/(a + b_b)$ or b_b/a from Monte Carlo calculations for waters characterized by b/a values ranging from 1.0 to 5.0. The spectral radiance reflectance evaluated just above the ocean surface is defined as the remote sensing reflectance $R_{rs}(\lambda)$. It should also be noted that remote sensing estimates of pigment (typically chlorophyll *a* or chlorophyll *a* + phaeopigments) concentrations use empirical relations involving ratios such as $L_{wn}(443\,nm)/L_{wn}(550\,nm)$ or $R_{rs}(490\,nm)/R_{rs}(555\,nm)$ or these ratios with other wavelength combinations.

One of the most commonly used instruments for measuring AOPs is a broadband scalar irradiance E_0 or PAR sensor, because of the need for determining the availability of light for phytoplankton and its relative simplicity. A spherical light collector made of diffusing plastic receives the light from approximately 4π sr. The light is transmitted via a fiberoptic connector or quartz light conducting rod to a photodetector which records an output voltage. The instrument is calibrated with a standard lamp. Analogous sensors use flat plate cosine or hemispherical collectors. Spectroradiometers are irradiance or scalar irradiance meters which use a variable monochromator placed between a light collector and the photodetector. Light separation can be achieved using sets of interference filters (usually 10 nm in bandwidth) selected for particular purposes (e.g., absorption peaks and hinge points for pigments). Alternatively, higher spectral resolution (~ 3 nm from 400 to 800 nm) can be achieved using grating monochromators. Cosine (flat-plate), spherical, or hemispherical collectors are used for several different measurements or calculations of the AOPs described above (e.g., E_d, L_u, K_d, etc.). Irradiance cosine collectors are designed such that their responses to parallel radiant flux should be proportional to the angle between the normal to the collector surface and the direction of the radiant flux. This is necessary as the angle between the incoming radiant flux and the collector is variable, as is the area of the projection. To account for the differences in refractive indices of air and water, an immersion correction factor (typically 1.3–1.4) is applied. It should be noted that radiance sensors are designed to accept light in a small solid angle (typically a viewing angle of a few degrees). Irradiance and radiance instruments are calibrated using well-characterized, standard light sources. Because of the growing concern about UV radiation, a number of instruments are designed to measure into the UV portion of the electromagnetic spectrum as well.

A grand challenge of oceanography is to greatly increase the variety and quantity of ocean measurements. Optical and other ocean measurements are expensive. Thus, a major goal is to develop new sensors and systems, which can be efficiently deployed from a host of available ocean platforms including ships, moorings, drifters, floats, and autonomous underwater vehicles (AUVs). A further need is to telemeter the data in near real-time. Ship-based observations are useful for detailed profiling (high vertical resolution), but moorings are better suited for high temporal resolution, long-term measurements. Drifters, floats, and AUVs can provide horizontal coverage unattainable from the other *in situ* platforms. Ultimately, all of these platforms, along with satellite- and plane-based systems, are needed to fill in the time–space continuum. Several optical systems, which have been used for mooring-based time series studies, are illustrated in **Figure 1**. These collective instrument systems include most of the sensors described earlier. The sampling for these instruments is typically done every few minutes (in some cases once per hour) for periods of several months. At this point in time, biofouling, rather than power or data storage, is the limiting factor. However, even this aspect is becoming less problematic with several new antibiofouling methods. In addition, more optical sensors are being deployed from other emerging platforms such as AUVs. The optimal utilization of optical data from the various platforms will require the use of advanced data merging methods and data assimilation models for predictions.

Optical Experiment Data Sets

Finally, a few examples of data, which have been collected with some of the optical instruments described above (**Figure 1**), are presented. An interdisciplinary experiment devoted to the understanding of relations between mixing processes and optical variability was conducted south of Cape Cod, Massachusetts on the continental shelf. Several spectral optical instruments collected IOP and AOP data sets using ship-based profiling and towed systems, moorings, and bottom tripods. The period of the experiment covered almost one year (July 1996–June 1997). Some instruments sampled at several minute to hourly intervals whereas others sampled with vertical resolution on the scale of centimeters for a few weeks during two intensive field campaigns. Several interesting observations were enabled. These included: sediment resuspension forced by two passing hurricanes (hurricane Edouard and hurricane Hortense), spring and fall phytoplankton blooms, water mass intrusions, and internal solitary waves. Time series data collected using the BIOP system (e.g. BIOPS and MORS in **Figure 1**) which was deployed from a mooring and a bottom tripod located near each other in 70 m waters, are shown in **Figure 2**. **Figure 2(A)** shows the 37-m time series of total spectral absorption (water absorption has been subtracted) using a nine-wavelength ac-meter. The major feature is related to hurricane Edouard. The spectral absorption contributions due to phytoplankton are shown for two days in the summer in **Figure 2(B)**; the peaks at 440 and 683 nm are caused by phytoplankton. **Figure 2(C, D)** shows ac-meter time series of spectral (nine wavelengths) scattering and attenuation coefficients. The record is dominated by sediment resuspension caused by mixing and waves created by hurricane Edouard and the more distant hurricane Hortense. The complete mooring data set showed that sediments were lifted more than 30 m above the ocean floor.

The second example highlights optical data collected from a ship-based profiling system (similar to the system shown in **Figure 1A**). The purpose of the measurements was to study the dispersion of contaminants (treated wastewaters) discharged about 4 km offshore (70 m water depth) into Mamala Bay, off the coast of Honolulu, Hawaii in the fall of 1994. Optical and physical measurements were made from shipboard using both profile and towing modes in the vicinity of the outfall plume. Optical instrumentation included a beam transmissometer

(660 nm), spectral ac-meters (nine wavelengths for a and c), PAR sensor, chlorophyll fluorometer, a particle size analyzer (laser diffraction method), and a spectral absorption and fluorescence instrument. Physical measurements of temperature, salinity and pressure were also done. Sampling was designed to track both the horizontal and vertical structure of effluent as manifest in the optical signals. The water contributions to a, b, and c were removed for the analyses. The collective optical and physical measurements enabled the partitioning of particle types into categories: particulate versus dissolved components, phytoplankton opposed to detrital components,

shallow layer versus deep layer phytoplankton, and old versus newly discharged sewage plume waters. Briefly, a profile (see **Figure 3**) taken near the end of the outfall diffuser (water depth ~ 80 m) displayed the following features: (1) sewage plume waters centered near 60 m as characterized by low salinity and very high values of spectral attenuation and absorption (greater in the shorter wavelengths), and (2) a shallow phytoplankton layer near 20 m with modest relative maxima in chlorophyll fluorescence and spectral absorption and attenuation coefficients in the blue. The profile of spectral single-scattering albedo, $\omega_0(\lambda) = b(\lambda)/c(\lambda)$, shows a general trend of increased

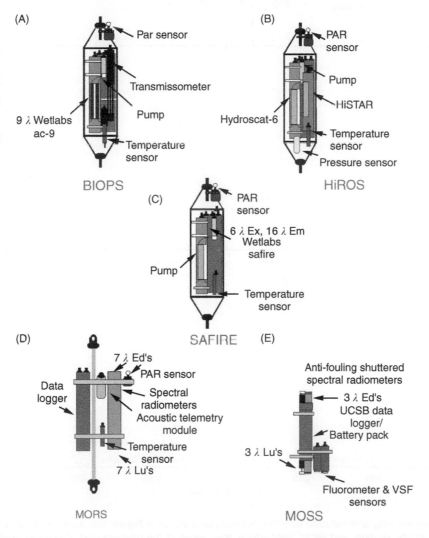

Figure 1 Schematic showing a variety of optical systems for mooring applications. (A) System used for measuring inherent optical properties (IOPs) including beam c (660 nm), spectral attenuation and absorption coefficients at nine wavelengths, along with PAR and temperature. (B) System used for measuring inherent optical properties including spectral attenuation and absorption coefficients at 100 wavelengths, spectral backscatter at six wavelengths, along with PAR, temperature, and pressure. (C) System measuring spectral fluorescence with 6 excitation wavelengths and 16 emission wavelengths; also PAR and temperature sensors are included. (D) System for measuring apparent optical properties (AOPs) including spectral downwelling irradiance and upwelling radiance at seven wavelengths and PAR along with temperature. A telemetry module is also included. (E) System for measuring apparent optical properties (AOPs) including spectral downwelling irradiance and upwelling radiance at three wavelengths along with instruments for measuring chlorophyll fluorescence, volume scattering function. An antifouling shutter system is also utilized.

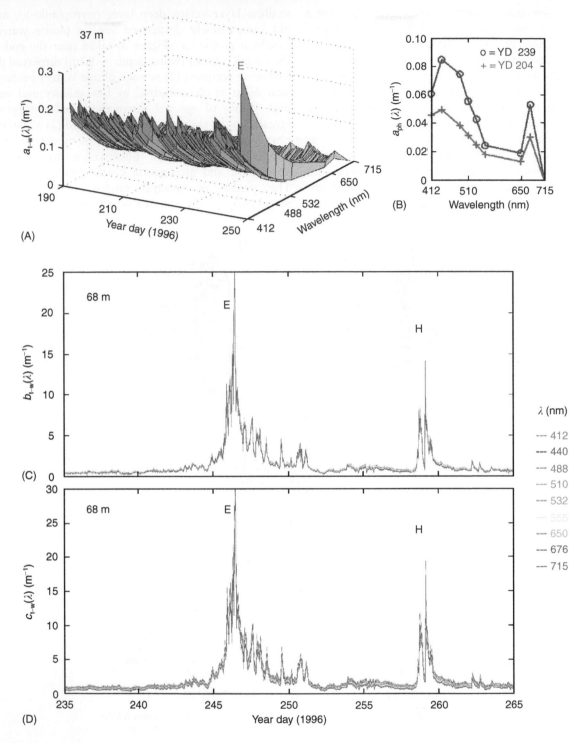

Figure 2 Illustrations showing data collected from the BIOPS system shown in **Figure 1**(A) on the continental shelf south of Cape Cod Massachusetts in the summer and fall of 1996. (A) A time series of the total spectral absorption coefficient of light (after subtracting the clear water component) at a depth of 37 m (total water depth is 70 m). (B) The spectral absorption on YD 204 and 239. (C) Time series of spectral scattering coefficient computed by differencing the total spectral attenuation and absorption coefficients (again, the clear water coefficient has been subtracted) at 68 m. The large peaks are attributed to passages of hurricanes Edouard (E) and Hortense (H). (D) as for (C) except for spectral beam attenuation coefficients. (Figures based on Chang and Dickey, 1999.)

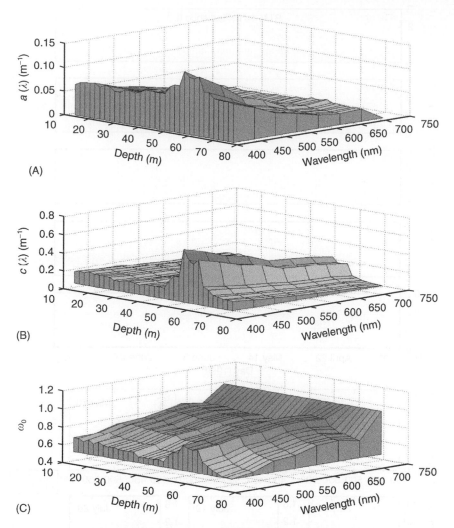

Figure 3 Data collected near an ocean sewage outfall in Mamala Bay off Honolulu, Hawaii in the fall of 1994. Vertical profiles of (A) spectral absorption coefficient, (B) spectral attenuation coefficient, and (C) single scattering albedo. Data were obtained from an ac-meter system similar to the one shown in **Figure 1(A)**. Eight wavelengths are displayed and the major peak near 60 m is caused by sewage plume waters with high scattering and absorption coefficents. (Figures based on Petrenko *et al.*, 1997.)

importance of scattering for greater wavelengths with the most pronounced effect for the sewage plume waters centered at 60 m (**Figure 3C**).

The final example describes an important linkage between *in situ* and remote sensing observations of ocean color. The Bermuda Testbed Mooring (BTM) is used to test new oceanographic instrumentation (including systems shown in **Figure 1**), for scientific studies devoted to biogeochemical cycling and climate change, and for groundtruthing (verification using *in situ* data) and algorithm development for ocean color satellites such as SeaWiFS. The latter aspect is the focus here. Two BTM optical systems utilizing radiometers for measurements of spectral downwelling irradiance and spectral upwelling radiance are illustrated in **Figure 1(D)** and **(E)**. These systems measure

at wavelengths, which are coincident with those of the SeaWiFS ocean color satellite, and sample at hourly intervals. Two or more systems are deployed at different depths and another radiometer system is deployed from the surface buoy to collect incident spectral downwelling irradiance. Derived quantities include time series of several of the AOP quantities introduced earlier (e.g., spectral diffuse attenuation coefficients, spectral reflectances including remote sensing reflectance, and water-leaving radiance). Time series of spectral upwelled radiance from the BTM radiometer system (**Figure 1D**) located at 14 m are shown for the period April through July, 1999 in the top set of panels of **Figure 4**. The variability is primarily caused by the daily cycle of solar insolation, cloud cover, and optical properties associated with

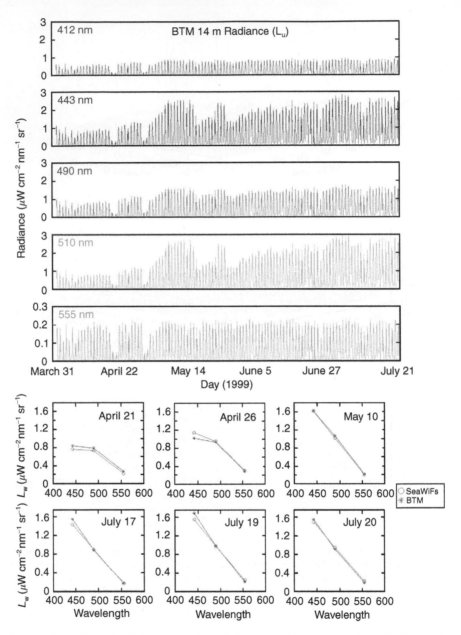

Figure 4 BTM radiometer data collected with the system shown in **Figure 1(D)**. The top set of panels shows a time series of spectral upwelling radiance (five wavelengths) from the 14-m spectral radiometer system (April–July, 1999). The bottom set of panels shows spectral water-leaving radiance derived from the BTM radiometers and from the SeaWiFS ocean color satellite for six days of the sampling period.

phytoplankton and their products. The bottom panels of **Figure 4** show a subset of the water-leaving radiance data as determined from the BTM radiometers (extrapolation to surface) and SeaWiFS satellite sensors for six days during the particular sampling period. These two types of data are critical for quantifying both the temporal and horizontal spatial variability of water clarity, PAR, phytoplankton biomass, and primary productivity.

Toward the Future

The last two decades have been marked by the emergence of a host of new optical instruments and systems; their applications are growing at a rapid rate. Looking toward the future, it is anticipated that optical instrumentation will be (1) more capable in spectral resolution, (2) smaller and require less power, (3) suitable for deployment from a variety of autonomous platforms, (4) designed for ease in telemetering of data for real-time applications, and (5) less costly as demand increases enabling higher volumes of data collection.

See also

Absorbance Spectroscopy for Chemical Sensors. IR Radiometers. Optical Particle Characterization. Satellite Remote Sensing Microwave Scatterometers.

Further Reading

Agrawal YC and Pottsmith HC (1994) Laser diffraction particle sizing in STRESS. *Continental Shelf Research* 14: 1101–1121.

Bartz R, Zaneveld JRV, and Pak H (1978) A transmissometer for profiling and moored observations in water. *Ocean Optics V, SPIE* 160: 102–108.

Booth CR (1976) The design and evaluation of a measurement system for photsynthetically active quantum scalar irradiance. *Limnology and Oceanography* 19: 326–335.

Chang GC and Dickey TD (1999) Partitioning *in situ* total spectral absorption by use of moored spectral absorption and attenuation meters. *Applied Optics* 38: 3876–3887.

Chang GC and Dickey TD (2001) Optical and physical variability on time-scales from minutes to the seasonal cycle on the New England continental shelf: July 1996–June 1997. *Journal of Geophysical Research* 106: 9435–9453.

Chang GC, Dickey TD, and Williams AJ 3rd (2001) Sediment resuspension over a continental shelf during Hurricanes Edouard and Hortense. *Journal of Geophysical Research* 106: 9517–9531.

Dana DR, Maffione RA, and Coenen PE (1998) A new *in situ* instrument for measuring the backward scattering and absorption coefficients simultaneously. *Ocean Optics XIV* 1: 1–8.

Dickey T (1991) Concurrent high resolution physical and bio-optical measurements in the upper ocean and their applications. *Reviews in Geophysics* 29: 383–413.

Dickey T, Granata T, Marra J, *et al.* (1993) Seasonal variability of bio-optical and physical properties in the Sargasso Sea. *Journal of Geophysical Research* 98: 865–898.

Dickey T, Marra J, Stramska M, *et al.* (1994) Bio-optical and physical variability in the sub-arctic North Atlantic Ocean during the spring of 1989. *Journal of Geophysical Research* 99: 22541–22556.

Dickey T, Frye D, Jannasch H, *et al.* (1998) Initial results from the Bermuda Testbed Mooring Program. *Deep-Sea Research I* 45: 771–794.

Dickey T, Marra J, Weller R, *et al.* (1998) Time-series of bio-optical and physical properties in the Arabian Sea: October 1994–October 1995. *Deep-Sea Research II* 45: 2001–2025.

Dickey TD, Chang GC, Agrawal YC, Williams AJ 3rd, and Hill PS (1998) Sediment resuspension in the wakes of Hurricanes Edouard and Hortense. *Geophysical Research Letters* 25: 3533–3536.

Dickey T, Zedler S, Frye D, *et al.* (2001) Physical and biogeochemical variability from hours to years at the Bermuda Testbed mooring site: June 1994–March 1998. *Deep-Sea Research*.

Foley D, Dickey T, McPhaden M, *et al.* (1998) Longwaves and primary productivity variations in the equatorial Pacific at 0°, 140°W February 1992–March 1993. *Deep-Sea Research II* 44: 1801–1826.

Gentien P, Lunven M, Lehaitre M, and Duvent JL (1995) In situ depth profiling of particle sizes. *Deep-Sea Research* 42: 1297–1312.

Griffiths G, Knap A, and Dickey T (1999) Autosub experiment near Bermuda. *Sea Technology*, December.

Jerlov NG (1976) *Marine Optics*. Amsterdam: Elsevier.

Kirk JTO (1994) *Light and Photosynthesis in Aquatic Ecosystems*. Cambridge: Cambridge University Press.

Mobley CD (1994) *Light and Water: Radiative Transfer in Natural Waters*. San Diego: Academic Press.

Moore CC, Zaneveld JRV, and Kitchen JC (1992) Preliminary results from *in situ* spectral absorption meter data. *Ocean Optics XI, SPIE* 1750: 330–337.

O'Reilly JE, Maritorena S, Mitchell BG, *et al.* (1998) Ocean color chlorophyll algorithms for SeaWiFS. *Journal of Geophysical Research* 103: 24937–24953.

Petrenko AA, Jones BH, Dickey TD, Le Haitre M, and Moore C (1997) Effects of a sewage plume on the biology, optical characteristics, and particle size distributions of coastal waters. *Journal of Geophysical Research* 102: 25 061–25 071.

Petrenko AA, Jones BH, and Dickey TD (1998) Shape and near-field dilution of the Sand Island sewage plume: observations compared to model results. *Journal of Hydraulic Engineering* 124: 565–571.

Petzold TJ (1972) *Volume scattering functions for selected waters, Scripps Institution of Oceanography Reference 72–78*. La Jolla, California: Scripps Institution of Oceanography.

Smith RC, Booth CR, and Star JL (1984) Oceanographic bio-optical profiling system. *Applied Optics* 23: 2791–2797.

Spinrad RW, Carder KL, and Perry MJ (1994) *Ocean Optics*. Oxford: Oxford University Press.

NEPHELOID LAYERS

I. N. McCave, University of Cambridge, Cambridge, UK

Introduction

A remarkable feature of the lower water column in most deep parts of the World Ocean is a large increase in light scattering and attenuation conferred by the presence of increased amounts of particulate material. This part of the water column is termed the bottom nepheloid layer (BNL). Another class of nepheloid layers found especially at continental margins are intermediate nepheloid layers (INLs) (**Figures 1** and **2**). These occur frequently at high levels off the upper continental slope and at the depth of the shelf edge. From here, they spread out across the continental margin. These INLs are similar to the inversions observed in the BNL on some profiles (**Figure 1**). (The surface nepheloid layer (SNL), not treated here, is simply the upper ocean layer in which particles are produced by biological activity, and which may have material from river plumes close to shore.)

The increase in light scattering is perceived relative to minimum values found at mid-water depths of 2000–4000 m (shallower on continental margins). The increased scattering is due to fine particles. This has been determined by particle-size measurements and filtration of seawater with determination of concentration by weight and volume. Most data on the distribution and character of nepheloid layers have been acquired by optical techniques, principally by the Lamont photographic nephelometer and the SeaTech transmissometer, and more recently the WetLabs transmissometer and light scattering sensor (LSS).

The optical work has revealed that the BNL is up to 2000-m thick (can be more in trenches) and generally has a basal uniform region, the bottom mixed nepheloid layer (BMNL), corresponding quite closely to the bottom mixed layer defined by uniform potential temperature (**Figure 1**). Above the BMNL there is a more or less exponential fall-off in intensity of light scattering up to the clear-water minimum marking the top of the BNL.

Both bottom and INLs are principally produced by resuspension of bottom sediments. Their distribution indicates the dispersal of resuspended sediment in the ocean basins and is thus a signature of both material and water transport away from boundaries (some of which may be internal such as ridges and seamounts). Most concentrated nepheloid layers occur on the continental shelf, upper slope, or deep continental margin. They indicate the locus of active resuspension and redeposition by strong bottom currents and internal waves.

Optics of Nephelometers: What They 'See'

Detection of deep-ocean nepheloid layers has been mainly through measurement of light scattering. The Lamont nephelometer has made the largest number of profiles in all oceans but is no longer in use. It used an incandescent bulb as the source and photographic film as the detector of the light scattered from angles between $\theta = 8°$ and $24°$ from the forward axis of the light beam. The film was continuously wound on as the instrument was lowered, resulting in an averaging of the received signal over about 25-m depth. The short-lived Geochemical Ocean Section Study (GEOSECS) nephelometer used a red ($\lambda = 633$ nm) laser source and a photoelectric cell to detect light scattered from $\theta = 3–15°$ off the axis of the beam. The SeaTech (now WetLabs, Inc.) LSS measures infrared light (880 nm) backscattered (180°) from particles in the sample volume using a solar-blind silicon detector.

Because of multiple scattering, nephelometers do not yield precise optical parameters. Optical transmission with a narrow beam can yield the attenuation coefficient (c). The most commonly used SeaTech and WetLabs transmissometers have a red light source ($\lambda = 660$ or 670 nm) and usually a 0.20–0.25-m path length.

Most of the contribution to the total scattering b comes from near-forward angles (low values of θ). Jerlov (1976) shows that, for surface waters, 47% of b occurs between $\theta = 0°$ and 3°, 79% between 0° and 15°, and 90% between 0° and 30°. The GEOSECS instrument records about 32% of b and the Lamont nephelometer about 16%. The total scattering is given by Mie theory (assumed spherical particles) as a function of particle size d and relative refractive index (relative RI) n, and wavelength of light λ. The relative indices of refraction of suspended material are dominated by components with $n = 1.05$ and 1.15, values probably characteristic of organic and mineral matter, respectively (e.g., RI of seawater 1.34, quartz 1.55, ratio $n = 1.15$).

Particles from clear ocean waters and weak nepheloid layers (concentration $C < 40$ mg m^{-3}) tend

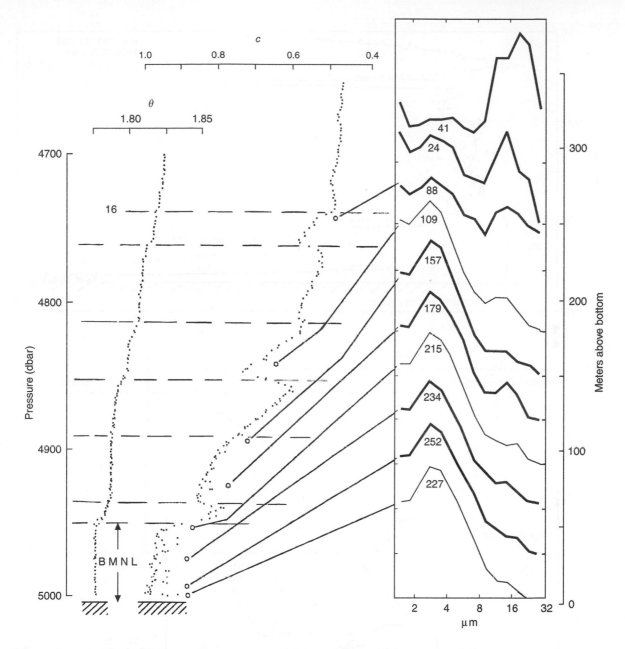

Figure 1 Data from the SeaTech transmissometer in the Atlantic showing the BMNL and a nepheloid layer comprising multiple steps in temperature and turbidity. Turbidity is given as c the attenuation coefficient, and θ is potential temperature in °C. Also shown on the right are particle-size spectra determined by Coulter counter. Reproduced from McCave IN (1983) Particulate size spectra, behavior and origin of nepheloid layers over the Nova Scotian Continental Rise. *Journal of Geophysical Research* 88: 7647–7666.

to have particle-size distributions by volume which are flat, equivalent to $k = 3$ in a particle number distribution of Junge type, $N = Kd^{-k}$ where N is the cumulative number of particles larger than diameter d and K and k are constants. However, this distribution does not appear to be maintained at sizes finer than about 2 µm where k decreases toward 1.5. In concentrated nepheloid layers, this distribution does not occur at all and a peaked distribution with a peak between 3 and 10 µm is encountered.

Morel has calculated scattering according to Mie theory for suspensions with Junge distributions and several indices of refraction. Recalculation into cumulative curves of percentage scattering in **Figure 3** illustrates the fact that most recorded scattering is produced by fine particles. The cases shown are for values of k of 2.1, 3.2, and 4.0 and a two-component (peaked) distribution with $k = 2.1$ up to $\alpha = \pi dn/\lambda = 32$ and $k = 4.0$ for larger sizes ($\alpha = 32$ is equivalent to $d = 5.6$ µm for $\lambda = 633$ nm). In each case, three

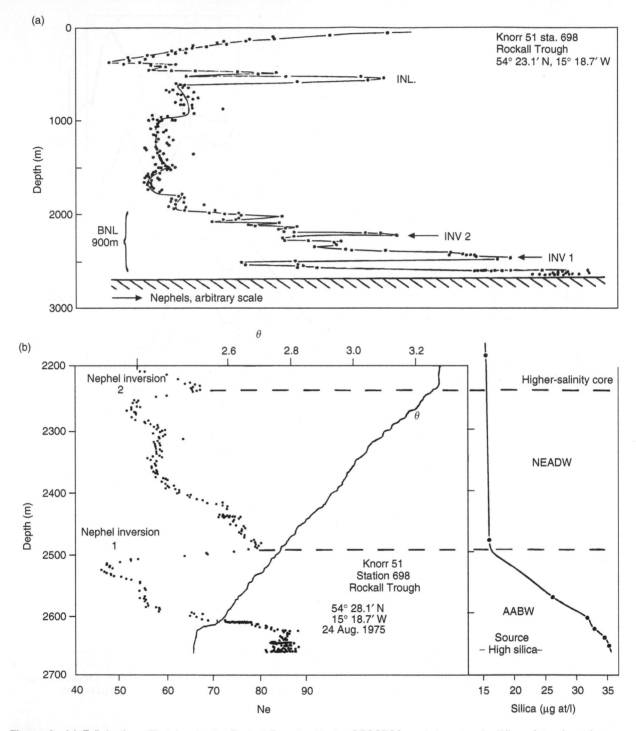

Figure 2 (a) Full-depth profile taken in the Rockall Trough with the GEOSECS nephelometer. An INL and two inversions are apparent. (b) Detail of the lower 500 m of the profile in (a) showing the relationship between the nephel (turbidity) inversions and hydrography. Reproduced from McCave IN (1986) Local and global aspects of the bottom nepheloid layers in the world ocean. *Netherlands Journal of Sea Research* 20: 167–181.

forward-scattering angles are given. It is clear that scattering close to the beam is more sensitive to large particles than that at 20°. (At $\theta < 0.5°$ we are essentially dealing with a transmissometer.) In the case of $k = 2.1$ only about 22% of the scattering is from smaller sizes ($\alpha < 32$) at $\theta = 20°$. Thus the curves for $\theta = 10°$ and 20° are generally representative, and the distribution for both $k = 3.2$ and the composite case show that 95% of the scattering is by particles $< 5\,\mu m$ for $\lambda = 633\,nm$.

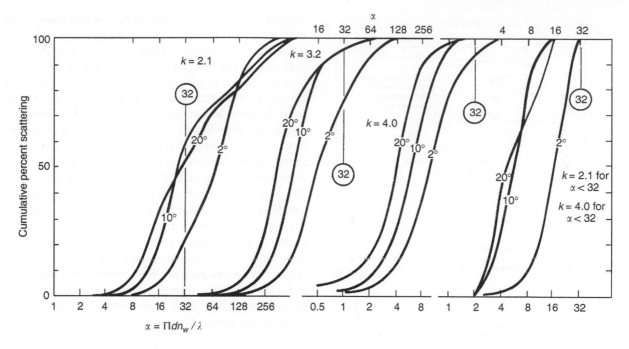

Figure 3 Cumulative percentage of scattering calculated from the data of Morel (1973) by McCave (1986). The right-hand case is for a peaked distribution with the peak at $\alpha = 32$, equivalent to $5.6\,\mu m$ for $\lambda = 633\,nm$ and $n = 1.15$. Note that material larger than the peak contributes virtually nothing to the scattering in this case. Reproduced from McCave IN (1986) Local and global aspects of the bottom nepheloid layers in the world ocean. *Netherlands Journal of Sea Research* 20: 167–181.

So it is dominantly the fine fraction of particles in nepheloid layers that is seen and recorded by nephelometers. These particles have very low settling velocities, less than $\sim 5 \times 10^{-6}\,m\,s^{-1}$. Although larger particles are present, they are rare and their properties and behavior cannot be invoked to explain features of distribution shown by nephelometers.

The SeaTech 0.25 m path length transmissometer has been used for most of the modern work on the structure and behavior of nepheloid layers. The transmission T is related to beam attenuation coefficient c over path length l as $T = e^{-cl}$. The major control of c is due to particles. Attenuation is due to absorption a and scattering b, thus $c = a + b$. The value of c for pure seawater is about $0.36\,m^{-1}$ for this instrument operating at $\lambda = 660\,nm$, and any excess is due to particulate effects. Because scattering very close to the beam is more sensitive to large particles, the transmissometer is more sensitive to larger particles and the nephelometer to smaller ones (**Figure 4**).

Nepheloid Layer Features

The principal features of nepheloid layers that must be accounted for are the facts that the concentration is generally highest close to the bed, decreasing upward, but also that this is not universally the case, because: inversions, upward increases of concentration, are also found; steeper gradients in concentration as well as inversions are often found at the boundaries of distinct density (temperature and/or salinity) changes, but their frequency decreases upward.

The thickness of the BNL is generally in the region of 500–1500 m, and exceptionally up to 2000 m. This is clearly greater than the thickness of the bottom mixed layer (**Figures 2** and **4**), a fact which rules out the possibility of simple mixing by boundary turbulence being a sufficient mechanism for BNL generation.

In several cases, the nepheloid layer is seen to transcend water masses. That is to say, the nepheloid layer shows a general decline in turbidity upward through interleaved water masses of differing sources and temperature/salinity characteristics though there may be a steeper turbidity gradient at the boundary between water masses.

The highest suspended sediment concentrations occur in the BMNL in regions of strong bottom currents where they are typically $100–500\,mg\,m^{-3}$ (this is the same as $\mu g\,l^{-1}$). In general, deep western boundary currents and regions of recirculation carry high particulate loads. However, high turbidity is also found beneath regions of high surface eddy kinetic energy (variance of current speed) when located over strong thermohaline bottom currents. High surface eddy kinetic energy is connected with high bottom eddy kinetic energy; thus, intermittent variability, when added to a strong steady

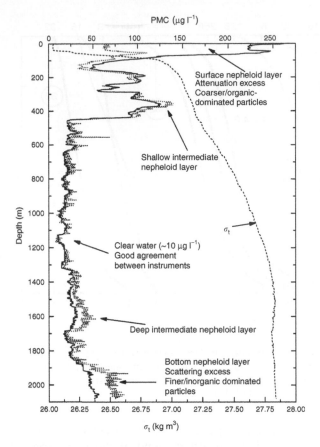

PMC (µg l⁻¹)

Figure 4 Profiles of particulate matter concentration calculated from beam attenuation (solid line), and light scattering (dotted line) against depth, together with the density structure (σ_t) of the water column (dashed line). Reproduced from Hall IR, Schmidt S, McCave IN, and Reyss JL (2000) Particulate matter distribution and Th-234/U-238 disequilibrium along the Northern Iberian Margin: Implications for particulate organic carbon export. *Deep-Sea Research I* 47: 557–582.

component, may be responsible for very high current speeds which produce intense sediment resuspension.

Separated Mixed-Layer Model

Nepheloid layer structure is consistent with a quasi-vertical transport mechanism involving turbulent mixing in bottom layers of ~10–50-m thickness (*see*), followed by their detachment and lateral advection along isopycnal (equal density) surfaces The detachment occurs in areas of steep topography as well as in areas of lower gradient at benthic fronts where sloping isopycnals intersect the bottom. In many nepheloid layers, there are sharp upward increases in sediment content associated with steps in other properties such as temperature and salinity (**Figure 1**). Both the step structure and inversions in particulate matter concentration are incompatible with vertical turbulent mixing (which occurs mainly in the BMNL)

up to a kilometer above the bed. It is not possible to mix sediment across sharp density steps without breaking them down. However, these features are explicable if the layers are recently separated from the bottom. Some layers marked by steps in potential temperature contain excess radon-222 (originating from bottom sediments) with a 3.8 day half-life, suggesting detachment of bottom layers within 2–3 weeks before sampling. It is anticipated that with time these layers become thinner by mixing at their boundaries and by lateral spreading to yield, eventually, a uniform stratification. In this, the upper part of the BNL, sheared-out mixed layers that have, on average, come further from the sloping sides of the basins and from regions with less frequent resuspension, have lower concentrations. The basal layers are on average more recently resuspended and also gain material by fall-out from above; thus, there is an overall decreasing particulate concentration, and increasing age upward.

Decay of Concentration: Aging of Particulate Populations

The particles composing the nepheloid layer may be modified due to aggregation with similar-sized particles and scavenging by larger rapidly settling ones. Aggregation may also be caused through biological activity, although little is known about such processes at great depths. Particles tend to settle and to be deposited onto the bed, from the bottom mixed layer. The larger particles should be deposited in a few weeks to months, 10–20 µm particles taking 50–20 days to settle from a 60-m-thick layer. This will not affect the layer perceived by nephelometers so quickly because the timescale of fine particle removal initially involves Brownian aggregation with a 'half-life' of several months to years.

The direct rate of deposition of very fine particles (0.5–1 µm) from a layer which remained in contact with the bed would be very slow. Concentration would halve in about 8 years. Thus the rate of decrease in concentration of 0.5–1-µm particles is due more to their being moved to another part of the size spectrum by aggregation (and then deposited) than to their being deposited directly. The fine material in dilute nepheloid layers has a mean residence time measured in years, demonstrated by the residence time of particle-reactive short half-life radionuclides such as [210]Pb ($t_{1/2} = 22.3$ years). In more concentrated nepheloid layers, a large proportion of this material will be removed in under a year, and in the BMNL residence times are tens to a hundred or so days estimated via [234]Th ($t_{1/2} = 24.1$ days). The dilute nepheloid layers in tranquil parts of the oceans could thus contain material that was resuspended very far away. The contribution of this material to the net sedimentation rate of these tranquil regions may not be negligible. The rate of deposition in the central South Pacific of only $0.5–2\,mm\,ky^{-1}$ could include up to $1\,mm\,ky^{-1}$ of fine material from the nepheloid layer.

With aging, the individual detached layers comprising the nepheloid layer lose material by aggregation and settling and lose their identity by being thinned through shearing. An originally discontinuous vertical profile of concentration with inversions is converted to one of relatively smooth upward decline in concentration. Present understanding of particle aggregation and sinking rates suggests that this takes a few years to achieve.

Chemical Scavenging by Particles in Nepheloid Layers

Many chemical species are particle-reactive and rapidly become adsorbed onto surfaces. This is why a number of elements are present in only trace quantities in seawater as outlined by Robert Anderson in 2004. The phenomenon of 'boundary scavenging', preferential removal of particle active species at

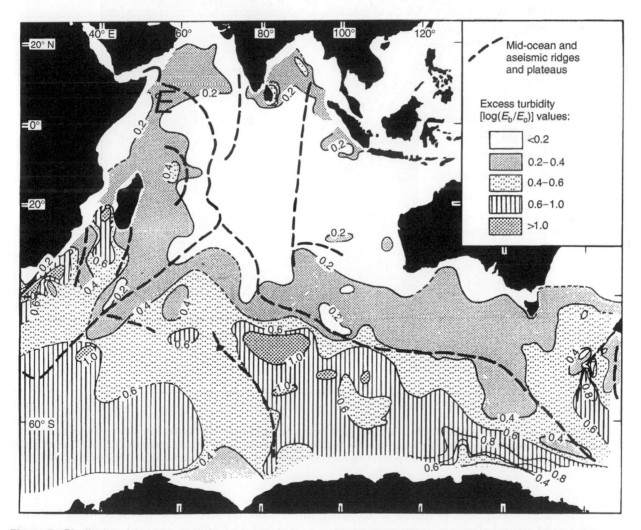

Figure 5 Distribution of excess turbidity for the Indian Ocean expressed as $\log(E/E_c)$ where E is the maximum light scattering near the bed and E_c is the value at the clear water minimum. A value of 1 thus represents a factor of 10 increase from the clear-water value. Reproduced from McCave IN (1986) Local and global aspects of the bottom nepheloid layers in the world ocean. *Netherlands Journal of Sea Research* 20: 167–181; based on Lamont nephelometer data presented by Kolla V, Sullivan L, Streeter SS, and Langseth MG (1976) Spreading of Antarctic bottom water and its effects on the floor of the Indian Ocean inferred from bottom water potential temperature, turbidity and sea-floor photography. *Marine Geology* 21: 171–189; and Kolla V, Henderson L, Sullivan L, and Biscaye PE (1978) Recent sedimentation in the southeast Indian Ocean with special reference to the effects of Antarctic bottom Water circulation. *Marine Geology* 27: 1–17.

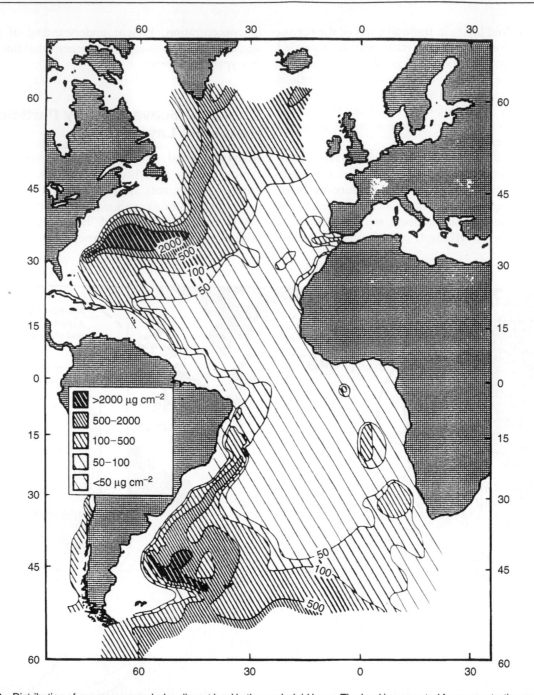

Figure 6 Distribution of excess suspended sediment load in the nepheloid layer. The load is computed for concentration excess over the value of the clear water minimum and integrated over the height of the nepheloid layer. Reproduced from Biscaye PE and Eittreim SL (1977) Suspended particulate loads and transports in the nepheloid layer of the abyssal Atlantic Ocean. *Marine Geology* 23: 155–172.

continental margins, has been proposed to account for the deficit of several elements in the open ocean. Short half-life radionuclides have shown that the particles in the BNL are responsible for trace element scavenging, and that the age of the suspensions in the BMNL is rather short, a few tens of days. The fact that the BMNL is almost always present means that this intense scavenging activity is continuous. Most

of the material involved in this scavenging becomes aggregated and falls to the bed on the continental slope and rise in the model of Amin and Huthnance. Turnewitsch and Springer observed that the most likely reason for the difference between measured and calculated thorium fluxes over the Porcupine Abyssal Plain (NE Atlantic) is lateral advection of [234]Th-depleted water, a loss likely due to particle scavenging in the BNL on the nearby continental margin. The $^{234}Th_t$ ('t' means total, i.e., dissolved and particle-associated) profile showed a $^{234}Th_t/^{238}U$ disequilibrium (the difference between $^{234}Th_t$ and ^{238}U activity) up to at least 500 mab (meters above bottom), and that secular $^{234}Th_t/^{238}U$ equilibrium was reached between 500 and 1000 mab. That is to say radiochemical evidence suggests lateral advection on the 100-day timescale (4–5 ^{234}Th half-lives) below 500 mab at a site quite close to the continental margin (about 250 km away). Above that, the timescale is longer.

The Turbidity Minimum

The source of nepheloid layers at heights over a few hundred meters above the bed is believed to be the sides of the ocean basins and protrusions from the bottom of the oceans such as seamounts, ridges, and rises. Major source regions are (1) deep zones of stronger currents, which in most areas are at depths greater than 3000 m; and (2) shallow areas where material resuspended at the shelf edge and upper slope by surface and internal wave motions, spreads seaward. Tidal motion on the rough topography of mid-ocean ridges is known to yield enhanced mixing as explained by Kurt Polzin and colleagues in 1997, and is also probably responsible for resuspension. Between these zones are depths of 1000–3000 m where stratification and currents are weak, there is no primary production of organic matter, and the ocean is often undersaturated with respect to calcite, aragonite, and opal. There is thus little particle supply from the side, and a decrease in downward transport due to dissolution and bacterial consumption of $CaCO_3$, SiO_2, and organic carbon, leading to a nepheloid minimum where (in the Atlantic) concentrations are 5–20 mg m^{-3}.

The concentrations given by nephelometers at the clear-water minimum differ by a factor of about 3 between areas under high surface productivity (high) and mid-gyre regions (low). The flux of large biogenic particles from the surface in these areas appears to provide a net supply of smaller particles to mid-depth (rather than scavenging them). This is also seen to be the case in temporal variability at a point, such that higher mid-water turbidity occurs in summer under higher productivity, and is lower in winter.

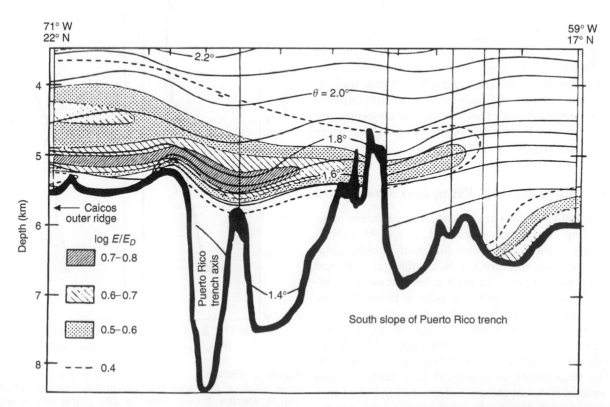

Figure 7 Interflow of suspended sediment generated in the Western Boundary Undercurrent, having detached from the bed on Caicos drift, flowing over the Puerto Rico trench. Reproduced from Tucholke BE and Eittreim SL (1974) The western boundary undercurrent as a turbidity maximum over the Puerto Rico Trench. *Journal of Geophysical Research* 79: 4115–4118.

Concentration and Spreading in the Atlantic and Indian Oceans

The broad picture of high-speed inflows on the western sides of ocean basins with a distributed return flow proposed by Henry Stommel has been confirmed by many hydrographic studies and current measurements. These broad patterns of the inflow of cold bottom water are very similar to the excess BNL concentration (excess over the clear-water minimum) for the Indian Ocean, and net BNL particulate load (mass per unit area) for the Atlantic based on Lamont nephelometer data (**Figures 5** and **6**). It is also apparent that the concentration or load of sediment generally declines going northward along the paths of the major inflows of Antarctic bottom water.

However, there are several points of very high concentration or load which are not obviously related to likely increases in mean bottom flow velocity, for example, the high in the center of the Argentine Basin (**Figure 6**). Richardson and colleagues showed in 1993 that these turbidity highs are sometimes caused by intermittently high velocities under regions where high surface eddy kinetic energy was propagated downward, resulting in high abyssal eddy kinetic energy, but at other times and places there was no correlation with current speeds. Thus some uncertainty remains over the role of high abyssal eddy kinetic energy versus locally accelerated thermohaline flow or deep recirculation loops in producing high concentration. Nevertheless, the major feature of concentrated BNLs is that they do delineate the paths of high-velocity bottom currents extremely well and they decline in concentration away from high-velocity boundary regions and away from the energetic Southern Ocean.

Boundary Mixing, INLs, and Inversions

Steps and inversions are best shown by modern high-sampling-rate sensors (**Figures 1, 2,** and **4**), though striking examples may also be found in older data. INLs are most common off the continental slope and submarine canyons. In both cases, focusing of internal waves (*see* and) or tides is believed to be responsible, due to the amplification of bottom flow velocities at points where the slope of the seabed S matches the wave characteristic slope C This causes both resuspension and boundary mixing (*see* and) The suspensions so generated have both fine and coarse particles at the time of generation but as they spread seaward they rain out the larger ones over the slope, leaving only finer material with slow sinking speeds.

Not all intermediate layers detach on slopes. Perhaps the biggest INL in the ocean occurs where the turbid plume of the North Atlantic Western Boundary Undercurrent leaves the bottom at the end of Caicos Drift off the Bahamas at about 5200 m depth and proceeds at that level over the 8000-m-deep Puerto Rico trench (**Figure 7**).

Pronounced inversions are seen at depths between 1 and 4 km along the continental margin close to the Congo (Zaire) River. A comparable situation is found over Nitinat Fan off the NW USA. This area lies off two submarine canyons which are the most likely sources of the intermediate layers. Thus,

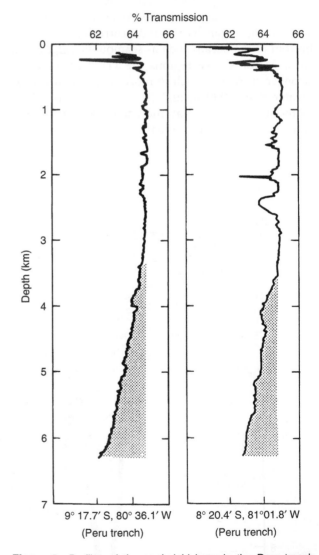

Figure 8 Profiles of the nepheloid layer in the Peru trench made with a 1 m path length transmissometer. Reproduced from Pak H, Menzies D, and Zaneveld JRV (1979) *Optical and Hydrographical Observations off the Coast of Peru during May–June 1977*, Data Report 77, Ref. 79-14, pp. 1–93. Corvallis, OR: Oregon State University.

canyons may also act as point sources for supply of turbid layers which mix out into the ocean interior.

Trenches and Channels

A final example of the influence of the sides of a basin on its nepheloid layers comes from the extremely thick nepheloid layers found in trenches and some deep-sea passages. The nepheloid layer thickness is that of the trench depth plus the thickness of the nepheloid layer over the adjacent seafloor for the Kuril–Japan trench. The result is a layer 2600-m thick. Nepheloid layers 2700-m thick have been recorded in the Peru trench between 8° and 10°S (**Figure 8**). Flow through Vema Channel between the Argentine and Brazil basins yielded a well-mixed water column in temperature and turbidity with excess radon-222 up to 400 m above the bed. Models of boundary layer development show that this cannot occur by vertical turbulent diffusion. Turbid layers must have spread quickly from the sides to the center of the channel, giving added support to the detached mixed-layer model.

The fact that there is more suspended material at depths greater than about 4000 m partly reflects the fact that waters at those depths are in contact with a much greater area of seabed in proportion to their volume than the shallower parts of the oceans. For 4–5 km depth the value is $0.83 \, km^2 \, km^{-3}$, whereas for 2–3 km it is $0.11 \, km^2 \, km^{-3}$. One might say that deeper waters feel more bed. The global distribution of nepheloid layers is, for instrumental reasons, the global distribution of fine particles ($< 2 \, \mu m$), because that is what the instruments that map them 'see', but they also contain larger particles which play an important role in sediment dynamics, especially in the lowest 10 m of the BMNL.

Further Reading

Amin M and Huthnance JM (1999) The pattern of cross-slope depositional fluxes. *Deep-Sea Research I* 46: 1565–1591.

Anderson RF (2003) Chemical tracers of particle transport. In: Elderfield H (ed.) *Treatise on Geochemistry, Vol. 6: The Oceans and Marine Geochemistry*, pp. 247–273. Oxford, UK: Pergamon.

Anderson RF, Bacon MP, and Brewer PG (1983) Removal of [230]Th and [231]Pa at ocean margins. *Earth and Planetary Science Letters* 66: 73–90.

Armi L and D'Asaro E (1980) Flow structures of the benthic ocean. *Journal of Geophysical Research* 85: 469–484.

Bacon MP and Rutgers van der Loeff MM (1989) Removal of thorium-234 by scavenging in the bottom nepheloid layer of the ocean. *Earth and Planetary Science Letters* 92: 157–164.

Baker ET and Lavelle JW (1984) The effect of particle size on the light attenuation coefficient of natural suspensions. *Journal of Geophysical Research* 89: 8197–8203.

Biscaye PE and Eittreim SL (1977) Suspended particulate loads and transports in the nepheloid layer of the abyssal Atlantic Ocean. *Marine Geology* 23: 155–172.

Dickson RR and McCave IN (1986) Nepheloid layers on the continental slope west of Porcupine Bank. *Deep-Sea Research* 33: 791–818.

Gardner WD (1989) Periodic resuspension in Baltimore Canyon by focusing of internal waves. *Journal of Geophysical Research* 94: 185–194.

Hall IR, Schmidt S, McCave IN, and Reyss JL (2000) Particulate matter distribution and Th-234/U-238 disequilibrium along the Northern Iberian Margin. Implications for particulate organic carbon export. *Deep-Sea Research I* 47: 557–582.

Jerlov NG (1976) *Marine Optics*. Amsterdam: Elsevier.

Johnson DA, McDowell SE, Sullivan LG, and Biscaye PE (1976) Abyssal hydrography, nephelometry, currents and benthic boundary layer structure in Vema Channel. *Journal of Geophysical Research* 81: 5771–5786.

Kolla V, Henderson L, Sullivan L, and Biscaye PE (1978) Recent sedimentation in the southeast Indian Ocean with special reference to the effects of Antarctic bottom water circulation. *Marine Geology* 27: 1–17.

Kolla V, Sullivan L, Streeter SS, and Langseth MG (1976) Spreading of Antarctic bottom water and its effects on the floor of the Indian Ocean inferred from bottom water potential temperature, turbidity and sea-floor photography. *Marine Geology* 21: 171–189.

McCave IN (1983) Particulate size spectra, behavior and origin of nepheloid layers over the Nova Scotian continental rise. *Journal of Geophysical Research* 88: 7647–7666.

McCave IN (1986) Local and global aspects of the bottom nepheloid layers in the world ocean. *Netherlands Journal of Sea Research* 20: 167–181.

McPhee-Shaw E (2006) Boundary–interior exchange: Reviewing the idea that internal-wave mixing enhances lateral dispersal near continental margins. *Deep-Sea Research II* 53: 42–59.

Menard HW and Smith SM (1966) Hypsometry of ocean basin provinces. *Journal of Geophysical Research* 71: 4305–4325.

Morel A (1973) *Indicatrices de Diffusion Calculées par la Théorie de Mie pour les Systèmes Polydisperses, en Vue de l'Application aux Particules Marines*, Report 10, pp. 1–75. Paris: Laboratoire d'Océanographie Physique, University of Paris VI.

Pak H, Menzies D, and Zaneveld JRV (1979) *Optical and Hydrographical Observations off the Coast of Peru during May–June 1977*, Data Report 77, Ref. 79-14, pp. 1–93. Corvallis, OR: Oregon State University.

Polzin KL, Toole JM, Ledwell JR, and Schmitt RW (1997) Spatial variability of turbulent mixing in the abyssal ocean. *Science* 276: 93–96.

Richardson MJ, Weatherly GL, and Gardner WD (1993) Benthic storms in the Argentine Basin. *Deep-Sea Research* 40: 957–987.

Thorndike EM (1975) A deep sea, photographic nephelometer. *Ocean Engineering* 3: 1–15.

Tucholke BE and Eittreim SL (1974) The western boundary undercurrent as a turbidity maximum over the Puerto Rico Trench. *Journal of Geophysical Research* 79: 4115–4118.

Turnewitsch R and Springer BM (2001) Do bottom mixed layers influence [234]Th dynamics in the abyssal near-bottom water column? *Deep-Sea Research I* 48: 1279–1307.

Zaneveld JRV, Roach DM, and Pak H (1974) The determination of the index of refraction of oceanic particulates. *Journal of Geophysical Research* 79: 4091–4095.

OPTICAL PARTICLE CHARACTERIZATION

P. H. Burkill and C. P. Gallienne,
Plymouth Marine Laboratory, West Hoe,
Plymouth, UK

Particles and Their Properties

Particles are ubiquitous in ocean waters, where they are intimately involved in defining the optical properties, productivity, and biogeochemistry of our seas. Marine particles exist in a wide range of sizes and concentrations, and exhibit an inverse relationship between size and concentration, and a positive relationship between concentration and ambient nutrient concentration (the 'trophic status') in surface waters of the ocean.

Particles range in size from the largest marine organisms (blue whales, *c.* 70 m length) down to the size that arbitrarily divides particles from dissolved materials. In biological oceanography, this is defined operationally as 0.2 μm. But here we will focus on particles that fall within the size range of the plankton. Plankton organisms range from viruses (*c.* 0.05 μm) up to larger zooplankton such as euphausiids (*c.* 2 cm). However, even within this size range, many particles are not living but instead contribute to the large pools of detritus that often predominate over living particles in the ocean.

Particle Characterization

A wide array of techniques is available for the characterization of marine particles. Although most are based on optical properties, nonoptical techniques, such as the acoustic doppler current profiler (ADCP) and the multifrequency echosounder, can also be used to quantify and characterize particles such as large zooplankton and fish in sea water. Optical characterization techniques vary considerably in their resolution. At one extreme, satellite-based remote sensing can be used to quantify and characterize the marine phytoplankton across whole ocean basins. At the other extreme, microscope-based techniques and analytical flow cytometry resolve single particles. For the biologist, microscopy is the benchmark procedure for identification of plankton. This is true whether the particles of interest are viruses, which are typically analyzed by electron microscopy, or bacteria, protozoa, or larger zooplankton, which are analyzed by light microscopy. As an adjunct to light microscopy, fluorescence-based techniques are used increasingly to characterize, and sometimes quantify, the chemical properties of cells. Such approaches can be extremely powerful, particularly when used in conjunction with fluorescently labeled molecular probes. Such probes can be tailored to target specific taxonomic groups. Although microscopy remains the benchmark, for the simple reason that 'what you see, you believe,' it is time consuming and costly. A wide range of techniques offer rapid analysis of particles. However, these tend to be 'black box' techniques and should always be used with appropriate controls. No single technique provides a panacea in particle analysis, and it is often useful to combine two or more complementry techniques.

Rapid optical techniques for analyzing plankton-sized particles may be based on scattered, fluorescent or transmitted light. Scattering and fluorescence methods are applicable to smaller particles (< 500 μm equivalent spherical diameter (ESD)), where as larger particles, such as zooplankton, are usually analyzed by transmission techniques. Two techniques that have been developed rapidly in the last decade, are analytical flow cytometry (AFC) and optical plankton counting (OPC). AFC and OPC are particularly suitable for the analysis of smaller particles (viruses to protozoa) and larger particles (metazoa), respectively.

Analytical Flow Cytometry

Technique Analytical flow cytometry (AFC) is a generic technique based on the multiparametric analysis of single particles at high speed. Originally developed for medical hematology and oncology, AFC is used increasingly in biological oceanography. Its strengths are derived from its quantitative capability, versatility, sensitivity, speed, statistical precision, and ability to identify and, in many instruments, sort particle subsets from heterogeneous populations. Its drawbacks are its cost and, for commercial instruments, the small volume of sample (*c.* 0.5 cm^3) analyzed. Particle characterization and quantification in flow cytometry relies on cellular fluorescence and light scatter, and the power of the technique derives from the ability to make multiple measurements

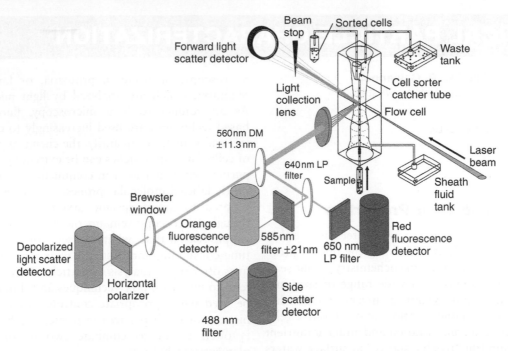

Figure 1 Operating principles of AFC in which samples containing the particles of interest are passed singly across a laser beam. Each particle scatters light and this is collected by forward and side light scatter detectors. Birefringent particles will tend to depolarize the vertically polarized laser light and this is measured at the appropriate detector. Fluorescence from each particle is collected and spectrally filtered so the wavelength of interest is detected by photomultiplier tubes. Output from each sensor is digitized and the data are transferred to a computer. Particle size and refractive index are determined by the light scatter and the chemical properties are determined by fluorescence. Particles exhibiting appropriate properties can be collected by sorting, whereas other particles pass to waste. (Figure produced by Glen Tarran, Plymouth Marine Laboratory.)

simultaneously on each cell at high speed. Typically, up to 5000 cells can be analyzed per second and sorting rates of $>10\,000\ \mathrm{s}^{-1}$ with $>98\%$ purity, can be achieved.

The principles of AFC (**Figure 1**) are based on hydrodynamically focusing a suspension that is streamed coaxially through a flow chamber so that individual particles pass singly through the focus of a high intensity light source. Suitable light sources include coherent wave lasers since these provide a very stable light beam that can easily be focused to small dimensions. Light flux from the highly focused light source generates enough fluorescence from individual cells for this to be measured in the few microseconds taken to traverse the light beam. As the particles traverse the beam, they scatter light and may also produce fluorescence. Cellular fluorescence arises from autofluorescent cells or cells stained with a fluorochrome, and this is collected by a high numerical aperture lens located orthogonally to the irradiation source and the sample stream. The light collected is spectrally filtered sequentially by dichroic mirrors which reflect specific wavelengths into photomultiplier tubes (PMT). The PMTs are optically screened by band-pass filters. The quantity of light incident upon each PMT, responding within a given color band, is then proportionally converted

into an electrical signal. The signal is amplified, digitized, and stored transiently in computer memory. The data are then displayed on a computer screen and stored onto disk as 'list mode' data. The list mode data can be considered to be analogous to a spreadsheet in which each row represents a particle, each column represents a different AFC sensor with values that represent quantitative optical signatures of each particle. The advantage of list mode data, as with data in any spreadsheet, is that it can be replayed, reanalyzed, and redisplayed.

Commercial cytometers are usually equipped with light-scatter detectors that are situated in the narrow forward and an orthogonal angle, as well as two or three fluorescence PMTs. Although a single laser is normal, AFC instruments can be equipped with two or more lasers to increase the number of fluorochrome excitation wavelengths. Some cytometers may use arc-lamp excitation particularly if UV irradiation is required. There is also a move to the use of diode lasers for applications that demand low power use. Flow chambers may vary in their hydrodynamic, optical, mechanical, and electrical characteristics to achieve high sensitivity and good stream stability. AFC instruments often have quartz cuvette sensing zones to improve sensitivity and to allow the application of UV irradiation. Specialized cytometers can also measure particle volumes based on the Coulter principle of electrical impedance alteration as particles flow through a restricted orifice. Other specialized instruments may generate images of particles in the sensing zone. Data processing and display procedures have been developed which handle fully crossed-correlated multidimensional data and this is achieved by microcomputers. Considerable developments have taken place in the last few years to apply sophisticated procedures such as multiparametric statistics or neural net generation, to identify and characterize particles from within heterogeneous mixtures.

Many AFC instruments are able to sort cells and this is invaluable for identification, manipulation or as a gateway to other analysis procedures. High speed 'sorting-in-air' is based on developments in ink-jet printing. Two populations may be sorted from the sample stream that undergoes oscillation, driven by a piezo-electric crystal that is mechanically coupled to the flow chamber. The crystal, driven at 30–40 kHz, produces uniform liquid droplets of which a small percentage contain single cells. The 'sort logic' circuitry compares processed signals from the sensors with pre-set, operator-defined ranges. When the amplitude falls within the pre-set range, an electronic time delay of a few microseconds is activated. This triggers an electrical droplet-charging pulse at the moment the cell arrives at the droplet formation break-off point. The droplet-charging pulse causes a group of droplets to be charged, and subsequently, deflected by a static electric field into a collection vessel. Cells failing the pre-set sort criteria do not trigger droplet-charging, and so pass undetected into the waste collector. Other sorting procedures include one in which a collecting arm moves into the sample stream to pick up particles that meet the programmed sort criteria. Sorting is an essential adjunct to AFC and is crucial to verifying both satisfactory instrument and analytical protocol operation.

Applications Oceanographic applications of AFC are now diverse and continue to expand rapidly. Although the fundamental principle of AFC remain constant, recent developments in optical sensitivity and design of AFC instruments have aided new applications. Fluorochrome chemistry and molecular biology are both richly endowed fields and developments in fluorescent assays of biochemical constituents, coupled with the ability to target individual taxa have proved invaluable for AFC applications in marine biology. Detection limits are adequate to measure cellular attributes of many planktonic cells.

Cellular fluorescence may be derived from two basic categories:

1. autofluorescence in which the fluorescent molecule of interest occurs naturally in the cell;
2. applied fluorescence in which the fluorescent dye is applied, or otherwise generated, and fluorescence is accumulated within the cell.

Phytoplankton. Analysis of phytoplankton by AFC is based on the presence of chlorophyll, a highly autofluorescent compound that is found in all viable plants. Chlorophyll is the phytoplankton's principal light-harvesting pigment, and absorbs light strongly in the blue and red regions of the visible spectrum. Blue light cellular absorption coincides with the emission of the argon ion laser at 488 nm that is commonly used in flow cytometers. Chlorophyll fluorescence is emitted in the far red ($\lambda_{em} = 680$ nm), thereby offering a useful Stokes' shift of some 200 nm. This window means that phytoplankton can be readily characterized and quantified by flow cytometers equipped with an argon laser (or other blue light source) and suitable spectral filtration (such as a 650 nm longpass filter) of fluorescent light emitted by cells onto a sensitive photomultiplier tube.

As well as chlorophyll, some phycobiliproteins are also autofluorescent. One of these is phycoerythrin which is found in cyanobacteria and cryptophytes. Although phycoerythrin absorbs light in the green–blue end of the spectrum, the 488 nm emission of the argon is sufficiently close to excite this compound. In studies of phytoplankton, fluorescence from phycoerythrin is measured by a separate photomultiplier tube that is spectrally filtered to collect emissions at 585 nm. Based on this differentiation and coupled with light scatter measurements (the magnitude of which is roughly proportional to cell size), AFC can readily differentiate and quantify the phytoplankton groups shown in **Table 1**.

In recent years, the application of powerful multivariate statistical and neural net procedures have been applied to further characterize algal taxa

Table 1 Routine AFC analysis of phytoplankton based on optical characteristics

Differentiation	AFC criteria
Phytoplankton	Chlorophyll autofluorescence
Prochlorococcus	Low chlorophyll and light scatter
Synechococcus	Low phycoerythrin and light scatter
Cryptophytes	Phycoerythrin and light scatter
Coccolithophores	High orthogonal light scatter and laser depolarization

from within the complex mixtures that are typical of sea water. Multivariate statistics that have been used include quadratic discriminant analysis and canonical variate analysis. The latter is a useful graphical technique for analyzing and displaying data, whereas quadratic discriminant analysis can discriminate over two-thirds of mixtures of 22 algal taxa, with classification rates >70%. Such approaches are more than two orders of magnitude faster than conventional flow cytometric analyses for discriminating and enumerating phytoplankton species.

Artificial neural nets (ANN) have proved to be extremely powerful in increasing AFC capability for differentiating algal taxa. This approach is based on training an ANN to recognize the optical characteristics of individual taxa. This is achieved by presenting the net with AFC data derived from unialgal cultures. The unknown samples are then analyzed by the AFC under the same conditions and the data passed through the trained ANN. The net outputs identification probabilities for each cell analyzed. Several types of ANN have been used and it is now possible for nets to differentiate and recognize >70 taxa with high accuracy. Considerable developments are anticipated in this field in the coming years.

As well as providing procedures for differentiation of phytoplankton from other particles, there are AFC protocols for quantifying cellular attributes of phytoplankton. These include the cellular concentrations of chlorophyll, phycoerythrin, protein and DNA as well as enzymes such as ribulose-1,5-bisphosphate carboxylase. AFC instruments are capable of great sensitivity and are able to quantify concentrations of cellular chlorophyll in phytoplankton in the range of about $1–2000$ fg cell^{-1}.

In practice, marine phytoplankton are typically analyzed using a fresh sample of sea water without pretreatment. The sample is analyzed at a constant rate so sample volume can be determined from analysis time. Chlorophyll-containing phytoplankton and those containing phycoerythrin are registered by the red and orange fluorescence emitted from single cells as they traverse the laser beam. Typical sample analysis time is generally 4–5 min. Examples of data generated by AFC protocols for the analyses of natural waters are shown in **Figure 2**.

Bacteria. Bacteria, traditionally quantified by epifluorescence microscopy, can now be differentiated from other particles and analyzed by AFC (**Figure 2**). Both approaches are based on the intercalation of a fluorochrome with the cell's nucleic acid. However, such intercalation is universal and often does not differentiate between autotrophic and heterotrophic bacteria. A range of fluorochromes have been used including 488 nm absorbing YOYO-1, YO-PRO-1, PicoGreen and SYBR Green as well as the more traditional UV excited bis-benzimide Hoechst 33342 or 4',6-diamidino-2-phenylindole (DAPI). Of these,

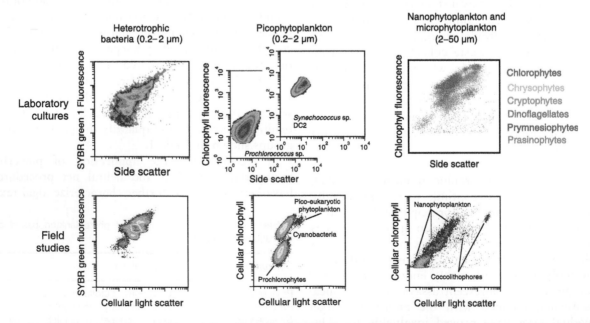

Figure 2 AFC characterization and differentiation of bacteria and phytoplankton in lab culture and in natural communities carried out at the Plymouth Marine Laboratory. Bacteria are measured using SYBR Green fluorescence and the different phytoplankton are measured by chlorophyll autofluorescence. Analysis can be verified by microscopic analysis of flow-sorted particles. (Figure produced by Glen Tarran, Plymouth Marine Laboratory.)

SYBR Green offers the practical advantage that autotrophic and heterotrophic bacteria can be differentiated readily.

It is now possible to Quantify the cellular protein and DNA content of bacteria in natural waters. Such AFC techniques use the intensity of SYPRO-protein or DAPI–DNA fluorescence of individual marine bacteria. Cultures of various marine bacteria have been measured in the range 60–330 fg protein cell^{-1}, but the amount of natural bacterioplankton from the North Sea in August 1998 was shown to be only 24 fg protein cell^{-1}. The total DNA of natural bacteria has been estimated to be about 3 fg cell^{-1} by AFC techniques.

Changes in bacterioplankton community composition have also been assessed by molecular biological AFC techniques. The combination of AFC analysis and sorting combined with denaturing gradient gel electrophoresis of polymerase chain reaction (PCR)-amplified 18S rDNA fragments and fluorescence *in situ* hybridization has been shown to be a rapid method of analyzing the taxonomic composition of bacterioplankton. Experimental manipulation of natural water samples resulted in a bacterial succession from members of a *Cytophaga flavobacterium* cluster, through gamma-proteobacteria and finally alpha-proteobacteria.

Protozoa. AFC-based techniques for the analysis of protozoa have, so far, been based on molecular probes. Ribosomal RNA species-specific probes to various members of the common heterotrophic flagellate genus, *Paraphysomonas*, have been developed (**Figure 3**). However, they have been restricted to laboratory applications since naturally occurring organisms exhibit cellular fluorescence levels that are often too low to distinguish from background. This observation may either be a reflection of poor probing efficiency or it may be due to the organisms low growth rates *in situ*.

Viruses. Viruses are now thought to be one of the most abundant types of particles in the ocean. AFC protocols are now available for enumerating natural marine viruses based on staining with the nucleic acid-specific dye SYBR Green-I. Interestingly AFC-based counts are often higher than those obtained by microscopy, suggesting that further development work is needed. However, this AFC protocol reveals two, and sometimes three, virus populations in natural samples, whereas microscopy would only differentiate one pool of viruses. Cultures of several different marine virus families (Baculoviridae, Herpesviridae, Myoviridae, Phycodnaviridae, Picornaviridae, Podoviridae, Retroviridae, and

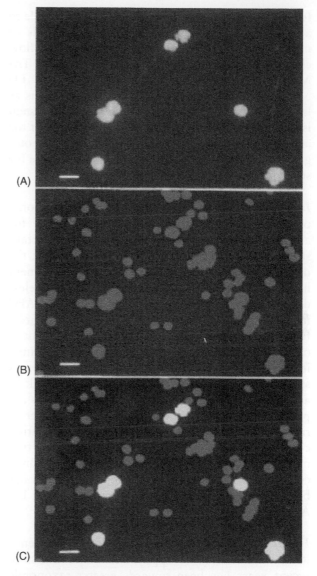

(A)

(B)

(C)

Figure 3 Photomicrographs showing a mixture of four *Paraphysomonas* species after hybridization with a mixture of PV1 and EUK probes tagged with fluorescein and rhodamine, respectively. PV1 is specific for *P. vestita* whereas EUK labels all eukaryotes. The mixture was irradiated at (A) 488 nm to show *P. vestita* labeled with PV1, (B) 568 nm to show organisms labeled with EUK, and (C) both 488 and 568 nm to reveal both probes. Scale bar is 10 μm. (Reproduced from Rice *et al.* (1997) with permission from the Society for General Microbiology.)

Siphoviridae) have also been stained with a variety of highly fluorescent nucleic acid-specific dyes. Highest fluorescence is achieved using SYBR Green I, allowing DNA viruses with genome sizes between 48.5 and 300 kb (kilobases) to be detected. Small genome-sized RNA viruses (7.4–14.5 kb) are at the current limit of detection by AFC.

Zooplankton and larval fish. Although commercially available AFC instruments are directly applicable for the analysis of microbial cells, it is also possible to adapt the generic AFC concept for the analysis of metazoan organisms. This involves a scaling up of flow chambers and the associated fluidics system. The Macro Flow Planktometer built within the EU MAROPT project has been applied to organisms as large as larval fish. An inherent property of this system is that it incorporates 'imaging in flow' as part of the analysis. Images may be stored either photographically or electronically and then be available for subsequent image analysis.

In-situ AFC. As we move towards operational oceanography, there is an increasing need for autonomous, *in situ* instrumentation. An important step towards this has been made recently with the development of CytoBuoy, an AFC instrument housed in a moored buoy and capable of wireless data transfer. CytoBuoy is one of the very few AFC instruments to have been designed and built purely for aquatic use. Its characteristics include enhanced optics and electronics designed to obtain maximal information on particle characteristics. Whereas standard cytometers reduce these to single peak or area 'list-mode' numbers, time resolved signals are preserved fully and transferred to the computer as raw data. Pulse shape signals aid identification considerably and allow, for the first time, a true measurement of particle length. The CytoBuoy concept has also been taken a step further and has been redesigned as a functional module of the UK Autosub autonomous underwater vehicle.

Future trends. The generic capability of AFC lends itself to a variety of applications. The thrust in recent years has been towards greater taxonomic resolution and this has been aided by the application of molecular techniques particularly involving oligonucleotide probes. These have many advantages including the ability to be tailored to target particular groups of organisms. Here the level of taxonomic targeting may range from general (e.g. differentiation of classes of phytoplankton) down to individual species. Molecular probes may also be coupled with chemotaxonomic capability that can analyze cellular function such as specific enzyme production.

Flow cytometry has opened up our ability to characterize marine particles with a greater degree of taxonomic resolution and further development of techniques such as ANN will increase this capability considerably. This should allow characterization of natural populations objectively in close to real-time. The quest for greater analytical resolution will undoubtedly continue. However, it is also possible that a taxonomic identification watershed may soon be reached. So far, the focus of AFC protocols has generally conformed to traditional taxonomic criteria. But there may be another route that remains to be explored. This involves an approach to classifying particles based directly on flow cytometry variables of light scatter, fluorescence, time of flight etc. Such an approach might prove worthwhile because of its direct simplicity. How such an approach would compare to traditional taxonomic identification remains to be addressed, and that remains an exciting challenge for the future.

Optical Plankton Counting

The Optical Plankton Counter (OPC) has been designed to analyze those zooplankton called the mesozooplankton whose size range conventionally spans 0.2–20 mm.

Technique The OPC uses a collimated beam of light through an enclosed volume, received by a photosensor. When a particle interrupts this beam of light the sensor produces an electronic response proportional to the cross-sectional area of the particle (**Figure 4**). This response is digitized and this digital size is converted into ESD using a semiempirical formula. This is the diameter of a sphere having the same cross-sectional area as the particle being measured, and can be simply converted into volume. The OPC comes in two versions: a towed instrument for *in situ* use and a benchtop laboratory version. The towed version has a sampling tunnel of 22 cm × 2 cm. The laboratory version has a glass flow cell 2 cm square.

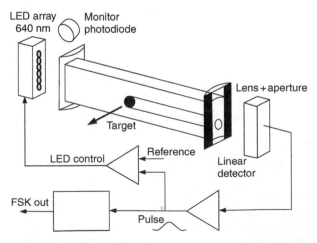

Figure 4 Schematic of the operating principle of the optical plankton counter. LED, light-emitting diode; FSK; frequency shift keying.

Applications The OPC is capable of large-scale, rapid and continuous sampling of zooplankton, providing a reliable measure of size distributed abundance and biovolume, between 0.25 and 20 mm ESD, at data rates up to 200 s^{-1}. The use of the *in situ* OPC on various towed platforms is now well established. The laboratory version is intended for characterizing preserved samples. It has also been deployed at sea in pump-through mode producing continuous real-time data on surface zooplankton abundance and size distribution, permitting near continuous sampling of epipelagic zooplankton across ocean basin scales. **Figure 5** shows data from the OPC in this mode on the Atlantic Meridional Transect. The data in **Figure 5** show one of 11 transects completed to date, each comprising a near-continuous 13 000 km transect of size distributed biovolume in the North and South Atlantic, illustrating the power of this kind of instrument. Data at this level of detail and spatial resolution acquired autonomously and continuously could not have been gathered over such large spatial scales by any other means.

Operational considerations Bias and coincidence require calibration of the instrument against some other sampling device. Initial calibration uses spherical glass beads of known size. Several researchers have noted nonlinearity in this calibration at the extremes of the size range, and have suggested that the operational size range should be reduced. Sensor response time and coincidence limit the densities at which the OPC can operate. Coincidence occurs when more than one particle is in the beam at the same time. They register as one larger particle and abundance will, therefore, be underestimated, and biomass overestimated. Coincidence in the towed OPC can be reduced in areas of high abundance by inserting into the sampling tunnel a transparent plate, reducing the sampling volume to one-fifth. There is a finite probability of coincidence occurring at all concentrations, increasing with abundance and flow rate through the instrument. It has been determined experimentally that at a count rate of 30 s^{-1}, more than 90% of particles will be counted. Smaller particles far outnumber larger ones, so coincidence will result in a loss of these smaller particles, and biovolume is underestimated to a lesser degree than abundance. Towed at 4 m s^{-1}, the standard OPC will pass 17.6 l s^{-1} through the sampling tunnel. Coincidence will therefore begin to have a significant effect on estimated abundance above concentrations of 1700 m^{-3}, or 8500 m^{-3} with the flow insert.

Object orientation can also present problems in this type of counter. Elongated organisms present a very different cross-sectional area depending upon whether they are side-on or end-on. Biovolume may be considerably underestimated in the latter case. It has been shown that on average the cross-sectional area measured for a randomly oriented object will be greater than 70% of the true value. Biovolume may also be overestimated by the spherical model assumed in the OPC calibration – most zooplankton have a shape closer to that of an oblate spheroid. We use an ellipsoidal model based on cross-sectional area of the particle. From cross-sectional area the length and width of the ellipsoid can be calculated (assuming a length to width ratio typical for copepods).

Several cases have been reported of OPCs producing counts many times higher than those from concurrent net samples. Comparison between OPC and a Longhurst Hardy Plankton Recorder (LHPR)

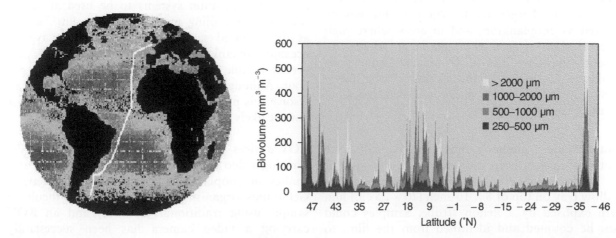

Figure 5 A 13 000 km transect of size-distributed epipelagic zooplankton biovolume in the North and South Atlantic from the OPC in continuous surface sampling mode on the Atlantic Meridional Transect.

showed that abundance and biovolume recorded by the OPC were consistently four times higher than for the LHPR (200 μm mesh). Using a 53 μm mesh showed that there had been significant undersampling of zooplankton <350 μm ESD by the LHPR. Studies like these show that care is required when comparing any two sampling methods. When comparing the OPC to net systems, careful selection of mesh size or OPC lower size threshold is required. Our own investigations indicate that the most suitable mesh size for a net used in comparison with the standard OPC is 125 μm (250 μm ESD, 3:1 ellipsoidal model, mesh size 75% of width of smallest animal to be quantitatively sampled).

It must be emphasized that OPC cannot discriminate living from nonliving particles. All particles within its operating range will be detected, including detritus, marine aggregates, air bubbles, etc. A recent study found average abundance in the Faroe–Shetland Channel and north–western North Sea to be 2.5 times higher for an OPC than for net samples and up to 40 times higher at extremely low concentrations. This was put down to detrital particles/marine aggregates, which may in some cases be the most abundant particle type in the water, and are often too fragile to be sampled by nets. Detrital particles may be considered part of the plankton, being significant in marine food webs, but their presence in large numbers will bias comparisons between OPC and net derived estimates of abundance.

Particle Imaging Instruments

Automated particle counters (electronic, acoustic, and optical) can give reliable data on zooplankton abundance and size distribution, but tell us little or nothing about the species present, except where dominant species are already known and well separated in size. Larger autotrophic particles may be counted as zooplankton, and in areas where high concentrations of detritus or marine aggregates are present, they are not discriminated from zooplankton. To make these discriminations, information on shape as well as size is needed, requiring the use of imaging devices.

Photographic methods High-speed silhouette photography has been undertaken at sea without the need for sample preservation. Samples in a shallow tray on top of 8 × 10-inch (20 × 25 cm) film were exposed by a xenon strobe. Samples could then be counted and identified from the film, to provide a permanent record. A towed version of the system using concentrating nets ahead of a camera system was subsequently developed. Although this system avoided some of the problems associated with the deployment and use of net systems, and preservation of samples, a human investigator still carried out counting and identification of the samples. Automated image processing of digitized photographic images could alleviate this problem, but duration and spatial resolution is limited by the amount of film that can be carried.

Video methods The most advanced video method for producing abundance and size distribution indicators of good temporal and spatial resolution, together with near-real-time classification to taxonomic groups, is the video plankton recorder. This consists of a towed frame 3 m long with four cameras each having concentric fields of view (5–100 mm) covering the size range of interest (0.5–20 mm. ESD). The imaged volume is defined by the oblique intersection of the cone defining the camera field of view, and that produced by a collimated, strobed 80-watt xenon light beam pulse 1 μs duration, producing dark field illumination. Sample volume varies from 1 ml to 1 liter. Fiberoptic telemetry is used to send video data to the surface, where information is recorded to tape. Subsequent upgrading of this system is expected to permit preprocessing to be done in real time, and near real time production of size and species distributions. Concurrent calibration of the device is provided by an integrated LHPR system. The device can differentiate detrital material and zooplankton, unlike optical and acoustic systems. The video plankton recorder has been used extensively around George's Bank in the north-west Atlantic.

Video cameras used in scientific imaging typically produce 10–100 million bytes s^{-1}, and image-processing algorithms are highly computer intensive. For real-time acquisition systems to be used at sea in continuous sampling mode, assuming 1 m^{-3} min^{-1} as the required sampling volume, and at typical oceanic zooplankton abundance of 50–5000 m^{-3}, we need to be able to process 3000–300 000 animals per hour. Although currently available technology can resolve this problem to some degree, the solution is not yet likely to be economical in size or cost.

ROV devices Remotely operated vehicles (ROVs) carrying video cameras have been used for *in situ* studies of zooplankton abundance and behavior. Gelatinous organisms are notoriously difficult to sample using traditional methods, and an ROV carrying a video camera has been successfully deployed for their study. ROVs are usually restricted to small-scale, observational studies, but may be useful for the location of zooplankton patches. Small imaging volume may preclude the study of rarer taxa, and poor image quality may preclude the study of smaller taxa. Zooplankton also exhibit attraction/avoidance responses to the presence of ROV systems.

See also

Acoustic Scattering by Marine Organisms. Fluorometry for Biological Sensing. Inherent Optical Properties and Irradiance. Remotely Operated Vehicles (ROVs).

Further Reading

Boddy L, Morris CW, and Wilkins MF (2000) Identification of 72 phytoplankton species by radial basis function neural network analysis of flow cytometric data. *Marine Ecology Progress Series* 195: 47–59.

Brussaard CPD, Marie D, and Bratbak G (2000) Flow cytometric detection of viruses. *Journal of Virological Methods* 85: 175–182.

Davis CS, Gallager SM, Marra M, and Stewart WK (1996) Rapid visualisation of plankton abundance and taxonomic composition using the video plankton recorder. *Deep-Sea Research II* 43: 1947–1970.

Dubelaar GBJ and Gerritzen PL (2000) CytoBuoy: a step forward towards using flow cytometry in operational oceanography. *Scientia Marina* 64: 255–265.

Gallienne CP and Robins DB (1998) Trans-oceanic characterisation of zooplankton community size structure using an Optical Plankton Counter. *Fisheries Oceanography* 7: 147–158.

Herman AW (1992) Design and calibration of a new optical plankton counter capable of sizing small zooplankton. *Deep Sea Research* 39(3/4): 395–415.

Jonker R, Groben R, Tarran G, *et al.* (2000) Automated identification and characterisation of microbial populations using flow cytometry: the AIMS project. *Scientia Marina* 64: 225–234.

Kachel V and Wietzorrek J (2000) Flow cytometry and integrated imaging. *Scientia Marina* 64: 247–254.

Olson RJ, Zettler ER, and DuRand MD (1993) Phytoplankton analysis using flow cytometry. In: Kemp, *et al.* (eds.) *Handbook of Methods in Aquatic Microbial Ecology*, pp. 175–186. Boca Raton: Lewis.

Reckermann M and Colijn F (2000) Aquatic flow cytometry: achievements and prospects. *Scientia Marina* 64(2): 119–268.

Rice J, Sleigh MA, Burkill PH, *et al.* (1997) Flow cytometric analysis of characteristics of hybridization of species-specific fluorescent oligonucleotide probes to rRNA of marine nanoflagellates. *Applied and Environmental Microbiology* 63: 938–944.

Rice J, O'Connor CD, Sleigh MA, *et al.* (1997) Fluorescent oligonucleotide rDNA probes that specifically bind to a common nanoflagellate. *Paraphysomonas vestita. Microbiology* 143: 1717–1727.

Schultze PC, Williamson CE, and Hargreaves BR (1995) Evaluation of a remotely operated vehicle (ROV) as a tool for studying the distribution and abundance of zooplankton. *Journal of Plankton Research* 17: 1233–1243.

Wood-Walker RS, Gallienne CP, and Robins DB (2000) A test model for optical plankton counter (OPC) coincidence and a comparison of OPC derived and conventional measures of plankton abundance. *Journal of Plankton Research* 22: 473–484.

Zubkov MV, Fuchs BM, Sturmeyer H, Burkill PH, and Amann R (1999) Determination of total protein content of bacterial cells using SYPRO staining and flow cytometry. *Applied and Environmental Microbiology* 65: 3251–3257.

PHOTOCHEMICAL PROCESSES

N. V. Blough, University of Maryland, College Park, MD, USA

Introduction

Life on Earth is critically dependent on the spectral quality and quantity of radiation received from the sun. The absorption of visible light (wavelengths from 400 to 700 nm) by pigments within terrestrial and marine plants initiates a series of reactions that ultimately transforms the light energy to chemical energy, which is stored as reduced forms of carbon. This complex photochemical process, known as photosynthesis, not only provides all of the chemical energy required for life on Earth's surface, but also acts to decrease the level of a major greenhouse gas, CO_2, in the atmosphere. By contrast, the absorption of ultraviolet light in the UV-B (wavelengths from 280 to 320 nm) and UV-A (wavelengths from 320 to 400 nm) by plants (as well as other organisms) can produce seriously deleterious effects (e.g. photoinhibition), leading to a decrease in the efficiency of photosynthesis and direct DNA damage (UV-B), as well as impairing or destroying other important physiological processes. The level of UV-B radiation received at the Earth's surface depends on the concentration of ozone (O_3) in the stratosphere where it is formed photochemically. The destruction of O_3 in polar regions, leading to increased levels of surface UV-B radiation in these locales, has been enhanced by the release of man-made chlorofluorocarbons (CFCs), but may also be influenced in part by the natural production of halogenated compounds by biota.

These biotic photoprocesses have long been recognized as critical components of marine ecosystems and air–sea gas exchange, and have been studied extensively. However, only within the last decade or so has the impact of abiotic photoreactions on the chemistry and biology of marine waters and their possible coupling with atmospheric processes been fully appreciated. Light is absorbed in the oceans not only by phytoplankton and water, but also by colored dissolved organic matter (CDOM), particulate detrital matter (PDM), and other numerous trace light-absorbing species. Light absorption by these constituents, primarily the CDOM, can have a number of important chemical and biological consequences including: (1) reduction of potentially harmful UV-B and UV-A radiation within the water column; (2) photo-oxidative degradation of organic matter through the photochemical production of reactive oxygen species (ROS) such as superoxide (O_2^-), hydrogen peroxide (H_2O_2), the hydroxyl radical (OH) and peroxy radicals (RO_2); (3) changes in metal ion speciation through reactions with the ROS or through direct photochemistry, resulting in the altered biological availability of some metals; (4) photochemical production of a number of trace gases of importance in the atmosphere such as CO_2, CO, and carbonyl sulfide (COS), and the destruction of others such as dimethyl sulfide (DMS); (5) the photochemical production of biologically available low molecular weight (LMW) organic compounds and the release of available forms of nitrogen, thus potentially fueling the growth of microorganisms from a biologically resistant source material (the CDOM). These processes provide the focus of this article.

Optical Properties of the Abiotic Constituents of Sea Waters

CDOM is a chemically complex material produced by the decay of plants and algae. This material, commonly referred to as gelbstoff, yellow substance, gilvin or humic substances, can be transported from land to the oceans by rivers or be formed directly in marine waters by as yet poorly understood processes. CDOM is the principal light-absorbing component of the dissolved organic matter (DOM) pool in sea waters, far exceeding the contributions of discrete dissolved organic or inorganic light-absorbing compounds. CDOM absorption spectra are broad and unstructured, and typically increase with decreasing wavelength in an approximately exponential fashion (**Figure 1**). Spectra have thus been parameterized using the expression [1].

$$a(\lambda) = a(\lambda_0) \cdot e^{-S(\lambda - \lambda_0)} \qquad [1]$$

$a(\lambda)$ and $a(\lambda_0)$ are the absorption coefficients at wavelength λ and reference wavelength λ_0, respectively, and S defines how rapidly the absorption increases with decreasing wavelength. Absorption coefficients are calculated from relation [2], where A is the absorbance measured across pathlength, r.

$$a(\lambda) = \frac{2.303 \cdot A(\lambda)}{r} \qquad [2]$$

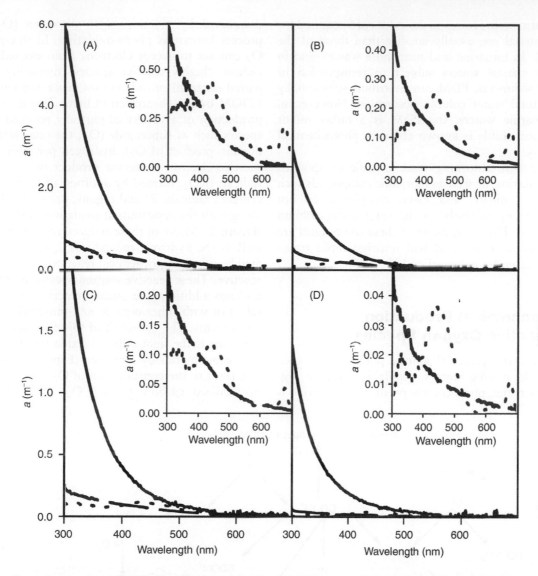

Figure 1 Absorption spectra of CDOM (—), PDM (– – –) and phytoplankton (– – –) from surface waters in the Delaware Bay and Middle Atlantic Bight off the east coast of the USA in July 1998: (A) mid-Delaware Bay at 39° 9.07′ N, 75° 14.29′ W; (B) Mouth of the Delaware Bay at 38° 48.61′ N, 75° 5.07′ W; (C) Mid-shelf at 38° 45.55′ N, 74° 46.60′ W; (D) Outer shelf at 38° 5.89′ N, 74° 9.07′ W.

Due to the exponential increase of $a(\lambda)$ with decreasing λ, CDOM absorbs light strongly in the UV-A and UV-B, and thus is usually the principal constituent within marine waters that controls the penetration depth of radiation potentially harmful to organisms (**Figure 1**). Moreover, for estuarine waters and for coastal waters strongly influenced by river inputs, light absorption by CDOM can extend well into the visible wavelength regime, often dominating the absorption by phytoplankton in the blue portion of the visible spectrum. In this situation, the amount and quality of the photosynthetically active radiation available to phytoplankton is reduced, thus decreasing primary productivity and potentially affecting ecosystem structure. High levels of absorption by CDOM in these regions can also seriously

compromise the determination of phytoplankton biomass through satellite ocean color measurements.

As described below, the absorption of sunlight by CDOM also initiates the formation of a variety of photochemical intermediates and products. The photochemical reactions producing these species ultimately lead to the degradation of the CDOM and the loss, or bleaching, of its absorption. This process can act as a feedback to alter the aquatic light field.

Particulate detrital material (PDM), operationally defined as that light-absorbing material retained on a GFF filter and not extractable with methanol, is a composite of suspended plant degradation products and sediment that also exhibits an exponentially rising absorption with decreasing wavelength (**Figure 1**); eqn [2] has thus been used to parameterize this material as well. However, the values of S acquired for this material are usually smaller than those of the CDOM. In estuarine and near-shore waters, and in shallow coastal waters subject to resuspension of bottom sediments, PDM can contribute substantially to the total water column absorption. However, in most marine waters, the PDM is a rather minor constituent. Little is known about its photochemical reactivity.

Other light-absorbing trace organic compounds such as flavins, as well as inorganic compounds such as nitrate, nitrite, and metal complexes, do not contribute significantly to the total water column absorption. However, many of these compounds are quite photoreactive and will undergo rapid transformation under appropriate light fields.

Photochemical Production of Reactive Oxygen Species

CDOM is the principal abiotic photoreactive constituent in marine waters. Available evidence suggests that the photochemistry of this material is dominated by reactions with dioxygen (O_2) in a process known as photo-oxidation. In this process, O_2 can act to accept electrons from excited states, radicals (highly reactive species containing an unpaired electron) or radical ions generated within the CDOM by the absorption of light. This leads to the production of a variety of partially reduced oxygen species such as superoxide (O_2^-, the one-electron reduction product of O_2), hydrogen peroxide (H_2O_2, the two-electron reduction product of O_2), peroxy radicals (RO_2, formed by addition of O_2 to carbon-centered radicals, R) and organic peroxides (RO_2H), along with the concomitant oxidation of the CDOM (**Figure 2**). Many of these reduced oxygen species as well as the hydroxyl radical (OH), which is generated by other photochemical reactions, are also quite reactive. These reactive oxygen species or ROS can undergo additional secondary reactions with themselves or with other organic and inorganic seawater constituents. The net result of this complex series of reactions is the light-induced oxidative degradation of organic matter by dioxygen (**Figure 2**). This process leads to the consumption of O_2, the production of oxidized carbon gases (CO_2, CO, COS), the

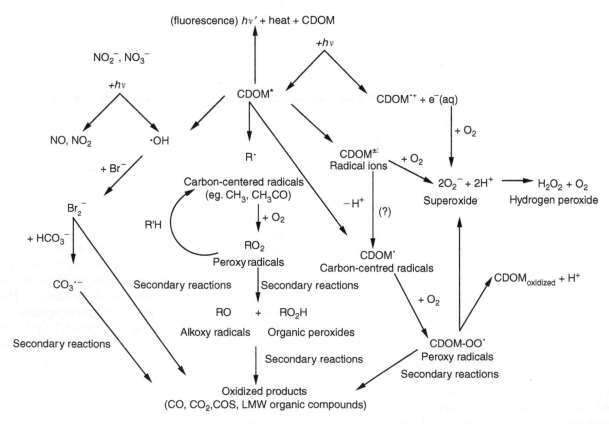

Figure 2 Schematic representation of the photochemical and secondary reactions known or thought to occur following light absorption by CDOM. For a more detailed description of these reactions see the text, Blough and Zepp (1995), and Blough (1997). Not shown in this diagram are primary and secondary reactions of metal species; for a description of these processes, see Helz *et al.* (1994) and Blough and Zepp (1995).

formation of a variety of LMW organic compounds, the release of biologically available forms of nitrogen, and the loss of CDOM absorption. Through direct photochemical reactions and reactions with the ROS, the speciation of metal ions is also affected.

These photochemical intermediates and products are produced at relatively low efficiencies. About 98–99% of the photons absorbed by CDOM are released as heat, while another $\sim 1\%$ are re-emitted as fluorescence. These percentages (or fractions) of absorbed photons giving rise to particular photoresponses are known as quantum yields (Φ). The Φ for the production of H_2O_2 and O_2^- (the two reduced oxygen species produced with highest efficiency), are approximately one to two orders of magnitude smaller than those for fluorescence, ranging from $\sim 0.1\%$ at 300 nm to $\sim 0.01\%$ at 400 nm. The Φ for other intermediates and products range even lower, from $\sim 0.01\%$ to 0.000 0001% (see below). The Φ for most of the intermediates and products created from the CDOM are highest in the UV-B and UV-A, and fall off rapidly with increasing wavelength; yields at visible wavelengths are usually negligible (see for example, **Figure 3**).

The hydroxyl radical, a very powerful oxidant, can be produced by the direct photolysis of nitrate and nitrite (eqns [I]–[III]).

$$NO_3^- + h\upsilon \rightarrow O^- + NO_2 \qquad [I]$$

$$NO_2^- + h\upsilon \rightarrow O^- + NO \qquad [II]$$

$$O^- + H_2O \rightarrow OH + OH^- \qquad [III]$$

The Φ values for these reactions are relatively high, about 7% for nitrite and about 1–2% for nitrate. However, because of the relatively low concentrations of these compounds in most marine surface waters, as well as their low molar absorptivities in the ultraviolet, the fraction of light absorbed is generally small and thus fluxes of OH from these sources also tend to be small. Recent evidence suggests that OH, or a species exhibiting very similar reactivity, is produced through a direct photoreaction of the CDOM; quinoid moieties within the CDOM may be responsible for this production. Quantum yields are low, $\sim 0.01\%$, and restricted primarily to the ultraviolet. In estuarine and near-shore waters containing higher levels of iron, the production of OH may also occur through the direct photolysis of iron–hydroxy complexes or through the Fenton reaction (eqn [IV]).

$$Fe^{2+} + H_2O_2 \rightarrow Fe^{3+} + OH + OH^- \qquad [IV]$$

Compounds that do not absorb light within the surface solar spectrum are also subject to photochemical modification through indirect or 'sensitized' photoreactions. In this case, the ROS or intermediates produced by direct photoreactions of a light-absorbing constituent such as CDOM can react secondarily with the nonabsorbing compounds. DMS and COS, two trace gases of some importance to the atmosphere, are thought to be destroyed and created, respectively, by sensitized photoreactions in marine surface waters.

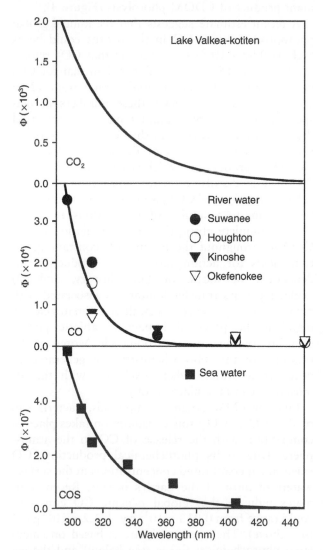

Figure 3 Wavelength dependence of the quantum yields (Φ) for the photochemical production of CO_2, CO, and COS. Data have been replotted from those dependencies originally reported in Vähätalo *et al.* (2000), Valentine and Zepp (1993), and Weiss *et al.* (1995) for CO_2, CO, and COS, respectively.

Photochemical Production and Consumption of LMW Organic Compounds and Trace Gases

The photolysis of CDOM produces a suite of LMW organic compounds and a number of trace gases. The production of these species presumably occurs through radical and fragmentation reactions arising from the net oxidative flow of electrons from CDOM to O_2 (see above), although the exact mechanism(s) have yet to be established. Most LMW organic compounds produced contain three or fewer carbon atoms and include such species as acetaldehyde, acetate, acetone, formaldehyde, formate, glyoxal, glyoxalate, methylglyoxal, propanal, and pyruvate. The Φ values for the production of individual compounds are low, ~ 0.001–0.0001%, with wavelengths in the UV-B the most effective; efficiencies decrease rapidly with increasing wavelength. Available evidence indicates that the Φ for O_2^- and H_2O_2 production are about one to two orders of magnitude larger than those for the LMW organic compounds, so it appears that the sum of the production rates for the known LMW organic compounds is small with respect to the flux of photochemical equivalents from CDOM to O_2.

Most, if not all, of these products are rapidly taken up and respired by bacteria to CO_2. Numerous investigators have presented evidence supporting enhanced microbial activity in waters exposed to sunlight, with bacterial activities increasing from 1.5- to almost 6-fold depending presumably on the length and type of light exposure, and the concentration and source of the CDOM. Recently, biologically labile nitrogen-containing compounds such as ammonia and amino acids have also been reported to be produced photochemically from CDOM. Because CDOM is normally considered to be biologically refractive, this recent work highlights the important role that abiotic photochemistry plays in the degradation of CDOM, not only through direct photoreactions, but also through the formation of biologically available products that can be respired to CO_2 or used as nutrients by biota. A recent estimate suggests that the utilization of biologically labile photoproducts could account for as much as 21% of the bacterial production in some near-surface waters.

Carbon dioxide and carbon monoxide are major products of the direct photolysis of CDOM (Figure 3). Quantum yields for CO production are about an order of magnitude smaller than those for O_2^- and H_2O_2 production, ranging from $\sim 0.01\%$ at 300 nm to $\sim 0.001\%$ at 400 nm. Available data indicate that the Φ for CO_2 range even higher, perhaps as much as 15–20-fold. The yields for CO_2 production must thus approach, if not exceed, those for O_2^- and H_2O_2. This result is somewhat surprising, since it implies that about one CO_2 is produced for each electron transferred from the CDOM to O_2, further implying a high average redox state for CDOM. Although CDOM (i.e. humic substances) is known to contain significant numbers of carboxyl moieties that could serve as the source of the CO_2, the yield for CO_2 production, relative to O_2^- and H_2O_2, would be expected to fall rapidly as these groups were removed photochemically; available evidence suggests that this does not occur. An alternative explanation is that other species, perhaps the CDOM itself, is acting as an electron acceptor. Regardless of mechanism, existing information indicates that CO_2 is the dominant product of CDOM photolysis (Figure 3).

A recent estimate suggests that the annual global photoproduction of CO in the oceans could be as high as 0.82×10^{15} g C. Assuming that CO_2 photoproduction is 15–20 times higher than that for CO, values for CO_2 formation could reach from 12 to 16×10^{15} g C y^{-1}. To place these numbers in perspective, the estimated annual input of terrestrial dissolved organic carbon to the oceans (0.2×10^{15} g C y^{-1}) is only 1.3–1.7% of the calculated annual CO_2 photoproduction, which is itself about 2–3% of the oceanic dissolved organic carbon pool. These calculated CO_2 (and CO) photoproduction rates may be high due to a number of assumptions, including (1) the complete absorption of UV radiation by the CDOM throughout the oceans, (2) constant quantum yields (or action spectra) for production independent of locale or light history, and (3) neglecting mass transfer limitations associated with physical mixing. Nevertheless, these estimates clearly highlight the potential impact of abiotic photochemistry on the oceanic carbon cycle. Moreover, the products of this photochemistry are generated in near-surface waters where exchange with the atmosphere can take place readily.

Like the LMW organic compounds, bacteria can oxidize CO to CO_2; this consumption takes place in competition with the release of CO to the atmosphere. Due to its photochemical production, CO exists at supersaturated concentrations in the surface waters of most of the Earth's oceans. Recent estimates indicate that global oceanic CO emissions could range from 0.013×10^{15} g y^{-1}–1.2×10^{15} g y^{-1} (see above). The upper estimate is based on calculated photochemical fluxes (see below) and the assumption that all CO produced is emitted to the atmosphere. The lower estimate was calculated using air–sea gas exchange equations and extensive measurements of CO concentrations in the surface waters and atmosphere of the Pacific Ocean. The source of the significant discrepancy between these two estimates has yet to be resolved. Depending on the answer, CO emitted to the atmosphere from the oceans could play a significant role in controlling OH levels in the marine troposphere.

Carbonyl sulfide (COS) is produced primarily in coastal/shelf waters, apparently by the CDOM-photosensitized oxidation of organosulfur compounds. UV-B light is the most effective in its formation, with Φ decreasing rapidly from $\sim 6 \times 10^{-7}$ at 300 nm to $\sim 1 \times 10^{-8}$ by 400 nm (**Figure 3**). The principal sinks of seawater COS are release to the atmosphere and hydrolysis to CO_2 and H_2S. Accounting for perhaps as much as one-third of the total source strength, the photochemical production of COS in the oceans is probably the single largest source of COS to the atmosphere, although more recent work has revised this estimate downward. Smaller amounts of carbon disulfide (CS_2) are also generated photochemically in surface waters through CDOM sensitized reaction(s); Φ values decrease from $\sim 1 \times 10^{-7}$ at 313 nm to 5×10^{-9} at 366 nm. The CS_2 emitted to the atmosphere can react with OH to form additional COS in the troposphere. Although it was previously thought that the oxidation of COS in the stratosphere to form sulfate aerosol could be important in determining Earth's radiation budget and perhaps in regulating stratospheric ozone concentrations, more recent work suggests that other sources contribute more significantly to the background sulfate in the stratosphere.

Dimethyl sulfide (DMS), through its oxidation to sulfate in the troposphere, acts as a source of cloud condensation nuclei, thus potentially influencing the radiative balance of the atmosphere. DMS is formed in sea water through the microbial decomposition of dimethyl sulfonioproprionate (DMSP), a compound believed to act as an osmolyte in certain species of marine phytoplankton. The flux of DMS to the atmosphere is controlled by its concentration in surface sea waters, which is controlled in turn by the rate of its decomposition. Estimates indicate that 7–40% of the total turnover of DMS in the surface waters of the Pacific Ocean is due to the photosensitized destruction of this compound, illustrating the potential importance of this pathway in controlling the flux of DMS to the atmosphere.

In addition to these compounds, the photochemical production of small amounts of non-methane hydrocarbons (NMHC) such as ethene, propene, ethane, and propane has also been reported. Production of these compounds appears to result from the photolysis of the CDOM, with Φ values of the order of 10^{-7}–10^{-9}. The overall emission rates of these compounds to the atmosphere via this source are negligible with respect to global volatile organic carbon emissions, although this production may play some role in certain restricted locales exhibiting stronger source strengths, or in the marine environment remote from the dominant terrestrial sources.

The photolysis of nitrate and nitrite in sea water produces nitrogen dioxide (NO_2) and nitric oxide (NO), respectively (eqns [I] and [II]). Previous work indicated that the photolysis of nitrite could act as a small net source of NO to the marine atmosphere under some conditions. However, this conclusion seems to be at odds with estimates of the steady-state concentrations of superoxide and the now known rate constant for the reaction of superoxide with nitric oxide ($6.7 \times 10^9 \, M^{-1} \, s^{-1}$) to form peroxynitrite in aqueous phases (eqn [V]).

$$O_2^- + NO \rightarrow {}^-OONO \qquad [V]$$

The peroxynitrite subsequently rearranges in part to form nitrate (eqn [VI]).

$$^-OONO \rightarrow NO_3^- \qquad [VI]$$

Even assuming a steady-state concentration of O_2^- (10^{-12} M) that is about two orders of magnitude lower than that expected for surface sea waters ($\sim 10^{-10}$ M), the lifetime of NO in surface sea waters would be only ~ 150 s, a timescale too short for significant exchange with the atmosphere except for a thin surface layer. Moreover, even in this situation, the atmospheric deposition of additional HO_2 radicals to this surface layer (to form O_2^-) would be expected to act as an additional sink of the NO (flux capping). It appears that most if not all water bodies exhibiting significant steady-state levels of O_2^-, produced either photochemically or thermally, should act as a net sink of atmospheric NO and probably of NO_2 as well. Further, although less is known about the steady-state levels of peroxy radicals in sea waters due largely to their unknown decomposition routes, their high rate constants for reaction with NO (1–$3 \times 10^9 \, M^{-1} \, s^{-1}$) indicate that they should also act as a sink of NO. In fact, methyl nitrate, a trace species found in sea waters, may in part be produced through the aqueous phase reactions (eqns [VII] and [VIII]) with the methylperoxy radical (CH_3OO) generated through a known photochemical reaction of CDOM (or through atmospheric deposition) and the NO arising from the photolysis of nitrite (or through atmospheric deposition).

$$CH_3OO + NO \rightarrow CH_3OONO \qquad [VII]$$

$$CH_3OONO \rightarrow CH_3ONO_2 \qquad [VIII]$$

The concentrations of NO and NO_2 in the troposphere are important because of the involvement of these gases in the formation of ozone.

The atmospheric deposition of ozone to the sea surface can cause the release of volatile iodine compounds to the atmosphere. There is also evidence that methyl iodide can be produced (as well as destroyed) by photochemical processes in surface sea waters. The release of these volatile iodine species from the sea surface or from atmospheric aqueous phases (aerosols) by these processes may act as a control on the level of ozone in the marine troposphere via iodine-catalyzed ozone destruction.

Trace metal photochemistry

A lack of available iron is now thought to limit primary productivity in certain ocean waters containing high nutrient, but low chlorophyll concentrations (the HNLC regions). This idea has spurred interest in the transport and photochemical reactions of iron in both seawaters and atmospheric aerosols. Very little soluble Fe(II) is expected to be available at the pH and dioxygen concentration of surface seawaters due to the high stability of the colloidal iron (hydr)oxides. The photoreductive dissolution of colloidal iron oxides by CDOM is known to occur at low pH; this process is also thought to occur in seawaters at high pH, but the reduced iron appears to be oxidized more rapidly than its detachment from the oxide surface. However, some workers have found that CDOM-driven cycles of reduction followed by oxidation increases the chemical availability, which was strongly correlated with the growth rate of phytoplankton. Significant levels of Fe(II) are also known to be produced photochemically in atmospheric aqueous phases (at lower pH) and could serve as a source of biologically available iron upon deposition to the sea surface.

Manganese oxides are also subject to reductive dissolution by light in surface seawaters. This process produces Mn(II), which is kinetically stable to oxidation in the absence of bacteria that are subject to photoinhibition. These two effects lead to the formation of a surface maximum in soluble Mn(II), in contrast to most metals which are depleted in surface waters due to biological removal processes. Other examples of the impact of photochemical reactions on trace metal chemistry are provided in Further Reading.

Photochemical Calculations

Global and regional estimates for the direct photochemical production (or consumption) of a particular photoproduct (or photoreactant) can be acquired with knowledge of the temporal and spatial variation of the solar irradiance reaching the Earth's surface combined with a simple photochemical model (eqn [3]).

$$F(\lambda, z) = E_D(\lambda, z) \cdot \Phi_i(\lambda) \cdot a_{Di}(\lambda) \qquad [3]$$

Here $F(\lambda, z)$ is the photochemical production (or consumption) rate; $E_D(\lambda, z)$ is the downwelling irradiance at wavelength, λ, and depth, z, within the water column; a_{Di} is the diffuse absorption coefficient for photoreactive constituent i; $\Phi_i(\lambda)$ is the quantum yield of this ith constituent. $E_D(\lambda, z)$ is well approximated by eqn [4].

$$E_D(\lambda, z) = E_{D0}(\lambda) \cdot e^{-K_d(\lambda) \cdot z} \qquad [4]$$

$E_{D0}(\lambda)$ is the downwelling irradiance just below the sea surface and $K_d(\lambda)$ is the vertical diffuse attenuation coefficient of downwelling irradiance. $K_d(\lambda)$ can be approximated by eqn [5].

$$K_d(\lambda) \approx \frac{\sum a_i(\lambda) + \sum b_{bi}(\lambda)}{\mu_D} \qquad [5]$$

where $\sum a_i(\lambda)$ and $\sum b_{bi}(\lambda)$ are the total absorption and backscattering coefficients, respectively, of all absorbing and scattering constituents within the water column, and μ_D is the average cosine of the angular distribution of the downwelling light. This factor accounts for the average pathlength of light in the water column, and for direct solar light is approximately equal to $\cos \theta$, where θ is the solar zenith angle (e.g. $\mu_D \sim 1$ when the sun is directly overhead). The diffuse absorption coefficient, a_{Di}, is given by eqn [6].

$$a_{Di} = \frac{a_i}{\mu_D} \qquad [6]$$

This model assumes that the water column is homogeneous, that $K_d(\lambda)$ is constant with depth, and that upwelling irradiance is negligible relative to $E_D(\lambda, z)$. Combining eqn [3,4 and 6] gives eqn [7].

$$F(\lambda, z) = \frac{E_{D0}(\lambda) \cdot e^{-K_d(\lambda) \cdot z} \cdot \Phi_i(\lambda) \cdot a_i(\lambda)}{\mu_D} \qquad [7]$$

This equation allows calculation of the spectral dependence of the production (consumption) rate as a function of depth in the water column, assuming knowledge of $E_{D0}(\lambda)$, $K_d(\lambda)$, $a_i(\lambda)$ and $\Phi_i(\lambda)$, all of which can be measured or estimated (**Figure 4**).

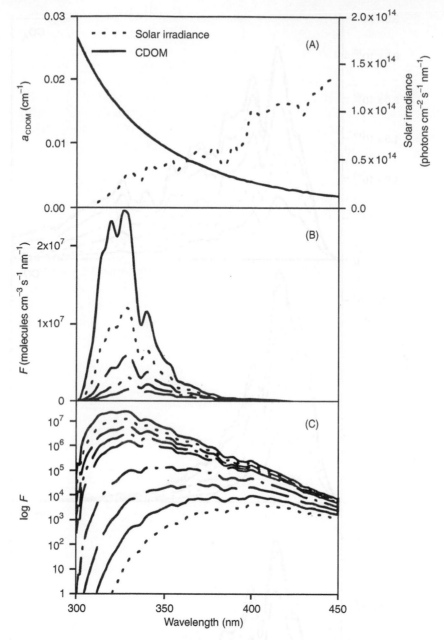

Figure 4 Spectral dependence of CO photoproduction rates with depth, plotted on a linear (B) and logarithmic (C) scale. Depths in (B) are (from top to bottom): surface, 0.5, 1, 1.5, and 2 m. Depths in (C) are (from top to bottom): surface, 0.5, 1, 1.5, 2, 4, 6, 8, and 10 m. These spectral dependencies were calculated using eqn [7], the wavelength dependence of the quantum yield for CO shown in **Figure 3**, and the CDOM absorption spectrum and surface solar irradiance shown in (A). The attenuation of irradiance down the water column in this spectral region was assumed to be only due to CDOM absorption, a reasonable assumption for coastal waters (see **Figure 1**). Note the rapid attenuation in production rates with depth in the UV-B, due to the greater light absorption by CDOM in this spectral region.

Integrating over wavelength provides the total production (consumption) rate at each depth.

Integration of eqn [7] from the surface to depth z provides the spectral dependence of the photochemical flux (Y) over this interval

$$Y(\lambda, z) = \frac{E_{D0}(\lambda) \cdot \left(1 - e^{-K_d(\lambda) \cdot z}\right) \cdot \Phi_i(\lambda) \cdot a_i(\lambda)/K_d(\lambda)}{\mu_D} \quad [8]$$

which upon substitution of eqn [4] becomes,

$$Y(\lambda, z) = E_{D0}(\lambda) \cdot \left(1 - e^{-K_d(\lambda) \cdot z}\right) \cdot \Phi_i(\lambda) \cdot \frac{a_i(\lambda)}{\sum a_i(\lambda) + \sum b_{bi}(\lambda)} \quad [9]$$

In most, but not all seawaters, the total absorption will be much greater than the total backscatter,

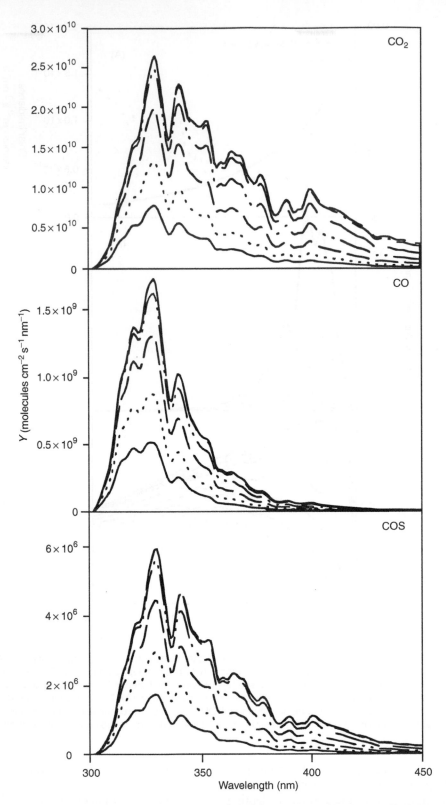

Figure 5 Spectral dependence of the photochemical flux with depth for CO_2, CO, and COS. Fluxes with depth are from the surface to 0.25, 0.5, 1.0, 2.0, and 4 m, respectively (bottom spectrum to top spectrum). Below 4 m, increases in the flux are nominal. These spectral dependencies were calculated using [10], the wavelength dependence of the quantum yields for CO_2, CO and COS shown in **Figure 3**, and the surface solar irradiance shown in **Figure 4A**. CDOM is assumed to absorb all photons in this spectral region (see **Figures 1** and **4**).

$\sum a_i(\lambda) \gg \sum b_{bi}(\lambda)$, and thus the backscatter can be ignored; this approximation is not valid for most estuarine waters and some coastal waters, where a more sophisticated treatment would have to be applied. This approximation leads to the final expression for the variation of the spectral dependence of the flux with depth (**Figure 5**),

$$F(\lambda, z) = E_{D0}(\lambda) \cdot \left(1 - e^{-K_d(\lambda) \cdot z}\right) \cdot \Phi_i(\lambda) \cdot \frac{a_i(\lambda)}{\sum a_i(\lambda)} \quad [10]$$

The spectral dependence of the total water column flux ($z \to \infty$) is then given by,

$$F(\lambda) = E_{D0}(\lambda) \cdot \Phi_i(\lambda) \cdot \frac{a_i(\lambda)}{\sum a_i(\lambda)} \quad [11]$$

with the total flux obtained by integrating over wavelength,

$$F \int_{\lambda} E_{D0}(\lambda) . \Phi_i(\lambda) \cdot \frac{a_i(\lambda)}{\sum a_i(\lambda)} d\lambda \quad [12]$$

To obtain global estimates of photochemical fluxes, many investigators assume that the absorption due to CDOM, a_{CDOM}, dominates the absorption of all other seawater constituents in the ultraviolet, and thus that $a_{CDOM}(\lambda) / \sum a_i(\lambda) \approx 1$. While this approximation is reasonable for many coastal waters, it is not clear that this approximation is valid for all oligotrophic waters. This approximation leads to the final expression for flux,

$$Y \int_{\lambda} E_{D0}(\lambda) . \Phi_i(\lambda) d\lambda \quad [13]$$

which relies only on the surface downwelling irradiance and the wavelength dependence of the quantum yield for the photoreaction of interest. Uncertainties in the use of this equation for estimating global photochemical fluxes include (1) the (usual) assumption that $\Phi(\lambda)$ acquired for a limited number of samples is representative of all ocean waters, independent of locale or light history, and (2) differences in the spatially and temporally averaged values of $E_{D0}(\lambda)$ utilized by different investigators.

Conclusions

The absorption of solar radiation by abiotic sea water constituents initiates a cascade of reactions leading to the photo-oxidative degradation of organic matter and the concomitant production (or consumption) of a variety of trace gases and LMW organic compounds (**Figure 2**), as well as affecting trace metal speciation. The magnitude and impact of these processes on upper ocean biogeochemical cycles and their coupling with atmospheric processes are just beginning to be fully quantified and understood. There remains the need to examine possible couplings between atmospheric gas phase reactions and photochemical reactions in atmospheric aqueous phases.

Further Reading

Blough NV (1997) Photochemistry in the sea-surface microlayer. In: Liss PS and Duce R (eds.) *The Sea Surface and Global Change*, pp. 383–424. Cambridge: Cambrige University Press.

Blough NV and Green SA (1995) Spectroscopic characterization and remote sensing of non-living organic matter. In: Zepp RG and Sonntag C (eds.) *The role of Non-living Organic Matter in the Earth's Carbon Cycle*, pp. 23–45. New York: John Wiley.

Blough NV and Zepp RG (1995) Reactive oxygen species in natural waters. In: Foote CS, Valentine JS, Greenberg A, and Liebman JF (eds.) *Reactive Oxygen Species in Chemistry*, pp. 280–333. New York: Chapman & Hall.

de Mora S, Demers S, and Vernet M (eds.) (2000) *The Effects of UV Radiation in the Marine Environment*. Cambridge: Cambridge University Press.

Häder D-P, Kumar HD, Smith RC, and Worrest RC (1998) Effects of UV-B radiation on aquatic ecosystems. *Journal of Photochemistry and Photobiology B* 46: 53–68.

Helz GR, Zepp RG, and Crosby DG (eds.) (1994) *Aquatic and Surface Photochemistry*. Ann Arbor, MI: Lewis Publishers.

Huie RE (1995) Free radical chemistry of the atmospheric aqueous phase. In: Barker JR (ed.) *Progress and Problems in Atmospheric Chemistry*, pp. 374–419. Singapore: World Scientific Publishing Co.

Kirk JTO (1994) *Light and Photosynthesis in Aquatic Ecosystems*. Cambridge: Cambridge University Press.

Moran MA and Zepp RG (1997) Role of photoreactions in the formation of biologically labile compounds from dissolved organic matter. *Limnology and Oceanography* 42: 1307–1316.

Thompson AM and Zafiriou OC (1983) Air–sea fluxes of transient atmospheric species. *Journal of Geophysical Research* 88: 6696–6708.

Vähätalo AV, Salkinoja-Salonen M, Taalas P, and Salonen K (2000) Spectrum of the quantum yield for photochemical mineralization of dissolved organic carbon in a humic lake. *Limnology and Oceanography* 45: 664–676.

Valentine RL and Zepp RG (1993) Formation of carbon monoxide from the photodegradation of terrestrial dissolved organic carbon in natural waters. *Environmental Science Technology* 27: 409–412.

Weiss EW, Andrews SS, Johnson JE, and Zafiriou OC (1995) Photoproduction of carbonyl sulfide in south Pacific Ocean waters as a function of irradiation wavelength. *Geophysical Research Letters* 22: 215–218.

Zafiriou OC, Blough NV, Micinski E, *et al.* (1990) Molecular probe systems for reactive transients in natural waters. *Marine Chemistry* 30: 45–70.

Zepp RG, Callaghan TV, and Erickson DJ (1998) Effects of enhanced solar ultraviolet radiation on biogeochemical cycles. *Journal of Photochemistry and Photobiology B* 46: 69–82.

RADIATIVE TRANSFER IN THE OCEAN

C. D. Mobley, Sequoia Scientific Inc., WA, USA

Introduction

Understanding how light interacts with sea water is a fascinating problem in itself, as well as being fundamental to fields as diverse as biological primary production, mixed-layer thermodynamics, photochemistry, lidar bathymetry, ocean-color remote sensing, and visual searching for submerged objects. For these reasons, optics is one of the fastest growing oceanographic research areas.

Radiative transfer theory provides the theoretical framework for understanding light propagation in the ocean, just as hydrodynamics provides the framework for physical oceanography. The article begins with an overview of the definitions and terminology of radiative transfer as used in oceanography. Various ways of quantifying the optical properties of a water body and the light within the water are described. The chapter closes with examples of the absorption and scattering properties of two hypothetical water bodies, which are characteristic of the open ocean and a turbid estuary, and a comparison of their underwater light fields.

Terminology

The optical properties of sea water are sometimes grouped into inherent and apparent properties.

- Inherent optical properties (IOPs) are those properties that depend only upon the medium and therefore are independent of the ambient light field. The two fundamental IOPs are the absorption coefficient and the volume scattering function. (These quantities are defined below.)
- Apparent optical properties (AOPs) are those properties that depend both on the medium (the IOPs) and on the directional structure of the ambient light field, and that display enough regular features and stability to be useful descriptors of a water body. Commonly used AOPs are the irradiance reflectance, the remote-sensing reflectance, and various diffuse attenuation functions.

'Case 1 waters' are those in which the contribution by phytoplankton to the total absorption and scattering is high compared to that by other substances. Absorption by chlorophyll and related pigments therefore plays the dominant role in determining the total absorption in such waters, although covarying detritus and dissolved organic matter derived from the phytoplankton also contribute to absorption and scattering in case 1 waters. Case 1 water can range from very clear (oligotrophic) to very productive (eutrophic) water, depending on the phytoplankton concentration.

'Case 2 waters' are 'everything else,' namely, waters where inorganic particles or dissolved organic matter from land drainage contribute significantly to the IOPs, so that absorption by pigments is relatively less important in determining the total absorption. Roughly 98% of the world's open ocean and coastal waters fall into the case 1 category, but near-shore and estuarine case 2 waters are disproportionately important to human interests such as recreation, fisheries, and military operations.

Table 1 summarizes the terms, units, and symbols for various quantities frequently used in optical oceanography.

Radiometric Quantities

Consider an amount ΔQ of radiant energy incident in a time interval Δt centered on time t, onto a surface of area ΔA located at position (x,y,z), and arriving through a set of directions contained in a solid angle $\Delta \Omega$ about the direction (θ, φ) normal to the area ΔA, as produced by photons in a wavelength interval $\Delta \lambda$ centered on wavelength λ. The geometry of this situation is illustrated in **Figure 1**. Then an operational definition of the spectral radiance is

$$L(x, y, z, t, \theta, \varphi, \lambda) \equiv \frac{\Delta Q}{\Delta t \Delta A \Delta \Omega \Delta \lambda}$$
$$[\text{Js}^{-1}\text{m}^{-2}\text{sr}^{-1}\text{nm}^{-1}] \quad [1]$$

In the conceptual limit of infinitesimal parameter intervals, the spectral radiance is defined as

$$L(x, y, z, t, \theta, \varphi, \lambda) \equiv \frac{\partial^4 Q}{\partial t\, \partial A\, \partial \Omega\, \partial \lambda} \quad [2]$$

Spectral radiance is the fundamental radiometric quantity of interest in optical oceanography: it completely specifies the positional (x,y,z), temporal (t), directional (θ, φ), and spectral (λ) structure of the light field. In many oceanic environments, horizontal

Table 1   Quantities commonly used in optical oceanography

Quantity	SI units	Symbol
Radiometric quantities		
Quantity of radiant energy	$J\,nm^{-1}$	Q
Power	$W\,nm^{-1}$	Φ
Intensity	$W\,sr^{-1}\,nm^{-1}$	I
Radiance	$W\,m^{-2}\,sr^{-1}\,nm^{-1}$	L
Downwelling plane irradiance	$W\,m^{-2}\,nm^{-1}$	E_d
Upwelling plane irradiance	$W\,m^{-2}\,nm^{-1}$	E_u
Net irradiance	$W\,m^{-2}\,nm^{-1}$	E
Scalar irradiance	$W\,m^{-2}\,nm^{-1}$	E_o
Downwelling scalar irradiance	$W\,m^{-2}\,nm^{-1}$	E_{ou}
Upwelling scalar irradiance	$W\,m^{-2}\,nm^{-1}$	E_{ou}
Photosynthetic available radiation	$Photons\,s^{-1}\,m^{-2}$	PAR
Inherent optical properties		
Absorption coefficient	m^{-1}	a
Volume scattering function	$m^{-1}\,sr^{-1}$	β
Scattering phase function	sr^{-1}	$\tilde{\beta}$
Scattering coefficient	m^{-1}	b
Backscatter coefficient	m^{-1}	b_b
Beam attenuation coefficient	m^{-1}	c
Single-scattering albedo	–	ω_o
Apparent optical properties		
Irradiance reflectance (ratio)	–	R
Remote-sensing reflectance	sr^{-1}	R_{rs}
Attenuation coefficients	m^{-1}	
of radiance $L(z, \theta, \varphi)$	m^{-1}	$K(\theta, \varphi)$
of downwelling irradiance $E_d(z)$	m^{-1}	K_d
of upwelling irradiance $E_u(z)$	m^{-1}	K_u
of PAR	m^{-1}	K_{PAR}

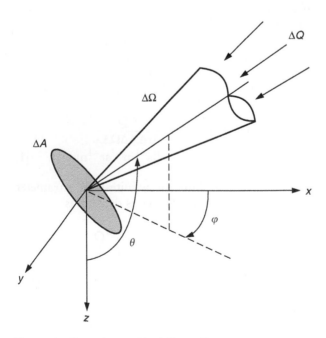

Figure 1   Geometry used to define radiance.

variations (on a scale of tens to thousands of meters) of the IOPs and the radiance are much less than variations with depth, in which case it can be assumed that these quantities vary only with depth z. (An exception would be the light field due to a single light source imbedded in the ocean; such a radiance distribution is inherently three-dimensional.) Moreover, since the timescales for changes in IOPs or in the environment (seconds to seasons) are much greater than the time required for the radiance to reach steady state (microseconds) after a change in IOPs or boundary conditions, time-independent radiative transfer theory is adequate for most oceanographic studies. (An exception is time-of-flight lidar bathymetry.) When the assumptions of horizontal homogeneity and time independence are valid, the spectral radiance can be written as $L(z, \theta, \varphi, \lambda)$.

Although the spectral radiance completely specifies the light field, it is seldom measured in all directions, both because of instrumental difficulties and because such complete information often is not needed. The most commonly measured radiometric quantities are various irradiances. Suppose the light detector is equally sensitive to photons of a given wavelength λ traveling in any direction (θ, φ) within a hemisphere of directions. If the detector is located at depth z and is oriented facing upward, so as to collect photons traveling downward, then the detector output is a measure of the spectral downwelling scalar irradiance at depth z, $E_{od}(z, \lambda)$. Such an instrument is summing radiance over all the directions (elements of solid angle) in the downward hemisphere; thus $E_{od}(z, \lambda)$ is related to $L(z, \theta, \varphi, \lambda)$ by

$$E_{od}(z,\lambda) = \int_{2\pi_d} L(z,\theta,\varphi,\lambda)d\Omega \ [Wm^{-2}nm^{-1}] \quad [3]$$

Here $2\pi_d$ denotes the hemisphere of downward directions (i.e., the set of directions (θ, φ) such that $0 \leq \theta \leq \pi/2$ and $0 \leq \varphi < 2\pi$, if θ is measured from the $+z$ or nadir direction). The integral over $2\pi_d$ can be evaluated as a double integral over θ and φ after a specific coordinate system is chosen.

If the same instrument is oriented facing downward, so as to detect photons traveling upward, then the quantity measured is the spectral upwelling scalar irradiance $E_{ou}(z, \lambda)$. The spectral scalar irradiance $E_o(z, \lambda)$ is the sum of the downwelling and upwelling components:

$$E_O(z,\lambda) \equiv E_{od}(z,\lambda) + E_{ou}(z,\lambda)$$
$$= \int_{4\pi} L(z,\theta,\varphi,\lambda)d\Omega \quad [4]$$

$E_o(z, \lambda)$ is proportional to the spectral radiant energy density ($J m^{-3} nm^{-1}$) and therefore quantifies how much radiant energy is available for photosynthesis or heating the water.

Now consider a detector designed so that its sensitivity is proportional to $|\cos \theta|$, where θ is the angle between the photon direction and the normal to the surface of the detector. This is the ideal response of a 'flat plate' collector of area ΔA, which when viewed at an angle θ to its normal appears to have an area of $\Delta A |\cos \theta|$. If such a detector is located at depth z and is oriented facing upward, so as to detect photons traveling downward, then its output is proportional to the spectral downwelling plane irradiance $E_d(z, \lambda)$. This instrument is summing the downwelling radiance weighted by the cosine of the photon direction, thus

$$E_d(z,\lambda) = \int_{2\pi_d} L(z,\theta,\varphi,\lambda)|\cos\theta|d\Omega$$
$$[Wm^{-2}nm^{-1}] \quad [5]$$

Turning this instrument upside down gives the spectral upwelling plane irradiance $E_u(z, \lambda)$. E_d and E_u are useful because they give the energy flux (power per unit area) across the horizontal surface at depth z owing to downwelling and upwelling photons, respectively. The difference $E_d - E_u$ is called the net (or vector) irradiance.

Photosynthesis is a quantum phenomenon, i.e., it is the number of available photons rather than the amount of radiant energy that is relevant to the chemical transformations. This is because a photon of, say, $\lambda = 400$ nm, if absorbed by a chlorophyll molecule, induces the same chemical change as does a photon of $\lambda = 600$ nm, even though the 400 nm photon has 50% more energy than the 600 nm photon. Only a part of the photon energy goes into photosynthesis; the excess is converted to heat or is re-radiated. Moreover, chlorophyll is equally able to absorb and utilize a photon regardless of the photon's direction of travel. Therefore, in studies of phytoplankton biology, the relevant measure of the light field is the photosynthetic available radiation, PAR, defined by

$$PAR(z) \equiv \int_{350\,nm}^{700\,nm} \frac{\lambda Z}{hc} E_o(z,\lambda)d\lambda$$
$$[photons\,s^{-1}m^{-2}] \quad [6]$$

where $h = 6.6255 \times 10^{-34}$ J s is the Planck constant and $c = 3.0 \times 10^{17}$ nm s^{-1} is the speed of light. The factor λ/hc converts the energy units of E_o to quantum units (photons per second). Bio-optical literature often states PAR values in units of mol photons s^{-1} m^{-2} or einst s^{-1} m^{-2} (where one einstein is one mole of photons).

Inherent Optical Properties

Consider a small volume ΔV of water, of thickness Δr as illuminated by a collimated beam of monochromatic light of wavelength λ and spectral radiant power $\Phi_i(\lambda)$ (W nm^{-1}), as schematically illustrated in **Figure 2**. Some part $\Phi_a(\lambda)$ of the incident power $\Phi_i(\lambda)$ is absorbed within the volume of water. Some part $\Phi_i(\psi, \lambda)$ is scattered out of the beam at an angle ψ, and the remaining power $\Phi_t(\lambda)$ is transmitted through the volume with no change in direction. Let $\Phi_s(\lambda)$ be the total power that is scattered into all directions.

The inherent optical properties usually employed in radiative transfer theory are the absorption and scattering coefficients. In the geometry of **Figure 2**, the absorption coefficient $a(\lambda)$ is defined as the limit of the fraction of the incident power that is absorbed within the volume, as the thickness becomes small

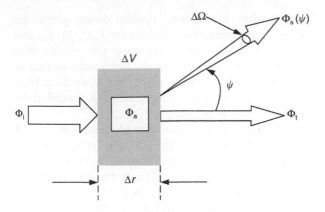

Figure 2 Geometry used to define inherent optical properties.

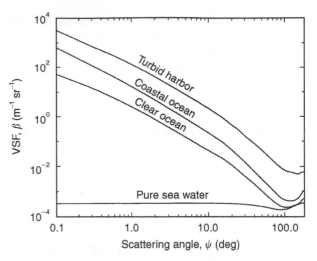

Figure 3 Volume scattering functions (VSF) measured in three different oceanic waters. The VSF of pure sea water is shown for comparison.

$$a(\lambda) \equiv \lim_{\Delta r \to 0} \frac{1}{\Phi_i(\lambda)} \frac{\Phi_a(\lambda)}{\Delta r} \quad [\mathrm{m}^{-1}] \qquad [7]$$

The scattering coefficient $b(\lambda)$ has a corresponding definition using $\Phi_s(\lambda)$. The beam attenuation coefficient $c(\lambda)$ is defined as $c(\lambda) = a(\lambda) + b(\lambda)$.

Now take into account the angular distribution of the scattered power, with $\Phi_s(\psi, \lambda)/\Phi_i(\lambda)$ being the fraction of incident power scattered out of the beam through an angle ψ into a solid angle $\Delta\Omega$ centered on ψ, as shown in **Figure 2**. Then the fraction of scattered power per unit distance and unit solid angle, $\beta(\psi, \lambda)$, is

$$\beta(\psi,\lambda) \equiv \lim_{\Delta r \to 0} \lim_{\Delta\Omega \to 0} \frac{\Phi_s(\psi,\lambda)}{\Phi_i(\lambda)\Delta r \Delta\Omega} \quad [\mathrm{m}^{-1}\mathrm{sr}^{-1}] \quad [8]$$

The spectral power scattered into the given solid angle $\Delta\Omega$ is just the spectral radiant intensity scattered into direction ψ times the solid angle: $\Phi_s(\psi, \lambda) = I_s(\psi, \lambda)\,\Delta\Omega$. Moreover, if the incident power $\Phi_i(\lambda)$ falls on an area ΔA, then the corresponding incident irradiance is $E_i(\lambda) = \Phi_i(\lambda)/\Delta A$. Noting that $\Delta V = \Delta r \Delta A$ is the volume of water that is illuminated by the incident beam gives

$$\beta(\psi,\lambda) = \lim_{\Delta V \to 0} \frac{I_s(\psi,\lambda)}{E_i(\lambda)\Delta V} \qquad [9]$$

This form of $\beta(\psi, \lambda)$ suggests the name volume scattering function (VSF) and the physical interpretation of scattered intensity per unit incident irradiance per unit volume of water. **Figure 3** shows measured VSFs (at 514 nm) from three greatly different water bodies; the VSF of pure water is shown for comparison. VSFs of sea water typically increase by five or six orders of magnitude in going from $\psi = 90°$ to $\psi = 0.1°$ for a given water sample, and scattering at a given angle ψ can vary by two orders of magnitude among water samples.

Integrating $\beta(\psi, \lambda)$ over all directions (solid angles) gives the total scattered power per unit incident irradiance and unit volume of water, in other words the spectral scattering coefficient:

$$b(\lambda) = \int_{4\pi} \beta(\psi,\lambda)\mathrm{d}\Omega = 2\pi \int_0^\pi \beta(\psi,\lambda)\sin\psi\,\mathrm{d}\psi \quad [10]$$

Eqn. [10] follows because scattering in natural waters is azimuthally symmetric about the incident direction (for unpolarized light sources and randomly oriented scatterers). This integration is often divided into forward scattering, $0 \le \psi \le \pi/2$, and backward scattering, $\pi/2 \le \psi \le \pi$, parts. Thus the backscatter coefficient is

$$b_b(\lambda) \equiv 2\pi \int_{\pi/2}^\pi \beta(\psi,\lambda)\sin\psi\,\mathrm{d}\psi \qquad [11]$$

The VSFs of **Figure 3** have b values ranging from 0.037 to 1.824 m^{-1} and backscatter fractions b_b/b of 0.013 to 0.044.

The preceding discussion assumed that no inelastic-scattering processes are present. However, inelastic scattering does occur owing to fluorescence by dissolved matter or chlorophyll, and to Raman scattering by the water molecules themselves. Power lost from wavelength λ by scattering into wavelength $\lambda' \neq \lambda$ appears as an increase in the absorption $a(\lambda)$. The gain in power at λ' appears as a source term in the radiative transfer equation.

Two more inherent optical properties are commonly used in optical oceanography. The single-scattering albedo is $\omega_o(\lambda) = b(\lambda)/c(\lambda)$. The single-scattering albedo is the probability that a photon will be scattered (rather than absorbed) in any given interaction, hence $\omega_o(\lambda)$ is also known as the probability of photon survival. The volume scattering phase function, $\tilde{\beta}(\psi, \lambda)$ is defined by

$$\beta(\psi, \lambda) \equiv \frac{\beta(\psi, \lambda)}{b(\lambda)} \quad [\text{sr}^{-1}] \qquad [12]$$

Writing the volume scattering function $\beta(\psi, \lambda)$ as the product of the scattering coefficient $b(\lambda)$ and the phase function $\tilde{\beta}(\psi, \lambda)$ partitions $\beta(\psi, \lambda)$ into a factor giving the strength of the scattering, $b(\lambda)$ with units of m^{-1}, and a factor giving the angular distribution of the scattered photons, $\tilde{\beta}(\psi, \lambda)$ with units of sr^{-1}. A striking feature of the sea water VSFs of **Figure 3** is that their phase functions are all similar in shape, with the main differences being in the detailed shape of the functions in the backscatter directions ($\psi > 90°$).

The IOPs are additive. This means, for example, that the total absorption coefficient of a water body is the sum of the absorption coefficients of water, phytoplankton, dissolved substances, mineral particles, etc. This additivity allows the development of separate models for the absorption and scattering properties of the various constituents of sea water.

The Radiative Transfer Equation

The equation that connects the IOPs and the radiance is called the radiative transfer equation (RTE). Even in the simplest situation of horizontally homogeneous water and time independence, the RTE is a formidable integro-differential equation:

$$\cos\theta \frac{\text{d}L(z, \theta, \varphi, \lambda)}{\text{d}z} = -c(z, \lambda)L(z, \theta, \varphi, \lambda)$$
$$+ \int_{4\pi} L(z, \theta', \varphi', \lambda)$$
$$\times \beta(z; \theta', \varphi' \to \theta, \varphi; \lambda)\text{d}\Omega'$$
$$+ S(z, \theta, \varphi, \lambda) \qquad [13]$$

The scattering angle ψ in the VSF is the angle between the incident direction (θ', φ') and the scattered direction (θ, φ). The source term $S(z, \theta, \varphi, \lambda)$ can describe either an internal light source such as bioluminescence, or inelastically scattered light from other wavelengths. The physical environment of a water body – waves on its surface, the character of its bottom, the incident radiance from the sky – enters the theory via the boundary conditions necessary to solve the RTE. Given the IOPs and suitable boundary conditions, the RTE can be solved numerically for the radiance distribution $L(z, \theta, \varphi, \lambda)$. Unfortunately, there are no shortcuts to computing other radiometric quantities. For example, it is not possible to write down an equation that can be solved directly for the irradiance E_d; one must first solve the RTE for the radiance and then compute E_d by integrating the radiance over direction.

Apparent Optical Properties

Apparent optical properties are always a ratio of two radiometric variables. This ratioing removes effects of the magnitude of the incident sky radiance onto the sea surface. For example, if the sun goes behind a cloud, the downwelling and upwelling irradiances within the water can change by an order of magnitude within a few seconds, but their ratio will be almost unchanged. (There will still be some change because the directional structure of the underwater radiance will change when the sun's direct beam is removed from the radiance incident onto the sea surface.)

The ratio just mentioned,

$$R(z, \lambda) \equiv \frac{E_u(z, \lambda)}{E_d(z, \lambda)} \qquad [14]$$

is called the irradiance reflectance (or irradiance ratio). The remote-sensing reflectance $R_{rs}(\theta, \varphi, \lambda)$ is defined as

$$R_{rs}(\theta, \varphi, \lambda) \equiv \frac{L_w(\theta, \varphi, \lambda)}{E_d(\lambda)} \quad [\text{sr}^{-1}] \qquad [15]$$

where L_w is the water-leaving radiance, i.e., the total upward radiance minus the sky and solar radiance that was reflected upward by the sea surface. L_w and E_d are evaluated just above the sea surface. Both $R_{rs}(\theta, \varphi, \lambda)$ and $R(z, \lambda)$ just beneath the sea surface are of great importance in remote sensing, and both can be regarded as a measure of 'ocean color.' R and R_{rs} are proportional (to a first-order approximation) to $b_b/(a + b_b)$, and measurements of R_{rs} above the surface or of R within the water can be used to estimate water quality parameters such as the chlorophyll concentration.

Under typical oceanic conditions, for which the incident lighting is provided by the sun and sky, the radiance and various irradiances all decrease approximately exponentially with depth, at least when far enough below the surface (and far enough above the bottom, in shallow water) to be free of boundary effects. It is therefore convenient to write the depth dependence of, say, $E_d(z, \lambda)$ as

$$E_d(z, \lambda) \equiv E_d(0, \lambda)\exp\left[-\int_0^z K_d(z', \lambda)\text{d}z'\right] \qquad [16]$$

where $K_d(z, \lambda)$ is the spectral diffuse attenuation coefficient for spectral downwelling plane irradiance. Solving for $K_d(z, \lambda)$ gives

$$K_d(z, \lambda) = -\frac{d\ln E_d(z, \lambda)}{dz}$$
$$= -\frac{1}{E_d(z, \lambda)} \frac{dE_d(z, \lambda)}{dz} \quad [m^{-1}] \quad [17]$$

The beam attenuation coefficient $c(\lambda)$ is defined in terms of the radiant power lost from a collimated beam of photons. The diffuse attenuation coefficient $K_d(z, \lambda)$ is defined in terms of the decrease with depth of the ambient downwelling irradiance $E_d(z, \lambda)$, which comprises photons heading in all downward directions (a diffuse, or uncollimated, light field). $K_d(z, \lambda)$ clearly depends on the directional structure of the ambient light field, hence its classification as an apparent optical property. Other diffuse attenuation coefficients, e.g., K_u, K_{od}, or K_{PAR}, are defined in an analogous manner, using the corresponding radiometric quantities. In most waters, these K functions are strongly correlated with the absorption coefficient a and therefore can serve as convenient, if imperfect, descriptors of a water body. However, AOPs are not additive, which complicates their interpretation in terms of water constituents.

Optical Constituents of Seawater

Oceanic waters are a witch's brew of dissolved and particulate matter whose concentrations and optical properties vary by many orders of magnitude, so that ocean waters vary in color from the deep blue of the open ocean, where sunlight can penetrate to depths of several hundred meters, to yellowish-brown in a turbid estuary, where sunlight may penetrate less than a meter. The most important optical constituents of sea water can be briefly described as follows.

Sea Water

Water itself is highly absorbing at wavelengths below 250 nm and above 700 nm, which limits the wavelength range of interest in optical oceanography to the near-ultraviolet to the near infrared.

Dissolved Organic Compounds

These compounds are produced during the decay of plant matter. In sufficient concentrations these compounds can color the water yellowish brown; they are therefore generally called yellow matter or colored dissolved organic matter (CDOM). CDOM absorbs very little in the red, but absorption increases rapidly with decreasing wavelength, and CDOM can be the dominant absorber at the blue end of the spectrum, especially in coastal waters influenced by river runoff.

Organic Particles

Biogenic particles occur in many forms.

Bacteria Living bacteria in the size range 0.2–1.0 μm can be significant scatterers and absorbers of light, especially at blue wavelengths and in clean oceanic waters, where the larger phytoplankton are relatively scarce.

Phytoplankton These ubiquitous microscopic plants occur with incredible diversity of species, size (from less than 1 μm to more than 200 μm), shape, and concentration. Phytoplankton are responsible for determining the optical properties of most oceanic waters. Their chlorophyll and related pigments strongly absorb light in the blue and red and thus, when concentrations are high, determine the spectral absorption of sea water. Phytoplankton are generally much larger than the wavelength of visible light and can scatter light strongly.

Detritus Nonliving organic particles of various sizes are produced, for example, when phytoplankton die and their cells break apart, and when zooplankton graze on phytoplankton and leave cell fragments and fecal pellets. Detritus can be rapidly photooxidized and lose the characteristic absorption spectrum of living phytoplankton, leaving significant absorption only at blue wavelengths. However, detritus can contribute significantly to scattering, especially in the open ocean.

Inorganic Particles

Particles created by weathering of terrestrial rocks can enter the water as wind-blown dust settles on the sea surface, as rivers carry eroded soil to the sea, or as currents resuspend bottom sediments. Such particles range in size from less than 0.1 μm to tens of micrometers and can dominate water optical properties when present in sufficient concentrations.

Particulate matter is usually the major determinant of the absorption and scattering properties of sea water and is responsible for most of the temporal and spatial variability in these optical properties. A central goal of research in optical oceanography is to understand how the absorption and scattering properties of these various constituents relate to the particle type (e.g., microbial species or mineral composition), present conditions (e.g., the physiological state of a living microbe, which in turn depends on nutrient supply and ambient lighting), and history (e.g., photo-oxidation of pigments in dead cells). Bio-geo-optical models have been developed that attempt (with varying degrees of success) to predict the IOPs in terms of the chlorophyll concentration or other simplified measures of the composition of a water body.

Examples of Underwater Light Fields

Solving the radiative transfer equation requires mathematically sophisticated and computationally intensive numerical methods. *Hydrolight* is a widely used software package for numerical solution of oceanographic radiative transfer problems. The input to *Hydrolight* consists of the absorption and scattering coefficients of each constituent of the water body (microbial particles, dissolved substances, mineral particles, etc.) as functions of depth and wavelength, the corresponding scattering phase functions, the sea state, the sky radiance incident onto the sea surface, and the reflectance properties of the bottom boundary (if the water is not assumed infinitely deep). *Hydrolight* solves the one-dimensional, time-independent radiative transfer equation, including inelastic scattering effects, to obtain the radiance distribution $L(z, \theta, \varphi, \lambda)$. Other quantities of interest such as irradiances or reflectances are then computed using their definitions and the solution radiance distribution.

To illustrate the range of behavior of underwater light fields, *Hydrolight* was run for two greatly different water bodies. The first simulation used a

chlorophyll profile measured in the Atlantic Ocean north of the Azores in winter. The water was well mixed to a depth of over 100 m. The chlorophyll concentration *Chl* varied between 0.2 and 0.3 mg m^{-3} between the surface and 116 m depth; it then dropped to less than 0.05 mg m^{-3} below 150 m depth. The water was oligotrophic, case 1 water, and commonly used bio-optical models for case 1 water were used to convert the chlorophyll concentration to absorption and scattering coefficients (which were not measured). A scattering phase function similar in shape to those seen in **Figure 3** was used for the particles; this phase function had a backscatter fraction of $b_b/b = 0.018$. The second simulation was for an idealized, case 2 coastal water body containing 5 mg m^{-3} of chlorophyll and 2 g m^{-3} of brown-colored mineral particles representing resuspended sediments. Bio-optical models and measured mass-specific absorption and scattering coefficients were used to convert the chlorophyll and mineral concentrations to absorption and scattering coefficients. The large microbial particles of low index of refraction were assumed to have a phase function with $b_b/b = 0.005$, and the small mineral particles of high index of refraction had $b_b/b = 0.03$. The water was assumed to be well mixed and to have a brown mud bottom at a depth of 10 m. Both simulations used a clear sky radiance distribution appropriate for midday in January at the Azores location. The sea surface was covered by capillary waves corresponding to a 5 m s^{-1} wind speed.

Figure 4 shows the component and total absorption coefficients just beneath the sea surface for these two hypothetical water bodies, and **Figure 5**

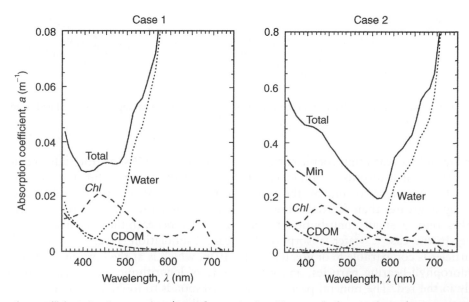

Figure 4   Absorption coefficients for the case 1 and case 2 water bodies. The contributions by the various components are labeled.

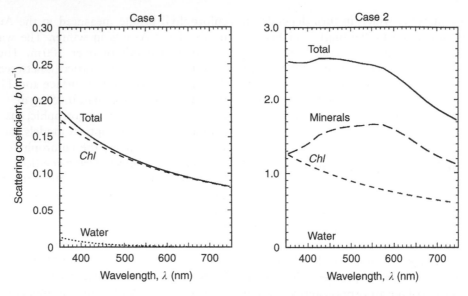

Figure 5 Scattering coefficients for the case 1 and case 2 water bodies. The contributions by the various components are labeled (CDOM is nonscattering).

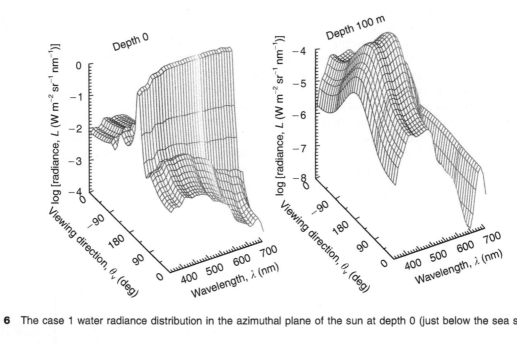

Figure 6 The case 1 water radiance distribution in the azimuthal plane of the sun at depth 0 (just below the sea surface) and at 100 m.

shows the corresponding scattering coefficients. For the case 1 water, the total absorption is dominated by chlorophyll at blue wavelengths and by the water itself at wavelengths greater than 500 nm. However, the water makes only a small contribution to the total scattering. In the case 2 water, absorption by the mineral particles is comparable to or greater than that by the chlorophyll-bearing particles, and water dominates only in the red. The mineral particles are the primary scatterers.

Figure 6 and 7 show the radiance in the azimuthal plane of the sun as a function of polar viewing direction and wavelength, for selected depths. For the case 1 simulation (**Figure 6**), the depths shown are zero, just beneath the sea surface, and 100 m; for the case 2 simulation (**Figure 7**), the depths are zero and 10 m, which is at the bottom. Note that the radiance axis is logarithmic. A viewing direction of $\theta_v = 0$ corresponds to looking straight down and seeing the upwelling radiance (photons traveling straight up).

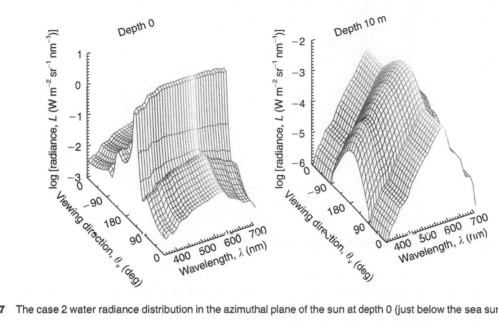

Figure 7   The case 2 water radiance distribution in the azimuthal plane of the sun at depth 0 (just below the sea surface) and at 10 m.

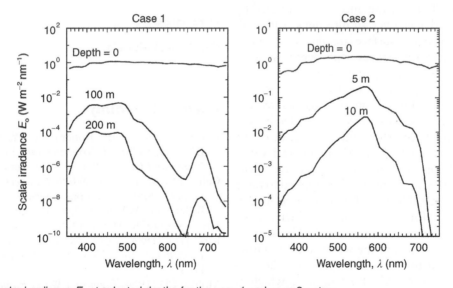

Figure 8   The scalar irradiance E_o at selected depths for the case 1 and case 2 waters.

Near the sea surface, the angular dependence of the radiance distribution is complicated because of boundary effects such as internal reflection (the bumps near $\theta_v = 90°$, which is radiance traveling horizontally) and refraction of the sun's direct beam (the large spike near $\theta_v = 140°$). As the depth increases, the angular shape of the radiance distribution smooths out as a result of multiple scattering. By 100 m in the case 1 simulation, the shape of the radiance distribution is approaching its asymptotic shape, which is determined only by the IOPs. In the case 2 simulation, the upwelling radiance $(-90° \leq \theta_v \leq 90°)$ at the bottom is isotropic; this is a consequence of having assumed the mud bottom to be a Lambertian reflecting surface. As the depth increases, the color of the radiance becomes blue for the case 1 water and greenish-yellow for the case 2 water. In the case 1 simulation at 100 m, there is a prominent peak in the radiance near 685 nm, even though the solar radiance has been filtered out by the strong absorption by water at red wavelengths. This peak is due to chlorophyll fluorescence, which is transferring energy from blue to red wavelengths, where it is emitted isotropically.

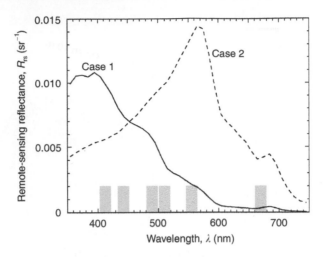

Figure 9 The remote-sensing reflectance R_{rs} for the case 1 and case 2 waters.

As already noted, the extensive information contained in the full radiance distribution is seldom needed. A biologist would probably be interested only in the scalar irradiance E_o, which is shown at selected depths in **Figure 8**. This irradiance was computed by integrating the radiance over all directions. Although the irradiances near the surface are almost identical, the decay of these irradiances with depth is much different in the case 1 and case 2 waters.

The remote-sensing reflectance R_{rs}, the quantity of interest for 'ocean color' remote sensing, is shown in **Figure 9** for the two water bodies. The shaded bars at the bottom of the figure show the nominal SeaWiFS sensor bands. The SeaWiFS algorithm for retrieval of the chlorophyll concentration uses a function of the ratio $R_{rs}(490 \, nm)/R_{rs}(555 \, nm)$. When applied to these R_{rs} spectra, the SeaWiFS algorithm retrieves a value of $Chl = 0.24 \, mg \, m^{-3}$ for the case 1 water, which is close to the average value of the measured profile over the upper few tens of meters of the water column. However, when applied to the case 2 spectrum, the Sea-WiFS algorithm gives $Chl = 8.88 \, mg \, m^{-3}$, which is almost twice the value of $5.0 \, mg \, m^{-3}$ used in the simulation. This error results from the presence of the mineral particles, which are not accounted for in the SeaWiFS chlorophyll retrieval algorithm.

These *Hydrolight* simulations highlight the fact that it is now possible to compute accurate underwater radiance distributions given the IOPs and boundary conditions. The difficult science lies in learning how to predict the IOPs for the incredible variety of water constituents and environmental conditions found in the world's oceans, and in learning how to interpret measurements such as R_{rs}. The development of bio-geo-optical models for case 2 waters, in particular, is a research topic for the next decades.

See also

Bio-Optical Models.

Further Reading

Bukata RP, Jerome JH, Kondratyev KY, and Pozdnyakov DV (1995) *Optical Properties and Remote Sensing of Inland and Coastal Waters.* New York: CRC Press.

Caimi FM (ed.) (1995) *Selected Papers on Underwater Optics.* SPIE Milestone Series, vol. MS 118. Bellingham, WA: SPIE Optical Engineering Press.

Jerlov NG (1976) *Marine Optics.* Amsterdam: Elsevier.

Kirk JTO (1994) *Light and Photosynthesis in Aquatic Ecosystems* 2nd. New York: Cambridge University Press.

Mobley CD (1994) *Light and Water Radiative Transfer in Natural Waters.* San Diego: Academic Press.

Mobley CD (1995) The optical properties of water. In Bass M (ed.) *Handbook of Optics*, 2nd edn, vol. I. New York: McGraw Hill.

Mobley CD and Sundman LK (2000) *Hydrolight 4.1 Users' Guide.* Redmond, WA: Sequoia Scientific. [See also www.sequoiasci.com/hydrolight.html].

Shifrin KS (1988) *Physical Optics of Ocean Water.* AIP Translation Series. New York: American Institute of Physics.

Spinrad RW, Carder KL, and Perry MJ (1994) *Ocean Optics.* New York: Oxford University Press.

Walker RE (1994) *Marine Light Field Statistics.* New York: Wiley.

TRANSMISSOMETRY AND NEPHELOMETRY

C. Moore, WET Labs Inc., Philomath, Oregon, USA

Introduction

Transmissometry and nephelometry are two of the most common optical metrics used in research and monitoring of the Earth's oceans, lakes, and streams. Both of these measurements relate to what we perceive as the clarity of the water, and both provide vital information in numerous studies of natural processes and human activities' impact upon water bodies. Applications involving these measurements range from monitoring drinking water suitability to understanding how carbon is transferred into and transported within ocean waters.

Transmissometry refers to measurements made by transmissometers or beam attenuation meters. These sensors infer the total light lost from a beam of light passing through the water. These losses are caused by two primary mechanisms. Suspended particles and the molecules of the water itself scatter the light away from its original path; the water, and dissolved and particulate matter contained within, absorb the light and convert it into heat, photosynthetic activity, fluorescence, and other forms of energy. Larger concentrations of scattering and absorbing substances therefore result in greater losses in signal.

Nephelometry refers to measurements made by optical scattering sensors, often referred to as turbidity sensors or nephelometers. These sensors project a beam of light into the water and measure the radiant flux of light scattered into the direction of a receiver. Since the receiver signal increases with greater numbers of particles, the device infers the concentration of suspended particles in the water.

Scattering sensors are used more commonly in environmental monitoring applications, especially in highly turbid waters with large concentrations of particles; transmissometers see more use in general scientific studies. However, the uses for which they are employed broadly overlap. Nevertheless, transmissometers perform quite different measurements from those of scattering sensors and the quantities they measure are independent of one another and typically offer no direct comparison. In fact, while the data products they provide may covary, the relationship between the values most certainly will differ depending upon the composition of the materials in the water.

Using a transmissometer one can derive an attenuation coefficient that mathematically describes the ability of the water to transmit light. This coefficient is a fundamental optical characteristic and an absolute quantity for a given medium. The scattering sensor, on the other hand, collects a very small portion of the scattered light and is usually calibrated to some secondary standard. The units of measurement are themselves relative to that standard. Other differences also prove crucial in defining these measurements. Limitations imposed by the instruments themselves, application-specific requirements, sensor sizes, and cost all play roles in determining the possible suitability of one measurement versus another. Thus, in order to best fit these two methods to potential applications, it is necessary to understand the measurements, the design of the sensors performing them, and the products that the sensors provide.

Measurements and Fundamental Values

In the realm of water sciences, transparency and turbidity are two of the most commonly used terms in describing optical clarity. These are general terms and typically not tied to absolute physical quantities other than through the use of secondary standards. However, the set of underlying optical processes that describe the impact of water-based media upon light propagating through them are well defined, if not completely understood. In the study of the transmission of light energy through water, the inherent optical properties (IOPs) refer to the set of intrinsic optical characteristics of the water and components contained therein. The IOPs define how light propagates through the water. In comparison to apparent optical properties (AOPs), the other general class of in-water optical measurements, the IOPs are not affected by changes in the radiance distribution from sunlight or other sources. The IOPs include coefficients for the attenuation, absorption, and scattering of light as well as the volume scattering function.

The coefficients of attenuation (c), absorption (a), and scattering (b) determine radiance losses of a ray of light propagating through the water. Light is either lost to absorption by the water and material contained within or it is scattered by the same. The attenuation coefficient accounts for losses attributed to both the absorption and the scattering and is equal to the sum of these coefficients eqn [1].

$$c = a + b \qquad [1]$$

One determines the beam attenuation coefficient by comparing the radiant flux of a collimated beam of light at source (F_s) with the radiant flux of the beam at a receiver detector (F_d), a finite distance (r) away. This ratio is known as the beam transmittance (T), given by eqn[2] or equivalently by eqn[3].

$$F_d/F_s = T = e^{-cr} \qquad [2]$$

$$c = -\ln(T)/r \qquad [3]$$

Here r is the path length between the source and the receiver. This coefficient is the value ultimately determined by a transmissometer. The attenuation coefficient is expressed in units of inverse meters (m^{-1}). Thus, when one refers to water with an attenuation coefficient of $1 \, m^{-1}$, the implication is that within a 1 m path the available light within a collimated beam is reduced to 1/e or approximately 37% of its original energy.

Within the visible light spectrum the scattering and absorption losses from the water itself remain effectively constant, and thus variability found in field measurements results from non-water particulate and dissolved matter. The extent of absorption-based losses compared to scattering-based losses depend both on the materials being measured and on the spectral configuration of the meters. Both the scattering and absorbing properties of water-based components are prone to variation with the wavelength of light at which measurements are conducted. Variations in the absorption depend heavily upon the amount of colored dissolved organic matter (CDOM) and chlorophyll content. CDOM absorbs very strongly in the blue wavelengths; chlorophyll absorbs heavily in the blue and in addition has a pronounced absorption peak in the deep red portion of the spectrum (676 nm). Absorption by these materials provides the appearance of color to the water. Visually, CDOM laden waters tend to appear brown, and chlorophyll-rich waters appear green. A deep blue cast to the water indicates very low levels of both of these substances. The spectral dependency of the scattering signals is largely due to the size of the particles from which the light is scattered (**Figure 1**).

In addition to the optical loss coefficients, the volume scattering function (VSF) forms another important component of the IOPs in describing the fate of light in water. The VSF describes optical scattering as a function of the angle, θ, away from the direction of propagation of the incident beam of light. The

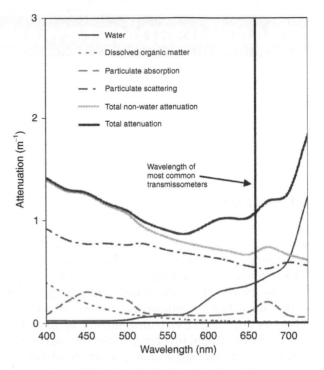

Figure 1 Relative contributions of water and non-water scattering and absorbing components are seen in formulation of the attenuation coefficient within 'typical' waters.

VSF coefficient, $\beta(\theta)$, defines the radiant energy lost into a given angular region of the light scattering and is expressed in terms of inverse meters per steradian. The VSF integrated over the entire spherical volume into which light is scattered provides b, the total scattering coefficient (eqn [4]).

$$b = 2\pi \int_{0}^{\pi} \beta(\theta)\sin(\theta)\mathrm{d}\theta \qquad [4]$$

The actual shape of the VSF depends upon the particle field being measured. Specific properties that define this shape include the particle size and shape and the index of refraction. Particle size is probably the single most pronounced factor in defining the VSF in that it dictates the regime of light interaction with the particles themselves. Very small particles that fall within the wavelength of the light impinging upon the particles are subject to molecular or Rayleigh scattering. This interaction is relatively weak, and creates a VSF that is relatively constant with angle. While Rayleigh scatterers are by far the most prevalent in most waters, most of the scattering signal seen by sensors is attributed to particles ranging from 1 μm to > 50 μm. The scattering behavior of these particles is typically modeled using Mie theory. Mie theory uses Maxwell's equations to predict perturbations of an incident planar wave by spherical particles in its path. In general, larger particles will create a greater degree of near-forward scattering.

Most scattering sensors are not considered tools for determination of in-water optical properties, but all scattering sensors including turbidity sensors measure the VSF within a given angular region, typically somewhere in the region of 90–160° with respect to the incident direction of the light. It is perhaps ironic that while these sensors are among the most ubiquitous of in-water optical tools, the VSF is one of the least-characterized of all the IOPs. This is because no single angle measurement can account for the shape of the entire function. This in turn points to a major source of error in all turbidity-based measurements. Different materials dictate different VSFs and a single angle measurement will vary with concentration from one type of material to the next. In actual fact a diverse amalgam of organic and inorganic particulates reside within most waters. This ultimately tends to homogenize the VSFs such that the variability in the VSF of the composite is less than the variability of individual components (**Figure 2**).

Most scattering measurements are based upon some standard such as formazin, diatomaceous earth, or more recently spherical styrene bead suspensions. These standards are used because they tend to be reproducible and easy to mix into various concentrations for calibrations. Units of quantity are expressed in form of turbidity units such as NTU (nephelometric turbidity units). Because of the disparate VSFs of these standards and natural waters, total attenuation (or particle concentration) cannot be obtained from turbidity measurements without

intercalibrating with transmissometers (or by filtering and weighing) in natural waters.

Sensors

Transmissometers

A basic transmissometer consists of a collimated light source projected through an in-water beam path and then refocused upon a receiver detector. Typically single-wavelength transmissometers employ a light-emitting diode coupled with an optical bandpass filter as the source. Source light is often split so that a portion of the beam impinges upon a reference or compensation detector that is either used in numerical processing of the data or integrated into a source stabilization feedback circuit. The source output is often modulated and the lamp and receiver detector samples are in phase with the source modulation. This greatly reduces ambient light detection by the receiver from the sun or other unwanted sources. Path lengths are fixed with distances typically ranging from 5 cm to 25 cm depending upon the waters in which the sensors are used (**Figure 3**).

The receiver detector converts radiant flux into current and its output is thus proportional to the radiant energy passed through the water. Electronics subsequent to the detector amplify and rectify the signal for digitization or direct output as a DC voltage level. This signal is known as the instrument transmittance (T_i) (eqn [5]).

$$T_i = S \times T \tag{5}$$

S represents the instrument transmittance scaling constant. This constant is a combined term that

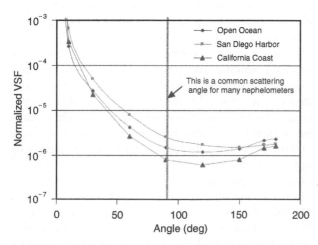

Figure 2 Normalized VSF data for three representative ocean water types. Note that at 90°, the most common nephelometer scattering angle, significant differences exist for the respective coefficients. Data collected by Theodore Petzold and Seibert Duntley of Scripps Institute of Oceanography.

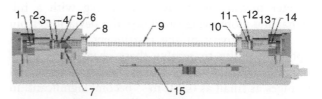

Figure 3 Cutaway view showing the primary optical components found in a modern transmissometer. A transmitter assembly and receiver assembly are mounted and aligned within a rigid frame. The transmitter assembly consists of (1) a source lamp; (2) a pinhole aperture; (3) a collimating lens; (4) field aperture; (5) an interference filter; (6) a beam splitter; (7) a reference detector; and (8) a pressure window. The beam (9) then passes through a fixed-path volume of water and enters the receiver assembly. The receiver consists of (10) a pressure window, (11) field aperture, (12) a refocus lens, (13) a pinhole aperture, and (14) the receiver detector. Signals from the detector are then fed to the electronics for processing and output (15).

includes signal amplification, losses through windows and lenses, and other sensor gain factors. From eqn[5] and assuming a 25 cm pathlength, we obtain eqn[6] or equivalently eqn[7].

$$T_i/S = e^{-c(0.25)} \qquad [6]$$

$$c = 4 \ln T_i - Q \qquad [7]$$

The constant $Q = 4 \ln S$ is a general scaling term that is removed, or compensated for, during the calibration process.

An ideal transmissometer would reject all but the parallel incident light into its receiver. This implies that there is no error associated with near-forward scattered light getting into the receiver. However, limitations in real-world optics make this a near impossibility. Transmissometers thus provide a value for a system attenuation coefficient that has a finite scattering error and is defined primarily by the acceptance angle of the receiver optics. These values range from around 0.5° to 1° in water for most commercial instruments. Because that VSF for in-water particles is highly peaked at these angles, this can result in underestimation of the attenuation coefficient and can also lead to sensor-to-sensor discrepancies in measurement. It thus becomes important to know this angle in treating data carefully. While it is possible to build sensors with narrower acceptance angles than 0.5°, scattering in the very near-forward direction becomes dominated by turbulent fluctuations in the density of the water itself. This turbulence-induced scattering is irrelevant to particulate studies and, depending upon the distances and receiver sizes involved, to most signal transmission applications.

The conceptual framework for the transmissometer measurement involves starting with a full signal and monitoring small negative deviations from it. The sensitivity of the instrument thus depends upon its ability to resolve these changes. In many oceanic and other clear water investigations, signal changes as small as $0.001 \, \text{m}^{-1}$ become significant. In a 25 cm instrument this implies a requirement for transmittance resolution on the order of 0.025%. At the other end of the environmental spectrum, many inland waterways and some harbor areas would render a 25 cm path instrument ineffective due to loss of all signal. Therefore, range and resolution become the two critical factors in determining a transmissometer's effectiveness in a given application. While it is easy to imagine using arbitrarily long path lengths to obtain increased sensitivity, the instrument path begins to impose other limitations upon its utility. Size and mechanical stability both

reduce utility of the longer path instruments. On the other hand, shorter paths impose more demands than just high levels of precision in measurement. Cleaning of optical surfaces also becomes a major issue in maintaining sensor reproducibility and accuracy. Again using the 25 cm path length instrument as an example, maintaining signal reproducibility of $0.01 \, \text{m}^{-1}$ over time requires a cleaning technique that gives results that repeat within 0.25% transmittance. For a 10 cm path length instrument, repeatability would need to be within 0.10% transmittance. Likewise, internal correction mechanisms such as compensation of temperature-related drift impose stringent requirements upon the sensor's electronics as well as the subsequent characterization process. Long-term drift and general mechanical stability also must be tightly constrained for the instrument to provide accurate results over time. The requirements prove challenging in light of the forty degree (centigrade) temperature swings and the 6000 meter depth excursions to which the instruments potentially get exposed.

While the calculation of the attenuation coefficient from raw transmittance is independent of the cross-sectional area of the beam, the beam size does play an important role in the transmissometer's ability to measure. Accurate transmittance measurements rely upon the water and the materials it contains acting as a homogenous medium. This model starts to break down in two important cases: when the number concentration of particulates becomes significantly low compared to the total volume of the illuminated sample area; and when the particle sizes become significantly large in comparison to the cross-sectional area of the beam. Taken in the extreme, one can easily imagine a very narrow beam providing a binary response at the receiver depending upon whether a particle occludes its path. Practically speaking, most transmissometers need to show minimal spiking for particle sizes up to $100 \, \mu\text{m}$ diameter. Particles more than a few micrometers in diameter are 'seen' by the receiver at about two times their actual size as a result of diffraction. This means for a beam of 5 mm nominal width that a single $100 \, \mu\text{m}$ particle could reduce signal at the receiver by approximately 0.08% or on the order of $0.0032 \, \text{m}^{-1}$ in a 25 cm path (or $0.008 \, \text{m}^{-1}$ in 10 cm path). This proves acceptable for most operational conditions. On the other hand, a 1 mm particle could create an 8% deviation in sensor output, creating a noticeable spike. Fortunately, 1 mm particles are extremely rare except in active erosion zones.

There are presently two primary methods used in calibrating transmissometers. The first uses fundamental principles of beam optics and knowledge of the index of refraction difference between air and water to directly estimate the sensor output. Electro-optical linearity in response to signal changes is assumed or verified. The sensor's gain level is set near full scale for transmission in air and the sensor is checked to ensure that if the source output is completely blocked it provides a real zero output. Accounting for the differences in reflection and transmission of the air–glass interfaces compared to the water–glass interfaces, one can then assume that, upon immersion, any further deviations in signal are due to the attenuation of the water and materials contained therein. This measurement is then verified by immersion in clean water and subsequent comparison to clean water values. Error terms in this method usually include deviations of the modeled optics from the real world. These errors include lens-induced focusing aberrations, alignment issues, spectral content of the source, and any dust or film on any of the optical components. The primary advantages of this method are that the calibration process relies only upon the air value measured by the meter, and that the attenuation due to the water is included in the water-based measurements.

The second method involves blanking the meter directly with clean water. More akin to calibration approaches used in spectrophotometry, this method involves immersion of the instrument into optically clean water, measuring the value, and setting that value as full-scale transmittance or, conversely, $0.000\,\mathrm{m}^{-1}$ attenuation (clean water values for the attenuation can then be added back in accordance with published values). The chief disadvantage of this method lies in the difficulty of creating and verifying optically clean water. While various levels of filtering can remove most of the particulates from the water, filters can also introduce bubbles. These bubbles are seen as particles by the sensor. Assuming that one achieves filtration without introducing any bubbles, bubble creation is still a concern in that any partial pressure imbalances between the gases contained within the water and the surrounding environment will result in subsequent bubble formation. Added to that is the possibility that the containers and the sensors themselves may also act as sources of particulate contamination. The chief advantage of this method is that it accommodates for small deviations in the real instrument with respect to the ideal.

The overriding issue with calibration of transmissometers is the same as in the discussion of the need for and difficulty of proper cleaning. In order to calibrate an instrument to operate accurately in cleaner waters, the calibrations must achieve accuracy to within 0.25% of full-scale measurement.

Ultimately, reproducibility of results becomes the best check for calibration. That said, this level of accuracy is really only required in conditions where particle concentrations are approaching minimal levels. Relative changes of transmittance will still be precisely reflected in the instrument's measurements.

Scattering Sensors

A simple scattering sensor consists of a source element projecting a beam of light in the water and a receiver detector positioned at a fixed angle with respect to the source. The source beam is sometimes stabilized by inclusion of a second receiver that receives a portion of the light coming directly out of the lamp. This signal is then fed back into the lamp driver circuitry to compensate for fluctuations in the source with time and temperature. The source beam has a defined primary projection angle and a distribution of light about that angle. Conversely, the receiver is placed at a specific angle and maintains a defined field of view about that angle. These factors combine to form the distribution of angular response for the scattered light (**Figure 4**).

As with transmissometers, it is necessary to reject ambient light from the sun and other non-sensor sources during measurement. With scattering sensors this is achieved both through the use of synchronously modulated light and detector amplification and also through the use of direct optical rejection. Direct optical rejection is employed at the source through the use of relatively narrow spectral band sources that emit light in the infrared away from the water-penetrating wavelengths of sunlight. Accordingly the receiver incorporates narrowband optical filters that reject wavelengths away from the primary emission bands of the source.

Figure 4 Typical scatter sensors and transmissometers.

Specific angular configurations used in modern scattering sensors vary widely. Some sensors are designed to operate within a highly constrained, narrowband, angular relationship, and some are designed to collect as much scattered light as possible and thus encompass a very wide angular range. In general two truths hold for all the designs: they will all provide a roughly linear response that is proportional to the particle concentration (at least in low to moderate concentrations); and different optical configurations will demonstrate different absolute response curves with respect to each other even when calibrated with the same standard (**Figure 5**).

A scattering sensor works by the simple principle that when particles are present they will scatter light and the receiver will collect some of that light. Using Beer's law, which states that increasing concentrations will result in a linear increase in output signal, the sensor's output varies from a zero value in clean water to a full-scale value at the upper end of its range. While it is convenient to assume a linear response with concentration, this is not strictly true. Light reaching the volume of interaction and the light scattered back into the detector is subject to secondary losses due to attenuation. As the concentration of scattering components in the water increases, so does the attenuation. This produces a nonlinearity in the output signal. In sensors with large interaction volumes and a wide angular response, this becomes a particularly messy analytical problem in that the light is subject to a large range of effective path lengths in propagation from the source and back to the receiver. In the extreme case, sensors exist that position a near-isotropic source next to a wide-angle detector such that they both project out, perpendicular to the same plane. In these sensors the effective volume of interaction is strictly a function the attenuation coefficient in that it is infinite other than for induced losses of light. As with transmissometers, the volume of interaction also affects a scattering sensor's sensitivity and the effect of larger particles upon the signal. Small volumes show less sensitivity and measure larger particles as signal spikes. The combined issues of long-path attenuation coupling and volumetric sensitivity point to the preference of designs incorporating larger beams with greater interaction volumes for measurements of cleaner waters and narrower beams with interaction volumes close to the sensor surface for use in highly turbid waters.

The response of a given scattering sensor is very highly dependent upon its specific optical configuration. Angle of interaction, angular distribution, wavelength at which the source emits, and the relative path distance from the source and back to the receiver are all factors in how a sensor will behave. As mentioned earlier, it should be expected that two different designs will provide two different responses. In studies in which researchers require only relative responses with space or time, this is not a major issue. A twofold change in a given concentration of particles will generate an associated response in the instrument output. However, many studies require some form of reproducible results. It is not enough that two sensors are calibrated to the same medium. They must also respond in the same way to any other medium that they might mutually measure. Standards such as ISO 7027 have been published. These standards impose constraints on the angle of interaction between the source and the receiver (90°), the angular distribution of the source, and its wavelength of operation, as well as other design parameters. The goal is to ensure that all sensors built within the constraints imposed by the standard will provide similar results in similar waters. This is a very important step toward achieving consistent results amenable to intercomparison.

Straightforward in concept, sensor calibration employing a standard suspension, provides several pitfalls in practice. First and foremost, no calibration can be achieved to better accuracy than the standard solutions themselves. Secondly, it is critical to ensure that the container in which the calibration takes place is not a cause of secondary reflections of light that can get back to the receiver. Care must also be taken to ensure that the suspension is not settling or flocculating during the measurement. Finally, one variation of this technique is to use arbitrary concentration of the calibration media and calibrate against another 'standard' precalibrated sensor.

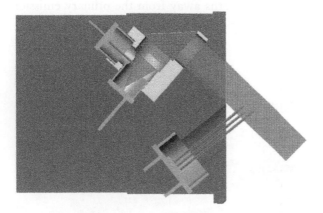

Figure 5 One of many possible optical configurations for a scattering sensor. A source assembly consisting of a LED lamp, reference detector, lens, and right angle prism projects light into the water. The receiver is placed to receive light at 90° with respect to projected source beam.

Great care should be applied when using this method. Standard sensors often already incorporate compensation schemes for linearizing the data. These schemes in turn are developed for use with a specific type of suspension. This can create dramatic and surprising results when using another suspension.

While scattering sensors are predominantly used to determine relative concentrations of particulates, another very important set of applications involve characterization of the volume scattering function itself. One of the important goals in observational oceanography involves the use of remotely sensed data from satellites and other airborne platforms to rapidly characterize large areas of surface and near surface waters. Of particular interest are the emerging methodologies associated with using ocean color data captured from airborne and space-borne platforms to provide information about the biology and chemistry of waters. In the United States, NASA projects such as the Coastal Zone Color Scanner and the more recent SeaWiFs satellite program stimulated this interest, and in the case of SeaWiFs continue to contribute a growing body of information. The light that these platforms receive is a function of the sea surface state and the resultant reflections and the water-leaving radiance. This radiance in turn is defined by the absorption and scattering characteristics of the water. Scattering in the region of 90–180° is specifically important because it represents incoming light from the sun that is scattered back into the atmosphere. To quantify this, a class of sensors called optical backscattering sensors have been developed and calibrated specifically for this purpose. In many respects these sensors are very similar to other scattering sensors in that they use the same basic optical configurations and respond similarly to variations in the particle field. The major differences involve design constraints upon the wavelengths of the source emitters and the angles of interaction. Equally importantly, the calibration of these sensors involves tying the sensor response directly to the volume scattering function.

Calibration of scattering sensors for radiometric measurements involves detailed knowledge of the sensor optics geometry and some known scattering agent. The prevalent method for single-angle measurements incorporates a sheet of highly reflective diffuse material and maps the sensor response as a function of the distance between the target and the sensor. This information is then applied to derive the angular weighting function of the interaction volume. Finally, this weighting function is applied to a typical ocean water VSF. More recently, researchers have begun to apply a calibration technique that incorporates known concentrations of scattering agents with well-defined VSFs. These two techniques address different elements of a sensor calibration and may well find optimum effectiveness when used in conjunction with one another (**Figure 5**).

Applications

Domains of Use

The use of transmissometers and nephelometers falls broadly into two categories. We want to study the water's optical properties and how they might relate to ongoing processes occurring in the water, and we want to determine how much foreign matter is in it. While, ultimately, both thrusts of study lead to measurement of the same media within a given body of water, the products that the instruments provide differ, and the requirements surrounding the given areas of study tend to drive the development of the different technologies. The factors ultimately determining the appropriateness of one sensor versus another do not always pertain to the data products provided. Size, cost, ease of deployment, ease of maintenance, and researcher's experiences all contribute to decisions on which type of sensor is the best to use.

Optical oceanographic research motivated much of the development of modern transmissometers. This arena also stimulated development of scattering sensors that are specifically designed and calibrated for providing coefficients related to the VSF. Much of this work in the United States revolves around Naval research needs, and primary development of sensors now available commercially was in large part funded through Naval research dollars. Naval applications include mine hunting, underwater tactical assessment for diving operations, and sea truthing for laser communications and imaging research. The US National Aeronautics and Space Administration (NASA) has also played a major role in developing underwater tools for optical characterization. These tools help calibrate the airborne sensors. Similarly, numerous other governments foster the development and use of these tools through their respective Naval, space and other scientific agencies. While not engaged in the study of ocean optical properties *per se*, many other ocean scientists working under aegis of funds supplied by these agencies use transmissometers and optical backscattering sensors in ongoing efforts to understand physical, biological, and chemical distributions and processes in the water.

Scattering sensors remain the dominant optical tools used by environmental researchers. These sensors' size and cost make them widely affordable and easily used, and the newer sensors incorporate fouling-retardant features such as shutters and biocidal exposed surfaces. As such they are becoming increasingly subscribed to as the sensor of choice in compliance-driven monitoring applications developed by various governmental agencies throughout the world. Naturally, the more attractive size and costs of scattering sensors also make them favorable choices in many larger-scale applications.

It is likely that remote sensing will to some degree change preferences for sensors among fresh water researchers over the next ten years. Presently there is relatively little airborne color data available for fresh water bodies, and thus many limnology researchers have not yet been compelled to measure optical properties of lakes directly. With the next generation color airborne sensors and new governmental mandates driving more effective broader-scale sampling strategies, the need and desire for transmissometer measurements and scattering measurements for VSF determination will undoubtedly grow.

How Sensors are Deployed

One major constraint in an underwater sampling is how to use the instrumentation effectively in the environment for which it is intended. Researchers often want to measure the water in places they cannot easily get to, or over timescales that make personal attendance of equipment an unappealing proposition. To these considerations must be added the requirement that the data gathered must truly reflect changes at the time and space scales of the governing processes within the water column, and the constraint imposed by doing this sampling at a reasonable cost. The sampling challenge becomes formidable. As a result, the development of effective sampling platforms has become as challenging and competitive a discipline of research as instrumentation design itself.

Transmissometers and scattering sensors are typically integrated into multiparameter sampling packages for acquiring and storing data (CTDs, data sondes, loggers). The packages are then deployed from boats or other platforms and lowered through the water column, travel on or are towed by a vessel, or are placed on buoys or mooring lines in order to log measurements over an extended period. Many variations of these basic methods exist but virtually all entail these basic concepts.

A new class of autonomous deployment platforms will serve to revolutionize underwater sampling. These range from miniature programmable underwater vehicles, to freely drifting ocean profilers that can continuously move through the water column, and to rapidly deployable profiling moorings. Many flavors of these various platforms are now emerging. Some will find important niches for acquisition of data over space and time.

Some Current Applications

There are many different applications engaging the use of transmissometers and scattering sensors. Table 1 represents only a sampling across numerous disciplines.

Extending Capabilities

As mankind's need to understand and monitor the Earth's waters has increased, they have driven the development of more rugged, more reliable, smaller, and cost-effective technologies for transmissometry and nephelometry as measurement techniques. These resultant technologies have not only carved greater roles for optical measurement methods but have also proved seminal in the development of entirely new sensors. Recently, a new generation of IOP tools has been made available to the oceanographic community. They include sensors for the determination of the in-water absorption coefficient, multiangle scattering sensors, and a set of IOP tools with spectral capabilities. Transmissometers and simple scattering sensors have laid the foundation for the optical techniques and data methods of these new devices. In turn, these new sensors promise to significantly enhance the role of IOP measurements in modern observing platforms.

One of the more significant recent breakthroughs in optical measurement techniques lies in the development of the absorption meter. This sensor uses a measurement method and optical geometry similar to a transmissometer except that it encompasses the sensor's beam path with a reflective tube and incorporates a large-area detector at the receiver end of the path. The reflective tube and large-area detector combine to collect the bulk of the light scattered from the source beam. Thus the light not detected is primarily due to absorption by the water and its constituents.

The wide-band spectral nature of sunlight coupled with the selective filtering capabilities of water and the absorption characteristics of phytoplankton and dissolved organic material make spectral optical characterization of the water highly desirable. Likewise, the spectral information from the scattering of particles provides more direct correlation with remote color data as well as a more complete description of the type of particles scattering. New tools encompassing spectral attenuation, absorption, and scattering are now commercially available.

Table 1 Applications of transmissometry and nephelometry

Application	Description
Monitoring terrestrial runoff and impact of industrial inflows on water quality	Scattering sensors stationed in rivers and streams allow researchers to determine impacts of inflows upon water quality. Inflows might be created by logging, agriculture, mining, land development, controlled and uncontrolled outflows from water treatment plants, natural runoff and other events that introduce new matter into the monitored bodies.
Compliance monitoring	United States' compliance monitoring of fresh water bodies is soon likely to include turbidity as a required parameter for ongoing measurement.
Determining biological distribution in the water	Both transmissometers and scattering sensors are deployed in viewing the biological variability in space and time through the water column.
Radiative transfer studies – optical closure	In verifying the optical relationships between the inherent and apparent optical properties, researchers seek to test the relationships through direct measure and comparison of values from the disparate instrument types. Scientists also seek to reconcile measurements of the inherent properties among themselves in validation of IOP theory.
Remote sensing validation	Satellite and other airborne remote imaging systems require in-water transmissometry, scattering, and absorption measurements to calibrate these sensors to water-borne optical properties.
Studying the benthic layer processes	In understanding the processes effecting the settling and re-suspension of particles near the bottom of the water column both scattering sensors and transmissometers can provide relative indications of particle flux.
Frazil ice formation	Transmissometers have been shown to 'see' signal fluctuations associated with the formation of frazil or supercooled ice. These studies are imperative in understanding how polar ice sheets are formed.
Diver visibility	Navies require better tactical assessment of waters for determining operational risk for divers and other visibility-related operations.
Small-scale structure in the water column	In coastal regimes many physical and ecological processes take place on smaller time and space scales than previously thought. The speed of acquisition and sensitivity of modern scattering sensors and transmissometers allow accurate particulate mapping within the water column, which in turn serves as a tracer for these processes.
Tracking particulate organic carbon	Data from transmissometers has been shown to accurately reflect total particulate organic carbon within the water column. Understanding in-water carbon transport processes is, in turn, vital to understanding carbon flux between the water and atmosphere through the uptake and output of CO_2.
Tracking bloom cycles	Transmissometers on moorings located both in open ocean and in coastal areas track seasonal bloom cycles as well as event-driven changes from major storms or other potential system disturbances.
Monitoring activity around thermal vents and underwater volcanoes	Scattering sensors on moorings and underwater vehicles track plumes from underwater vents and eruptions.

These tools are playing increasingly important roles in various applications.

Despite the plethora of scattering sensor data available, very little information exists concerning the range of variability of the VSF, and how it relates to different water masses and the processes within them. One of the chief constraints in fully characterizing the VSF is that it requires a multiangle scattering measurement encompassing in excess of 4 orders of magnitude of scattered light intensity. After some seminal work performed by researchers at the Scripps Institute of Oceanography during the late 1960s and early 1970s, very little has since been done to add to this body of data. In fact, VSF functions measured then remain *de facto* calibration standards for instruments being built today. In recent years researchers in Europe and the United States have refocused attention upon this issue. As a result,

a new set of multiangle scattering sensors is now coming into commercial availability.

Other development efforts and new instrumentation incorporate scattering and transmittance measurements in unique ways to obtain specific underwater chemical and biological components. One example of these includes an underwater transmissometer that uses polarized light to determine concentrations of particulate inorganic carbon. These instruments promise to fill a vital niche in understanding the fate of carbon in the seas. Another example in development are underwater flow cytometers. While the prevalence of IOP measurements look at bulk phase phenomena, new instruments are now available as ship-board and dock mounted units that couple scattering and fluorescence measurements of individual cells and organisms to provide identifying signatures. Patterned after laboratory flow-cytometers, the in water devices will offer break-through capability in typing specific organisms in their natural environment.

One of the most exciting aspects of the recent advancements in IOP-related technologies lies in the opportunities offered by their combined use. One marked example lies in the characterization of particle aggregations in the water. While the attenuation or scattering at one wavelength will provide data about relative concentrations of particles within the water column, spectral data from these sensors combined with absorption measurements can move us a long way toward characterizing the aggregation into various biological and inorganic components.

Summary

Transmissometry and nephelometry provide increasingly valuable information relating to the light-transmitting characteristics of water as well as an idea of the relative concentration of suspended material within lakes and oceans. While sometimes viewed as near-synonymous techniques, these methods use different measurement methods, provide different products, and have different strengths and weaknesses in considering the applications to which they are applied. Applications vary widely and across numerous disciplines, but tend to be divided into two major classes: those that attempt to characterize the fundamental optical properties of the water; and those that seek the relative concentrations of foreign particulate matter in the water. In general, nephelometry is the preferred technique in environmental and fresh water applications and transmissometry is more common in oceanographic research. Although transmissometry and nephelometry differ as measurement techniques, in their application domains, and in subsequent calibration and handling, all of these sensors are capable of providing outputs in terms of absolute coefficients that describe the fate of light passing through water. These coefficients of light transfer are collectively known as the inherent optical properties or IOPs. Their values are related through the volume scattering function that describes scattering as a function of angle into which light is deflected. While these sensors play an increasing role in observing in water processes, they also provide a technological foundation for a new generation of

sensors that extend IOP capabilities. These new sensors hold the ability to determine absorption coefficients, to determine coefficients as a function of wavelength, and to characterize the volume scattering function at more than one angle. These improvements not only allow more complete characterization of natural waters but also provide a tangible means of relating remotely sensed data from air and space to in-water processes.

See also

Optical Particle Characterization. Radiative Transfer in the Ocean. Satellite Remote Sensing Microwave Scatterometers. Turbulence Sensors.

Further Reading

Bogucki DJ, Domaradzki JA, Stramski D, and Zaneveld JRV (1998) Comparison of near-forward light scattering on oceanic turbulence and particles. *Applied Optics* 37: 4669–4677.

Bricaud A, Morel A, and Prieur L (1981) Absorption by dissolved organic matter of the sea (yellow substance) in the UV and visible domains. *Limnology and Oceanography* 26: 43–53.

Greenberg AE, Clescerl LS, and Eaton AD (eds.) (1992) *Standard Methods for the Examination of Water and Wastewater* 18th edn. Washington, DC: American Public Health Association, AWWA, WEF.

Jerlov NG (1976) *Marine Optics*. Amsterdam: Elsevier.

Kirk JTO (1994) *Light and Photosynthesis in Aquatic Ecosystems*. Cambridge: Cambridge University Press.

Mobley CD (1994) *Light and Water: Radiative Transfer in Natural Waters*. New York: Academic Press.

Pegau WS, Paulson CA, and Zaneveld JRV (1996) Optical measurements of frazil concentration. *Cold Regions Science and Technology* 24: 341–353.

Petzold TJ (1972) *Volume Scattering Functions for Selected Ocean Waters*. Reference Publication 72–28. La Jolla, CA: Scripps Institute of Oceanography.

Tyler JE, Austin RW, and Petzold TJ (1974) Beam transmissometers for oceanographic measurements. In: Gibbs RJ (ed.) *Suspended Solids in Water*. New York: Plenum Press.

Zaneveld JRV, Bartz R, and Kitchen JC (1990) A reflective-tube absorption meter. *Ocean Optics X, Proceeding of the Society for Photo-Optical Instrumentation and Engineering* 1302: 124–136.

SENSORS: CHEMICAL

WET CHEMICAL ANALYZERS

A. R. J. David, Bere Alston, Devon, UK

Introduction

Since the early 1960s there has been a requirement for seawater laboratories to carry out increasing numbers of routine analyses, many of which were performed by traditional manual methods. The performance of standard manual methods was generally variable due to human error and the efficiency was poor. Method automation has since enabled increased numbers of samples to be analyzed with improved efficiency and reduced the risk of human error. Air-segmented continuous flow analyzers (CFA) and flow injection analyzers (FIA) have handled the bulk of this automation. Instrument manufacturers have continued to improve both hardware and software over the years, which has resulted in better reliability and analytical performance.

Air-Segmented Continuous Flow Analyzers

These instruments are in widespread use for the determination of nutrient concentrations in natural waters. The technique is based on the fundamental principles developed in 1957 and converts a series of discrete samples into a continuous flowing carrier stream by a pumping system. Reagents are added by continuous pumping and merging of the sample carrier and reagent streams. The sample carrier stream is segmented with air before reagent addition, which typically allows between 20 and 80 samples to be processed in an hour. The insertion of standards in the sample carrier stream provides regular datum points during a particular analysis. There is usually no problem with distinguishing between the samples at the detection stage as the regular timing between stages is controlled. However, unless precautions are taken to prevent carryover, interaction can occur in a continuous system causing loss in discrimination between successive samples at the detection stage.

Figure 1 shows a typical air-segmented CFA manifold arrangement for total oxidized nitrogen (TON) determinations. It can be seen that the manifold is relatively complex, with eight separate streams pumped at fast flow rates. The sample stream is air-segmented prior to merging with the ammonium chloride ($10 \ g \ l^{-1}$) carrier stream and the bubbles are removed by a debubbler before entering the cadmium wire reduction coil. Air-segmentation is then re-introduced into the sample stream before merging with the separate sulfanilamide and N1NED streams. A second debubbler finally removes the bubbles before entering the flow cell.

Sophisticated laboratory CFA systems are used in shipborne laboratories for the routine determination of nitrate, nitrite, silicate, phosphate, and ammonia

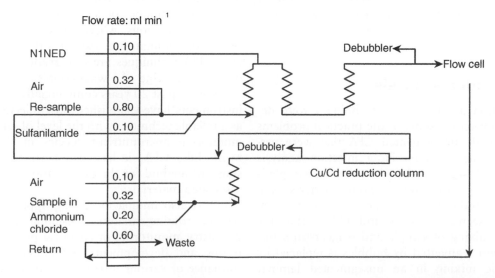

Figure 1 Schematic diagram of a typical air-segmented continuous flow analyzer manifold (for total oxidized nitrogen).

in sea water simultaneously. Although the technology is now very mature and the instruments perform extremely well in true laboratory conditions, the same cannot be said for shipborne applications. The main drawback with air-segmented systems is the need for a reproducible bubble pattern, which can be difficult to achieve when at sea in rough conditions. The air bubbles are compressible and therefore will create pulsations in the system and for most detector designs the bubbles have to be removed to avoid flow problems within the cell.

Multichannel systems are complex and consequently require specialist knowledge of the system in order to achieve optimum performance in adverse conditions, which is not always possible when operating a watch system on a research cruise. However, these instruments have been in use for many years with standard validated methods and generally they are the standard by which other techniques are judged.

In addition to the standard methods applied to CFA, various techniques have been developed over the years in order to eliminate some of the problems associated with air-segmented continuous flow techniques, e.g. the use of EDTA (ethylene diamine tetraacetic acid) to segment the carrier stream instead of air. The technique is also very flexible, allowing customized methods to be developed for a variety of additional chemical species and more sensitive methods for the normal nutrient species. For example, methods for the determination of nanomolar concentrations of nitrate and nitrite in sea water using an air-segmented CFA system have been developed.

Air-segmented CFA has been widely used at sea for the analysis of all major nutrient species and until the mid-1980s was the only technique that was available at a reasonable cost for automated analysis.

Flow Injection Analysis

Flow injection analysis (FIA) techniques were developed to overcome some of the practical problems associated with air-segmented CFA that were perceived by some workers. Flow injection analysis differs from air-segmented CFA in that the sample is injected directly into a moving liquid carrier stream without the addition of air. The main distinction between air-segmented CFA and FIA is that the continuous mixing of sample and reagents in a turbulent stream segmented by bubbles is replaced by the periodic mixing in an unsegmented laminar stream. The periodic mixing in a FIA system is achieved by injecting the sample (or reagent in the case of reverse flow injection) into a liquid carrier stream flowing to the detector, where the analyte forms a colored species in the reaction zone, which contains both sample and reagent. The reaction does not have to go to completion, and as the flow is incompressible, the extent of the reaction will be similar in all samples. This gives significant advantages over air-segmented CFA, including faster analysis rates and less complicated equipment, which has resulted in FIA being readily adapted to the analysis of seawater nutrients. In the late 1970s, methods were described for the simultaneous determination of nitrite and nitrate by FIA. This allowed up to 30 samples per hour to be analyzed with a relative precision of 1% in the range 0–0.05 μM for Nitrite-N and 0–0.1 μM for Nitrate-N. Similar work was carried out later that year, where up to 90 samples were analyzed with a relative precision of 0.5 and 1.5% for nitrite and nitrate respectively; in the range of 0.1–0.5 mg l^{-1} for Nitrite-N and 1–5 mg l^{-1} for Nitrate-N. Further developments of this method, which utilized copperized cadmium wire in a pre-valve in-valve reduction technique, allowed synchronous determinations of nitrite and (nitrite + nitrate) using one manifold and detector. In the early 1980s, methods were described which were modifications to previous methods that utilized reverse flow injection analysis, whereby the sample was the carrier stream and reagents were injected into it. This method gave a limit of detection (LOD) of 0.1 μM and allowed up to 70 samples per hour to be analyzed with a relative precision of 1%. FIA is finding increased use in the water industry, where laboratories that had previously used air-segmented CFA have introduced FIA to complement their working practices. FIA is the preferred technique for small batch sizes and low-level concentrations, where speed of analysis is essential to eliminate the risk of airborne contamination. FIA techniques are also readily adaptable to online monitoring of a watercourse.

Over the past 20 years commercially available FIA instruments have established flow injection analysis as a reliable technique with the level of sensitivity for monitoring micronutrient species in the environment. Like CFA, it has also been accepted as a standard method for the examination of waters and associated materials.

FIA is also finding increasing applications in research, routine analysis, teaching of analytical chemistry, monitoring of chemical processes, sensor testing and development, and enhancing the performance of various instruments. It is also used for the measurement of diffusion coefficients, reaction rates, stability constants, composition of complexes and extraction constants and solubility products. The versatility of FIA has allowed the technique to be adapted to different detection systems such as electrochemistry, molecular spectroscopy, and atomic spectroscopy, using numerous manifold configurations. FIA systems can also be designed to dilute or to preconcentrate the analyte; to perform separations based on solvent extraction, ion exchange, gas diffusion or dialysis; and to prepare unstable reagents *in situ*.

Figure 2 shows a typical FIA manifold arrangement for the determination of total oxidized nitrogen (TON).

The basic principle of FIA is the injection of a liquid sample into a moving, unsegmented carrier stream of a suitable liquid. The injected sample forms a zone, which is transported towards a detector that continuously records the absorbance, electrode potential or other parameter as the sample passes through the detector.

Optimization and design of the flow channels to achieve maximum sampling frequency, best reagent and sample economies, and proper exploitation of the chemistries is possible through understanding of the physical and chemical processes taking place during the movement of the fluids through the FIA manifold.

The simplest FIA analyzer, shown schematically in **Figure 3**, consists of a pump, which is used to propel the carrier stream through a narrow tube; an injection port, by means of which a well-defined volume of a sample solution is injected into the carrier stream in a reproducible manner; and a microreactor in which the sample zone reacts with the components of the sample stream, forming a species which is sensed by a flow-through detector and recorded.

The height, width, and areas of a typical peak output from a simple FIA system are all related to the concentration of the analyte. The time span between the sample injection and the peak height is the residence time during which the chemical reaction takes place. With rapid response times, typically in the range of 5–20 s, a sampling frequency of two samples per minute can be achieved. The injected sample volumes may be between 1 and 300 μl, which in turn requires typically between 0.5 and 5 ml of reagent per sampling cycle. This makes FIA a simple, automated micro-chemical technique, which is capable of a high sampling rate and a low sample and reagent consumption.

FIA is based on the combination of three principles: sample injection, controlled dispersion of the injected sample zone, and reproducible timing of its movement from the injection point to the detector. The chemical reactions take place whilst the sample material is dispersing within the reagent. The physical dispersion processes form the concentration gradient of the sample zone. The sample zone broadens as it moves downstream and changes from the original asymmetrical shape to a more symmetrical and eventually Gaussian form. For standard

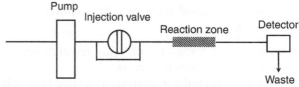

Figure 3 Schematic diagram of a simple FIA system.

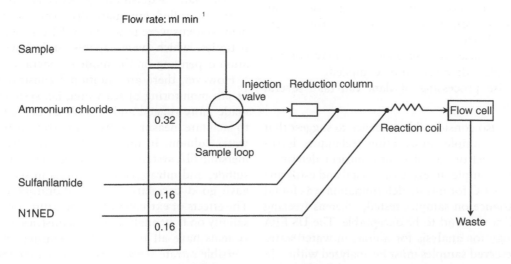

Figure 2 Schematic diagram showing a typical FI manifold (for total oxidized nitrogen).

conditions, the procedure is totally reproducible in that one injected sample behaves in the same way as all other subsequently injected samples.

In Situ Monitoring

As greater pressure is placed upon the environment through anthropogenic activities and climatic changes, effective monitoring and control of nutrient enrichment is vital to protect what is becoming a very delicately balanced marine environment. For example, quantitative knowledge of nutrients and primary production is essential for investigating the ecology and biogeochemistry of aquatic ecosystems. Until recently, the only way to monitor nutrient levels has been to collect discrete samples or use research vessels with onboard laboratory facilities to steam through a particular area of study. Collected samples are either analyzed at the collection site if convenient or transported to a central laboratory to be analyzed later. Monitoring schemes such as these may not detect short-term changes such as algal bloom conditions, storm events, or point discharges between sampling events. In addition, weather conditions and the high cost and logistics of using research vessels for routine studies may not allow complete data sets to be compiled. Therefore, the current problems associated with compiling long-term or continuous data can be summarized as follows:

- High cost and logistics of using research vessels for long-term routine surveys.
- Samples collected for analysis at a later time rely on good sample preservation techniques. This can lead to erroneous results with no means of replacing samples.
- Existing methods require some degree of human input to perform tasks, which can introduce experimental error.
- Research cruise weather conditions may not permit complete data sets to be compiled.
- Post-cruise processing of data can take several months.

There is also considerable evidence to suggest that many of the sample preservation techniques introduce some level of variance into nutrient determinations. For example, freezing of coastal and estuarine water at $-10°C$ for nitrate determination was found to give variance on samples tested, whereas freezing at $-20°C$ was found to be acceptable. The US EPA methodology for analysis for anions in water states that unpreserved samples must be analyzed within 48 hours otherwise they can be preserved with sulfuric acid at $pH < 2$ for 28 days. However, this has been shown to have an effect on environmental samples where nitrite is converted to nitrate by further microbial activity, whereas pH 12 is thought to eliminate the conversion-causing bacteria. Consequently, significant errors may be introduced into preserved samples and in both cases the cost and logistics of continuously monitoring a particular environment would generally be prohibitive.

In Situ Instruments

Advances in analytical chemistry have made it feasible to perform a wide range of chemical determinations *in situ*. The development of field automated methods, e.g. flow injection, has been particularly important in this respect.

The application of *in situ* automated FIA techniques can produce low-cost, rapid analysis with high sampling frequency analytical systems that are simple and easily maintained. Early examples of FIA based field instruments successfully completed field trials on the River Frome in Dorset. *In situ* FIA techniques have also been developed for environmental monitoring of phosphate, ammonia, and aluminum, all using solid-state LED/photodiode detectors. These early systems were based on mains-powered microcomputers which are unsuitable for use in portable battery-powered systems. Since then, advances in microchip technology have resulted in the availability of specialized microcontroller devices for control and automation of a variety of everyday uses. This technology has been exploited in the development of *in situ* FIA monitoring systems for the analysis of nitrate and phosphate in natural waters. All functions required for field-based operation, i.e. control of peristaltic pumps, injection and switching valves, data acquisition, processing, and logging were controlled automatically. The early development systems were powered by 12 V sealed lead-acid batteries, which were capable of 2–3 week's operation depending on the mode of operation.

However, there are additional constraints for the *in situ* monitoring of sea water, i.e. systems must be made more rugged and capable of being submersed to facilitate measurements at a particular point in the water column. In the late 1980s, the use of a submersible FIA system was reported to monitor silicate, sulfide, and nitrate concentrations in sea water that gave good correlation with laboratory techniques. The effects of extremes of temperature, pressure, and salinity on flow analysis and chemistries used in these systems have all been studied. During 1996 a submersible nitrate sensor based on FIA techniques was successfully tested in estuarine and coastal waters, over complete tidal cycles, and to depths of 40 m. Commercially available submersible FIA instruments permit laboratory FIA methods to be used *in situ*.

Other commercially available wet chemistry field instruments, which utilize the same basic proven chemistries as the CFA and FIA instruments, are available. For example, some instruments utilize rugged microprocessor-controlled syringe systems and a unique design of manifold for the collection of the sample, reagent addition, and colorimetric determination of the resultant colored species.

Field instruments enable a wide range of chemical determinations to be performed *in situ* and as the technology matures greater use of these will be made in the years to come.

Further Reading

Crompton TR (1989) *Analysis of Seawater.* Sevenoaks: Butterworths.

David ARJ, McCormack T, Morris AW, and Worsfold PJ (1998) *Anal. Chim. Acta* 361: 63.

Grasshoff K, Ehrhardt M, and Kremling K (1976) *Methods of Seawater Analysis.* New York: Verlag Chemie.

HMSO (1981) *Oxidised Nitrogen in Waters: Methods of Examination of Waters and Associated Materials.* London: HMSO.

HMSO (1988) *Discrete and Air Segmented Automated Methods of Analysis including Robots, An Essay Review,* 2nd edn: *Methods for the Examination of Waters and Associated Materials.* London: HMSO.

HMSO (1990) *Flow Injection Analysis, An Essay Review and Analytical Methods: Methods for the Examination of Waters and Associated Materials.* London: HMSO.

Karlberg B and Pacey GE (1989) *Flow Injection Analysis – A Practical Guide.* Amsterdam: Elsevier.

Ruzicka J and Hansen EH (1988) *Flow Injection Analysis,* 2nd edn. Chichester: Wiley Interscience.

Strickland JDH and Parsons TRA (1972) *Practical Handbook of Seawater Analysis,* 2nd edn.

US EPA No. 353.2 (1979) *Methods for the Chemical Analysis of Water and Wastes.* Washington:

Valcarcel MD and Luque de Castro MD (1987) *Flow Injection Analysis – Principles and Applications.* Chichester: Ellis Horwood.

ACOUSTIC METHODS

ACOUSTIC MEASUREMENT OF NEAR-BED SEDIMENT TRANSPORT PROCESSES

P. D. Thorne and P. S. Bell, Proudman Oceanographic Laboratory, Liverpool, UK

Introduction: Sediments and Why Sound Is Used

Marine sediment systems are complex, frequently comprising mixtures of different particles with non-cohesive (sands) and cohesive (clays and muds) properties. The movement of sediments in coastal waters impacts on many marine processes. Through the actions of accretion, erosion, and transport, sediments define most of our coastline. Their deposition and resuspension by waves and tidal currents in estuarine and nearshore environments control seabed morphology. Fine sediments, which act as reservoirs for nutrients and contaminants and as regulators of light transmission through the water column, have significant impact on water chemistry and on primary production. Therefore, an improved understanding of sediment dynamics in coastal waters has relevance to a broad spectrum of marine science ranging from physical and chemical processes, to the complex biological and ecological structures supported by sedimentary environments. However, it is commonly acknowledged that our capability to describe the coupled system of the bed, the hydrodynamics, and the sediments themselves is still relatively primitive and often based on empiricism. Recent advances in observational technologies now allow sediment processes to be investigated with greater detail and precision than has previously been the case. The combined use of acoustics, laser and radar, both at large and small scales, is facilitating exciting measurement opportunities. The enhancement of computing capabilities also allows us to make use of more complex coupled sediment–hydrodynamic models, which, linked with the emerging observations, provide new openings for model development.

It is readily acknowledged by sedimentologists that the presently available commercial instrumentation does not satisfy a number of requirements for near-bed sediment transport processes studies. Here we focus on the development of acoustics to fulfill some of these needs. The question could be asked: Why use acoustics for such studies? Sediment transport can be thought of as dynamic interactions between: (1) the seabed morphology, (2) the sediment field, and (3) the hydrodynamics. These three components interrelate with each other in complex ways, being mutually interactive and interdependent as illustrated schematically as a triad in **Figure 1**. The objective therefore is to measure this interacting triad with sufficient resolution to study the dominate mechanisms. Acoustics uniquely offers the prospect of being able to nonintrusively provide profiles of the flow, the suspension field, and the bed topography. This exclusive combination of being able to measure all three components of the sediment triad, co-located and simultaneously, has been and is the driving force for applying acoustics to sediment transport processes.

The idea of using sound to study fundamental sediment processes in the marine environment is attractive, and, in concept, straightforward. A pulse of high-frequency sound, typically in the range 0.5–5 MHz in frequency, and centimetric in length, is transmitted downward from a directional sound source usually mounted a meter or two above the bed. As the pulse propagates down toward the bed, sediment in suspension backscatters a proportion of the sound and the bed generally returns a strong echo. The amplitude of the signal backscattered from the suspended sediments can be used to obtain vertical profiles of the suspended concentration and particle size. Utilizing the rate of change of phase of the backscattered signal provides profiles of the three orthogonal components of flow. The strong echo from the bed can be used to measure the bed forms.

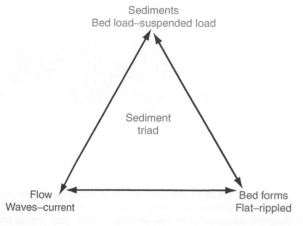

Figure 1 Illustration of the sediment processes triad and their interactions.

Acoustics therefore has the potentiality to provide profile measurements of near-bed sediment processes, with sufficient spatial and temporal resolution to allow turbulence and intrawave processes to be probed; this coupled with the bedform morphology observations provides sedimentologists and coastal engineers with an extremely powerful tool to advance understanding of sediment entrainment and transport. All of this is delivered with almost no influence on the processes being measured, because sound is the instrument of measurement.

Some Historical Background

For over two decades the vision of a number of people involved in studying small-scale sediment processes in the coastal zone has been to attempt to utilize the potential of acoustics to simultaneously and nonintrusively measure seabed morphology, suspended sediment particle size and concentration profiles, and profiles of the three components of flow, with the required resolution to observe the perceived dominant sediment processes. The capabilities to measure the three components have developed at different rates, and it is only in the past few years that the potentiality of an integrated acoustic approach for measuring the triad has become realizable. A schematic of the vision is shown in **Figure 2**.

The figure shows a visualization of the application of acoustics to sediment transport processes. 'A' is a multifrequency acoustic backscatter system (ABS), consisting, in this case, of three downward-looking narrow beam transceivers. The differential scattering characteristic of the suspended particles with frequency is used to obtain profiles of suspended sediment particle size and concentration profiles. 'B' is a three-axis coherent Doppler velocity profiler (CDVP) for measuring co-located profiles of the three orthogonal components of flow velocity; two horizontal and one vertical. It consists of an active narrow beam transceiver pointing vertically downward and two passive receivers having a wide beam width in the vertical and a narrow beam width in the horizontal. This system uses the rate of change of phase from the backscattered signal to obtain the three velocity components. 'C' is a pencil beam transceiver which rotates about a horizontal axis and functions as an acoustic ripple profiler (ARP). This is used to extract the bed echo and provide profiles of the bed morphology along a transect. These measurements are used to obtain, for example, ripple height and wavelength, and assess bed roughness. 'D' is a high-resolution acoustic bed (ripple) sector scanner (ARS) for imaging the local bed features. Although the ARS does not provide quantitative measurements of bed form height, it does provide the spatial distribution and this can be very useful when used in conjunction with the ARP. 'E' is a rapid backscatter ripple profiler system (BSARP) for measuring the instantaneous relationship between bed forms and the suspended sediments above it.

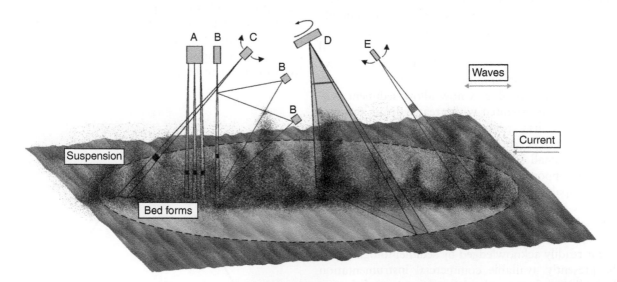

Figure 2 A vision of the application of acoustics to sediment transport processes. A, multifrequency acoustic backscatter for measuring suspended sediment particle size and concentration profiles. B, coherent Doppler velocity profiler for measuring the three orthogonal components of flow velocity. C, bed ripple profiler for measuring the bed morphology along a transect. D, high-resolution sector scanner for imaging the local bed features. E, backscatter scanning system for measuring the relationship between bed form morphology and suspended sediments.

What Can Be Measured?

The Bed

Whether the bed is rippled or flat has a profound influence on the mechanism of sediment entrainment into the water column. Steep ripples are associated with vortices lifting sediment well away from the bed, while for flat beds sediments primarily move in a confined thin layer within several grain diameters of the bed. Therefore knowing the form of the bed is a central component in understanding sediment transport processes. The development of the ARP and the ARS has had a significant impact on how we interpret sediment transport observations. These specifically designed systems typically either provide quantitative measurements of the evolution of a bed profile with time, the ARP, or generate an image of the local bed features over an area, the ARS.

Figure 3 shows data collected with a 2-MHz narrow pencil beam ARP in a marine setting. The figure shows the variability of a bed form profile, over nominally a 3-m transect, covering a 24-h period. Over this period the bed was subject to both tidal currents and waves and the figure shows the complex evolution of the bed with periods of ripples and less regular bed forms. The figure clearly shows the detailed quantitative measurements of the bed that can be obtained with the ARP.

Acoustic ripple scanners, ARSs, are based on sector-scanning technology, which has been specifically adapted for high-resolution images of bed form morphology. They typically have a frequency of around 1–2 MHz with beam widths of about 1° in the horizontal and 30° in the vertical. As the pulse is backscattered from the bed, the envelope of the signal is measured and usually displayed as image intensity.

An example of the data collected by an ARS is shown in Figure 4. As can be seen, this provides an aerial image of the bed, clearly showing the main bed features. The advantage of the ARS is the area coverage that is obtained, as opposed to a single line profile with the ARP; however, direct information on the height of features within the image cannot readily be extracted.

Ideally one would like to combine the two instruments and recently such systems have become available. This is essentially an ARP which also rotates horizontally through 180° and therefore allows a three-dimensional (3-D) measurements of the bed. The sonar gathers a single swath of data in the vertical plane and then rotates the transducer around the vertical axis and repeats the process until a circular area underneath the sonar has been scanned in a sequence of radial spokes. An example of data collected by a 3-D ARP operating at a frequency of 1.1 MHz is shown in Figure 5.

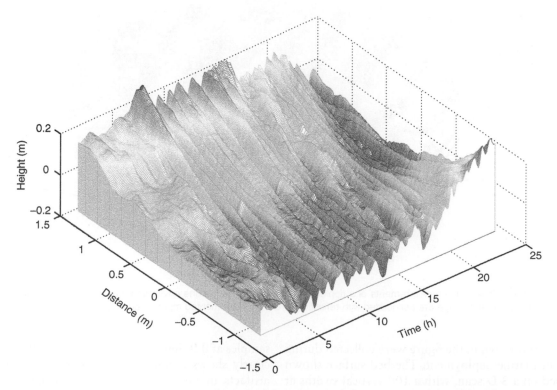

Figure 3 ARP measurements of the temporal evolution of a 3-m profile on a rippled bed, covering a 24-h period.

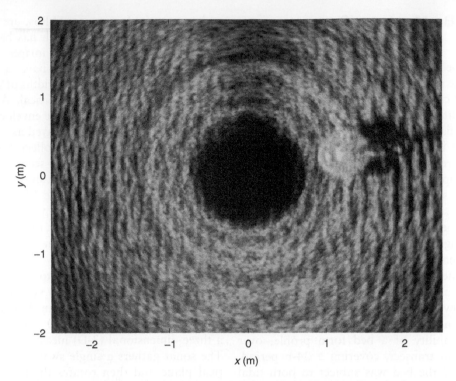

Figure 4 Image of ripples on the bed in a large-scale flume facility, the Delta Flume, collected using an acoustic ripple sector scanner, ARS. The 0.5-m-diameter circle at the center right of the image is one of the feet of the instrumented tripod used to collect the measurements, while the dark area at the center of the image is a blind spot directly beneath the sonar.

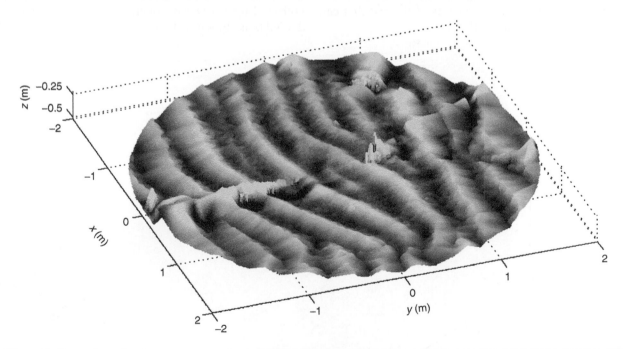

Figure 5 Three-dimensional measurements of a rippled bed collected in the marine environment using a 3-D ripple scanner, 3-D ARP. The artifacts in the image are associated with reflections from the instrumented tripod used to collect the measurements.

The data shown in the figure were collected during a recent marine deployment. The bed surface shown is based on a 3-D scan, with a 100 vertical swaths at 1.8° intervals, each of which comprised 200 acoustic samples at 0.9° intervals and spanning 180°. The plot clearly shows a rippled bed. There are one or two artifacts in the plot; these are associated with reflection from the rig on which the 3-D ARP was mounted. However, the figure plainly contains information on both the horizontal and vertical dimensions of the ripples and can therefore be used to precisely define quantitatively the features of the bed over an area. The 3-D ARP is a substantial advance on both the ARP and the ARS.

The Flow

The success of acoustic Doppler current profilers (ADCPs), which typically provide mean current profiles with decimeter spatial resolution, and more recently the acoustic Doppler velocimeter (ADV), which measures, subcentimetric, subsecond, three velocity components at a single height, has stimulated interest in using acoustics to measure near-bed velocity profiles. The objective is to use the same backscattered signal as used by the ABSs, but process the rate of change of phase of the signal (rather than the amplitude as used by ABSs) to obtain velocity profiles with comparable spatial and temporal resolution to ABSs. The phase technique is utilized in CDVPs and the phase approach has been the preferred method for obtaining high spatial and temporal resolution velocity profiles. An illustration of a three-axis CDVP is shown in **Figure 6**. A narrow-beam, downwardly pointing transceiver, Tz, transmits a pulse of sound. The scattered signal is picked up by Tz, and two passive receivers Rx and Ry which are orthogonal to each other and have a wide beam in the vertical and narrow in the horizontal.

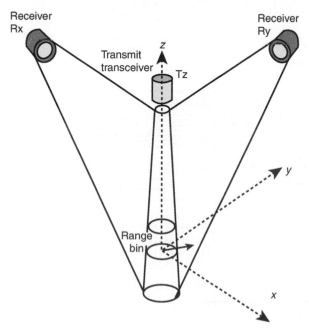

Figure 6 Schematic of the transducer arrangement for a three-axis CDVP.

To examine the capability of such a system, measurements from a three-axis CDVP have been compared with a commercially available ADV. The system had a spatial resolution of 0.04 m, operated over a range of 1.28 m, and provided 16-Hz velocity measurements of the vertical and two horizontal components of the flow. **Figure 7** shows a typical example of CDVP and ADV time series, power spectral density, and probability density function plots for **u**, the streamwise flow. The velocities presented in **Figure 7** show CDVP results that compare very favorably with the ADV measurements; having time series, spectra, and probability distributions in general agreement. There are differences in the spectrum; the CDVP spectra begins to depart from the ADV above about 4 Hz, with the CDVP measuring larger spectral components at the higher frequencies. This trend was common to all the records and is a limitation of the CDVP system used to obtain the data, rather than an intrinsic limitation to the technique. **Figure 8** illustrates the capability of the CDVP for flow visualization in the marine environment. The figure shows mean zeroed velocity vectors, **u–w**, **v–w**, and **u–v**, plotted over a 5-s time period, between 0.05 and 0.7 m above the bed. The length of the velocity vectors is indicated in the figure. A single-point measurement instrument such as an ADV can provide the time-varying velocity vectors at a single height above the bed; however, the spatial profiling which is achievable with the three-axis CDVP provides a capability to visualize structures in the flow. The structures seen in **Figure 8** are associated with combined turbulent and wave flows. This type of plot exemplifies the value of developing a three-axis CDVP with co-located measurement volumes, since it clearly illustrates the fine-scale temporal and spatial flow structures which can be measured in the near-bed flow regime. Linking such measurements with ABS profiles of particle size and concentration will provide a very powerful tool for studying near-bed fluxes and sediment transport processes.

The Suspended Sediments

Multifrequency acoustic backscattering, ABS, can be used to obtain profiles of mean particle size and concentration. The ABS is the only system available that profiles both parameters rapidly and simultaneously. Also the bed echo references the profile to the local bed position. This is important because all sediment transport formulas use the bed as the reference point and predict profiles of suspension parameters relative to the bed location. Examples of the results that can be obtained are shown

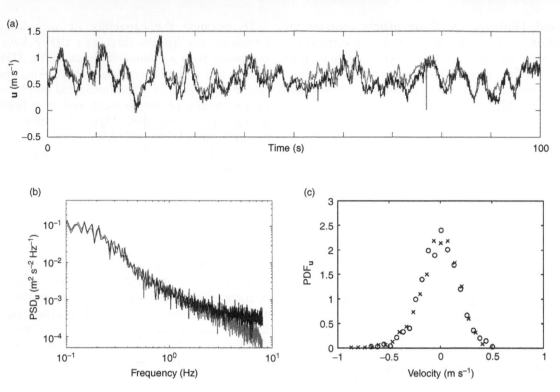

Figure 7 Comparison of the streamwise flow, **u**, measured by an ADV (red) and a CDVP (black). (a) The velocities measured at 16 Hz over 100 s. (b) The power spectra of the zero-mean velocities. (c) The probability density functions of the zero-mean velocities for the ADV (open circles) and the CDVP (crosses).

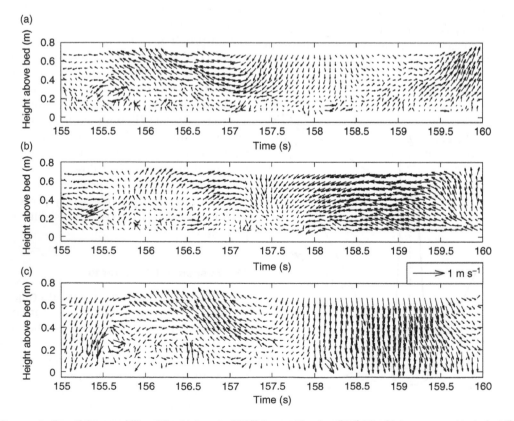

Figure 8 Demonstration of the capability of the triple-axis CDVP to provide visualizations of intrawave and turbulent flow. Plots in (a)–(c) show a time series over a 5 s period of the zero-mean velocities displayed as vectors **u**–**w**, **v**–**w**, and **u**–**v**, respectively.

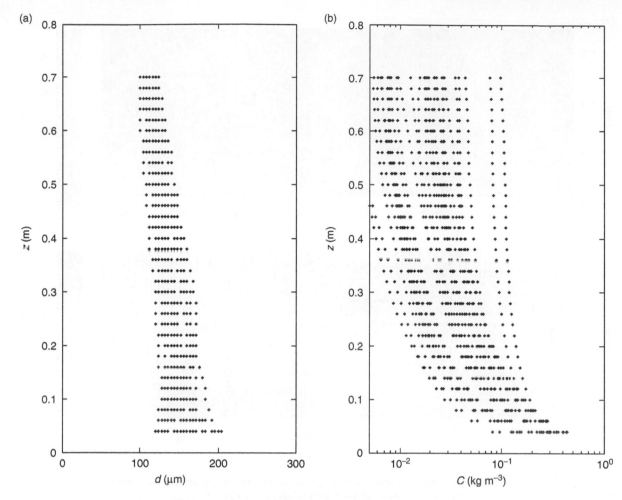

Figure 9 Profiles of suspended sediments: (a) particle size and (b) concentration.

in **Figures 9** and **10**. These observations were collected off Santa Cruz Pier, California, as a seasonal storm passed through the area over the period of a couple of days. **Figure 9** shows profiles of the suspended sediment particle size and concentration. The profiles of particle size are seen to be relatively consistent with a mean diameter of c. 180 μm near the bed and reducing to c. 120 μm at 0.7 m above the bed. The variability in size is relatively limited and due to changing bed and hydrodynamic conditions as the storm passed by.

In **Figure 9(b)** the suspended concentrations are comparable in their form, although they have absolute values that vary by greater than an order of magnitude. This variation in suspended concentration is associated with the changing conditions as the storm passed through the observational area. It is interesting to note that over the period even though there is a large variation in concentration, the particle size remains nominally consistent. **Figure 10** shows the temporal variation in particle size and

concentration with height above the bed. It can be clearly seen that some of the periods of increased particle size are associated with substantial suspended sediment events, as one might expect, however, there are one or two events where the correlation is not as clear. **Figures 9** and **10** clearly illustrate the capability of ABS to simultaneously measure profiles of concentration and particle size and the combination of both significantly adds to the assessment and development of sediment transport formulas.

A Case Study of Waves over a Rippled Bed

Here the use of acoustics is illustrated by application to a specific experimental study. Over large areas of the continental shelf outside the surf zone, sandy seabeds are covered with wave-formed ripples. If the ripples are steep, the entrainment of sediments into the water column, due to the waves, is considered

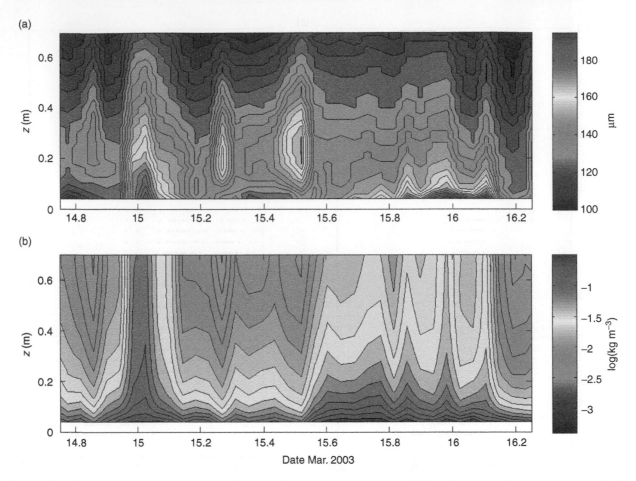

Figure 10 Measurements of the temporal variations with height above the bed, of: (a) particle size with the color bar scaled in microns and (b) logarithmic concentration with the color bar scaled relative to $1.0 \, \text{kg m}^{-3}$.

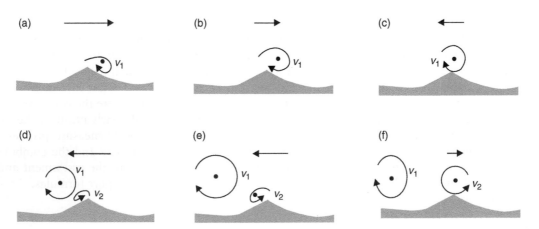

Figure 11 A schematic of vortex sediment entrainment by waves over a steeply rippled bed. The arrows show the direction and relative magnitude of the near-bed wave velocity. v_1 and v_2 are the lee slope-generated vortices.

to be primarily associated with the generation of vortices. This process is illustrated in **Figure 11**. As shown in **Figures 11(a)** and **11(b)**, a spinning parcel of sediment-laden water, v_1, is formed on the leeside of the ripple at the peak positive velocity in the wave cycle. This sediment-rich vortex is then thrown up into the water column at around flow reversal (**Figures 11(c)** and **11(d)**), carrying sediment well away from the bed and allowing it to be transported by the flow. At the same time, a sediment-rich vortex,

v_2, is being formed on the opposite side of the ripple due to the reversed flow. As shown in **Figures 11(d)–11(f)**, v_2 grows, entrains sediment, becomes detached, and moves over the crest at the next flow reversal carrying sediments into suspension. The main feature of the vortex mechanism is that sediment is carried up into the water column twice per wave cycle at flow reversal. This mechanism is completely different to the flat bed case where maximum near-bed concentration is at about the time of maximum flow velocity.

To study this fundamental process of sediment entrainment, experiments were conducted in one of the world's largest man-made channels, specifically constructed for such sediment transport studies, the Delta Flume; this is located at the De Voorst Laboratory of the Delft Hydraulics in the north of the Netherlands. The flume is shown in **Figure 12(a)**; it is 230 m in length, 5 m in width, and 7 m in depth, and it allows waves and sediment transport to be studied at full scale. A large wave generator at one end of the flume produced waves that propagated along the flume, over a sandy bed, and dissipated on a beach at the opposite end. The bed was comprised of coarse sand, with a mean grain diameter by weight of 330 μm, and this was located approximately halfway along the flume in a layer of thickness 0.5 m and length 30 m. In order to make the acoustic and other auxiliary measurements, an instrumented tripod platform was developed and is shown in **Figure 12(b)**. The tripod, STABLE II (Sediment Transport and Boundary Layer Equipment II), used an ABS to measure profiles of particle size and concentration, a pencil beam ARP to measure the bed forms, and, in this case, electromagnetic current meters (ECMs) to measure the horizontal and vertical flow components. **Figure 12(c)** shows a wave propagating along the flume with STABLE II submerged in water of depth 4.5 m, typical of coastal zone conditions.

To investigate and then model the vortex entrainment process it was necessary to establish at the outset whether or not the surface waves were generating ripples on the bed in the Delta Flume. Using an ARP a 3-m transect of the bed was measured over time. The results of the observations over a 90-min recording period are shown in **Figure 13**. Clearly ripples were formed on the bed and the ripples were mobile. To obtain flow separation and hence vortex formation requires a ripple steepness (ripple height/

(a)   (b)   (c)

Figure 12 (a) Photograph of the Delta Flume showing the sand bed at approximately midway along the flume and the wave generator at the far end of the flume. (b) The instrumented tripod platform, STABLE II, used to make the measurements. (c) A surface wave propagating along the flume.

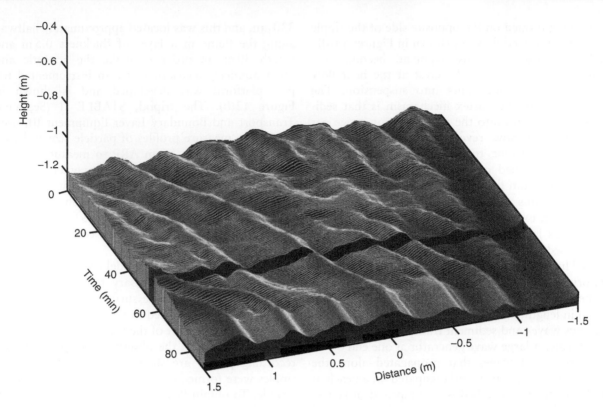

Figure 13   Acoustic ripple profiler measurements of a transect of the bed, over time, in the Delta Flume.

ripple wavelength) of the order of 0.1 or greater; an analysis of the observations showed that this was indeed the case.

Using the ABS, some of the most detailed full scale measurements of sediment transport over a rippled bed under waves were captured. These measurements from the Delta Flume are shown in **Figure 14**. The images shown were constructed over a 20-min period as a ripple passed beneath the ABS. The suspended concentrations over a ripple, at the same velocity instants during the wave cycle, were combined to generate a sequence of images of the concentration over the ripple with the phase of the wave. Four images from the sequence have been shown to illustrate the measured vortex entrainment. The length and direction of the arrows in the figure give the magnitude and direction of the wave velocity, respectively. Comparison of **Figure 14** with **Figure 11** shows substantial similarities. In **Figure 14(a)**, there can be observed the development of a high-concentration event at high flow velocity above the lee slope of the ripple, v_1. In **Figure 14(b)**, as the flow reduced in strength, the near-bed sediment-laden parcel of fluid travels up the leeside of the ripple toward the crest. As the flow reverses, this sediment-laden fluid parcel, v_1, travels over the crest and expands. As the reverse flow increases in strength (**Figure 14(d)**), the parcel v_1 begins to lift

away from the bed and a new sediment-laden lee vortex, v_2, is initiated on the lee slope of the ripple.

In order to capture the essential features of these data within a relatively simple, and hence practical, 1-DV (one-dimensional in the vertical) model, the data has first been horizontally averaged over one ripple wavelength at each phase instant during the wave cycle. The resulting pattern of sediment suspension contours is shown in the central panel of **Figure 15**, while the upper panel shows the oscillating velocity field measured at a height of 0.3 m above the bed. The concentration contours shown here are relative to the ripple crest level, the mean (undisturbed) bed level being at height $z = 0$. The measured concentration contours presented in **Figure 15** show two high concentration peaks near the bed that propagate rapidly upward through a layer of thickness corresponding to several ripple heights. The first, and the strongest, of these peaks occurs slightly ahead of flow reversal, while the second, weaker and more dispersed peak, is centered on flow reversal. The difference in the strengths of the two peaks reflects the greater positive velocity that can be seen to occur beneath the wave crest (time $= 0$ s) than beneath the wave trough (time $= 2.5$ s). Between the two concentration peaks the sediment settles rapidly to the bed. Maybe rather unexpectedly this settling effect occurs at the times of strong forward and backward velocity at measurement

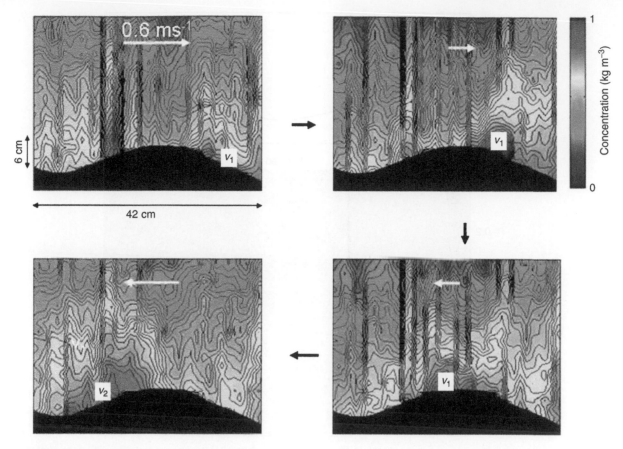

Figure 14   Acoustic imaging of suspended sand entrainment over a rippled bed due to waves, at four phases of the wave velocity. The length of the white arrow in each plot gives the magnitude and direction of the near-bed wave velocity.

levels well above the bed. The underlying mechanism of sediment entrainment by vortices shed at or near flow reversal is clearly evident in the spatially averaged measurements shown in **Figure 15**.

Any conventional 'flat rough bed' model that attempts to represent the above sequence of events in the suspension layer runs into immediate and severe difficulties, since such models predict maximum near-bed concentration at about the time of maximum flow velocity, and not at flow reversal. Here therefore, for the first time in a 1-DV model, it has been attempted to capture these effects realistically through the use of a strongly time-varying eddy viscosity that represents the timing and strength of the upward mixing events due to vortex shedding. The model initially predicts the size of the wave-induced ripples and the size of the grains found in suspension, and then goes on to solve numerically the equations governing the upward diffusion and downward settling of the suspended sediment. The resulting concentration contours in the present case are shown on the lower panel of **Figure 15**. The essential two-peak structure of the eddy shedding process can be seen to be represented rather well, with the initial

concentration peak being dominant. The decay rate of the concentration peaks as they go upward is also represented quite well, though a phase lag develops with height that is not seen to the same extent in the data. Essentially, the detailed acoustic observations of sediment entrainment under waves over ripples of moderate steepness have begun to establish a new type of 1-DV modeling, thereby allowing the model to go on to be used for practical prediction purposes in the rippled regime, which is the bed form regime of most importance over wide offshore areas in the coastal seas.

Discussion and Conclusions

The aim of this article has been to illustrate the application of acoustics in the study of near-bed sediment processes. It was not to detail the theoretical aspects of the work, which can be found elsewhere. To this end, measurements of bed forms, the hydrodynamics, and the movement of sediments have been used. These results show that acoustics is progressively approaching the stage where it

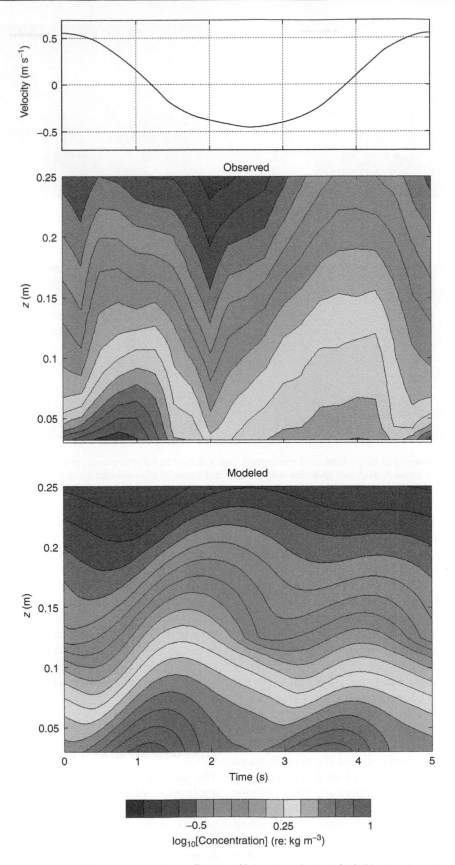

Figure 15 Measurement and modeling of suspended sediments with height z above a rippled bed under a 5-s period wave.

can measure nonintrusively, co-located and simultaneously, with high temporal and spatial resolution, all three components of the interacting sediment triad. **Figure 16** shows the instrumentation used in a further recent deployment in the Delta Flume. This shows the convergence of the instrumentation and also its use in conjunctions with instruments such as laser *in situ* scattering transmissometry (LISST).

Although substantial advances have been made in the past two decades, the application of acoustics to sediment transport processes is still in an ongoing developmental phase and there are limitations and shortcomings that need to be overcome, and further applications explored. Although there have been few reports to date on data collected using 3-D ARP, such systems are now becoming available. The 3-D ARP is a substantial development over the single-line ARP and the ARS, and should make a considerable contribution to the measurement and understanding of the formation and development of bed forms. There have been a number of reports on single-axis CDVP; however, it is the three-axis CDVP which is the way ahead. Again, these instruments are coming online, though they are still very much a research tool. However, the concept has now essentially been

proven and its use with a vengeance in sediment studies will begin to make an important impact in the next couple of years. It was with the concept of using acoustics to measure suspended sediment concentration that its application to the other components of the small-scale sedimentation processes triad followed. To date the use of sound to measure suspended sediment concentration and particle size has been successful when systems have been deployed over nominally homogeneous sandy beds. However, all who use acoustics recognize that the marine sedimentary environment is frequently much more complicated, and suspensions of cohesive sediments and combined cohesive and noncohesive sediments are common. To employ acoustics quantitatively in mixed and cohesive environments requires the development of a description of the scattering properties of suspensions of cohesive sediments and sediment mixtures. This would be interesting and very valuable work, and should significantly extend the deployment regime over which acoustic backscattering can be employed quantitatively.

In conclusion, the objective of this article has been to describe the role of acoustics, in near-bed sediment transport studies. It is clearly acknowledged

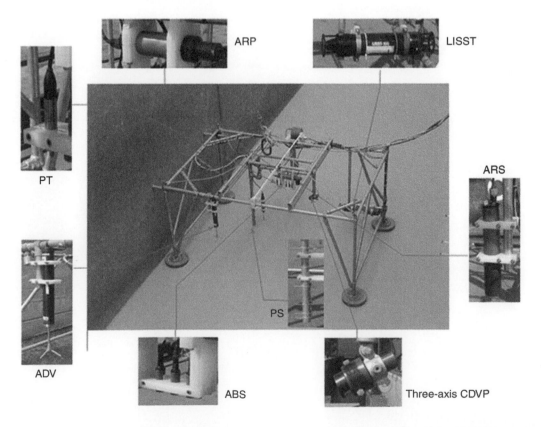

Figure 16 The instrumentation used in a recent Delta Flume experiment. LISST, laser *in situ* scattering transmissometry, PS, pumped samples, PT, pressure transducer.

that acoustics is one of a number of technologies advancing our capabilities to probe sediment processes. However, its nonintrusive profiling ability, coupled with its capability to measure all three components of the sediment dynamics triad, make it a unique and very powerful tool for studying the fundamental mechanisms of sediment transport.

See also

Acoustic Scattering by Marine Organisms. Acoustics in Marine Sediments.

Further Reading

Crawford AM and Hay AE (1993) Determining suspended sand size and concentration from multifrequency acoustic backscatter. *Journal of the Acoustical Society of America* 94(6): 3312–3324.

Davies AG and Thorne PD (2005) Modelling and measurement of sediment transport by waves in the vortex ripple regime. *Journal of Geophysical Research* 110: C05017 (doi:1029/2004JC002468).

Hay AE and Mudge T (2005) Principal bed states during SandyDuck97: Occurrence, spectral anisotropy, and the bed storm cycle. *Journal of Geophysical Research* 110: C03013 (doi:10.1029/2004JC002451).

Thorne PD and Hanes DM (2002) A review of acoustic measurements of small-scale sediment processes. *Continental Shelf Research* 22: 603–632.

Vincent CE and Hanes DM (2002) The accumulation and decay of near-bed suspended sand concentration due to waves and wave groups. *Continental Shelf Research* 22: 1987–2000.

Zedel L and Hay AE (1999) A coherent Doppler profiler for high resolution particle velocimetry in the ocean: Laboratory measurements of turbulence and particle flux. *Journal of Atmosphere and Ocean Technology* 16: 1102–1117.

Relevant Websites

http://www.aquatecgroup.com
– Aquatec Group Ltd.
http://www.marine-electronics.co.uk
– Marine Electronics Ltd.
http://www.pol.ac.uk
– POL Research, Proudman Oceanographic Laboratory.

ACOUSTIC NOISE

I. Dyer, Marblehead, MA, USA

Introduction

Some ocean scientists consider ambient noise to be a fairly simple and well-behaved property of the ocean. Ambient noise, after all, is often reported and summarized in highly averaged form, its naturally large variance mostly unstated. Other ocean scientists consider the variational complexity of ambient noise a richly colored portrait carrying images of basic ocean processes, including the physics of various noise sources and the acoustics of multiple noise propagation paths. Space and incomplete knowledge precludes a description here that can fully satisfy all ocean scientists or technologists. Instead, the objective is to summarize those aspects of ocean ambient noise that convey the more important recent research results and the more significant remaining research questions.

A 1962 summary of ambient noise measurements in the ocean (see **Figure 1**) is still useful today, at least to classify the various noise sources and their average levels and smooth frequency spectra. Prevailing noises (those observed almost always) are caused by wave–wave interactions at the sea surface, by distributed seismic activity in the earth, by atmospheric or oceanic turbulence, by distant shipping, by wind-induced sea surface agitation, and by thermally induced molecular agitation. According to Wenz, wave–wave interaction effects, seismic background, and/or turbulence dominates the noise at VLF (very low frequency band: $1 < f < 20\,\text{Hz}$), with power spectral density of the pressure field $S(f) \propto f^{-4}$. Distant shipping noise dominates at LF (low frequency band: $20 < f < 200\,\text{Hz}$), has a broad spectral peak around $50\,\text{Hz}$, and falls off sharply for $f > 200\,\text{Hz}$ as f^{-6}. At MF (midfrequency band: $200\,\text{Hz} < f < 50\,\text{kHz}$), noise caused by sea surface agitation typically dominates, with a broad peak within $200\,\text{Hz} < f < 2\,\text{kHz}$ and, beyond $f \approx 2\,\text{kHz}$, with $S(f) \propto f^{-1.7}$. Finally, molecular agitation typically dominates the noise at HF (high frequency band: $f > 100\,\text{kHz}$), with $S(f) \propto f^2$.

Other noise sources are classified as temporally intermittent or spatially discrete, rather than prevailing, and can often dominate. These include sounds from marine earthquakes, from marine animals, from nearby ships or other nearby commercial activities in the ocean, from rain/hail/snow striking the sea surface, and from fractures of ice in the north or south polar oceans. With such a large number of prevailing and other noise sources, the band designations given in the previous paragraph are unlikely to be associated unequivocally with just one noise source or, for that matter, adopted fully by most ambient noise ·researchers or practitioners. They are of use, however, to help present the material to follow.

The spectral summaries used in this Introduction are based on Wenz, and although still useful, modifications and additions are needed in the light of new knowledge. Urick published an excellent summary of ambient noise data acquired in various measurement programs through about 1980. Practitioners commonly use these data, plus the Wenz results, for prediction. Nevertheless, basic understanding of many ambient noise mechanisms through about 1980 was meager and, indeed, some suggested mechanisms were considered speculative. Fortunately, mechanisms for prevailing ambient noises have received considerable research attention since then, particularly from 1985 or so. The other noises have also been researched, in general to a lesser degree. Two volumes edited by Kerman and one by Buckingham and Potter are conference proceedings of recent ambient noise research, and are extraordinary seminal contributions to the understanding of ambient noise mechanisms in the ocean. The continuing flow of research results in archival journals and books, and the aforementioned volumes, provide important modifications and additions to the classical summary of ambient noise by Wenz. In what follows the more important new knowledge, or lack thereof, is summarized.

ULF Band: Wave–Wave Interaction Noise

Measurements within $0.1 < f < 2\,\text{Hz}$, which has come to be called the ultralow frequency band (ULF), extended the Wenzian picture one decade lower in frequency[1], and showed ULF noise to be a function of wind speed. The data (**Figure 2**), have a strong peak f_0 located between about 0.2 and 0.7 Hz, with

---

[1] To accommodate the ULF band, the band scheme described in the Introduction is redefined to be VLF2–20 Hz.

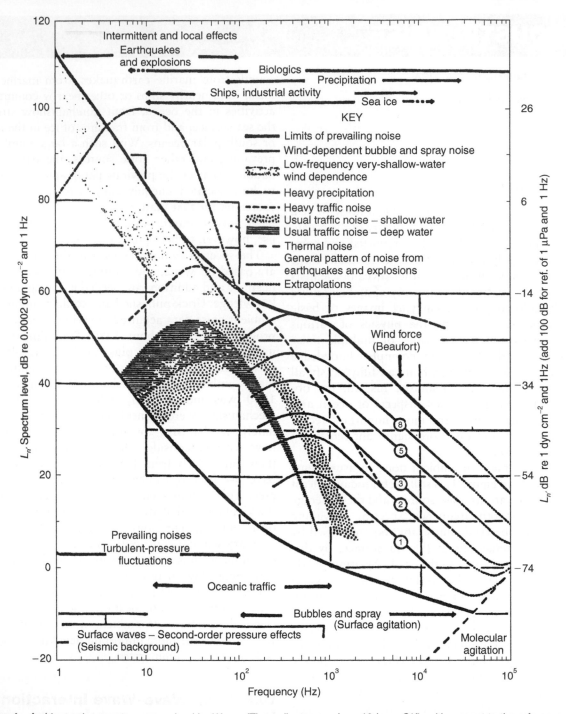

Figure 1 Ambient noise spectra summarized by Wenz. (The ordinates are $L_n = 10\ log_{10}\ S(f)$, with respect to the reference value. Add 100 dB to the right-hand scale to obtain L_n in dB re 1 μPa and 1 Hz.) The Beaufort Force translates to wind speeds, in ms^{-1}, as follows: 1, 0.5–1.5; 2, 2–3; 3, 3.5–5; 5, 8.5–10.5; 8, 17–20. (Reproduced from Wenz, 1962.)

$S(f)$ at higher f proportional to about $f^{-3}-f^{-5}$, dependent upon wind speed. A long history of measurements, as well as theoretical surface wave interaction studies, presaged this result. The appearance of systematic data such as in **Figure 2** apparently sparked even more research efforts that ultimately confirmed the basic aspects of ULF noise.

Pressure Spectral Density

Wave–wave interaction noise is caused by opposing wind-driven surface gravity waves, each at frequency f_w, that to second-order create a pressure field in the water at $f = 2f_w$. (Orders higher than the second have been shown to be negligible.) In its simplest

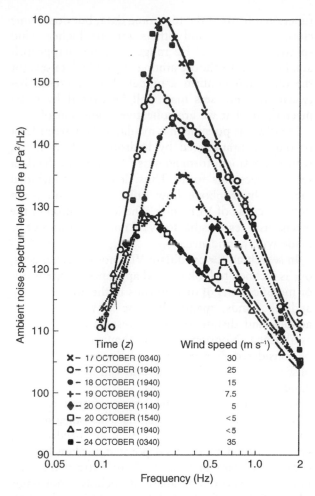

Figure 2 Ambient noise pressure derived from vertical displacement (seismometer) data at the water/bottom interface (depth about 100 m). (Reproduced from Kibblewhite and Evans, 1985.)

form, the spectral density S of the pressure field is

$$S(f,z) = (\pi^3 \rho^2 g^2/2c^2)[S_w(f/2)]^2 f^3 \Phi(f/2) T_p(f,z) \quad [1]$$

where ρ is water density, c is water sound speed, g is acceleration of gravity, S_w is the spectral density of the surface wave elevation, Φ is an integral over all azimuth angles ϕ of the product of the normalized azimuthal directivity of the opposing surface waves (for ϕ and $\phi + \pi$), and T_p is a function normalized by its value at depth $z = 0$ that relates the pressure spectral density at the surface to that at z, which importantly includes acoustic coupling to the sea bottom. For simplicity, eqn [1] includes contributions to the noise field only for horizontal phase speeds given by $c_x \geq c$. Because of its exponential decay in z, the remaining regime $c_x < c$ is important mostly for depths approaching $z = 0$, especially for lower f, and is included in fuller analyses.

The noise spectral density $S(f)$ is proportional to $[S_w(f/2)]^2$, the surface elevation density squared and shifted in frequency. Since the frequency dependence of the last three terms in eqn [1] is relatively weak around the peak frequency f_{wo} of S_w, the peak of S is essentially $f_o \approx 2f_{wo}$. Both f_{wo} and S_w are functions of wind speed U, as affected by other sea conditions (fetch, sea age, etc.), and similarly lead to the dependence of f_o and S on U.

Microseism Spectral Density

Displacement, rather than pressure, is often measured on the seafloor. Termed microseism noise, it can be obtained from eqn [1] with the substitution of a modified transfer function $T_d(f,z)$ for $T_p(f,z)$, each in general given for all z, to yield the displacement spectral density $S_d(f,z)$ rather than $S(f,z)$. These transfer functions incorporate the acoustics of the medium, including the seismoacoustics of the bottom. It is no surprise, therefore, that noise pressure or displacement data may show subsidiary peaks in the frequency domain, additional to the major peak at f_o, associated with acoustic modes of the oceanic waveguide.

With modifications as outlined in the foregoing and as detailed in the literature, and with site-specific seismic–acoustic bottom properties and time- and site-specific sea surface elevations, data such as in **Figure 2** are remarkably well predicted with use of the theory symbolized by eqn [1]. The theory may also be extended to cover interaction of swells with wind waves, with use of $S_w S_s$ instead of S_w^2, where S_s is the swell spectral density, and with Φ similarly including the swell azimuthal directivity.

VLF Band: Atmospheric Turbulence Forcing as a Noise Source

The heading of this section is in reality a question. In the VLF band (2–20 Hz), and to somewhat higher frequencies in seas with very low shipping density, noise measurements do not match direct extrapolation of wave–wave interaction data as in **Figure 2**. Other mechanisms may become important in this band, and one has attracted considerable attention, namely turbulent pressure fluctuations in the atmosphere that drive the ocean surface. Is the latter the responsible mechanism? And if not, what is? A robust answer is not yet available.

In one of the more extensive data summaries dating from the 1960s, Crouch and Burt showed that in the absence of shipping noise, and at low wind speeds, S is $\propto f^{-1}U^0$ between about 10–50 Hz. This holds until a crossover wind speed U_c is reached

($\approx 16 \, \mathrm{m\ s^{-1}}$ at 11 Hz), beyond which S is roughly $\propto f^{-3} U^4$. The crossover speed U_c tends to increase with increasing frequency up to about 50 Hz.

Nichols reported some data that also included frequencies lower than those of Crouch/Burt. These data show that for a small spread of wind speeds around $U \approx 9 \, \mathrm{m\ s^{-1}}$, $S \propto f^{-2.5}$ from 3 to 8 Hz, and $S \propto f^0$ from 8 to 20 Hz. These results compare well with those of Crouch/Burt at the overlapping frequencies, and add a functional form below 8 Hz not covered by Crouch/Burt. Nichols also summarized unpublished data for U varying from about 3 to 9 m s^{-1}, that show $S \propto f^{-5} U^5$ from 2 to 5 Hz, and $S \propto f^0 U^5$ from 5 to 10 Hz. (The $f^0 U^5$ functional form could be argued from the data shown to be $f^0 U^0$ at lower U, and $f^0 U^{>5}$ at higher U.) This form at the lowest frequencies has a much sharper falloff with frequency than the Nichols' form, but preference should be given to it because the frequency resolution was much finer. Accordingly, the overall result, labeled 'Nichols', is stated here as $S \propto f^{-5} U^5$ from 2 to 5 Hz, and $f^0 U^5$ from 5 to 20 Hz.

More recently, as part of LF noise measurement programs, data in the upper part of the VLF band were published. Deep ocean noise versus U at $f = 13$ and 50 Hz has been reported. For $5 \le U \le 15 \, \mathrm{m\ s^{-1}}$, these data have a functional form approximately $S \propto f^{-0.8} U^{1.3}$ (perhaps more like $f^0 U^{1.6}$ at the highest U). A report of shallow water noise versus U in third-octave bands for $10 \le f \le 20$ Hz, with form $S \propto f^{-3.2} U^{3.4}$ for $U = 3$ and 5 m s^{-1} in late fall, had data that differed so widely for $U = 5$ and 9 m s^{-1} in late spring that a functional form could not be stated. One report of deep ocean noise at $f = 15$ and 25 Hz, for $2 \le U \le 12 \, \mathrm{m\ s^{-1}}$, suggests $S \propto f^0 U^{-0.5}$ for $U < U_c$, and $S \propto f^0 U^3$ for $U > U_c$, with the crossover speed $U_c \approx 8 \, \mathrm{m\ s^{-1}}$.

What noise mechanism could account for all the foregoing observations? An extrapolation of wave–wave interaction noise to the VLF band, from data such as in **Figure 2**, suggests that $S \propto f^{-4} U^{1/2}$, give or take one integer in the exponent of f, and one-half integer in the exponent of U. However, this is unacceptably far from the data. The Crouch/Burt data suggest $f^{-1} U^0$ and $f^{-3} U^4$ for low and high U, respectively. The overall Nichols result is $f^{-5} U^5$ from 2 to 5 Hz, and $f^0 U^5$ from 5 to 20 Hz. Other data give $f^{-0.8} U^{1.3}$, $f^{-3.2} U^{3.4}$ or an indefinite form, and $f^0 U^{-0.5}$ and $f^0 U^3$ for low and high U, respectively. Without significant modifications applicable to the VLF band, it seems that the wave–wave interaction possibility must be set aside.

Next, consider the atmospheric turbulence model. It has evolved as most theories do, but is contentious. It predicts $S \propto f^0 U^4$. This, too, is mostly far from the

functional form of the foregoing data, but does come close to Nichols and others (at the higher wind speeds) for the 5–25 Hz range. It seems inappropriate, however, to choose among available data sets for the ones that confirm a model. The difference between the data sets may well be caused by some mechanism that we are collectively ignorant of.

Finally, it is possible that available data are at least partially contaminated by hydrophone flow noise, whose functional form goes as f^{-4}. None of the data sets matches this form. Thus, it can be concluded that the hydrophone flow noise mechanism is an unlikely cause of VLF noise. The identification of the mechanism responsible for VLF noise can thus not be made with confidence.

In searching for candidate VLF noise mechanisms, one is inclined to look toward appropriate extensions or modifications of mechanisms in the adjacent ULF and LF bands, mainly because wave–wave interactions and distant shipping, respectively, are well established. Nonprevailing mechanisms should also be considered. For example, whale vocalizations are observed for $15 < f < 35$ Hz and can affect the VLF band.

Noise data sets beyond those referred to here, supported by environmental data as suggested by candidate mechanisms, may well be needed. The Crouch/Burt data set incorporated a plausible but convoluted data analysis path to extract the VLF noise. The reported database of Nichols is not large. The VLF data of other workers could have been affected, as the authors acknowledged, by distant shipping noise. Perhaps because many of these research efforts were aimed at other objectives, environmental data provided with the noise data are generally too fragmentary to aid the search for VLF mechanisms.

LF Band: Distant Shipping Noise

Evolving technology has altered the view of distant shipping noise. Increasing use of large aperture acoustic arrays, with attendant high-resolution beamwidths and focused scanning in range and use of high-resolution frequency filters, blurs the distinction once sharp between distant and local ships. That is, ambient noise at LF can be observed with high-resolution technology as a countable number of discrete ship noise sources, rather than as a sum of noise from a very large number of widely distributed ships.

Frequency Spectra

Figure 3 shows the noise radiated by a contemporary cargo ship. The radiation is largely tonal, as has long

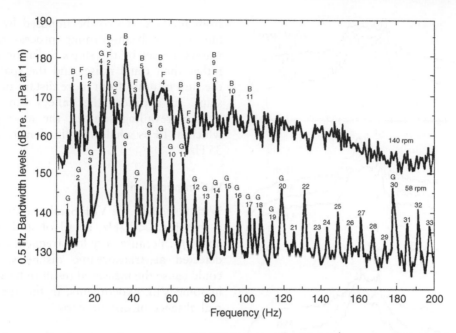

Figure 3 On-axis source level spectra of a cargo ship at 8 and 16 knots (4 and 8 m s^{-1}) measured directly below the ship. Noise levels at distances beyond 1 m may be obtained by subtracting the transmission loss. Levels for bandwidths other than 0.5 Hz can not be determined from this figure because the bandwidths of the tones are not given. (B, F, and G in the figure identify, respectively, the harmonics of the (propeller) blade rate, the (diesel engine) firing rate, and the (ship's service) generator rate.) (Reproduced from Arveson and Vendittis, 2000.)

been known. However, this is not obvious from the spectra shown in **Figure 1**, because they entail sums over many ships. The tonal envelopes in **Figure 3** maximize between about 20 and 80 Hz, in good agreement with the summation spectra shown in **Figure 1**. Acoustic propagation losses in the ocean change the shape of the source spectrum shown in **Figure 3**; above about 80 Hz, the spectrum observed distantly is increasingly reduced with increasing f and with increasing distance from the ship.

Directional Spectra

Noise radiated by a ship is a function of azimuth ϕ_s, and vertical angle θ_s, in a cylindrical coordinate system attached to the ship. The azimuthal spectral shape $S_\phi(\phi; f, z)$ observed for a single ship at longer ranges is close to that measured near the ship. However, propagation of the noise to large ranges fundamentally affects the vertical directional spectrum $S_\theta(\theta; f, z)$. In these spectra, ϕ and θ are in a coordinate system attached to the observer ($\theta = 0$ is the local horizontal plane). **Figure 4** shows S_θ as measured in deep water by a vertical line array. It sums over the directional spectrum in azimuth S_ϕ, and therefore over the areal distribution of ships. A prominent feature of S_θ is a pedestal of high noise around the horizontal, which is weakly dependent on f and z, and varies in half-width θ_w from about ± 15 to $20°$. These values are consistent with

$cos\theta_w \approx c_z/c_b$, where c_z and c_b are the sound speed at the observation depth and at the bottom, respectively. Distant shipping noise in deep water thus arrives mostly from source radiation at the surface near $|\theta_s| = 0$, and then propagates to the observer via refraction and surface reflection paths in the water. Bottom reflection or transmission losses are relatively high, so that these paths are less important. Other effects influence the pedestal including, but not limited to, surface waves that modulate the source amplitude, scattering rather than specular reflection from the rough sea surface, and scattering from a seamount or continental margin. Because of these oceanographic and topographical complexities, the shape of the noise pedestal in **Figure 4** is not general but instead suggestive of the main features of S_θ in the deep ocean.

In shallow waters, if $c_z/c_b > 1$ (downward refracting profile), then S_θ is governed by path or mode losses, including those attributable to the bottom. If $c_z/c_b < 1$ (upward refracting profile), then S_θ would have a pedestal, but with θ_w typically an order of magnitude smaller than that observed in the deep ocean.

Summation Issues

With several thousand ships underway in each of the heavily traveled oceans, taking account of all at the same time to determine distant shipping noise is

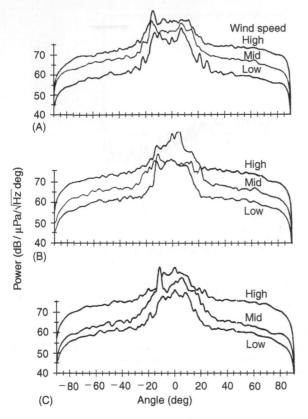

Figure 4 Vertical angle directional spectrum S_Θ summed in azimuth in deep water for $f = 75\,\text{Hz}$ and $U = 3, 7$, and $11\,\text{m s}^{-1}$. Positive angles are upward looking. (A) Shallow, (B) mid, and (C) deep depth refers to the vertical line array (VLA) centered at $z = 850, 1750$, and $2650\,\text{m}$. The half-power beamwidth of the VLA at this frequency is about $1.3°$ when steered to $\theta = 0$. (Reproduced from Sotrin and Hodgkiss, 1990.)

neither feasible nor necessary. At one extreme, low-resolution data in the LF band (20–200 Hz) are relatively insensitive to the detailed noise source characteristics of individual ships. Because summation from a large number of ships merges the details, only the broad and slowly evolving trends in shipping lane location, in shipping density, and in shipping composition will affect the level and horizontal directionality of the noise. At the other extreme, high-resolution data resolve the frequency and spatial spectra associated with distant ships, thus giving the experimenter the *in situ* noise field in relevant detail. The experimenter can avoid such noise in the spectral valleys between tones or the spatial valleys between high-noise beams[2]. Aside

---

[2] Many ocean processes broaden tonal bandwidths and spatial beamwidths as the sound propagates from source to receiver. Such broadening is typically large enough to be a candidate ocean monitoring tool, but not so much that it completely fills the valleys.

from limits that may be imposed by the focusing, filtering, or beam-forming processor, the spectral valleys are set by the ship's radiation of continuous rather than tonal noise, and the spatial valleys by the sum of noise from more distant or less powerful ships. In addition, the valleys can be influenced by nonprevailing sources, the most prominent of which at LF is whale vocalization from about 15 to 35 Hz.

VLF Implications

With respect to the VLF band, the tonal envelope of a ship for $f < 20\,\text{Hz}$, at all of its higher speeds is about f^0. Because very low frequency sounds can be detected at transoceanic distances, distant ships could cause the measured result to be $S \propto f^0 U^0$ found in some VLF experiments, in the apparent but not real absence of distant ships.

MF Band: Wind-driven Sea Surface Noise

Bubbles created by wind-driven surface waves have long been thought to be the dominant source of prevailing noise in the MF band (0.2–50 kHz). Many basic physical details, however, have only recently become better understood, and some relevant additional questions only recently posed. Various wave-breaking processes of wind-driven surface waves entrain air in the upper part of the ocean. Air-filled bubbles in the water are pinched off from the entrained air, which in turn oscillate and radiate noise as acoustic monopoles. Such noise thus entails wave-breaking and bubble hydrodynamics, both of which are addressed elsewhere in this encyclopedia. The acoustical aspects are addressed here.

Vertical Directional Spectrum

In its simplest form, the theory for noise generated by a uniform distribution of sources on the surface is, from ray acoustics,

$$S_\theta^+(\theta; f, z) = (c_s/c_z)\Sigma D\{\sin\theta_s(1 - R_b R_s)\}^{-1} \qquad [2]$$
$$\theta_w \le \theta \le \pi/2$$

$$S_\theta^-(\theta; f, z) = R_b S_\theta^+, \; -\pi/2 \le \theta \le -\theta_w \qquad [3]$$

where

$$\cos\theta_w = c_z/c_s \quad \text{for} \quad c_z/c_s < 1, \qquad [4]$$
$$\text{or} \quad \theta_w = 0 \quad \text{for} \quad c_z/c_s > 1$$

and where the directional spectrum in vertical angle S_θ, per unit solid angle, is a function of f (at least through Σ, the pressure spectral density of the source per unit surface area) and of observation depth z (through sound speed c_z at z). The superscripts + and – refer to positive (upward looking) and negative (downward looking) θ obtained from cos $\cos\theta = (c_z/c_s)\cos\theta_s$ (the grazing angle θ_s and sound speed c_s pertain to z = 0). $D(\theta_s, f)$ is the directivity of the elemental bubble noise sources distributed just below the surface. For present purposes, the surface source distribution Σ, and the acoustic waveguide, is taken as uniform and independent of range and azimuth, contributions from propagation in the bottom are neglected, and volumetric absorption in the seawater is neglected. (These simplifications are adopted to keep the main ideas clear but, for more precise needs, can readily be replaced by assumptions that are more realistic.)

For a downward refracting sound speed profile to depth $z(c_z/c_s < 1)$, the ray theory of [3][4] predicts a refractive shadow zone or notch of width $2\theta_w$ around $\theta = 0$, the horizontal plane. Wave theory must be used to properly predict the field in the notch, which also can be partially filled by scattering of the noise from midwater depths by fish schools and by ocean inhomogeneities. For an upward refracting profile $(c_z/c_s > 1)$, the field around $\theta = 0$ is directly due to the surface-generated noise, plus possible scattering contributions.

In eqns [2] and [3], R_b and R_s are, respectively, the coefficients of bottom and surface specular reflection. Terms involving these parameters can be important in the directional spectrum (but since perfect reflection is not likely for an acoustic waveguide in the ocean, they do not lead to singularities as eqns [2]–[4] might appear to suggest). For example, consider that R_b and R_s approach unity (but do not reach it) as the grazing angles at the bottom and the surface, respectively, approach zero. Then, for θ within about $\pm\pi/4$, S_θ can be increased in typical situations by about 10 dB. In addition, the bottom propagation paths neglected here can actually contribute, especially at the lower MF frequencies. Thus details of the acoustic waveguide affect S_θ^+ and S_θ^- and, along with the sound speed profile $c(z)$), could account for the plethora of somewhat dissimilar measured MF vertical directional spectra in the literature.

Eqn [2] contains the bubble source directivity $D(\theta_s, f)$ that, unfortunately, is not known with confidence. At least two models for directive radiation from aggregated bubbles have been considered. One assumes an exponential decrease of uncorrelated monopoles below a horizontal perfectly reflecting surface, and the other assumes a similarly situated monopole distribution concentrated on a submerged plane. Then, respectively,

$$D = 2\{1 - \mathrm{sinc}(2k_s d\sin\theta_s)\} \qquad [5]$$

$$D = 4\sin^2(k_s d\sin\theta_s) \qquad [6]$$

where k_s is the acoustic wavenumber at the surface, d is the effective depth (the e-folding depth and the δ-function depth, respectively), and sinc $(x) \equiv (\sin x)/x$. In the limit $k_s d \sin\theta_s \ll 1$, these functions have the same shape, and close to the same magnitude (≈ 1 dB different). Data, however, show that the two are distinct. For eqn [5], the data suggest $k_s d \approx \pi$, whereas for eqn [6] $k_s d \approx \pi$ In either case, the idealized states assumed in eqn [6] might not represent the relevant complexity of the radiating bubbles beneath a breaking wave. For example, the exponential decay of bubble density with depth may well be a good model for horizontally isotropic bubbles quasistatically present as a result of previous wave breaking events, but a poor model for radiating bubbles immediately caused by a new event.

Integration of eqns [2]–[4] over θ to obtain the noise spectral density $S(f, z)$ also depends sensitively on R_b and R_s (and on possible bottom propagation paths). This emphasizes the need to compare experimentally derived values of $S(f, z)$ with appropriate knowledge of the acoustic waveguide. Alternatively, with use of eqns [2]–[4], Σ may be extracted from vertical line array (VLA) data. When a VLA is steered to $\theta = \pi/2$, the specular reflection and the bottom propagation paths will contribute at most

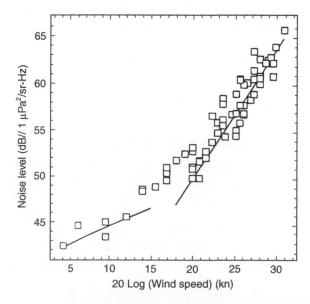

Figure 5 Level of the source spectral density Σ, in dB re μPa^2 per m^2 Hz, for $f = 110$ Hz. The 10 m wind speed U is in kn (1 kn $\approx 1/2$ m s^{-1}). (Reproduced from Chapman and Cornish, 1993.)

weakly. Such a measurement is thus dominated by local surface sources, so that Σ may be compared among measurements with less concern for waveguide properties.

Source Spectral Density

Chapman and Cornish measured Σ in deep water with an upward-looking VLA. They apparently assumed eqn [6] for D, with $k_s d \approx \pi/2$. Their data at $f = 110\,\text{Hz}$, and for the wind speed interval $2 < U < 15\,\text{m s}^{-1}$, are reproduced in **Figure 5**, and show a crossover wind speed $U_c \approx 4.3\,\text{m s}^{-1}$. The frequency interval for their measurements is $13 < f < 300\,\text{Hz}$, within which they found that U_c is about $4.3\,\text{m s}^{-1}$ for $f \leq 110\,\text{Hz}$, and that U_c is somewhat smaller ($\approx 3.5\,\text{m s}^{-1}$) for $f > 110\,\text{Hz}$. Chapman/Cornish attribute the crossover speed to a transition in source mechanism physics. Furthermore, by regression analyses, their data show that for $U < U_c$, $\Sigma \propto f^{-2.1} U^{0.6}$, and for $U > U_c$, $\Sigma \propto f^{-2.1} U^{2.7}$. These results hold on average within the speed and frequency intervals measured.

Kewley, Browning, and Carey reviewed and compared several data sets, mostly deep water VLA measurements, to extract Σ. They also used eqn [6] and $k_s d \approx \pi/2$, and concluded that for $30 < f < 1300\,\text{Hz}$ and $1 < U < 15\,\text{m s}^{-1}$, $U_c \approx 6\,\text{m s}^{-1}$ with $\Sigma \propto U^1$ for $U < U_c$ and $\Sigma \propto U^3$ for $U > U_c$. The Kewley et al. wind speed exponents of 1 and 3 are not too different from those of Chapman/Cornish. When one considers that the former tilted their exponent choices somewhat to agree with extant physical models proposed for the below and above U_c regimes, the agreement can be considered quite satisfactory.[3] What is more relevant, however, is that a universal spectral shape is not evident for either regime in the Kewley et al. comparisons. More likely than not Σ, f, and D need to be scaled by hydrodynamic parameters other than or additional to U, as shown below.

Basic Wave-Breaking Correlates

Research results on hydrodynamically based scaling of noise from breaking waves have been reported. Kerman has proposed that at $u_*/u_c \approx 1$, where u_* is the friction velocity and u_c is the minimum phase speed of gravity/capillary surface waves, the wave-breaking process transitions from one that has an aerodynamically smooth sea surface to another that is rough. Kennedy analyzed VLA data in a deep, acoustically isolated bay ($40 < f < 4000\,\text{Hz}$, $2 < U < 15\,\text{m s}^{-1}$), with unlimited wind fetch but limited wave fetch. It was found that $u_*/u_c 0.9$ defined a rough surface regime. (It may therefore be presumed that the crossover speed discussed in the foregoing section is $U_c \approx 0.9 u_c$) **Figure 6** shows that the spectral data for $u_*/u_c > 0.9$ aggregate to an almost universal scalable spectrum. What garners the caveat of 'almost' is that frequency is scaled by f_p, the observed peak frequency. Both Kerman and Kennedy point out that f_p does not vary strongly. It ranges from about 300 to 800 Hz in the Kennedy data, and is not unlike that sketched by Wenz (**Figure 1**). But experimental interest does not always include measurement of f_p, in which case a user of **Figure 6** must slide the frequency scale without benefit of Kennedy's judgement. Neither, however, can properly be accused of intellectual sloth. Kerman provides a model for f_p, which contains wave-breaking parameters that unfortunately are poorly known. Kennedy's collapsed spectral spread although acceptably small, is large enough, and the frequency dependence for $1/3 < f/f_p < 10$ is weak enough, to forego fine attention to f_p. Although apparently not used, f_p is related semi-empirically to breaking-wave whitecap size $f_p (\text{in Hz}) \approx 1400/(\sqrt{L}W)$, where L and W are, respectively, the whitecap along-crest length and cross-crest width, both in meters].

The source spectral density Σ in **Figure 6** is obtained by Kennedy from a dipole directivity model. In effect, eqn [6] was used with $k_s d \ll 1$, in this limit known as a compact dipole. With this assumption, the integral of $D(f, \theta_s)$ over a hemispherical surface

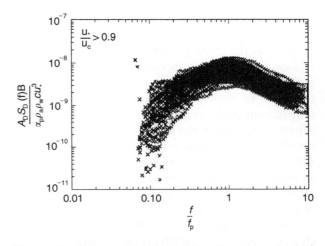

Figure 6 Source spectral density Σ versus f, both nondimensional as described in the text. The source directivity model used is presumably eqn [6] with $k_s d \ll 1$ (the compact dipole model). Data are for the aerodynamically rough regime ($u_* > u_c$). (Reproduced from Kennedy, 1992.)

---

[3] When compared at the same U and f, Σ is about 3 dB higher in the Chapman/Cornish data set than in the data reviewed by Kewley et al.

yields $A_D = 2\pi/3$ and this appears as one of the Σ scaling terms in **Figure 6**.[4] Another term is $B = 1000\,\text{Hz}$, the nominally observed bandwidth of the noise; its role is simply to create an integral measure ΣB for dimensional clarity. In the term $E = 5\rho_a u_*^3$, ρ_a is air density, and E is the major scaling variable, the average rate of energy dissipated per unit surface area by the breaking waves. Finally, $A_D \Sigma B/(\rho_w c_s)$, where ρ_w is water density, is the average rate of acoustic energy radiated per unit surface area.

Virtually simultaneously and independently, others have researched in greater depth some concepts that are related to the Kerman/Kennedy result. Noise spectral density $S(f, z)$ has been correlated with dissipation E, in deep water under steady wind and wave conditions, for the intervals $4.3 < f < 14\,\text{kHz}$ and $2 < U < 12\,\text{m s}^{-1}$. The data on average show $S \propto \Sigma \propto f^{-0.4} E^{0.74}$, with the exponent of E varying from 0.86 to 0.67 from the low to the high end of the frequency interval. At constant E, the frequency dependence agrees reasonably with an extrapolation of **Figure 6**. But the dissipation dependence can not be compared without scaling the peak frequency f_p, which was not observed. Thus, for a range of E one can seek the range of f_p to satisfy linear scaling in E. The peak frequency f_p would then need to decrease about a factor of 4 from low to high E, a factor so large as to suggest that a major change in noise physics occurs at these higher frequencies. Does the quasistatic bubble layer below the sea surface increasingly attenuate the noise, or increasingly inhibit its generation, at these frequencies? Bubbles are known to attenuate sound as a function of frequency and size distribution, but data analyses do not consider this.

With use of the Fresnel field of an array of hydrophones, sound radiated by individual breaking waves has been measured in deep water ($0.35 < f < 4\,\text{kHz}$, $4 < U < 15\,\text{m s}^{-1}$). The on-axis source levels of individual breaking events, were obtained and modeled as spatially and temporally discrete compact dipoles eqn ([6] with $k_s d \ll 1$). The individual source levels were correlated with U and c_b, the latter being the speed of a breaking wave event, a measure closely connected to breaking wave dissipation E. The correlation with c_b was found to be significantly better than that with U and, via physical arguments it was concluded that the source levels are well correlated with E. This measurement technique is also important as it determined the probability density of the dipole source levels, and the spatial density of discrete breaking wave events. It was also concluded, again via physical arguments, that the source spectral density for the frequencies measured are on average $\propto E^{0.83}$, which in view of the lower frequencies observed might be taken as reasonably consistent with the $E^{0.74}$ obtained by other workers. Thus the question remains on a possible frequency-dependent bubble layer effect.

The foregoing results clearly have not answered all questions on MF noise caused by breaking waves. They do, however, provide more general predictive tools than those previously available, and identify at least some of the more important physical attributes of noise from breaking waves.

HF Band: Molecular Noise

Molecules impinging on the surface of a pressure sensor cause noise, as estimated from physical principles and as plotted in **Figure 1**. Molecular motion, and thus momentum reversal on the sensor (i. e., force per unit area) is a function of molecular kinetic energy, and thus seawater temperature. On an absolute temperature scale, all oceans may be considered at a constant temperature. Hence, one line in **Figure 1** is sufficient to estimate the noise.

See also

Acoustics, Arctic. Acoustics, Shallow Water. Ships.

Further Reading

Arveson PT and Vendittis DJ (2000) Radiated noise characteristics of a modern cargo ship. *Journal of the Acoustical Society of America* 107: 118–129.

Buckingham MJ and Potter JR (eds.) (1995) *Sea Surface Sound '94*, vol. 3. Singapore: World Scientific, 494pp.

Chapman NR and Cornish JW (1993) Wind dependence of deep ocean ambient noise at low frequencies. *Journal of the Acoustical Society of America* 93: 782–789.

Crouch WW and Burt PJ (1972) The logarithmic dependence of surface-generated ambient-sea-noise spectrum level on wind speed. *Journal of the Acoustical Society of America* 51: 1066–1072.

Kennedy RM (1992) Sea surface sound source dependence on wave-breaking variables. *Journal of the Acoustical Society of America* 91: 1974–1982.

---

[4] A spherical surface for the integral would seem more appropriate, since the only way a monopole can become a dipole is by including the negative image above the free surface, in which case $A_D = 4\pi/3$. Had a noncompact dipole been assumed with $k_s d = \pi/2$ (eqn [6]), then $A_D = 2\pi$. There is as much as 5 dB difference in these values compared to the one used by Kennedy.

Kerman BR (1984) Underwater sound generation by breaking wind waves. *Journal of the Acoustical Society of America* 75: 149–165.

Kerman BR (ed.) (1988) *Sea Surface Sound*, vol. 1. Dordrecht: Kluwer Academic Publishers, 639pp.

Kerman BR (ed.) (1993) *Sea Surface Sound*, vol. 2. Dordrecht: Kluwer Academic Publishers, 750pp.

Kewley DJ, Browning DG, and Carey WM (1990) Low-frequency wind-generated ambient noise source levels. *Journal of the Acoustical Society of America* 88: 1894–1902.

Kibblewhite AC and Evans KC (1985) Wave–wave interactions, microseisms, and infrasonic ambient noise in the ocean. *Journal of the Acoustical Society of America* 78: 981–994.

Nichols RH (1981) Infrasonic ambient ocean noise measurements: Eleuthera. *Journal of the Acoustical Society of America* 69: 974–981.

Sotrin BJ and Hodgkiss WS (1990) Fine-scale measurements of the vertical ambient noise field. *Journal of the Acoustical Society of America* 87: 2052–2063.

Urick RJ (1986) *Ambient Noise in the Sea*. Los Altos, CA, Peninsula Publishing.

Wenz GM (1962) Acoustic ambient noise in the ocean: spectra and sources. *Journal of the Acoustical Society of America* 34: 1936–1955.

ACOUSTIC SCATTERING BY MARINE ORGANISMS

K. G. Foote, Woods Hole Oceanographic Institution, Woods Hole, Massachusetts, USA

Historical Overview

Development of underwater sonar as a tool for navigation and military operations, following sinking of the *Titanic* in 1912, led inevitably to applications to marine organisms. By the 1930s, echoes from fish schools had been detected. In the 1940s, the deep sound-scattering layer was observed. Its biological origin in mesopelagic fish was identified in the 1950s. At the same time, applications to commercial fish were pursued with vigor, and both scientific echo sounders and fishery echo sounders began to be manufactured.

Steady improvements in transduction enabled individual fish of certain species and sizes to be detected at ranges of hundreds of meters. The ultrasonic frequency of 38 kHz was becoming a standard at this time; it was subsequently shown to be near the optimum for achieving detection of commercially important fish in the presence of attenuation due to spherical spreading and absorption. Parallel to studies of single-fish scattering at ultrasonic frequencies were studies of scattering at sonic frequencies, especially to determine the resonance frequency in swimbladder-bearing fish, which is a measure of size.

Echo integration was introduced in 1965 as a tool for quantifying fish aggregations at essentially arbitrary conditions of numerical density. This was rapidly developed, and it has been used routinely in surveys of fish stock abundance since about 1975. Introduction of standard-target calibration in the early 1980s served the cause of quantification by providing a rapid, high-accuracy method of enabling the results of echo integration to be expressed in absolute physical units. With few exceptions, standard-target calibration has become the method of choice.

Sonar, with one or more obliquely oriented or steerable beams, began to find common application in the 1970s for counting fish schools that might be missed by a vertical echo sounder beam. This was a significant development for acknowledging the narrowness of the sampling volume of vertically oriented directional echo sounder beams and the possibility of fish avoidance reactions to the transducer platform, typically a research vessel.

In another parallel development, the Doppler principle was exploited to measure the rate of approach or recession of fish targets. Both horizontally oriented echo sounder beams and sonar beams were used. Early applications determined the swimming speeds of schools of small pelagic fish and individual salmon in rivers.

Applications of acoustics to fish in the 1970s were accompanied by notable applications to zooplankton, if pursued less intensively owing to differences in commercial importance. Because of the enormous diversity of zooplankton species in size, shape, and composition, it was recognized early that insonification over a band of frequencies is required, even for routine observation. This has usually been achieved by the use of multiple resonant transducers, but genuinely broadband sonars are also proving successful in yielding spectra of individual euphausiids and copepods.

Recognition of the importance of bandwidth in scattering by zooplankton was accompanied by appreciation of the role of interpretive models. Acoustic scattering models have been developed and applied to fish since the 1950s and to zooplankton since the 1970s.

The transition from analog to digital technologies in the 1970s facilitated processing of echo data. This has become steadily more automated and sophisticated, but always with operator control of important decisions through the man–machine interface.

Other developments in technology since the 1970s have extended the range of applications of acoustic scattering by marine organisms. Multiple-element transducers have been used to determine the three-dimensional locations and movement, as well as the target strength, of individual animals. Compact, high-frequency sonars have been mounted on fish capture gear to observe the behavior of fish during catching operations. Steerable high-frequency sonars have been used to track fish schools during capture and to map their three-dimensional shapes.

Physical Basis for Scattering

Acoustic scattering by a marine organism is, in principle, no different from that of any other kind of scattering. Differences in the physical properties of the causative bodies with respect to the surrounding medium are accompanied by reflection and refraction, or more generally diffraction, of incident waves. Organisms, with contrasts in mass density or elasticity relative to sea water, are thus sources of scattering.

The processes of reflection, refraction, and diffraction occur at surfaces, both external and internal, marking discrete changes in physical properties and throughout the volume or inside embedded inhomogeneities, as characterized by continuous changes in properties. The net result of the individual processes is a redistribution in space of the incident energy field. Changes in direction and amplitude characterize the scattering.

Classification of Marine Organisms as Scatterers

Marine organisms are conveniently divided into groups based on considerations of taxonomy and anatomy. Two major groups are those of fish and zooplankton, but others are also treated.

Fish may be distinguished as cartilaginous or bony. Bony fish may be acoustically distinguished because the fish possesses or lacks a gas-filled swimbladder. Swimbladders may be closed, with gas exchange effected by the *rete mirabile*, or open, with gas exchange effected by gulping air at the surface or by releasing a sphincter muscle on a duct leading to the exterior. The respective swimbladder types are called physoclists and physostomes. They are illustrated by cod (*Gadus morhua*) and herring (*Clupea harengus*), respectively. Some mesopelagic fish possess gas-filled swimbladders, including a number of myctophid species. Some other myctophids, as well as the deepwater fish orange roughy (*Hoplostethus atlanticus*), possess swimbladders that are invested with wax esters. The whiptail (*Coryphaenoides subserrulatus*), a macrurid, possesses a swimbladder that contains gas in a spongy matrix of tissue. Swimbladderless fish are illustrated by mackerel (*Scomber scombrus*). Cartilaginous fish lack a swimbladder, but their liver is large and presents a marked density contrast with the surrounding fish flesh.

Zooplankton come in many shapes and sizes, but acoustically their variable physical composition admits of a severe reduction. Three prominent classes have been identified: the liquidlike, the hard-shelled, and the gas-bearing. These are illustrated by, respectively, euphausiids, pteropods, and siphonophores.

Other marine organisms have also been detected by scattering. These include squid, gelatinous zooplankton, algae, benthos, marine mammals, and even diving birds. The first five groups are considered in a separate section in the following.

Dependences of Scattering

In general, scattering by marine organisms is affected by a number of factors. Some are listed here.

Intrinsic factors. Intrinsic to the scatterers are size, shape, internal composition, and condition. Condition may be affected by the stage of development, presence of reproductive products, and degree of stomach filling. Behavior is another intrinsic factor, if often directly affected or determined by the external environment. It is typically quantified through the attitude, or orientation, of the organism and its velocity of movement.

Extrinsic factors. Scattering is affected by the insonification signal, hence by its spectral composition. For impulsive signals, the spectrum may be broadly continuous. For a typical pulsed sinusoid containing many wavelengths, the spectrum will be narrow, and the signal can be characterized by the center frequency, pulse duration, and amplitude. Depth and history of depth excursion may also influence the scattering, as in the case of rapid depth changes for physoclists. For swimbladdered fish lacking *rete mirabile*, depth excursions will necessarily affect the swimbladder form, with the volume changing in accordance with Boyle's law, thus inversely with the ambient pressure.

Quantification of Scattering

Nomenclature

Scattering properties of organisms are distinguished as belonging to individual organisms or to aggregations of organisms. The fundamental scattering property of a single organism is the scattering amplitude. This is described through the idealization of a plane harmonic wave incident on a finite scattering body. At a great distance r from the body, the scattered pressure field or amplitude p_{sc} is related to the incident pressure amplitude p_{inc} by eqn [1].

$$p_{sc} = p_{inc} f \exp(ikr)/r \qquad [1]$$

In eqn [1] f is the far-field scattering amplitude, r is the distance from the scatterer, k is the wavenumber $2\pi/\lambda$, and λ is the acoustic wavelength. The scattering amplitude f describes the angular characteristics of the scattered field. The differential or bistatic scattering cross-section is $|f|^2$. In the backscattered direction $f = f_b$, and the backscattering cross-section is given by eqn [2], where the dual convention of using both σ_{bs} and σ is shown.

$$\sigma_{bs} = |f_b|^2 = \frac{\sigma}{4\pi} \qquad [2]$$

The target strength TS is a logarithmic measure (eqn [3]; where r_0 is the reference distance, typically 1 m).

$$TS = 10 \log \frac{\sigma_{bs}}{r_0^2} \qquad [3]$$

When many scatterers are concentrated in a volume in which individual scatterers cannot be distinguished by their echoes, a collective standard measure of scattering is used. This is the volume scattering coefficient. In the backscattered direction, the volume backscattering coefficient s_v is given by eqn [4], where $f_{b,i}$ is the backscattering amplitude for the ith scatterer of N, and V is the volume.

$$s_v = V^{-1} \sum_{i=1}^{N} |f_{b,i}|^2 \qquad [4]$$

The volume backscattering strength is given by eqn [5].

$$S_v = 10 \log(r_0 s_v) \qquad [5]$$

A quantity useful in echo integration is the area or column backscattering coefficient s_a, (eqn [6]), where the integration is performed over the range interval $[r_1, r_2]$.

$$s_a = \int_{r_1}^{r_2} s_v \, dr \qquad [6]$$

In scattering by fish, a numerically more convenient measure of s_a is eqn [7], which refers the backscattering to the reference area of one square nautical mile.

$$s_A = 4\pi 1852^2 s_a \qquad [7]$$

This form is particularly useful, for the fundamental equation of echo integration is simply eqn [8], where ρ_A is the numerical density of fish referred to the same area of one square nautical mile, and σ is the characteristic or mean backscattering cross section.

$$s_A = \rho_A \sigma \qquad [8]$$

Another measure of scattering is the extinction cross-section. This measures the relative loss of energy due to scattering and internal absorption. It may be defined for an individual scatterer, but is generally applied to aggregations of organisms if they are sufficiently numerous.

With few exceptions, the issue of calibration must be addressed when making measurements. Standard methods are available for this, the aim being to define the system characteristics so that the result of a measurement, a voltage signal for instance, can be expressed as a pressure-wave amplitude in the water medium.

Measurement

There are dozens of techniques for measuring the scattering properties of individual organisms and aggregations of organisms. These are commonly distinguished as being *in situ*, without constraint in the natural environment of the organisms, or *ex situ*, hence constrained in some way, wherever this might be.

Target strength is a key quantity in many investigations. It may be determined with a single-beam echo sounder; for example, by repeated measurement of similar organisms that are acoustically resolved and by appropriate statistical reduction of these measurements. Alternatively, it may be measured directly with a dual- or split-beam echo sounder, in which the beam pattern can be determined in the direction of the organism, enabling the backscattering cross-section to be extracted from each individual echo.

Similar measurements can be performed on single organisms *ex situ* with greater control and hence knowledge of their state during measurement. Measurements on tethered organisms, constrained to maintain a given orientation during insonification, are popular.

Aggregations of organisms are frequently quantified acoustically through the volume backscattering coefficient. If the number and occupied volume of the organisms are known, then the characteristic target strength can be inferred through eqn [9].

$$S_v = 10 \log n + TS \qquad [9]$$

Here n is the numerical density of organisms, and TS is the so-called mean target strength corresponding to a single organism, but derived as the logarithmic measure of the mean backscattering cross-section.

Cages are often employed to confine a known or knowable number of organisms to a fixed volume. Measurement of S_v can then yield a value for TS.

Modeling

The importance of target strength in many studies involving scattering by marine organisms is so great that recourse is frequently made to theoretical models. On the basis of assumptions about the shape and internal composition of subject organisms, mathematical expressions may be derived that can be evaluated for particular conditions of concentration or frequency that might not be realistically explored through measurement. Ultimately, measurements may be used to refine models, and models to interpret measurements.

Fish as Scatterers

Swimbladder-bearing Fish

The swimbladder shape varies with species and with condition of the individual specimen. An example of a swimbladder *in corpus* is shown in **Figure 1**. Here the swimbladder of an Atlantic herring (*Clupea harengus*) has been exposed by careful dissection.

Low frequencies   At low frequencies, with acoustic wavelengths much greater than characteristic swimbladder dimensions, the effect of a pressure wave on the swimbladder is essentially that of uniform compression and rarefaction. Consequently, a spherical model can be used. In fact, some swimbladder-bearing fishes have been modelled successfully as a spherical gas bubble surrounded by a finite layer of fish flesh that acts as a viscous fluid medium supporting surface tension on the interface between the shell and fish flesh. The volume of a bubble of radius a is equivalent to that of the swimbladder. Equation [10] gives the resonance frequency v_0 of an immersed spherical gas bubble,

$$v_0 = \frac{1}{2\pi}\left(\frac{3\gamma P}{\rho a^2}\right)^{1/2} \qquad [10]$$

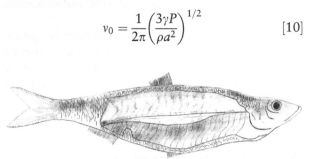

Figure 1   Drawing of a specimen of Atlantic herring (*Clupea harengus*), female, 36.0 cm long, 453 g, with exposed swimbladder. (Drawing by H. T. Kinacigil, used with permission.)

where γ is the ratio of specific heats at constant pressure and volume, P is the ambient pressure at depth, ρ is the mass density of fish flesh, and a is the equivalent spherical radius. For elongated bubbles or swimbladder shapes, the resonance frequency is modified.

The backscattering cross-section σ at frequency v is given by eqn [11].

$$\sigma = \frac{4\pi a^2}{[v_0/(vH)]^2 + \left[(v_0/v)^2 - 1\right]^2} \qquad [11]$$

H is the damping factor given by eqn [12], where c is the speed of sound in water, and ξ is the viscosity of fish flesh.

$$H^{-1} = \frac{2\pi a v}{v_0 c} + \frac{\xi}{\pi a^2 v_0 \rho}, \qquad [12]$$

Some numerical values for the various parameters are $\rho = 1050 \, \text{kg m}^{-3}$ and $\xi = 50 \, \text{Pa s}$. The speed of sound in sea water varies over the range 1450–1550 m s^{-1}, depending on temperature, salinity, and pressure.

For gadoids and clupeoids in the size range 8–30 cm, v_0 varies over 2.2–0.3 kHz. Given the inverse relationship of resonance frequency and size in eqn [10], smaller fish will have higher resonance frequencies. Thus, mesopelagic fish with partially wax-invested swimbladders may have resonance frequencies in the low ultrasonic range. Very large swimbladdered fish, say with a total fish length exceeding 1 m, will have resonance frequencies of the order of hundreds of hertz. The corresponding backscattering cross-section, hence target strength, can be computed from eqs [10]–[12]. It is important to note that the quality factor of the resonance condition, eqn [13], where Δv describes the range in frequency over which σ decreases to one-half its maximum value, may be of the order of 1.5–3.

$$Q = v_0/\Delta v \qquad [13]$$

Implicit in the low-frequency condition of the model is that σ is independent of orientation. Averages of σ with respect to arbitrary orientation distributions will be identical to σ itself.

When computing average values of σ for aggregations of swimbladdered fish of varying size, σ must be averaged with respect to the size distribution. The characteristic target strength is determined from the definition in eqn [3].

Intermediate   frequencies   As   the   acoustic wavelength   decreases   toward   characteristic swimbladder dimensions, the scattering becomes markedly directional, and the backscattering begins to depend sensitively on the orientation of the fish. From measurements made both *in situ* and *ex situ*, the empirical relationship of eqn [14] between mean target strength *TS* at 38 kHz and total fish length *l* in centimeters has been derived for a number of gadoids.

$$TS = 20 \log l - 67.5 \qquad [14]$$

Equation [15] applies for clupeoids

$$TS = 20 \log l - 71.9 \qquad [15]$$

The average backscattering cross-section σ may be determined immediately from eqn [3]. For a cod of length $l = 50$ cm, $TS = -33.5$ dB and $\sigma = 56$ cm^2. For a herring of length $l = 30$ cm, $TS = -42.4$ dB and $\sigma = 7.2$ cm^2.

Blue whiting is an important commercial stock in both hemispheres, and it is routinely surveyed by acoustics. To convert measurements of acoustic density at 38 kHz to numerical density in accordance with the echo integration equation [8], eqn [16], where *l* is the fork length in centimeters. is used for the northern-hemisphere blue whiting (*Micromesistius poutassou*):

$$TS = 21.7 \log l - 72.8 \qquad [16]$$

Equation [17] applies for the southern-hemisphere southern blue whiting (*Micromesistius australis*), where *l* is again the fork length in centimeters.

$$TS = 25.0 \log l - 81.4 \qquad [17]$$

Coincidentally, perhaps, the target strength of yellowfin tuna (*Thunnus albacares*) at 38 kHz is nearly identical to that of *Micromesistius australis* and is given by eqn [18].

$$TS = 25.3 \log l - 80.6 \qquad [18]$$

The target strength of bigeye tuna (*Thunnus obesus*) under similar conditions is given by eqn [19].

$$TS = 24.3 \log l - 73.3 \qquad [19]$$

These relations were established from specimens in the approximate size range 50–130 cm and 3–50 kg.

The whiptail (*Coryphaenoides subserrulatus*), with a swimbladder containing gas-filled spongy tissue, seems to have a mean *in situ* target strength at 38 kHz that is consistent with the equation developed for another macrurid, the blue grenadier or hoki (*Macruronus novaezelandie*) (eqn [20], where *l*

is the total fish length in centimeters).

$$TS = 20 \log l - 72.7 \qquad [20]$$

Some stocks of orange roughy (*Hoplostethus atlanticus*) are being surveyed about their seamount habitats. Determination of the target strength of this deepwater fish with fat-invested swimbladder is admittedly problematical. Some work suggests convergence of the mean target strength of a 35 cm long orange roughy at 38 kHz to about -48 dB. If the standard equation for mean target strength–length were used, namely eqn [21],

$$TS = 20 \log l + b \qquad [21]$$

the coefficient *b* would be -79 dB.

For modeling scattering by swimbladdered fish at these frequencies, the Kirchhoff approximation model can be used. This assumes that the fish is represented by the swimbladder, which acts as a pressure-release surface where it is directly insonified, and as a surface without response otherwise.

A more general scattering model is that of the boundary-element method. The swimbladder is represented by a mesh of points, called nodes, spanning the surface, illustrated in **Figure 2**. The harmonic wave equation is solved numerically, assuming continuity of pressure and normal component of velocity at each node. It is thus possible to model the effects of internal gas density and pressure.

To convert modeled values for σ as a function of orientation to an average value, an orientation distribution is required. Ideally, this is done on the basis of *in situ* observations, but often such data are lacking and an orientation distribution must be assumed. Some orientation distributions are described in the literature. In some special circumstances it has

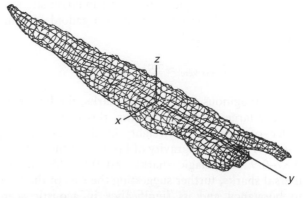

Figure 2 Boundary element model of the swimbladder of a specimen of pollack (*Pollachius pollachius*), 34.5 cm in length, with anterior end to the lower right (*y* direction). (Model by D. T. I. Francis, used with permission.)

been possible to infer the orientation distribution by a combination of acoustic measurement and modeling.

The relationship of maximum and average measures of σ is given approximately by eqn [22].

$$\sigma_{max} \approx 7\sigma_{ave} \qquad [22]$$

Alternatively, eqn [23] can be used.

$$TS_{max} \approx TS_{ave} + 5 \text{ dB} \qquad [23]$$

Measures of the extinction cross-section are relatively rare, there being few occasions when it is necessary to compensate for scattering losses. However, measurement or inference suggest that the extinction cross-section is very roughly 1–3 times the backscattering cross-section at intermediate frequencies. Ultimately, the cross-sections and their ratio must depend on the behavior of the organism, as quantified through the orientation distribution.

High frequencies When the acoustic wavelength becomes very small compared to the swimbladder size, scattering by other tissues may become important. The contributions of head structure, vertebrae, and even scales at very high frequencies have been established through *ex situ* measurement. Modeling of scattering by such structures can be computationally excessive, suggesting the advantages of stochastic modeling if direct measurement is not possible or convenient.

Swimbladderless Fish

The mackerel is a prominent example of a swimbladderless fish. Its target strength must be attributed to the non-swimbladder structures and hence is intrinsically complicated at nearly all frequencies. At intermediate frequencies, the mean target strength is roughly 10 dB less than that of a gadoid of comparable size (eqn [24]).

$$TS_{mackerel} \approx TS_{gadoid} - 10 \text{ dB} \qquad [24]$$

For cartilaginous fish, such as sharks, the liver may be very large. In pelagic sharks, this may be of the order of 7–23% by weight; in demersal sharks, 3–6%. The specific gravity of lipids is of the order of 0.87–0.92 in pelagic sharks and 0.93–0.94 in demersal sharks, further suggesting the role of the liver in buoyancy and its significance in acoustic scattering. At least for the pelagic sharks, the size and difference in mass density may explain much of the target strength. Were a model to be constructed,

a pelagic shark might be represented by a body with the size, shape, and physical properties of the liver.

Zooplankton as Scatterers

Liquid-like Bodies

A number of prominent and abundant zooplankton can be classified as liquidlike in their acoustic properties. Extensive modeling and measurement have demonstrated that internal shear waves have negligible influence in scattering by such organisms. The animals are thus generally fluidlike in their properties. If the same animals lack sizable organs or other tissue presenting large contrasts in mass density or compressibility relative to the sea water immersion medium, then the acoustic properties of the organisms are more particularly liquidlike, and their acoustic scattering is consequently relatively weak. Two examples of zooplankton with liquidlike properties are euphausiids and copepods. These are also representative of homogeneous and inhomogeneous scatterers, respectively.

Homogeneous liquidlike bodies The expectation of relatively weak scattering by euphausiids has been confirmed by measurement. For example, the target strength of Antarctic krill (*Euphausia superba*) of mean lengths 30–39 mm is in the range from − 88 to − 83 dB at 38 kHz and from − 81 to − 74 dB at 120 kHz. The respective acoustic wavelengths are 39 and 12.5 mm.

For a scattering body that is relatively long compared to the wavelength, the scattering will be inherently directional. Laboratory measurement has demonstrated strong effects of orientation on scattering by euphausiids in the size range 30–42 mm at frequencies of 120 kHz and higher.

In modeling scattering by homogeneous liquidlike zooplankton, there are just two significant material properties, the mass density and compressibility, or longitudinal-wave sound speed. A variety of models can be used to represent shape. At low frequencies, a single euphausiid can be represented by a finite circular cylinder or even a sphere, with volume equal to that of the animal. At higher frequencies, the same animal might be represented as a finite, bent, tapered cylinder or, better, by the actual shape of the exoskeleton.

Scattering models for euphausiids have demonstrated the sensitive dependence of target strength on both the material properties and orientation of the organism. Given the rarity of measurements of material properties, their seasonal and individual variability, and the generally unknown orientation, there has been little systematization of measured values of target strength.

Theoretical understanding of scattering by euphausiids has succeeded in associating large lobes with the echo spectrum at rather short acoustic wavelengths. When these are combined with knowledge of the target strength to within about an order of magnitude, it is possible to classify euphausiids by their acoustic signature.

Inhomogeneous liquidlike bodies Copepods, like euphausiids, also display relatively weak acoustic scattering. Unlike euphausiids, however, their internal structure is acoustically distinct, being composed of two dominant scatterers, a prosome and an embedded oil sac. Because of the low density of lipids in the oil sac, of the order of 900 kg m^{-3}, the prosome must be correspondingly more massive. Because the copepod body as a whole is close to neutral buoyancy in sea water, the target strength is due to the internal contrast in mass density and compressibility, or longitudinal-wave sound speed, between the prosome and oil sac.

Measurement has shown that the target strength of a 2 mm long copepod, *Calanus finmarchicus*, is in the approximate range from −95 to −90 dB over the frequency range 1600–2400 kHz.

Copepods have been modeled as composite two-liquid-body structures. Numerical values for the mass density and longitudinal-wave sound speed have been derived from measurements or have been assumed. The shapes of embedded oil sac and encompassing prosome, illustrated in **Figure 3**, have been determined from videomicroscopic cross-sections in dorsal and lateral views. Results of modeling of copepods have shown the expected weak dependence on orientation at low or moderate frequencies, and an overall mean target strength that is in line with measured values.

Hard-shelled Bodies

An example of a hard-shelled zooplankton is the pteropod *Limacina retroversa*, a marine snail with a spiral shell, opercular opening, and wings that propel it through the mid-water column. The target strength of specimens of shell length 2 mm has been measured over the approximate frequency range from 350 to 750 kHz. The target strength varies between −80 and −60 dB, depending on both frequency and orientation.

The pteropod has been modeled as a rough spherical shell with a circular opening. Predictions of scattering have been in reasonable agreement with

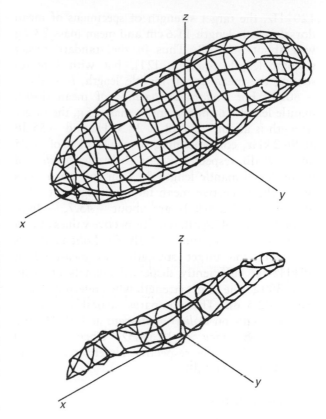

Figure 3 Boundary element models of the prosome and oil sac of a specimen of *Calanus finmarchicus*, stage 6 female, 2.74 mm in length, with anterior end to the lower left (*x* direction). (Models by D. T. I. Francis, used with permission.)

measurements at wavelengths roughly comparable to the maximum shell dimension.

Gas-bearing Bodies

Siphonophores are representatives of gas-bearing zooplankton, with gas inclusions in the pneumatophores. These are generally small compared to overall dimensions of specimens, and the target strength varies widely over the frequency range 350–750 kHz. In particular, the target strength varies over the range from −90 to −60 dB, but with no apparent systematic dependence on frequency. This wide range is suggestive of interference between echoes from the gas inclusions and the nongaseous tissue, the basis of an acoustic model.

Other Organisms as Scatterers

Squid

A number of specimens of squid have been observed by acoustics. These include *Todarodes pacifica*, *Loligo opalescens*, and *Loligo vulgaris reynaudii*. In a survey of the second species, performed at

120 kHz, the target strength of specimens of mean dorsal mantle length 11.6 cm and mean mass 23.7 g was about − 59 dB. Thus in the standard target strength–length equation [21], but with *l* representing the mean dorsal mantle length, *b* is about − 80 dB. For *Todarodes pacifica* of mean dorsal mantle length 16 cm and mean mass 95 g, the target strength is about − 51 dB at 28.5 kHz and − 55 dB at 96.2 kHz, corresponding to values of *b* of − 75 and − 79 dB, respectively. For *Todarodes pacifica* of mean dorsal mantle length 23.7 cm and mean mass 340 g, the respective mean target strengths at 28.5, 50, 96.2, and 200 kHz are about − 45.7, − 46.5, − 48.0, and − 47.6 dB, with respective values of *b* of − 75, − 74, − 76, and − 76 dB. For *Loligo vulgaris reynaudii*, the target strength was measured at 38 kHz for sufficiently dispersed animals of mean mass 300 g. The target strength when referred to 1 kg was − 42.5 dB. This compares favorably with the measurements on *Loligo opalescens* at 120 kHz and *Todarodes pacifica* at 28.5 kHz. When expressed relative to 1 kg, the respective target strengths are − 42.3 and − 41.1 dB.

Common Jellyfish

In anticipation of acoustic surveying of the ctenophore *Mnemiopsis leidyi* and other gelatinous zooplankton, namely *Aurelia aurita* and *Pleurobrachia pileus*, in the Black Sea, measurements have been made of the target strength of the common jellyfish *Aurelia aurita*. Functional regression equations have related the mean target strength in decibels to the disk diameter *d* in centimeters. At 120 kHz, the relation is eqn [25].

$$TS = 14.7 \log d - 74.6 \qquad [25]$$

At 200 kHz it is eqn [26].

$$TS = 39.6 \log d - 104.4 \qquad [26]$$

Thus for a specimen with mean diameter 10 cm, $TS = -59.9$ and − 64.8 dB at the respective frequencies.

Algae

Algae, such as kelp, are being surveyed by acoustics. For purposes of quantification, the acoustic properties of the plants themselves are being studied, both by experiment and by theoretical modeling. Measurements have been performed on leaves of *Laminaria saccarina* and *L. digitata* at three ultrasonic frequencies. The lengths of these span the range 0.7–2 m; the widths 0.4–0.9 m; the thicknesses 1–5 mm;

and the masses 0.33–0.8 kg. Target strengths expressed relative to 1 kg of biomass vary from − 35 to − 28 dB at 50 kHz, from − 33 to − 24 dB at 70 kHz, and from − 29 to − 22 dB at 200 kHz.

Smaller algae, the phytoplankton *Prorocentrum micans*, *Peridinium triquetrum*, *Olistodiscus luteus*, *Dunaliella salina*, *Platimous viridis*, and *Phaeodactilum tricornutum*, are also being studied by acoustics. Measurements of reverberation, in particular, are being used in attempts to quantify the volume of gas vacuoles.

Clams

Both the razor clam (*Tagelus dombeii*) and the surf clam (*Mesodesma donacium*) have been surveyed by acoustics. Beds of the razor clam have been surveyed in shallow water over a flat bottom. Echograms that show the bottom–surface–bottom reflection in addition to the first bottom reflection show an enhanced registration above the so-called second bottom echo. Counting of its characteristic serrations provides a quantitative measure of clam density.

Marine Mammals

A few measurements have been reported on the target strength of the sperm whale (*Physeter catodon*) and the humpback whale (*Megaptera novaeangliae*) *in situ*. Measurements have been made of the Atlantic bottlenose dolphin (*Tursiops truncatus*) in captivity. Measurements made on a 2.2 m long 126 kg female dolphin in broadside aspect at the surface revealed a mean target strength that decreased from about − 10 dB at the lowest measurement frequency of 23 kHz to about − 24 dB at 45 kHz, rising to about − 20 dB at 65 kHz, then falling to − 25 dB at 80 kHz. The observed degree of variability about these nominal values due to repeated insonification was 4–11 dB to within the first standard deviation to either side.

Challenges

For all of the instances and applications of acoustic scattering by marine organisms, there is an enormous demand for enhanced imaging capability and more quantitative understanding, including both improved measurement methods and models. In addition to refinement of current measurement methods, including those for quantifying concentrations of marine organisms, instruments are being developed or adapted for application. These include high-frequency sonars, multibeam sonars, and continuously broadband echo sounders, operating at both low and high frequencies.

In general, the addition of bandwidth to acoustic devices, whether achieved by multiple frequencies or a continuous spectrum, is a firm objective of many development efforts. Its usefulness in classification is appreciated from certain studies in zooplankton scattering, but it would aid studies of nekton scattering if successful.

Recognition of the importance of understanding the acoustic properties of individual organisms is similarly influential in promoting developments and applications. Determining the properties of single organisms when found *en masse* remains a challenge, as does quantifying avoidance reactions or avoiding inducing them. While there are many techniques for determining target strength, their application requires ingenuity to elucidate some of the principal dependences. The general lack of information on the depth dependence of target strength for gas-bearing organisms is a particular, prominent example.

Modeling of scattering by marine organisms offers much potential for resolving physically intractable problems, such as those involving separation of echoes from individual organisms in the midst of their social aggregations or inferring the acoustic properties of organisms that are very fragile or that occur in extreme environments. Both analytical and numerical models, however, require knowledge of the physical properties, shape, and behavioral characteristics, such as the orientation distribution, of the subject organisms. Acoustic inference of the *in situ* properties of organisms, by special measurement techniques and aided by models, appears very attractive if generally difficult.

An enhanced imaging capability based on acoustic scattering is also valuable. If realized in a compact device, this could aid fishing practice, as in providing fishers with information on the species and size of organisms present in the water column or on the bottom without actually having to capture the organism to make the determination. For the researcher, being able to distinguish different organisms with overlapping distributions would be invaluable in aiding the study of relationships, ultimately to advance the goals of ecosystem analysis and understanding.

Acknowledgments

This is Woods Hole Oceanographic Institution contribution number 10271.

See also

Bioacoustics. Sonar Systems.

Further Reading

Craig RE (ed.) (1984) *Fisheries Acoustics* A symposium held in Bergen, 21–24 June 1982, *Rapports et Proces-Verbaux des Reunions*, vol. 184. Copenhagen: International Council for the Exploration of the Sea.

Foote KG (1997) Target strength of fish. In: Crocker MJ (ed.) *Encyclopedia of Acoustics*, vol. 1, pp. 493–500. New York: Wiley.

Foote KG and Stanton TK (2000) Acoustical methods. In: Harris RP, *et al.* (ed.) *ICES Zooplankton Methodology Manual*, pp. 223–258. London: Academic Press.

Freon P and Misund OA (1999) *Dynamics of Pelagic Fish Distribution and Behaviour: Effects on Fisheries and Stock Assessment*. Oxford: Fishing New Books.

Karp WA (ed.) (1990) *Developments in Fisheries Acoustics*, A symposium held in Seattle, 22–26 June 1987, *Rapports et Proces-Verbaux des Reunions*, vol. 189. Copenhagen: International Council for the Exploration of the Sea.

Margetts AR (ed.) (1977) *Hydro-acoustics in Fisheries Research*, A symposium held in Bergen, 19–22 June 1973, *Rapports et Proces-Verbaux des Reunions*, vol. 170. Copenhagen: International Council for the Exploration of the Sea.

Medwin H and Clay CS (1998) *Fundamentals of Acoustical Oceanography*. San Diego, CA: Academic Press.

Nakken O and Venema SC (eds.) (1983) *Symposium on Fisheries Acoustics*, Selected papers of the ICES/FAO Symposium on Fisheries Acoustics, Bergen, Norway, 21–24 June 1982, FAO *Fisheries Report* no. 300. Rome: Food and Agriculture Organization of the United Nations.

Ona E (1990) Physiological factors causing natural variations in acoustic target strength of fish. *Journal of the Marine Biological Association of the United Kingdom* 70: 107–127.

Physics of Sound in the Sea (1969) Reprint of the 1946 edition. Washington, DC: Department of the Navy.

Progress in Fisheries Acoustics (1989) *Proceedings of the Institute of Acoustics*, vol. 11, no. 3. St. Albans: Institute of Acoustics.

Simmonds EJ and MacLennan DN (eds.) (1996) *Fisheries and Plankton Acoustics*, Proceedings of an ICES international symposium held in Aberdeen, Scotland, 12–16 June 1995. *ICES Journal of Marine Science*, vol. 53, no. 2. London: Academic Press.

Urick RJ (1983) *Principles of Underwater Sound*, 3rd edn. New York: McGraw-Hill.

ACOUSTIC SCINTILLATION THERMOGRAPHY

P. A. Rona, Rutgers University, New Brunswick, NJ, USA

C. D. Jones, University of Washington, Seattle, WA, USA

Introduction: Mapping Diffuse Flow at Seafloor Hydrothermal Sites

Acoustic scintillation thermography or AST is a method to detect and map diffuse flow at seafloor hydrothermal sites. Diffuse flow is the discharge of low-temperature hydrothermal solutions (up to tens of degrees Celsius) as seepage through areas of the seafloor. It is considered widespread at low intensity (temperature less than $1\,^\circ C$; flow rate less than $1\,cm\,s^{-1}$) in ocean basins and at higher intensity (temperature less than $100\,^\circ C$; flow rate less than $1\,m\,s^{-1}$) in seafloor hydrothermal fields. Quantitative assessment of diffuse flow is important because the cumulative thermal flux of diffuse flow through areas of the seafloor may equal or exceed that of focused flow from associated high-temperature (up to $405\,^\circ C$) higher-flow-velocity (flow rate tens to hundreds of centimeters per second) black smoker vents. Chemical fluxes in diffuse flow may be selectively significant. Measuring diffuse flow is difficult. Occurrence is patchy and temperatures and flow velocities are low. Because it is clear, lacking the suspended particulate matter of black smoker plumes, diffuse flow cannot be detected by measuring attenuation or backscatter of light and sound.

Method

The AST method uses the phase-coherent correlation of acoustic backscatter from consecutive sonar scans of the seafloor to detect weak fluctuation in the index of refraction of the water near the seafloor. Random index of refraction changes result from temporal variations in the water temperature caused by turbulent mixing, which produce detectable changes in travel time of an acoustic ray as the ray propagates from a stationary acoustic transducer through the diffuse flow to the seafloor and is scattered back through the diffuse flow. These changes in travel time increase from the nearer to the farther portion of the sonar footprint, causing the associated echo waveform to change in shape with each transmission and resulting in decorrelation of successive sound pulses. In essence, the scintillation of the acoustic wave as it passes through a turbulent flow field and scatters off the underlying seafloor makes the seafloor appear to shimmer, much as the eye would detect hot water or hot air shimmering against a stationary backdrop. For certain types of seafloor hydrothermal flow (such as the case with diffuse flow where buoyant turbulent microplumes are concentrated near the seafloor), the AST method can provide a means of remotely detecting areas of flow and a potential method of measuring the scales of temperature and velocity fluctuations in the near-bottom boundary layer.

Consider the scattering geometry illustrated in **Figure 1**, where a turbulent boundary layer overlays a rough seafloor and the incident acoustic field propagates forward and back between the transducer and the backscattering seafloor through the boundary layer. Assume that the seafloor is not changing in time and there is no motion of the acoustic transducer, but the turbulent flow is evolving temporally. Between consecutive and rapid sonar returns from the same spot on the seafloor, turbulent mixing will cause slight spatial and temporal fluctuations of the index of refraction of the water near the seafloor. These fluctuations will cause weak random forward scattering of the acoustic field as it propagates

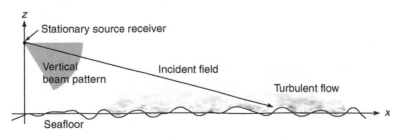

Figure 1 Propagation of an acoustic pulse through a turbulent boundary layer near the seafloor produced by diffuse hydrothermal flow and backscatter of the pulse from the seafloor.

through the turbulent flow, resulting in temporal fluctuations in the amplitude and phase of the acoustic field measured at the receiver. The temporal changes in backscatter can be measured by estimating the phase-coherent temporal decorrelation of backscatter between successive echoes, whereas the backscatter intensity from the turbulence itself is too weak to measure. In this case, the rate and magnitude of temporal decorrelation can be used to detect and potentially measure scales of mixing in the boundary layer.

Application

The method was first applied to detect and map diffuse flow in seafloor hydrothermal fields using a human occupied vehicle (HOV). Several later experiments have shown that deep ocean diffuse flow can be detected over relatively large areas of the seafloor using a remotely operated vehicle (ROV) platform. In these experiments, backscatter was recorded using a 200 kHz multibeam sonar mounted on an ROV as it hovered above the seafloor at intervals along a track line. **Figure 2** shows an area of detected diffuse flow (as verified by video) and the corresponding backscattered intensity image. The two images are derived from the same sonar scans of the seafloor taken at 0.1 s intervals. Future applications will extend the AST method to seafloor observatories to record changes in area of diffuse flow at a seafloor site on timescales up to years.

An informative representation is to drape areas of diffuse flow detected by the AST method over a corresponding bathymetric map to show the relation between diffuse flow and seafloor morphology.

Volume and heat fluxes from areas of diffuse flow can be estimated when *in situ* measurements of water temperature and vertical flow velocity are made in the areas of diffuse flow. For example, *in situ* measurements in diffuse flow of temperature using thermistors and of vertical flow velocity using video to record rise rate against a calibrated rod in an acoustically imaged area of diffuse flow and black smokers in the Main Endeavour Field indicated that heat flux of the diffuse flow was a multiple of that of the associated black smokers. **Figure 3** shows areas of decorrelation intensity (draped onto bathymetry) for a relatively large area of seafloor surveyed at the Clam Bed hydrothermal field on the Juan de Fuca Ridge (near the Main Endeavour Field site). The AST method was applied using a sonar mounted on a hovering ROV to map the diffuse flow over a 3500 m × 900 m area of the axial valley of the Endeavour segment of the Juan de Fuca Ridge. Using *in situ* sensors to simultaneously measure temperature and flow velocity, a diffuse heat flux of 150 MW was integrated over the areas of the AST decorrelation anomalies, indicating that the heat transferred by diffuse flow is a significant component of total heat flux in the study area.

The temporal correlation of the backscattered field is found by correlating collocated returns from the seafloor as a function of transmission time. Consider the monostatic geometry for backscatter as illustrated in **Figure 1**, where a source/receiver system is fixed in space and measures backscatter at discrete pulse transmission times τ_n. The temporal correlation of backscatter signals between a transmission at time τ_m and a later transmission at τ_n is defined as

$$C(\mathbf{r}; \tau_m, \tau_n) = \langle u(\mathbf{r}; \tau_m) u^*(\mathbf{r}; \tau_n) \rangle \qquad [1]$$

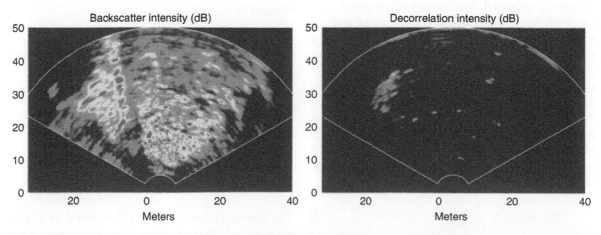

Figure 2 An AST image of diffuse flow near Hulk vent in the Main Endeavour Field on the northern Juan de Fuca Ridge is shown in the right panel. The areas of diffuse flow (light patches) are detected as an increase in the decorrelation intensity. A conventional sonar image of the same area is shown in the left panel. Colors indicate level of backscatter intensity (red is highest intensity) related to bottom roughness.

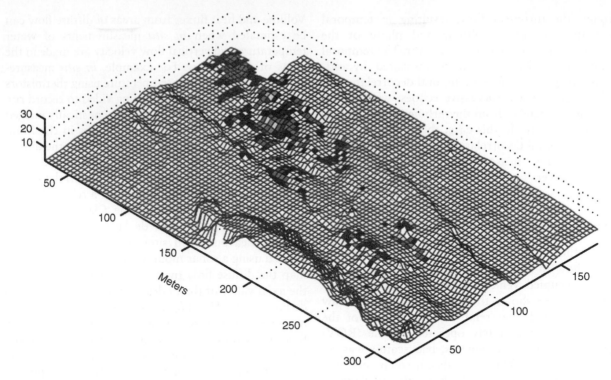

Figure 3 Areas of diffuse hydrothermal flow from the seafloor (Clam Bed site on the northern Juan de Fuca Ridge) detected by backscatter decorrelation (area in meters; Jones *et al.*, 2000). Colors indicate level of decorrelation intensity (yellow is higher intensity) corresponding to intensity of diffuse flow.

where $u(\mathbf{r}; \tau_n) = A(\mathbf{r}; \tau_n)\exp[i\phi(\mathbf{r}; \tau_n)]$ is the complex envelope of the backscattered field as a function of range $\mathbf{r}$ along the seafloor. Range is determined by round-trip acoustic time-of-flight as is usual in sonar. The complex envelope, also known as the baseband signal, incorporates the amplitude (A) and phase (ϕ) of the echo signal, and the asterisk in eqn [1] denotes complex conjugation. For $m = n$, the correlation function is real and proportional to the signal intensity. With $m \neq n$, the correlation function will be complex due to signal fluctuations in amplitude and phase, with phase dominating. If the area of seafloor observed has not changed because of motion of the source/receiver, fluctuations in the signal will be due to changes in the water above the seafloor as a function of time τ_n. With $m = 0$ as the reference time and $n = 1, 2, 3, \ldots$ increasing monotonically, eqn [1] is the cumulative temporal correlation function. For a single realization and finite number of discrete data points, an estimate of the ideal correlation function is found by windowing the backscattered signals in range and summing over range bins,

$$\hat{C}(\mathbf{r}_i;\ \tau_0, \tau_n) = \frac{1}{M}\sum_{i=1}^{M} u(\mathbf{r}_i;\ \tau_0)u^*(\mathbf{r}_i;\ \tau_n) \quad [2]$$

where $\mathbf{r}_i$ is a set of M discrete range values that correspond to a finite area (or patch) of the seafloor centered at range $\mathbf{r}$.

The correlation function is a measure of the change in the scattering medium, but noise and motion will also cause signal decorrelation. To correct for the effects of low signal-to-noise levels when the return from the seafloor is weak, it is convenient to define the decorrelation intensity as

$$I_\rho(\mathbf{r}; \tau_0, \tau_n) = I(\mathbf{r}; \tau_0, \tau_n)[1 - \hat{\rho}(\mathbf{r}; \tau_0, \tau_n)] \quad [3]$$

where the normalized temporal correlation coefficient is

$$\hat{\rho}(\mathbf{r}; \tau_0, \tau_n) = \left|\hat{C}(\mathbf{r}; \tau_0, \tau_n)\right| \bigg/ \sqrt{\hat{C}(\mathbf{r}; \tau_0, \tau_0)\ \hat{C}(\mathbf{r}; \tau_n, \tau_n)}$$

$$[4]$$

The scalar intensity I is the geometric mean of the intensity in the range window, defined as

$$I(\mathbf{r}; \tau_0, \tau_n) = \sqrt{\hat{C}(\mathbf{r}; \tau_0, \tau_0)\ \hat{C}(\mathbf{r}; \tau_n, \tau_n)} \quad [5]$$

The decorrelation intensity [3] is useful as a relative measure of change when the scattering strength over an area of the seafloor is nearly uniform and helps avoid the comparison of areas where the scattered signal level is relatively low. Motion of the source/receiver is more difficult to compensate and will depend on the randomness in the seafloor. Angular motion of the source/receiver between recordings should be much smaller than the angular correlation scale of the scattering from the seafloor. If there is motion the rate of signal decorrelation due to motion must be less than the rate of signal decorrelation due to the diffuse flow.

The size of the area of seafloor (or patch) used to estimate the correlation is defined by the length of the transmitted pulse and the number of samples needed to form an accurate estimate. For a bandwidth-limited Gaussian signal with bandwidth B, it is well known that the normalized mean-square error of the correlation estimate is given as

$$\varepsilon^2\left[\hat{C}(\mathbf{r};\tau_0,\tau_n)\right] = \frac{1}{2BT}\left[1 + \rho^{-2}(\mathbf{r};\tau_0,\tau_n)(1 + N/S)^2\right]$$
[6]

where ρ is the desired true correlation coefficient and T is the length of the record used to form the estimate (corresponding to the number of points M or patch size). The signal-to-noise ratio (S/N) is assumed constant and uncorrelated between transmission times τ_0 and τ_n.

See also

Acoustic Measurement of Near-Bed Sediment Transport Processes. Acoustic Noise. Acoustic Scattering by Marine Organisms. Acoustics, Deep Water. Acoustics in Marine Sediments. Acoustics, Shallow Water. Bioacoustics.

Further Reading

Bendat JS and Piersol AG (2000) *Random Data: Analysis and Measurement Procedures*, 3rd edn., pp. 291–296. New York: Wiley-Interscience.

Johnson HP, Hautala SL, Tivey MA, *et al.* (2002) Survey studies hydrothermal circulation on the northern Juan de Fuca Ridge. *EOS, Transactions, American Geophysical Union* 83(73): 78–79.

Jones CD, Jackson DR, Rona PA, and Bemis KG (2000) Observations of hydrothermal flow (abstract). *Journal of the Acoustical Society of America* 108(5): 2544–2545.

Rona PA, Jackson DR, Wen T, Jones CD, Mitsuzawa K, and Bemis KG (1997) Acoustic mapping of diffuse flow at a seafloor hydrothermal site: Monolith Vent, Juan de Fuca Ridge. *Geophysical Research Letters* 24(19): 2351–2354.

ACOUSTICS IN MARINE SEDIMENTS

T. Akal, NATO SACLANT Undersea Research Centre, La Spezia, Italy

Introduction

Because of the ease with which sound can be transmitted in sea water, acoustic techniques have provided a very powerful means for accumulating knowledge of the environment below the ocean surface. Consequently, the fields of underwater acoustics and marine seismology have both used sound (seismo-acoustic) waves for research purposes.

The ocean and its boundaries form a composite medium, which support the propagation of acoustic energy. In the course of this propagation there is often interaction with ocean bottom. As the lower boundary, the ocean bottom is a multilayered structure composed of sediments, where acoustic energy can be reflected from the interface formed by the bottom and subbottom layers or transmitted and absorbed. At low grazing angles, wave guide phenomena become significant and the ocean bottom, covered with sediments of different physical characteristics, becomes effectively part of the wave guide. Depending on the frequency of the acoustic energy, there is a need to know the acoustically relevant physical properties of the sediments from a few centimeters to hundreds of meters below the water/sediment interface.

Underwater acousticians and civil engineers are continuously searching for practical and economical means of determining the physical parameters of the marine sediments for applications in environmental and geological research, engineering, and underwater acoustics. Over the past three decades much effort has been put into this field both theoretically and experimentally, to determine the physical properties of the marine sediments. Experimental and forward/inverse modeling techniques indicate that the acoustic wave field in the water column and seismo-acoustic wave field in the seafloor can be utilized for remote sensing of the physical characteristics of the marine sediments.

Sediment Structure as an Acoustic medium

Much of the floor of the oceans is covered with a mixture of particles of sediments range in size from boulder, gravels, coarse and fine sand to silt and clay, including materials deposited from chemical and biological products of the ocean, all being saturated with sea water. Marine sediments are generally a combination of several components, most of them coming from the particles eroded from the land and the biological and chemical processes taking place in sea water. Most of the mineral particles found in shallow and deep-water areas, have been transported by runoff, wind, and ice and subsequently distributed by waves and currents.

After these particles have been formed, transported, and transferred, they are deposited to form the marine sediments where the physical factors such as currents, dimensions and shapes of particles and deposition rate influence the spatial arrangements and especially sediment layering. Particles settle to the ocean floor and remain in place when physical forces are not sufficiently strong to move them. In areas with strong physical forces (tidal and ocean currents, surf zones etc.) large particles dominate, whereas in low motion energy areas (ocean basins, enclosed bays) small particles dominate.

During the sedimentation process these particles, based on the physical and chemical interparticle forces between them, form the sedimentary acoustic medium: larger particles (e.g., sands) by direct contact forces; small particles (e.g., clays and fine silts) by attractive electrochemical forces; and silts, remaining between sands and clays are formed by the combination of these two forces. The amount of the space between these particles is the result of different factors, mainly size, shape, mineral content and the packing of the particles determined by currents and the overburden pressure present on the ocean bottom.

Figure 1 is an example of a core taken very carefully by divers, to ensure an undisturbed internal structure of the sediment sample. Sediment structures have been quantified by using X-ray computed tomography to obtain values of density with a millimeter resolution for the full three-dimensional volume. The image shown in **Figure 1** is a false color 3D reconstruction of a core sample at a site where sediments consist of sandy silt (75% sand, 15% silt, and 10% clay) and shell pieces.

The results of the X-ray tomography of the same core, can also be shown on an X-ray cross-section slice along the center of the core (**Figure 2A**) and the corresponding two-dimensional spectral density levels for that cross-section (**Figure 2B**). The complex

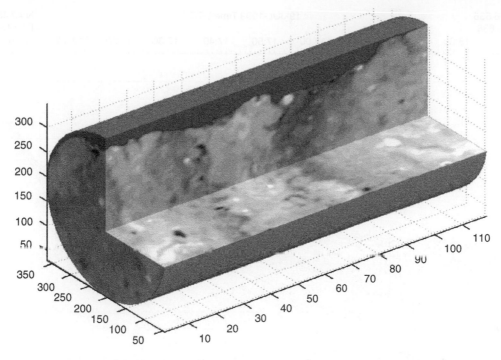

Figure 1 Three-dimensional reconstruction of a sediment core sample containing 75% sand, 15% silt and 10% clay. Scale in mm.

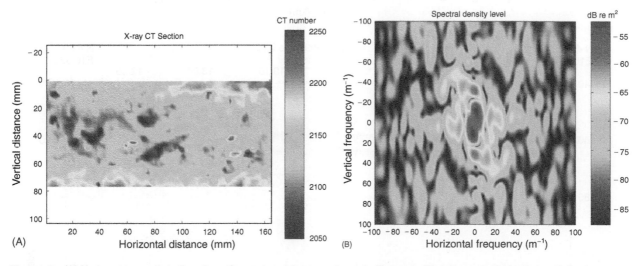

Figure 2 (A) X-ray cross-section slice along the center of the core shown in **Figure 1**. (B) 2-D spectral density levels for the cross-section shown in (A).

structure of the sediments can be with seen strong heterogeneity (local density fluctuation) of the medium that controls the interaction of the seismo-acoustic energy. In addition to the complex fine structure described above, the seafloor can also show complex layering (**Figure 3A**) or a simpler structure (**Figure 3B**). These structures result from the lowering of sea levels during the glaciation of the Pleistocene epoch during which sand was deposited over wide areas of the continental shelves. Unconsolidated

sediments subsequently covered the shelves as the sea level are in postglacial times.

Biot–Stoll Model

Various theories have been developed to describe the geoacoustic response of marine sediments. The most comprehensive theory is based on the Biot model as elaborated by Stoll. This model takes into account various loss mechanisms that affect the response of

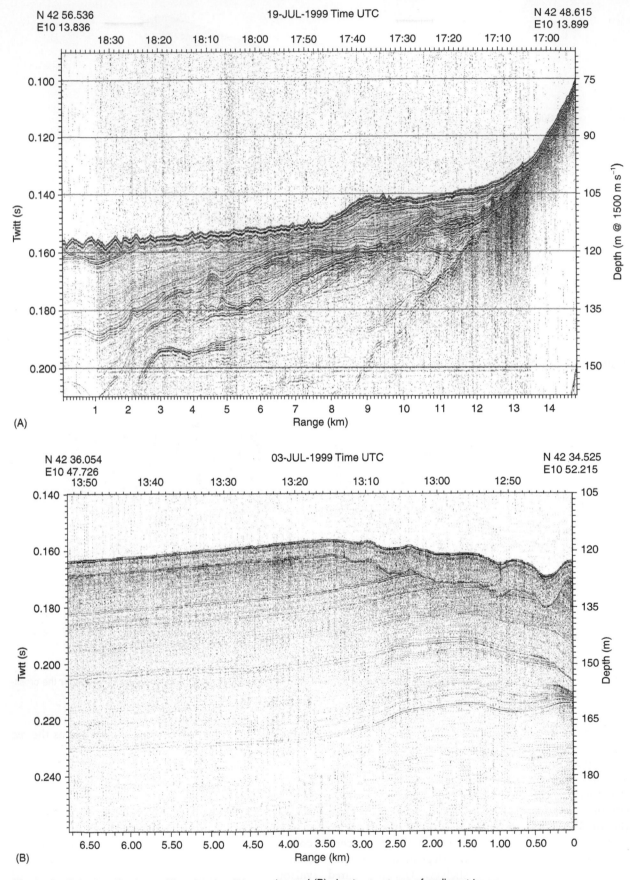

Figure 3 Seismic reflection profiles showing (A) complex and (B) simple structures of sediment layers.

porous sediments that are saturated with fluid. The Biot–Stoll theory shows that acoustic wave velocity and attenuation in porous, fluid-saturated sediments depend on a number of parameters including porosity, mean grain size, permeability, and the properties of the skeletal frame.

According to the Biot–Stoll theory, in an unbounded, fluid-saturated porous medium, there are three types of body wave. Two of these are dilational (compressional) and one is rotational (shear). One of the compressional waves ('first kind') and the shear wave are similar to body waves in an elastic medium. In a compressional wave of the 'first kind', the skeletal frame and the sea water filling the pore space move nearly in phase so that the attenuation due to viscous losses becomes relatively small. In contrast, due to out-of-phase movement of the frame and the pore fluid, the compressional wave of the 'second kind' becomes highly attenuated. The Biot theory and its extensions by Stoll have been used by many researchers for detailed description of the acoustic wave–sediment interaction when basic input parameters such as those shown in **Table 1** are available.

Seismo-acoustic Waves in the Vicinity of the Water–Sediment Interface

As mentioned above, when acoustic energy interacts with the seafloor, the energy creates two basic types of deformation: translational (compressional) and rotational (shear). Solution of the equations of the wave motion shows that each of these types of deformation travels outward from the source with its own velocity. Wave type, velocity, and propagation direction vary in accordance with the physical properties and dimensions of the medium.

The ability of seafloor sediments to support the seismo-acoustic energy depends on the elastic properties of the sediment, mainly the bulk modulus (incompressibility, K) and the shear modulus (rigidity, G). These parameters are related to the compressional and shear velocities C_P and C_S respectively, by:

$$C_P = [(K + 4G/3)\rho]^{1/2}$$

$$C_S = (G/\rho)^{1/2}$$

where, ρ is the bulk density. **Table 2** shows basic seismic–acoustic wave types and their velocities related to elastic parameters

These two types of deformations (compressional and shear) belong to a group of waves (body waves) that propagate in an unbounded homogeneous medium. However, in nature the seafloor is bounded and stratified with layers of different physical properties. Under these conditions, propagating energy undergoes characteristic conversions every time it interacts with an interface: propagation velocity,

Table 1 Basic input parameters to Biot–Stoll model

Frequency-independent Variables	
Porosity (%)	P
Mass density of grains	ρ_r
Mass density of pore fluid	ρ_f
Bulk modulus of sediment grains	K_r
Bulk modulus of pore fluid	K_f
Variables affecting global fluid motion	
Permeability	k
Viscosity of pore fluid	η
Pore-size parameter	a
Structure factor	α
Variables controlling frequency-dependent response of frame	
Shear modulus of skeletal frame	$\bar{\mu} = \mu_r(\omega) + i\mu_i(\omega)$
Bulk modulus of skeletal frame	$\bar{K}_b = K_{br}(\omega) + iK_{bi}(\omega)$

Table 2 Basic seismo-acoustic wave types and elastic parameters

Basic wave type		Wave velocity
Body wave	Compressional	$C_P = [(K + 4G/3)/\rho]^{1/2}$
	Shear	$C_S = (G/\rho)^{1/2}$
Ducted wave	Love	$C_L = (G/\rho)^{1/2}$
Surface wave	Scholte	$C_{SCH} = (G/\rho)^{1/2}$
Elastic parameters in terms of wave velocities (C) and bulk density (ρ)		
Bulk modulus (incompressibility)	$K = \rho(C_P^2 - 4C_S^2/3)$	
Compressibility	$\beta = 1/K$	
Young's modulus	$E = 2C_S^2\rho(1 + \sigma)$	
Poisson's ratio (transv./long. strain)	$\sigma = (3E - \rho C_P^2)/(3E + 2\rho C_P^2)$	
Shear modulus (rigidity)	$G = \rho C_S^2$	
Lame's constant	$\lambda = \rho(C_P^2 - 2C_S^2)$	

energy content, and spectral structure and propagation direction changes. In addition, other types of waves, i.e., ducted waves and surface waves, may be generated. These basic types of waves and their characteristics together with their arrival structure as synthetic seismograms for orthogonal directions are illustrated in **Figure 4**.

Body Waves

These waves propagate within the body of the material, as opposed to surface waves. External forces can distort solids in two different ways. The first involves the compression of the material without changing its shape; the second implies a change in shape without changing its volume (distortion). From earthquake seismology, these compressional and distortion waves are called primus (P) and secoundus (S), respectively, for their arrival sequence on earthquake records. However, the distortional waves are very often called shear.

Compressional waves Compressional waves involve compression of the material in such a manner that the particles move in the direction of propagation.

Shear waves Shear waves are those in which the particle motion is perpendicular to the direction of propagation. These waves can be generated at a layer interface by the incidence of compressional waves at other than normal incidence. Shear wave energy is polarized in the vertical or horizontal planes, resulting in vertically polarized shear waves (SV) and horizontally polarized shear waves (SH). However, if the interfaces are close (relative to a wavelength), one cannot distinguish between body and surface waves.

Ducted Waves

Love waves Love waves are seismic surface waves associated with layering; they are characterized by horizontal motion perpendicular to the direction of propagation (SH wave). These waves can be considered as ducted shear waves traveling within the duct of the upper sedimentary layer where total reflection occurs at the boundaries; thus the waves represent energy traveling by multiple reflections.

Interface Waves

Seismic interface waves travel along or near an interface. The existence of these waves demands the combined action of compressional and shear waves. Thus at least one of the media must be solid, whereas

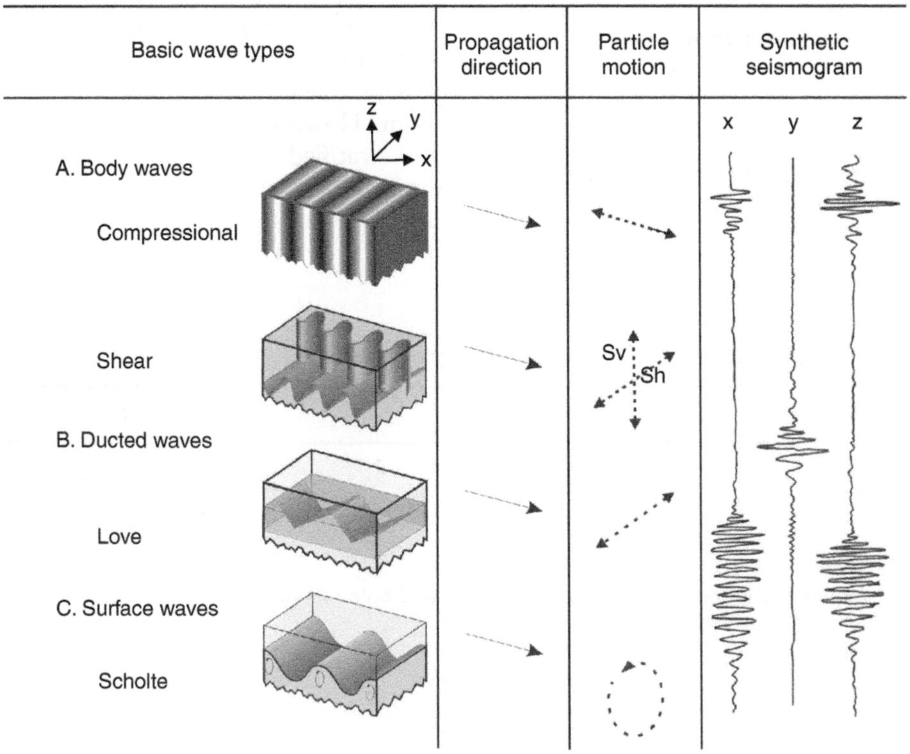

Figure 4 Basic seismo-acoustic waves in the vicinity of a water–sediment interface.

the other may be a solid (Stoneley wave), a liquid (Scholte wave), or a vacuum (Rayleigh wave). For two homogeneous half-spaces, interface waves are characterized by elliptical particle motion confined to the radial/vertical plane and by a constant velocity that is always smaller than the shear wave.

When different types of seismic waves propagate and interact with the layered sediments they are partly converted into each other and their coupling may create mixed wave types in the vicinity of the interface. **Figure 4** shows the basic seismo-acoustic waves in the vicinity of a water/sediment interface. Basic characteristics of these waves together with their particle motion and synthetic seismograms at three orthogonal directions (x, y and z) are also illustrated in the same Figure. Under realistic conditions, in which the seafloor cannot be considered to be homogeneous, isotropic, nor a half-space, some of these waves become highly attenuated or travel together, making identification very difficult. In fact, the interface waves shown in **Figure 4** are for a layered seafloor, where the dispersion of the signal is evident (homogeneous half-space would not give any dispersion).

Under realistic conditions, i.e., for an inhomogeneous, bounded and anisotropic seafloor, some of these waves convert from one to another. The different wave types may travel with different speeds or together, and they generally have different attenuation. As an example, **Figure 5** shows signals from an explosive source (0.5 kg trinitrotoluene (TNT))

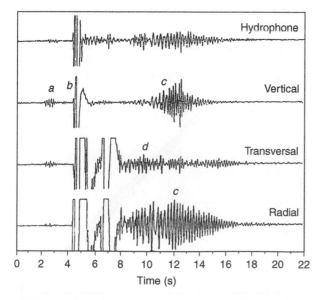

Figure 5 Signals from an explosive source of 0.5 kg trinitrotoluene received by a hydrophone and three orthogonal geophones at 1.5 km distance (a, Head wave; b, Compressional wave; c, Interface wave; d, Love wave).

received by a hydrophone and three orthogonal geophones placed on the seafloor, at a distance of 1.5 km from the source. The broadband signal of a few milliseconds duration generated by the explosive source is dispersed over nearly 18 s demonstrating the arrival structure of different types of waves. The characteristics of these waves are indicated in the figure for four different sensors. They can be identified in order of their arrival time as: (a) head wave; (b) water arrival (compressional wave); (c) interface wave; (d) Love waves.

Seafloor Roughness

The roughness of the water–sediment interface and layers below is another important parameter that needs to be considered in sediment acoustics. The seafloor contains a wide spectrum of topographic roughness, from features of the order of tens of kilometers, to those of the order of millimeters. The shape of the seafloor and its scattering effect on acoustic signals is be covered here.

Techniques to Measure Geoacoustic Parameters of Marine Sediments

The geoacoustic properties of the seafloor defined by the compressional and shear wave velocities, their attenuation, together with the knowledge of the material bulk density, and their variation as a function of depth, are the main parameters needed to solve the acoustic wave equation. To be able to determine these properties of the seabed, different techniques have been developed using samples taken from the seafloor, instruments and divers conducting measurements *in situ*, and remote techniques measuring seismo-acoustic waves and inverting this information with realistic models into sediment properties. Some of the current methods of obtaining geoacoustic parameters of the marine sediments are briefly described here.

Laboratory Measurements on Sediment Core Samples

Most of our knowledge of the physical properties of sediments is acquired through core sampling. A large number of measurements on marine sediments have been made in the past. In undisturbed sediment core samples, under laboratory conditions, density and compressional velocity can be measured with accuracy, and having measured values of density and compressional velocity, the bulk modulus can be selected as the third parameter, where it is can be calculated (**Table 1**).

There are several laboratory techniques available to measure some of the sediment properties. However, the reliability of such measurements can be degraded by sample disturbance and temperature and pressure changes. In particular, the acoustic properties are highly affected by the deterioration of the chemical and mechanical bindings caused by the differences in temperature and pressure between the sampling and the laboratory measurements. Controlling the relationships between various physical parameters can be used to check the accuracy of measured parameters of sediment properties. It has been shown that the density and porosity of sediments have a relationship with compressional velocity, and different empirical equations between them have been established.

Over the past three decades, at the NATO Undersea Research Centre (SACLANTCEN), a large number of laboratory and *in situ* measurements have been made of the physical properties of the seafloor sediments. These measurements have been conducted on the samples with the same techniques as when the same laboratory methods were applied. This data set with a great consistency of the hardware and measurement technique, has been used to demonstrate the physical characteristics of the sediment that affect acoustic waves.

The relationship between measured physical parameters From 300 available cores, 20 000 measured data samples for the density and porosity, and 10 000 samples for the compressional velocity were obtained. To be able to handle this large data set taken from different oceans at different water depths, all bulk density and compressional velocity data were converted into relative density (ρ) and relative velocity (C) with respect to the *in situ* water values:

$$\rho = \rho_S / \rho_W$$
$$C = C_S / C_W$$

where, ρ_S is sediment bulk density, ρ_W is water density, C_S is sediment compressional velocity and C_W is water compressional velocity. The data are not only from the water–sediment interface but cover sedimentary layers of up to 10 m deep.

Relative density and porosity The density and porosity of the marine sediment are least affected during coring and laboratory handling of the samples. Porosity is given by the percentage volume of the porous space and sediment bulk density by the weight of the sample per unit volume. The relationship between porosity and bulk density has been investigated by many authors with fewer data than used here and shown to have a strong linear correlation. Theoretically this linearity only exists if the dry densities of the mineral particles are the same for all marine sediments. The density of the sediment would then be the same as the density of the solid material at zero porosity, and the same as the density of the water at 100% porosity. **Figure 6** shows this relationship.

Porosity and relative compressional wave velocity The relationship between porosity and compressional wave velocity has received much attention in the literature because porosity can be measured easily and accurately. Data from the SACLANTCEN sediment cores giving the relationship between porosity and compressional wave velocity are shown in **Figure 7**. As shown in **Figure 6** due to the linear relation between density and porosity, the relationship between density and compressional wave velocity is similar to the porosity compressional wave velocity relation.

In situ Techniques

There are several *in situ* techniques available that use instruments lowered on to the seafloor mainly by means of submersibles, remotely operated vehicles (ROVs), and autonomous underwater vehicles AUVs and divers. The first deep-water, *in situ* measurements of sediment properties were made from the bathyscaph *Trieste* in 1962. These measurements provided accurate results due to the minimum disturbance of temperature and pressure changes compared to bringing the sample to the surface and for

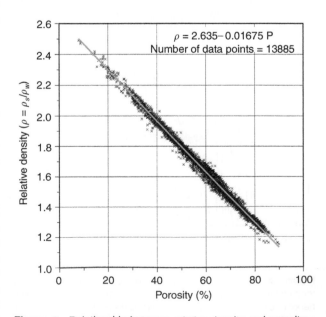

Figure 6 Relationship between relative density and porosity.

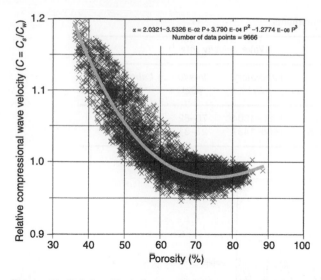

$$\alpha = 2.0321 - 3.5326\ \text{E-02}\ P + 3.790\ \text{E-04}\ P^2 - 1.2774\ \text{E-06}\ P^3$$
Number of data points = 9666

Figure 7 Relationship between relative compressional wave velocity and porosity.

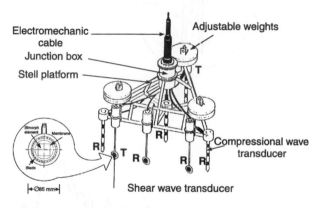

Figure 8 *In Situ* Sediment Acoustic Measurement System (ISSAMS) for near-surface *in situ* geoacoustic measurements.

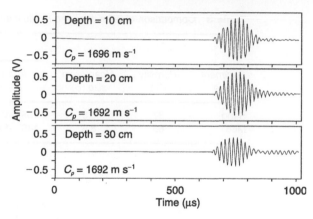

Figure 9 Examples of received compressional wave signals from fine sand sediment.

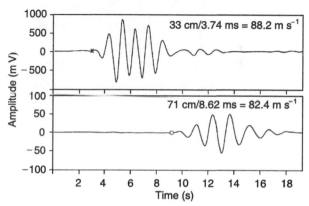

Figure 10 Examples of signals recorded from two shear wave receivers in hard-packed fine sand sediments.

analysis in the laboratory. The most reliable direct geoacoustic measurement techniques for marine sediments are *in situ* techniques. Some of the approaches used in recent years are discussed below.

Near-surface method A system has been developed to measure sediment geoacoustic parameters, including compressional and shear wave velocities and their attenuation at tens of centimeters below the sediment–water interface. **Figure 8** shows the main features of the *In situ* Sediment Acoustic Measurement System (ISSAMS). Shear and compressional wave probes are attached to a triangular frame that uses weights to force the probes into the sediment. In very shallow water, divers can be used to insert the probes into the sediment, whereas in deeper water, a sleeve system (not shown) allows the ISSAMS to penetrate into the seafloor.

Using the compressional and shear wave transducers measurements are made with a continuous wave (cw) pulse technique where the ratio of measured transducer separation and pulse arrival time yields the wave velocity. Samples of compressional and shear wave data are shown in **Figure 9** and **10** respectively. **Table 3** gives a comparison of laboratory and *in situ* values of compressional and shear wave velocities from two different types of Adriatic Sea sediment.

Cross-hole method Measurements as a function of depth in sediments can be made with boreholes, using either single or cross-hole techniques. Boreholes are made by divers using water–air jets to penetrate thin-walled plastic tubes for cross-hole measurements. **Figure 11** shows the experimental set up for the cross-hole measurements. The source is in the form of an electromagnetic mallet securely coupled to the inner wall of one of the plastic tubes with a hydraulic clamping device.

Table 3 Comparison of laboratory and *in situ* measurements

Sediment type	Porosity (%)	Mean grain size (ϕ)	Wave velocity (ms^{-1})			
			In situ (C_P)	Laboratory (C_P)	In situ (C_S)	Laboratory (C_S)
Sand	37	3.5	1557–1568	1580–1604	78–82	50
Mud	68	8.6	1467–1488	1468–1487	27–31	15

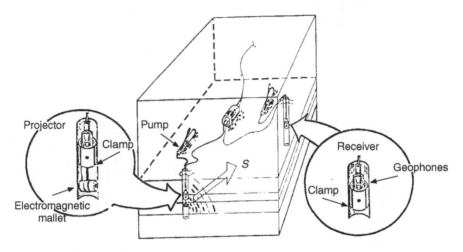

Figure 11 The experimental setup for cross-hole measurements.

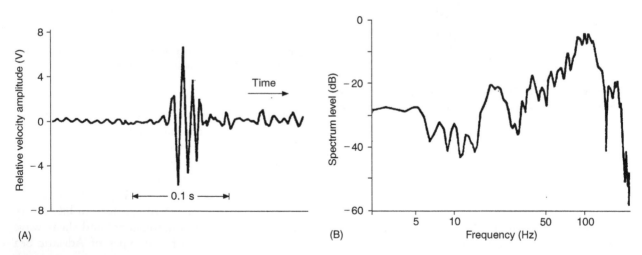

Figure 12 Cross-hole shear wave signal (A) and its spectrum (B) for a silty-clay bottom in the Ligurian Sea.

With a separation range of 2.9 m, moving-coil geophone receivers are also coupled to the inner wall of the second plastic tube with a hydraulic clamping device. The electromagnetic mallet generates a point source on the thin plastic tube wall, which results in a multipolarized transient signal to the sediment. Depending on the orientation of receiving sensors, vector components of the propagating compressional and shear waves are received and analyzed for velocity and attenuation parameters. Examples of a time series and a frequency spectrum for a shear wave signal received by a geophone with a natural frequency of 4.5 Hz are shown in **Figure 12**.

Figure 13 shows shear wave velocity as a function of depth obtained in the Ligurian Sea. It can be seen that the shear wave velocities are around 60 m^{-1} s at the sediment interface and increase with depth.

Remote Sensing Techniques

Even though *in situ* techniques provide the most reliable data, they are usually more time consuming and expensive to make and they are limited to small areas. Remote sensing and inversion to obtain geoacoustic parameters can cover larger areas in less time and provide reliable information. These techniques are based on the use of a seismo-acoustic

signal received by sensors on the seafloor and/or in the water column. **Figure 14** illustrates a characteristic shallow water signal from an explosion received by a hydrophone close to the sea bottom. Three different techniques to extract information relative to bottom parameters from these signals are described briefly below.

Reflected waves

The half-space seafloor. When the seafloor consists of soft unconsolidated sediments, due to its very low shear modulus it can be treated as fluid.

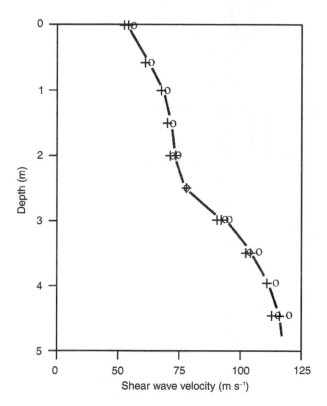

Figure 13 Shear wave velocity profile obtained from cross-hole measurements.

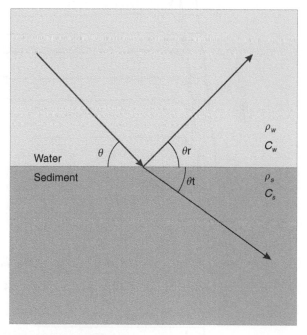

Figure 15 Geometry and notations for a simple half-space water–sediment interface.

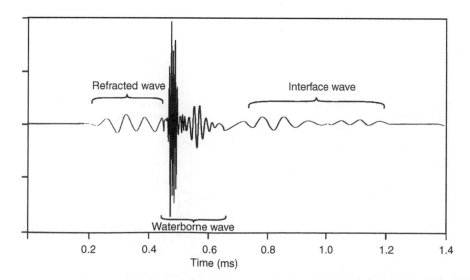

Figure 14 A characteristic shallow-water signal.

Reflection may occur whenever an acoustic wave is incident on an interface between two media. The amount of energy reflected, and its phase relative to the incident wave, depend on the ratio between the physical properties on the opposite sides of the interface. It is possible to calculate frequency independent reflection loss as the ratio between the amplitude of the shock pulse and the peak of the first reflection (after correcting for phase shift, absorption in the water column, and differences in spreading loss). If the relative density $\rho = \rho_S/\rho_W$, and the relative compressional wave velocity

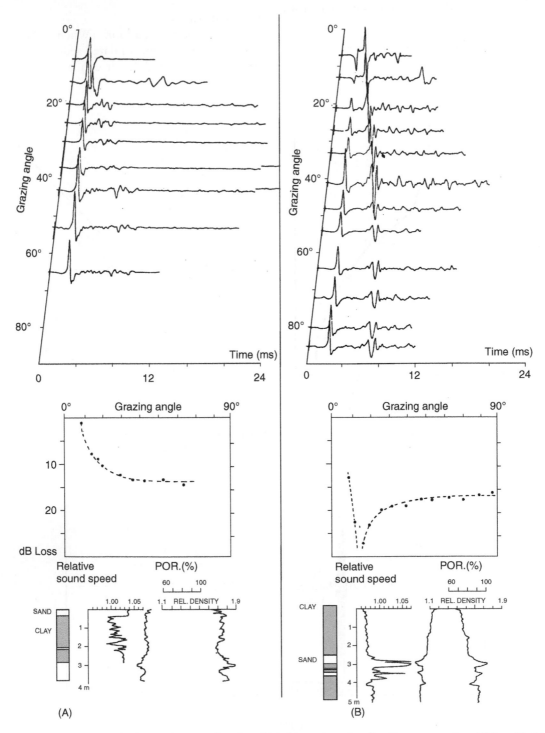

Figure 16   Acoustic signals and reflection loss as a function of grazing angle and sediment core properties: (A) for critical angle; (B) for angle of incidence case.

$C = C_S/C_W$, are used to present the contrast between the two media, reflection coefficient for such a simple environmental condition can be written as:

$$R = \frac{\rho\sin\Theta - \sqrt{(1/C^2 - \cos^2\Theta)}}{\rho\sin\Theta - \sqrt{(1/C^2 - \cos^2\Theta)}}$$

Where Θ is the grazing angle with respect to the interface as shown in **Figure 15**. For an incident path normal to a reflecting horizon, i.e., $\Theta = 90°$, the reflection coefficient is:

$$R = \rho C - 1/\rho C + 1$$

Critical angle case. When the velocity of compressional wave velocity is greater in the sediment layer $(C>1)$, as the grazing angle is decreased, a unique value is reached at which the acoustic energy totally reflects back to the water column. This is known as the critical angle and is given by:

$$\Theta_{cr} = \arccos(1/C)$$

When the grazing angle is less than this critical angle, all the incident acoustic energy is reflected.

However, the phase of the reflected wave is then shifted relative to the phase of the incidence wave by an angle varying from 0° to 180° and is given as:

$$\Phi = -2\arctan\frac{\sqrt{(\cos^2\Theta - 1/C^2)}}{\rho\sin\Theta}$$

Figure 16 shows measured and calculated reflection losses (20 log R) for this simple condition together with the basic physical properties of the core sample taken in the same area for explosive signals at different grazing angles.

Angle of intromission case. Especially in deep-water sediments the sound velocity in the top layer of the bottom is generally less than in the water above $(C<1)$. In such conditions there is an angle of incidence at which all of the incident energy is transmitted into the sedimentary layer and the reflection coefficient becomes zero:

$$\cos\Theta_i = \sqrt{\left(\frac{\rho^2 - 1/C^2}{\rho^2 - 1}\right)}$$

The phase shift is 0° when the ray angle is greater than the intromission angle and 180° when it is smaller. Thus, acoustical characteristics of the bottom, such as the critical angle, the angle of incidence, the phase shift,

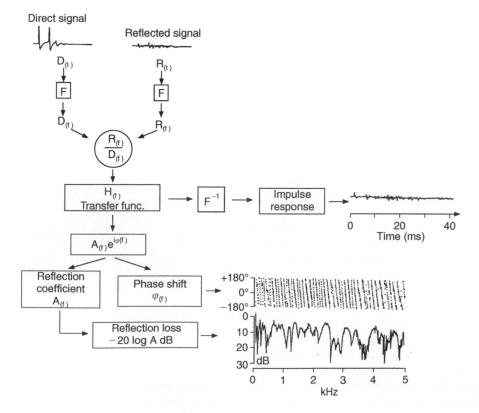

Figure 17 Acoustic reflection data and analysis technique for a layered seafloor.

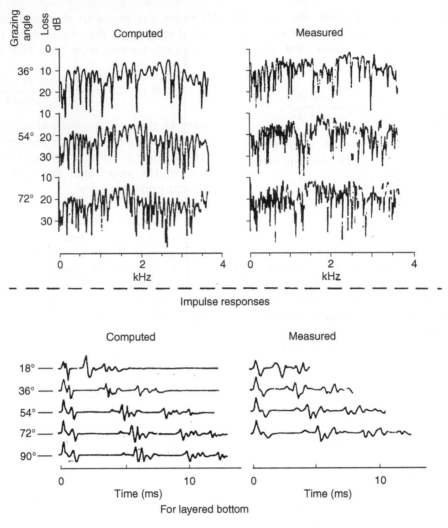

Impulse responses

For layered bottom

Figure 18 Measured and calculated reflection losses for a layered seafloor.

and the reflection coefficient, are primarily influenced by the relative density and relative sound velocity of the environment as a function of the ray angle.

The layered seafloor. Since the seafloor is generally layered, a simple peak amplitude approximation cannot be implemented because of the frequency dependence. In this case one calculates the transfer function (or reflection coefficient) from the convolution of the direct reference signal with the reflected signal. Examples of the phase shift and reflection loss as a function of frequency are shown in **Figure 17**. The reflectivity can also be described in the time domain by the impulse response, which is the inverse Fourier transform of the transfer function as shown in the same figure. This type of data can be utilized for inverse modeling to obtain the unknown parameters of the sediments. **Figure 18** is an example of measured and calculated reflection losses and impulse responses for a layered seafloor.

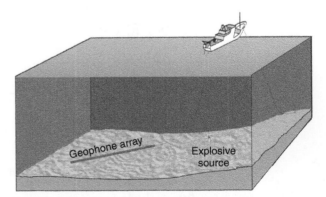

Figure 19 Experimental setup to measure seismo-acoustic waves on the seafloor.

Refracted waves Techniques developed for remote sensing of the uppermost sediments (25–50 m below the seafloor) utilize broad-band sources (small

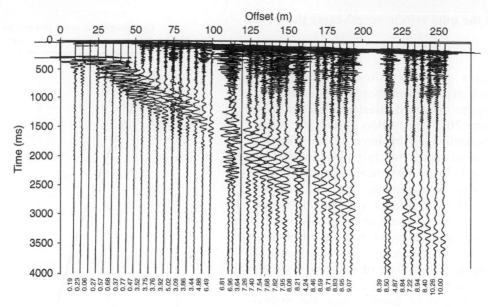

Figure 20   Signals received by geophone array.

explosives) and an array of geophones deployed on the seafloor. To obtain estimates of the bottom properties as a function of depth, both refracted compressional and shear waves as well as interface waves are analyzed. Inversion of the data is carried out using modified versions of techniques developed by earthquake seismologists to study dispersed Rayleigh-waves and refracted waves. **Figure 19** illustrates the basic experimental setup and **Figure 20** the signals received by an array of 24 geophones that permit studies of both interface and refracted waves.

These data are analyzed and inverted to obtain both compressional and shear wave velocities as a function of depth in the seafloor. Studies of attenuation and lateral variability are also possible using the same data set.

Figure 21 shows the expanded early portion of the geophone data shown in **Figure 20**, where, the first arriving energy out to a range of about 250 m has been fitted with a curve showing that compressional waves refracted through the sediments just beneath the seafloor travel faster as they penetrate more deeply into the sediment. At zero offset, the slope indicates a velocity of 1505 m s^{-1} whereas at a range offset at 250 m the slope corresponds to a velocity of 1573 m s^{-1}. At ranges over 250 m, a strong head wave becomes the first arrival and, the interpretation would be that there is an underlying rock layer with compressional wave velocity of about 4577 m s^{-1}.

A compressional wave velocity–depth curve for the upper part of the seafloor can be derived from the first arrivals shown in **Figure 21** using the classical Herglotz–Bateman–Wieckert integration method.

The slope of the travel-time curve (fitted parabola) gives the rate of change of the range with respect to time that is also the velocity of propagation of the diving compressional wave at the level of its deepest penetration (turning point) into the sediment. At each range Δ, the depth corresponding to the deepest penetration is then calculated using the following integral

$$z(V) = 1/\pi \int_0^\Delta \cosh^{-1}(V(\mathrm{d}t/\mathrm{d}x))\mathrm{d}x$$

where

$$1/V = (\mathrm{d}t/\mathrm{d}x)_{x=\Delta}$$

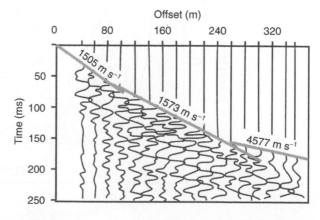

Figure 21   Expanded early portion of the data shown in **Figure 20**.

The result is the solid velocity–depth curve shown in **Figure 22**.

Interface waves In order to obtain a shear wave velocity–depth profile from the data, later arrivals corresponding to dispersed interface waves may be utilized (**Figure 20**). The portion of each individual signal corresponding to the interface-wave arrival can be processed using multiple filter analysis to create a group velocity dispersion diagram (Gabor diagram). The result of applying this technique to a dispersed signal is a filtered time signal whose envelope reaches a maximum at the group velocity arrival time for a selected frequency. The envelope is computed by taking the quadrature components of the inverse Fourier transform of the filtered signal. Filtering is carried out at many discrete frequencies over selected frequency bands. Once the arrival times are converted in to velocity, the envelopes are arranged in a matrix and contoured and dispersion curves are obtained by connecting the maximum values of the contour diagram (**Figure 23**).

Having obtained the dispersion characteristics of the interface waves, the geoacoustic model, made of a stack of homogeneous layers with different compressional and shear wave velocities for each layer that predict the measured dispersion curve is determined. **Figure 24** illustrates a number of examples from the Mediterranean sea covering data from soft clays to hard sands.

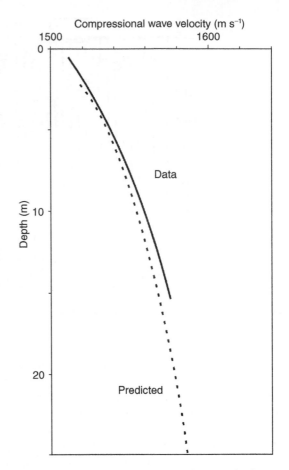

Figure 22 Compressional wave velocity versus depth curves derived from data and predicted (dashed line) by the model.

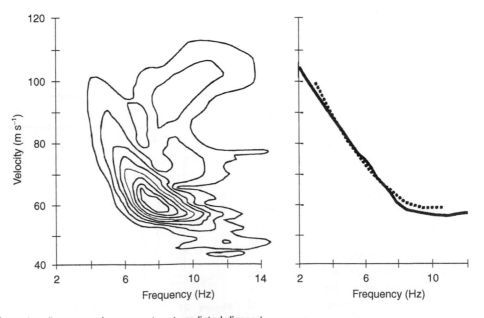

Figure 23 Dispersion diagram and measured and predicted dispersion curves.

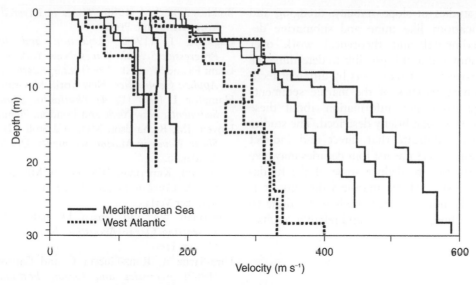

Figure 24 Summary of shear wave velocity profiles from the Mediterranean Sea.

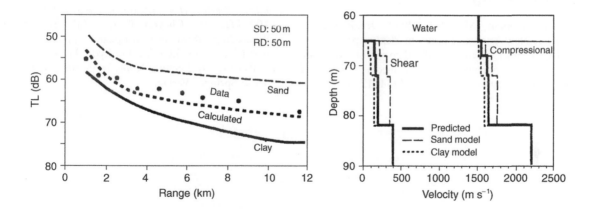

Figure 25 Comparison of transmission loss (TL) data and SAFARI prediction. Effects of changing bottom parameters from sand to clay is also shown together with the input profiles utilized for SAFARI predictions.

Transmission Loss Technique

The seafloor is known to be the controlling factor in low-frequency shallow water acoustic propagation. Forward modeling is performed with models giving exact solutions to the wave equation, i.e., SAFARI where, compressional and shear wave velocities, the attenuation factors associated with these waves, and the sediment density as a function of depth are the main input parameters. Acoustic energy propagating through a shallow water channel interacts with the seafloor causing partitioning of waterborne energy into different types of seismic and acoustic waves. The propagation and attenuation of these waves observed in such an environment are strongly dependent on the physical characteristics of the sea bottom. Transmission loss (TL), representing the amount of energy lost along an acoustic propagation path, carries the information relative to the environment through which the wave is propagating. **Figure 25** shows a comparison of TL data and model predictions together with the input parameters used at 400 Hz. The effects of changes in bottom parameters are also shown in the figure. This technique becomes extremely useful when seafloor information is sparse.

Conclusions

Acoustic/seismic characteristics of the marine sediments have been of interest to a wide range of activities covering commercial operations involving trenching and cable lying, construction of offshore foundations, studies of slope stability, dredging and military applications like mine and submarine detection. Experimental and theoretical work over the years has shown that it is possible to determine the geoacoustic properties of sediments by different techniques. The characteristics of the marine sediments and techniques to obtain information about these characteristics have been briefly described. The studies so far conducted indicate that direct and indirect methods developed over the last four decades may give sufficient information to deduce some of the fundamental characteristics of the marine sediments. It is evident that still more research needs to be done to develop these techniques for fast and reliable results.

See also

Acoustics, Shallow Water.

Further Reading

Akal T and Berkson JM (eds.) (1986) *Ocean Seismo-Acoustics*. New York and London: Plenum press.

Biot MA (1962) Generalized theory of acoustic propagation in porous dissipative media. *Journal of Acoustical Society of America* 34: 1254–1264.

Brekhovskikh LM (1980) *Waves in Layered Media*. New York: Academic Press.

Cagniard L (1962) *Reflection and Refraction of Progressive Seismic Waves*. New York: McGraw-Hill.

Grant FS and West GF (1965) *Interpretation and Theory in Applied Geophysics*. New York: McGraw Hill.

Hampton L (ed.) (1974) *Physics of Sound in Marine Sediments*. New York and London: Plenum Press.

Hovem JM, Richardson MD, and Stoll RD (eds.) (1992) *Shear Waves in Marine Sediments*. Dordrecht: Kluwer Academic.

Jensen FB, Kuperman WA, Porter MB, and Schmidt H (1999) *Computational Ocean Acoustics*. New York: Springer-Verlag.

Kuperman WA and Jensen FB (eds.) (1980) *Bottom Interacting Ocean Acoustics*. New York and London: Plenum Press.

Lara-Saenz A, Ranz-Guerra C and Carbo-Fite C (eds) (1987) *Acoustics and Ocean Bottom*. II F.A.S.E. Specialized Conference. Inst. De Acustica, Madrid.

Pace NG (ed.) (1983) *Acoustics and the Sea-Bed*. Bath: Bath University Press.

Pouliquen E, Lyons AP, Pace NG *et al.* (2000) *Backscattering from unconsolidated sediments above 100 kHz*. In: Chevret P and Zakhario ME (ed.) Proceedings of the fifth European Conference on Under water Acoustics, ECUA 2000 Lyon, France.

Stoll RD (1989) *Lecture Notes in Earth Sciences*. New York: Springer-Verlag.

ACOUSTICS, ARCTIC

P. N. Mikhalevsky, Science Applications
International Corporation, McLean, VA, USA

Introduction

The Arctic Ocean is an isolated mediterranean basin
with only limited communication with the world's
oceans, principally the Atlantic Ocean via the Fram
Strait and the Barents Sea, and the Pacific Ocean via
the Bering Strait. The ubiquitous feature of the Arctic
Ocean is the sea ice that covers the entire Arctic
basin during the winter months and only retreats off
the shallow water shelf areas in the summer months,
creating a permanent cap over most of the central
Arctic basin (**Figure 1**). The presence of the year-
round sea ice cover determines the unique character of
acoustic propagation and ambient noise in the Arctic
Ocean. The sea ice nsulates the Arctic Ocean from
solar heating in the summer months, creating a year-
round upward refracting sound speed profile with the
sound speed minimum at the water–ice interface.
Sound, therefore, is refracted upward and is continuously reflected from the ice as it propagates,
causing attenuation by scattering, mode conversion,
and absorption that increases rapidly with frequency.
The lack of solar forcing and the Arctic Ocean's restricted communication with the other oceans of the
world creates a very stable acoustic channel with
significantly reduced fluctuations of acoustic signals in
comparison with the temperate oceans. In contrast to
the central basin, acoustic propagation on the Arctic
shelves and in the marginal ice zones (MIZs, those
areas between the average ice minimum and maximum) (**Figure 1**), is quite complex and variable owing
to the seasonal retreat of the sea ice, river run-off, and
bottom interaction (*see* Acoustics, Shallow Water,
Acoustics in Marine Sediments).

Over the last half-century Arctic acoustics research
and development has largely supported submarine
operations. The importance of the Northern Sea
Route to the Soviet Union, and the prospect of Soviet
nuclear ballistic missile submarines exploiting the
unique Arctic acoustic environment to remain undetected provided the need for this research. Since
the end of the Cold War and the beginning of concern about 'global warming' there has been a new
focus for Arctic acoustics on acoustic thermometry
and acoustic remote sensing (*see* Tomography). The
Arctic Ocean is the world's 'air-conditioner', maintaining the surface heat balance, and it provides fresh
water to the world's oceans, principally in the form
of sea ice discharged from the Fram Strait. The latter
regulates convective overturning in the Greenland
and Norwegian Seas that in turn drives the global
thermohaline circulation with significant impact on
climate. Monitoring changes in the temperature and
stratification of the Arctic Ocean and sea ice thickness using acoustics is an important capability that
will improve our understanding of the Arctic Ocean
and its role in global climate change.

History

The first large program dedicated to acoustics research in the Arctic Ocean was undertaken by the US
Navy Underwater Sound Laboratory, as it was then
known, in 1958 in connection with the International
Geophysical Year. Also in that year the US nuclear
submarine *Nautilus* (SSN 571) made its historic
voyage to the North Pole, marking the beginning of
regular submarine operations in the Arctic Ocean.
From 1958 to 1975 the USA conducted much of its
acoustics research from manned ice islands (thick
tabular sections of land-fast ice that occasionally
break away from the Ellesmere ice shelf) that remained within the polar pack ice (*see* Cryosphere:
Sea Ice). The most famous of these was Fletcher's Ice
Island, also known as T3, which was discovered in
June 1950. T3, originally 14.5 km long, 6.4 km wide
and 52 m thick, provided an ideal platform for Arctic
research including acoustics. The first experiments in
1958 were aimed at investigating the feasibility of
using the RAFOS (Ranging and Fixing of Sound) for
submarine navigation in Arctic waters. Explosive
signals were deployed from T3 and other ice stations
while the *USS Skate* (SSN 578) attempted to receive
the signals for acoustic crossfixing. While RAFOS
was never operationally deployed, a significant
amount was learned about the upward refracting
sound speed profile, propagation, and resulting
scattering losses, as well as ambient noise and
reverberation.

The Soviet Union operated at least two manned
year-round ice stations in the Arctic continuously
from 1937 through 1992. In 1971 they began
POLEX (Polar Experiment), which was an intense
set of oceanographic, ice, and atmospheric measurements over the entire Arctic Ocean basin that

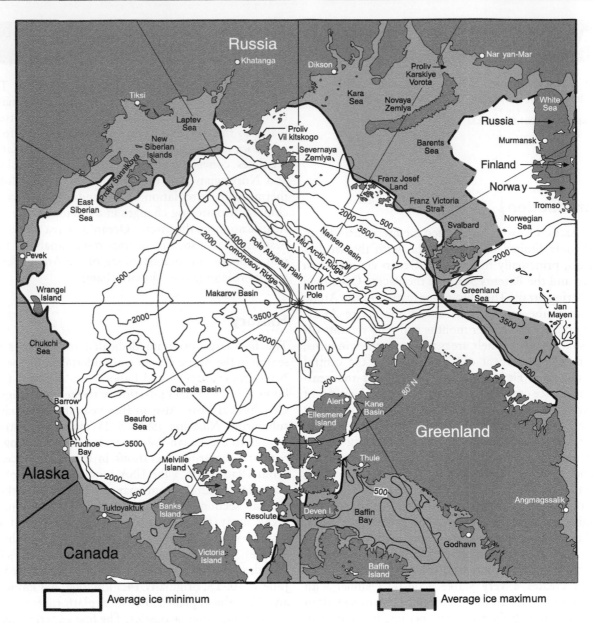

Figure 1 Map of the Arctic Ocean showing the average minimum and maximum sea ice extent, as well as the major bathymetric features, and other significant geographic locations.

continued for a decade. From 1971 through 1994 the US Office of Naval Research sponsored or co-sponsored many international acoustic-related research science programs based from seasonal ice camps in the Arctic, staged out of facilities in Alaska, Canada, Greenland, and Svalbard. These included the Arctic Ice Dynamics Experiments (AIDJEX), 1971–76; the FRAM and MIZEX series in the central Arctic basin and marginal ice zones respectively, 1979–87; the Coordinated East Arctic Experiment (CEAREX), 1988–89; the Sea Ice Mechanics Initiative (SIMI), 1993–94; and the Transarctic Acoustic Propagation Experiment (TAP) in 1994. These programs

quantified sea ice properties and ice scattering processes, bottom and surface reverberation, Arctic plate tectonics with seismic reflection and refraction, identified the mechanisms of ice generated ambient noise, discovered the exceptional phase stability of low frequency propagation ($\sim$20–30 Hz), and demonstrated the feasibility of basin scale acoustic thermometry in the Arctic Ocean.

The US, UK, and Soviet Union (now Russia) have operated submarines in the Arctic Ocean and conducted a significant amount of acoustics research. In particular, the US Navy's Submarine Ice Exercises (SUBICEXs) were conducted over almost three decades by the Arctic Submarine Laboratory (ASL) under the leadership of Dr Waldo Lyon, often in conjunction with seasonal ice camps. Dr Lyon was one of the first to investigate propagation in polar stratified waters, beginning shortly after the end of World War II, working with Canadian researchers. From the Naval Electronics Laboratory (NEL) in San Diego, Dr Lyon led the development of upward-looking ice profiling sonar, and the forward-looking ice avoidance sonar, that proved critical for safe under-ice submarine operations. SUBICEXs were designed to test all aspects of submarine operations including surfacing through ice, and weapons effectiveness (see Under Ice in Further Reading). In 1993 the US Navy supported a dedicated 60 day science cruise in the Arctic with the USS Pargo (SSN 650). This started the Submarine Science Ice Expeditions (SCICEX) that were conducted each year from 1995 to 1999. These multidisciplinary cruises have confirmed the large changes in the Arctic Ocean thermohaline structure, including the warming of the Arctic Intermediate Water (AIW, see below) that was measured acoustically during the TAP experiment in 1994. Other data and modeling have shown strong evidence of decadal Arctic oscillations, but longer observational time series are needed. A trans-basin acoustic section was started in October 1998 from a source located in the Franz Victoria Strait to a receiving array in the Lincoln Sea. This 3-year US/Russian collaborative effort is expected to be followed by more permanent long-term monitoring of the Arctic Ocean using acoustics.

Arctic Ocean Sound Speed Structure

The structure of the typical sound speed profile in the Arctic consists of components that correspond closely to the primary stratified water masses of the Arctic Ocean (Figure 2). The uppermost layer is Polar Water (PW) defined by temperatures $< 0°C$ and salinities < 34.5 ppt and extends from the surface to depths of 100–200 m. There is often a mixed layer of Polar Water 30–60 m thick just below the ice with a nearly constant temperature near the freezing point and nearly constant salinity at around 33.6 ppt. Below the mixed layer the temperature and salinity increase uniformly, attaining 0°C and 34.5 ppt respectively, at the base of the Polar Water layer. Arctic Intermediate Water (AIW) – also known as the Atlantic Layer reflecting its origin (and sometimes referred to as Atlantic Intermediate Water) – is defined by temperatures $> 0°C$ and salinities > 34.5 ppt and extends from the base of the Polar Water to a depth of 1000 m. Temperature increases with depth through the AIW layer to a maximum at approximately 200–500 m depth and then decreases with depth to 0°C at about 1000 m. AIW enters the Arctic Ocean via the West Spitzbergen Current from the Fram Strait and via the Barents Sea east of Franz Josef Land (see Ocean Currents: Arctic Basin Circulation). The shallower depths (200 m) of the temperature maximum of the AIW occur in the eastern Arctic and approaches 2°C. This temperature maximum deepens to 500 m the western Arctic in the Canada Basin and is approximately 0.4°C. Salinity

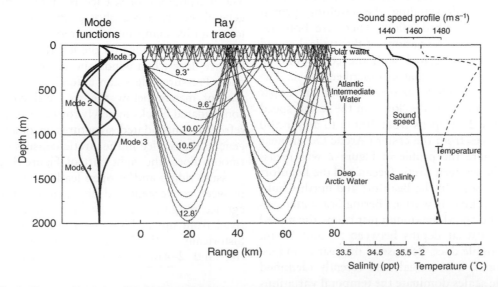

Figure 2 The Arctic sound speed profile shown on the right was computed from the measured temperature and salinity, also shown, from an ice camp in the eastern Arctic Ocean in April 1994. A ray trace for this sound speed profile is plotted for a source at a depth of 100 m. The mode shapes computed at 20 Hz using this profile are shown on the left. The major Arctic water masses (see text) are indicated.

increases with depth in the AIW layer from 34.5 ppt to 34.9 ppt at about the same depth of the temperature maximum and remains constant at 34.9 ppt below this depth. Deep Arctic Water (DW) is a relatively homogeneous water mass from a depth of 1000 m to the bottom, with temperatures $<0°C$ and a nearly constant salinity of 34.9 ppt.

The resulting sound speed profile (**Figure 2**) for these typical deep Arctic basin conditions has a minimum at the ice–water interface of 1435–1440 m s^{-1} and increases with depth to the bottom. In the near-surface mixed layer of the PW the sound speed gradient is $+0.016$ s^{-1}, due entirely to the increase of pressure with depth, as temperature and salinity are constant. Below the mixed layer the sound speed increases rapidly with the increasing temperature and salinity (temperature having the far dominant effect) to the depth of the maximum temperature in the AIW layer with gradients of $+0.1$ s^{-1} or more, reaching a sound speed between 1455 and 1465 m s^{-1}. Below the depth of the AIW temperature maximum the sound speed continues to increase, but with reduced gradients of $+0.01$ s^{-1} or less as the temperature decreases, but the pressure increases and the salinity is constant. Below 1000 m in the DW both temperature and salinity are nearly constant and the sound speed continues to increase with the $+0.016$ s^{-1} gradient to the bottom associated with the increasing pressure.

In general the Arctic or polar profile can be characterized (and is often approximated) roughly as a bilinear upward refracting profile with the large positive gradient at the near surface creating a strong near-surface duct and a smaller positive gradient below this 'knee' to the bottom. This structure is important to understanding the resulting propagation effects described below. There are both regional and temporal variations of this typical sound speed structure associated with the corresponding variations in the water masses described above. The strong positive gradient associated with the upper AIW layer weakens as the 'knee' in the sound speed profile (**Figure 2**) deepens from 200 m to about 500 m as one moves from the eastern Arctic Ocean near Svalbard (where the profile in **Figure 2** was measured) into the western Arctic Ocean and the Beaufort and Chukchi Seas. In the Beaufort and north Chukchi Seas the influx of warmer Bering Sea water can create a sound speed maximum just below the mixed layer of the PW at depths between 50 and 80 m, creating a double duct. Because of the year-round ice cover, seasonal, interannual, and recently identified decadal time scales dominate the temporal variations of the sound speed profile. The seasonal and interannual variability is confined almost entirely to the PW layer. The decadal variability is seen in the PW

Figure 3 A photograph of the pack ice in the central Arctic Ocean taken at the FRAM II Ice Camp in May 1980. A weathered pressure ridge is shown on the left bounding a relatively flat multi-year ice floe on the right that is typical of the central arctic. The blocks of ice in the foreground are 3–4 m high. (Photo by P. Mikhalevsky.).

and AIW layers. Mesoscale and internal wave dynamics are approximately one order of magnitude less energetic in the Arctic Ocean than in the temperate seas and so have a small influence on the sound speed profile and this results in the exceptional signal stability at low frequencies (discussed below).

Propagation

Acoustic propagation in the Arctic is dominated by the repeated reflection of the sound from the sea ice as illustrated in **Figure 2**. The ray trace shown was computed using the sound speed profile of **Figure 2**. The morphology of sea ice is quite complicated in general. **Figure 3** is a photograph taken in the pack ice at an ice camp in the eastern Arctic. It shows a typical pressure ridge on the left that forms between two ice floes when winds and currents compress the pack. Pressure ridges can extend as much as several tens of meters below the ice, and consist of unconsolidated blocks in newly formed ridges as well as refrozen consolidated structures in older ridges. In general the sea ice in the Arctic consists of relatively smooth floes (the right part of **Figure 3**) laced with pressure ridges and/or open leads as the pack works between convergent and divergent conditions. Average ice roughness and thickness measurements derived from upward-looking submarine sonars have historically (1958–76) been in the range of 1–5 m rms and 2–4 m respectively.[1] Recent analysis of

[1] LeSchack LA (1980) *Arctic Ocean Sea-ice Statistics Derived from Upward-Looking Sonar Data Recorded during Five Nuclear Submarine Cruises.* Technical Report to ONR under Contract

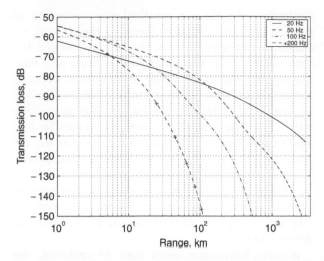

Figure 4 The transmission loss plotted as a function of range for 20, 50, 100, and 200 Hz. The source and receiver were located at 60 m depth. The ice was modeled as a rough elastic plate including statistical parameters for the upper and lower side of the ice representative of the ice floe and ridge structure typical of the central Arctic.

SCICEX data (1993–97), however, has shown that since the late 1970s the ice has thinned by as much as 40%,[2] which would also imply a reduction in ice roughness as well. The thicker ice and correspondingly greater roughness is typically found in the eastern Arctic and just north of the eastern Canadian Archipelago and Greenland, where the transpolar drift pushes the ice.

The acoustic energy is reflected and scattered from the rough ice and converted to both shear waves and compressional waves within the ice, resulting in significant frequency-dependent attenuation loss. Because of this, trans-basin propagation in the Arctic is limited to frequencies typically below 30 Hz or wavelengths exceeding about 50 m. **Figure 4** shows the propagation loss plotted as a function of range for several frequencies. The curves in **Figure 4** were generated using a model that includes a rough elastic plate representing the ice cover, with a two scale roughness spectrum that models pressure ridges and the smoother intervening floes.[3] The model includes the statistics for both the underside and surface of the ice since at lower frequencies as the wavelength

(*footnote continued*)
N00014-76-C-0757/NR 307-374. Maryland, LeSchack Associates, Ltd.

[2]Rothrock DA, Yu Y and Maykut GA (1999) Thinning of the arctic sea-ice cover. *Geophysical Research Letters* 26(23): 3469–3472.

[3]Kudryashov VM (1996) Calculation of the acoustic field in an arctic waveguide. *Physical Acoustics*. 42: 386–389.

becomes much greater than the ice thickness the reflection occurs at the ice–air interface. Submarine ice draft statistics from the eastern Arctic over the Nansen Basin (**Figure 1**) were used for **Figure 4**. As **Figure 4** shows the Arctic acoustic waveguide is a low pass filter. So much so in fact that in comparison with propagation in the temperate oceans which scatter from surface waves that have a similar roughness spectrum, at a given range, equivalent propagation loss in the Arctic requires transmitting at a frequency as much a factor of ten less.

On the left side of **Figure 2** the acoustic mode shapes for modes 1–4 at 20 Hz are plotted for the profile shown. The acoustic modes can be thought of as the interference pattern of upward and downward going acoustic rays (**Figure 2**) whose turning depths (the depth where the rays are horizontal) correspond with the depth of the deepest peak of the mode amplitude. At 20 Hz, mode 1 consists of those paths that are trapped in the strong near-surface duct created by the thermocline in the transition from the PW to AIW. The higher modes correspond to rays with greater launch angles that turn at successively deeper depths in the Arctic Ocean. Rough surface scattering theory tells us that the loss per bounce or reflection increases with grazing angle and frequency (i.e., the steeper rays, or higher modes have greater loss per bounce). However, even though the lower modes (lower grazing angles) have lower per bounce loss, those that are trapped in the upper duct experience many more interactions with the ice, as clearly seen in **Figure 2**, and consequently their net loss per kilometer is higher. At shorter ranges for a source and receiver in the upper duct (**Figure 4**) the propagation is dominated by the lower modes (or equivalently the lower grazing angle rays that are trapped), but these get stripped away as the range increases and at longer ranges (>100–200 km) the higher modes (or equivalently the higher grazing angle rays that are not trapped) dominate. Because the source and receiver are shallow (at 60 m depth) for the model run in **Figure 4**, there is a less efficient excitation of mode 1 at 20 Hz (i.e., destructive interference with the surface reflected energy) than at the higher frequencies. Thus less of the energy is trapped in the near-surface duct. This results in lower energy being received on the shallow receiver at closer ranges and hence the higher initial propagation loss at 20 Hz. As the frequency increases the mode shapes tend to compress toward the surface (because the wavelength is getting smaller). For a shallow source more of the energy is trapped in this upper duct at higher frequencies (see discussion of frequency dispersion and **Figure 6** below). Given the higher per bounce losses at the higher frequencies,

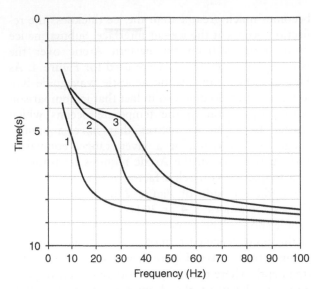

Figure 6 A composite of various measurements of Arctic ambient noise. (1) Central Arctic pack ice Nansen Basin, first- and multi-year floes, 2–3 m thick, with ridging, April 1982. (Reproduced with permission from Dyer, 1984.) (2) Beaufort Sea, summer and fall conditions: (a) highest levels observed, (b) typical cold weather situation following rapid temperature drop and thermal-induced cracking, (c) quietest conditions during warmer stable temperature periods with low winds September–October, 1961 and May–September 1962. (Reproduced with permission from DiNapoli *et al.*, 1978, in Von Winkle, 1984.) (3) Shore-fast ice in the Canadian Archipelago, winter conditions, thermal cracking, February 1963. (Reproduced with permission from Milne AR and Ganton JH (1964) Ambient noise under Arctic-Sea ice. *Journal of the Acoustic Society of America* 36(5): 855–863.) (4) Shore-fast ice in the Canadian Archipelago, spring conditions: (a) noisiest conditions observed during diurnal cooling, (b) quietest conditions observed during diurnal warming, April 1961 (Milne and Ganton 1964). (5) Chukchi Sea, spring conditions, cold stable temperatures, low winds, April 1999 (Mikhalevsky, APLIS Ice Camp).

and the greater number of bounces experienced by these trapped modes, it can be seen why attenuation in the Arctic increases rapidly with frequency. For deeper sources the near-surface duct is not as important and higher grazing angle (higher mode) propagation and loss are dominant at all ranges; however, the higher loss per bounce at higher frequencies still results in more rapid attenuation as the frequency increases.

There is a significant body of research on the scattering and reverberation of acoustic energy from the sea ice. The earliest models relied on rough free surface scattering theory with empirical fits to important parameters such as rms ice roughness and average ridge spacing. The free surface scattering models that included ridge-like morphology performed reasonably well at frequencies from 250 Hz to 2000 Hz. At lower frequencies they tended to underpredict the loss. When elastic coupling and

scattering into the ice (particularly conversion to shear waves and the attenuation in the ice) are included, better agreement at the low frequencies has been achieved. In particular the ice keels and their spacing are important in the conversion of acoustic compressional waves into flexural modes in the ice. Le Page and Schmidt (*see* Further Reading) have shown that the intensity of acoustic scattering into the flexural modes of a rough ice plate strongly depends on the width of the roughness spatial spectra. In order to achieve consistent agreement with data the propagation loss models have become more complex, demanding additional input data, particularly about the structure of the sea ice and its constitutive properties.

Bottom interaction must also be included, for those paths which cross major features like the Lomonosov Ridge, and propagation on the arctic shelves (**Figure 1**). The Lomonosov Ridge can rise to 1500 m below the surface. **Figure 2** shows that modes 5 and higher (rays >11–$12°$) will start to interact with this feature and will begin to get stripped away for long-range propagation. As the frequency decreases below 20 Hz the modes tend to expand away from the surface and finally all modes will begin to interact with the bottom except in the very deepest parts of the Arctic basin. This leads to a lower frequency bound at 5–10 Hz for efficient long-range propagation. However, for high source levels at these frequencies there is significant backscatter reverberation from these bathymetric features that can be detected and mapped. Excellent correlation with known Arctic bathymetry as well as the discovery of uncharted features at basin scale ranges (1000 km) has been achieved using large explosive sources deployed from the ice. Over the shelves the shallow water propagation dictates bottom and surface/ice interaction at all frequencies; however, there are optimal propagation frequencies similar to deep water that depend upon the depth and boundary properties.

The stable upward refracting Arctic sound speed profile causes a very predictable modal and frequency dispersion. As can be seen in **Figure 2**, successively higher modes are propagating at higher sound speeds corresponding to higher group velocities. As a consequence at a given range, mode 1 arrives last, preceded in order by modes 2, 3, 4, etc. This can be observed by transmitting a wideband waveform such as an impulsive or explosive source and plotting the signal intensity as a function of frequency and time. **Figure 5** shows the arrival time from the computed group velocities of modes 1–3 for an explosive shot at a range of approximately 580 km in the Beaufort Sea. Although not plotted,

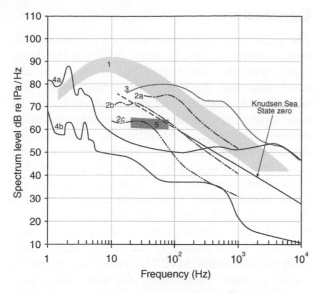

Figure 6 The arrival time of a broadband acoustic pulse plotted as a function of frequency showing the modal dispersion with mode 1 arriving last, preceded by modes 2 and 3 respectively. The absolute travel time can be obtained by adding 396.5 s to the time shown on the vertical axis. This modeled result agreed very closely with the received data. (Adapted with permission from DiNapoli *et al.*, 1978, in Von Winkle, 1984.)

this calculation agreed very closely with the measured data. Note also from **Figure 5** that as the frequency increases (as described above), the modes compress towards the surface and first mode 1, then mode 2 and finally mode 3 become trapped in the slower near-surface duct, as evidenced by the longer travel times. This modal dispersion and arrival pattern is a classic Arctic acoustic result. The data upon which **Figure 5** is based were taken in 1962; 32 years later in 1994 the TAP experiment measured the same arrival pattern, attesting to the very long-term stable water mass structure in the Arctic Ocean. Observations of the changes in the arrival times of these modes are directly related to the temperature changes of these water masses (since sound speed increases approximately 5 m/s per 1 °C). In particular, at 20 Hz mode 2 is most sensitive to temperature changes in the AIW, as can be seen from **Figure 2**. A decrease in the travel time of mode 2 by 2 s was the indicator of the warming in the AIW observed in the 1994 TAP experiment. In addition to the long-term stability of the Arctic sound speed profile, the short-term stability results in exceptional amplitude (~ 1 dB rms) and phase stability (~ 0.01 cycles rms) of acoustic signals at low frequencies even over ranges approaching 3000 km for up to 1 h (the longest continuous observations made to date). This applies to essentially fixed terminals. This was first observed in 1980 at 15–30 Hz and 300 km and then

again in 1994 and 1999 at 20 Hz and 2700 km. This implies that coherent integration of up to 1 h with optimal gain is achievable (and has been demonstrated).

At very high frequencies (more than tens of kHz) reflection occurs at the ice–water interface. Attenuation is rapid, due not only to the severe scattering from the ice, but to volumetric absorption as well. These frequencies are used for short-range applications including the upward-looking ice profiling sonar, forward-looking ice avoidance sonar, and downward-looking depth sounders and bottom-mapping sonar. Torpedoes also operate at these frequencies and ice capture can be a problem, but Doppler processing can distinguish moving targets from the stationary ice (*see* Sonar Systems).

Ambient Noise

The ambient noise in the Arctic is highly variable, exhibiting some of the quietest as well as the noisiest ocean noise conditions of all the world's oceans. A composite of various measurements of Arctic ambient noise is shown in **Figure 6**. In **Figure 6**, Knudsen Sea State Zero refers to the ambient noise level in the temperate oceans at sea state zero (the quietest conditions) for comparison (*see* Acoustic Noise). Ice-generated ambient noise is the dominant mechanism contributing to the general character of ambient noise in the Arctic Ocean from a few tenths of Hz up to 10 000 Hz. Episodic noise is also present in the form of seismic events, such as earthquakes along the Mid Arctic Ridge (**Figure 1**), biologics (mostly marine mammals in the marginal ice zones), and man-made noise from ice breakers, and seismic exploration.

Ice noise is generated when the sea ice deforms, fractures, and breaks in response to environmental forcing such as wind, current, thermal, and internal and surface wave-induced stresses. In general, during stable or warming temperature conditions with low winds the quietest conditions obtain, particularly under shore-fast ice. During periods of rapid cooling ice-fracturing events resulting from thermal-induced tensile stresses can lead to higher noise levels. Higher noise levels also occur in the pack ice when the ice is in motion due to nonthermal forcing such as high winds causing bumping, grinding and rubbing of the ice flows. In the marginal ice zone where the ice concentrations are typically less than in the central Arctic pack, ice concentration and surface gravity wave-induced flexural floe failure are the primary correlates with the ambient noise. The actual noise-generating mechanics are complicated but fall into at least two general categories. In the low frequency range (5–100 Hz) the important noise-generating mechanism is the unloading motion of the ice immediately following breaking. This has been shown to have a dipole radiation characteristic. The breaking process itself is important at intermediate frequencies (100–2000 Hz) and has an octopole radiation characteristic likely resulting from a slip–dislocation process.

Noise from earthquakes and other seismic events occur with some regularity within the Arctic Ocean, concentrated along the Mid Arctic Ridge (**Figure 1**). These events have been recorded on acoustic arrays suspended from the ice within approximately 300 km of the Mid Arctic Ridge. The received frequencies are typically 20 Hz and below. Earthborne acoustic pressure and shear waves emanate from the source and couple through the ocean floor above the epicenter to compressional waterborne waves that propagate vertically and are reflected from the ice canopy into the Arctic sound channel. Most of the energy from these events arrives via this path, but there are weaker precursors associated with crustal propagating compressional and shear waves that couple to the water directly below the receive array. These arrivals are easily identified by their vertical arrival angle at the array as well as their arrival time.

Biological noise in the Arctic is concentrated in the marginal ice zones and near the edge of the pack ice. The bowhead is the most numerous of the baleen whales in Arctic waters, migrating along the coast of Alaska in the Chukchi and Beaufort Seas in the spring and then exiting the Arctic waters in the fall. They can vocalize from 25 to 3500 Hz, but their dominant frequencies are between 100 and 400 Hz. Of the toothed whales the beluga and narwhal are the most common, with calls ranging from a few hundred Hz to as high as 20 kHz. Many species of pinnipeds, including hair seals such as the bearded, hooded, harp, ringed, and ribbon seal, and the walrus frequent the marginal ice zones. As a group their calls typically range from a few hundred Hz to 10 kHz.

Unlike the temperate oceans where shipping typically dominates the ambient noise spectrum, man-made noise in the Arctic is a small contributor except for specific events that are isolated in time and space. Such events include icebreakers, seismic exploration, and some military and experimental activities. Icebreaker noise peaks in the 50–100 Hz range, but with broadband contributions up to 1000 Hz. Seismic exploration occurs during the summer and fall seasons when the ice extent is minimum (**Figure 1**), allowing easier access to the shelves. Most of the activity has been confined to the Beaufort Sea off the North Slope of Alaska near Prudhoe Bay.

Conclusions

The presence of the year-round ice cap creates the upward refracting sound speed profile in the Arctic Ocean with the sound speed minimum at the ice–water interface. Reflection, scattering, mode conversion, and absorption by the rough elastic sea ice cover causes high attenuation as frequency increases, limiting long-range propagation to very low frequencies. Bathymetric effects are important near the major ocean ridges, basin margins, and on the shelf areas where significant mode coupling can occur. The Arctic sound channel is very stable and predictable in the central Arctic basins and there is a close correspondence of propagating acoustic modes with the major water masses of the Arctic Ocean, especially the important AIW. This latter fact makes the use of acoustic thermometry for monitoring long-term Arctic Ocean temperature change particularly suitable. Ongoing research is exploring ways to relate changes in acoustic travel time and intensity to monitor other important variables in the Arctic Ocean including changes in the PW, DW, and the halocline, and average sea-ice thickness and roughness. The latter measurements, when combined with sea ice extent from satellite remote sensing, could provide an estimate of sea ice mass in the Arctic.

The role of the Arctic Ocean in shaping and responding to global climate change is only beginning to be explored. Cost-effective, long-term, year-round synoptic observations in the Arctic Ocean require new measurement strategies. The year-round ice cover in the Arctic prevents the use of satellites for direct ocean observations common in the ice-free oceans. Shore-cabled mooring-based observations using advanced biogeochemical sensors and acoustic sources and hydrophone arrays, as well as instrumented autonomous underwater vehicles (AUVs) and under-ice drifters, represent new approaches for observing the Arctic Ocean. Interestingly the RAFOS concept (using nonexplosive sources) is being evaluated anew as a way to track AUVs and drifters in the Arctic as well as for acoustic communication of data. It is clear that Arctic acoustics will have as large a role to play in this important new endeavor in the future, as it has had in the submarine and military operations of the past.

See also

Acoustics, Arctic. Acoustics in Marine Sediments. Acoustic Noise. Acoustics, Shallow Water. Bioacoustics. Nepheloid Layers. Tomography.

Further Reading

Dyer I (1984) Song of sea ice and other Arctic Ocean melodies. In: Dyer I and Chryssostomidis C (eds.) *Arctic Technology and Policy*, pp. 11–37. Washington, DC: Hemisphere Publishing.

Dyer I (1993) Source mechanisms of Arctic Ocean ambient noise. In: Kerman BR (ed.) *Natural Physical Sources of Underwater Sound*, pp. 537–551. Netherlands: Kluwer Academic.

Leary WM (1999) *Under Ice*. College Station: Texas A&M University Press.

LePage K and Schmidt H (1994) Modeling of low-frequency transmission loss in the central Arctic. *Journal of the Acoustic Society of America* 96(3): 1783–1795.

Mikhalevsky PN, Gavrilov AN, and Baggeroer AB (1999) The transarctic acoustic propagation experiment and climate monitoring in the Arctic. *IEEE Journal of Oceanic Engineering* 24(2): 183–201.

Newton JL (1989) Sound speed structure of the Arctic Ocean including some effects on acoustic propagation. *US Navy Journal of Underwater Acoustics* 39(4): 363–384.

Richardson WJ, Greene CR Jr, Malme CI, and Thomson DH (1995) *Marine Mammals and Noise*. San Diego: Academic Press.

Urick RJ (1975) *Principles of Underwater Sound*. New York: McGraw-Hill.

Von Winkle WA (1984) *Naval Underwater Systems Center (NUSC) Scientific and Engineering Studies: Underwater Acoustics in the Arctic*. New London: Naval Underwater Systems Center Publisher.

ACOUSTICS, DEEP OCEAN

W. A. Kuperman, Scripps Institution of
Oceanography, University of California,
San Diego, CA, USA

Introduction

The acoustic properties of the ocean, such as the
paths along which sound from a localized source will
travel, are mainly dependent on its sound speed
structure. The sound speed structure is dependent on
the oceanographic environment described by variations in temperature, salinity, and density with
depth or horizontal position. This article will review
the ocean acoustic environment, sound propagation,
ambient noise, scattering and reverberation, and the
passive and active sonar equation.

Ocean Acoustic Environment

Sound propagation in the ocean is governed by the
spatial structure of the sound speed and the sound
speed in the ocean is a function of temperature, salinity, and ambient pressure. Since the ambient pressure is a function of depth, it is customary to express
the sound speed (c) in meters per second as an empirical function of temperature (T) in degrees celsius,
salinity (S) in parts per thousand and depth (z) in
meters, e.g. eqn [1].

$$c = 14449.2 + 4.6T - 0.055T^2 + 0.00029T^3$$
$$+ (1.34 - 0.01T)(S - 35) + 0.016z \qquad [1]$$

There exist more accurate formulas, if needed.

Figure 1 shows a typical set of sound speed profiles, indicating greatest variability near the surface.
In a warmer season (or warmer part of the day,
sometimes referred to as the 'afternoon effect'), the
temperature increases near the surface and hence the
sound speed increases toward the sea surface. In
nonpolar regions where mixing near the surface due
to wind and wave activity is important, a mixed layer
of almost constant temperature is often created. In
this isothermal layer, sound speed increases with
depth because of the increasing ambient pressure, the
last term in eqn [1]. This is the surface duct region.
Below the mixed layer is the thermocline where the
temperature and hence the sound speed decrease

with depth. Below the thermocline, the temperature
is constant and the sound speed increases because of
increasing ambient pressure. Therefore, between the
deep isothermal region and the mixed layer, there is a
minimum sound speed; the depth at which this
minimum takes place is referred to as the axis of the
deep sound channel. However, in polar regions, the
water is coldest near the surface so that the minimum
sound speed is at the surface.

Figure 2 is a contour display of the sound speed
structure of the North and South Atlantic with the
deep sound channel axis indicated by the heavy dashed
line. Note that the deep sound channel becomes
shallower toward the poles. Aside from sound speed
effects, the ocean volume is absorbtive and will cause
attenuation that increases with acoustic frequency.

The ocean surface and bottom also have a strong
influence on sound propagation. The ocean surface,
though a perfect reflector when flat, causes scattering
when its roughness becomes comparable in size with
the acoustic wavelength. The ocean bottom, depending on its local structure will scatter and also
attenuate the acoustic field.

Units

The decibel (dB) denotes a ratio of intensities (see
Section 3.3) expressed in terms of a logarithmic (base

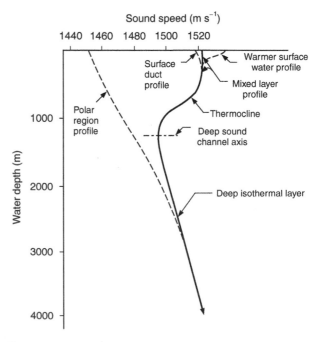

Figure 1 Generic sounds speed profiles.

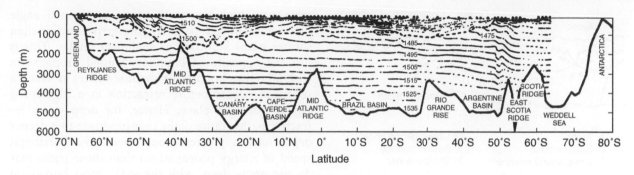

Figure 2 Sound speed contours of the North and South Atlantic along 30.50°W. Dashed line indicates axis of deep sound channel. (From Northrop and Colborn (1974).)

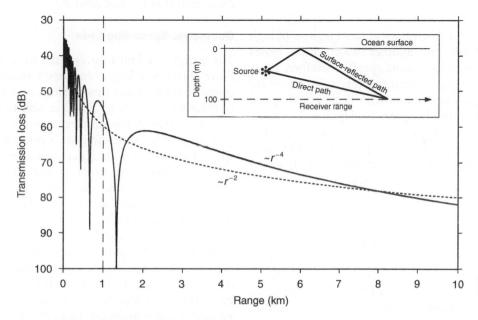

Figure 3 The insert shows the geometry of the Lloyd mirror effect. The plots show a comparison of Lloyd mirror to spherical spreading. Transmission losses are plotted in decibels corresponding to losses of 10 log r^2 and 10 log r^4, respectively, as explained in the test.

10) scale. The ratio of two intensities, I_1/I_2 is 10 log (I_1/I_2) in dB units. Absolute intensities are expressed using an accepted reference intensity of a plane wave having an rms pressure equal to 10^{-5} dyn cm^{-2} or, equivalently, 1 µPa. Transmission loss is a decibel measure of relative intensity, the latter being proportional to the square of the acoustic amplitude.

Sound Propagation

Very Short-range Propagation

The pressure amplitude from a point source in free space falls off with range r as r^{-1}; this geometric loss is called spherical spreading. Most sources of interest in the deep ocean are nearer the surface than the bottom. Hence, the two main short-range paths are the direct path and the surface-reflected path. When

these two paths interfere, they produce a spatial distribution of sound often referred to as a 'Lloyd mirror pattern' as shown in the insert of **Figure 3**.

Basic Long-range Propagation Paths

Figure 4 is a schematic of propagation paths in the ocean resulting from the sound speed profiles (indicated by the dashed line) described above in **Figure 1**. These paths can be understood from Snell's law eqn [2], which relates the ray angle $\theta(z)$, with respect to the horizontal, to the local sound speed $c(z)$ at depth z.

$$\frac{\cos\theta(z)}{c(z)} = \text{constant} \qquad [2]$$

The equation requires that the higher the sound speed, the smaller the angle with the horizontal,

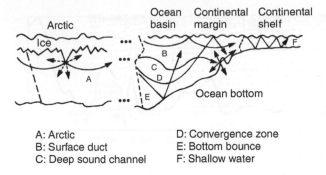

A: Arctic
B: Surface duct
C: Deep sound channel
D: Convergence zone
E: Bottom bounce
F: Shallow water

Figure 4 Schematic representation of sound propagations paths in the ocean.

meaning that sound bends away from regions of high sound speed or, put another way, sound bends toward regions of low sound speed. Therefore, paths A, B, and C are the simplest to explain since they are paths that oscillate about the local sound speed minima. For example, path C depicted by a ray leaving a source near the deep sound channel axis at a small horizontal angle propagates in the deep sound channel. This path, in temperate latitudes where the sound speed minimum is far from the surface, permits propagation over distances of thousands of kilometers. Path D, which is at slightly steeper angles and is usually excited by a near surface source, is convergence zone propagation, a spatially periodic (35–65 km) refocusing phenomenon producing zones of high intensity near the surface due to the upward refracting nature of the deep sound speed profile. Regions between these zones are referred to as shadow regions. Referring back to **Figure 1**, there may be a depth in the deep isothermal layer at which the sound speed is the same as it is at the surface; this depth is called the critical depth and is the lower limit of the deep sound channel. A positive critical depth specifies that the environment supports long-distance propagation without bottom interaction, whereas a negative critical depth specifies that the ocean bottom is the lower boundary of the deep sound channel. The bottom bounce path E is also a periodic phenomenon but with a shorter cycle distance and shorter propagation distance because of losses when sound is reflected from the ocean bottom.

An alternative way of describing paths is by denoting that they are composed of combinations of refraction (R), surface reflection (SR) and bottom reflection (BR) processes. Thus, **Figure 4** also represents some of these paths. Note from Snell's law that refractive paths involve a path turning around at the highest speed in its duct of confinement. Because such ray paths spend much of their propagation in

regions of high sound speed, larger launch angle paths with longer path lengths arrive earlier than shorter paths launched at shallower angles. This is just the opposite of boundary-limited propagation such a bottom bounce (or shallow water) in which a reflection occurs before refraction in a high-speed region can take place. Hence, for deep water refractive paths, those paths that penetrate to a deeper depth have a greater group speed (the horizontal speed of energy propagation) than those paths that do not go as deep, with the axial, most horizontal path along the deep sound channel axis having the slowest group speed. For pulse propagation, this axial arrival is the last arrival.

Geometric Spreading Loss

The energy per unit time emitted by a sound source flows through a larger area with increasing range. Intensity is the power flux through a unit area, which translates to the energy flow per unit time through a unit area. Hence, the simplest example of geometric loss is spherical spreading for a point source in free space, for which the area increases as $4\pi r^2$, where r is the range from the point source. Thus spherical spreading results in an intensity decay proportional to r^{-2}. Since intensity is proportional to the square of the pressure amplitude, the fluctuations in pressure p induced by the sound decay as r^{-1}. For range-independent ducted propagation, that is, where rays are refracted or reflected back toward the horizontal direction (which is the case for most long-range propagation), there is no loss associated with the vertical dimension. In this case, the spreading surface is the area of a cylinder whose axis is in the vertical direction passing through the source: $2\pi r H$ where H is the depth of the duct and is constant. Geometric loss in the near-field Lloyd mirror regime requires consideration of interfering beams from direct and surface reflected paths. To summarize, the geometric spreading laws for the pressure field (recall that intensity is proportional to the square of the pressure) are

- Spherical spreading loss: $p \propto r^{-1}$
- Cylindrical spreading loss: $p \propto r^{-1/2}$
- Lloyd mirror loss: $p \propto r^{-2}$.

Volume Attenuation

Volume attenuation increases with frequency. In **Figure 3**, the losses associated with path C include only volume attenuation and scattering because this path does not involve boundary interactions. The volume scattering can be biological in origin or can arise from interaction with large internal wave activity in the vicinity of the upper part of the deep sound channel where paths are refracted before they would interact with the surface. Both of these effects are small for low frequencies. This same internal wave region is also on the lower boundary of the surface duct, allowing scattering out of the surface duct and thereby also constituting a loss mechanism for the surface duct. This mechanism also leaks sound into the deep sound channel, a region that without scattering would be a shadow zone for a surface duct source. This type of scattering from internal waves is also thought to be a major source of fluctuation of the sound field.

Attenuation is characterized by an exponential decay of the sound field. If A_0 is the rms amplitude of the sound field at unit distance from the source, then the attenuation of the sound field causes the amplitude to decay with distance r along the path according to eqn [3], where the unit of α is nepers/distance (nepers is a unitless quantity).

$$A = A_0 \exp(-\alpha r) \qquad [3]$$

This attenuation coefficient can be expressed in decibels per unit distance by the conversion $\alpha' = 0.686\alpha$. The frequency dependence of attenuation can be roughly divided into four regimes as displayed in **Figure 5**. In region I, leakage out of the sound channel is believed to be the main cause of attenuation. The main mechanisms associated with regions II and III are boric acid and magnesium sulfate chemical relaxation. Region IV is dominated by the shear and bulk viscosity associated with fresh

water. A summary of the approximate frequency dependence (f in kHz) of attenuation (in units of dB km^{-1}) is given in eqn [4] with the terms sequentially associated with regions I–IV in **Figure 5**.

$$\alpha' (dBkm^{-1}) = 3.3 \times 10^{-3} + \frac{0.11f^2}{1+f^2} + \frac{43f^2}{4100 + f^2}$$
$$+ 2.98 \times 10^{-4}f^2 \qquad [4]$$

Bottom Loss

The structure of the ocean bottom affects those acoustic paths that interact with the ocean bottom. This bottom interaction is summarized by bottom reflectivity, the amplitude ratio of reflected and incident plane waves at the ocean-bottom interface as a function of grazing angle, θ (see **Figure 6A**). For a simple bottom that can be represented by a semi-infinite half-space with constant sound speed c_b and density ρ_b, the reflectivity is given by eqn [5] with the subscript w denoting water

$$\mathcal{R}(\theta) = \frac{\rho_b k_{wz} - \rho_w k_{bz}}{\rho_b k_{wz} + \rho_w k_{bz}}, \qquad [5]$$

The wavenumbers are given by eqn [6].

$$k_{iz} = (\omega/c_i)\sin\theta_i = k\sin\theta_i; \quad i = w, b \qquad [6]$$

The incident and transmitted grazing angles are related by Snell's law according to eqn [7] and the incident grazing angle θ_w is also equal to the angle of the reflected plane wave.

$$c_b \cos\theta_w = c_w \cos\theta_b \qquad [7]$$

For this simple water-bottom interface for which we take $c_b > c_w$, there exists a critical grazing angle θ_c below which there is perfect reflection (eqn [8]).

$$\cos\theta_c = \frac{c_w}{c_b} \qquad [8]$$

For a lossy bottom, there is no perfect reflection, as also indicated in a typical reflection curve in **Figure 6B**. These results are approximately frequency independent. However, for a layered bottom, the reflectivity has a complicated frequency dependence as shown in the example in **Figure 6C**, where the contours are in decibels. This example shows the simple reflectivity result below 200 Hz and then a more complicated frequency dependence at higher frequencies. It should be pointed out that if the density of the second medium vanishes, the reflectivity reduces to the pressure release case of $\mathcal{R}(\theta) = -1$.

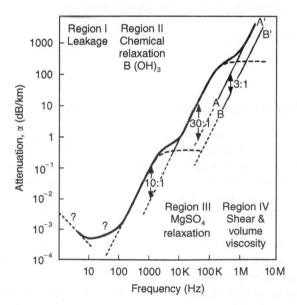

Figure 5 Regions of different dominant processes of attenuation of sound in sea water. (From Urick (1979).) The attenuation α is given in dB per 1000 yards.

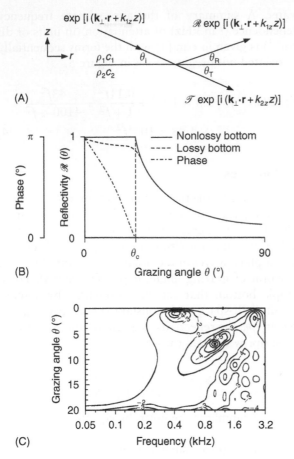

(A)

(B)

(C)

Figure 6 The reflection and transmission processes. Grazing angles are defined relative to the horizontal. (A) A plane wave is incident on an interface separating two media with densities and sound speeds ρc. $\mathscr{R}(\theta)$ and $\mathscr{T}(\theta)$ are reflection and transmission coefficients. Snell's law is a statement that $k_\perp$, the horizontal component of the wavevector, is the same for all three waves. (B) Rayleigh reflection curve (eqn [5]) as a function of the grazing angle (θ in (A)) indicating critical θ_c. The dashed curve shows that if the second medium is lossy, there is less than perfect reflection below the critical angle. (C) Examples of contour of reflection loss ($20 \log \mathscr{R}$) for a layered bottom, showing frequency and grazing angle dependence. The simpler reflectivity curve for each frequency is obtained from a vertical slice.

Propagation Models

An ocean acoustic environment is often very complex, with range- and depth-dependent properties. Such an environment does not in general lend itself to simple analytic predictions of sound propagation. Even in range-independent environments there are many paths (multipaths) and these paths combine to form a complex interference pattern. For example, the convergence zones are an example of a more complex structure that cannot be described by a monotonic geometric spreading law. Acoustic models play an important role in predicting sound propagation; the inputs to these models are oceanographic quantities ultimately translated into the acoustically relevant parameters of sound speed, density, and attenuation.

Sound propagation in the ocean is mathematically described by the wave equation whose coefficients and boundary conditions are derived from the ocean environment. There are essentially four types of models (computer solutions to the wave equation) to describe sound propagation in the sea: ray, spectral or fast field program (FFP), normal mode (NM), and the parabolic equation (PE). Ray theory is an asymptotic high-frequency approximation to the wave equation, whereas the latter three models are more or less direct solutions to the wave equations under an assortment of milder restrictions. The high-frequency limit does not include diffraction phenomena. All of these models can handle depth variation of the ocean acoustic environment. A model that also takes into account horizontal variations in the environment (i.e., sloping bottom or spatially variable oceanography) is termed range-dependent. For high frequencies (a few kHz or above), ray theory is the most practical. The other three model types are more applicable and usable at lower frequencies (below 1 kHz). The hierarchy of underwater acoustic models is shown in schematic form in **Figure 7**. The output of these models is typically propagation loss, which is the intensity relative to a unit source at unit distance, expressed in decibels. Transmission loss is the negative of propagation loss, and hence, a positive quantity.

An example of the output of propagation models is shown in **Figure 8**, indicating agreement between the models. However, we also see a difference among the models in that ray theory predicts a sharper shadow zone than the wave theory model (i.e., the 10–30 km region in **Figure 8B**); this is an expected result from the infinite-frequency ray approximation.

Scattering and Reverberation

Scattering caused by rough boundaries or volume inhomogeneities is a mechanism for loss (attenuation), reverberant interference, and fluctuation. Attenuation from volume scattering was addressed above. In most cases, it is the mean or coherent (or specular) part of the acoustic field that is of interest for a sonar or communications application, and scattering causes part of the acoustic field to be randomized. Rough surface scattering out of the 'specular direction' can be thought of as an attenuation of the mean acoustic field and typically increases with increasing frequency. A formula often used to describe reflectivity from a rough boundary is eqn [9], where $\mathscr{R}(\theta)$ is the reflection coefficient of the

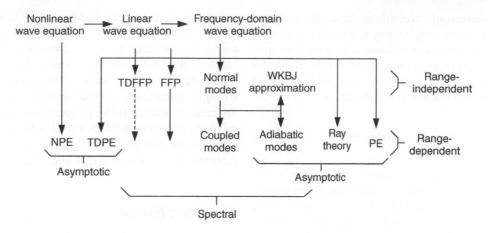

Figure 7 Heirarchy of underwater acoustic models. TD refers to time domain. NPE is the nonlinear parabolic equation that describes high-amplitude (e.g., shockwave) propagation. The arrows are directed toward the flow of derivation of the model

smooth interface and Γ is the Rayleigh roughness parameter defined as $\Gamma \equiv 2k\sigma \sin \theta$ where $k = 2\pi/\lambda$, λ is the acoustic wavelength, and σ is the rms roughness (height).

$$\mathscr{R}'(\theta) = \mathscr{R}(\theta) \exp - \frac{\Gamma^2}{2} \qquad [9]$$

The scattered field is often referred to as reverberation. Surface, bottom, or volume scattering strength, $S_{S,B,V}$ is a simple parametrization of the production of reverberation and is defined as the ratio in decibels of the sound scattered by a unit surface area or volume referenced to a unit distance, (Insert equqtion), to the incident plane wave intensity, I_{inc} (eqn [10]).

$$S_{S,B,V} = 10 \log \frac{I_{scat}}{I_{inc}} \qquad [10]$$

The Chapman–Harris curves predicts the ocean surface scattering strength in the 400–6400 Hz region; eqn [11], where θ is the grazing angle in degrees, w the wind speed in $m\,s^{-1}$ and f is the frequency in Hz.

$$S_S = 3.3\beta \log \frac{\theta}{30} - 42.4 \log \beta + 2.6;$$

$$\beta = 107 (wf^{1/3})^{-0.58} \qquad [11]$$

A more elaborate formula exists that is more accurate for lower wind speeds and lower frequencies.

The simple characterization of bottom backscattering strength utilizes Lambert's rule for diffuse scattering, given by eqn [12] where the first term is determined empirically.

$$S_B = A + 10 \log \sin^2 \theta \qquad [12]$$

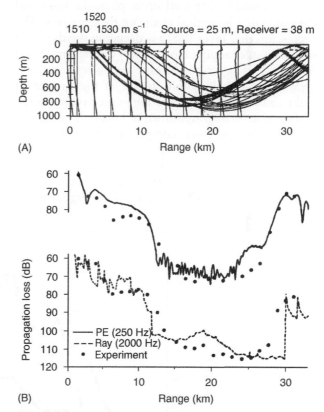

Figure 8 Model and data comparison for a range-dependent deep water case. (A) Sound speed profiles as a function of range together with a ray trace showing the breakdown of surface duct propagation. (B) Parabolic equation comparison with data at 250 Hz and ray theory comparison with data at 2000 Hz.

Under the assumption that all incident energy is scattered into the water column with no transmission in to the bottom, A is -5 dB. Typical realistic values for A that have been measured are -17 dB for big Basalt Mid-Atlantic Ridge cliffs and -27 dB for sediment ponds.

Volume scattering strength is typically reduced to a surface scattering strength by taking (Insert equqtion) as an average volume scattering strength within some layer at a particular depth; then the corresponding surface scattering strength is given by eqn [13], where H is the layer thickness.

$$S_S = S_v + 10 \log H \qquad [13]$$

The column or integrated scattering strength is defined as the case for which H is the total water depth.

Volume scattering usually decreases with depth (about 5 dB per 300 m) with the exception of the deep scattering layer. For frequencies less than 10 kHz, fish with air-filled swimbladders are the main scatterers. Above 20 kHz, zooplankton or smaller animals that feed upon phytoplankton and the associated biological chain are the scatterers. The deep scattering layer (DSL) is deeper in the day than in the night, changing most rapidly during sunset and sunrise. This layer produces a strong scattering increase of 5–15 dB within 100 m of the surface at night and virtually no scattering in the daytime at the surface since it migrates down to hundreds of meters. Since higher pressure compresses the fish swimbladder, the backscattering acoustic resonance tend to be at a higher frequency during the day when the DSL migrates to greater depths. Examples of day and night scattering strengths are shown in **Figure 9**.

Finally, near-surface bubbles and bubble clouds can be thought of as either volume or surface scattering mechanisms acting in concert with the rough surface. Bubbles have resonances (typically greater than 10 kHz) and at these resonances, scattering is strongly enhanced. Bubble clouds have collective properties; among these properties is that a bubbly mixture, as specified by its void fraction (total bubble gas volume divided by water volume) has a considerable lower sound speed than water.

Ambient Noise

There are essentially two types of ocean acoustic noise: man-made and natural. Generally, shipping is the most important source of man-made noise, though noise from offshore oil rigs is becoming more and more prevalent. Typically, natural noise dominates at low frequencies (below 10 Hz) and high frequencies (above a few hundred hertz). Shipping fills in the region between ten and a few hundred hertz. A summary of the spectrum of noise is shown in **Figure 10**. The higher-frequency noise is usually parametrized according to sea state (also Beaufort number) and/or wind. **Table 1** summarizes the description of sea state.

The sound speed profile affects the vertical and angular distribution of noise in the deep ocean. When there is a positive critical depth, sound from

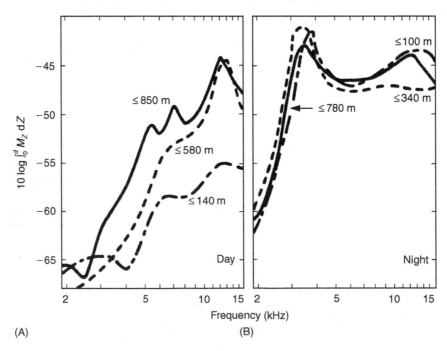

(A)   (B)

Figure 9 Day and night scattering strength measurements using an explosive source as a function of frequency. The spectra measured at various times after the explosion are labeled with the depth of the nearest scatterer that could have contributed to the reverberation. The ordinate corresponds to S_v in eqn [13]. (From Chapman and Marchall (1966))

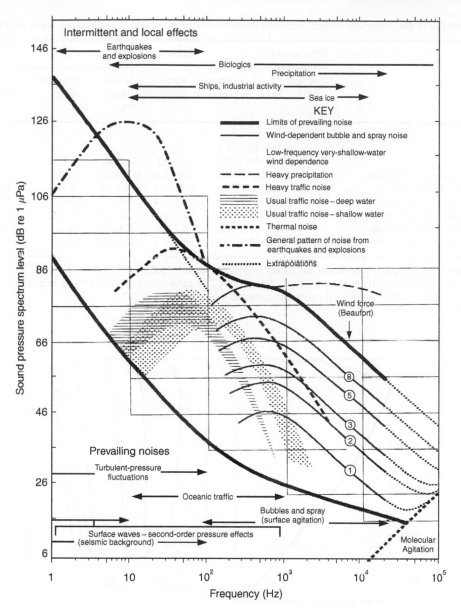

Figure 10 Composite of ambient noise spectra. (From Wenz (1962).)

surface sources can travel long distances without interacting with the ocean bottom, but a receiver below this critical depth should sense less surface noise because propagation involves interaction with lossy boundaries, surface and/or bottom. This is illustrated in **Figure 11**, which shows a deep water environment with measured ambient noise. **Figure 12** is an example of vertical directivity of noise that also follows the propagation physics discussed above. The shallower depth is at the axis of the deep sound channel, while the other is at the critical depth. The pattern is narrower at the critical depth where the sound paths tend to be horizontal since the rays are turning around at the lower boundary of the deep sound channel.

In a range-independent ocean, Snell's law predicts a horizontal noise notch at depths where the speed of sound is less than the near-surface sound speed. Returning to eqn [2] and reading off the sound speeds from **Figure 11** at the surface ($c = 1530 \, \mathrm{m \, s^{-1}}$) and say, $300 \, \mathrm{m}$ ($1500 \, \mathrm{m \, s^{-1}}$), a horizontal ray ($\theta = 0$) launched from ocean surface would have an angle with respect to the horizontal of about $11°$ at $300 \, \mathrm{m}$ depth. All other rays would arrive with greater vertical angles. Hence we expect this horizontal notch. However, the horizontal notch is often not seen at shipping noise frequencies. This is because shipping tends to be concentrated in continental shelf regions; range-dependent propagation couples such noise sources to the deep ocean. Thus, for example,

Table 1   Descriptions of the ocean sea surface. Approximate relation between scales of wind speed, wave height, and sea state

Sea criteria	Wind speed			12-h wind	Fully risen sea			
	Beaufort scale	Range (m s^{-1})	Mean (m s^{-1})	Wave height[a, b] (m)	Wave height[a, b] (m)	Duration[b, c] (h)	Fetch[b, c] (km)	Seastate scale
Mirrorlike	0	<0.5						0
Ripples	1	0.5–1.7	1.1					1/2
Small wavelets	2	1.8–3.3	2.5	<0.30	<0.30			1
Large wavelets, scattered whitecaps	3	3.4–5.4	4.4	0.30–0.61	0.30–0.61	<2.5	<19	2
Small waves, frequent whitecaps	4	5.5–8.4	6.9	0.61–1.5	0.61–1.8	2.5–6.5	19–74	3
Moderate waves, many whitecaps	5	8.5–11.1	9.8	1.5–2.4	1.8–3.0	6.5–11	74–185	4
Large waves, whitecaps everywhere, spray	6	11.2–14.1	12.6	2.4–3.7	3.0–5.2	11–18	185–370	5
Heaped-up sea, blown spray, streaks	7	14.2–17.2	15.7	3.7–5.2	5.2–7.9	18–29	370–740	6
Moderately high, longwaves, spindrift	8	17.3–20.8	19.0	5.2–7.3	7.9–11.9	29–42	740–1300	7

[a]The average height of the highest one-third of the waves (significant wave height).
[b]Estimated from data given in US Hydrographic Office (Washington, DC) publications HO 604 (1951) and HO 603 (1955).
[c]The minimum fetch and duration of the wind needed to generate a fully risen sea.
From Wenz (1962).

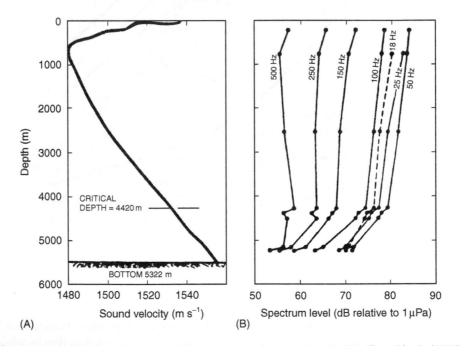

Figure 11   (A) Sound speed profile and (B) noise level as a function of depth in the Pacific. (From Morris (1978))

propagation down a continental slope converts high-angle rays to lower angles at each bounce. There are also deep sound channel shoaling effects that result in the same trend in angle conversion.

Sonar Equation

A major application of underwater acoustics is sonar technology. The performance of a sonar is often described simply in terms of the sonar equation. The methodology of the sonar equation is analogous to an accounting procedure involving acoustic signal, interference, and system characteristics.

It is instructive, beyond the specific application to conventional sonars, to understand this accounting methodology and below is a simplified summary.

Passive Sonar Equation

A passive sonar system uses the radiated sound from a target to detect and locate the target. A radiating object of source level SL (all units are in decibels) is received at a hydrophone of a sonar system at a lower signal level S because of the transmission loss 'TL' it suffers (e.g., cylindrical spreading plus attenuation or a TL computed from one of the propagation models) (eqn [14]).

$$S = SL - TL \qquad [14]$$

The noise, N, at a single hydrophone is subtracted from eqn [14] to obtain the signal-to-noise ratio (SNR) at a single hydrophone (eqn [15]).

$$SNR = SL - TL - N \qquad [15]$$

Typically a sonar system consists of an array or antenna of hydrophones that provides signal-to-noise enhancement through a beamforming process; this process is quantified in decibels by array gain AG that is therefore added to the single hydrophone SNR to give the SNR at the output of the beamformer (eqn [16]).

$$SNR_{BF} = SL - TL - N + AG \qquad [16]$$

Because detection involves additional factors including sonar operator ability, it is necessary to specify a detection threshold (DT) level above the SNR_{BF} at which there is a 50% (by convention) probability of detection. The difference between these two quantities is called signal excess (SE) (eqn [17]).

$$SE = SL - TL - N + AG - DT \qquad [17]$$

This decibel bookkeeping leads to an important sonar engineering descriptor called the figure of merit, FOM, which is the transmission loss that gives a zero signal excess (eqn [18]).

$$FOM = SL - N + AG - DT \qquad [18]$$

The FOM encompasses the various parameters a sonar engineer must deal with: expected source level, the noise environment, array gain and the detection threshold. Conversely, since the FOM is a transmission loss, one can use the output of a propagation model (or, if appropriate, a simple geometric loss plus attenuation) to estimate the minimum range at which a 50% probability of detection can be expected. This range changes with oceanographic conditions and is often referred to as the 'range of the day' in navy sonar applications.

Active Sonar Equation

A monostatic active sonar transmits a pulse to a target and its echo is detected at a receiver co-located with the transmitter. A bistatic active sonar has the receiver in a different location from the transmitter. The main differences between the passive and active cases is that the source level is replaced by a target strength, TS; reverberation and hence reverberation level, RL, is usually the dominant source of interference as opposed noise; and the transmission loss is over two paths: transmitter to target and target to receiver. In the monostatic case the transmission loss is $2TL$, where TL is the one-way transmission loss; and in the bistatic case the transmission loss is the sum (in dB) over paths from the transmitter to the target and the target to the receiver, $TL_1 + TL_2$. The concept of the detection threshold is useful for both passive and active sonars. Hence, for signal excess, we have eqn [19].

$$S = SL - TL_1 + TS - TL_2$$
$$- (RL + N) + AG - DT \qquad [19]$$

The corresponding FOM for an active system is defined for the maximum allowable two-way transmission loss with $TS = 0\,\mathrm{dB}$.

See also

Acoustic Noise. Acoustic Scattering by Marine Organisms. Acoustics, Arctic. Acoustics, Shallow Water. Acoustics in Marine Sediments. Bioacoustics. Deep-Sea Drilling Methodology. Seismology Sensors. Sonar Systems. Tomography.

Further Reading

Anderson VC (1979) Variations of the vertical directivity of noise with depth in the North Pacific. *Journal of the Acoustical Society of America* 66: 1446–1452.

Brekhovskikh LM and Lysanov YP (1991) *Fundamentals of Ocean Acoustics*. Berlin: Springer-Verlag.

Chapman RP and Harris HH (1962) Surface backscattering strengths measured with explosive sound sources.

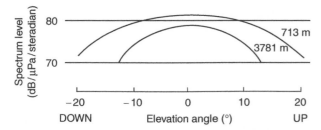

Figure 12 The vertical directionality of noise at the axis of the deep sound channel and at the critical depth in the Pacific. (From Anderson (1979))

Journal of the Acoustical Society of America 34: 1592–1597.

Chapman RP and Marchall JR (1966) Reverberation from deep scattering layers in the Western North Atlantic. *Journal of the Acoustical Society of America* 40: 405–411.

Collins MD and Siegmann WL (2001) *Parabolic Wave Equations with Applications*. New York: Springer-AIP.

Dushaw BD, Worcester PF, Cornuelle BD, and Howe BM (1993) On equations for the speed of sound in seawater. *Journal of the Acoustical Society of America* 93: 255–275.

Jensen FB, Kuperman WA, Porter MB, and Schmidt H (1994) *Computational Ocean Acoustics*. Woodbury: AIP Press.

Keller and Papadakis JS (eds.) (1977) *Wave Propagation in Underwater Acoustics*. New York: Springer-Verlag.

Makris NC, Chia SC, and Fialkowski LT (1999) The bi-azimuthal scattering distribution of an abyssal hill. *Journal of the Acoustical Society of America* 106: 2491–2512.

Medwin H and Clay CS (1997) *Fundamentals of Acoustical Oceanography*. Boston: Academic Press.

Morris GB (1978) Depth dependence of ambient noise in the Northeastern Pacific Ocean. *Journal of the Acoustical Society of America* 64: 581–590.

Munk W, Worcester P, and Wunsch C (1995) *Acoustic Tomography*. Cambridge: Cambridge University Press.

Nicholas M, Ogden PM, and Erskine FT (1998) Improved empirical descriptions for acoustic surface backscatter in the ocean. *IEEE-JOE* 23: 81–95.

Northrup J and Colborn JG (1974) Sofar channel axial sound speed and depth in the Atlantic Ocean. *Journal of Geophysical Research* 79: 5633–5641.

Ross D (1976) *Mechanics of Underwater Noise*. New York: Pergamon.

Ogilvy JA (1987) Wave scattering from rough surfaces. *Reports on Progress in Physics* 50: 1553–1608.

Urick RJ (1979) *Sound Propagation in the Sea*. Washington, DC: US GPO.

Urick RJ (1983) *Principles of Underwater Sound*. New York: McGraw Hill.

Spiesberger JL and Metzger K (1991) A new algorithm for sound speed in seawater. *Journal of the Acoustical Society of America* 89: 2677–2688.

Wenz GM (1962) Acoustics ambient noise in the ocean: spectra and sources. *Journal of the Acoustical Society of America* 34: 1936–1956.

ACOUSTICS, SHALLOW WATER

F. B. Jensen, SACLANT Undersea Research Centre, La Spezia, Italy

Introduction

Using the commonly accepted definition of shallow water to mean coastal waters with depth up to 200 m, the shallow-water regions of the world constitute around 8% of all oceans and seas. These regions are particularly important since they are national economic zones and also more assessible.

Sound waves in the sea play the role of light in the atmosphere, i.e. acoustics is the only means of 'seeing' objects at distances beyond a few hundred meters in seawater. All forms of electromagnetic waves (light, radar) are rapidly attenuated in seawater. Low-frequency acoustic signals, on the other hand, propagate with little attenuation and can be heard over thousands of kilometers in the deep ocean.

The use of sound in the sea is ubiquitous. It is employed by the military to detect mines and submarines, and ship-mounted sonars measure water depth, ship speed and the presence of fish shoals. Side-scan systems are used to map bottom topography, sub-bottom profilers for getting information about the deeper layering, and other sonar systems for locating pipelines and cables on and beneath the seafloor. Sound is also used for navigating submerged vehicles, for underwater communications and for tracking marine mammals. In an inverse sense sound is used for measuring physical parameters of the ocean environment and for monitoring oceanic processes through the techniques of acoustical oceanography and ocean acoustic tomography.

Optimal sonar design for this great variety of applications demands using a wide range of acoustic frequencies. Practical shallow-water systems cover a frequency range from 50 Hz to 500 kHz, which, with a mean sound speed of 1500 m s^{-1}, correspond to acoustic wavelengths from 30 m down to 3 mm.

The principal characteristic of shallow-water propagation is that the sound-speed profile is nearly constant over depth or downward refracting, meaning that long-range propagation takes place exclusively via lossy bottom-interacting paths. This is very different from deep-water scenarios, where the sound-speed structure is such that sound is refracted away from the bottom and therefore can propagate to long ranges with little attenuation. Moreover, the environmental variability is much higher in coastal regions than in the deep ocean, with the result that there is much more acoustic variability in shallow water than in deep water.

The Ocean Acoustic Environment

The ocean is an acoustic waveguide limited above by the sea surface and below by the seafloor. The speed of sound in the waveguide plays the same role as the index of refraction does in optics. Sound speed is normally related to density and compressibility. In the ocean, density is related to static pressure, salinity and temperature. The sound speed in the ocean is an increasing function of temperature, salinity, and pressure, the latter being a function of depth. It is customary to express sound speed (c) as an empirical function of three independent variables: temperature (T) in degrees centigrade, salinity (S) in parts per thousand (‰), and depth (D) in meters. A simplified expression for this dependence is

$$c = 1449.2 + 4.6T - 0.055T^2 + 0.00029T^3 \\ + (1.34 - 0.010T)(S - 35) + 0.016D \quad [1]$$

In shallow water, where the depth effect on sound speed is small, the primary contributor to sound speed variations is the temperature. Thus, for a salinity of 35‰, the sound speed in seawater varies between 1450 m s^{-1} at 0°C and 1545 m s^{-1} at 30°C.

Seasonal and diurnal changes affect the oceanographic parameters in the upper ocean. In addition, all of these parameters are a function of geography. In a warmer season (or warmer part of the day) in shallow seas where tidal mixing is weak, the temperature increases near the surface and hence the sound speed increases toward the sea surface. This near-surface heating (and subsequent cooling) has a profound effect on surface-ship sonars. Thus the diurnal heating causes poorer sonar performance in the afternoon – a phenomenon known as the afternoon effect. The seasonal variability, however, is much greater and therefore more important acoustically.

A ray picture of propagation in a 100-m deep shallow water duct is shown in **Figure 1**. The sound-speed profile in the upper panel is typical of the Mediterranean in the summer. There is a warm surface layer causing downward refraction and hence repeated bottom interaction for all ray paths. Since

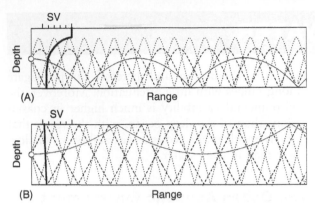

Figure 1 Ray paths in shallow water for typical Mediterranean summer and winter profiles. (A) In summer sound interacts repeatedly with the seabed but not with the sea surface. (B) In winter sound interacts with both the sea surface and the seabed, except for shallow rays emitted near the horizontal.

the seafloor is a lossy boundary, propagation in shallow water is dominated by bottom reflection loss at low and intermediate frequencies ($< 1\,\text{kHz}$) and scattering losses at high frequencies. The seasonal variation in sound-speed structure is significant with winter conditions being nearly iso-speed (**Figure 1B**). The result is that there is less bottom interaction in winter than in summer, which again means that propagation conditions are generally better in winter than in summer.

Of course, the ocean sound-speed structure is neither frozen in time nor space. On the contrary, the ocean has its own weather system. There are currents, internal waves and thermal microstructure present in most shallow-water areas. **Figure 2** illustrates the sound speed variability along a 15 km-long track in the Mediterranean Sea. The data were recorded on a towed thermistor chain covering depths between 5 and 90 m. In general, this type of time-varying oceanographic structure has an effect on sound propagation, both as a source of attenuation (acoustic energy being scattered into steeperangle

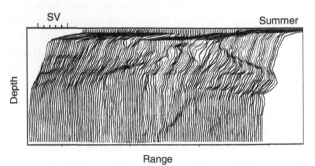

Figure 2 Spatial variability of sound speed in shallow-water area of the Mediterranean. The depth covered is around 100 m and the range 15 km.

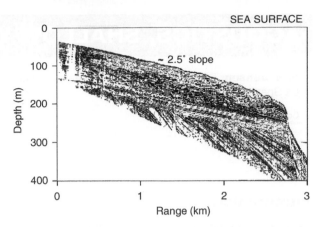

Figure 3 Seismic profile of bottom layering in coastal-water area of the Mediterranean.

propagation paths suffers increased bottom reflection loss) and of acoustic signal fluctuations with time.

Turning to the upper and lower boundaries of the ocean waveguide, the sea surface is a simple horizontal boundary and a nearly perfect reflector. The seafloor, on the other hand, is a lossy boundary with varying topography. Both boundaries have small-scale roughness associated with them which causes scattering and hence attenuation of sound due to the increased bottom reflection loss associated with steep-angle propagation paths. In terms of propagation physics, the seafloor is definitely the most complex boundary, exhibiting vastly different reflectivity characteristics in different geographical locations.

The structure of the ocean bottom in shallow water generally consists of a thin stratification of sediments overlying the continental crust. The nature of the stratification is dependent on many factors, including geological age and local geological activity. Thus, relatively recent sediments will be characterized by plane stratification parallel to the sea bed, whereas older sediments and sediments close to the crustal plate boundaries may have undergone significant deformation. An example of a complicated bottom layering is given in **Figure 3**, which displays a seismic section from the coastal Mediterranean. The upper stratification here is almost parallel to the seafloor, whereas deeper layers are strongly inclined.

Transmission Loss

The decibel (dB) is the dominant unit in ocean acoustics and denotes a ratio of intensities (not pressures) expressed on a $\log_{10}$ scale.

An acoustic signal traveling through the ocean becomes distorted due to multipath effects and weakened due to various loss mechanisms. The standard measure in underwater acoustics of the change in signal strength with range is transmission loss defined as the ratio in decibels between the acoustic intensity $I(r,z)$ at a field point and the intensity I_0 at 1 m distance from the source, i.e.

$$
\begin{aligned}
\mathrm{TL} &= -10\log\frac{I(r,z)}{I_0} \\
&= -20\log\frac{|p(r,z)|}{|p_0|} \quad [\mathrm{dB\ re\ 1\,m}].
\end{aligned}
\quad [2]
$$

Here use has been made of the fact that the intensity of a plane wave is proportional to the square of the pressure amplitude. The major contributors to transmission loss in shallow water are: geometrical spreading loss, water volume attenuation, bottom reflection loss, and various scattering losses.

Geometrical Spreading

The spreading loss is simply a measure of the signal weakening as it propagates outward from the source. **Figure 4** shows the two geometries of importance in underwater acoustics. First consider a point source in an unbounded homogeneous medium (**Figure 4A**). For this simple case the power radiated by the source is equally distributed over the surface area of a sphere surrounding the source. If the medium is assumed to be lossless, the intensity is inversely proportional to the surface of the sphere, i.e. $I \propto 1/(4\pi R^2)$. Then from eqn [2] the spherical spreading loss is given by

$$
\mathrm{TL} = 20\log r \quad [\mathrm{dB\ re\ 1\,m}] \quad [3]
$$

where r is the horizontal range in meters.

When the medium has plane upper and lower boundaries as in the waveguide case in **Figure 4B**, the farfield intensity change with horizontal range becomes inversely proportional to the surface of a cylinder of radius R and depth D, i.e. $I \propto 1/(2\pi RD)$.

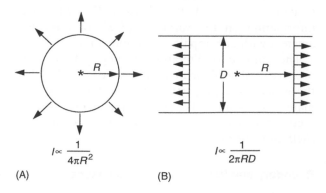

(A)   (B)

Figure 4   Geometrical spreading laws. (A) Spherical spreading; (B) Cylindrical spreading.

The cylindrical spreading loss is therefore given by

$$
\mathrm{TL} = 10\log r \quad [\mathrm{dB\ re\ 1\,m}] \quad [4]
$$

Note that for a point source in a waveguide, there is spherical spreading in the nearfield ($r \leq D$) followed by a transition region toward cylindrical spreading which applies only at longer ranges ($r \gg D$).

As an example consider propagation in a shallow-water waveguide to a range of 20 km with spherical spreading applying on the first 100 m. The total propagation loss (neglecting attenuation) then becomes: $40\,\mathrm{dB} + 23\,\mathrm{dB} = 63\,\mathrm{dB}$. This figure represents the minimum loss to be expected at 20 km. In practice, the total loss will be higher due both to the attenuation of sound in seawater, and to various reflection and scattering losses.

Sound Attenuation in Seawater

When sound propagates in the ocean, part of the acoustic energy is continuously absorbed, i.e. the energy is transformed into heat. Moreover, sound is scattered by different kinds of inhomogeneities, also resulting in a decay of sound intensity with range. As a rule, it is not possible in real ocean experiments to distinguish between absorption and scattering effects; they both contribute to sound attenuation in seawater.

A simplified expression for the frequency dependence (f in kHz) of the attenuation is given by Thorp's formula,

$$
\alpha = \frac{0.11f^2}{1+f^2} + \frac{44f^2}{4100+f^2} \quad [\mathrm{dB\ km^{-1}}], \quad [5]
$$

where the two terms describe absorption due to chemical relaxations of boric acid, $B(OH)_3$, and magnesium sulphate, $MgSO_4$, respectively.

According to eqn [5] the attenuation of low-frequency sound in seawater is indeed very small. For instance, at 100 Hz a tenfold reduction in sound intensity ($-10\,\mathrm{dB}$) occurs over a distance of around 8300 km. Even though attenuation increases with frequency ($r_{-10\,\mathrm{dB}} \simeq 150\,\mathrm{km}$ at 1 kHz and $\simeq 9\,\mathrm{km}$ at 10 kHz), no other kind of radiation can compete with sound waves for long-range propagation in the ocean.

Bottom Reflection Loss

Reflectivity, the ratio of the amplitudes of a reflected plane wave to a plane wave incident on an interface separating two media, is an important measure of the effect of the bottom on sound propagation. Ocean bottom sediments are often modeled as fluids which means that they support only one type of sound wave – a compressional wave.

The expression for reflectivity at an interface separating two homogeneous fluid media with density ρ_i and sound speed c_i, $i = 1, 2$, was first worked out by Rayleigh as

$$R(\theta_1) = \frac{(\rho_2/\rho_1)\sin\theta_1 - \sqrt{\left((c_1/c_2)^2 - \cos^2\theta_1\right)}}{(\rho_2/\rho_1)\sin\theta_1 + \sqrt{\left((c_1/c_2)^2 - \cos^2\theta_1\right)}} \qquad [6]$$

where θ_1 denotes the grazing angle of the incident plane wave of unit amplitude.

The reflection coefficient has unit magnitude, meaning perfect reflection, when the numerator and denominator of eqn [6] are complex conjugates. This can only occur when the square root is purely imaginary, i.e. for $\cos\theta_1 > c_1/c_2$ (total internal reflection). The associated critical grazing angle below which there is perfect reflection is found to be

$$\theta_c = \arccos\left(\frac{c_1}{c_2}\right) \qquad [7]$$

Note that a critical angle only exists when the sound speed of the second medium is higher than that of the first.

A closer look at eqn [6] shows that the reflection coefficient for lossless media is real for $\theta_c > \theta_c$, which means that there is loss ($|R| = 1$) but no phase shift associated with the reflection process. On the other hand, for $\theta_c < \theta_c$ we have perfect reflection ($|R| = 1$) but with an angle-dependent phase shift. In the general case of lossy media (c_i complex), the reflection coefficient is complex, and, consequently, there is both a loss and a phase shift associated with each reflection.

The critical-angle concept is very important for understanding the waveguide nature of shallow-water propagation. **Figure 5** shows bottom loss curves ($BL = -10\log|R|^2$) for a few simple fluid bottoms with different compressional wave speeds (C_p), densities and attenuations. Note that for a lossy bottom we never get perfect reflection. However, there is in all cases an apparent critical angle ($\theta_C \simeq 33°$ for $c_p = 1800\,\mathrm{m\,s^{-1}}$ in **Figure 5**), below which the reflection loss is much smaller than for supercritical incidence. With paths involving many bottom bounces such as in shallow-water propagation, bottom losses even as small as a few tenths of a decibel per bounce accumulate to significant total losses since the propagation path may involve many tens or even hundreds of bounces.

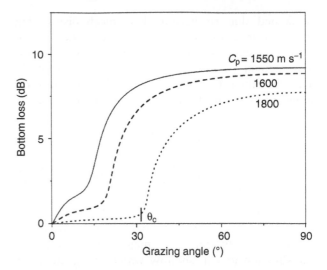

Figure 5 Bottom reflection loss curves for different bottom types. Note that low-speed bottoms (clay, silt) are more lossy than high-speed bottoms (sand, gravel). C_p is the compressional wave speed.

Real ocean bottoms are complex layered structures of spatially varying material composition. A geoacoustic model is defined as a model of the real seafloor with emphasis on measured, extrapolated, and predicted values of those material properties important for the modeling of sound transmission. In general, a geoacoustic model details the true thicknesses and properties of sediment and rock layers within the seabed to a depth termed the effective acoustic penetration depth. Thus, at high frequencies ($> 1\,\mathrm{kHz}$), details of the bottom composition are required only in the upper few meters of sediment, whereas at low frequencies ($< 100\,\mathrm{Hz}$) information must be provided on the whole sediment column and on properties of the underlying rocks.

The information required for a complete geoacoustic model should include the following depth-dependent material properties: the compressional wave speed, C_p; the shear wave speed, C_p; the compressional wave attenuation, C_p; the shear wave attenuation, C_p; and the density, C_p. Moreover, information on the variation of all of these parameters with geographical position is required.

The amount of literature dealing with acoustic properties of seafloor materials is vast. **Table 1** lists the geoacoustic properties of some typical seafloor materials, as an indication of the many different types of materials encountered just in continental shelf and slope environments.

Boundary and Volume Scattering Losses

Scattering is a mechanism for loss, interference and fluctuation. A rough sea surface or seafloor causes

Table 1 Geoacoustic properties of continental shelf environments

Bottom type	p (%)	ρ_b/ρ_w –	C_p/C_w –	C_p (m s^{-1})	C_s (m s^{-1})	α_p ($\alpha B \lambda_p^{-1}$)	α_s ($\alpha B \lambda_s^{-1}$)
Clay	70	1.5	1.00	1500	<100	0.2	1.0
Silt	55	1.7	1.05	1575	$C_s{}^a$	1.0	1.5
Sand	45	1.9	1.1	1650	$C_s{}^b$	0.8	2.5
Gravel	35	2.0	1.2	1800	$C_s{}^c$	0.6	1.5
Moraine	25	2.1	1.3	1950	600	0.4	1.0
Chalk	–	2.2	1.6	2400	1000	0.2	0.5
Limestone	–	2.4	2.0	3000	1500	0.1	0.2
Basalt	–	2.7	3.5	5250	2500	0.1	0.2

$^a C_s = 80\ \bar{z}^{0.3}$
$^b c_s = 110\ \bar{z}^{0.3}$
$^c c_s = 180\ \bar{z}^{0.3}$
$C_w = 1500\ \text{ms}^{-1}$, $\rho_w - 1000\ \text{kg m}^{-3}$.

attenuation of the mean acoustic field propagating in the ocean waveguide. The attenuation increases with increasing frequency. The field scattered away from the specular direction, and, in particular, the back-scattered field (called reverberation) acts as interference for active sonar systems. Because the ocean surface moves, it will also generate acoustic fluctuations. Bottom roughness can also generate fluctuations when the sound source or receiver is moving. The importance of boundary roughness depends on the sound-speed profile which determines the degree of interaction of sound with the rough boundaries.

Often the effect of scattering from a rough surface is thought of as simply an additional loss to the specularly reflected (coherent) component resulting from the scattering of energy away from the specular direction. If the ocean bottom or surface can be modeled as a randomly rough surface, and if the roughness is small with respect to the acoustic wavelength, the reflection loss can be considered to be modified in a simple fashion by the scattering process. A formula often used to describe reflectivity from a rough boundary is:

$$R'(\theta) = R(\theta)e^{-0.5\Gamma^2} \qquad [8]$$

where $R'(\theta)$ is the new reflection coefficient, reduced because of scattering at the randomly rough interface. Γ is the Rayleigh roughness parameter defined as

$$\Gamma = 2k\sigma\sin\theta \qquad [9]$$

where $k = 2\pi/\lambda$ is the acoustic wavenumber and σ is the *rms* roughness. Note that the reflection coefficient for the smooth ocean surface is simply -1 (the pressure-release condition is obtained from eqn [6] by setting $\rho_2 = 0$) so that the rough-sea-surface reflection coefficient for the coherent field is $R'(\theta) = -\exp(-0.5\Gamma^2)$. For the ocean bottom, the

appropriate geoacoustic parameters (see **Table 1**) are used for evaluating $R'(\theta)$, and the rough-bottom reflection coefficient is then obtained from eqn [8].

Volume scattering is thought to arise primarily from biological organisms. For lower frequencies (less than 10 kHz), fish with air-filled swim bladders are the main scatterers whereas above 20 kHz, zooplankton or smaller animals that feed on the phytoplankton, and the associated biological food chain, are the scatterers. Many of the organisms undergo a diurnal migration rising towards the sea surface at sunset and descending to depth at sunrise. Since the composition and density of the populations vary with the environmental conditions, the scattering characteristics depend on geographical location, time of day, season and frequency. As an example, data from the Mediterranean Sea for volume scattering losses due to fish shoals show excess losses of 10–15 dB for a propagation range of 12 km and frequencies between 1 and 3 kHz.

Finally, scattering off bubbles near the surface is sometimes referred to as either a volume- or surface-scattering mechanism. These bubbles arise not only from sea surface action, but also from biological origins and from ship wakes. Furthermore, bubbles are not the only scattering mechanism, but bubble clouds may have significantly different sound speed than plain seawater thereby altering local refraction conditions. At the sea surface, the relative importance of roughness versus bubble effects is not yet resolved.

Transmission-loss Data

Figure 6 gives an example of transmission-loss variability in shallow water. The graph displays a collection of experimental data from different shallow-water areas (100–200 m deep) all over the

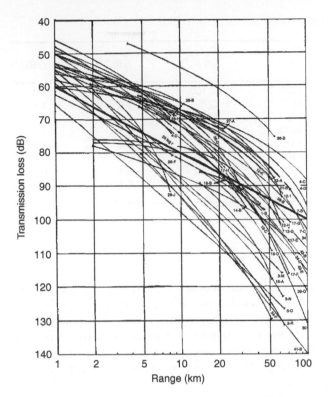

Figure 6 Transmission loss variability in shallow water.

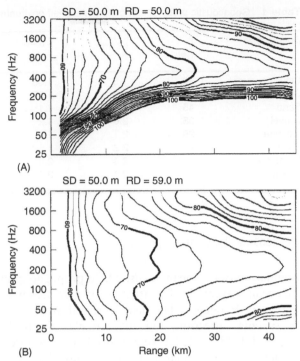

Figure 7 Examples of frequency-dependent propagation losses measured in two shallow-water areas: (A) Barents Sea, (B) English Channel. Note the presence of an optimum frequency of propagation between 200 and 400 Hz.

world. The data refer to downward-refracting summer conditions in the frequency band 0.5–1.5 kHz. Two features are of immediate interest. One is the spread of the data amounting to around 50 dB at 100 km and caused primarily by the varying bottom-loss conditions in different areas of the world. The second feature is the fact that transmission is generally better than free-field propagation (20 log r) at short and intermediate ranges but worse at longer ranges. This peculiarity is due to the trapping of energy in the shallow-water duct, which improves transmission at shorter ranges (cylindrical versus spherical spreading), but, at the same time, causes increased boundary interaction, which degrades transmission at longer ranges.

A second example of transmission-loss variability in shallow water is given in **Figure 7**, where broadband data from two different geographical areas are compared. The data set in **Figure 7A**, was collected in the Barents Sea in 60 m water depth. Note the high transmission losses recorded below 200 Hz, where energy levels fall off rapidly indicating that most of the acoustic energy emitted by the source is lost to the seabed. It is believed that this excess attenuation is caused by the coupling of acoustic energy into shear waves in the seabed. In contrast to the high-loss environment in the Barents Sea, **Figure 7B** shows a data set from the English Channel in 90 m water depth. Here propagation conditions are excellent

over the entire frequency band. This second data set represents typical propagation conditions for thick sandy sediments with negligible shear-wave effects.

Cutoff Frequency and Optimum Frequency

A common feature of all acoustic ducts is the existence of a low-frequency cutoff. Hence, there is a critical frequency below which the shallow-water channel ceases to act as a waveguide, causing energy radiated by the source to propagate directly into the bottom. The cutoff frequency is given by,

$$f_0 = \frac{c_w}{4D\sqrt{\left(1 - (c_w/c_b)^2\right)}} \quad [10]$$

This expression is exact only for a homogeneous water column of depth D and sound speed c_w overlying a homogeneous bottom of sound speed c_b. As an example, let us take $D = 100$ m, $c_w = 1500$ m s^{-1}, and $c_b = 1600$ m s^{-1} (sand–silt), which yields $f_0 \simeq 11$ Hz.

Sound transmission in shallow water has the characteristic frequency-dependent behavior shown in **Figure 7**, i.e. there is an optimum frequency of propagation at longer ranges. Thus the 80 dB contour line extends farthest in range for frequencies around 400 Hz in **Figure 7A** and around 200 Hz in

Figure 7B, implying that transmission is best at these frequencies – the optimum frequencies of propagation for the two sites.

Optimum frequency is a general feature of ducted propagation in the ocean. It occurs as a result of competing propagation and attenuation mechanisms at high and low frequencies. In the high-frequency regime we have increasing volume and scattering loss with increasing frequency. At lower frequencies the efficiency of the duct to confine sound decreases (the cutoff phenomenon). Hence propagation and attenuation mechanisms outside the duct (in the seabed) become important. In fact, the increased penetration of sound into a lossy seabed with decreasing frequency causes the overall attenuation of waterborne sound to increase with decreasing frequency. Thus we get high attenuation at both high and low frequencies, whereas intermediate frequencies have the lowest attenuation. It can be shown that the optimum frequency for shallow-water propagation is strongly dependent on water depth ($f_{opt} \propto D^{-1}$), has some dependence on the sound-speed profile, but is only weakly dependent on the bottom type. Typically, the optimum frequency is in the range 200–800 Hz for a water depth of 100 m.

Signal Transmission in the Time Domain

Even though underwater acousticians have traditionally favored spectral analysis techniques for gaining information about the band-averaged energy distribution within a shallow-water waveguide, additional insight into the complication of multipath propagation can be obtained by looking at signal transmission in the time domain.

Figure 8 indicates that the signal structure measured downrange will consist of a number of arrivals with time delays determined by the pathlength

differences, and individual pulse shapes being modified due to frequency-dependent amplitude and phase changes associated with each boundary reflection. From simple geometrical considerations, the time dispersion is found to be

$$\Delta_\tau \simeq \frac{R}{\bar{c}} \left(\frac{1}{\cos\theta} - 1 \right) \quad [11]$$

where R is the range between source and receiver, $\bar{c}$ is the mean sound speed in the channel, and θ is the maximum propagation angle with respect to the horizontal. This angle will be determined either by the source beamwidth or by the critical angle at the bottom (the smaller of the two). Since the dispersion considered here is solely due to the geometry of the waveguide, it is called geometrical dispersion.

An example of measured pulse arrivals over a 10 h period in the Mediterranean is given in **Figure 9**. Note that the time-varying ocean (internal waves, currents, tides) causes strong signal fluctuations with time, particularly in the earlier part of the signal. The time dispersion is 15–20 ms and at least four main energy packets, each consisting of several ray arrivals can be identified.

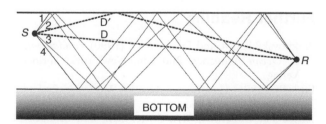

Figure 8 Schematic of ray arrivals in shallow-water waveguide. A series of multipath arrivals are expected, represented by eigenrays connecting source (S) and receiver (R). The shortest path is the direct arrival D followed by the surface-reflected arrival D'. Next comes a series of four arrivals all with a single bottom bounce; then four rays with two bottom bounces as illustrated in the figure, followed by four rays with three bottom bounces, etc.

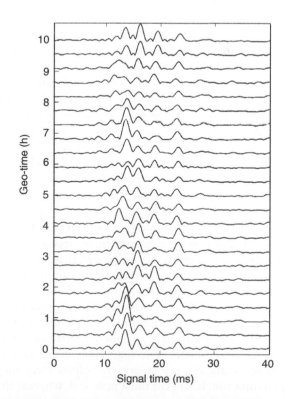

Figure 9 Measured pulse arrivals versus geo-time over a 10 km shallow-water propagation track in the Mediterranean Sea. The bandwidth is 200–800 Hz.

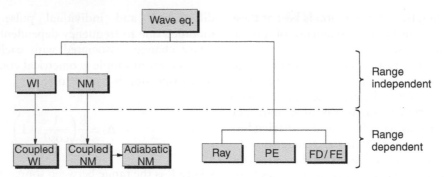

Figure 10 Hierarchy of numerical models in underwater acoustics. WI, wavenumber integration; NM, normal modes; PE, parabolic equation; FD, finite difference; FE, finite element.

Numerical Modeling

The advent of computers has resulted in an explosive growth in the development and use of numerical models since the mid-1970s. Numerical models have become standard research tools in acoustic laboratories, and computational acoustics is becoming an evermore important branch of the ocean acoustic science. Only the numerical approach permits an analysis of the full complexity of the acoustic problem.

An assortment of models has been developed over the past 25 years to compute the acoustic field in shallow-water environments in both the frequency and time domains. Entire textbooks are dedicated to the development of theoretical and numerical formalisms which can provide quantitative acoustic predictions for arbitrary ocean environments. Sound propagation is mathematically described by the wave equation, whose parameters and boundary conditions are descriptive of the ocean environment. As shown in **Figure 10**, there are essentially five types of models (computer solutions to the wave equation) to describe sound propagation in the sea: wavenumber integration (WI); normal mode (NM); ray; parabolic equation (PE) and direct finite-difference (FD) or finite-element (FE) solutions of the full wave equation. All of these models permit the ocean environment to vary with depth. A model that also permits horizontal variations in the environment, i.e. sloping bottom or spatially varying oceanography, is termed range dependent.

As shown in **Figure 10**, an a priori assumption about the environment being range independent, leads to solutions based on spectral techniques (WI) or normal modes (NM); both of these techniques can, however, be extended to treat range dependence. Ray, PE and FD/FE solutions are applied directly to range-varying environments. For high frequencies (a few kilohertz or above), ray theory, the infinite frequency approximation, is still the most practical, whereas the other five model types become more and more applicable below, say, a kilohertz in shallow water.

Models that handle ocean variability in three spatial dimensions have also been developed, but these models are used less frequently than two-dimensional versions because of the computational cost involved.

Conclusions

The acoustics of shallow water has been thoroughly studied both experimentally and theoretically since World War II. Today the propagation physics is well understood and sophisticated numerical models permit accurate simulations of all processes (reflection, refraction, scattering) that contribute to the complexity of the shallow-water problem. Sonar performance predictability, however, is limited by knowledge of the controlling environmental inputs. The current challenge is therefore how best to collect relevant environmental data from the world's enormously variable shallow-water areas.

See also

Acoustic Scattering by Marine Organisms. Acoustics in Marine Sediments. Acoustics, Arctic. Acoustic Noise. Sonar Systems. Tomography.

Further Reading

Brekhovskikh LM and Lysanov YP (1990) *Fundamentals of Ocean Acoustics*, 2nd edn. New York: Springer-Verlag.

Etter PC (1996) *Underwater Acoustic Modeling*, 2nd edn. London: E & FN Spon.

Jensen FB, Kuperman WA, Porter MB, and Schmidt H (2000) *Computational Ocean Acoustics*. New York: Springer-Verlag.

Medwin H and Clay CS (1998) *Fundamentals of Acoustical Oceanography*. San Diego: Academic Press.

Urick RJ (1996) *Principles of Underwater Sound*, 3rd edn. Los Altos: Peninsula Publishing.

BIOACOUSTICS

P. L. Tyack, Woods Hole Oceanographic Institution,
Woods Hole, USA

Introduction

The term 'bioacoustics' has two different usages in
ocean sciences. Biological oceanographers use active
sonars to map organisms in the sea. Since they use
sound to detect marine life, they often call this approach 'bioacoustics'. The other sense of bioacoustics involves studying how animals use sound
themselves in the ocean. This is the kind of bioacoustics covered in this article.

Humans are visual animals, and we think of vision
as a primary distance sense because light carries so
well in terrestrial environments. However, light is
useful for vision under the sea only over ranges of
tens of meters at best. Sound, on the other hand,
propagates extremely well in water – that is why
oceanographers so often select sound as a medium
for exploring the sea or for communicating under the
sea. Sound propagates so well under water that a
depth charge exploded off Australia can be heard in
Bermuda. Just as we can hear well but emphasize
vision, so many marine mammal species see well but
emphasize hearing. It is possible to gauge the relative
importance of audition versus vision in animals by
comparing the number of nerve fibers in the auditory
versus the optic nerves. Of all marine mammals, the
cetaceans are the most specialized to use sound.
Most cetaceans have auditory:optic ratios of fiber
counts that are 2–3 times those of land mammals,
suggesting that audition is more important than vision. Some cetaceans also use sound to echolocate.
Dolphins have a large repertoire of vocalizations
spanning frequencies from below 100 Hz to over 100
kHz, and dolphins have evolved high-frequency
echolocation similar to some human-made sonars
and to the biosonar used by bats.

Marine mammals not only hear well, they are also
very vocal animals. The sounds of marine mammals
are now well known, but the first recordings identified from a marine mammal species were only made
in the late 1940s. In the 1950s and 1960s, there was
rapid growth in studies of how dolphins echolocate
using high-frequency click sounds and of field studies
associating different sounds with different species of

marine mammal. Marine mammal bioacoustics during this period was concerned primarily with identifying which species produced which sounds heard
under water. Much of this research was funded by
naval research organizations because biological
sources of noise can interfere with military use of
sound in the sea.

Elementary Acoustics

Sound consists of mechanical vibrations that propagate through a medium. Sound induces movements
or displacements of the particles in the medium.
Imagine a small sphere that expands to create a
denser area. This compression will propagate as
particles are displaced in the direction of propagation. If the sphere then contracts, it can create an
area of rarefaction, or lower density, and this also
can propagate outward. These compressions or rarefactions can be expressed in terms of particle displacement or as a pressure differential.

Now imagine a sound source that creates a series
of compressions and rarefactions that propagate
through the medium. A source with a purely sinusoidal pattern of compression and rarefaction would
produce energy at only one frequency. The frequency
of this sound is measured in cycles per second. A
sound that takes t seconds to make a full cycle has a
frequency $f = t^{-1}$. Older references may refer to frequency in cycles per second, but the modern unit of
frequency is the Hertz (Hz) and a frequency of 1000
Hz is expressed as one kiloHertz (1 kHz). If a sound
took 1 s for a full cycle, it would have a frequency of
1 Hz. The wavelength of a tonal sound is the distance
from one measurement of the maximum pressure to
the next maximum. The speed of sound is approximately $1500 \, \text{m s}^{-1}$ in water, roughly five times the
value in air, $340 \, \text{m s}^{-1}$. The speed of a sound c is
related in a simple way to the frequency f and the
wavelength λ by $c = \lambda f$. An under-water sound with
$f = 1$ Hz would have $\lambda = 1500$ m; for $f = 1500$ Hz,
$\lambda = 1$ m. Not all sounds have energy limited to one
frequency. Sounds that have energy in a range of
frequencies, say in the frequency range between 2000
and 3000 Hz (2 and 3 kHz), would be described as
having a bandwidth of 1 kHz.

One can imagine a sound wave as a growing sphere
propagating outward from a compression or rarefaction generated by a point source. The initial
movement of the source will have transmitted a certain amount of energy to the medium. If none of this

energy is lost as the sound propagates, then it will be evenly diluted over the growing sphere. The acoustic intensity is defined as the amount of energy flowing through an area over a unit of time. As the sphere increases in radius from 1 to r, the surface area increases to $4\pi r^2$. The intensity of a sound thus declines as the inverse of the square of the range from the source (r^{-2}). A sound in the middle of the ocean can be thought of as spreading in this way until it encounters a boundary such as the surface or seafloor that might cause reflection, or an inhomogeneity in the medium that might cause refraction. One fascinating acoustic feature of the deep ocean is that sound rays propagating upward may refract downward as they encounter warmer water near the surface, and downward-propagating rays will refract upward as they encounter denser water at depth. When one is far from a sound source compared to the ocean depth, the sound energy may be concentrated by refraction in the deep ocean sound channel. This sound can be thought of as spreading in a plane, to a first approximation. In this case, sound intensity would decline as the inverse of the first power of the range, or r^{-1}. This involves much lower loss than the inverse square spreading loss in an unbounded medium.

Sound spreading is a 'dilution' factor and is not a true loss of sound energy. Absorption, on the other hand, is conversion of acoustic energy to heat. The attenuation of sound due to absorption is a constant per unit distance, but this constant is dependent upon signal frequency. While absorption yields trivial effects at frequencies below 100 Hz, it can significantly limit the range of higher frequencies, particularly above 40 kHz or so. A 100 Hz sound can travel over a whole ocean basin with little absorption loss, while a 100 kHz sound would lose half its energy just traveling about 100 m.

Relating Acoustic Structure to Biological Function of Marine Mammal Calls

Understanding the physics of sound in the sea can help us understand why animals make the kinds of sound they do. For example, the calls of baleen whales are low-frequency because they are adapted for long-range propagation in the deep sea. Large baleen whales have evolved abilities to produce and to hear low-frequency calls well-suited for long-range communication. Blue whales and fin whales produce the lowest-frequency signals of all marine mammals, so low that humans can barely hear them. The long moans of blue whales, *Balaenoptera musculus*, have fundamental frequencies in the 14–

36 Hz band and they last several tens of seconds. The pulses of finback whales, *Balaenoptera physalus*, range roughly between 15 and 30 Hz and last on the order of 1 s. Particularly during the breeding season in mid-latitudes, finbacks produce series of pulses in a regularly repeating pattern in bouts that may last many days.

These loud low-frequency sounds appear to be specialized for long-range propagation in the sea. Absorption is negligible at the frequencies of these sounds. While acoustic models predicted that these sounds could be detected at ranges of hundreds of kilometers, it is only recently that this has been confirmed empirically. During the Cold War, the US Navy developed bottom-mounted hydrophones to locate ships and to track them. After the end of the Cold War, these sophisticated systems were made available to biologists, who have worked with Navy personnel to locate and track whales over long ranges, including one whale tracked for more than 1700 km over 43 days (**Figure 1**). These arrays have proven capable of detecting whales at ranges of hundreds to thousands of kilometers, as was predicted by the earlier acoustic models.

The physics of sound can also help explain why dolphins specialize in high-frequency sounds. Dolphins can detect distant objects acoustically by producing loud clicks and then listening for echoes. The clicks used by dolphins for echolocation have been well described. The echolocation clicks of bottlenose dolphins are very short (< 100 μs), with a rapid rise-time and a relatively broad bandwidth from several tens of kilohertz up to near 150 kHz (**Figure 2A**).

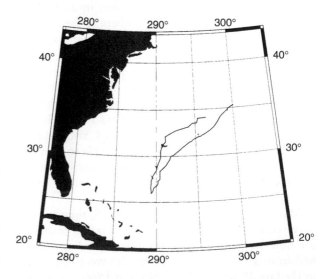

Figure 1 Track of a calling blue whale, *Balaenoptera musculus*, as it swam 1700 km over 43 days. The whale was tracked using the Integrated Underwater Sound Surveillance System (IUSS) of the US Navy. (From Figure 4.17 of Au *et al.* (2000).)

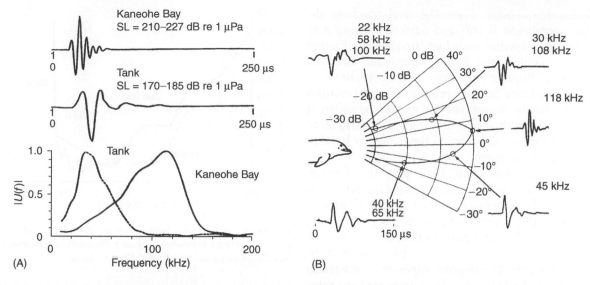

Figure 2 (A) Waveform and spectrum of echolocation clicks of bottlenose dolphins, *Tursiops truncatus*, in open ocean (Kaneohe Bay) and in a tank. The spectrum of the click from the tank (indicated with a dashed line) has a lower frequency peak at 40 kHz. (B) Beam pattern of *Tursiops* echolocation clicks. (μPa = micropascal, reference for sound pressure measurements. SL = source level.) ((A) from Figure 9.1, (B) from Figure 9.5 of Au *et al.* (2000).)

Captive dolphins in a reverberant pool make clicks that are less loud and lower in frequency than dolphins working on long-range echolocation in an open bay. The high-frequency components of these clicks are highly directional. If one moves 10 degrees off the axis of the beam, the click energy is halved and the click contains energy at lower frequencies (**Figure 2B**). The detection abilities of echolocating dolphins are truly remarkable. For example, trained bottlenose dolphins can detect a 2.54 cm solid steel sphere at 72 m, nearly a football field away.

The optimal frequency of a sound used for echolocation depends upon the size of the expected target. Absorption imposes a penalty for higher frequencies, but small targets can best be detected by short-wavelength, or high-frequency, signals. In the nineteenth century, Lord Rayleigh solved the frequency dependence of sound scattering from rigid spherical targets; this is called Rayleigh scattering. A spherical target of radius r reflects maximum energy when the wavelength of the sound impinging on it equals the circumference of the sphere, or when $\lambda = 2\pi r$. The echo strength drops off rapidly from signals with wavelength $\lambda > 2\pi r$. Since $\lambda = c/f$, one can equate the two λ terms to get $c/f = 2\pi r$. The relationship $f = c/2\pi r$ can be found by rearranging terms to calculate the optimal frequency for reflecting sound energy off a spherical target of radius r. Higher frequencies than this would still be effective sonar signals, but frequencies below f would show a strong decrease in effectiveness with decreasing frequency. A dolphin echolocating on rigid targets with a 'radius' of 0.5 cm should use a frequency $f \geq c/$

$2\pi r = 1500/(2\pi \times 0.005) \sim 50 \mathrm{kHz}$. This is within the frequency range of dolphin echolocation clicks, which include energy up to about 150 kHz. This upper frequency is appropriate for detecting spherical targets with radii as small as 1.5 mm. The hearing of dolphins is also most sensitive at frequencies of roughly 50–100 kHz. If dolphins have a need to echolocate on rigid targets with sizes in the 1 cm range, that helps explain why their echolocation system emphasizes these high frequencies.

Marine Mammal Hearing

In order to detect sound, animals require a receptor that can transduce the forces of particle motion or pressure changes into neural signals. Most mechanoreceptors in animals involve cells with hairlike cilia on their surfaces. As these cilia move, the electric potential between the inside and the outside of the receptor cells changes, and this potential difference modifies the rate of nerve impulses that signal other parts of the nervous system.

Terrestrial mammals evolved an ear that is divided into three sections: the outer, middle, and inner ear. The outer ear and middle ear function in terrestrial mammals to transduce airborne sound into vibrations of a fluid the inner ear of mammals which contains the cochlea, the organ in which sound energy is converted into neural signals. Sound enters the cochlea via the oval window and causes a membrane, called the basilar membrane, to vibrate. This membrane is mechanically tuned to vibrate at different frequencies. Near the oval window, the basilar membrane is stiff and narrow, causing it to vibrate when excited with high frequencies. Farther into the cochlea, the basilar membrane becomes wider and 'floppier', making it more sensitive to lower frequencies. Sensory cells at different positions along the basilar membrane are excited by different frequencies, and their rate of firing is proportional to the amount of sound energy in the frequency band to which they are sensitive.

Marine mammals share basic patterns of mammalian hearing but also have varying adaptations for listening under water as opposed to in air. All marine mammals other than sirenians, the sea otter, and cetaceans spend critical parts of their lives on land or ice and some phocid seals communicate both in air and under water. The relative importance of hearing in air and under water has been compared for three pinniped species whose hearing has been tested in both environments. The California sea lion (*Zalophus californianus*) is adapted to hear best in air; the harbor seal (*Phoca vitulina*) can hear equally well in air and under water; and the northern elephant seal (*Mirounga angustirostris*) has an auditory system adapted for under water sensitivity at the expense of aerial hearing.

The eardrum and middle ear in terrestrial mammals functions to efficiently transmit airborne sound to the inner ear where the sound is detected in a fluid. No such matching is required for an animal living in the water, and cetaceans, which are adapted exclusively for listening under water, do not have an air-filled external ear canal. The problem for cetaceans is isolating the ears acoustically, and the inner ear is surrounded by an extremely dense bone that is isolated from the skull. High-frequency sound is thought to enter the dolphin head through a thin section of bone in the lower jaw and is conducted to the inner ear via fatty tissue that acts as a waveguide.

Hearing abilities have been tested for those species of marine mammals that can be held in captivity. **Figure 3** shows audiograms from a dolphin, porpoise, and several pinnipeds. As discussed above, dolphins have hearing specialized to hear very high frequencies up to ten times the upper limit of human hearing. Seals have less acute hearing than do dolphins and they are less able to hear the highest frequencies. The frequency range of hearing has never been tested in baleen whales. Hearing is usually tested by training an animal, and baleen whales are so big that only a few have been kept for short periods in captivity. However, both their low-frequency vocalizations and the frequency tuning of their cochlea suggest they are specialized for low-frequency hearing.

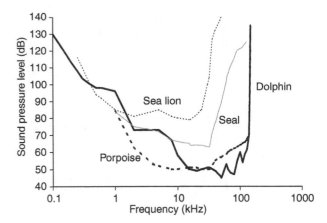

Figure 3 Audiograms from a variety of marine mammals: dolphin *Tursiops truncatus*; porpoise *Phocoena phocoena*; sea lion *Zalophus californianus*; seal *Phoca vitulina*.

Mammalian hearing is designed to analyze the frequency content of sound. Among mammals, dolphins have extraordinarily good abilities of discriminating different frequencies. They can detect a change of as little as 0.2% in frequency, which is close to the resolution of human hearing.

Vocalizations of Marine Mammals

When terrestrial carnivores and ungulates invaded the sea, they encountered new constraints and opportunities for sensing signals. The sirenians, cetaceans, phocid seals, and the walrus (*Odobenus rosmarus*) evolved specializations for using sound to communicate under water and to explore the marine environment; other taxa, including the otariid pinnipeds, sea otter (*Enhydra lutra*) and polar bear (*Ursus maritimus*), vocalize mainly in air. As with hearing, cetaceans show the most elaborate and extreme specializations for acoustic communication under water.

The best-known acoustic displays of marine mammals are the reproductive advertisement displays called songs. The songs of humpback whales are the best known advertisement display in the cetaceans, but bowhead whales also sing. Male seals of some species repeat acoustically complex songs during the breeding season. Songs are particularly common among seals that inhabit polar waters and that haul out on ice. The songs of bearded seals, *Erignathus barbatus*, are produced by sexually mature adult males and are heard most frequently during the peak of the breeding season. Male walruses, *Odobenus rosmarus*, also produce complex visual and acoustic displays near herds of females during their breeding season. They use their lips to whistle, and also produce loud sounds of breathing that are audible in air when they surface during these displays. When they dive, displaying males produce a series of pulses under water followed by bell-like sounds. Antarctic Weddell seal males repeat under water trills (rapid alternations of notes) during the breeding season.

Marine mammals also produce a broad variety of displays, including threat displays and recognition displays used for individual or group recognition.

Mechanisms of Sound Production

Most terrestrial mammals produce vocal sounds by vibrating vocal cords in the larynx. It is thought that the polar bear and most pinnipeds make sounds using similar mechanisms. Some adaptations for diving may affect vocalization mechanisms in pinnipeds. Pinnipeds have a more flexible trachea than do terrestrial mammals, so that air inside can compress during a dive, and they have a wider trachea to allow higher rates of air flow. Most pinnipeds can vocalize under water without emitting bubbles; some species have sacs attached to the trachea or upper respiratory sac, but the role of these in vocalization has not been determined. Walruses have

many ways of producing sounds. They produce gonglike impulse sounds using specialized pharyngeal sacs, and can even use their lips to whistle in air.

Odontocetes have well developed vocal folds in the larynx, but most biologists argue that odontocetes produce sounds as air flows past the nasal plugs or phonic lips in the upper nasal passages (**Figure 4A**). Mechanisms for sound production must also match the acoustic impedance to the medium of air or sea water, and they may function to direct some sounds in a beam. The beam pattern of dolphin clicks (shown in **Figure 2B**) stems from a complex interaction of reflection from the skull and air sacs, coupled with refraction in soft tissues (**Figure 4A**).

There is a more detailed model of sound production for sperm whales (*Physeter macrocephalus*) than for other cetacean species. Sperm whales have a large organ called the spermaceti organ, which lies dorsal and anterior to the skull (**Figure 4B**). Below the spermaceti organ is the 'junk', which is composed of a series of fatty structures separated by dense connective tissue. The primary vocalizations of sperm whales are distinctive clicks comprising a burst of pulses with equally spaced interpulse intervals (IPIs). Bioacousticians suggest that these regular IPIs may result from reverberation within the

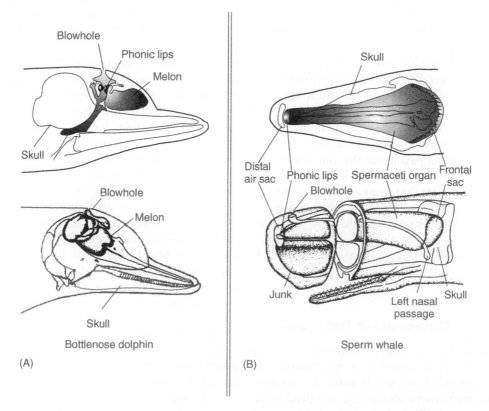

(A)

(B)

Figure 4 Functional anatomy of sound production in two odontocete cetaceans: (A) bottlenose dolphin *Tursiops truncatus*; (B) sperm whale *Physeter macrocephalus*. (Adapted from Figures 1.4 and 3.1 of Au *et al.* (2000).)

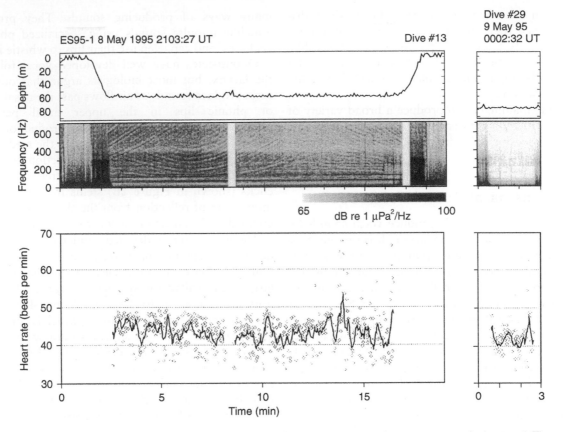

Figure 5 Data on dive profile, acoustic record, and acoustically determined heart rate from a tag on an elephant seal. The acoustic record shows a vessel passing. The closest point of approach occurs at the minimum frequency of the 'U' shaped pattern in the spectrogram at about 6 min. The heart rate differs little from the surrounding times or from a quiet period in a later dive from the same seal. (μPa = micropascal, reference for sound pressure measurements.) (Adapted from Figure 8 of Burgess *et al.* (1998).)

spermaceti organ. The frontal sac at the posterior end of the spermaceti organ has been suggested as a potential reflector of sound and the distal sac as a partial reflector of sound at the anterior end (**Figure 4B**). The source of the sound energy in the click is thought to come from a strong valve (phonic lips) in the right nasal passage at the anterior end of the spermaceti organ (**Figure 4B**). This sound production model suggests that some of the energy from the first pulse within the click is transmitted directly into the water. The remaining pulses are hypothesized to occur as some of the sound energy passes through the anterior reflector into the ocean at each reflection there.

Methods for Bioacoustic Research

It has been difficult to integrate visual observation of social behavior with patterns of vocalization in submerged mammals because it is difficult to identify which animal within an interacting group produces a sound under water. Biologists studying terrestrial animals take it for granted that they can identify

which animal is vocalizing by using their own ears to locate the source of a sound and then looking for movements associated with sound production. Humans cannot locate sounds under water in the same way that they locate airborne sounds. Furthermore, marine mammals seldom produce visible motions coordinated with sound production under water. It is even more difficult to attempt behavioral observations on marine mammals during a dive when they are out of sight. The need for some technique to track behavior during a dive and to identify which cetacean produces which sound during normal social interaction has been discussed for over three decades. Two different approaches have emerged: (1) passive acoustic location of sound sources using an array of hydrophones; (2) recording information about behavior and sound production by attaching a tag onto the animal.

Acoustic location of vocalizing animals is a useful method for identifying which animal is producing a sound. It involves no manipulation of the animals, merely placement of hydrophones near them. In some applications, animals may vocalize frequently enough and be sufficiently separated that source location data may suffice to indicate which animal produces a sound. Tracks of continuously vocalizing finback and blue whales have been made using bottom-mounted hydrophones. **Figure 1** shows a 1700 km track of a blue whale that was tracked in the early 1990s by US Navy personnel using arrays of hydrophones initially developed to track submarines. Bottom-mounted recording devices are proving cheaper alternatives for biologists today.

Bioacousticians have also developed smaller, portable hydrophone arrays that can be deployed rapidly from a ship or from shore. These arrays have been used to locate vocalizing finback whales, right whales (*Eubalaena glacialis*), sperm whales, and several species of dolphins. Vertical hydrophone arrays can in some settings be used to calculate the range and depth of vocalizing whales. One classic configuration involves a linear horizontal array of hydrophones that is towed behind a ship. Signal processing techniques allow one to determine what bearing a sound is coming from, and to reconstruct the signal from that bearing. Bioacousticians are only just beginning to explore how to use these techniques in behavioral studies of whales.

The second technique does not require locating each animal within a group. If an animal carries a telemetry device that transmits acoustic data recorded at the animal, then the device can record all vocalizations of the animals along with most everything else it hears. This kind of tag can also record depth of dive, movement and orientation of the tagged animal. However, it is difficult to telemeter information through sea water, and marine mammals might sense many of the signals one might want to use for telemetry. These problems with telemetry have led biologists to develop recoverable tags that record data while on an animal, but that need to be recovered from the animal in order for the data to be downloaded. Recently, biologists have had successful programs recovering such tags from many different kinds of marine mammal. Recoverable acoustic tags may have scientific uses well beyond identifying vocalizations. **Figure 5** shows acoustic and dive data sampled from an elephant seal. The tag was able to monitor both the acoustic stimuli heard by the whale, and orientation sensors monitored not just the depth of the dive but also movement patterns such as the fluke beat and physiological parameters such as heart rate. This information is useful to determine reactions of marine mammals to man-made noise, an issue of growing concern.

See also

Acoustic Scattering by Marine Organisms. Sonar Systems.

Further Reading

Au WWL (1993) *The Sonar of Dolphins.* New York: Springer Verlag.

Au WWL, Popper AS and Fay R (eds.) (2000) *Hearing by Whales and Dolphins.* Springer Handbook of Auditory Research Series. New York: Springer Verlag.

Burgess WC, Tyack PL, LeBoeuf BJ, and Costa DP (1998) A programmable acoustic recording tag and first results from free-ranging northern elephant seals. *Deep-Sea Research* 45: 1327–1351.

Kastak D and Schusterman RJ (1998) Low-frequency amphibious hearing in pinnipeds: Methods, measurements, noise, and ecology. *Journal of the Acoustical Society of America* 103: 2216–2228.

Medwin H and Clay CS (1998) *Fundamentals of Acoustical Oceanography.* New York: Academic Press.

Miller P and Tyack PL (1998) A small towed beamforming array to identify vocalizing resident killer whales (*Orcinus orca*) concurrent with focal behavioral observations. *Deep-Sea Research* 45: 1389–1405.

Rayleigh, Lord (1945) *The Theory of Sound.* New York: Dover.

Tyack P (1998) Acoustic communication under the sea. In: Hopp SL, Owren MJ, and Evans CS (eds.) *Animal Acoustic Communication: Recent Technical Advances,* pp. 163–220. Heidelberg: Springer Verlag.

Tyack PL (2000) Functional aspects of cetacean communication. In: Mann J, Connor R, Tyack PL, and Whitehead H (eds.) *Cetacean Societies: Field Studies of Dolphins and Whales,* pp. 70–307. Chicago: University of Chicago Press.

Wartzok D and Ketten DR (1999) Marine mammal sensory systems. In: Reynolds JE III and Rommel SA (eds.) *Biology of Marine Mammals,* vol. 1, pp. 117–175. Washington, DC: Smithsonian Press.

SEISMIC REFLECTION METHODS FOR STUDY OF THE WATER COLUMN

I. Fer, University of Bergen, Bergen, Norway
W. S. Holbrook, University of Wyoming, Laramie, WY, USA

Introduction

Recent work has shown that marine seismic reflection profiling, a technique commonly used by geophysicists and geologists to image the Earth's crust beneath the seafloor, can produce surprisingly detailed images of thermohaline fine structure in the ocean. Seismic reflection profiling produces images by mapping the locations of 'bounce points' where acoustic waves expanding out from the shot location have reflected from subsurface interfaces where changes in material density and/or sound speed occur. Similar to sonar, the arrival time of a reflection at a receiver, together with a measurement or estimate of the sound-speed profile, is used to estimate the depth of the feature that generated the reflection.

Within the ocean, density-compensating, fine-scale (1–10-m thick) temperature-salinity contrasts (e.g., temperature and salinity having opposing contributions to the density to keep the density constant) result in small changes in sound speed that produce weak, but distinct, reflections. Recent studies image returns that are specular reflections from laterally continuous steps or sheets in the ocean's temperature-depth structure. The reflectors imaged in the ocean are 100–1000 times weaker than those from the solid Earth below. However, by increasing the gain of the processing system, oceanic fine structure can be made visible. Until recently, seismologists have been largely unaware of these weak reflections in their data and have not routinely processed returns from the water column, since their focus is on the structure of the Earth. The discovery of seismic reflections from the water column and the ability to image large volumes of the ocean at full depth and at high lateral resolution opens up new possibilities for probing the structure of the ocean with 'seismic oceanography'. Spectacular images of thermohaline fine structure in the ocean have been produced from features such as intrusions, internal waves, and mesoscale eddies.

Primer on the Method

Marine seismic reflection data are acquired by firing, at constant distance intervals, a sound source of relatively low-frequency (10–150 Hz) energy produced by an array of air guns towed several meters beneath the sea surface. The data are recorded on hydrophone cables, or streamers, which are typically several kilometers long, contain 40–100 hydrophone channels per kilometer, and are towed 3–8 m below the sea surface. The pressure variations of the reflected sound signals are recorded by hydrophones installed at regular intervals (usually 12.5 m) along the streamer. For a flat reflector, the point in the ocean's interior at which a reflecting ray reaches its maximum depth before reflecting (the 'bottoming' or 'reflecting' point) occurs halfway between the source and the receiver. Therefore, the subsurface sampling interval is half the hydrophone spacing (i.e., usually 6.25 m).

Reflections appear on the resulting records as curved arrivals whose travel time increases hyperbolically with distance from the source point. With the exception of the direct arrival, refractions are not returned to the streamer from within the water column – only reflections. Shots are fired at intervals of 50–150 m, creating a large volume of overlapping reflections on successive shot records. The vessel maintains a constant speed along a straight track allowing for redundant sampling of a fixed point. Traces on the shot records are re-sorted into common midpoint (CMP) records, which gather together all traces that have a common source–receiver midpoint. The hyperbolic curvature of the reflections in each CMP ensemble is then removed, and the resulting flattened reflections are summed (or stacked) together to create a single trace at each CMP. Summed traces for each CMP are then joined into a stacked section. Each trace in the stack can be thought of as the response that would be generated by a simple, single-channel geometry, that is, the wave field from a single shot that travels downward, reflects off a boundary at normal incidence, and is returned and recorded on a single hydrophone located at the source point. The final processing step is migration, which accounts for any distortions in the stacked section caused by the presence of dipping (nonhorizontal) reflectors. The resulting migrated image resembles a snapshot cross section of the acoustic impedance of the ocean and underlying Earth at comparable horizontal and vertical sampling of about 5–10 m.

Comparison with Other Marine Acoustic Technology

Seismic reflection profiling differs substantially from, and thus complements, the high-frequency acoustic imaging commonly used in oceanography. High-frequency (typically above 10 kHz) ocean acoustics image scatterers in the water column – that is, diffracting bodies with spatial dimensions on the order of the dominant wavelength of the sound source, such as zooplankton, bubble clouds or microstructure of temperature, salinity, or velocity. Scattering theories have been developed to quantify relationship between microstructure intensity or zooplankton concentration to the backscatter strength. Many observational programs have imaged internal waves in shallow water, often two-layer flows in straits that show dramatic patterns of wave growth and breaking;

however, high-frequency techniques are limited to ranges of a few hundred meters. For the low-frequency sound (10–150 Hz) of the seismic profiles, scattering theory is inappropriate, because the wavelength of the sound is much larger than the length scale of common scatterers in the ocean.

Unlike the long-range refracted sound typically used in acoustic tomography in the ocean, the reflections imaged with seismic reflection methods are sensitive to the vertical derivative of the sound speed, rather than the value of the sound speed itself. The travel times of the reflections are sensitive to the average sound speed, but the reflection amplitudes are sensitive to the vertical gradients. The frequencies generated by air gun sources are typically in the 10–150-Hz range, corresponding to vertical wavelengths of 10–150 m. The marine seismic reflection method can resolve layers with thickness of about

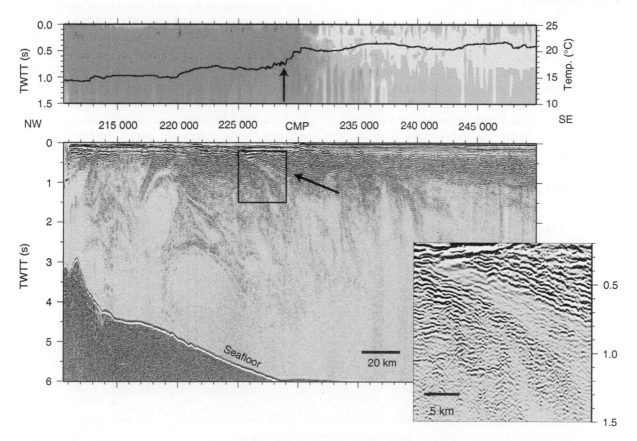

Figure 1 (Top) Stacking sound speed, approximately equal to root-mean-square sound speed, in the ocean (color) and sea surface temperature (SST, black) measured during a seismic survey in the Newfoundland Basin. Cold colors correspond to low sound speed (minimum of $\sim$1440 m s^{-1}); warm colors reflect higher sound speed (maximum of $\sim$1530 m s^{-1}). Horizontal axis is the common-mid-point (CMP) with 6.25-m spacing. The arrow marks the front between the Labrador current (NW of the front) and the North Atlantic current (SE of the front), visible as an abrupt $\sim$5 °C increase in temperature at CMP 229 000. (Bottom) Stacked seismic section of water column with a vertical exaggeration of 27. Vertical axis is two-way-travel time (TWTT) in seconds; the base of the section at 6 s corresponds to a depth of $\sim$4500 m in the ocean. An intrusive feature is marked by an arrow. Box denotes the portion of the profile depicted in the inset which shows slabs losing coherency at depths of $\sim$1000 m. Reproduced from Holbrook WS, Páramo P, Pearse S, and Schmitt RW (2003) Thermohaline fine structure in an oceanographic front from seismic reflection profiling. *Science* 301: 821–824, with permission from AAAS.

one-quarter wavelength, that is, from several meters to several tens of meter thick, provided that the layers are separated by relatively sharp vertical gradients in sound speed as a result of predominant temperature contrasts. For example, at surface pressure, the sound speed increases by $4 \, \mathrm{m \, s^{-1}}$ by increasing the temperature of seawater of 35 psu salinity from 5 to 6 °C. The oceanic fine structure is characterized by vertical scales of meters to tens of meters which are not directly affected by molecular diffusion. These scales are comparable to the dominant wavelength of the sound source. Hence, marine seismic reflection data is ideally tuned to detect oceanic fine structure and seismic theory developed for reflections from a laterally coherent layered medium can be confidently applied.

Observations

To date, published observations using seismic oceanography come from three areas: the North Atlantic Front, the Kuroshio Front, and the Norwegian Sea.

Seismic reflection data acquired in August 2000 across the oceanographic front between the Labrador current (LC) and the North Atlantic current (NAC) in Newfoundland basin yielded images showing the two-dimensional thermohaline structure of the ocean in unprecedented detail (**Figure 1**). LC and NAC are characterized by cold-fresh and warm-salty water masses, respectively, separated by a front (marked by an arrow in the upper panel of **Figure 1**, LC is on the NW side of the front) where strong thermohaline intrusions with 20- to 30-m-thick fine structure were previously reported. Wedges of inclined reflections in the vicinity of the front are representative of this phenomenon. A plausible mechanism that creates intrusive layers is double diffusion driven solely by the factor 100 difference between the molecular diffusivity of heat and salt. Temperature-salinity structure in the intrusive layers is modified by the fluxes induced by double diffusion when both salinity and temperature increase or decrease with depth. The strong tilt of the reflections matches the expected inclination of density surfaces affected by double-diffusive processes. Nearly horizontal, 0.5- to 5-km-long reflections in the warm waters of the NAC are consistent with the weak tilt of lateral intrusions as a result of double-diffusive mixing observed previously in the NAC, whereas the isopycnal surfaces tilt more strongly in the vicinity of the front as a result of the geostrophic dynamical balance. Coherent slabs of 100- to 300-m-thick reflections inclined at 1–4° from the horizontal become progressively weaker and lose coherence with increasing depth, reaching down to

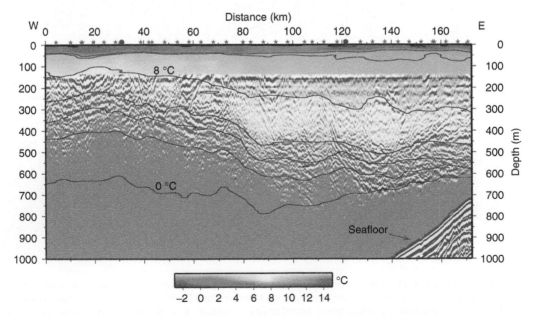

Figure 2 Ocean temperature (color) overlain by seismic reflection data (black-and-white image). Thin solid lines are selected isotherms contoured every 2 °C, derived from XBT and XCTD casts. Red stars and blue circles at top show locations of XBTs and XCTDs, respectively. The warm (7–14 °C) AW is separated from the cold (− 0.5–2 °C) NSDW by a boundary layer delineated by rapidly changing temperatures and strong seismic reflections. The top 140 m of the seismic profile, containing interference from the direct arrival, has been muted. Reproduced from Nandi P, Holbrook WS, Pearse S, Páramo P, and Schmitt RW (2004) Seismic reflection imaging of water mass boundaries in the Norwegian Sea. *Geophysical Research Letters* 31: L23311 (doi:10.1029/2004GL021325). Copyright (2004) American Geophysical Union. Reproduced by permission of American Geophysical Union.

about 1.5 s two-way-travel time (TWTT, approximately 1100-m depth).

Soon after the discovery of the ability to seismically image oceanic thermohaline fine-structure in such detail, the first joint seismic reflection/physical oceanography study was conducted in the Norwegian Sea: a dense array of expendable bathythermographs (XBTs) and expendable conductivity-temperature depth (XCTD) probes were deployed on several lines during the acquisition of seismic reflection data. The resulting data showed conclusively that the marine seismic reflection method can map distinct water masses. At the survey site, the major water masses are the warm and saline Atlantic Water (AW) carried by the Norwegian Atlantic current and underlying cold and less-saline Norwegian Sea Deep Water (NSDW). Relatively nonreflective zones in the upper ~400 m and below the 0 °C isotherm in **Figure 2** correspond to AW and NSDW, respectively. Fine-scale temperature variability measured by the XBT/XCTDs in the boundary between the two water masses affects the sound speed and results in acoustic impedance contrasts manifested by the strong reflective zone in **Figure 2**.

Reflections generally follow isotherms derived from the XBT/XCTD survey (**Figure 2**). A comparison of a temperature profile measured by an XCTD to the coincident seismic reflection profile reveals the sensitivity of the reflections to fine-scale variability in temperature. Temperature fluctuations derived by high-pass filtering the temperature profile for lengths smaller than 35 m, the dominant wavelength of the range of the sound source used in this study, match the seismic reflection signal (**Figure 3**). Reflector amplitudes are enhanced where temperature anomalies are large. Even temperature fluctuations as small as 0.04 °C, comparable to the measurement accuracy of ± 0.02 °C, can be imaged with seismic reflection techniques.

Seismic profiles across the Kuroshio current, off the Muroto peninsula southwest of Japan, revealed fine structure continuous at a horizontal scale over 40 km between 300- and 1000-m depth (**Figure 4**). Successive seismic profiles at 2–3-day intervals, corresponding to about cross section snapshots every 180–450 km of the Kuroshio current moving with 1 m s^{-1}, suggest that the lateral continuity of the fine structure at the Kuroshio current is not a transient feature and can persist at least for 20 days.

The observations across the Kuroshio current motivated a joint seismic reflection and physical oceanographic survey in summer 2005 within the Kuroshio extension front east of Japan, where warm Kuroshio current meets cold Oyashio water. The hydrography and ocean currents were sampled using XBT, XCTD, expendable current profiler (XCP), and

vessel-mounted acoustic Doppler current profiler (ADCP). Isotherms derived from XCTD measurements across the Kuroshio current show the temperature characteristics of the frontal system with intrusive features (**Figure 5(a)**). Two lines were acquired, one to the north (Line 1) and another to the south (Line 2) of the Kuroshio current core. The first-order difference in the seismic images is that the reflectors become nearly horizontal in Line 2 with increasing distance from the front, whereas more fine structure and steeply inclined reflectors are common in Line 1, consistent with the temperature section (**Figure 5(a)**). With the purpose of determining the best seismic source configuration to image the oceanic fine structure, the seismic lines of this survey were repeated with different air gun chamber volumes (20 l, 9 l, and a 3.4 l air gun of generator–injector, GI, type). Upper portions of panels (b) to (c) of **Figure 5** show Line 1 sampled with decreasing air gun source strength over a duration of 54 h. The

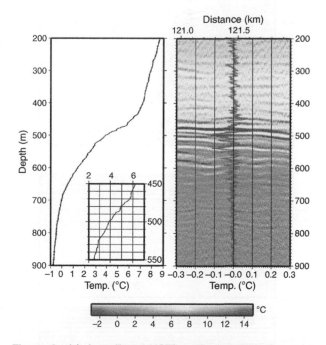

Figure 3 (a) An unfiltered XCTD profile located at km 121.5 on the seismic profile (**Figure 2**) showing temperature from 200–900-m depth. (b) Short-wavelength temperature variations (red), produced by removing wavelengths greater than 35 m from the XCTD temperature profile, plotted with a 5-km-wide section of the reflection image surrounding the XCTD location (black and white image). Background color scheme is ocean temperature, plotted as in **Figure 2**. The seismic image has been shifted upward by 14 m to reflect the lag between the onset of energy and peak amplitude in the seismic wavelet. Reproduced from Nandi P, Holbrook WS, Pearse S, Páramo P, and Schmitt RW (2004) Seismic reflection imaging of water mass boundaries in the Norwegian Sea. *Geophysical Research Letters* 31: L23311 (doi:10.1029/2004GL021325). Copyright (2004) American Geophysical Union. Reproduced by permission of American Geophysical Union.

reflection patterns are consistent between consequent transects and correlate with hydrographic property changes in the water column. Depth-distance maps of the ADCP backscatter intensity collected in concert with the seismic reflection data show some similarities to the seismic images regarding the inclined reflection patterns and their locations. Similar to the observations in the Norwegian Sea (**Figure 2**),

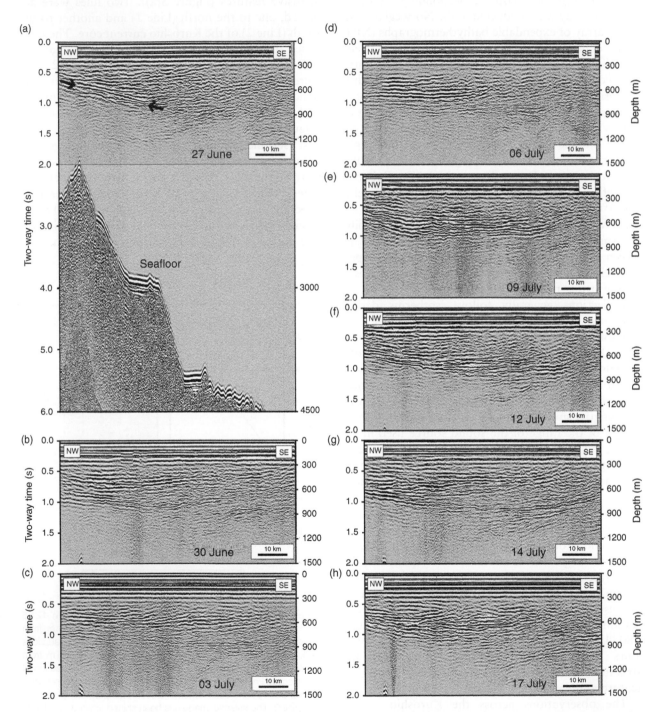

Figure 4 Selected seismic profiles acquired across the Kuroshio current at 2–3-day intervals with date indicated on each panel. The orientation of the line is northwest (left) to southeast (right). The vertical axis to the left is the TWTT and that on the right is the depth calculated assuming an acoustic velocity of 1500 m s^{-1}. The image for (a) 27 June is shown for TWTT 0–6 s, whereas other panels are shown for 0–2 s. The reflection marked in (a) with arrows has a slope of $\sim$1/120. Reproduced from Tsuji T, Noguchi T, Niino H, et al. (2005) Two-dimensional mapping of fine structures in the Kuroshio Current using seismic reflection data. *Geophysical Research Letters* 32: L14609 (doi:10.1029/ 2005GL023095). Copyright (2005) American Geophysical Union. Reproduced by permission of American Geophysical Union.

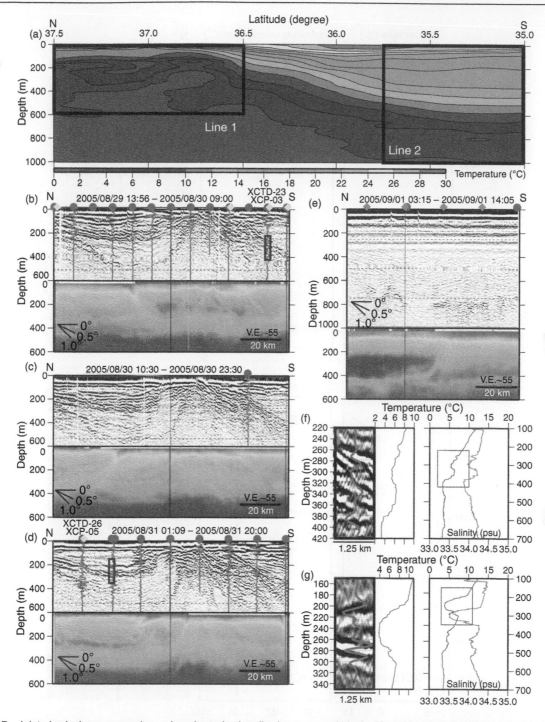

Figure 5 Joint physical oceanography and marine seismic reflection survey within the Kuroshio extension front east of Japan. Two seismic lines, indicated by boxes in (a) were sampled: Line 1 to the north and Line 2 to the south of the core of the Kuroshio current. (a) Temperature section derived from XCTD measurements. Panels (b) to (e) show seismic reflection profile (gray scale, top) and ADCP intensity (color, bottom) for (b–d) Line 1 and (e) Line 2. The air guns used are (b) 20 l, (c) 9 l, (d) 3.4 l, and (e) 20 l. Time period of observation is indicated above each seismic profile. Red circles, green diamonds, and yellow diamonds are XCTD, XBT, and XCP locations, respectively. Red waveforms on the seismic reflection profiles are synthetic seismograms calculated from XCTD profiles. Expanded data from two blue rectangles marked in (b) and (d) are shown in (f) and (g), respectively. (f) Comparison of Line 1 seismic profile (20 l air gun) and XCTD data marked with blue rectangle in (b). (left) Seismic profile overlain by a synthetic seismogram (red), (right) temperature (blue) and salinity (red) profiles derived from XCTD measurements. (middle) Blow-up of temperature from the rectangle to the right, which corresponds to the depth range of seismic profile to the left. (g) Same as (f) except using 3.4 l air gun at location shown by blue rectangle in (d). Reproduced from Nakamura Y, Noguchi T, Tsuji T, Itoh S, Niino H, and Matsuoka T (2006) Simultaneous seismic reflection and physical oceanographic observations of oceanic fine structure in the Kuroshio extension front. *Geophysical Research Letters* 33: L23605 (doi:10.1029/ 2006GL027437). Copyright (2006) American Geophysical Union. Reproduced by permission of American Geophysical Union.

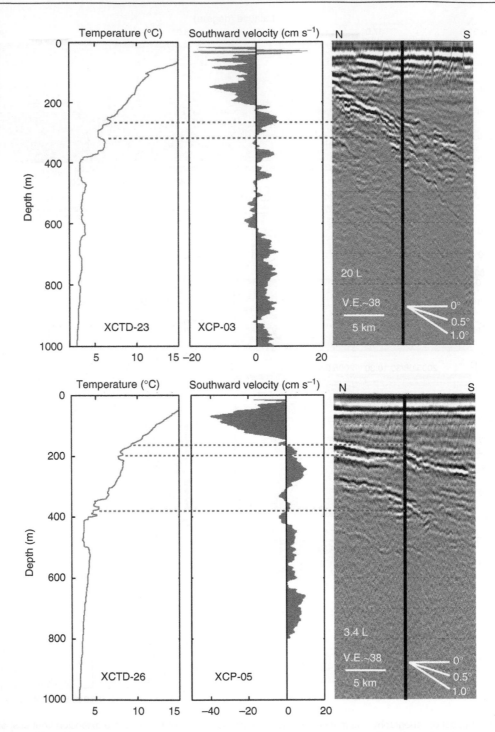

Figure 6 Comparison of (left) XCTD temperature, (middle) southward velocity recorded by XCP and (right) seismic reflection data. XCTD and XCP station numbers indicated on panels are marked in **Figures 5(b)** and **5(d)**. The dashed blue horizontal lines indicate abrupt temperature and velocity changes correlated with seismic reflections. Location of the XCP and XCTD measurements relative to the seismic profile is shown by vertical black lines. All data are collected at Line 1 (see **Figure 5(a)**) with top panels using a 20 l air gun and bottom panels using a 3.4 l air gun. These two measurements were separated in time by ~19 h. Reproduced from Nakamura Y, Noguchi T, Tsuji T, Itoh S, Niino H, and Matsuoka T (2006) Simultaneous seismic reflection and physical oceanographic observations of oceanic fine structure in the Kuroshio extension front. *Geophysical Research Letters* 33: L23605 (doi:10.1029/ 2006GL027437). Copyright (2006) American Geophysical Union. Reproduced by permission of American Geophysical Union.

reflections are enhanced at the base of the thermocline (temperature range 4–8 °C).

Given an observed CTD profile a synthetic reflection pattern can be constructed to check whether the seismic interpretation of thermohaline fine structure is credible or not. The sound-speed profile in seawater (calculated as a function of temperature, salinity, and depth) is used to calculate acoustic impedance contrasts. Acoustic impedance is relatively more sensitive to changes in sound speed than changes in density. Reflection coefficients derived from acoustic impedance are used to generate synthetic seismograms. Red waveforms on the seismic reflection profiles of **Figure 5** are such synthetic

seismograms constructed from the XCTD profiles. The synthetic profiles match the seismic observations well, lending confidence to the physical interpretation of the reflectors (**Figure 6**).

Common in all seismic reflection images are undulations of reflectors which cannot be attributed to processing artifacts. It is well known that the velocity shear of the internal wave field of the stratified ocean displaces the isopycnals, isotherms, and irreversible fine structure caused by mixing events. The undulating reflectors can be interpreted as the acoustic snapshots of the fine-structure displacements due to the internal waves. This assertion was recently tested by digitizing selected reflectors where reflectors

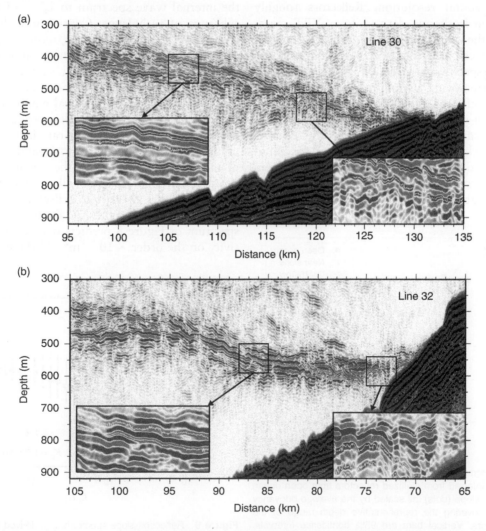

Figure 7 Seismic sections for Lines (a) 30 and (b) 32 acquired across the isobaths in the Norwegian Sea. Data are displayed so that peaks and troughs of reflections are red and blue, respectively, in the water column, and gray and black in the solid earth. Yellow lines show reflectors picked and digitized by auto-tracking for spectral analysis shown in **Figure 8**. (a) Stacked seismic section of line 30 shows a clear progression toward the slope from smooth, continuous fine structure (inset left) to highly disrupted fine structure (inset right). (b) Stacked seismic section of Line 32, similar to Line 30, shows fine structure changing from smoothly undulating to more discontinuous pattern as the continental slope is approached. Reproduced from Holbrook WS and Fer I (2005) Ocean internal wave spectra inferred from seismic reflection transects. *Geophysical Research Letters* 32: L15604 (doi:10.1029/ 2005GL023733). Copyright (2005) American Geophysical Union. Reproduced by permission of American Geophysical Union.

roughly parallel isotherms (hence representative of isotherm displacements) and comparing the wavenumber spectrum to the isopycnal displacement expected from oceanic internal wave field. In the open ocean, frequency and wavenumber domain representations of internal wave energy show a remarkably uniform energy spectrum that can be modeled by the so-called Garrett–Munk spectrum, which has a frequency, ω, and wavenumber, k, decay of energy with a power law near ω^{-2} and k^{-2}. This spectrum is thought to be maintained by a cascade of energy from large scales and low frequencies to small scales and high frequencies due to interaction of different wave packets. Two seismic lines presented in **Figure 7** were acquired in the Norwegian Sea at 6.25-m horizontal resolution. Reflectors roughly following the isotherms derived from XBT measurements in the vicinity are digitized (yellow traces in **Figure 7**) to calculate the horizontal wavenumber (k_x) power spectra of reflection displacements, ζ, obtained by removing a straight line fit over the extent of each reflector (typically 5–30 km).

It is known that the internal wave characteristics differ significantly in proximity to the sloping boundaries where internal wave–boundary interactions lead to a variety of processes that can enhance the internal wave energy, shear, and mixing. Consistently, the seismic reflection images show a marked change from smooth, continuous patterns to more choppy reflections with high vertical displacements as the continental slope is approached. The spectra calculated away from the seafloor and horizontally within 10 km of the continental slope are compared to the Garrett–Munk spectrum in **Figure 8**. For the open ocean reflectors, the agreement with the oceanic internal wave horizontal wave number spectrum is within a factor of 2 down to wavelengths of about 30 m. Near the continental slope reflector displacement spectrum suggests enhanced internal wave energy which could be caused by wave reflection and generation processes at the sloping boundary. At a length scale of about 300 m, there is an apparent change of slope from k_x^{-2} of the internal wave spectrum to $k_x^{-5/3}$ of the inertial subrange of turbulence. Consistent with this inference from the seismic profiles, recent measurements of joint isopycnal displacement and turbulence-dissipation rates indicate that turbulence subrange extends to surprisingly large horizontal wavelengths (>100 m). A fit of the spectral amplitude to a turbulence-model spectrum yields reasonable estimates of the turbulence-dissipation rate in the water column, even when applied to horizontal scales of tens of meters, which could not be possible in vertical profiles. The method is equally applicable to horizontal transects of seismic reflectors which follow isopycnal displacements. A fit to the near-slope spectrum of **Figure 8** predicts diapycnal eddy diffusivity on the order $\sim 10^{-3}\,\text{m}^2\,\text{s}^{-1}$ (**Figure 9**).

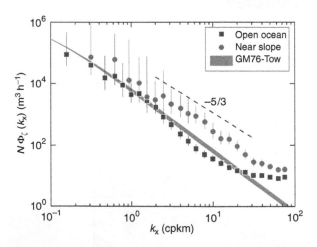

Figure 8 Horizontal wavenumber spectra of vertical displacements inferred from digitized reflectors from open ocean (squares), near slope (dots), all scaled by the average buoyancy frequency, N, covering the representative depth range of the chosen reflectors. Vertical bars are 95% confidence intervals. Garrett–Munk tow spectrum is shown as a band for the observed range of $N = (1.7 - 4.4) \times 10^{-3}\,\text{s}^{-1}$. The dashed line shows the $-5/3$ slope of the inertial subrange of turbulence, for reference. Reproduced from Holbrook WS and Fer I (2005) Ocean internal wave spectra inferred from seismic reflection transects. *Geophysical Research Letters* 32: L15604 (doi:10.1029/2005GL023733). Copyright (2005) American Geophysical Union. Reproduced by permission of American Geophysical Union.

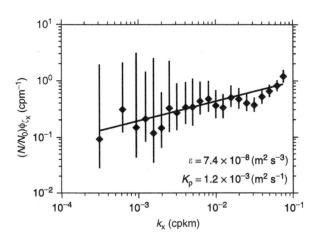

Figure 9 Reflection-slope spectrum, $\phi_{\zeta x}$, derived from the near-slope spectrum shown in **Figure 8**. Spectrum is normalized by the buoyancy frequency, $N = 3.5 \times 10^{-3}\,\text{s}^{-1}$, relative to $N_0 = 5.3 \times 10^{-3}\,\text{s}^{-1}$. Dissipation rate of turbulent kinetic energy, ε, is estimated by a fit of the turbulent spectrum in the $4 \times 10^{-3} - 2 \times 10^{-2}$ cpm wavenumber range. Eddy diffusivity is estimated using $K_\rho = 0.2\varepsilon/N^2$. Reproduced with permission from Klymak JM and Moum JN (2007) Oceanic isopycnal slope spectra: Part II: Turbulence. *Journal of Physical Oceanography* 37: 1232–1245.

Conclusions

Standard multichannel marine seismic reflection profiling has the capability to image order 10-m vertical scale sound speed and density fine structure in the ocean, if processing gains are sufficiently high. Seismic reflection images visualize ocean fronts, water mass boundaries, eddies, intrathermocline lenses, and thermohaline intrusions. With the very dense lateral sampling of reflection profiling (order 10 m), exceptional detail can be seen in the structure of oceanic reflectors throughout portions of the water column where fine structure is present. Interpreting marine seismic reflection data in terms of ocean fine structure and internal motions is presently being investigated to develop this technique into a tool that can produce useful and trusted information on properties of dynamical interest to the physical oceanographers. A basic physical understanding is achieved on the origin of low-frequency reflections in the water column. Carefully analyzed horizontal transects of reflectors suggest that undulating patterns common in most seismic transects are a proxy to the isopycnal displacements and can be used to remotely estimate the internal wave energy as well as turbulent kinetic energy dissipation rate and mixing where ocean fine structure allows. In the near future, joint seismic-physical oceanography surveys are anticipated, which will provide ground-truth and further improvements in interpreting and analyzing seismic reflection sections and finally allow for exploitation of the extensive global archive of marine seismic reflection data for interpreting the ocean dynamics.

Nomenclature

k	wavenumber (cycles per meter, cpm)
k_x	horizontal wavenumber (cycles per meter, cpm; or cycles per kilometer, cpkm)
K_ρ	diapycnal eddy diffusivity ($m^2 s^{-1}$)
N	buoyancy frequency (s^{-1})
ε	dissipation of turbulent kinetic energy, per unit mass ($m^2 s^{-3}$)
ζ	vertical displacement (m)
ζ_x	horizontal derivative of ζ $(-)$
ω	frequency (Hz)

See also

Sonar Systems.

Further Reading

Holbrook WS and Fer I (2005) Ocean internal wave spectra inferred from seismic reflection transects. *Geophysical Research Letters* 32: L15604 (doi:10.1029/ 2005GL 023733).

Holbrook WS, Páramo P, Pearse S, and Schmitt RW (2003) Thermohaline fine structure in an oceanographic front from seismic reflection profiling. *Science* 301: 821–824.

Klymak JM and Moum JN (2007) Oceanic isopycnal slope spectra: Part II: Turbulence. *Journal of Physical Oceanography* 37: 1232–1245.

Nakamura Y, Noguchi T, Tsuji T, Itoh S, Niino H, and Matsuoka T (2006) Simultaneous seismic reflection and physical oceanographic observations of oceanic fine structure in the Kuroshio extension front. *Geophysical Research Letters* 33: L23605 (doi:10.1029/ 2006GL027 437).

Nandi P, Holbrook WS, Pearse S, Páramo P, and Schmitt RW (2004) Seismic reflection imaging of water mass boundaries in the Norwegian Sea. *Geophysical Research Letters* 31: L23311 (doi:10.1029/ 2004GL021325).

Páramo P and Holbrook WS (2005) Temperature contrasts in the water column inferred from amplitude-versus-offset analysis of acoustic reflections. *Geophysical Research Letters* 32: L24611 (doi:10.1029/ 2005GL024533).

Tsuji T, Noguchi T, Niino H, *et al.* (2005) Two-dimensional mapping of fine structures in the Kuroshio current using seismic reflection data. *Geophysical Research Letters* 32: L14609 (doi:10.1029/ 2005GL023095).

SONAR SYSTEMS

A. B. Baggeroer, Massachusetts Institute of
Technology, Cambridge, MA, USA

Introduction and Short History

Sonar (Sound Navigation and Ranging) systems
are the primary method of imaging and communicating within the ocean. Electromagnetic energy does
not propagate very far since it is attenuated by either
absorption or scattering – visibility beyond 100 m is
exceptional. Conversely, sound propagates very well
in the ocean especially at low frequencies; consequently, sonars are by far the most important systems used by both man and marine life within the
ocean for imaging and communication.

Sonars are classified as being either active or passive. In active systems an acoustic pulse, or more
typically a sequence of pulses, is transmitted and a
receiver processes them to form an 'image' or to
decode a data message if operating as a communication system. The image can be as simple as the
presence of a discrete echo or as complex as a visual
picture. The receiver may be coincident with the
transmitter – a monostatic system, or separate – a
bistatic system. Both the waveform of the acoustic
pulse and the beamwidths of both the transmitter
and receiver are important and determine the performance of an active system. One typically associates an active sonar with the popular perception of
sonar systems. Many marine mammals use active
sonar for navigation and prey localization, as well as
communication in ways which we are still attempting to understand. Many of the signals used by
modern sonars have some of the same features as
those of marine mammals.

Passive systems only receive. They sense ambient
sound made by a myriad of sources in the ocean such
as ships, submarines, marine mammals, volcanoes.
These systems have been, and still are, especially
important in anti-submarine warfare (ASW) where
stealth is an important issue, and an active ping
would reveal the location of the source.

The use of sound for detecting underwater
objects was first introduced in a patent by
Richardson in June 1912 for the 'sonic detection
of icebergs,' 2 months after the sinking of the
Titanic.[1] This was soon followed by the

development of the Fessenden oscillator in 1914
which eventually led to the development of fathometers, an acoustic system for measuring the
depth to the seabed. The French physicist/chemist
Paul Langevin was the first to detect a submarine
using sonar in 1918, motivated by the extensive
damage of German U-boats. Between World Wars I
and II both Britain and the US sponsored sonar
research, especially on transducers. The former was
conducted under the Antisubmarine Detection Investigation Committee, or ASDIC as sonar is still
often referred to within the British military, and the
latter was performed at the Naval Research
Laboratory.

The re-emergence of the German U-boat stimulated the modern era of sonars and the physics of
sound propagation in the ocean where major research
programs were chartered in the USA (Columbia,
Harvard, Scripps Institution of Oceanography,
Woods Hole Oceanographic Institution), UK, and
Russia. A very comprehensive summary was compiled by the US National Defense Research Council
after World War II, which still remains a valuable
reference (*see* Further Reading Section).

The development of the nuclear submarine, both as
an attack boat (SSN) or as a missile carrier (SSBN)
provided a major emphasis for sonar throughout the
cold war. The USA, UK, Russia, and France all had
substantial research programs on sonar for many
applications, but ASW certainly had a major priority.
The nuclear submarine could deny use of the oceans
but could also unleash massive destruction with
nuclear missiles. With the end of the cold war, ASW
now has a lower priority; however, the submarine still
remains the platform of choice for many countries
since modern diesel/electric submarines operating on
batteries are extremely hard to detect and localize.
Undoubtedly, the most extensively used reference was
compiled by Urick (1975), which is frequently referenced as a handbook for sonar engineers.

While military operations have dominated the
development of sonars, they are now used extensively for both scientific and commercial applications. The use of fathometers and closely related
seismic methods provided much of the important
data validating plate tectonics. There is also a lot
of overlap between geophysical exploration for

---

[1] Much of this material in the history has been extracted from
Beyer, 1999.

hydrocarbons and modern sonars. High resolution and multibeam systems are extensively used for charting the seabed and its sub-bottom characteristics, fish finding, current measurements exploiting Doppler, as well as archaeological investigations.

Active Sonar systems

The major components of an active system are indicated in **Figure 1**. A waveform generator forms a pulse or 'ping', which is then modulated, or frequency shifted, to an operating frequency, f_o which may be as low as tens of Hertz for very long-range systems, or as high as 1 MHz, for high resolution short-range imaging sonars. Next, the signal is often 'beamformed' by an array of transducers, that focuses the signal in specific directions either by mechanically rotating the array or by introducing appropriate time delays or phase shifts. The signals are amplified and then converted from an electrical signal to a sound wave by the transmit transducers. Efficient transduction, the conversion of electric power to sound power, and even a modest amount of directivity of the transmitter requires that the transducer have dimensions on the scale of the wavelength of the operating frequency; hence, low frequency transmitters are typically large and not very efficient, whereas high frequency transmitters are smaller and very efficient.

The pinging rate, usually termed the pulse repetition frequency (PRF) is determined by the duration over which strong echos (called 'returns') from the previously transmitted pulse can be expected, so that one return does not overlap and become confused with another. With some systems with well confined response durations, several pulses may be in transit at the same time.

The ocean introduces three important components before it is detected by a receiver.

- There is the desired echo from the target itself. This may be a simple echo, especially if the target is close, but it may also include many multipaths and/or modes as a result of reflections of the ocean surface and bottom as well as paths refracted completely within the ocean itself.
- The ocean is filled with spurious, or unwanted reflectors which produce reverberation. The dominant source of this is the sea bottom, but the sea surface and objects (e.g. fish) can be important as well. Typically, the bottom is characterized in terms of a scattering strength per unit area insonified.
- Finally, the ocean is filled with ambient noise which is created by both natural and man-made sources. At low frequencies, 50–500 Hz, shipping tends to dominate the noise in the Northern Hemisphere, especially near shipping lanes. Wind and wave processes as well as rain can also be important. In specific areas, marine life may be a very important component.

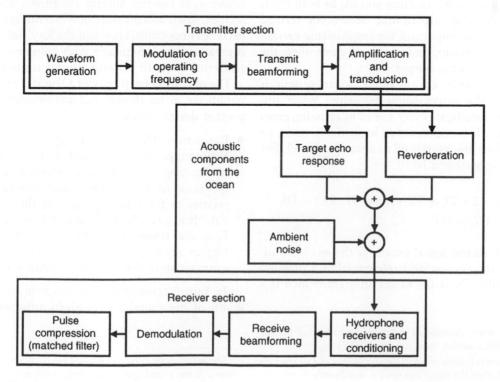

Figure 1 Active sonar system components.

The sonar receiver implements operations similar to the transmitter. Hydrophones convert the acoustic signal to an electric one whereupon it usually undergoes some 'signal conditioning' to amplify it to an appropriate level. In modern sonars the signal is digitized, since most of the subsequent operations are more easily implemented by digital signal processors. Next, a receiver beamformer, which may be quite different from the transmitter, focuses energy arriving from specific directions for spatial imaging. This is also done either by mechanically steering the array or by introducing time delays or phase shifts (if the processing is done in the frequency domain). The signal is then usually demodulated to a low frequency band which simplifies further electronics and signal processing steps. Finally it is 'pulse compressed,' or 'matched filtered,' which is a process that maximizes energy arriving at the travel time that corresponds to the range to a target. The matched filter is simply a correlation operation which seeks the best replica of the transmitted signal among all the signal components introduced by the ocean. In the simplest form of processing a sequence of 'pings' is rastered, i.e. the echo time-series are displayed one after the other, to construct an image. This is typical of a sidescan sonar system. In more sophisticated systems, especially those operating at low frequencies where phase coherence can be preserved or extracted, a sequence of outputs from the pulse compression filter is processed to form images. This is typical of synthetic aperture sonars. In both types of systems display algorithms, sometimes termed 'normalizers,' are important for emphasizing certain features by improving contrast and controlling the dynamic range of the output.

The performance of an active sonar system is captured in the active sonar equation; while imperfect in its details, it is very useful in assessing gross performance. It is expressed in logarithmic units, or decibels which are referenced to a standard level. For a monostatic system it is:

$$SE = SL - 2*TL + TS - \max(NL, RL) + DI_t$$
$$+ AG_r - DT$$

where[2,3] SE, is the signal excess at the receiver output; SL, is the source level referenced to a pressure level of 1 μPa; TS, is the target strength, which is a function of aspect angle; NL, is the level of the ambient noise at a single hydrophone; RL, is the reverberation which is determined by the area insonified, the scattering strength, and the signal level; DI_t, is the directivity index of the transmitter, which is a measure of the gain compared omnidirectional radiation; AG_r, is the array gain of the receiver in the direction of the target (often this is described as a receiver); directivity index, DI_r; DT, is the direction threshold for a target to be seen on the output display (this can be a complicated function of the complexity of the environment and the sophistication of an operator); TL, is the transmission loss, i.e. the loss in signal energy as it transmits to the target and returns. If the $SE > 0$, then a target is discernible on a display.

The notation $\max(NL, RL)$ distinguishes the two important regimes for an active sonar. When $NL > RL$, the sonar is operating in a noise-limited environment; conversely, when $RL > NL$ the environment is reverberation-limited which is the case in virtually all applications.

Reverberation is the result of unwanted echoes from the sea surface, seafloor, and volume. In the simplest formulation its level for surfaces both bottom and top is usually characterized by a scattering strength per unit area, so the level is given by the product of the resolved area multiplied by the scattering strength (or sum if using a decibel formulation). Often, the Rayleigh parameter $\frac{2\pi\sigma_s}{\lambda}\sin(\phi)$ is used, where σ_s is the rms surface roughness and ϕ is the incident angle as a measure of when surface roughness becomes important, i.e. when the Rayleigh parameter is greater than 1. Similarly with volume scattering, a scattering strength per unit volume is used.

The operating frequency of a sonar is an important parameter in its design and performance. The important design issues are:

- Resolution. There are two aspects of resolution – 'cross-range' resolution and 'in range' resolution. 'Cross-range' resolution is determined by the dimensions of the transmit and receive apertures relative to the wavelength, λ, of the acoustic signal.[4] It is given by $R\lambda/L$ where R is the range and L is the transmitter and/or receiving aperture. Higher resolution requires higher operating frequencies since these result in smaller $R\lambda/L$ ratios.

'In range' resolution is determined by the bandwidth of the signal and is given approximately by

---

[2]There are many versions of the sonar equation and the nomenclature differs among them (see Urick, 1975)

[3]If the system is bistatic wherein the transmitter and receiver are not colocated, then the sonar equation is significantly more complicated (Cox, 1989)

---

[4]The wavelength in uncomplicated media is given by $\lambda = c/f_o$, where c is the sound speed, nominally $1500\,\text{m s}^{-1}$ and f_o is the operating frequency.

$c/2W$, where W is the available bandwidth and c is the sonar speed. Since W is usually proportional to the operating frequency, f_o, one tends to try to use higher frequencies, which limits practical ranges. Also the sonar channel is often very band-limited. As a result in most sonar systems the 'in range' resolution is typically significantly smaller than the 'cross-range' resolution, so care needs to be taken in interpreting images.

- Maximum operating range. Acoustical signals can propagate over very long ranges, but there are a number of phenomena which can both enhance and attenuate the signal power. These include the geometrical spreading, the stratification of the sound speed versus depth, absorption, and scattering processes. The first two are essentially independent of frequency while the latter two have strong dependencies. **Figure 2** indicates the absorption loss in dB km^{-1} of sound at 20° and 35 apt salinity.[5] Essentially, the absorption loss factor increases quadratically with frequency. The two 'knees' in the figure relate to the onset of losses introduced by ionic relaxation phenomena. The net implication is that efforts are made to minimize the operating frequency so that it contributes <10 dB of loss for the desired range, i.e $\alpha(f_o)R < 10$ dBwhere $\alpha(f_o)R$ is attenuation per unit distance in dB.

The penalty of this, however, is a less directive signal, so it is more difficult to avoid contact with the ocean boundaries and consequent scattering losses, especially the bottom is shallow-water environments. As a result, the actual operating frequency of a sonar is a compromise based on the sound speed profile, the directivity of transmitter, receiver beamformers, and desired operating range and isolation.

- Target cross-section. The physics of sound reflecting from a target can be very complicated and there are few exact solutions, most of which involve long, complicated mathematical functions. In addition, the geometry of the target can introduce a significant aspect dependence. The important scaling number is $2\pi a/\lambda$ where a is a characteristic length scale presented to the incoming sound wave. Typically, if the number is less than unity, the reflected target strength depends upon the fourth power of frequency, so the target strength is quite small; this is the so-called Rayleigh region of scattering. Conversely, if this

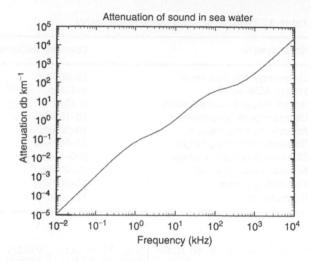

Figure 2 Attenuation of sound vs frequency.

number is larger than unity, the target strength normalized by the presented area is typically between 1 and 10. There are two additional features to consider: (i) large, flat surfaces lead to large returns, often termed specular, and (ii) in a bistatic sonar, the forward scattering, essentially the shadow, is determined almost solely by the intercepted target shape, often called Babinet's principle.

The net effect is a desire for higher operating frequencies in order to stay in the Rayleigh region where there is significant target strength.

Overall high frequencies produce better resolution and higher target strengths. However, at high frequencies acoustic propagation is more complicated and absorption limits the range.

Table 1 indicates the operating frequencies of some typical active sonars.

Sonar System Components

The components in **Figure 1** all have significant impact on the performance of an active sonar. The signal processing issues are complex and there is a large sonar-related literature as well as radar where the issues are similar. The essential problem is to separate a target amidst the reverberation and ambient noise in either of two operating realms – 'noise-limited' and 'reverberation'-limited environments. In a 'noise-limited' environment the ambient background noise limits the performance of a system, so increasing the transmitter output power improves performance. A 'reverberation-limited' environment is one where the noise is composed of mostly unwanted reflections from objects other than the target, so increasing the transmitted power simply increases both the target and reverberation returns simultaneously with no

---

[5] db km^{-1} represents 10 log of the fractional loss in power per kilometer representing an exponential decay versus range.

Table 1 Features of some typical active sonars

Sonar system	Operating frequency range	Wavelength	Nominal range
Long-range, low frequency	50–500 Hz	30–3 m	1000 km
Military ASW sonars	3–4 kHz	0.5–0.75 m	100 km
Bottom-mapping echosounders	3–4 kHz	0.5–0.75 m	vertical
High-resolution fathometers	10–15 kHz	15–10 cm	vertical
Acoustic communications	10–30 kHz	15–5 cm	10 km
Sidescan sonar (long-range)	50–100 kHz	6–1.5 cm	5 km
Sidescan sonar (short range)	500–1000 kHz	3–1.5 mm	100 m
Acoustic localization nets	10–20 kHz	15–5 cm	vertical
Fish-finding sonars	25–200 kHz	6–1 cm	1–5 km
Recreational	100–250 km	15–6 mm	vertical

net gain in signal to noise ping. Most active sonars operate in a reverberation limited environment. Effective design of an active sonar depends upon controlling reverberation through a combination of waveform design and beamforming.

Waveform design There are two basic approaches to waveform design for resolving targets – 'range gating' and 'Doppler gating.' The simplest approach to 'range gating' is a short, high powered pulse. This essentially resolves every reflector and an image is constructed by successive pulses and then the returns are rastered. While this is the simplest waveform it has limitations when operating in environments with high noise and reverberation levels, since the peak power of most sonars, both man-made and marine mammal, is limited. This shortcoming can often be mitigated by exploiting bandwidth (resolution α/β). This has led to a large literature on waveform design with the most popular being frequency modulated (FM) and coded (PRN) signals. With these signals[6] the center frequency is swept, or 'chriped' across a frequency band at the transmitter and correlated, or 'compressed' at the receiver. This class of signals is commonly used by marine mammals including whales and dolphins for target localization.

'Doppler gating' is based on differences in target motion. A moving target imparts a Doppler shift to the reflected signal which is proportional to operating frequency and the ratio v/c, where v is the target speed and c is the sound speed. 'Doppler gating' is particularly useful in some ASW contexts since it is difficult to keep a submarine stationary, thus a properly designed signal, one that resolves Doppler,

can distinguish it against a fixed reverberant background. The ability to resolve Doppler frequency depends upon the duration of a signal, with a dependence of 1/duration, so good 'Doppler gating' waveforms are long.

There has been a lot of research on the topic of optimal waveform design. Ideally, one wants a waveform which can resolve range and Doppler simultaneously which implies long duration, and wide bandwidth. These requirements are difficult to satisfy simultaneously.

Beamforming Both the transmitter and the receiver beamformers provide spatial resolution for the sonar system. The angular resolution in degrees is approximately $\Delta\theta \approx 60\lambda/L$, where λ is the wavelength and L is the aperture length. Since acoustic wavelengths are large when compared with optical wavelengths, the angular resolutions tend to be large especially at low frequencies.[7] Beamformers, often termed array processors in receivers, have been an important research topic for several decades with the advent of digital signal processing which permitted increasingly more sophistication, especially in the realm of adaptive methods. One of the simplest transmit beamformers consists of a line array of transducers each radiating the same signal. The simple receiver and also a line of transducers adds all the signals together. This resolves the paths perpendicular or broadside to the array. If one wants to 'steer' the array, or resolve another direction, the array must be mechanically rotated. This method of beamforming is still used by many systems since it is quite robust. Another simple beamformer is a planar array of transducers.

---

[6] Pseudo Random Noise (PRN) are coded signals which appear to be random noise. Well designed signals have useful mathematical constructs which led to good outputs at the output of the pulse compression, or matched filter, processor.

[7] Sonars with angular resolutions of 1° are generally considered to have high resolution. Compare this with that of the human eye with a nominal diameter of 4 mm and the wavelength of light in the visible region is 0.4 μm leading to a resolution scale of 0.1 ms.

Digital signal processing has led to more sophisticated array processing, especially for receivers. Beamformers which steer beams electronically by introducing delays, or phase shifts, shape beams to control sidelobes, place nulls to control strong reflectors, and reduce jamming are now practical because these features can be practical electronically rather than mechanically.

Examples of Active Sonar Images

This section describes two examples of sonars used for mapping seafloor bathymetry. In the first the sonar is carried on an unmanned underwater vehicle (UUV) close to the seafloor. The operating frequency is 675 kHz and the beam is mechanically steered from port to starboard as well as fore and aft as the UUV proceeds along its track, so the beams are steered forward and directly below the UUV (**Figure 3**). The onset time of the first echo return is the parameter of interest. It is converted to the depth of the seafloor after including the vehicle position, the direction of the beam and possibly refraction effects in the water itself. Usually straight-line acoustic propagation is assumed.

The signals are combined to generate a high resolution map of the seafloor. The processing to achieve this includes editing for spurious responses, registration of the rasters or images from successive transmission using the navigation sensors on the UUV (or more generally any vehicle) and normalization to improve the contrast so that weak features can be detected amidst strong ones.

The second example of an active sonar is a multibeam bathymetric mapper. Most of these systems for deep water operate at a 12 kHz center frequency. The transmit beam is produced by a linear array running fore to aft along the bottom of the ship, thwartships beam, which produces a swath which resolves the seafloor along track (**Figure 4**). The receiver array is oriented port to starboard. The signals from this array are beamformed electronically, so the seafloor is resolved port to starboard within the transmitted swath, since the patch is the product of the transmit and receive beamwidth. This configuration allows two-dimensional resolution with two linear arrays instead of a full planar array. The depths from each of the multibeams are measured by combining the travel time and the ray refraction from the sound speed profile to obtain a depth. Subsequent processing edits anomalous returns and interpolates all the data to generate the contour map. Active sonar systems with additional features have been developed for special applications, but they all use the basic principles described above.

Passive Sonars

Passive sonars that only listen and do not transmit are used in a variety of applications including the military for antisubmarine warfare (ASW), tracking and classification of marine mammals, earthquake detection, and nuclear test ban monitoring.

Since the signals are passive there is no pulse compression, or matched filtering, so a passive sonar design primarily focuses upon the 'short-term' frequency wavenumber spectrum, or the directional spectrum and the power density spectrum and how it evolves in time. The data are nonstationary and inhomogeneous, but many of the processing algorithms

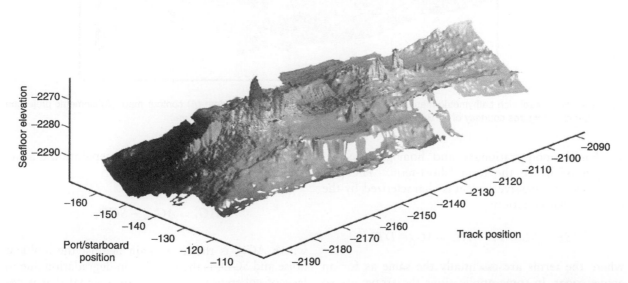

Figure 3 Image with forward-looking and down-looking sonar. (Figure courtesy of Dr Dana Yoerger, Woods Hole Oceanographic Institution.)

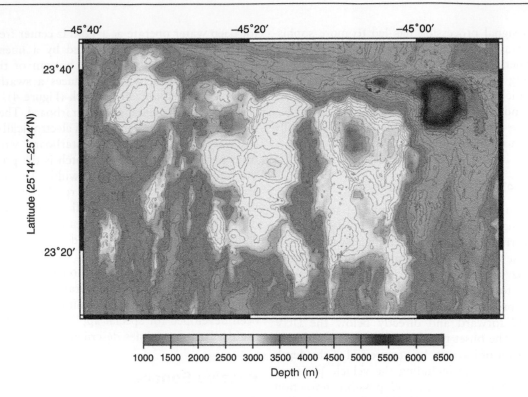

Figure 4 High resolution bathymetric map of the seafloor near the Mid-Atlantic Ridge: (A) contour map; (B) isometric projection (from top-right). (Figures courtesy of Dr Brian Tucholke, Woods Hole Oceanographic Institution.)

are based upon stationary and homogeneous assumptions; hence the term 'short-term.' The performance of a passive system is characterized by the passive sonar equation:

$$SE = SL - TL - NL + AG_r - DT$$

where the terms are essentially the same as for an active sonar. In some applications the arrays are so large that the coherence of the received signal is important, and it is necessary to separate the array gain, AG_r into two terms, or:

$$AG_r = AG_{r,n} - SGD_s$$

where $AG_{r,n}$ is the array gain against the ambient noise and SGD_s is the signal gain degradation due to lack of coherence. $SGD_s = 0$ for a signal that is coherent across the entire array.

Passive Sonar Beamforming

The signals received by the sonar's hydrophone are preconditioned, which might include editing bad data channels, calibration, and filtering. They are then beamformed, either in the time domain by introducing delays to compensate for the travel time across the array, or in the frequency domain. With digital signals the former usually requires upsampling or interpolation of the data to avoid distortion. The latter is accomplished by FFT (fast Fourier transforms), phase shifting to compensate for the delays, and then IFFT (inverse fast Fourier transforming). Frequency domain beamforming allows simpler implementation of adaptive techniques that are useful in cases where the ambient field has many discrete components. Adaptive algorithms form beams with notches, i.e. poor response in the direction of interferers, thereby suppressing them. Many algorithms have been designed to accomplish this, but the MVDR (minimum variance distortion filter – first introduced by Capon) and related algorithms have been used most extensively in practice.

Passive Sonar Display Formats

The output of the beamformer is a time-series for each beam. In certain applications the time-series itself may be of interest, however, in most cases the time-series is further processed to assist in extracting weak signals from the background noise. The parameter for signal processing schemes and display formats includes time (the epoch for data processing, T), angle (azimuth and elevation), and frequency (the spectral content of the data) (**Figure 5**).

Bearing-time Recording

Bearing time processing takes the beam outputs over a specified frequency band and plots the output versus time. Two modes of processing are often used: (i) energy detection, which forms an average of the beam outputs, or (ii) cross-correlation detection, where the array is split at the beamformer and then the two outputs are cross-correlated versus the direction. The processing is often classified according to the width of the band used, passive broadband (PBB) or passive narrowband (PNB).

The data for each time epoch are normalized to improve the contrast for signals of interest and each raster is plotted. Over a sequence of epochs, the directional components in the ambient field which are associated with shipping are observed. By maneuvering the array one can triangulate to obtain a range to each source as well.

Low Frequency Acoustic Recording and Analysis (LOFAR) grams   Once a bearing or direction of interest has been determined, spectral analysis of the selected beam is used to produce a LOFAR gram, which is a plot of the signal spectrum for each analysis epoch, T, versus time. By examining the

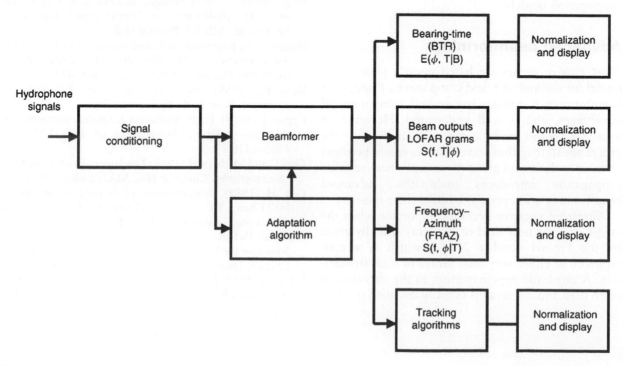

Figure 5   Passive sonar with modes of displaying output.

features of the LOFAR gram, such as frequency, peaks, harmonic signals, and their changes as a function of time, the source of the signal and some of its characteristics, such as speed, can be deduced. As in the case of the bearing-time display, the LOFAR gram is normalized to enhance features of interest.

FRAZ displays Frequency-azimuth, or FRAZ displays plot the spectral content as a function of frequency and azimuth for each epoch, T. Often a number of FRAZ outputs are averaged to improve signal to noise. FRAZ displays allow connection of the spectral content of a single source along a given bearing since the display contains a number of lines for each source at a given azimuth. As with the previous display, normalization algorithms to improve contrast are usually employed.

Trackers The objective of an ASW passive sonar is to detect, classify and track sources of radiating sound. Trackers are used to follow sources through direction and frequency space. There are a number of tracking algorithms build upon various signal models. Some separate the direction and frequency dimensions while others are coupled models. Since the ambient field can often have a number of sources, some targets and some interferers and since there are complicated propagation effects, the design of trackers is difficult. Most involve some form of Kalman filtering and some have integral propagation models.

Advanced beamforming

Most passive sonars are based upon a plane wave model for the ambient field components. Plane wave beamforming is robust, has several computational advantages, and is well understood. However, in many sonars this model is not adequate because the arrays are so long that wavefront curvature becomes an issue or the arrays are used vertical where acoustic propagation introduces multipaths. Advanced beamforming concepts can address these issues.

Wavefront curative becomes important when the target is in the near field of the array (usually given by the Fresnel number $2L^2/\lambda$) which is a consequence of either very long arrays or high frequencies. A quadratic approximation to the curvature is often used and the array is actually designed to focus it at specific range (rather than at infinite range which is the case for plane waves). When focused at short ranges, long-range targets are attenuated. This introduces the focal range as another parameter for displaying the sonar output.

At long ranges and low frequencies or in shallow water acoustic signals have complex multipath or multimode propagation which leads to coherent interference along a vertical array or a very long horizontal array. The appropriate array processing is to determine the full field Green's function for the signal and match the beamforming to it, a technique known as matched field processing (MFP). MFP requires knowledge of the sound speed profile along the propagation path, so its performance depends upon the accuracy of environmental data. It is a computationally intensive process, but has the powerful advantage of being able to resolve both target depth and range, as well as azimuth. MFP is an active subject of passive sonar research.

See also

Acoustic Scattering by Marine Organisms. Acoustics, Shallow Water.

Further Reading

Baggeroer AB (1978) Sonar signal processing. In: Oppenheim AV (ed.) *Applications of Digital Signal Processing.* Englewood Cliffs NJ: Prentice-Hall.

Baggeroer A, Kuperman WA, and Mikhalevsky PN (1993) An overview of Matched Field Processing. *IEEE Journal of Oceanic Engineering* 18: 401–424.

Beyer RT (1999) *Sounds of Our Times: 200 Years of Acoustics.* New York: Springer-Verlag.

Capon J (1969) High resolution frequency-wavenumber spectrum analysis. *Proceedings of the IEEE* 57: 1408–1418.

Clay C and Medwin H (1998) *Fundamentals of Acoustical Oceanography.* Chestnut Hill, MA: Academic Press.

Cox H (1989) Fundamentals of bistatic active sonar. In: Chan YT (ed.) *Underwater Acoustic Data Processing.* Kluwer Academic Publishers.

National Defense Research Council (1947) *Physics of Sound on the Sea: Parts 1–6.* National Defense Research Council, Division 6, Summary Technical Report.

Urick RJ (1975) *Principles of Underwater Sound for Engineers,* 2nd edn. McGraw-Hill Book Co.

TOMOGRAPHY

P. F. Worcester, University of California
at San Diego, La Jolla, CA, USA

Introduction

Ocean acoustic tomography is a method for acoustic remote sensing of the ocean interior that takes advantage of the facts that the propagation of sound through the ocean is sensitive to quantities of oceanographic interest, such as temperature and water velocity, and that the ocean is nearly transparent to low-frequency sound so that signals can be transmitted over long distances. The procedure is (i) to transmit acoustic signals through the ocean, (ii) to make precise measurements of the properties of the received signals, e.g., travel times, and (iii) to use inverse methods to infer the state of the ocean traversed by the sound field from the measured properties. The characteristics of the ocean between the sources and receivers are determined, rather than the characteristics of the ocean at the instruments as is the case for conventional thermometers and current meters.

Ocean acoustic tomography has a number of attractive attributes. It makes possible the rapid and repeated measurement of ocean properties over large areas, taking advantage of the speed with which sound travels in water ($\sim 1500\,\mathrm{m\,s^{-1}}$). It permits the monitoring of regions in which it is difficult to install instruments to make direct measurements, such as the Gulf Stream or the Strait of Gibraltar, using sources and receivers on the periphery of the region. Acoustic measurements are inherently spatially integrating, suppressing the small-scale variability that can contaminate point measurements and providing direct measurements of horizontal and vertical averages over large ranges. Finally, the amount of data grows as the product $(S \times R)$ of the number of acoustic sources S and receivers R, rather than linearly as the sum of the number of instruments $(S + R)$ as is the case for point measurements.

Ocean acoustic tomography was originally introduced by Munk and Wunsch in 1979 to address the difficult problem of observing the evolving ocean mesoscale. Mesoscale variability has spatial scales of order 100 km and timescales of order one month. The short timescales mean that ships move too slowly for ship-based measurements to be practical. The short spatial scales mean that moored sensors must be too closely spaced to be practical. Munk and Wunsch proposed that the travel times of acoustic signals propagating between a relatively small number of sources and receivers could be used to map the evolving temperature field in the intervening ocean. Their work led directly to the first 3D ocean acoustic tomography experiment, conducted in 1981. In spite of the marginal acoustic sources that were available at the time, the experiment showed that it was possible to use acoustic methods to map the evolving mesoscale field in a 300 km by 300 km region (**Figure 1**).

It was quickly realized, however, that the integral measures provided by acoustic methods are powerful tools for addressing certain types of problems, including the measurement of integral quantities such as heat content, mass transport, and circulation. Acoustic measurements of the integrated water velocity around a closed contour, for example, provide the circulation, which is directly related to the areal-average vorticity in the interior by Stokes' theorem. Vorticity is difficult to measure in other ways. The suppression of small-scale variability in the spatially integrating acoustic measurements also makes them well suited to measure large-scale phenomena, such as the barotropic and baroclinic tides. Finally, the integral measurements provided by the acoustic data can be used to test the skill of dynamic models and to provide strong model constraints.

Acoustic scattering due to small-scale oceanic variability (e.g., internal waves) causes the properties of the received acoustic signals to fluctuate. Although these fluctuations limit the precision with which the signal characteristics can be measured and with which oceanic parameters such as temperature and water velocity can be inferred, it was soon realized that measurements of the statistics of the fluctuations can be used to infer the statistical properties of the small-scale oceanic variability, such as internal-wave energy level, as a function of space and time.

Summarizing, tomographic methods can be used to map the evolving ocean, to provide integral measures of its properties, and to characterize the statistical behavior of small-scale oceanic variability.

Ocean Acoustics: The Forward Problem

The 'forward' problem in ocean acoustics is to compute the properties of the received signal given the sound-speed $C(x, y, z)$ and current $v(x, y, z)$ fields

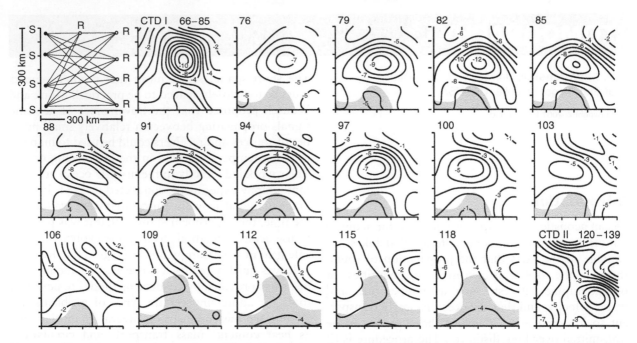

Figure 1 The 1981 tomography experiment. The first panel shows the geometry, with four source (S) and five receiver (R) moorings on the periphery of a 300 km by 300 km region in the north-west Atlantic Ocean. Subsequent panels show the sound-speed perturbations at 700 m depth derived from the acoustic data at 3-day intervals, with regions of high uncertainty shaded. The initial and final panels are derived from two ship-borne conductivity-temperature-depth (CTD) surveys, each of which required about 20 days to complete. The label on each panel is the year day in 1981. The contour interval is 1 m s^{-1} (0.2°C). Adapted from Cornuelle B, Wunsch C, Behringer D, *et al.* (1985) Tomographic maps of the ocean mesoscale. Part I: Pure acoustics. *Journal of Physical Oceanography* 15: 133–152.

between the source and receiver. Acoustic remote sensing of the ocean interior requires first a full understanding of the forward problem, i.e., of methods for finding solutions to the wave equation. A variety of approaches are available to do this, including geometric optics, normal mode, and parabolic equation methods. The appropriate method depends in part on the character of the sound-speed and current fields (e.g., range-independent or range-dependent) and in part on the choice of the observables in the received signal to use in the inverse problem.

The approach most commonly used in ocean acoustic tomography has been to transmit broadband signals designed to measure the impulse response of the ocean channel and to interpret the peaks in the impulse response in terms of geometric rays. Ray travel times are robust observables in the presence of internal-wave-induced scattering because of Fermat's principle, which states that ray travel times are not sensitive to first-order changes in the ray path. Other observables are possible, however. The peaks in the impulse response are in some cases more appropriately interpreted in terms of normal-mode arrivals, for example, and the observables are then modal group delays. Another possibility is to

perform full-field inversions that use the time series of intensity and phase for the entire received signal as observables. Unfortunately, neither normal modes nor the intensities and phases of the received signal are robust in the presence of internal-wave-induced scattering, and so tend to be useful only at short ranges and/or low frequencies where internal-wave-induced scattering is less important. A number of other possible observables have been proposed as well. In what follows the use of ray travel times as observables will be emphasized, in part because they have been the observable most commonly used to date and in part because they are robust to internal-wave-induced scattering.

Ocean Sound Channel

The speed with which sound travels in the ocean increases with increasing temperature, salinity, and pressure. As a result, over much of the temperate world ocean there is a subsurface minimum in sound speed at depths of roughly 1000 m. Sound speed increases toward the surface above the minimum because of increasing temperature and toward the bottom below the minimum because of increasing pressure. Salinity does not play a major role because its effect on sound speed is normally less than that of either temperature or pressure. The depth of the sound-speed minimum is called the sound channel axis. The axis shoals towards high latitudes where the surface waters are colder, actually reaching the surface during winter at sufficiently high latitudes.

The sound speed gradients above and below the sound channel axis refract acoustic rays toward the axis in accord with Snell's law. Near-horizontal rays propagating outward from an omnidirectional source on the axis will therefore tend to be trapped, cycling first above and then below the axis (**Figure 2**). Such rays are referred to as refracted–refracted (RR) rays. Steeper rays will interact with the surface and/or seafloor. Rays can reflect from the sea surface with relatively low loss. Rays that are refracted at depth

and reflected from the sea surface are referred to as refracted-surface-reflected (RSR) rays. Both RR and RSR rays can propagate to long distances and are commonly used in ocean acoustic tomography. Rays that interact with the seafloor tend to be strongly scattered, however. Rays with multiple bottom interactions therefore tend not to propagate to long ranges.

A receiver at a specified range from an omnidirectional acoustic source will detect a discrete set of ray arrivals (**Figure 2**), corresponding to the rays that are at the depth of the receiver at the appropriate range. These rays are called eigenrays and are designated $\pm p$, where $\pm$ indicates an upward/downward launch direction and p is the total number of ray turning points (including reflections). The ray

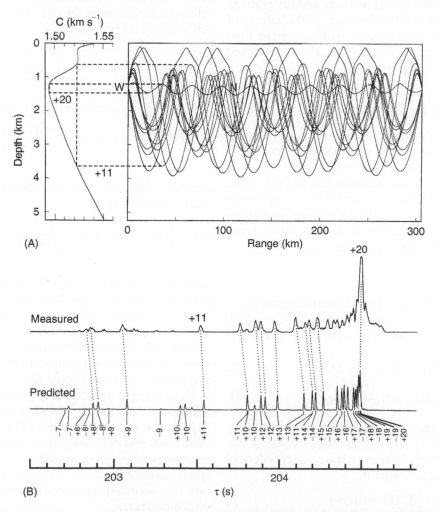

Figure 2 (A) Sound-speed profile in the western North Atlantic and the corresponding ray paths for source and receiver near the depth of the sound channel axis and about 300 km apart. The geometry is that of a reciprocal acoustic transmission experiment conducted in 1983 with transceivers designated W(est) and N(orth). (B) Measured and predicted acoustic amplitudes as a function of time for the 1983 experiment. The arrivals are labeled with their ray identifier. The earliest arrivals are from steep ray paths that cycle through nearly the entire water column. The latest arrivals are from flat ray paths that remain near the sound-channel axis. The differences between the measured and predicted arrival times are the data used in tomographic inversions. Adapted from Howe BM, Worcester PF and Spindel RC (1987) Ocean acoustic tomography: mesoscale velocity. *Journal of Geophysical Research* 92: 3785–3805.

geometry controls the vertical sampling properties of tomographic measurements. One can obtain significant vertical resolution even for the case of source and receiver both on the sound channel axis, because the eigenrays in general have a range of turning depths.

Ray Travel Time

To first order in $|\mathbf{v}|/C$, the travel time τ_i of ray i is

$$\tau_i = \int\limits_{\Gamma_i} \frac{ds}{C(\mathbf{r}) + \mathbf{v}(\mathbf{r}) \cdot \mathbf{r}'} \qquad [1]$$

where Γ_i is the ray path for ray i along which distance s is measured and $\mathbf{r}'$ is the tangent to the ray at position $\mathbf{r}$. The sign of $\mathbf{v} \cdot \mathbf{r}'$ depends on the direction of propagation, and the travel times and ray paths in opposite directions differ because of the effects of currents. (Sound travels faster with a current than against a current.) The eigenrays Γ_i are obtained using a numerical eigenray code.

The Inverse Problem

The 'inverse' problem is to compute the sound-speed $C(\mathbf{r})$ and current $\mathbf{v}(\mathbf{r})$ fields given the measured travel times. In fact, a great deal is normally known about $C(\mathbf{r})$ in the ocean from climatological or other data. The interesting problem is therefore to compute the perturbations from an assumed reference state, using the measured perturbations from the travel times computed for the reference state.

Data

Travel times are in general a nonlinear function of the sound-speed and current fields, because the ray path Γ_i depends on $C(\mathbf{r})$ and $\mathbf{v}(\mathbf{r})$. Linearize by setting

$$C(\mathbf{r}) = C(\mathbf{r}, -) + \Delta C(\mathbf{r}) \qquad [2]$$

$$\mathbf{v}(\mathbf{r}) = \mathbf{v}(\mathbf{r}, -) + \Delta \mathbf{v}(\mathbf{r}) \qquad [3]$$

where $C(\mathbf{r}, -), \mathbf{v}(\mathbf{r}, -)$ are the known reference states. The argument $(-)$ denotes the dependence of the variables only on the reference state, independent of the measurements. Normally,

$$|\Delta C(\mathbf{r})| \ll C(\mathbf{r}, -) \qquad [4]$$

$$|\Delta \mathbf{v}(\mathbf{r})| \ll \mathbf{v}(\mathbf{r}, -) \qquad [5]$$

In general, however,

$$|\Delta \mathbf{v}(\mathbf{r})| > |\mathbf{v}(\mathbf{r}, -)| \qquad [6]$$

because the fluctuations in current at a fixed location in the ocean are typically large compared to the time-mean current.

Setting $\mathbf{v}(\mathbf{r}, -) \equiv 0, \Delta \mathbf{v}(\mathbf{r}) \equiv \mathbf{v}(\mathbf{r})$, forming perturbation travel times, and linearizing to first order in $\Delta C/C$ and $|\mathbf{v}|/C$ gives

$$\Delta \tau_i^+ = \tau_i^+ - \tau_i(-) =$$
$$-\int\limits_{\Gamma_{i(-)}} \frac{[\Delta C(\mathbf{r}) + \mathbf{v}(\mathbf{r}) \cdot \mathbf{r}'(-)]}{C^2(\mathbf{r}, -)} ds \qquad [7]$$

$$\Delta \tau_i^- = \tau_i^- - \tau_i(-) =$$
$$-\int\limits_{\Gamma_{i(-)}} \frac{[\Delta C(\mathbf{r}) - \mathbf{v}(\mathbf{r}) \cdot \mathbf{r}'(-)]}{C^2(\mathbf{r}, -)} ds \qquad [8]$$

where $\Gamma_{i(-)}, \mathbf{r}'(-)$ are the ray path and tangent vector for the reference state. The superscript plus (minus) refers to propagation in the $+ (-)$ direction. The reference travel time is

$$\tau_i(-) = \int\limits_{\Gamma_{i(-)}} \frac{ds}{C(\mathbf{r}, -)} \qquad [9]$$

The sum of the travel time perturbations

$$\Delta s_i = \frac{1}{2}(\Delta \tau_i^+ + \Delta \tau_i^-) = -\int\limits_{\Gamma_{i(-)}} \frac{ds}{C^2(\mathbf{r}, -)} \Delta C(\mathbf{r}) \qquad [10]$$

depends only on the sound-speed perturbation $\Delta C(\mathbf{r})$. The difference

$$\Delta d_i = \frac{1}{2}(\Delta \tau_i^+ + \Delta \tau_i^-) = -\int\limits_{\Gamma_{i(-)}} \frac{ds}{C^2(\mathbf{r}, -)} \mathbf{v}(\mathbf{r}) \cdot \mathbf{r}'(-) \qquad [11]$$

depends only on the water velocity $\mathbf{v} \cdot \mathbf{r}'$ along the ray path. Forming sum and difference travel times separates the effects of ΔC and $\mathbf{v}$. This separation is crucial for measuring $\mathbf{v}$, because $|\mathbf{v}|$ is usually much smaller than ΔC. It is not crucial for measuring ΔC, however, and one-way, rather than sum, travel time perturbations are often used for this purpose.

The data used in the inverse problem can therefore either be the one-way travel time perturbations, e.g., $\Delta \tau_i^+$, or the sum and difference travel time perturbations, Δs_i and Δd_i. The use of one-way travel time measurements to estimate ΔC is sometimes given the special name of acoustic thermometry, reflecting the fact that sound-speed perturbations depend mostly on temperature.

Reference States and Perturbation Models

The perturbation field ΔC and therefore the data depend on the choice of reference state, $C(\mathbf{r}, -)$. Although there is some freedom in the choice, reference states that include the range and time dependence of the sound-speed field available from ocean climatologies, for example, usually yield reference ray paths that are acceptably close to the true ones. Accurate prior estimates of the ray paths help ensure that the sampling properties of the rays are included properly in the inverse procedure and that the nonlinearities associated with ray-path mismatches are minimized.

The continuous perturbation fields ΔC and $\mathbf{v}$ are parametrized with a finite number of discrete parameters using a model. Because the tomographic inverse problem is always underdetermined, it is important to use perturbation models that are oceanographically meaningful and that provide an efficient representation of ocean variability. Separable models using a linear combination of basis functions describing the horizontal (x, y) and vertical (z) structures,

$$\Delta C(x, y, z) = \sum_{k,l,m} a_{klm} X_k(x) Y_l(y) F_m(z) \quad [12]$$

have frequently been used for simplicity, although more general models are of course possible. The coefficients a_{klm} are the model parameters to be determined, and the X_k, Y_l, F_m are the basis functions.

Inverse Methods: Vertical Slice

The inverse problem is most simply described for the case of a single acoustic source–receiver pair. Neglecting currents and assuming that the sound-speed perturbation is a function of depth only,

$$\Delta \tau_i = - \int_{\Gamma_i(-)} \frac{\Delta C(z)}{C^2(\mathbf{r},-)} ds + \delta \tau_i, \quad i = 1, ..., M \quad [13]$$

where there are M rays. (Note that although the sound-speed perturbation has been assumed to be independent of range, the reference state can be a more general function of position.) The quantity $\delta \tau_i$ has been introduced to represent the noise contribution that is inevitably present. The noise term arises not only from observational errors but also from modeling errors associated with the representation of ΔC using a finite number of parameters and from nonlinearity errors associated with the use of the ray paths for the reference state rather than the true ray paths. (In the absence of currents the problem can be restated somewhat more simply in terms of sound slowness, $S = C^{-1}$, if desired.)

Substituting

$$\Delta C(z) = \sum_m a_m F_m(z) \quad [14]$$

gives

$$\Delta \tau_i = \sum_m a_m \left[- \int_{\Gamma_i(-)} \frac{F_m(z)}{C^2(\mathbf{r},-)} ds \right] + \delta \tau_i,$$
$$i = 1, ..., M \quad [15]$$

$$= \sum_m E_{im} a_m + \delta \tau_i, \quad i = 1, ..., M \quad [16]$$

The elements E_{im} depend only on prior information. This equation can be written in compact matrix notation as

$$\mathbf{y} = \mathbf{Ex} + \mathbf{n} \quad [17]$$

where

$$\mathbf{y} = [\Delta \tau_i], \quad \mathbf{E} = \{E_{im}\}, \quad \mathbf{x} = [a_m], \quad \mathbf{n} = [\delta \tau_i] \quad [18]$$

The inverse problem consists (i) of finding a particular solution $\hat{\mathbf{x}}$ and (ii) of determining the uncertainty and resolution of the particular solution. Writing the estimate $\hat{\mathbf{x}}$ as a weighted linear sum of the observations,

$$\hat{\mathbf{x}} = \mathbf{By} = \mathbf{B}(\mathbf{Ex} + \mathbf{n}) \quad [19]$$

For zero-mean noise, $\langle \mathbf{n} \rangle = 0$, the expected value is

$$\langle \hat{\mathbf{x}} \rangle = \mathbf{BE} \langle \mathbf{x} \rangle \quad [20]$$

The matrix $\mathbf{BE}$ is called the resolution matrix. It gives the particular solution as a weighted average of the true solution $\mathbf{x}$, with weights given by the row vectors of $\mathbf{BE}$. If the resolution matrix is the identity matrix $\mathbf{I}$, then the particular solution is the true solution. If the row vectors of $\mathbf{BE}$ are peaked on the diagonal with low values elsewhere, the particular solution is a smoothed version of the true solution. The solution uncertainty is described by the covariance matrix

$$\mathbf{P} = \left\langle (\hat{\mathbf{x}} - \mathbf{x})(\hat{\mathbf{x}} - \mathbf{x})^T \right\rangle \quad [21]$$

where superscript T denotes transpose.

There is an immense literature on inverse methods, and a variety of approaches are available to construct the inverse operator $\mathbf{B}$, including least squares, singular-value decomposition (SVD), and Gauss–Markov estimation. To provide an example of one approach that has been widely used, the Gauss–Markov estimate is discussed briefly here. (The Gauss–Markov estimate is sometimes known as the 'stochastic inverse' or as 'objective mapping'.) The Gauss–Markov estimate is derived by minimizing the expected uncertainty between the true value x_j and the estimate $\hat{x}_j$, i.e., by individually minimizing the diagonal elements of the uncertainty covariance matrix **P**. The result is the Gauss–Markov theorem,

$$\mathbf{B} = \boldsymbol{\Phi}_{xy}\boldsymbol{\Phi}_{xy}^{-1} \qquad [22]$$

where

$$\boldsymbol{\Phi}_{xy} \equiv \langle \mathbf{xy}^T \rangle, \boldsymbol{\Phi}_{yy} \equiv \langle \mathbf{yy}^T \rangle \qquad [23]$$

are the model–data and data–data covariance matrices, respectively. These covariances can be rewritten using $\mathbf{y} = \mathbf{Ex} + \mathbf{n}$. The model–data covariance matrix becomes

$$\boldsymbol{\Phi}_{xy} = \langle \mathbf{xx}^T\mathbf{E}^T \rangle = \boldsymbol{\Phi}_{xy}\mathbf{E}^T \qquad [24]$$

where it has been assumed that the model **x** and noise **n** are uncorrelated. The data–data covariance matrix becomes

$$\boldsymbol{\Phi}_{yy} = \langle (\mathbf{Ex}+\mathbf{n})(\mathbf{Ex}+\mathbf{n})^T \rangle = \mathbf{E}\boldsymbol{\Phi}_{xx}\mathbf{E}^T + \boldsymbol{\Phi}_{nn} \qquad [25]$$

Finally, the inverse estimate, $\hat{\mathbf{x}} = \mathbf{By}$, can be written in the familiar form

$$\hat{\mathbf{x}} = \mathbf{By} = \boldsymbol{\Phi}_{xx}\mathbf{E}^T\left(\mathbf{E}\boldsymbol{\Phi}_{xx}\mathbf{E}^T + \boldsymbol{\Phi}_{nn}\right)^{-1}\mathbf{y} \qquad [26]$$

The Gauss–Markov estimator requires that the perturbation model discussed above include the *a priori* specification of the statistics of the model parameters, i.e., of the covariance matrix $\boldsymbol{\Phi}_{xx}$.

The solution uncertainty **P** includes contributions due to data error and due to a lack of resolution. In most realistic cases the lack of resolution dominates the solution uncertainty estimate. A key, and unfamiliar, feature of the acoustic methods is that the solution uncertainty matrix is not diagonal, i.e., the uncertainties in the model parameters are correlated in a way that depends on the ray sampling properties. These correlated uncertainties often cancel in the computation of integral properties of the solution, such as the vertically averaged heat content.

Once a solution and its uncertainty have been found, the solution must be evaluated for consistency with the various assumptions made in its construction before it can be accepted. The statistics of the residuals $\hat{\mathbf{y}} - \mathbf{y}$), where $\hat{\mathbf{y}} = \mathbf{E}\hat{\mathbf{x}}$, need to be examined for consistency with the assumed noise statistics $\boldsymbol{\Phi}_{nn}$, for example. Further, ray trajectories should be recomputed for the field

$$C(\mathbf{r}) = C(\mathbf{r}, -) + \Delta C(\mathbf{r}) \qquad [27]$$

and the resulting ray travel times compared with the original data to test for consistency. Significant differences imply that the reference state is inadequate or the model is inadequately formulated. When nonlinearities are important, iterative or other methods are needed to find a solution consistent with the original data.

The linear inverse methods used in ocean acoustic tomography are well known and widely used in a variety of fields. The crucial problem in the application to tomography is the construction of the model used to describe oceanic variability, including the choice of parametrization and the specification of the (co)variances of the model parameters and noise.

Sampling Properties of Acoustic Rays

Vertical slice: range-independent The vertical sampling properties of acoustic rays are most easily understood for the range-independent case, in which sound speed is a function of depth z only. In that case, eqn [13] can be converted to an integral over depth

$$\Delta\tau_i = -\int_{\tilde{z}^-(-)}^{\tilde{z}^+(-)} \frac{\mathrm{d}z}{C^2(z,-)\sqrt{1-\left(C-(z,-)/\tilde{C}\right)^2}}\Delta C(z) + \delta\tau_i, \quad i = 1, \dots, M \qquad [28]$$

using Snell's law, $C(z)/cos(\theta) = \tilde{C}$, where θ is the ray angle relative to the horizontal and $\tilde{C}$ is the sound speed at the ray turning points $\tilde{z}^\pm$. The function

$$\frac{1}{C^2(z,-)\sqrt{1-\left(C(z,-)/\tilde{C}\right)^2}} \qquad [29]$$

gives the weighting with which $\Delta C(z)$ contributes to $\Delta\tau_i$. There are (integrable) singularities in the weighting function at both the upper and lower turning point depths, where $(\tilde{z}^\pm, -) = \tilde{C}$. The ray travel times are therefore most sensitive to sound-speed perturbations at the ray turning points (**Figure 3**).

The value of the weighting function is the same for depths $z^\pm$ above and below the sound-channel axis at which

$$C(z^+, -) = C(z^-, -) \qquad [30]$$

There is a fundamental up–down ambiguity for acoustic measurements in mid-latitudes. It is in principle impossible to distinguish from the acoustic data alone whether the observed travel-time perturbations are due to sound-speed perturbations located above or below the sound channel axis. This ambiguity has to be resolved from *a priori* information or from other data. This up–down ambiguity

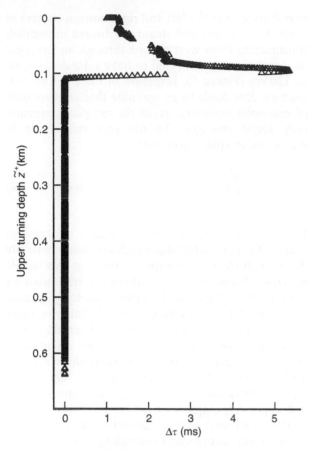

Figure 3 Travel-time perturbations computed at about 1000 km range in the North-east Pacific for a sound-speed perturbation with an amplitude of -1 m s^{-1} at 100 m depth, linearly decreasing to zero at 90 m and 110 m. The travel-time perturbations are zero for rays with upper turning depths below 110 m, because they do not sample the perturbed region. The perturbations are sharply peaked for rays with upper turning depths between 90 m and 110 m. Rays that have upper turning depths above 90 m have nonzero perturbations because they traverse the perturbed region, but the perturbations are relatively small because the ray weighting function falls off rapidly with distance from the turning point. Adapted from Cornuelle BD, Worcester PF, Hildebrand JA, *et al.* (1993) Ocean acoustic tomography at 1000-km range using wavefronts measured with a large-aperture vertical array. *Journal of Geophysical Research* 98: 16365–16377.

is not present in polar regions when the temperature profile is close to adiabatic, so that sound speed is a minimum at the surface and increases monotonically with depth (pressure).

Vertical slice: range-dependent A ray trapped in the sound channel, cycling between upper and lower turning points at regular intervals, samples the ocean periodically in space, so that its travel time is sensitive to some spatial frequencies but is unaffected by others. The key to understanding the horizontal sampling properties of acoustic travel times is to consider the wavenumber domain, rather

than physical space, using a truncated Fourier series in x as the model for the sound-speed perturbations,

$$\Delta C(x, z) = \sum_k \sum_m a_{km} \exp\left[\frac{2\pi i}{L}(kx)\right] F_m(z),$$
$$k = 0, \pm 1, \dots, \pm N \qquad [31]$$

The Fourier series is periodic over a domain of length L and is truncated at harmonic N. The domain is normally chosen sufficiently large (say twice the size of the source–receiver range) to avoid artifacts within the area of interest that might be caused by the periodicity.

The travel-time perturbations are then

$$\Delta \tau_i = \sum_k \sum_m a_{km} \left\{ - \int_{\Gamma_{i(-)}} \frac{ds}{C^2(\mathbf{r}, -)} \right.$$
$$\left. \times \exp\left[\frac{2\pi i}{L}(kx)\right] F_m(z) \right\} + \delta \tau_i \qquad [32]$$

where the integrals depend only on prior information. The problem has again reduced to the form $\mathbf{y} = \mathbf{Ex} + \mathbf{n}$, with the solution vector $\mathbf{x}$ containing an ordered set of the complex Fourier coefficients a_{km}.

The sensitivity of the travel-time inverse to various wavenumbers can be quantified by plotting the diagonal of the resolution matrix $\mathbf{BE}$ defined in eqn [20]. For the specific case of two moorings separated by 600 km, with a source and five widely-separated receivers on each mooring, the resolution matrix shows the sensitivity of tomographic measurements to the features that match the ray periodicity (i.e., have the same wavelength as the ray double loops) (**Figure 4**). Further, because the ray paths are somewhat distorted sinusoids in midlatitudes, the resolution matrix displays sensitivity to harmonics of the basic ray double loop length. Finally, as expected, the measurements are sensitive to the mean. There are obvious spectral gaps for wavenumbers between the mean and first harmonics of the ray paths, and again between the first and second harmonics. The harmonics extend over bands of wavenumbers because the eigenrays connecting the source and receiver have a range of double-loop lengths.

Horizontal slice The sampling issues present when the goal is to map the evolving ocean using integral data are most easily understood by considering the two-dimensional horizontal slice problem. In this case sound speed is assumed to be constant in the vertical, so that ray paths travel in straight lines in the horizontal plane containing the sources and receivers. Neglecting currents,

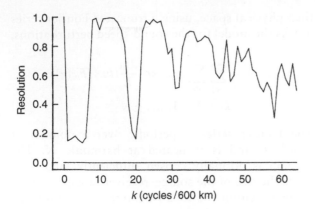

Figure 4 Diagonal elements of the resolution matrix ('transfer function') for tomographic measurements over a 600-km path in a range-dependent ocean. The plot is for the lowest baroclinic mode. Adapted from Cornuelle BD and Howe BM (1987) High spatial resolution in vertical slice ocean acoustic tomography. *Journal of Geophysical Research* 92: 11680–11692.

$$\Delta\tau_i = -\int_{\Gamma_{i(-)}} \frac{\Delta C(x,y)}{C^2(x,y,-)}\,\mathrm{d}s + \delta\tau_i,$$
$$i = 1, \ldots, M \qquad\qquad [33]$$

where there is one ray path per source–receiver pair and there are a total of M ray paths connecting the sources and receivers.

As was the case for the range-dependent vertical slice problem, the key to understanding the horizontal sampling properties of acoustic travel times is to consider the wavenumber domain, rather than physical space, using a truncated Fourier series in x and y as the model for the sound-speed perturbations,

$$\Delta C(x,y) = \sum_k \sum_l a_{kl}\exp\left[\frac{2\pi i}{L}(kx + ly)\right],$$
$$k, l = 0, \pm 1, \ldots, \pm N \qquad\qquad [34]$$

The Fourier series is doubly periodic over the square domain of size L and is truncated at harmonic N.

The travel-time perturbations are then

$$\Delta\tau_i = \sum_k \sum_l a_{kl}\left\{ -\int_{\Gamma_{i(-)}} \frac{\mathrm{d}s}{C^2(x,y,-)}\right.$$
$$\left.\exp\left[\frac{2\pi i}{L}(kx + ly)\right]\right\} + \delta\tau_i \qquad [35]$$

where the integrals depend only on prior information. The problem has again reduced to the form $\mathbf{y} = \mathbf{Ex} + \mathbf{n}$, with the solution vector $\mathbf{x}$ containing an ordered set of the complex Fourier coefficients a_{kl}.

To explore the horizontal sampling properties of integral data, consider a simple scenario in which

two ships start in the left and right bottom corners of a 1000-km square and steam northward in parallel, transmitting from west to east through an isotropic mesoscale field constructed to have a $1/e$ decay scale of 120 km (**Figure 5**). Inversion of the resulting travel-time data leads to an estimate that consists only of east–west contours, as all the ray paths measure only zonal averages. To interpret this result in wavenumber space, note that

$$\int_0^L \mathrm{d}x\Delta C(x,y) = 0 \quad \text{for} \quad k \neq 0 \qquad [36]$$

East–west transmissions therefore give information only on the parameters a_{0l}, which are nearly perfectly determined for the assumptions made in this simple scenario. Similarly, for north–south transmissions between two ships traveling from east to west, only the parameters a_{k0} are determined. Combining east–west and north–south transmissions determines both a_{0l} and a_{k0}, but nonetheless fails to give useful maps because the majority of the wavenumbers are still undetermined. Adding scans at $45°$ determines wavenumbers for which $k = l$, giving improved, but still imperfect, maps.

The conclusion is that generating accurate maps from integral data requires sampling geometries with ray paths at many different angles to provide adequate resolution in wavenumber space. This requirement must be independently satisfied in any region with dimensions comparable to the ocean correlation scale. These results are a direct consequence of the projection-slice theorem.

Time-dependent Inverse Methods

The discussion of inverse methods to this point has implicitly assumed that data from a single instant in time are used to estimate the state of the ocean at that instant. Observations from different times can be combined to generate improved estimates of the evolving ocean, however, using a time-dependent ocean model to connect the oceanic states at those times. One seeks to minimize the misfit between the estimate $\hat{\mathbf{x}}(t)$ and the true state $\mathbf{x}(t)$ over some finite time span, instead of at a single instant.

The practice of combining data with time-evolving ocean circulation models, referred to as 'assimilation' or 'state estimation', simultaneously tests and constrains the models. A variety of approaches are available to solve this problem, including, for example, Kalman filtering and the use of adjoint methods. Although the problem of combining integral tomographic data with time-evolving models does not differ in any fundamental way from the

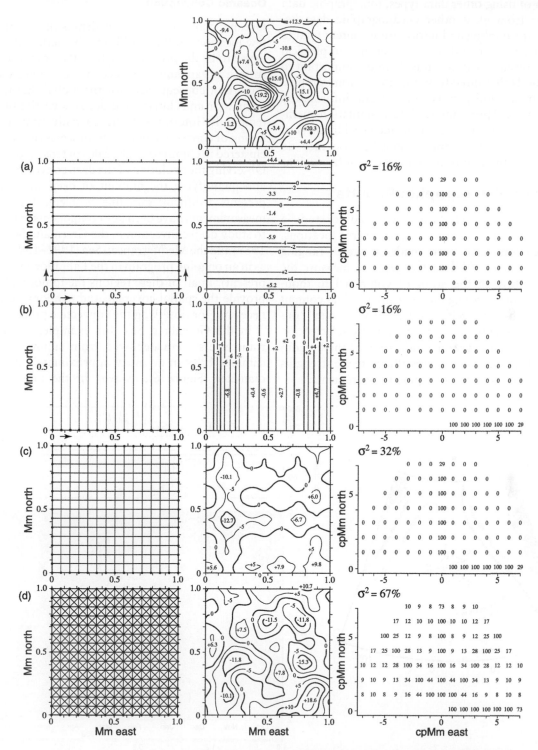

Figure 5 The top center panel is the 'true ocean,' constructed assuming a horizontally homogeneous and isotropic wavenumber spectrum, to be mapped using tomographic data. (A) W→E transmissions between two northward-traveling ships (left panel). Inversion of the travel time perturbations produces east–west contours in ΔC (middle) with only a faint relation to the 'true ocean.' Expected predicted variances in wavenumber space (right) are 0% (no skill) except for $(k,l) = (0,1),(0,2),\dots,(0,7)$, which account for $\sigma^2 = 16\%$ of the *a priori* ΔC variance. (B) S→N transmissions between two eastward traveling ships. (C) Combined W→E and S→N transmissions, accounting for 32% of the ΔC variance. (D) Combined W→E, S→N, SW→NE, and SE→NW transmissions, accounting for 67% of the variance and giving some resemblance to the true ocean. Distances are shown in magameters (Mm) and wavenumbers are shown in cycles per megameter (cpMm). Adapted from Cornuelle BD, Munk WH and Worcester PF (1989) Ocean acoustic tomography from ships. *Journal of Geophysical Research* 94: 6232–6250.

problem of using other data types, tomographic data do differ from most other oceanographic data because their sampling and information content tend to be localized in spectral space rather than in physical space, as discussed above. It is therefore important to use methods that directly assimilate the tomographic measurements and preserve the integral information they contain. Approximate data assimilation methods optimized for measurements localized in physical space are generally inappropriate because they do not preserve the nonlocal tomographic information.

Selected Tomographic Results

Tomographic methods have been used to study a wide range of ocean processes, at diverse locations. Measurements have been made at scales ranging from a few tens of kilometers (e.g., to measure the transport through the Strait of Gibraltar) to thousands of kilometers (e.g., to measure the heat content in the north-east Pacific Ocean). This review concludes by presenting results from a few selected experiments to provide some indication of the breadth of possible applications and to illustrate the strengths and weaknesses of tomographic measurements.

Oceanic Convection

Oceanic convection to great depths occurs at only a few locations in the world (*see*)Nonetheless, it is believed to be the process by which the properties of the surface ocean and deep ocean are connected, with important consequences for the global thermohaline circulation and climate. The deep convective process is temporally intermittent and spatially compact, consisting of convective plumes with scales of about 1 km clustered in chimneys with scales of tens of kilometers. Observing the evolution of the deep convective process and quantifying the amount of deep water formed presents a difficult sampling problem. Tomographic measurements have been key components in programs to study deep convection in the Greenland Sea (1988–1989) and the Mediterranean Sea (1991–1992), as well as in an ongoing program in the Labrador Sea (1996 to present). In all of these regions the tomographic data provide the spatial coverage and temporal resolution necessary for observing the convective process.

In the Greenland Sea, for example, six acoustic transceivers were deployed from summer 1988 to summer 1989 in an array approximately 210 km in diameter (**Figure 6**), as part of the intensive field

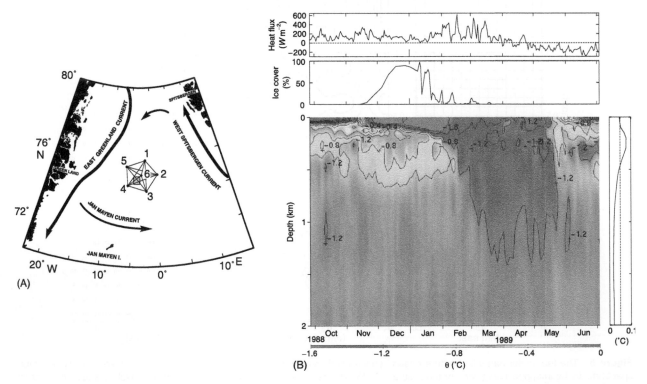

Figure 6 (A) Geometry of the tomographic transceiver array deployed in the Greenland Sea during 1988–1989. Mooring 2 failed about one month after deployment. A deep convective chimney was observed near the center of the array during March 1989 (shaded region). (B) Time-depth evolution of potential temperature averaged over the chimney region. Contour interval is 0.2°C. Typical rms uncertainty (°C) as a function of depth is shown to the right. Total heat flux (from the British Meteorological Office) and daily averaged ice cover (derived from satellite SSM/I measurements) are shown above. Adapted from Morawitz WML, Cornuelle BD and Worcester PF (1996) A case study in three-dimensional inverse methods: combining hydrographic, acoustic, and moored thermistor data in the Greenland Sea. *Journal of Atmospheric and Oceanic Techniques* 13: 659–679.

phase of the International Greenland Sea Project. The acoustic data were combined with moored thermistor data and hydrographic data to estimate the evolution of the three-dimensional temperature field $T(x, y, z)$ in the Greenland Sea during winter. (During the convective period, the hydrographic data were found to be contaminated by small-scale variability and were not useful for determining the chimney and gyre-scale structure.) A convective chimney reaching depths of about 1500 m was observed to the south west of the gyre center during March 1989. The chimney had a spatial scale of about 50 km and a timescale of about 10 days (**Figure 7**). The location of the chimney seemed to be sensitively linked to the distribution of the relatively warm, salty Arctic Intermediate Water found at intermediate depths. Potential temperature profiles

extracted from the three-dimensional inverse estimates were averaged over the chimney region to show the time-evolution of the chimney (**Figure 6**). A one-dimensional vertical heat balance adequately described changes in total heat content in the chimney region from autumn 1988 until the time of chimney break-up, when horizontal advection became important and warmer waters moved into the region. The average annual deep-water production rate in the Greenland Sea for 1988–1989 was estimated from the average temperature change over the region occupied by the tomographic array to be about $0.1 \times 10^6 \text{m}^3 \text{ s}^{-1}$.

Barotropic and Baroclinic Tides

Sum and difference travel times from long-range reciprocal transmissions provide precise measurements of the sound-speed (temperature) changes associated with baroclinic (internal) tidal displacements and of barotropic tidal currents, respectively.

The availability of global sea-surface elevation data from satellite altimeter measurements has made possible the development of improved global tidal models. Tomographic measurements of tidal currents made in both the central North Pacific and western North Atlantic Oceans have shown that the harmonic constants for current derived from a recent global tidal model (TPXO.2) are accurate to a fraction of a millimeter per second in amplitude and a few degrees in phase in open ocean regions (**Figure 8**). Small, spatially coherent differences between the modeled and measured harmonic constants are found in the western North Atlantic near complicated topography that is unresolved in the model. These differences are almost certainly due to errors in the TPXO.2 currents. The integrating nature of the tomographic measurements suppresses short-scale internal waves and internal tides, providing tidal current measurements that are substantially more accurate than those derived from current-meter data.

Tomographic measurements of sound-speed fluctuations at tidal frequencies from the same experiments revealed large-scale internal tides that are phase-locked to the barotropic tides. Prior to these measurements it had commonly been assumed that midocean internal tides are not phase-locked to the barotropic tides (except for locally forced internal tides) and have correlation length scales of order only 100–200 km. The measurements in the North Pacific were consistent with a large-scale, phase-locked internal tide that had been generated at the Hawaiian Ridge and then propagated to the tomographic array over 2000 km to the north (**Figure 9**).

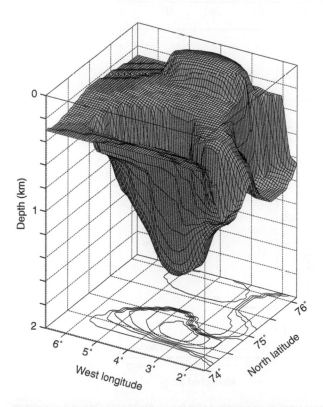

Figure 7 Mixed layer depth in the central Greenland Sea on 19 March 1989, as defined by the minimum depth of the −1.2°C isotherm. The ocean is colder than −1.2°C above this depth as a result of surface cooling and is warmer below. The main chimney reaches a maximum depth of about 1500 m in an area about 50 km in diameter centered on 74.75°N, 3.5°W, south west of the gyre center. A secondary chimney with a maximum depth of about 1000 m is evident to the north east of the gyre center, separated from the primary chimney by a ridge of warmer water. Contours of mixed layer depth are shown below. Adapted from Morawitz WML, Sutton PJ, Worcester PF *et al.* (1996) Three-dimensional observations of a deep convective chimney in the Greenland Sea during winter 1988/1989. *Journal of Physical Oceanography* 26: 2316–2343.

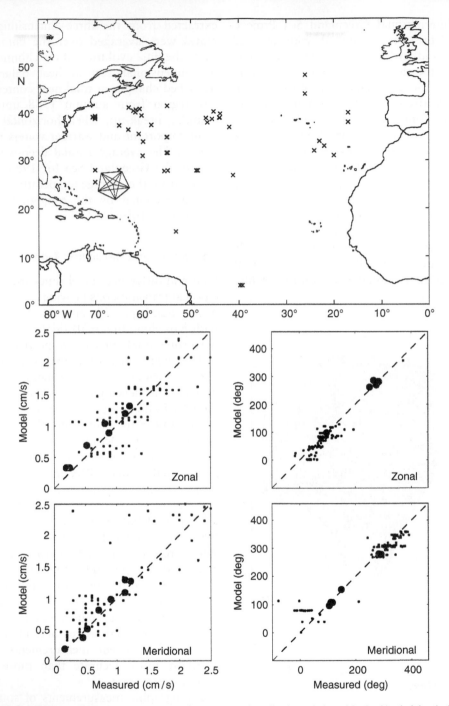

Figure 8 (Bottom) Comparison of the M$_2$ current harmonic constants (amplitude and phase) in the North Atlantic Ocean derived from reciprocal acoustic transmissions (filled circles) and from current meter data (dots) with those predicted by a global tidal model derived from satellite altimeter measurements (TPXO.2). (Top) The acoustic data are from the pentagonal tomographic transceiver array deployed in the western North Atlantic between Puerto Rico and Bermuda during 1991–1992. The current-meter mooring locations are indicated by crosses. Adapted from Dushaw BD, Egbert GD, Worcester PF *et al.* (1997) A TOPEX/POSEIDON global tidal model (TPXO.2) and barotropic tidal currents determined from long-range acoustic transmissions. *Progress in Oceanography* 40: 337–367.

These observations were subsequently confirmed from satellite altimeter data. The measurements in the western North Atlantic revealed a diurnal internal wave resonantly trapped between the shelf just north of Puerto Rico and the turning latitude for the diurnal K$_1$ internal tide, 1100 km distant at 30.0°N (**Figure 10**). In both cases the peak-to-peak temperature variations associated with the maximum displacement of the first baroclinic modes were only about 0.04°C. Once again, the acoustic observations

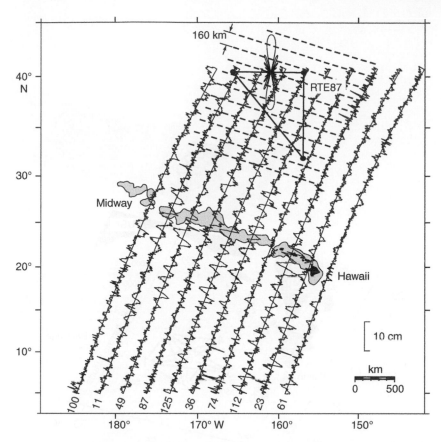

Figure 9 Schematic diagram showing the phase-locked internal tide generated at the Hawaiian Ridge and the triangular tomographic array deployed north of Hawaii during 1987 used to detect it. The dashed lines represent the crests of a wave with 160 km wavelength. Each leg of the tomographic array functions as a linear array with maximum sensitivity to an incident plane wave propagating perpendicular to the leg (i.e., with wave crests aligned parallel to the leg). The beam pattern (in dB) of the 750-km northern leg for a model-1 incident wave with a wavelength of 160 km is indicated. The satellite altimeter data that subsequently confirmed the tomographic observations are also shown. High-pass filtered M_2 surface elevations (cm) are plotted along ten ascending TOPEX/ POSEIDON ground tracks. Adapted from Dushaw BD, Cornuelle BD, Worcester PF, Howe BM and Luther DS (1995) Barotropic and baroclinic tides in the central North Pacific Ocean determined from long-range reciprocal acoustic transmissions. *Journal of Physical Oceanography* 25: 631–647; Ray RD and Mitchum GT (1996) Surface manifestation of internal tides generated near Hawaii. *Geophysical Research Letters* 23: 2101–2104.

of the baroclinic tide average in range and depth, suppressing internal-wave noise and providing enhanced estimates of the deterministic part of the internal-tide signal compared to measurements made at a point, such as by moored thermistors.

Heat Content

Acoustic methods have been used to measure the heat content of the ocean and its variability on basin scales, taking advantage of the integrating nature of acoustic transmissions to rapidly and repeatedly make range- and depth-averaged temperature measurements at ranges out to about 5000 km.

In the Mediterranean Sea, for example, a network of seven tomographic instruments was deployed for nine months during 1994 in the THETIS-2 experiment, including cross-basin

transmissions from Europe to Africa (**Figure 11**). Although it is normally difficult to obtain heat content measurements comparable to those provided by the acoustic data, in this case one of the transmission paths was intentionally aligned with the route of a commercial ship, from which expendable bathythermograph (XBT) measurements were made at two-week intervals. The acoustic average of potential temperature between 0 and 2000 m depth over the 600-km path and the corresponding XBT average between 0 and 800 m depth agreed within the expected uncertainty of 0.03°C (**Figure 11**). Further, the evolution of the three-dimensional heat content of the western Mediterranean estimated from the acoustic data was found to be consistent with the integral of the surface heat flux provided by European Centre for Medium-range Weather Forcasts (ECMWF), after

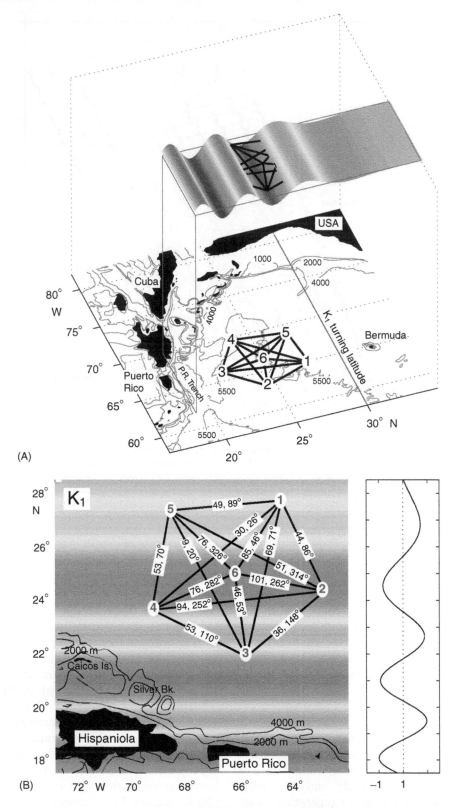

Figure 10 (A) Schematic diagram showing the predicted displacement of the lowest internal mode for the resonant diurnal (K_1) internal tide north of Puerto Rico and the six-element tomographic array deployed during 1991–1992 used to observe it. The tomographic array is about 670 km in diameter. (B) The predicted displacement of the diurnal (K_1) internal tide and the measured harmonic constants (amplitude and phase) for each acoustic path. Adapted from Dushaw BD and Worcester PF (1998) Resonant diurnal internal tides in the North Atlantic. *Geophysical Research Letters* 25: 2189–2192.

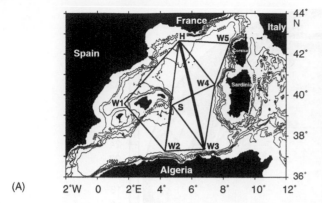

(A)

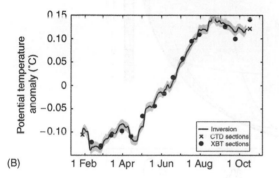

(B)

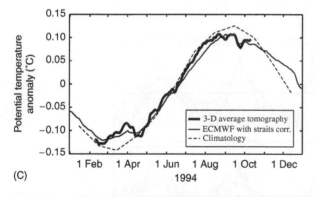

(C)

Figure 11 (A) Geometry of the THETIS-2 experiment in the western Mediterranean Sea, showing the instrument locations and acoustic transmission paths. The transmission path from source H to receiver W3 (heavy solid) coincided with an XBT section occupied every two weeks. (B) Range- and depth-averaged potential temperature (relative to 13.111°C) over 0–2000 m depth and over the 600 km path from source H to receiver W3 derived from the acoustic data, from CTD data, and from XBT data. The shaded band indicates the uncertainty in the temperature estimates derived from the acoustic data. (C) Evolution of the three-dimensional average heat content for the western Mediterranean during 1994 derived from the acoustic data, from the ECMWF surface heat fluxes corrected for heat transport through the Straits of Gibraltar and Sicily, and from climatology. Adapted from Send U, Krahmann G, Mauuary D *et al.* (1997) Acoustic observations of heat content across the Mediterranean Sea. *Nature* 385: 615–617.

correction for the heat flux through the Straits of Gibraltar and Sicily (**Figure 11**). The acoustic data were subsequently combined with satellite altimeter data and an ocean general circulation model to generate a consistent description of the basin-scale temperature and flow fields in the western Mediterranean and their evolution over time. Acoustic and altimetric data are complementary for this purpose, with the acoustic data providing information on the ocean interior with moderate vertical resolution and the altimetric data providing detailed horizontal coverage of the ocean surface.

Similar measurements have been made in the Arctic Ocean in the Transarctic Acoustic Propagation (TAP) experiment during 1994 and in the Arctic Climate Observations using Underwater Sound (ACOUS) project beginning in 1999. During the TAP experiment, ultralow-frequency (19.6 Hz) acoustic transmissions propagated across the entire Arctic basin from a source located north of Svalbard to a receiving array located in the Beaufort Sea at a range of about 2630 km. Modal travel time measurements yielded the surprising result that the Atlantic Intermediate Water layer was about 0.4°C warmer than expected from historical data. This result was subsequently confirmed by direct measurements made from icebreakers and submarines. Acoustic data collected on a similar path during April 1999 as part of the ACOUS project indicated further warming of about 0.5°C, which was again confirmed by direct measurements made from submarines. Acoustic methods can provide the long-term, continuous observations in ice-covered regions that are difficult to obtain using other approaches.

Finally, measurements of basin-scale heat content in the Northeast Pacific were made intermittently from 1983 through 1989 using transmissions from an acoustic source located near Kaneohe, Hawaii, and more recently from 1996 through 1999 during the Acoustic Thermometry of Ocean Climate (ATOC) project using sources located off central California and north of Kauai, Hawaii. Data from the ATOC project have shown that ray travel times may be used for acoustic thermometry at least out to ranges of about 5000 km. The estimated uncertainty in range- and depth-averaged temperature estimates made from the acoustic data at these ranges is only about 0.01 °C. Comparisons between sea-surface height measurements made with a satellite altimeter and sea-surface height estimates derived using the range-averaged temperatures computed from the acoustic data indicate that thermal expansion alone

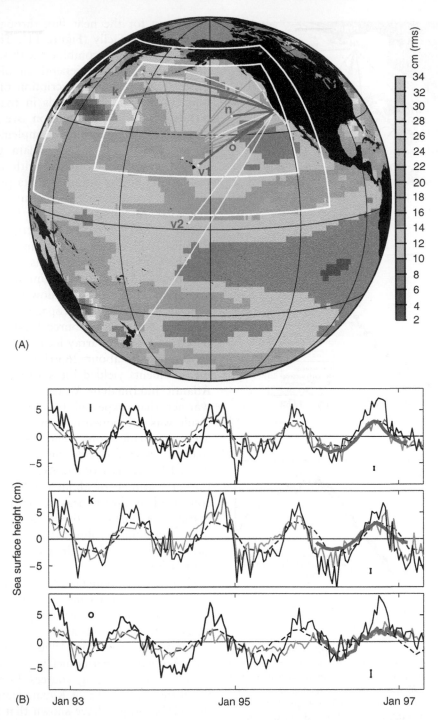

Figure 12 (A) The ATOC acoustic array is superimposed on a map of the root-mean-square (rms) sea level anomaly from altimetric measurements. Transmission paths from sources off central California and north of Kauai to a variety of receivers are shown. (B) The range-averaged sea-surface height anomaly along several of the acoustic sections from satellite altimeter data (solid black), inferred from the acoustic data (solid red), computed from climatological temperature fluctuations (dashed), and derived from an ocean general circulation model (solid blue). Uncertainties are indicated for the acoustic estimates. Adapted from ATOC Consortium (1998) Ocean climate change: comparison of acoustic tomography, satellite altimetry, and modeling. *Science* 281: 1327–1332.

is inadequate to account for all of the observed changes in sea level (**Figure 12**). Analysis of the results obtained when the acoustic and altimetric data were used to constrain an ocean general circulation model indicates that the differences result largely from a barotropic redistribution of mass, with variable salt anomalies a contributing, but smaller, factor.

Appendix: Conversion from Sound Speed to Temperature

Tomgraphic methods fundamentally provide information on the oceanic sound-speed and water-velocity fields. For most oceanographic purposes, however, temperature T and salinity Sa are of more interest than sound speed. Although sound speed C is a function of both T and Sa (as well as pressure), temperature perturbations are normally by far the most important contributor to sound-speed perturbations.

A simple nine-term equation for sound speed due to Mackenzie is

$$
\begin{aligned}
C(T,Sa,D) = {} & 1448.96 + 4.591T - 5.304 \times 10^{-2}T^2 \\
& + 2.374 \times 10^{-4}T^3 + 1.340(Sa - 35) \\
& + 1.630 \times 10^{-2}D + 1.675 \times 10^{-7}D^2 \\
& - 1.025 \times 10^{-2}T(Sa - 35) \\
& - 7.139 \times 10^{-13}TD^3 \qquad [37]
\end{aligned}
$$

where C is in ms^{-1}, T is in degrees Celsius, Sa is in parts per thousand (ppt), and D is the depth (positive down) in meters. Differentiating,

$$
\begin{aligned}
\frac{\partial C}{\partial T} = {} & 4.59 - 0.106T + 7.12 \times 10^{-4}T^2 \\
& -1.03 \times 10^{-2}(Sa - 35) \qquad [38]
\end{aligned}
$$

$$
\frac{\partial C}{\partial Sa} = 1.34 - 1.03 \times 10^{-2}T \qquad [39]
$$

where a slight depth dependence has been dropped. The derivative $\partial T/\partial C = 1/(\partial C/\partial T)$ varies significantly with temperature (**Figure 13**).

To first order, the fractional change in sound speed is then

$$
\Delta C/C = \alpha \Delta T + \beta \Delta Sa \qquad [40]
$$

where

$$
\alpha = \frac{1}{C}\frac{\partial C}{\partial T} \approx 2.4 \times 10^{-3}(^{\circ}C)^{-1} \qquad [41]
$$

$$
\beta = \frac{1}{C}\frac{\partial C}{\partial Sa} \approx 0.8 \times 10^{-3}(\mathrm{ppt})^{-1} \qquad [42]
$$

at 10°C. For a locally linear temperature–salinity relation,

$$
Sa = Sa(T(-)) + \mu \Delta T \qquad [43]
$$

where $T(-)$ is the reference temperature profile corresponding to the reference sound-speed profile $C(-)$ and $\Delta T = T - T(-)$. The fractional change in

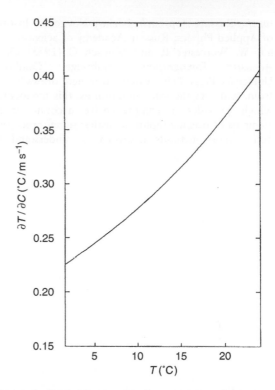

Figure 13 The derivative $\partial T/\partial C$ as a function of temperature. Adapted from Dushaw BD, Worcester PF, Cornuelle BD and Howe BM (1993) Variability of heat content in the central North-Pacific in summer 1987 determined from long-range acoustic transmissions. *Journal of Physical Oceanography* 23: 2650–2666.

sound speed is then

$$
\Delta C/C = \alpha \Delta T(1 + \mu\beta/\alpha) \qquad [44]
$$

In midlatitudes a typical value for μ might be $0.1(\mathrm{ppt})(^{\circ}C)^{-1}$, giving $\mu\beta/\alpha \approx 0.03$. Thus the sound-speed perturbation ΔC depends to first order only on the temperature perturbation ΔT.

The sound-speed perturbation profile $\Delta C(z)$ derived from the acoustic data can be easily converted to the corresponding temperature perturbation profile

$$
\Delta T(z) = \int_{C(-)}^{C(-)+\Delta C} \frac{\partial T}{\partial C} \, \mathrm{d}C, \qquad [45]
$$

where the integral allows for the dependence of $\partial T/\partial C$ on temperature.

See also

Acoustics, Arctic. Acoustics in Marine Sediments.

Further Reading

Khil'ko AI, Caruthers JW, and Sidorovskaia NA (1998) *Ocean Acoustic Tomography: A Review with Emphasis on the Russian Approach*. Nizhny Novgorod: Institute of Applied Physics, Russian Academy of Sciences.

Munk W, Worcester P, and Wunsch C (1995) *Ocean Acoustic Tomography*, Cambridge: Cambridge University Press. (The review given here draws heavily from, and uses the same notation as, this monograph, which provides a comprehensive account of the elements of oceanography, acoustics, signal processing, and inverse methods necessary to understand the application of tomographic methods to studying the ocean.)

Munk W and Wunsch C (1979) Ocean acoustic tomography: a scheme for large scale monitoring. *Deep-Sea Research* 26: 123–161.

Worcester PF (1977) Reciprocal acoustic transmission in a mid-ocean environment. *Journal of the Acoustical Society of America* 62: 895–905.

SATELLITE, AIRCRAFT, OR SHIP-BORNE REMOTE SENSING

AIRCRAFT REMOTE SENSING

L. W. Harding, Jr. and W. D. Miller, University of Maryland, College Park, MD, USA
R. N. Swift, and C. W. Wright, NASA Goddard Space Flight Center, Wallops Island, VA, USA

Introduction

The use of aircraft for remote sensing has steadily grown since the beginnings of aviation in the early twentieth century and today there are many applications in the Earth sciences. A diverse set of remote sensing uses in oceanography developed in parallel with advances in aviation, following increased aircraft capabilities and the development of instrumentation for studying ocean properties. Aircraft improvements include a greatly expanded range of operational altitudes, development of the Global Positioning System (GPS) enabling precision navigation, increased availability of power for instruments, and longer range and duration of missions. Instrumentation developments include new sensor technologies made possible by microelectronics, small, high-speed computers, improved optics, and increased accuracy of digital conversion of electronic signals. Advances in these areas have contributed significantly to the maturation of aircraft remote sensing as an oceanographic tool.

Many different types of aircraft are currently used for remote sensing of the oceans, ranging from balloons to helicopters, and from light, single engine piston-powered airplanes to jets. The data and information collected on these platforms are commonly used to enhance sampling by traditional oceanographic methods, giving increased spatial and temporal resolution for a number of important properties. Contemporary applications of aircraft remote sensing to oceanography can be grouped into several areas, among them ocean color, sea surface temperature (SST), sea surface salinity (SSS), wave properties, near-shore topography, and bathymetry. Prominent examples include thermal mapping using infrared (IR) sensors in both coastal and open ocean studies, lidar and visible radiometers for ocean color measurements of phytoplankton distributions and 'algal blooms', and passive microwave sensors to make observations of surface salinity structure of estuarine plumes in the coastal ocean. These topics will be discussed in more detail in subsequent sections. Other important uses of aircraft remote sensing are to test instruments slated for deployment on satellites, to calibrate and validate space-based sensors using aircraft-borne counterparts, and to make 'under-flights' of satellite instruments and assess the efficacy of atmospheric corrections applied to data from space-based observations.

Aircraft have some advantages over satellites for oceanography, including the ability to gather data under cloud cover, high spatial resolution, flexibility of operations that enables rapid responses to 'events', and less influence of atmospheric effects that complicate the processing of satellite data. Aircraft remote sensing provides nearly synoptic data and information on important oceanographic properties at higher spatial resolution than can be achieved by most satellite-borne instruments. Perhaps the greatest advantage of aircraft remote sensing is the ability to provide consistent, high-resolution coverage at larger spatial scales and more frequent intervals than are practical with ships, making it feasible to use aircraft for monitoring change.

Disadvantages of aircraft remote sensing include the relatively limited spatial coverage that can be obtained compared with the global coverage available from satellite instruments, the repeated expense of deploying multiple flights, weather restrictions on operations, and lack of synopticity over large scales. Combination of the large-scale, synoptic data that are accessible from space with higher resolution aircraft surveys of specific locations is increasingly recognized as an important and useful marriage that takes advantages of the strengths of both approaches.

This article begins with a discussion of sensors that use lasers (also called active sensors), including airborne laser fluorosensors that have been used to measure chlorophyll (chl-a) and other properties; continues with discussions of lidar sensors used for topographic and bathymetric mapping; describes passive (sensors that do not transmit or illuminate, but view naturally occurring reflections and emissions) ocean color remote sensing directed at quantifying phytoplankton biomass and productivity; moves to available or planned hyper-spectral aircraft instruments; briefly describes synthetic aperture radar applications for waves and wind, and closes with a discussion of passive microwave measurements of salinity. Readers are directed to the Further Reading section if they desire additional information on individual topics.

Active Systems

Airborne Laser Fluorosensing

The concentrations of certain waterborne constituents, such as chl-a, can be measured from their fluorescence, a relationship that is exploited in shipboard sensors such as standard fluorometers and flow cytometers that are discussed elsewhere in this encyclopedia. NASA first demonstrated the measurement of laser-induced chl-a fluorescence from a low-flying aircraft in the mid-1970s. Airborne laser fluorosensors were developed shortly thereafter in the USA, Canada, Germany, Italy, and Russia, and used for measuring laser-induced fluorescence of a number of marine constituents in addition to chl-a. Oceanic constituents amenable to laser fluorosensing include phycoerythrin (photosynthetic pigment in some phytoplankton taxa), chromophoric dissolved organic matter (CDOM), and oil films. Airborne laser fluorosensors have also been used to follow dyes such as fluorescein and rhodamine that are introduced into water masses to trace their movement.

The NASA Airborne Oceanographic Lidar (AOL) is the most advanced airborne laser fluorosensor. The transmitter portion features a dichroic optical device to spatially separate the temporally concurrent 355 and 532 nm pulsed-laser radiation, followed by individual steering mirrors to direct the separated beams to respective oceanic targets separated by ~1 m when flown at the AOL's nominal 150 m operational altitude. The receiver focal plane containing both laser-illuminated targets is focused onto the input slits of the monochromator. The monochromator output focal planes are viewed by custom-made optical fibers that transport signal photons from the focal planes to the photo-cathode of each photo-multiplier module (PMM) where the conversion from photons to electrons takes place with a substantial gain. Time-resolved waveforms are collected in channels centered at 404 nm (water Raman) and 450 nm (CDOM) from 355 nm laser excitation, and at 560 nm and 590 nm (phycoerythrin), 650 (water Raman), and 685 nm (chl-a) from the 532 nm laser excitation. The water Raman from the respective lasers is the red-shifted emission from the OH^- bonds of water molecules resulting from radiation with the laser pulse. The strength of the water Raman signal is directly proportional to the number of OH^- molecules accessed by the laser pulse. Thus, the water Raman signal is used to normalize the fluorescence signals to correct for variations in water attenuation properties in the surface layer of the ocean. The AOL is described in more detail on http:// lidar.wff.nasa.gov.

The AOL has supported major oceanographic studies throughout the 1980s and 1990s, extending the usefulness of shipboard measurements over wide areas to permit improved interpretation of the ship-derived results. Examples of data from the AOL show horizontal structure of laser-induced fluorescence converted to chl-a concentration (**Figure 1**). Prominent oceanographic expeditions that have benefited from aircraft coverage with the AOL include the North Atlantic Bloom Experiment (NABE) of the Joint Global Ocean Flux Study (JGOFS), and the Iron Enrichment Experiment (IRONEX) of the Equatorial Pacific near the Galapagos Islands. This system has been flown on NASA P-3B aircraft in open-ocean missions that often exceeded 6 h in duration and collected hundreds of thousands of spectra. Each 'experiment' is able to generate both active and passive data in 'pairs' that are used for determining ocean color and recovering chl-a and other constituents, and that are also useful in the development of algorithms for measuring these constituents from oceanic radiance spectra.

Airborne Lidar Coastal Mapping

The use of airborne lidar (light detection and ranging) sensors for meeting coastal mapping requirements is a relatively new and promising application of laser-ranging technology. These applications include high-density surveying of coastline and beach morphology and shallow water bathymetry. The capability to measure distance accurately with lidar sensors has been available since the early 1970s, but their application to airborne surveying of terrestrial features was seriously hampered by the lack of knowledge of the position of the aircraft from which the measurement was made. The implementation of the Department of Defense GPS constellation of satellites in the late 1980s, coupled with the development of GPS receiver technology, has resulted in the capability to provide the position of a GPS antenna located on an aircraft fuselage in flight to an accuracy approaching 5 cm using kinematic differential methodology. These methods involve the use of a fixed receiver (generally located at the staging airport) and a mobile receiver that is fixed to the aircraft fuselage. The distance between mobile and fixed receivers, referred to as the baseline, is typically on the order of tens of kilometers, and can be extended to hundreds of kilometers by using dual frequency survey grade GPS receivers aided by tracking the phase code of the carrier from each frequency.

Modern airborne lidars are capable of acquiring 5000 or more discrete range measurements per second. Aircraft attitude and heading information are

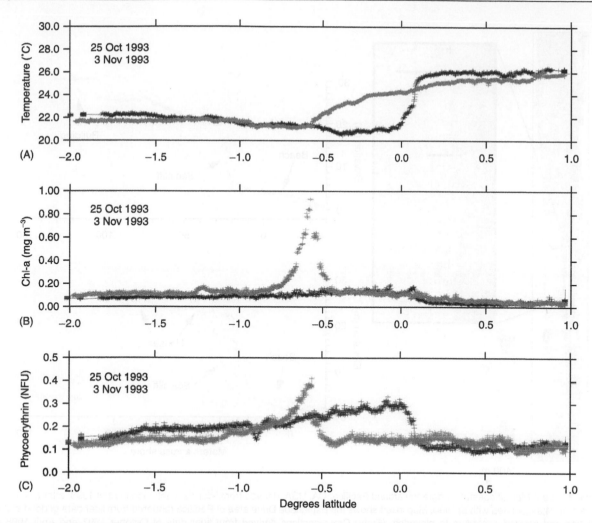

Figure 1 Cross-section profiles flown across a large oceanic front west of the Galapagos Islands on 25 October and 3 November 1993. (A) A dramatic horizontal displacement of the front as measured with an infrared radiometer; (B) and (C) show corresponding changes in laser-induced fluorescence of chl-a and phycoerythrin. This flight was made as part of the original IRONEX investigation in late 1993. NFU, normalized fluorescence units.

used along with the GPS-determined platform position to locate the position of the laser pulse on the Earth's surface to a vertical accuracy approaching 10 cm with some highly accurate systems and <30 cm for most of these sensors. The horizontal accuracy is generally 50–100 cm. Depending on the pulse repetition rate of the laser transmitter, the off-nadir pointing angle, and the speed of the aircraft platform, the density of survey points can exceed one sample per square meter.

NASA's Airborne Topographic Mapper (ATM) is an example of a topographic mapping lidar used for coastal surveying applications. An example of data from ATM shows shoreline features off the east coast of the USA (**Figure 2**). ATM was originally developed to measure changes in the elevation of Arctic ice sheets in response to global warming. The sensor was applied to measurements of changes in coastal morphology beginning in 1995. Presently, baseline

topographic surveys exist for most of the Atlantic and Gulf coasts between central Maine and Texas and for large sections of the Pacific coast. Affected sections of coastline are re-occupied following major coastal storms, such as hurricanes and 'Nor'easters', to determine the extent of erosion and depositional patterns resulting from the storms. Additional details on the ATM and some results of investigations on coastal morphology can be found on websites (http://lidar.wff.nasa.gov and http://aol.wff.nasa.gov/aoltm/projects/beachmap/98results/).

Other airborne lidar systems have been used to survey coastal morphology, including Optec lidar systems by Florida State University and the University of Texas at Austin. Beyond these airborne lidar sensors, there are considerably more instruments with this capability that are currently in use in the commercial sector for surveying metropolitan areas, flood plains, and for other terrestrial

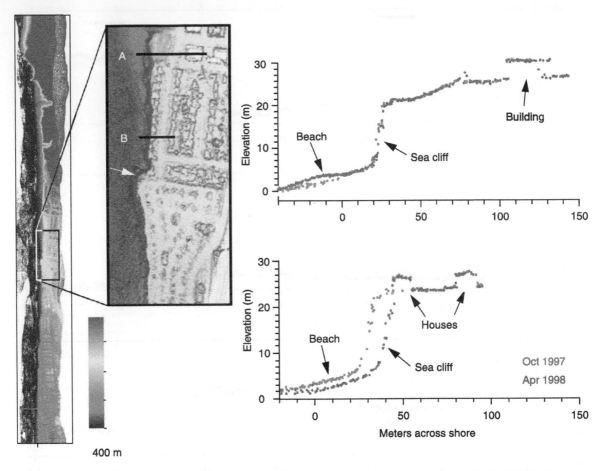

400 m

Figure 2 (Left) Map of coastal topography around Pacifica, CA, USA, derived from lidar data obtained in April 1998, after a winter of severe storms associated with El Niño. Map insert shows the Esplanade Drive area of Pacifica rendered from lidar data gridded at 2 m resolution and colored according to elevation. (Right) Cross-sections derived from lidar data of October 1997 and April 1998 at locations marked in the map inset. The profiles in (A) show a stable cliff and accreting beach, whereas about 200 m to the south the profiles in (B) show erosion of the sea cliff and adjacent beach resulting in undermining of houses. Each profile shows individual laser spot elevations that fall within a 2 m wide strip oriented approximately normal to shoreline. (Reproduced with permission from Sallenger *et al.*, 1999.)

applications. At the last count (early 2000) there were approximately 60 airborne lidars in operation worldwide, with most engaged in a variety of survey applications generally outside the field of coastal mapping.

Pump and Probe Fluorometry

Several sections in this chapter describe recoveries of phytoplankton biomass as chl-a by active and passive measurements. Another recent accomplishment is an active, airborne laser measurement intended to aid in remote detection of photosynthetic performance, an important ingredient of primary productivity computations. Fluorometric techniques, such as fast repetition rate (FRR) fluorometry (explained in another article of this encyclopedia), have provided an alternative approach to ^{14}C assimilation and O_2 evolution in the measurement of primary productivity. This

technology has matured with the commercial availability of FRR instruments that can give vertical profiles or operate in a continuous mode while underway. There have been several attempts to develop airborne lidar instruments to determine phytoplankton photosynthetic characteristics from aircraft in the past decade. NASA scientists have deployed a pump and probe fluorometer from aircraft, wherein the AOL laser (described above) acts as the pump and a second laser with variable power options and rapid pulsing capabilities (10 ns) functions as the probe.

Passive Systems

Multichannel Ocean Color Sensor (MOCS)

Passive ocean color measurements using visible radiometers to measure reflected natural sunlight from the ocean have been made with a number of instruments in the past two decades. These instruments include the Multichannel Ocean Color Sensor (MOCS) that was flown in studies of Nantucket Shoals in the early 1980s, the passive sensors of the AOL suite that have been used in many locations around the world, and more recently, simple radiometers that have been deployed on light aircraft in regional studies of Chesapeake Bay (see below). MOCS was one of the earliest ocean color sensors used on aircraft. It provided mesoscale data on shelf and slope chl-a in conjunction with shipboard studies of physical structure, nutrient inputs, and phytoplankton primary productivity.

Ocean Data Acquisition System (ODAS) and SeaWiFS Aircraft Simulator (SAS)

Few aircraft studies have obtained long time-series sufficient to quantify variability and detect secular trends. An example is ocean color measurements made from light aircraft in the Chesapeake Bay region for over a decade, providing data on chl-a and SST from >250 flights. Aircraft over-flights of the Bay using the Ocean Data Acquisition System (ODAS) developed at NASA's Goddard Space Flight Center commenced in 1989. ODAS was a nadir-viewing, line-of-flight, three-band radiometer with spectral coverage in the blue-green region of the visible spectrum (460–520 nm), a narrow $1.5°$ field-of-view, and a 10 Hz sampling rate. The ODAS instrument package included an IR temperature sensor (PRT-5, Pyrometrics, Inc.) for measuring SST. The system was flown for ~ 7 years over Chesapeake Bay on a regular set of tracks to determine chl-a and SST. Over 150 flights were made with ODAS between 1989 and 1996, coordinated with *in situ* observations from a multi-jurisdictional monitoring program and other cruises of opportunity.

ODAS was flown together with the SeaWiFS Aircraft Simulator (SAS II, III, Satlantic, Inc., Halifax, Canada) beginning in 1995 and was retired soon thereafter and replaced with the SAS units. SAS III is a multi-spectral (13-band, 380–865 nm), line-of-flight, nadir viewing, 10 Hz, passive radiometer with a $3.5°$ field-of-view that has the same wavebands as the SeaWiFS satellite instrument, and several additional bands in the visible, near IR, and UV. The SAS systems include an IR temperature sensor (Heimann Instruments, Inc.). Chl-a estimates are obtained using a curvature algorithm applied to water-leaving radiances at wavebands in the blue-green portion of the visible spectrum with validation from concurrent shipboard measurements. Flights are conducted at ~ 50–60 m s^{-1} (100–120 knots), giving an along-track profile with a resolution of 5–6 m averaged to

50 m in processing, and interpolated to 1 km^2 for visualization. Imagery derived from ODAS and SAS flights is available on a web site of the NOAA Chesapeake Bay Office for the main stem of the Bay (http://noaa.chesapeakebay.net), and for two contrasting tributaries, the Choptank and Patuxent Rivers on a web site of the Coastal Intensive Sites Network (CISNet) (http://www.cisnet-choptank.org).

Data from ODAS and SAS have provided detailed information on the timing, position, and magnitude of blooms in Chesapeake Bay, particularly the spring diatom bloom that dominates the annual phytoplankton cycle. This April–May peak of chl-a represents the largest accumulation of phytoplankton biomass in the Bay and is a proximal indicator of over-enrichment by nutrients. Data from SeaWiFS for spring 2000 show the coast-wide chl-a distribution for context, while SAS III data illustrate the high-resolution chl-a maps that are obtained regionally (**Figure** 3). A well-developed spring bloom corresponding to a year of relatively high freshwater flow from the Susquehanna River, the main tributary feeding the estuary, is apparent in the main stem Bay chl-a distribution. Estimates of primary productivity are now being derived from shipboard observations of key variables combined with high-resolution aircraft measurements of chl-a and SST for the Bay.

Hyper-spectral Systems

Airborne Visible/Infrared Imaging Spectrometer (AVIRIS)

The Airborne Visible/Infrared Imaging Spectrometer (AVIRIS) was originally designed in the late 1980s by NASA at the Jet Propulsion Laboratory (JPL) to collect data of high spectral and spatial resolution, anticipating a space-based high-resolution imaging spectrometer (HIRIS) that was planned for launch in the mid-1990s.

Because the sensor was designed to provide data similar to satellite data, flight specifications called for both high altitude and high speed. AVIRIS flies almost exclusively on a NASA ER-2 research aircraft at an altitude of 20 km and an airspeed of 732 km h^{-1}. At this altitude and a $30°$ field of view, the swath width is almost 11 km. The instantaneous field of view is 1 mrad, which creates individual pixels at a resolution of 20 m^2. The sensor samples in a whisk-broom fashion, so a mirror scans back and forth, perpendicular to the line-of-flight, at a rate of 12 times per second to provide continuous spatial coverage. Each pixel is then sent to four separate spectrometers by a fiber optic cable. The spectrometers are arranged so that they each cover a part of the

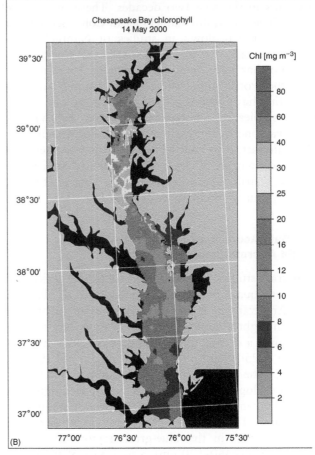

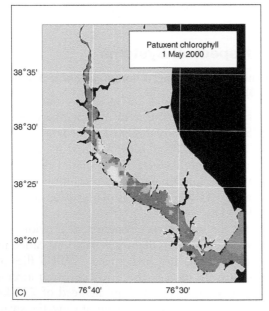

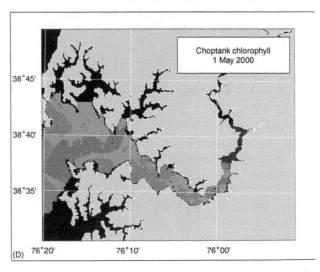

Figure 3   Spring chl-a (mg m^{-3}) in: (A) the mid-Atlantic region from SeaWiFS; (B) Chesapeake Bay; (C) Patuxent R; (D) Choptank R from SAS III.

spectrum from 0.40 to 2.4 μm, providing continuous spectral coverage at 10 nm intervals over the entire spectrum from visible to near IR. Data are recorded to tape cassettes for storage until rectification, atmospheric correction, and processing at JPL. A typical AVIRIS 'scene' is a 40 min flight line. At ER-2 flight parameters, this creates an image roughly 500 km long and 11 km wide. Data are encoded at 12-bits for a high degree of discrimination. The physical dimensions of AVIRIS are quite large, 84 cm wide × 160 cm long × 117 cm tall at a weight of 720 pounds.

Data collected with AVIRIS have been used for terrestrial, marine, and atmospheric applications. Accomplishments of AVIRIS include separation of the chl-a signature from bottom reflectance for clear lake waters of Lake Tahoe and turbid waters near Tampa Bay, interpretation of spectral signals from resuspended sediment and dissolved organic materials in W. Florida, and of suspended sediment and kelp beds in S. California. Recent efforts have focused on improving atmospheric correction procedures for both AVIRIS and satellite data, providing inputs for bio-optical models which determine inherent optical properties (IOPs) from reflectance, algorithm development, and sporadic attempts at water quality monitoring (e.g., chl-a, suspended sediment, diffuse attenuation coefficient, k_d). AVIRIS data have recently been used as an input variable to a neural network model developed to estimate water depth. The model was able to separate the contributions of different components to the total water-leaving radiance and to provide relatively accurate estimates of depth (rms error = 0.48 m).

Compact Airborne Spectrographic Imager (CASI)

The Compact Airborne Spectrographic Imager (CASI) is a relatively small, lightweight hyper-spectral sensor that has been used on a variety of light aircraft. CASI was developed by Itres Research Ltd (Alberta, Canada) in 1988 and was designed for a variety of remote sensing applications in forestry, agriculture, land-use planning, and aquatic monitoring. By allowing user-defined configurations, the 12-bit, push-broom-type sensor (333 scan lines per second) using a charge-coupled detector (CCD) can be adapted to maximize either spatial (37.8° across track field of view, 0.077° along-track, 512 pixels – pixel size varies with altitude) or spectral resolution (288 bands at 1.9 nm intervals between 400 and 1000 nm). Experiments have been conducted using CASI to determine bottom type, benthic cover, submerged aquatic vegetation, marsh type, and in-water constituents such as suspended sediments, chl-a, and other algal pigments.

Portable Hyper-spectral Imager for Low-Light Spectroscopy (PHILLS)

The Portable Hyper-spectral Imager for Low-Light Spectroscopy (PHILLS) has been constructed by the US Navy (Naval Research Laboratory) for imaging the coastal ocean. PHILLS uses a backside-illuminated CCD for high sensitivity, and an all-reflective spectrograph with a convex grating in an Offner configuration to produce a distortion-free image. The instrument benefits from improvements in large-format detector arrays that have enabled increased spectral resolution and higher signal-to-noise ratios for imaging spectrographs, extending the use of this technology in low-albedo coastal waters. The ocean PHILLS operates in a push-broom scanned mode whereby cross-track ground pixels are imaged with a camera lens onto the entrance slit of the spectrometer, and new lines of the along-track ground pixels are attained by aircraft motion. The Navy's interest in hyper-spectral imagers for coastal applications centers on the development of methods for determining shallow water bathymetry, topography, bottom type composition, underwater hazards, and visibility. PHILLS precedes a planned hyper-spectral satellite instrument, the Coastal Ocean Imaging Spectrometer (COIS) that is planned to launch on the Naval Earth Map Observer (NEMO) spacecraft.

Radar Altimetry

Ocean applications of airborne radar altimetry systems include several sensors that retrieve information on wave properties. Two examples are the Radar Ocean Wave Spectrometer (ROWS), and the Scanning Radar Altimeter (SRA), systems designed to measure long-wave directional spectra and near-surface wind speed. ROWS is a K_u-band system developed at NASA's Goddard Space Flight Center in support of present and future satellite radar missions. Data obtained from ROWS in a spectrometer mode are used to derive two-dimensional ocean spectral wave estimates and directional radar backscatter. Data from the pulse-limited altimeter mode radar yield estimates of significant wave height and surface wind speed.

Synthetic Aperture Radar (SAR)

Synthetic Aperture Radar (SAR) systems emit microwave radiation in several bands and collect the reflected radiation to gain information about sea surface conditions. Synthetic aperture is a technique that is used to synthesize a long antenna by combining signals, or echoes, received by the radar as it moves along a flight track. Aperture refers to the opening that is used to collect reflected energy and form an image. The analogous feature of a camera to the aperture would be the shutter opening. A synthetic aperture is constructed by moving a real aperture or antenna through a series of positions along a flight track.

NASA's Jet Propulsion Laboratory and Ames Research Center have operated the Airborne SAR (AIRSAR) on a DC-8 since the late 1980s. The radar of AIRSAR illuminates the ocean at three microwave wavelengths: C-band (6 cm), L-band (24 cm), and P-band (68 cm). Brightness of the ocean (the amount of energy reflected back to the antenna) depends on the roughness of the surface at the length scale of the microwave (Bragg scattering). The primary source of roughness, and hence brightness, at the wavelengths used is capillary waves associated with wind. Oceanographic applications derive from the responsiveness of capillary wave amplitude to factors that affect surface tension, such as swell, atmospheric stability, and the presence of biological films. For example, the backscatter characteristics of the ocean are affected by surface oil and slicks can be observed in SAR imagery as a decrease of radar backscatter; SAR imagery appears dark in an area affected by an oil spill, surface slick, or biofilm, as compared with areas without these constituents.

Microwave Salinometers

Passive microwave radiometry (L-band) has been tested for the recovery of SSS from aircraft, and it may be possible to make these measurements from space. Salinity affects the natural emission of EM radiation from the ocean, and the microwave signature can be used to quantify SSS. Two examples of aircraft instruments that have been used to measure SSS in the coastal ocean are the Scanning Low-Frequency

Microwave Radiometer (SLFMR), and the Electronically Thinned Array Radiometer (ESTAR).

SLFMR was used recently in the estuarine plume of Chesapeake Bay on the east coast of the USA to follow the buoyant outflow that dominates the nearshore density structure and constitutes an important tracer of water mass movement. SLFMR is able to recover SSS at an accuracy of about 1 PSU (**Figure 4**). This resolution is too coarse for the open ocean, but is quite suitable for coastal applications where significant gradients occur in regions influenced by freshwater inputs. SLFMR has a bandwidth of 25 MHz, a frequency of 1.413 GHz, and a single antenna with a beam width of approximately 16° and six across-track positions at ±6°, ±22° and ±39°. Tests of SLFMR off the Chesapeake Bay demonstrated its effectiveness as a 'salinity mapper' by characterizing the trajectory of the Bay plume from surveys using light aircraft in joint operations with ships. Flights were conducted at an altitude of 2.6 km, giving a resolution of about 1 km. The accuracy of SSS in this example is ~0.5 PSU.

ESTAR is an aircraft instrument that is the prototype of a proposed space instrument for measuring SSS. This instrument relies on an interferometric technique termed 'aperture synthesis' in the across-track dimension that can reduce the size of the antenna aperture needed to monitor SSS from space. It has been described as a 'hybrid of a real and a synthetic aperture radiometer.' Aircraft surveys of

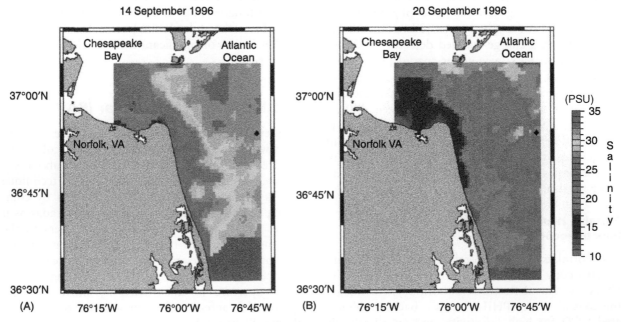

Figure 4 Sea surface salinity from an airborne microwave salinity instrument for (A) 14 September 1996; (B) 20 September 1996. Images reveal strong onshore-offshore gradients in salinity from the mouth of Chesapeake Bay to the plume and shelf, and the effect of high rainfall and freshwater input on the salinity distribution over a 1-week interval. (Adapted with permission from Miller *et al.*, 1998.)

SSS using ESTAR in the coastal current off Maryland and Delaware showed good agreement with thermosalinograph measurements from ships in the range of 29–31 PSU.

See also

Bio-Optical Models. Fluorometry for Biological Sensing. Inherent Optical Properties and Irradiance. IR Radiometers. Optical Particle Characterization. Satellite Oceanography, History and Introductory Concepts. Satellite Remote Sensing of Sea Surface Temperatures. Satellite Remote Sensing SAR.

Further Reading

Blume H-JC, Kendall BM, and Fedors JC (1978) Measurement of ocean temperature and salinity via microwave radiometry. *Boundary Layer Meteorology* 13: 295–308.

Campbell JW and Esaias WE (1985) Spatial patterns in temperature and chlorophyll on Nantucket Shoals from airborne remote sensing data, May 7–9, 1981. *Journal of Marine Research* 43: 139–161.

Harding LW Jr, Itsweire EC, and Esaias WE (1994) Estimates of phytoplankton biomass in the Chesapeake Bay from aircraft remote sensing of chlorophyll concentrations, 1989–92. *Remote Sensing Environment* 49: 41–56.

Le Vine DM, Zaitzeff JB, D'Sa EJ, *et al.* (2000) Sea surface salinity: toward an operational remote-sensing system. In: Halpern D (ed.) *Satellites, Oceanography and Society*, pp. 321–335. Elsevier Science.

Miller JL, Goodberlet MA, and Zaitzeff JB (1998) Airborne salinity mapper makes debut in coastal zone. *EOS Transactions of the American Geophysical Union* 79: 173–177.

Sallenger AH Jr, Krabill W, Brock J, *et al.* (1999) *EOS Transactions of the American Geophysical Union* 80: 89–93.

Sandidge JC and Holyer RJ (1998) Coastal bathymetry from hyper-spectral observations of water radiance. *Remote Sensing Environment* 65: 341–352.

IR RADIOMETERS

C. J. Donlon, Space Applications Institute, Ispra, Italy

Introduction

Measurements of sea surface temperature (SST) are most important for the investigation of the processes underlying heat and gas exchange across the air–sea interface, the surface energy balance, and the general circulation of both the atmosphere and the oceans. Complementing traditional subsurface contact temperature measurements, there is a wide variety of infrared radiometers, spectroradiometers, and thermal imaging systems that can be used to determine the SST by measuring thermal emissions from the sea surface. However, the SST determined from thermal emission can be significantly different from the subsurface temperature ($> \pm 1\,\mathrm{K}$) because the heat flux passing through the air–sea interface typically results in a strong temperature gradient. Radiometer systems deployed on satellite platforms provide daily global maps of SSST (sea surface temperature) at high spatial resolution ($\sim 1\,\mathrm{km}$) whereas those deployed from ships and aircraft provide data at small spatial scales of centimeters to meters. In particular, the development of satellite radiometer systems providing a truly synoptic view of surface ocean thermal features has been pivotal in the description and understanding of the global oceans.

This article reviews the infrared properties of water and some of the instruments developed to measure thermal emission from the sea surface. It focuses on *in situ* radiometers although the general principles described are applicable to satellite sensors treated elsewhere in this volume.

Infrared Measurement Theory

Infrared (IR) radiation is heat energy that is emitted from all objects that have a temperature above 0 K ($-273.16°C$). It includes all wavelengths of the electromagnetic spectrum between $0.75\,\mu\mathrm{m}$ and $\sim 100\,\mu\mathrm{m}$ (**Figure 1**) and has the same optical properties as visible light, being capable of reflection, refraction, and forming interference patterns.

The following total quantities, conventional symbols and units provide the theoretical foundation for the measurement of IR radiation and are schematically shown in **Figure 2**. Spectral quantities can be represented by restricting each to a specific waveband.

- Radiant energy Q, is the total energy radiated from a point source in all directions in units of joules (J).
- Radiant flux $\phi = \mathrm{d}Q/\mathrm{d}t$ is the flux of all energy radiated in all directions from a point source in units of watts (W).
- Emittance $M = \mathrm{d}\phi/\mathrm{d}A$ is the radiant flux density from a surface area A in units of $\mathrm{W\,m^{-2}}$. This is an integrated flux (i.e., independent of direction) and will therefore vary with orientation relative to a nonuniform source.
- Radiant intensity $I = \mathrm{d}\phi/\mathrm{d}\omega$ is the radiant flux of a point source per solid angle ω (steradian, sr) and is a directional flux in units of $\mathrm{W\,sr^{-1}}$.
- Radiance $L = \mathrm{d}I/\mathrm{d}(A\cos\theta)$ is the radiant intensity of an extended source per unit solid angle in a given direction θ, per unit area of the source projected in the same θ. It has units of $\mathrm{W\,sr^{-1}\,m^{-2}}$.

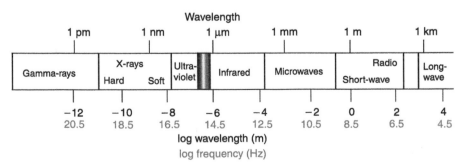

Figure 1 Schematic diagram of the electromagnetic spectrum showing the location and interval of the infrared waveband.

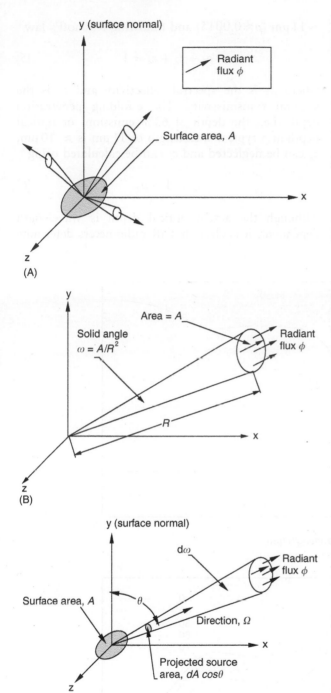

(A)

(B)

(C)

Figure 2 Schematic definition of (A) Emittance, E; (B) radiant intensity, I; (C) radiance, L.

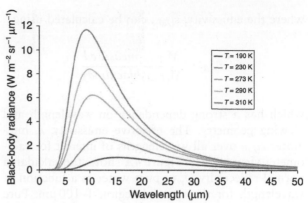

Figure 3 Black-body radiance as a function of wavelength computed using eqns [1] and [2] for different target temperatures.

Planck's law describes the emittance of a perfectly emitting surface (or black body) at a temperature T in Kelvin. It is the radiant flux (ϕ) per unit bandwidth centered at wavelength λ leaving a unit area of surface in any direction in units of W m^{-2} m^{-1}.

$$M_{\lambda,T} = \frac{2\pi h c^2}{\lambda^5 (e^{hc/\lambda kT} - 1)} \quad [1]$$

where $h = 6.626 \times 10^{-34}$ Js is Planck's constant, $c = 2.998 \times 10^8$ m s^{-1} is the speed of light and $k = 1.381 \times 10^{-23}$ JK^{-1} is Boltzmann's constant. The sea surface is considered a Lambertian source (i.e., uniform radiance in all directions) so that the spectral radiance L_λ is related to M_λ by

$$L_\lambda = \frac{M_\lambda}{\pi} \quad [2]$$

Figure 3 shows L_λ computed for several temperatures as a function of wavelength. Considering temperatures of 273–310 K as representative of the global ocean, maximum emission occurs at a wavelength of 9.3–10.7 μm. Atmospheric attenuation is minimal at ~3.5 μm, 9.0 μm and 11.0 μm that are the spectral intervals often termed atmospheric 'windows'. Instruments operating within these intervals are optimal for sea surface measurements – especially in the case of satellite deployment where atmospheric attenuation can be significant. In the 3–5 μm spectral region. L_λ is a strong function of temperature (**Figure 3**) highlighting the possibility to increase in radiometer sensitivity by utilizing this spectral interval.

By measuring L_λ using eqn [2] and inverting eqn [1], the spectral brightness temperature, $B_{(T,\lambda)}$, rather than the temperature is determined because these equations assume that sea water is a perfect emitter or black body. In practice, the sea surface does not behave as a black body (it is slightly reflective in the infrared) and therefore its spectral and geometric properties need to be considered. The emittance of a perfect emitter at the actual temperature T, wavelength λ, and view angle θ is given by

$$M_{(T,\lambda,\theta)} = \frac{\pi L_{(T,\lambda,\theta)}}{\varepsilon_{(\lambda,\theta)}} \quad [3]$$

where the emissivity, $\varepsilon_{(\lambda,\theta)}$, can be calculated using

$$\varepsilon_{(\lambda,\theta)} = \frac{M_{(T,\lambda,\theta)} measured}{M_{(T,\lambda,\theta)} blackbody} \qquad [4]$$

which has a strong dependence on wavelength and viewing geometry. The effective emissivity, ε, integrates $\varepsilon_{(\lambda,\theta)}$ over all wavelengths of interest for radiometer view angle θ. **Figure 4A** shows the calculated normal reflectivity, ρ, of pure water as a function of wavelength for the spectral region 1–100 µm. Pure water differs only slightly from sea water in this context. Note that ρ is minimal at a wavelength of

~11 µm ($\rho \approx 0.0015$) and following Kirchoff's law

$$\rho_\lambda + \tau_\lambda + \varepsilon_\lambda = 1 \qquad [5]$$

where ρ_λ is the spectral reflectivity and τ_λ is the spectral transmissivity. The e-folding penetration depth (i.e., the depth of 63% emission) or optical depth at a typical wavelength of 11 µm is $\approx$ 10 µm, τ_λ can be neglected and ε_λ can be calculated using

$$\varepsilon_\lambda = 1 - \rho_\lambda \qquad [6]$$

Although the actual optical depth is wavelength dependent, it is clear that IR radiometers determine

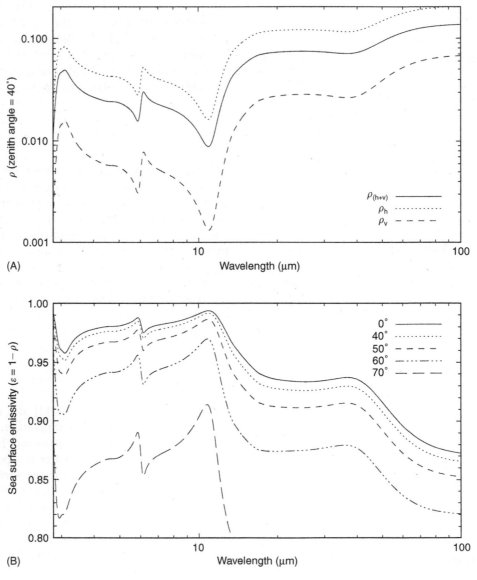

Figure 4 (A) The normal reflectivity, ρ, of the pure water as a function of wavelength (full line). Also shown are the horizontal polarized (ρ_h) and the vertically polarized (ρ_v) components of ρ. (B) The spectral emissivity of pure water as a function of viewing zenith angle. h, height above sea surface.

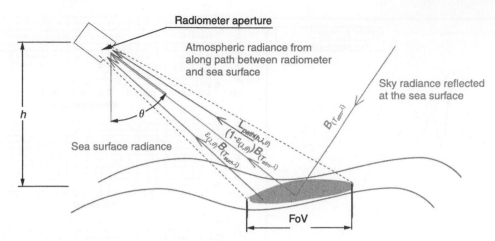

Figure 5 Schematic diagram showing the radiance components measured by an IR radiometer viewing the sea surface. Fov, Field of view.

the temperature of a very thin 'skin' layer of the ocean. This temperature is termed the sea surface skin temperature (SSST) and is distinct (although related) to the subsurface SST. Note that this is in contrast to the situation for short-wave solar radiation (having wavelengths of ~ 0.4–$0.7\,\mu m$) which, for clear water, penetrates to a depth of $\sim 100\,m$.

Figure 4B shows the ε_λ of pure water computed from eqn [6] as a function of both viewing zenith angle and spectral wavelength. Inspection of **Figure 4** reveals that the best radiant temperature measurement will be made when viewing a calm sea surface at an angle of 0–40° from nadir.

An IR radiometer is an optical instrument designed to measure L_λ entering an instrument aperture (**Figure 5**). The radiance measured by radiometer, $L_{(T,\lambda,\theta)}$, having a spectral bandwidth λ, viewing the sea surface at a zenith angle θ and temperature T is given by:

$$L_{(T,\lambda,\theta}=\int_0^\alpha \xi_\lambda[\varepsilon_{(\lambda,\theta)}B\left(T_{surf},\lambda\right)+\left(1-\varepsilon_{(\lambda,\theta)}\right)B(T_{atm},\lambda)$$
$$+L_{path(h,\lambda,\theta)}]d\lambda$$

[7]

where ξ_λ is the spectral response of the radiometer, $B(T_{surf},\lambda$ and $B(T_{atm},\lambda$ are the Planck function for surface temperature T_{surf}, and atmospheric temperature T_{atm}, and $L_{path(h,\lambda,\theta)}$ is the radiance emitted by the atmosphere between the radiometer at height h above the sea surface reflected into the radiometer field-of-view (FoV) at the sea surface. Note that the horizontal and vertical polarization components of reflectivity shown in **Figure 4** are unequal. It is important to consider the polarization of surface reflectance when making measurements of the sea

surface because diffuse downwelling sky radiance measured by a radiometer after reflection at the sea surface is polarized.

IR Radiometer Design

There are four fundamental components to all IR radiometer instruments described below.

Detector and Electronics System to Measure Radiance and Control the Radiometer

A detector system provides an output proportional to the target radiance incident on the detector. There are two main types of detector: thermal detectors that respond to direct heating and quantum detectors that respond to a photon flux. In general, thermal detectors have a response that is weakly dependent on wavelength and can be operated at ambient temperatures whereas rapid response quantum detectors require cooling and are wavelength dependent.

Fore-optics System to Filter, Direct and Focus Radiance

All optical components have an impact on radiometer reliability and accuracy. Mirrors should be free of aberration to minimize unwanted stray radiance reaching the detector. Several materials have good reflection characteristics in the IR including, gold, polished aluminum, and cadmium. Care should be exercised when choosing an appropriate mirror substrate and reflection coating to avoid decay in the marine atmosphere. A glass substrate having a 'hard' scratch-resistant polished gold surface provides >98% reflectance and good environmental wear.

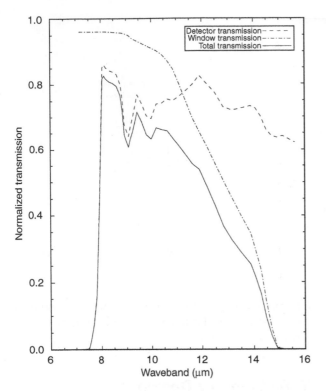

Figure 6 Normalized spectral transmission of an IR window and detector shown together with the combined total response.

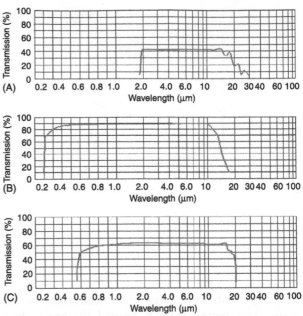

Figure 7 Spectral transmission for common IR window materials. (A) Germanium, (B) sodium chloride, (C) zinc selenide.

Spectral filter windows and lenses require spectral properties that, together with the detector characteristics, define the overall spectral characteristics of a radiometer. **Figure 6** shows the combined spectral response for a broadband radiometer together with the component spectral response of the window and detector.

Environmental System to Protect and Thermally Stabilize the Radiometer

For any optical instrument intended for use in the harsh marine environment, adequate environmental protection is critical. Rain, sea-water spray, and high humidity can destroy a poorly protected instrument rapidly and components such as electrical connections and fore-optics should be resistant to these effects. *In situ* radiometer windows are particularly important in this context. They should not significantly reduce the incoming signal or render it noisy, and be strong enough to resist mechanical, thermal and chemical degradation. **Figure 7** describes several common materials. Germanium (Ge) windows have good transmission characteristics but are very brittle. Sodium chloride (NaCl) is a low-cost, low-absorption material but is of little use in the marine environment because it is water-soluble. Zinc selenide (ZnSe) has high transmission, is nonhygroscopic

and resistant to thermal shock but is soft and requires a protective 'hard' surface finish coating.

Certain window materials achieve better performance when antireflection (AR) coatings are used to minimize reflection from the window. For example, when an AR coating is used on a ZnSe window the transmission increases from ~70% to ~90%. Other coatings provide windows that polarize the incident radiance signal such as optically thin interference coatings and wire grid diffraction polarizers.

Deposition of marine NaCl on all optical components (especially calibration targets) presents an unavoidable problem. Although NaCl itself has good infrared transmission properties (**Figure 7**), contaminated surfaces may become decoupled from temperature sensors and the noise introduced by the optical system will increase. Finally, adequate thermal control using reflective paint together with substantial instrument mass is required so that instruments are not sensitive to thermal shock. In higher latitudes, it may be necessary to provide an extensive antifreeze capability.

Calibration System to Quantify the Radiometer Output

The role of a calibration system is to quantify the instrument output in terms of the measured radiance incident on the detector. Calibration techniques are specific to the particular design of radiometer and vary considerably from simple bias corrections to systems providing automatic precision two-point blackbody calibrations. Proper calibration accounts for the following primary sources of error:

- the effect of fore-optics;
- unavoidable drifts in detector gain and bias;
- long-term degradation of components.

Finally, careful radiometer design and configuration can avoid many measurement errors. Examples include: poor focusing and optical alignment; filters having transmission above or below the stated spectral bandwidth (termed 'leaks'); inadequate protection against thermal shock, stability, and reliability of the calibration system; poor electronics and the general decay of optomechanical components.

Application of IR Radiometers

There are many different radiometer designs and deployment strategies ranging from simple single-channel hand-held devices to complex spectro-radiometers. Instrument accuracy, sensitivity and stability depends on both the deployment scenario and radiometer design. Accordingly, the following sections describe several different examples.

Broadband Pyrgeometers

A pyrgeometer is an integrating hemispheric radiometer which, by definition, measures the band-limited spectral emittance, E, so that the angle dependency in eqn [7] is redundant. They are used to determine the long-wave heat flux at the air–sea interface by measuring the difference between atmospheric and sea surface radiance either using two individual sensors (**Figure 8A**) or as a single combined sensor (**Figure 8B**). A thermopile detector is often used which is a collection of thermocouple detectors composed of two dissimilar metallic conductors connected together at two 'junctions.' The measurement junction is warmed by incident

radiance relative to a stable reference junction and a µV signal is produced. The response of a thermopile has little wavelength dependence and a hemispheric dome having a filter (~ 3–$50\,\mu m$) deposited on its inner surface typically defines the spectral response. Direct compensation for thermal drift using a temperature sensor located at the thermopile reference junction is sometimes used but regular calibration using an independent laboratory blackbody is mandatory. An accuracy of $<10\,W\,m^{-2}$ is possible after significant correction for instrument temperature drift and stray radiance contribution using additional on-board temperature sensors.

Narrow Beam Filter Radiometers

Narrow beam filter radiometers are often used to determine the SSST for air–sea interaction studies and for the validation of satellite derived SSST and there are several low-cost instruments that provide suitable accuracy and spectral characteristics. Many of these use a simple thermopile or thermistor detector together with a low-cost broadband-focusing lens. Typically, they have simple self-calibration techniques based on the temperature of the instrument and/or detector. Consequently they have poor resistance to thermal shock and have fore-optics that readily degrade in the marine atmosphere. However, handled with care, these devices are accurate to ± 0.1 K, albeit with limited sensitivity.

Precision narrow beam filter radiometers often use pyroelectric detectors that produce a small electrical current in response to changes in detector temperature forced by incident radiation. They have a fast response at ambient temperatures but require a modulated signal to operate. Modulation is accomplished by using an optical chopper having high reflectivity 'vanes' to alternately view a reference radiance source by reflection and a free path to the target radiance. The most common chopper systems are rotary systems driven by a small electric motor phase locked to the

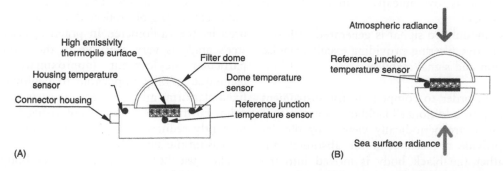

Figure 8 (A) A typical design of a long-wave pyrgeometer. (B) A net-radiation pyrgeometer for determination of the net long-wave flux at the sea surface.

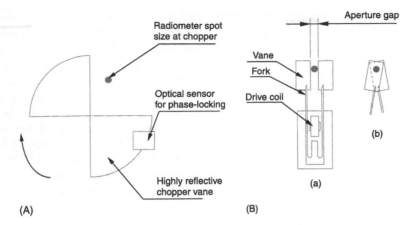

Figure 9 (A) Schematic layout of a rotary chopper; (B) schematic layout of a tuning fork chopper, (a) open, (b) closed.

detector output by an optoelectronic sensor **Figure 9A.** An alternative design driven by a small oscillating electromagnetic coil called a tuning fork chopper is shown in **Figure 9B.** As the coil resonates, the reflective vanes of the chopper oscillate alternately opening and closing an aperture 'gap.'

Dynamic detector bias compensation is inherent when using an optical chopper. The detector alternately measures radiance from the sea surface L_{src} and a reference blackbody, L_{bb} (sometimes this is the detector itself) reflected by the chopper vanes resulting in two signals

$$S_1 = L_{bb} + \delta \qquad [8]$$

$$S_2 = L_{src} + \delta \qquad [9]$$

Assuming L_{bb} remains constant during a short chopping cycle, the bias term δ in eqns [8] and [9] is eliminated

$$\Delta S = S_1 - S_2 = L_{bb} - L_{src} \qquad [10]$$

It is important to recognize the advantages to this technique, which is widely used:

- There is minimal thermal drift of the detector;
- The detector is dynamically compensated for thermal shock;
- A precise modulated signal is generated well suited to selective filtering providing excellent noise suppression and signal stability.

However, in order to compensate for instrument gain changes, an additional blackbody sources(s) is required. These are periodically viewed by the detector to provide a mechanism for absolute calibration. Either the black body is moved into the detector FoV or an adjustable mirror reflects radiance from the black body on to the detector.

Calibration cycles should be made at regular intervals so that gain changes can be accurately monitored and calibration sources need to be viewed using the same optical path as that used to view the sea surface.

A basic 'black-body' calibration strategy uses an external bath of sea water as a high ε (>0.95) reference as shown in **Figure 10.** In this scheme, the radiometer periodically views the water bath that is stirred vigorously to prevent the development of a thermal skin temperature deviation. The view geometry for the water bath and the sea surface are assumed to be identical and, by measuring the temperature of the water bath the radiometer can be absolutely calibrated. An advantage of this technique is that $\varepsilon_{(\lambda,\theta)}$ is not required to determine the SSST. However, in practice, it is difficult to continuously operate a water bath at sea and surface roughness differences between the bath and sea surface are ignored.

On reflection at the sea surface, diffuse sky radiance is polarized and, at Brewster's angle ($\sim 50°$ from nadir at a wavelength of 11 μm), the vertical v-polarization is negligible for a given wavelength (**Figure 11**). Only the horizontal h-polarization component remains so that if the radiometer filter response is v-polarized (i.e., only passes v-polarized radiance), negligible reflected sky radiance is measured by the radiometer. In practice, because Brewster's angle is very sensitive to the geometry of a particular deployment (approximately $\pm 2°$) this technique is only applicable to deployments from fixed platforms and when the sea surface is relatively calm. Further, the use of a polarizing filter will significantly reduce the signal falling on the detector increasing the signal-to-noise ratio.

The use of fabricated black-body cavities (**Figure 12A**) provides an accurate, versatile and, compact calibration system. Normally, two

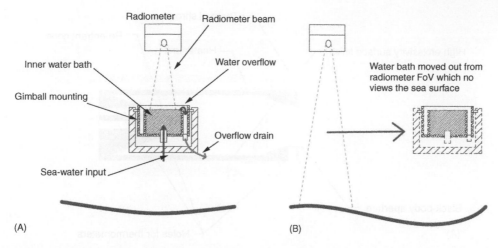

Figure 10 Schematic diagram showing the stirred water bath calibration scheme. (A) The radiometer in calibration mode, (B) the radiometer viewing the sea surface after the water bath has been moved out of the field of view (FoV).

black-body cavities are used, one of which follows the ambient temperature of the instrument and a second is heated to a nominal temperature above this. High ε (>0.99) is attained by a combination of specialized surface finish and black-body geometry. The cavity radiance is determined as a function of the black body temperature that is easily measured. **Figure 12B** shows a schematic outline of a typical black-body radiometer design using a rotary chopper and **Figure 12C** provides a schematic diagram of a typical output signal.

Note that for all calibration schemes, larger errors are expected beyond the calibrated temperature range which can be a problem for sky radiance measurements where clear sky temperatures of <200 K are common.

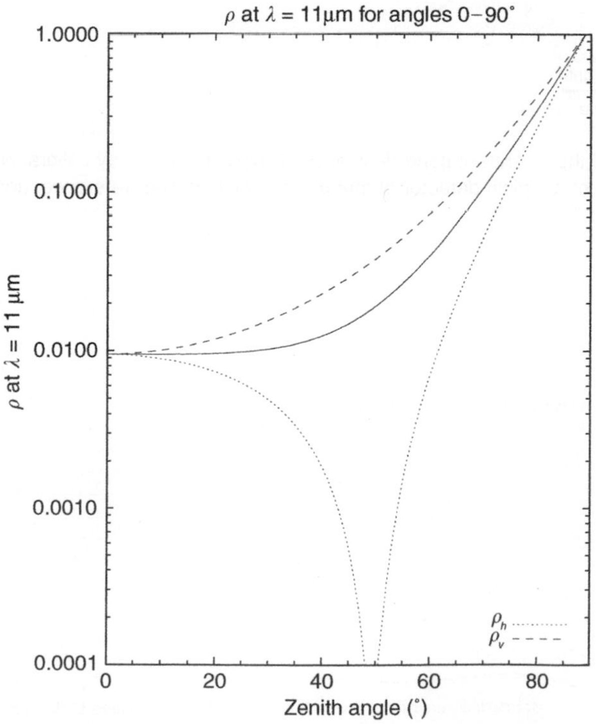

Multichannel Radiometers

The terms in eqn [7] are directly influenced by the height, h, of the radiometer above the sea surface and are different in magnitude for *in situ* and spacecraft deployments. In the case of a sea surface *in situ* radiometer deployment, $L_{path(h,\lambda,\theta)}$ is typically neglected because h is normally <10 m unless the atmosphere has a heavy water vapor loading (e.g., $>90\%$) or an aircraft deployment is considered. However, for a spacecraft deployment, this is a significant term requiring explicit correction. Conversely, the $B(T_{atm}, \lambda)$ term is critical to the accuracy of an *in situ* radiometer deployment but of little impact (except perhaps at the edge of clouds) for a satellite instrument deployment because $L_{path(h,\lambda,\theta)}$ dominates the signal. A multispectral capability can be used to explicitly account for $L_{path(h,\lambda,\theta)}$ in eqn [7] because of unequal atmospheric attenuation for different spectral wavebands. Multichannel radiometers are exclusively used on satellite platforms for this reason. Many *in situ* multichannel radiometers are designed primarily for the radiant calibration or validation of specific satellite radiometers and the development of satellite radiometer atmospheric correction algorithms. They often have several selectable filters matched to those of the satellite sensor.

ρ at $\lambda = 11\,\mu m$ for angles 0–90°

ρ at $\lambda = 11\,\mu m$

Zenith angle (°)

ρ_h ⋯⋯

ρ_v - - - -

Figure 11 Polarization of sea surface reflection at $11\,\mu m$ as a function of view angle. Total polarization is shown as a solid line.

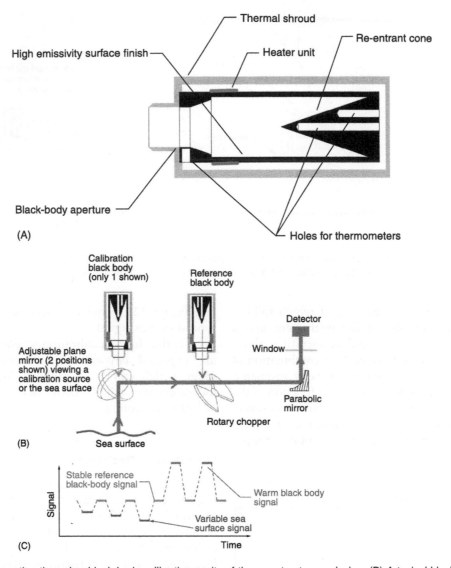

Figure 12 (A) A section through a black-body calibration cavity of the re-entrant cone design. (B) A typical black-body calibration radiometer using a rotary optical chopper. (C) A schematic diagram of a typical detector output signal, showing sea, reference, and calibration signals.

It is worth noting that multiangle view radiometers are also capable of providing an explicit correction for atmospheric attenuation. Often operated from satellite and aircraft, these instruments provide a direct measure of atmospheric attenuation by making two views of the same sea surface area at different angles using a geometry that doubles the atmospheric pathlength (**Figure 13**). The assumption is made that atmospheric and oceanic conditions are stationary in the time between each measurement.

Spectroradiometers

A recent development is the use of Fourier transform infrared spectrometers (FTIR) that are capable of accurate (~ 0.05 K). High spectral resolution

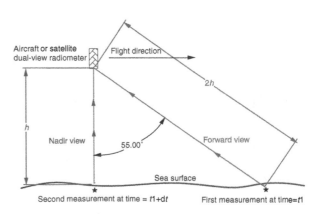

Figure 13 Schematic diagram of dual view, double atmospheric path radiometer deployment geometry.

($\sim 0.5\,cm^{-1}$) measurements over a broad spectral range (typically ~ 3–$18\,\mu m$) as shown in **Figure 14**. The FTIR provides a unique tool for the development of new IR measurement techniques and investigation of processes at the air–sea interface. For example, the Marine-Atmospheric Emitted Radiance Interferometer (M-AERI) has pioneered a SSST algorithm that uses a narrow spectral region centered at $7.7\,\mu m$ that is less susceptible to the influence of cloud cover and sky emissions that at 10–$12\,\mu m$. The FTIR can also be used to measure air temperatures by viewing the atmosphere at $\sim 15\,\mu m$ (a spectral region opaque due to CO_2 emission) that are accurate to $<0.1\,K$.

Of considerable interest is the ability of an FTIR provide an indirect estimate of $\rho_{(\lambda,\theta)}$ so that by using eqn [5], $\varepsilon_{(\lambda,\theta)}$ can be computed. The sky radiance spectrum has particular structures associated with atmospheric emission–absorbance lines (**Figure 14A**) that are physically uncorrelated with the smooth spectrum of $\rho_{(\lambda,\theta)}$ (**Figure 4**). The spectrum of $\rho_{(\lambda,\theta)}$ can be derived by subtracting a scaled $B(T_{arm},\lambda)$

spectrum to minimize the band-limited variance of the $B(T_{sea},\lambda)$ spectrum (**Figure 14B**).

Finally, direct measurement of the thermal gradient at the air–sea interface to obtain the net heat flux has been demonstrated using an FTIR in the laboratory. The FTIR uses the 3.3–$4.1\,\mu m$ spectral interval that has an effective optical depth (EOD) depending on the wavelength (EOD $= 0\,\mu m$ at $3.3\,\mu m$ whereas at $3.8\,\mu m$ EOD $= 65\,\mu m$) demonstrating the versatility of the FTIR. However, measurement integration times are long and further progress is required before this technique is applicable for normal field operations.

Thermal Imagers

Another recent development is the application of IR imagers and thermal cameras for high-resolution process studies such as fine-scale variability of SSST, wave breaking (**Figure 15**), and understanding air–sea gas and heat transfer. They are also used during

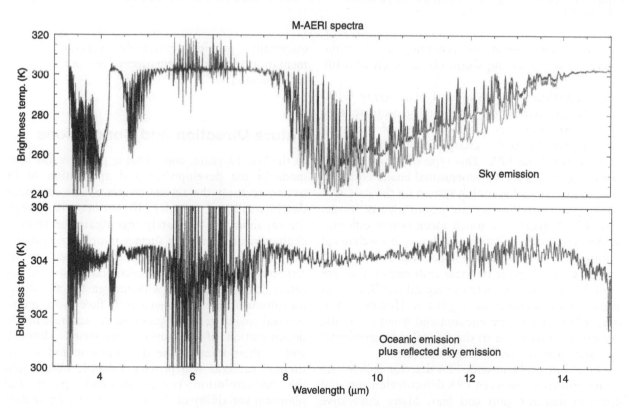

Figure 14 Spectra of emitted sky and sea view radiation measured by the M-AERI FTIR in the tropical Western Pacific Ocean on March 24, 1996. (A) Spectrum of sky radiance and (B) spectrum of corresponding sea radiance Sky measurements were made at 45° and zenith (red) above the horizon and ocean measurements were made at 45° below the horizon. The cold temperatures in the sky spectra show where the atmosphere is relatively transparent. The 'noise' in the 5.5–7 μm range is caused by the atmosphere being so opaque that the radiometer does not 'see' clearly the instrument internal black-body targets and calibration is void. The spectrum of upwelling radiation (B) consists of emission from the sea surface, reflected sky emission and emission from the atmospheric pathlength between the sea surface and the radiometer.

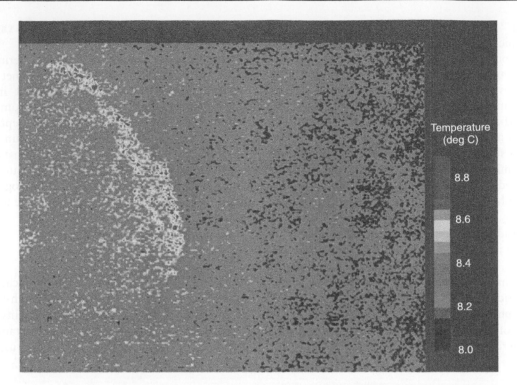

Figure 15   A thermal image of a breaking wave. Each pixel is ~25 × 25 cm and the wavelength of the camera is 8–12 μm. (Courtesy of D. Woolf.)

air–sea rescue operations providing a nighttime capability for detecting warm objects such as a life raft or survivors.

These instruments use a focal plane array (FPA) detector (a matrix of individual detectors, e.g., 256 × 256) located at the focal point of incoming radiance together with a charge couple device that is used to 'read' the FPA. This type of 'staring array' system generates a two-dimensional image either by using a mechanical scanning system (in the case of a small FPA array) or as an instantaneous image. Larger FPA arrays are much more power efficient, lighter and smaller than more elaborate mechanical scanning systems. Rapid image acquisition (> 15 frames s^{-1}) is typical of these instruments that are available in a wide variety of spectral configurations and a typical accuracy of ~ ± 0.1 K. However, considerable problems are encountered when obtaining sky radiance data due to the difficulty of geometrically matching sea and sky radiance data.

The major problem with FPA detector technology is nonuniformity between FPA detector elements and drifts in detector gain and bias. Many innovative self-calibration methods which range in quality are used to correct for these problems. For example, a small heated calibration plate assumed to be at an isothermal temperature is periodically viewed by the detector to provide an absolute calibration. However, further development of this technology will eventually provide extremely versatile instrumentation for the investigation of fine-scale sea surface emission.

Future Direction and Conclusions

In the last 10 years, considerable progress has been made in the development and application of IR sensors to study the air–sea interface. The continued development and use of FTIR sensors will provide the capability to accurately investigate the spectral characteristics of the sea surface in order to optimize the spectral intervals used by space sensors to determine SSST. It can be expected that in the near future, new algorithms will emerge for the direct measurement of the air–sea heat flux using multispectral sounding techniques and the accurate *in situ* determination of sea surface emissivity. Although still in their infancy, the development and use of thermal cameras will provide valuable insight into the fine-resolution two-dimensional spatial and temporal variability of the ocean surface. These data will be useful in developing and understanding the sampling limitations of large footprint satellite sensors and in the refinement of validation protocols. Finally, as satellite radiometers are now providing consistent and accurate observations of the SSST (e.g., ATSR), there is a need for autonomous operational *in situ* radiometer systems for ongoing validation of their data. Such intelligent systems that are extremely robust against the harsh realities of the marine environment are currently being developed.

See also

Radiative Transfer in the Ocean. Satellite Remote Sensing of Sea Surface Temperatures.

Further Reading

Bertie JE and Lan ZD (1996) Infrared intensities of liquids: the intensity of the OH stretching band revisited, and the best current values of the optical constants H_2O (1) at 25°C between 15,000 and $1 cm^{-1}$. *Applied Spectroscopy* 50: 1047–1057.

Donlon CJ, Keogh SJ, Baldwin DJ, *et al.* (1998) Solid state measurements of sea surface skin temperature. *Journal of Atmospherical Oceanic Technology* 15: 775–787.

Donlon CJ and Nightingale TJ (2000) The effect of atmospheric radiance errors in radiometric sea surface skin temperature measurements. *Applied Optics* 39: 2392–2397.

Jessup AT, Zappa CJ, and Yeh H (1997) Defining and quantifying micro-scale wave breaking with infrared imagery. *Journal of Geophysical Research* 102: 23145–23153.

McKeown W and Asher W (1997) A radiometric method to measure the concentration boundary layer thickness at an air–water interface. *Journal of Atmospheric and Oceanic Technology* 14: 1494–1501.

Shaw JA (1999) Degree of polarisation in spectral radiances from water viewing infrared radiometers. *Applied Optics* 15: 3157–3165.

Smith WL, Knuteson RO, Rivercombe HH, *et al.* (1996) Observations of the infrared radiative properties of the ocean – implications for the measurement of sea surface temperature via satellite remote sensing. *Bulletin of the American Meteorological Society* 77: 41–51.

Suarez MJ, Emery WJ, and Wick GA (1997) The multi-channel infrared sea truth radiometric calibrator (MISTRC). *Journal of Atmospheric and Oceanic Technology* 14: 243–253.

Thomas JP, Knight RJ, Roscoe HK, Turner J, and Symon C (1995) An evaluation of a self-calibrating infrared radiometer for measuring sea surface temperature. *Journal of Atmospheric and Oceanic Technology* 12: 301–316.

REMOTE SENSING OF COASTAL WATERS

N. Hoepffner and G. Zibordi, Institute for
Environment and Sustainability, Ispra, Italy

Introductory Comments

Coastal waters occupy at the most 8–10% of the ocean surface and only 0.5% of its volume, but represent an important fraction in terms of economic, social, and ecological value. With growing concerns about the rapid and negative changes of the coastal areas and marine resources, there has been over the last decade a pressing request for the development of quantitative and cost-effective methodologies to detect and characterize both long-term changes and short-term events in the coastal environment. The challenge, however, is that the properties of coastal waters are controlled by complex interactions and fluxes of material between land, ocean, and atmosphere. As a result, coastal zones are among the most changeable environment on Earth, typically involving phenomena which vary along time and space scales shorter than those in the open ocean.

In this view, remote sensing techniques have a key role to play, providing consistent products for a wide range of coastal applications over scales otherwise inaccessible from standard ship surveys. In spite of this advantage, remote sensing techniques have also their limitations and should not be seen as a substitute to field measurements but rather complementary to them in terms of both sampling scales and variables to be measured.

In this article, particular emphasis is given to the Space-borne sensors measuring radiances in the visible and near-infrared, enabling the observation of biogeochemical processes in the marine environment. Coastal waters are optically more complex than the open oceanic waters because of the large influence by the land system and catchment areas delivering significant amount of dissolved and particulate material. Thus, relative to other Earth-observing techniques, optical remote sensing of coastal waters requires more specificity in the treatment of the signal, and in the choice of the algorithms, to support any quantitative assessment of relevant biogeochemical variables and their associated processes.

Coastal Water Optics

The radiance leaving the water at a given wavelength, λ, $L_w(\lambda)$, in the direction of the remote sensor results from three major processes of unequal magnitude: (1) the reflected sunlight at the surface of the water; (2) the light that has entered the sea surface and retransmitted back to the atmosphere through scattering; and (3) the fluorescence of some specific material in suspension in the upper water column. In addition, in shallow and relatively clear waters, remote sensing techniques are often affected by the ocean bottom.

As the photons enter the water column, the spectral signature of the sunlight is altered depending on the optical properties of the medium itself and its constituents. Optically complex waters are commonly labeled as 'case 2' waters in a bipartite classification scheme (see Bio-Optical Models) where 'case 1' refers to marine waters in which phytoplankton and covarying materials are the principal components responsible for the changes in the optical properties. Optical properties of case 2 waters are influenced by substances that vary independently from phytoplankton. Such substances are inorganic particles in suspension (e.g., sediment), and some colored or chromophoric dissolved organic materials (CDOMs), also called yellow substances.

The water-type classification is based on the relative contribution of these three types of substances affecting the light field, irrespectively of possible changes in the concentration of each substance. For example, case 1 waters can range from clear, oligotrophic open ocean waters with chlorophyll concentration less than $0.1\,mg\,m^{-3}$, to some eutrophic conditions where chlorophyll can reach a concentration of $10\,mg\,m^{-3}$.

Accordingly, all coastal water samples can be optically charted in a triangular diagram (**Figure 1**) in which the axes represent the fractional contributions due to each of the three types of optically active components (AOCs). As the data point moves toward one of the three apices, the sample will be classified as either case 1 water, case 2 'yellow-substance' dominated water, or case 2 'suspended material' dominated water. This classification may appear rather restricted considering the huge variety of living and nonliving compounds that contributes to the optical variability of marine waters, but it is important in remote sensing of coastal waters to select the type of algorithms which are best suited to retrieve biogeochemical variables.

The Color of Coastal Waters

As for any water body, the color of coastal waters is defined by the spectral variations in reflectance, $R(\lambda)$, at the sea surface. This is an apparent optical

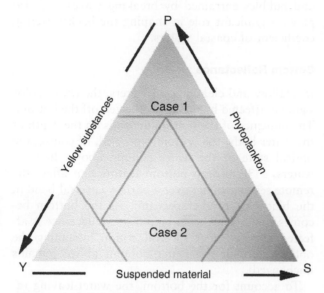

Figure 1 Ternary or triangular diagram illustrating the relative contribution of phytoplankton, dissolved organic matter (yellow substances) and suspended material (nonalgal particles), and the division between case 1 and case 2 waters. Copyright: IOCCG.

property (AOP) determined by the ratio between the upwelling and downwelling irradiances at a given wavelength (*see* Satellite Remote Sensing: Ocean Colour). This quantity is strictly related to the normalized water-leaving radiance, $L_{WN}(\lambda)$, which can be determined from the total radiance measured by a space-borne sensor. The reflectance varies depending on the illumination and viewing geometry, as well as on the inherent optical properties (IOPs) of the water constituents. In the context of remote sensing, the IOPs of relevance are the absorption coefficient, $a(\lambda)$, and the backscattering coefficient, $b_b(\lambda)$, which represents the integral of the volume scattering function, $\beta(\lambda,\varphi)$, in the backward direction from 90° to 180°. Accordingly, the reflectance at the sea surface is formulated as

$$R(\lambda) = f[b_b(\lambda)/(a(\lambda) + b_b(\lambda))] \qquad [1]$$

where f is a function of the solar zenith angle and depends also on the shape of the volume scattering function.

In coastal waters, the optically active constituents may vary in space and time, independently of each other, and consequently, this variation affects the color (i.e., reflectance) of the water (**Figure 2**).

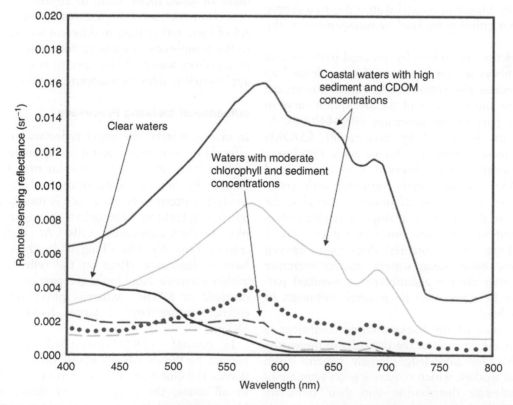

Figure 2 Examples of remote sensing reflectance spectra for different water types. Maximum reflectance shifts to higher wavelength in coastal waters with high concentration of nonalgal particles and CDOMs. Adapted from IOCCG.

Optical remote sensing in coastal waters thus necessitates a thorough assessment of the water bulk optical properties, quantifying the role of each individual component. To this end, the total spectral absorption and backscattering coefficients are divided into as many additive components:

$$a(\lambda) = a_w(\lambda) + a_{phy}(\lambda) + a_{NAP}(\lambda) + a_{CDOM}(\lambda) \quad [2]$$

$$b_b(\lambda) = b_{b\,w}(\lambda) + b_{b\,phy}(\lambda) + b_{b\,NAP}(\lambda) \quad [3]$$

where the subscript 'w' refers to water molecules (pure water), 'phy' to phytoplankton, 'NAP' to nonalgal particle (e.g., sediment), and 'CDOM' to colored dissolved organic matter (i.e., yellow substances, assumed to have negligible scattering properties).

The basic theory developed for open ocean (see Bio-Optical Models) to parametrize the wavelength dependence of each term in eqns [2] and [3] is directly applicable in coastal waters. The implementation of these models require, however, different strategies to account for unrelated distributions between the various optically active components as well as differences in the values of some model parameters. A notable difference, for example, is in the contribution of nonalgal particles which can be significantly higher in coastal waters due to a supply of sediment either from land or resuspension at the bottom.

The spectral absorption by nonalgal particles and CDOMs obeys similar exponential shape defined by their respective absorption coefficients at a reference wavelength and the slope of the absorption curve in the lower part of the spectrum (400–440 nm). In coastal areas influenced by river runoff, CDOMs mainly consist of terrestrial humic acids with lower slope values than those observed in open ocean and in coastal waters not directly connected with terrestrial sources. In the case of nonalgal particles, the variability in the exponential slope in marine waters is restricted to a narrow range of values (~ 0.01 to < 0.02 nm^{-1}). A potential distinction between coastal and open ocean waters remains therefore difficult, even though assemblage of nonalgal particles may be dominated by mineral sediments or organic debris.

The analysis of the backscattering coefficient is still advancing because of the difficulty in measuring it directly. Our knowledge largely benefits from theoretical studies, which require a good estimate of the particle-size distribution and their refractive index. In coastal waters, the large variety of particle type and size increases the complexity of the calculation. In addition, the presence of fine mineral particles with high refractive index represents there a crucial factor controlling the reflectance variability when compared with open ocean waters. Specific blooms of phytoplankton (e.g., coccolithophores) and bubbles entrained by breaking waves can also play a significant role in shaping the backscattering coefficient of coastal waters.

Bottom Reflectance

In shallow and clear coastal waters, the ocean color signal is affected by the light reflected off the bottom. The influence of the bottom depends on the depth of the water body, the bottom type, and, of course, the optical state of the water column above. In clear waters, a 30-m-deep bottom feature can affect the remote sensing signal over a narrow spectral band in the blue part of the spectrum. As the bottom becomes shallower, the surface water reflectance is affected over a wider spectral region (assuming the optical state of the water column above remains unchanged).

To account for the bottom, the water-leaving radiance in shallow waters is usually modeled using a two-flow system where the upwelled irradiance is made of a water column and a bottom component of known reflectance. The seabed, however, may be made of sand, rocks, mud, or covered with benthic organisms such as algae, seagrasses, and coral reefs. All of these reflect light in different ways, which add to the complexity in using remote sensing technique in nearshore waters, albeit open to new challenges to apply satellite data for mapping benthic features.

Influence of Inelastic Processes

In marine waters, the water reflectance can also be affected by some transpectral (or inelastic) processes such as those associated with solar-stimulated fluorescence by organic compounds, and Raman (molecular) scattering. Fluorescence is mostly generated by phytoplankton chlorophyll pigments, and by phycoerythrin-containing cells. At high concentration (typically, $Chl > 1$ mg m^{-3}), chlorophyll can have a significant effect on the reflectance signal within a narrow band in the red part of the spectrum centered at 685 nm, whereas phycoerythrin fluorescence occurs over a larger band centered at 585 nm.

In coastal 'CDOM-dominated' waters, fluorescence by the dissolved organic matter can also influence the blue part of the spectrum, c. 430–450 nm. In all cases, the integration of fluorescence into bio-optical models requires a good knowledge of the absorption coefficients of each component at the excitation wavelengths, as well as their fluorescence quantum yield. Note that in coastal waters, the effect of Raman scattering on the surface reflectance is reduced as the water turbidity increases.

Algorithms for Coastal Waters

Established algorithms are currently on a routine basis by the space agencies and others to generate multi-sensors archives of satellite biogeochemical products over the global ocean. These algorithms were developed to perform as efficiently as possible in the open ocean. The complexity of coastal waters, as well as the atmosphere above, requires more complex algorithms capable of handling nonlinear multivariate bio-optical systems. In addition, the large variability in time and space of the bulk properties of coastal waters hampers any implementation of a single and global algorithm that would work with the same efficiency in all regions.

Atmospheric Correction

The atmospheric correction process is applied to remove the effects of the atmosphere that contribute to the signal measured by a satellite sensor. The objective of this process is the discrimination, from top-of-atmosphere radiance, of the signal emerging from the sea carrying information on the materials suspended and dissolved in seawater. The atmospheric correction of coastal data is challenged by the presence of continental aerosols, bottom reflectance, and adjacency of land. Minimizing these perturbing effects, which generally are site-specific, requires knowledge of the regional aerosol and bottom optical properties. Specifically, effects of continental aerosols are minimized by properly accounting for their scattering phase function and single scattering albedo; the increase in radiance due to bottom reflectance can be removed through iterative processes knowing the water depth and spectral reflectance of the seabed; and the adjacency effects are minimized by determining the reduction in image contrast as a function of the aerosol and of the land reflectance.

Coastal Water Algorithms

In optical remote sensing of marine waters, the algorithms used to retrieve the concentrations of water constituents are either analytical (or semi-) or empirical. Empirical algorithms (so-called blue–green ratio algorithms; see Satellite Remote Sensing: Ocean Colour) have been implemented in the processing of satellite data over open ocean with various degrees of success. But the performance of these algorithms remains limited to the range of input values that have

been used for their development. In a highly changeable environment such as coastal waters, these algorithms are clearly restricted to specific regions, and even specific periods. Some improvements can be made to increase the efficiency of these algorithms in turbid waters, by using more than one band ratio, within spectral regions that are more sensitive to the optical conditions observed locally. For example, additional wavelengths in the blue part of the spectrum could be used to separate the dissolved material from the particulate matter, whereas other bands in the green and orange part may provide information on specific phytoplankton blooms often observed in coastal waters (see **Table 1**). In reality, the spectral characteristics of combined seawater constituents are not unique, suggesting that no single wavelength can really be representing information on just one constituent. Under these conditions, more complex modeling approaches, and optimization techniques are envisaged to solve the problem.

Analytical (or semi-) algorithms are based on the inversion of a forward radiance model describing both the relationship between water constituents and the water reflectance at the surface (typically as in eqn [1]), and the light propagation through the water and atmosphere. This approach makes use of known optical properties for the specific components, even though the spectral characteristics of these components are usually determined from empirical formulation, thus justifying the term 'semi-analytical' to describe these algorithms. In coastal waters, the number of unknowns in each set of equations may increase rapidly with the number of independent variables (i.e., water constituents), causing difficulties in solving the system without applying some

Table 1 Example of spectral range for coastal waters applications

Product name	Critical spectral range (nm)
Chlorophyll, and other pigment concentration	443–445 (max. abs.), 550–560 (min. abs.) 490, 548, 640 (for other pigments)
Red tides	510
Sediment concentration	530–650, NIR
CDOM	410–420
Seabed reflectance	490–580
Chlorophyll fluorescence	680–685
Aerosol optical properties	700–NIR
Atmospheric corrections	
Sea surface temperature (SST)	TIR

NIR, near-infrared; TIR, thermal infrared.

approximation that eventually leads to a lower accuracy of the algorithms.

Nonlinear optimization techniques and other machine learning methods such as artificial neural networks provide an interesting alternative to examine complex coastal waters and to handle multivariate systems. Models based on these methods determine the output values (e.g., the concentrations of optically active components) from input data (e.g., water reflectance at various wavelengths) through nonlinear multidimensional parametric functions (**Figure 3**). The determination of the model parameters, as well as the assessment of the model performance rely on a reference data set.

Remote sensing of coastal waters and algorithm development are still topics of ongoing research in spite of substantial effort and promising results that have taken place within the last few years. Improvements remain to be achieved on several aspects of both the atmospheric correction and in-water bio-optical modeling to ensure an operational retrieval of the products at a reasonable accuracy. These investigations rely on field experiments and the collection of high-quality optical data for a wide range of water bodies.

With the rapid changes occurring in coastal waters, a unique algorithm that will serve all optical conditions with the same accuracy is doubtful. On the other hand, the implementation of more than one regional algorithm into a unique processing scheme remains a challenge, as that approach could result in significant discontinuities at the region boundaries.

Data Validation and Measurement Protocols

The normalized water-leaving radiance, $L_{WN}(\lambda)$, is the primary ocean color remote sensing product obtained from top-of-atmosphere data corrected for the atmospheric perturbations. Higher-level marine products, like the concentration of phytoplankton pigments or seawater inherent optical properties (i.e., absorption, scattering, and backscattering coefficients), are determined from $L_{WN}(\lambda)$. The accuracy of coastal remote sensing products may be affected by the presence of turbid waters or absorbing aerosols. Because of this, the validation of remote sensing $L_{WN}(\lambda)$ and aerosol optical thickness data, resulting from the atmospheric correction process, is a first fundamental step in assessing the accuracy of ocean color products. Additional steps include the validation of derived products to assess the consistency of the whole process leading to the determination of higher-level products.

Validation measurements can result from oceanographic campaigns, or from continuous data collection by autonomous systems deployed on fixed platforms or ships of opportunity. Oceanographic campaigns ensure a wide spatial data collection with a comprehensive optical and biogeochemical characterization of each sampling site, but with a poor time resolution. Autonomous systems enable continuous a highly repeated collection of a restricted set of parameters at a specific site or along a given route. Obviously, measurement campaigns and

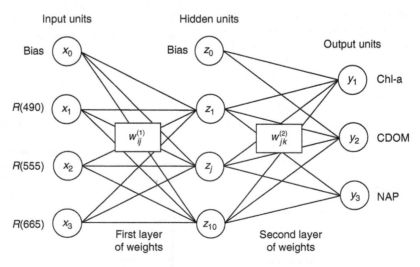

Figure 3 Representation of a multilayer perceptron (MLP) neural network adopted for the retrieval of the optically active seawater components. In this particular case, the network architecture consists of three inputs (reflectances at three satellite wavelengths), 10 hidden units, and three outputs (chlorophyll *a*, colored dissolved organic matter, and nonalgal particulate matter). An additional input is implemented for the bias parameter. Note that the MLP architecture presented here has only an illustrative purpose. Adapted from D'Alimonte D and Zibordi G (2003) Phytoplankton determination in an optically complex coastal region using a multilayer perceptron neural network. *IEEE Transactions on Geoscience and Remote Sensing* 41(12): 2861–2868.

autonomous systems are complementary and both contribute to the objective of validating remote sensing products.

Within the framework of *in situ* optical technologies commonly applied for the determination of the fundamental quantity $L_{WN}(\lambda)$, both in- and above-water radiometry can produce individual measurements or time series of optical data. In this context, various in-water autonomous systems have been widely exploited in open ocean regions where the biofouling perturbation is low. Differently, the use of autonomous above-water radiometry is more suitable to produce long-term measurements for the validation of ocean color data in optically complex coastal waters.

Offshore fixed platforms like lighthouses, oceanographic towers, and oil rigs, generally located at several nautical miles from the main land in coastal areas (i.e., regions permanently or occasionally affected by bottom resuspension, coastal erosion, river inputs, or by relevant anthropogenic impact), are suitable for the deployment of autonomous above-water radiometers provided that (1) their distance from the coast makes negligible the adjacency effects in the remote sensing data; and (2) the bottom structure and type have no effect (**Figure 4**).

Time, spatial, and spectral coincident pairs of satellite and *in situ* data, so-called matchups, are the basis for any validation analysis. Due to the intrinsic higher variability exhibited by coastal waters, when compared to oceanic waters, particular care must be given to the spatial variability of satellite data and to the temporal variability of *in situ* observations (**Figure 5**).

Remote Sensing Coastal Applications

Remote sensing of coastal waters satisfies on its own a large number of applications, owing to substantial progress in sensor technology, as well as algorithm performance, made over the last few years. Important contributions have been accomplished in various disciplines of marine sciences, like marine physics and biogeochemistry, process modeling, as well as fisheries and coastal management. More importantly, satellites provide continuous long time series of bio-optical and geophysical variables, thus representing a cost-effective tool to monitor changes in coastal waters that might occur as a direct or indirect result of external pressures (e.g., climate change and human activities). In parallel, the scientific and technical community is progressing in the search for

Figure 4 Lighthouse tower in the Baltic Sea used for the deployment of above-water radiometer systems supporting the validation of ocean color products in coastal areas.

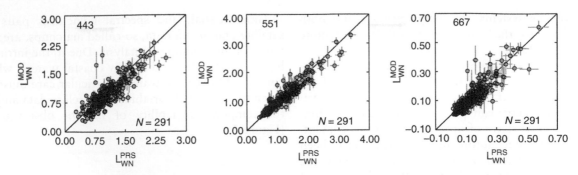

Figure 5 Scatter plots of satellite-derived L_{WN}^{MOD} vs. *in situ* L_{WN}^{MOD} normalized water-leaving radiances in units of mW cm^{-2} μm^{-1}s^{-1} at the 443, 551, and 667 nm MODIS center-wavelengths.

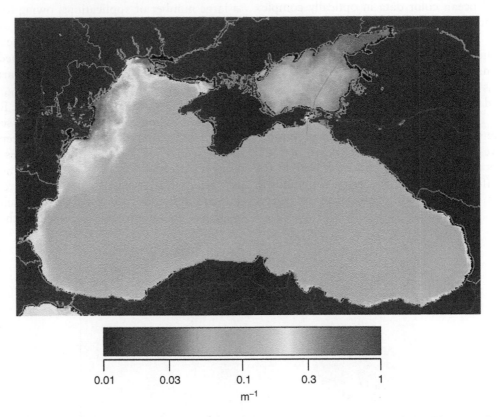

Figure 6 Light attenuation coefficient K_d in units of m^{-1} for the Black Sea derived from SeaWiFS monthly composite (Jan. 2003). The coefficient is calculated using a semi-analytical algorithm to retrieve the IOPs of the water constituents. Source: Joint Research Centre of the European Commission © Joint Research Centre, 2006.

the optimal model to be applied in these complex waters, taking benefit from advanced instruments with higher spectral resolution and better calibration. Recent development of hyperspectral sensor prototypes, providing the full spectral coverage over the visible wavelengths, offers new opportunities to discriminate optical disparities, and thus to classify turbid waters and quantify individual seawater components down to phytoplankton species.

Optical remote sensing in coastal waters typically provides quantitative estimates of phytoplankton biomass (indexed by chlorophyll concentrations), total suspended matter, and dissolved organic substances, with accuracies as good as in open ocean (*c.* 30–40% in biomass) when using regional dedicated algorithms. These components, in turn, influence the light attenuation coefficient in the water which is also directly retrieved from satellite (**Figure 6**), providing information on water transparency. The latter represents an important quality parameter to assess the ecological status of lakes and coastal waters. Remotely sensed chlorophyll concentration and light attenuation are also required to estimate water-column primary production, which represents a critical indicator to assess coastal eutrophication, and to analyze the carbon budget directly associated with the continental shelf pump.

The concentrations of suspended sediment and dissolved organic matter can be used to track river plumes and other coastal dynamics, and to calibrate and validate sediment transport models. Recently, the availability of IOPs, that is, absorption and backscattering coefficients, as direct products of satellite ocean color, has given an alternative approach to address changes in the water constituents, as well as water transparency, further extending satellite applications to particle size and composition characteristics.

Several algal species can be present in coastal waters as intense blooms (e.g., coccolithophorid species, red tides, *Trichodesmium*), which can be readily distinguished on the basis of their differences in pigment composition and optical properties (**Figure 7**). In that case, remote sensing acts as an efficient technique to monitor these anomalous blooms; some are associated with harmful consequences on the surrounding ecosystem, causing mass mortalities of marine organisms.

In shallow coastal waters, remote sensing is being used to monitor benthic structure and classify bottom types. At present, satellite data most commonly exploited for that purpose are equipped with radiometers, such as the SPOT-HRV and Landsat TM series with typically 10–30 m spatial resolution, or even IKONOS with spatial resolution better than 10 m. Even though these sensors have a limited set of wavelengths (three to four broad spectral bands in the visible), the spectral contrast between various bottom structures (e.g., corals and noncoral objects) can be used to construct classification algorithms based on differential reflectance (**Figure 8**).

Surface features simultaneously observed from ocean color and thermal imagery have assisted coastal fisheries to locate target areas with high fish densities. Many pelagic species aggregate along frontal structures and upwelling areas that bear distinctive signals on the sea surface temperature and primary production (or phytoplankton biomass) maps. Moreover, many aspects of the life cycle of fishes (e.g., reproduction and early-stage feeding) depends on physical and biological processes, and discerning these from remote sensing is important for marine conservation and sustainable management of the marine coastal resources.

Satellite instruments, other than optical sensors, can provide additional information of importance to coastal water management. Microwave sensors collect data either passively or actively in that part of the electromagnetic spectrum ranging from 3 mm to 1 m wavelength. These instruments provide significant advantages in viewing the Earth under all-weather conditions, even through clouds; but their spatial resolution, which depends on the physical dimension of the antenna, is often not adequate for coastal applications. Sophisticated synthetic aperture radars (SARs) are an exception to that rule and can retrieve surface features at extremely fine resolution of the order of meters. This remote sensing technique is used for fisheries applications to monitor fishing vessels in coastal waters, contributing thus to the management of the stocks through evaluation of the fishing effort. In addition to ship survey, SAR data are exploited in coastal waters to monitor sea surface roughness and to detect pollution at the surface. Oil spill detection is of particular importance to coastal management. The presence of oil slicks on the sea surface increases the surface tension of seawater, reducing its natural roughness which appears as a black signature on a SAR image. Multiple techniques using, for example, ocean color and SAR can be helpful to discriminate between dense algal blooms

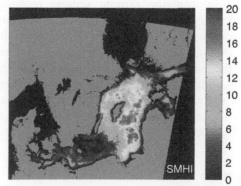

Figure 7 Cyanobacteria bloom in the Baltic Sea. Left: MODIS-Aqua 30 Jul. 2003. True color image of the Baltic Sea showing a bloom of *Nodularia spumigena*. Source: http://oceancolor.gsfc.nasa.gov. Right: Occurrence of cyanobacterial accumulations in the Baltic during 2003 expressed as number of days (data based on satellite images). Source: http://www.helcom.fi

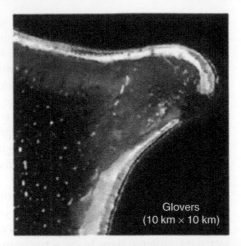

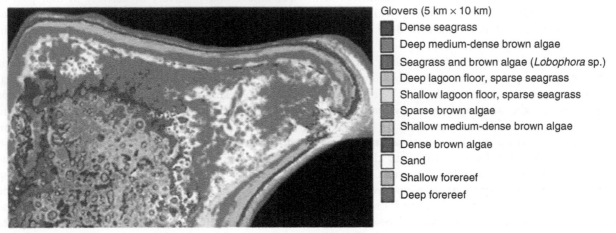

Glovers (5 km × 10 km)

- Dense seagrass
- Deep medium-dense brown algae
- Seagrass and brown algae (*Lobophora* sp.)
- Deep lagoon floor, sparse seagrass
- Shallow lagoon floor, sparse seagrass
- Sparse brown algae
- Shallow medium-dense brown algae
- Dense brown algae
- Sand
- Shallow forereef
- Deep forereef

Figure 8 Classification of tropical coral reef environment (Belize, Caribbean Sea). Top: RGB color composite based on the red, green, and blue bands of IKONOS sensor. Bottom: Eleven-class scheme classification of the benthic features. Reproduced from Andréfouët S, Kramer P, Torres-Pulliza D, *et al.* (2003) Multi-site evaluation of IKONOS data for classification of tropical coral reef environments. *Remote Sensing of Environment* 88: 128–143, with permission from Elsevier.

and surface slicks. The low revisit cycle (e.g., 35 days for the European Remote Sensing satellites ERS-1 and ERS-2 instruments) limits, however, the use of SAR to monitor rapidly changing events.

In conclusion, there is a large and ever-increasing interest in applying remote sensing technique in coastal waters. It offers a unique solution to monitor this environment at scales that are compatible with many biological and physical processes. At the same time, the number of satellite products is growing rapidly with the advent of new sensors, better algorithms and validation experiments, providing a range of key indicators to support scientific investigations, as well as operations for coastal management.

See also

Bio-Optical Models. Inherent Optical Properties and Irradiance. Satellite Remote Sensing: Ocean Colour.

Further Reading

Andréfouët S, Kramer P, Torres-Pulliza D, *et al.* (2003) Multi-site evaluation of IKONOS data for classification of tropical coral reef environments. *Remote Sensing of Environment* 88: 128–143.

Babin M, Stramski D, Ferrari GM, *et al.* (2003) Variations in the light absorption coefficients of phytoplankton, nonalgal particles, and dissolved organic matter in coastal waters around Europe. *Journal of Geophysical Research* 108(C7): 3211 (doi:10.1029/2001JC000882).

Berthon J-F, Mélin F, and Zibordi G (2007) Ocean colour remote sensing of the optically complex European seas. In: Barale V and Gade M (eds.) *Remote Sensing of European Seas*, pp. 35–52. Dordrecht: Springer.

Bukata RP, Jerome JH, Kondratyev KY, and Pozdnyakov DV (eds.) (1995) *Optical Properties and Remote Sensing of Inland and Coastal Waters*. Boca Raton, FL: CRC Press.

D'Alimonte D and Zibordi G (2003) Phytoplankton determination in an optically complex coastal region using a multilayer perceptron neural network. *IEEE*

Transactions on Geoscience and Remote Sensing 41(12): 2861–2868.

Gordon HR (2002) Inverse methods in hydrologic optics. *Oceanologia* 44: 9–58.

Hansson M (2004) Cyanobacterial blooms in the Baltic Sea. HELCOM Indicator Fact Sheets 2004. Online (2008-05-06) http://www.helcom.fi/environment2/ifs/archive/ifs2004/enGB/cyanobacteria (accessed July 2008).

IOCCG (2000) Remote sensing of ocean colour in coastal, and other optically-complex waters. In: Sathyendranath S (ed.) *Report of the International Ocean-Colour Coordinating Group, No. 3.* Dartmouth, NS: IOCCG.

IOCCG (2006) Remote sensing of inherent optical properties: Fundamentals, tests of algorithms and applications. In: Lee ZP (ed.) *Report of the International Ocean-Colour Coordinating Group, No. 5.* Dartmouth, NS: IOCCG.

Miller RL, Del Castillo CE, and McKee BA (eds.) (2005) *Remote Sensing of Coastal Aquatic Environments: Technologies, Techniques and Applications.* Dordrecht: Springer.

Oceanography Magazine (2004) *Special Issue: Coastal Ocean Optics and Dynamics. The Official Magazine of the Oceanography Society* 17(2).

Zibordi G, Holben B, Hooker SB, *et al.* (2006) A network for standardized ocean color validation measurements. *EOS Transactions, American Geophysical Union* 87(30): 293–297.

SATELLITE ALTIMETRY

R. E. Cheney, Laboratory for Satellite Altimetry, NOAA, Silver Spring, Maryland, USA

Introduction

Students of oceanography are usually surprised to learn that sea level is not very level at all and that the dominant force affecting ocean surface topography is not currents, wind, or tides; rather it is regional variations in the Earth's gravity. Beginning in the 1970s with the advent of satellite radar altimeters, the large-scale shape of the global ocean surface could be observed directly for the first time. What the data revealed came as a shock to most of the oceanographic community, which was more accustomed to observing the sea from ships. Profiles telemetered back from NASA's pioneering altimeter, Geos-3, showed that on horizontal scales of hundreds to thousands of kilometers, the sea surface is extremely complex and bumpy, full of undulating hills and valleys with vertical amplitudes of tens to hundreds of meters. None of this came as a surprise to geodesists and geophysicists who knew that the oceans must conform to these shapes owing to spatial variations in marine gravity. But for the oceanographic community, the concept of sea level was forever changed. During the following two decades, satellite altimetry would provide exciting and revolutionary new insights into a wide range of earth science topics including marine gravity, bathymetry, ocean tides, eddies, and El Niño, not to mention the marine wind and wave fields which can also be derived from the altimeter echo. This chapter briefly addresses the technique of satellite altimetry and provides examples of applications.

Measurement Method

In concept, radar altimetry is among the simplest of remote sensing techniques. Two basic geometric measurements are involved. In the first, the distance between the satellite and the sea surface is determined from the round-trip travel time of microwave pulses emitted downward by the satellite's radar and reflected back from the ocean. For the second measurement, independent tracking systems are used to compute the satellite's three-dimensional position relative to a fixed Earth coordinate system. Combining these two measurements yields profiles of sea surface topography, or sea level, with respect to the reference ellipsoid (a smooth geometric surface which approximates the shape of the Earth).

In practice, the various measurement systems are highly sophisticated and require expertise at the cutting edge of instrument and modeling capabilities. This is because accuracies of a few centimeters must be achieved to properly observe and describe the various oceanographic and geophysical phenomena of interest. **Figure 1** shows a schematic of the Topex/Poseidon (T/P) satellite altimeter system. Launched in 1992 as a joint mission of the American and French Space agencies (and still operating as of 2001), T/P is the most accurate altimeter flown to date. Its microwave radars measure the distance to the sea surface with a precision of 2 cm. Two different frequencies are used to solve for the path delay due to the ionosphere, and a downward-looking microwave radiometer provides measurements of the integrated water vapor content which must also be known. Meteorological models must be used to estimate the attenuation of the radar pulse by the atmosphere, and other models correct for biases created by ocean waves. Three different tracking systems (a laser reflector, a Global Positioning System receiver, and a 'DORIS' Doppler receiver) determine the satellite orbit to within 2 cm in the radial direction. The result of all these measurements is a set of global sea level observations with an absolute accuracy of 3–4 cm at intervals of 1 s, or about 6 km, along the satellite track. The altimeter footprint is exceedingly small – only 2–3 km – so regional maps or 'images' can only be derived by averaging data collected over a week or more.

Gravitational Sea Surface Topography

Sea surface topography associated with spatial variations in the Earth's gravity field has vertical amplitudes 100 times larger than sea level changes generated by tides and ocean currents. To first order, therefore, satellite altimeter data reveal information about marine gravity. Within 1–2% the ocean topography follows a surface of constant gravitational potential energy known as the geoid or the equipotential surface, shown schematically in **Figure 1**. Gravity can be considered to be constant in time for most purposes, even though slight changes do occur as the result of crustal motions,

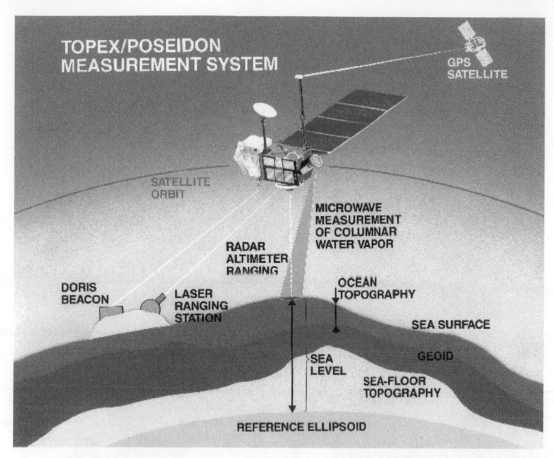

Figure 1 Schematic diagram of satellite radar altimeter system. Range to the sea surface together with independent determination of the satellite orbit yields profiles of sea surface topography (sea level) relative to the Earth's reference ellipsoid. Departures of sea level from the geoid drive surface geostrophic currents.

redistribution of terrestrial ice and water, and other slowly varying phenomena. An illustration of the gravitational component of sea surface topography is provided in **Figure 2**, which shows a T/P altimeter profile collected in December 1999 across the Marianas Trench in the western Pacific. The trench represents a deficit of mass, and therefore a negative gravity anomaly, so that the water is pulled away from the trench axis by positive anomalies on either side. Similarly, seamounts represent positive gravity anomalies and appear at the ocean surface as mounds of water. The sea level signal created by ocean bottom topography ranges from ∼1 m for seamounts to ∼10 m for pronounced features like the Marianas Trench, and the peak-to-peak amplitude for the large-scale gravity field is nearly 200 m.

Using altimeter data collected by several different satellites over a period of years, it is possible to create global maps of sea surface topography with extraordinary accuracy and resolution. When these maps are combined with surface gravity measurements, models of the Earth's crust, and bathymetric data collected by ships, it is possible to construct three-dimensional images of the ocean floor – as if all the water were drained away (**Figure 3**). For many oceanic regions, especially in the Southern Hemisphere, these data have provided the first reliable maps of bottom topography. This new data set has many scientific and commercial applications, from numerical ocean modeling, which requires realistic bottom topography, to fisheries, which have been able to take advantage of new fishing grounds over previously uncharted seamounts.

Dynamic Sea Surface Topography

Because of variations in the density of sea water over the globe, the geoid and the mean sea surface are not exactly coincident. Departures of the sea surface with respect to the geoid have amplitudes of about 1 m and constitute what is known as 'dynamic topography'. These sea surface slopes drive the geostrophic circulation: a balance between the surface slope (or surface pressure gradient) and the Coriolis force (created by the Earth's rotation). The

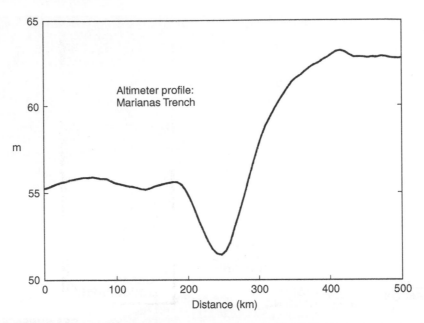

Figure 2 Sea surface topography across the Marianas Trench in the western Pacific measured by Topex/Poseidon. Heights are relative to the Earth's reference ellipsoid.

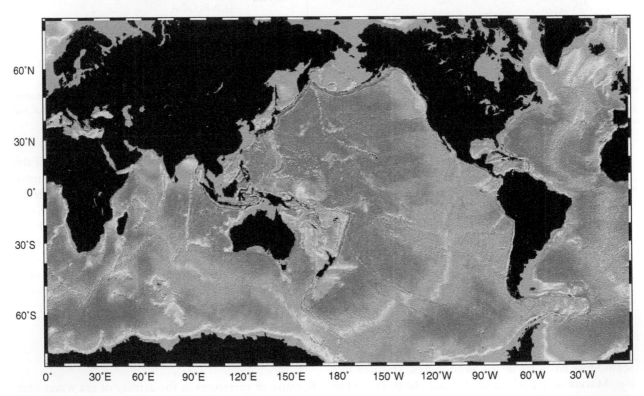

Figure 3 Topography of the ocean bottom determined from a combination of satellite altimetry, gravity anomalies, and bathymetric data collected by ships. (Courtesy of Walter H. F. Smith, NOAA, Silver Spring, MD, USA.)

illustration in **Figure 4** shows an estimate of the global geostrophic circulation derived by combining a mean altimeter-derived topography with a geoid computed from independent gravity measurements. Variations are between −110 cm (deep blue) and 110 cm (white). The surface flow is along lines of equal dynamic topography (red arrows). In the Northern Hemisphere, the flow is clockwise around

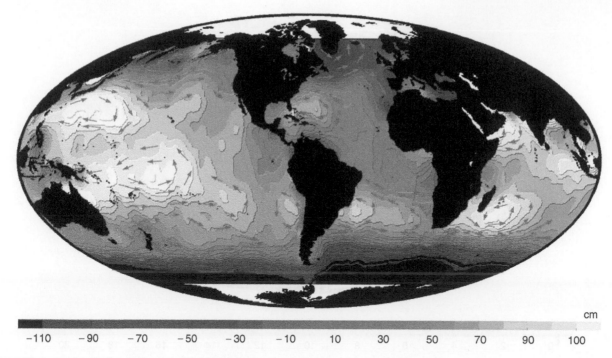

Figure 4 Surface geostrophic circulation determined from a combination of satellite altimetry and a model of the marine gravity field. (Courtesy of Space Oceanography Division, CLS, Toulouse, France.)

the topography highs, while in the Southern Hemisphere, the flow is counter-clockwise around the highs. The map shows all the features of the general circulation such as the ocean gyres and associated western boundary currents (e.g. Gulf Stream, Kuroshio, Brazil/Malvinas Confluence) and the Antarctic Circumpolar Current.

At the time of writing, global geoid models are not sufficiently accurate to reveal significant new information about the surface circulation of the ocean. However, extraordinary gravity fields will soon be available from dedicated satellite missions such as the Challenging Minisatellite Payload (CHAMP: 2000 launch), the Gravity Recovery and Climate Experiment (GRACE: 2002 launch), and the Gravity Field and Steady-state Ocean Circulation Explorer (GOCE: 2005 launch). These satellite missions, sponsored by various agencies in the USA and Europe, will employ accelerometers, gravity gradiometers, and the Global Positioning System to virtually eliminate error in marine geoid models at spatial scales larger than 300 km and will thereby have a dramatic impact on physical oceanography. Not only will it be possible to accurately compute global maps of dynamic topography and geostrophic surface circulation, but the new gravity models will also allow recomputation of orbits for past altimetric satellites back to 1978, permitting studies of long, global sea level time-series. Furthermore,

measurement of the change in gravity as a function of time will provide new information about the global hydrologic cycle and perhaps shed light on the factors contributing to global sea level rise. For example, how much of the rise is due simply to heating and how much to melting of glaciers? Together with complementary geophysical data, satellite gravity data represent a new frontier in studies of the Earth and its fluid envelope.

Sea Level Variability

At any given location in the ocean, sea level rises and falls over time owing to tides, variable geostrophic flow, wind stress, and changes in temperature and salinity. Of these, the tides have the largest signal amplitude, on the order of 1 m in mid-ocean. Satellite altimetry has enabled global tide models to be dramatically improved such that mid-ocean tides can now be predicted with an accuracy of a few cm (*see*)In studying ocean dynamics, the contribution of the tides is usually removed using these models so that other dynamic ocean phenomena can be isolated.

The map in **Figure 5** shows the variability of global sea level for the period 1992–98. It is derived from three satellite altimeter data sets: ERS-1, T/P, and ERS-2 (ERS is the European Space Agency Remote Sensing Satellite), from which the tidal signal has been removed. The map is dominated by mesoscale

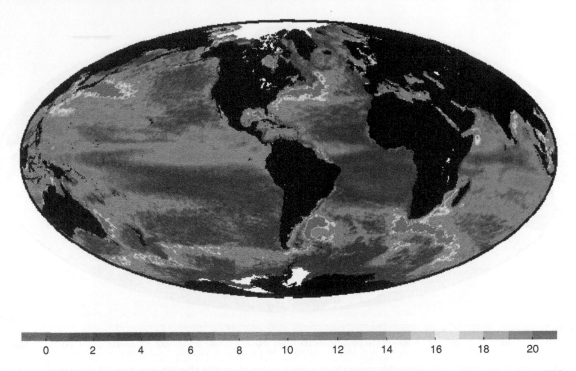

Figure 5　Variability of sea surface topography over the period 1992–98 from three satellite altimeters: Topex/Poseidon, ERS-1, and ERS-2. Highest values (cm) correspond to western boundary currents which meander and generate eddies. (Courtesy of Space Oceanography Division, CLS, Toulouse, France.)

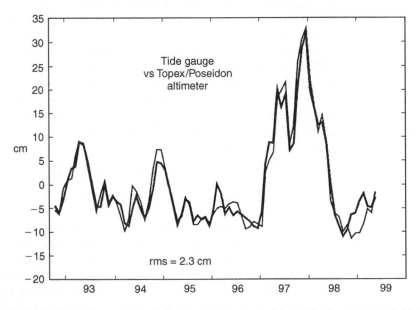

Figure 6　Monthly mean sea level deviation near the Galapagos Islands derived from tide gauge data and altimeter data. The ∼2 cm agreement demonstrates the accuracy of altimetry for observing sea level variability. The effect of the 1997–98 El Niño is apparent.

(100–300 km) variability associated with the western boundary currents, where the rms variability can be as high as 30 cm. This is due to a combination of current meandering, eddies, and seasonal heating and cooling. Other bands of relative maxima (10–15 cm) can be seen in the tropics where interannual signals such as El Niño are the dominant contributor. The smallest variability is found in the eastern portions of the major ocean basins where values are <5 cm rms.

To examine a sample of the sea level signal more closely, **Figure 6** shows the record from the region of

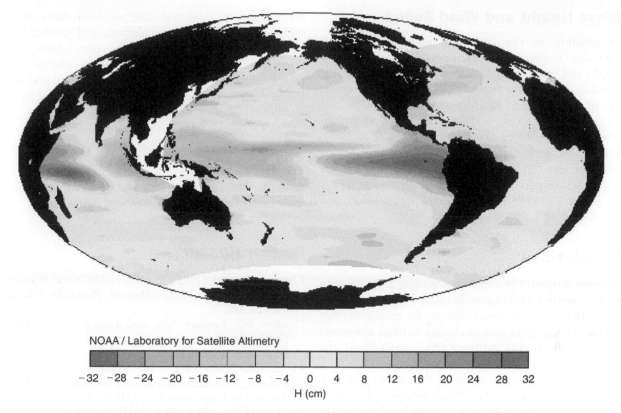

NOAA / Laboratory for Satellite Altimetry

−32 −28 −24 −20 −16 −12 −8 −4 0 4 8 12 16 20 24 28 32

H (cm)

Figure 7 Global sea level anomaly observed by the Topex/Poseidon altimeter at the height of the 1997–98 El Niño. High (low) sea level corresponds to areas of positive (negative) heat anomaly in the ocean's upper layers.

the Galapagos Islands in the eastern equatorial Pacific. The plot includes two time-series: one from the T/P altimeter and the other from an island tide gauge, both averaged over monthly time periods. These independent records agree at the level of 2 cm, an indication of the remarkable reliability of satellite altimetry. The plot also illustrates changes associated with the El Niño event which took place during 1997–98. During El Niño, relaxation of the Pacific trade winds cause a dramatic redistribution of heat in the tropical oceans. In the eastern Pacific, sea level during this event rose to 30 cm above normal by December 1977 and fell by a corresponding amount in the far western Pacific. The global picture of sea level deviations observed by the T/P altimeter at this time is shown in **Figure 7**. Because sea level changes can be interpreted as changes in heat (and to a lesser extent, salinity) in the upper layers, altimetry provides important information for operational ocean models which are used for long-range El Niño forecasts.

Global Sea Level Rise

Tide gauge data collected over the last century indicate that global sea level is rising at about 1.8 mm y^{-1}. Unfortunately, because these data are relatively sparse and contain large interdecadal fluctuations, the observations must be averaged over 50–75 years in order to obtain a stable mean value. It is therefore not possible to say whether sea level rise is accelerating in response to recent global warming. Satellite altimeter data have the advantage of dense, global coverage and may offer new insights on the problem in a relatively short period of time. Based on T/P data collected since 1992, it is thought that 15 years of continuous altimeter measurements may be sufficient to obtain a reliable estimate of the current rate of sea level rise. This will require careful calibration of the end-to-end altimetric system, not to mention cross-calibration of a series of two or three missions (which typically last only 5 years). Furthermore, in order to interpret and fully understand the sea level observations, the various components of the global hydrologic system must be taken into account, for example, polar and glacial ice, ground water, fresh water stored in man-made reservoirs, and the total atmospheric water content. It is a complicated issue, but one which may yield to the increasingly sophisticated observational systems that are being brought to bear on the problem.

Wave Height and Wind Speed

In addition to sea surface topography, altimetry provides indirect measurements of ocean wave height and wind speed (but not wind direction). This is made possible by analysis of the shape and intensity of the reflected radar signal: a calm sea sends the signal back almost perfectly, like a flat mirror, whereas a rough sea scatters and deforms it. Wave height measurements are accurate to about 0.5 m or 10% of the significant wave height, whichever is larger. Wind speed can be measured with an accuracy of about $2\,\mathrm{m\,s^{-1}}$. For additional information, see articles 69 and 129.

Conclusions

Satellite altimetry is somewhat unique among ocean remote sensing techniques because it provides much more than surface observations. By measuring sea surface topography and its change in time, altimeters provide information on the Earth's gravity field, the shape and structure of the ocean bottom, the integrated heat and salt content of the ocean, and geostrophic ocean currents. Much progress has been made in the development of operational ocean applications, and altimeter data are now routinely assimilated in near-real-time to help forecast El Niño, monitor coastal circulation, and predict hurricane intensity. Although past missions have been flown largely for research purposes, altimetry is rapidly moving into the operational domain and will become a routine component of international satellite systems during the twenty-first century.

See also

Satellite Remote Sensing Microwave Scatterometers. Satellite Remote Sensing SAR.

Further Reading

Cheney RE (ed.) (1995) TOPEX/POSEIDON: Scientific Results. *Journal of Geophysical Research* 100: 24 893–25 382

Douglas BC, Kearney MS, and Leatherman SP (eds.) (2001) *Sea Level Rise: History and Consequences*. London: Academic Press.

Fu LL and Cheney RE (1995) Application of satellite altimetry to ocean circulation studies, 1987–1994. *Reviews of Geophysics Suppl*: 213–223.

Fu LL and Cazenave A (eds.) (2001) *Satellite Altimetry and Earth Sciences*. London: Academic Press.

SATELLITE OCEANOGRAPHY, HISTORY AND INTRODUCTORY CONCEPTS

W. S. Wilson, NOAA/NESDIS, Silver Spring, MD, USA
E. J. Lindstrom, NASA Science Mission Directorate, Washington, DC, USA
J. R. Apel[†], Global Ocean Associates, Silver Spring, MD, USA

Published by Elsevier Ltd.

Oceanography from a satellite – the words themselves sound incongruous and, to a generation of scientists accustomed to Nansen bottles and reversing thermometers, the idea may seem absurd.

Gifford C. Ewing (1965)

Introduction: A Story of Two Communities

The history of oceanography from space is a story of the coming together of two communities – satellite remote sensing and traditional oceanography.

For over a century oceanographers have gone to sea in ships, learning how to sample beneath the surface, making detailed observations of the vertical distribution of properties. Gifford Ewing noted that oceanographers had been forced to consider "the class of problems that derive from the vertical distribution of properties at stations widely separated in space and time."

With the introduction of satellite remote sensing in the 1970s, traditional oceanographers were provided with a new tool to collect synoptic observations of conditions at or near the surface of the global ocean. Since that time, there has been dramatic progress; satellites are revolutionizing oceanography. (Appendix 1 provides a brief overview of the principles of satellite remote sensing.)

Yet much remains to be done. Traditional subsurface observations and satellite-derived observations of the sea surface – collected as an integrated set of observations and combined with state-of-the-art models – have the potential to yield estimates of the three-dimensional, time-varying distribution of properties for the global ocean. Neither a satellite nor an *in situ* observing system can do this on its own. Furthermore, if such observations can be collected over the long term, they can provide oceanographers with an observational capability conceptually similar to that which meteorologists use on a daily basis to forecast atmospheric weather.

Our ability to understand and forecast oceanic variability, how the oceans and atmosphere interact, critically depends on an ability to observe the three-dimensional global oceans on a long-term basis. Indeed, the increasing recognition of the role of the ocean in weather and climate variability compels us to implement an integrated, operational satellite and *in situ* observing system for the ocean now – so that it may complement the system which already exists for the atmosphere.

The Early Era

The origins of satellite oceanography can be traced back to World War II – radar, photogrammetry, and the V-2 rocket. By the early 1960s, a few scientists had recognized the possibility of deriving useful

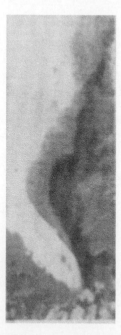

Figure 1 Thermal infrared image of the US southeast coast showing warmer waters of the Gulf Stream and cooler slope waters closer to shore taken in the early 1960s. While the resolution and accuracy of the TV on *Tiros* were not ideal, they were sufficient to convince oceanographers of the potential usefulness of infrared imagery. The advanced very high resolution radiometer (AVHRR) scanner (see text) has improved images considerably. Courtesy of NASA.

[†]Deceased

oceanic information from the existing aerial sensors. These included (1) the polar-orbiting meteorological satellites, especially in the 10–12-μm thermal infrared band; and (2) color photography taken by astronauts in the Mercury, Gemini, and Apollo manned spaceflight programs. Examples of the kinds of data obtained from the National Aeronautics and Space Administration (NASA) flights collected in the 1960s are shown in **Figures 1** and **2**.

Such early imagery held the promise of deriving interesting and useful oceanic information from space, and led to three important conferences on space oceanography during the same time period.

In 1964, NASA sponsored a conference at the Woods Hole Oceanographic Institution (WHOI) to examine the possibilities of conducting scientific research from space. The report from the conference, entitled *Oceanography from Space*, summarized findings to that time; it clearly helped to stimulate a number of NASA projects in ocean observations and sensor development. Moreover, with the exception of the synthetic aperture radar (SAR), all instruments flown through the 1980s used techniques described in

this report. Dr. Ewing has since come to be justifiably regarded as the father of oceanography from space.

A second important step occurred in 1969 when the Williamstown Conference was held at Williams College in Massachusetts. The ensuing Kaula report set forth the possibilities for a space-based geodesy mission to determine the equipotential figure of the Earth using a combination of (1) accurate tracking of satellites and (2) the precision measurement of satellite elevation above the sea surface using radar altimeters. Dr. William Von Arx of WHOI realized the possibilities for determining large-scale oceanic currents with precision altimeters in space. The requirements for measurement precision of 10-cm height error in the elevation of the sea surface with respect to the geoid were articulated. NASA scientists and engineers felt that such accuracy could be achieved in the long run, and the agency initiated the Earth and Ocean Physics Applications Program, the first formal oceans-oriented program to be established within the organization. The required accuracy was not to be realized until 1992 with *TOPEX/ Poseidon*, which was reached only over a 25-year

Figure 2 Color photograph of the North Carolina barrier islands taken during the Apollo-Soyuz Mission (AS9-20-3128). Capes Hatteras and Lookout, shoals, sediment- and chlorophyll-bearing flows emanating from the coastal inlets are visible, and to the right, the blue waters of the Gulf Stream. Cloud streets developing offshore the warm current suggest that a recent passage of a cold polar front has occurred, with elevated air–sea evaporative fluxes. Later instruments, such as the coastal zone color scanner (CZCS) on *Nimbus-7* and the SeaWiFS imager have advanced the state of the art considerably. Courtesy of NASA.

period of incremental progress that saw the flights of five US altimetric satellites of steadily increasing capabilities: *Skylab*, *Geos-3*, *Seasat*, *Geosat*, and *TOPEX/Poseidon* (see **Figure 3** for representative satellites).

A third conference, focused on sea surface topography from space, was convened by the National Oceanic and Atmospheric Administration (NOAA), NASA, and the US Navy in Miami in 1972, with 'sea surface topography' being defined as undulations of the ocean surface with scales ranging from

approximately 5000 km down to 1 cm. The conference identified several data requirements in oceanography that could be addressed with space-based radar and radiometers. These included determination of surface currents, Earth and ocean tides, the shape of the marine geoid, wind velocity, wave refraction patterns and spectra, and wave height. The conference established a broad scientific justification for space-based radar and microwave radiometers, and it helped to shape subsequent national programs in space oceanography.

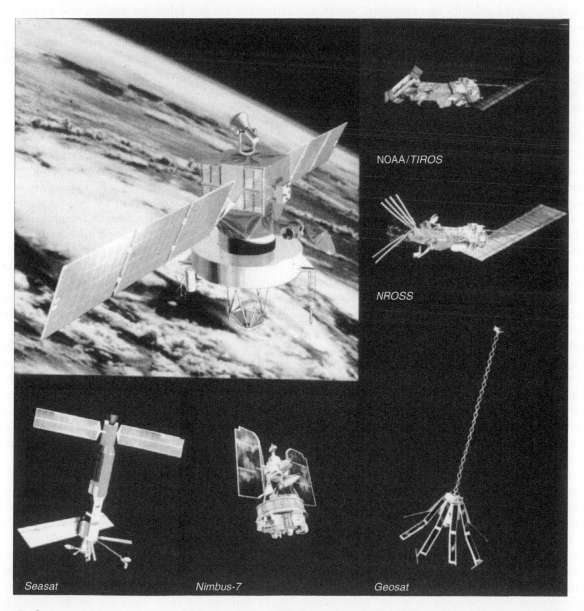

Figure 3 Some representative satellites: (1) *Seasat*, the first dedicated oceanographic satellite, was the first of three major launches in 1978; (2) the *Tiros* series of operational meteorological satellites carried the advanced very high resolution radiometer (AVHRR) surface temperature sensor; *Tiros-N*, the first of this series, was the second major launch in 1978; (3) *Nimbus-7*, carrying the CZCS color scanner, was the third major launch in 1978; (4) *NROSS*, an oceanographic satellite approved as an operational demonstration in 1985, was later cancelled; (5) *Geosat*, an operational altimetric satellite, was launched in 1985; and (6) this early version of *TOPEX* was reconfigured to include the French *Poseidon*; the joint mission *TOPEX/Poseidon* was launched in 1992. Courtesy of NASA.

The First Generation

Two first-generation ocean-viewing satellites, *Skylab* in 1973 and *Geos-3* in 1975, had partially responded to concepts resulting from the first two of these conferences. *Skylab* carried not only several astronauts, but a series of sensors that included the S-193, a radar-altimeter/wind-scatterometer, a long-wavelength microwave radiometer, a visible/infrared scanner, and cameras. S-193, the so-called Rad/Scatt, was advanced by Drs. Richard Moore and Willard Pierson. These scientists held that the scatterometer could return wind velocity measurements whose accuracy, density, and frequency would revolutionize marine meteorology. Later aircraft data gathered by NASA showed that there was merit to their assertions. *Skylab*'s scatterometer was damaged during the opening of the solar cell panels, and as a consequence returned indeterminate results (except for passage over a hurricane), but the altimeter made observations of the geoid anomaly due to the Puerto Rico Trench.

Geos-3 was a small satellite carrying a dual-pulse radar altimeter whose mission was to improve the knowledge of the Earth's marine geoid, and coincidentally to determine the height of ocean waves via the broadening of the short transmitted radar pulse upon reflection from the rough sea surface. Before the end of its 4-year lifetime, *Geos-3* was returning routine wave height measurements to the National Weather Service for inclusion in its Marine Waves Forecast. Altimetry from space had become a clear possibility, with practical uses of the sensor immediately forthcoming. The successes of *Skylab* and *Geos-3* reinforced the case for a second generation of radar-bearing satellites to follow.

The meteorological satellite program also provided measurements of sea surface temperature using far-infrared sensors, such as the visible and infrared scanning radiometer (VISR), which operated at wavelengths near 10 µm, the portion of the terrestrial spectrum wherein thermal radiation at terrestrial temperatures is at its peak, and where coincidentally the atmosphere has a broad passband. The coarse, 5-km resolution of the VISR gave blurred temperature images of the sea, but the promise was clearly there. **Figure 1** is an early 1960s TV image of the southeastern USA taken by the NASA *TIROS* program, showing the Gulf Stream as a dark signal. While doubts were initially held by some oceanographers as to whether such data actually represented the Gulf Stream, nevertheless the repeatability of the phenomenon, the verisimilitude of the positions and temperatures with respect to conventional wisdom, and their own objective judgment finally convinced most workers of the validity of the data. Today, higher-resolution, temperature-calibrated infrared imagery constitutes a valuable data source used frequently by ocean scientists around the world.

During the same period, spacecraft and aircraft programs taking ocean color imagery were delineating the possibilities and difficulties of determining sediment and chlorophyll concentrations remotely. **Figure 2** is a color photograph of the North Carolina barrier islands taken with a hand-held camera, with Cape Hatteras in the center. Shoals and sediment- and chlorophyll-bearing flows emanating from the coastal inlets are visible, and to the right, the blue waters of the Gulf Stream. Cloud streets developing offshore the warm stream suggest a recent passage of a cold polar front and attendant increases in air–sea evaporative fluxes.

The Second Generation

The combination of the early data and advances in scientific understanding that permitted the exploitation of those data resulted in spacecraft sensors explicitly designed to look at the sea surface. Information returned from altimeters and microwave radiometers gave credence and impetus to dedicated microwave spacecraft. Color measurements of the sea made from aircraft had indicated the efficacy of optical sensors for measurement of near-surface chlorophyll concentrations. Infrared radiometers returned useful sea surface temperature measurements. These diverse capabilities came together when, during a 4-month interval in 1978, the USA launched a triad of spacecraft that would profoundly change the way ocean scientists would observe the sea in the future. On 26 June, the first dedicated oceanographic satellite, *Seasat*, was launched; on 13 October, *TIROS-N* was launched immediately after the catastrophic failure of *Seasat* on 10 October; and on 24 October, *Nimbus-7* was lofted. Collectively they carried sensor suites whose capabilities covered virtually all known ways of observing the oceans remotely from space. This second generation of satellites would prove to be extraordinarily successful. They returned data that vindicated their proponents' positions on the measurement capabilities and utility, and they set the direction for almost all subsequent efforts in satellite oceanography.

In spite of its very short life of 99 days, *Seasat* demonstrated the great utility of altimetry by measuring the marine geoid to within a very few meters, by inferring the variability of large-scale ocean surface currents, and by determining wave heights. The wind scatterometer could yield oceanic surface wind velocities equivalent to 20 000 ship observations per day. The scanning multifrequency radiometer also provided wind speed and atmospheric water content data; and the SAR penetrated clouds to show features on the surface of the sea, including surface and internal waves, current boundaries, upwellings, and rainfall patterns. All of these measurements could be extended to basin-wide scales, allowing oceanographers a view of the sea never dreamed of before. *Seasat* stimulated several subsequent generations of ocean-viewing satellites, which outline the chronologies and heritage for the world's ocean-viewing spacecraft. Similarly, the early temperature and color observations have led to successor programs that provide large quantities of quantitative data to oceanographers around the world.

The Third Generation

The second generation of spacecraft would demonstrate that variables of importance to oceanography could be observed from space with scientifically useful accuracy. As such, they would be characterized as successful concept demonstrations. And while both first- and second-generation spacecraft had been exclusively US, international participation in demonstrating the utility of their data would lead to the entry of Canada, the European Space Agency (ESA), France, and Japan into the satellite program during this period. This article focuses on the US effort. Additional background on US third-generation missions covering the period 1980–87 can be found in the series of *Annual Reports for the Oceans Program* (NASA Technical Memoranda 80233, 84467, 85632, 86248, 87565, 88987, and 4025).

Partnership with Oceanography

Up to 1978, the remote sensing community had been the prime driver of oceanography from space and there were overly optimistic expectations. Indeed, the case had not yet been made that these observational techniques were ready to be exploited for ocean science. Consequently, in early 1979, the central task was establishing a partnership with the traditional oceanographic community. This meant involving them in the process of evaluating the performance of *Seasat* and *Nimbus-7*, as well as building an ocean science program at NASA headquarters to complement the ongoing remote sensing effort.

National Oceanographic Satellite System

This partnership with the oceanographic community was lacking in a notable and early false start on the part of NASA, the US Navy, and NOAA – the National Oceanographic Satellite System (NOSS). This was to be an operational system, with a primary and a backup satellite, along with a fully redundant ground data system. NOSS was proposed shortly after the failure of *Seasat*, with a first launch expected in 1986. NASA formed a 'science working group' (SWG) in 1980 under Francis Bretherton to define the potential that NOSS offered the oceanographic community, as well as to recommend sensors to constitute the 25% of its payload allocated for research. However, with oceanographers essentially brought in as junior partners, the job of securing a new start for NOSS fell to the operational community – which it proved unable to do. NOSS was canceled in early 1981. The prevailing and realistic view was that the greater community was not ready to implement such an operational system.

Science Working Groups

During this period, SWGs were formed to look at each promising satellite sensing technique, assess its potential contribution to oceanographic research, and define the requirements for its future flight. The notable early groups were the TOPEX SWG formed in 1980 under Carl Wunsch for altimetry, Satellite Surface Stress SWG in 1981 under James O'Brien for scatterometry, and Satellite Ocean Color SWG in 1981 under John Walsh for color scanners. These SWGs were true partnerships between the remote sensing and oceanographic communities, developing consensus for what would become the third generation of satellites.

Partnership with Field Centers

Up to this time, NASA's Oceans Program had been a collection of relatively autonomous, in-house activities run by NASA field centers. In 1981, an overrun in the space shuttle program forced a significant budget cut at NASA headquarters, including the Oceans Program. This in turn forced a reprioritization and refocusing of NASA programs. This was a blessing in disguise, as it provided an opportunity to initiate a comprehensive, centrally led program – which would ultimately result in significant funding for the oceanographic as well as remote-sensing communities. Outstanding relationships with individuals like Mous Chahine in senior management at the Jet Propulsion Laboratory (JPL) enabled the partnership between NASA headquarters and the two prime ocean-related field centers (JPL and the Goddard Space Flight Center) to flourish.

Partnerships in Implementation

A milestone policy-level meeting occurred on 13 July 1982 when James Beggs, then Administrator of NASA, hosted a meeting of the Ocean Principals Group – an informal group of leaders of the ocean-related agencies. A NASA presentation on opportunities and prospects for oceanography from space was received with much enthusiasm. However, when asked how NASA intended to proceed, Beggs told the group that – while NASA was the sole funding agency for space science and its missions – numerous agencies were involved in and support oceanography. Beggs said that NASA was willing to work with other agencies to implement an ocean satellite program, but that it would not do so on its own. Beggs' statement defined the approach to be pursued in implementing oceanography from space, namely, a joint approach based on partnerships.

Research Strategy for the Decade

As a further step in strengthening its partnership with the oceanographic community, NASA collaborated with the Joint Oceanographic Institutions Incorporated (JOI), a consortium of the oceanographic institutions with a deep-sea-going capability. At the time, JOI was the only organization in a position to represent and speak for the major academic oceanographic institutions. A JOI satellite planning committee (1984) under Jim Baker examined SWG reports, as well as the potential synergy between the variety of oceanic variables which could be measured from space; this led to the idea of understanding the ocean as a system. (From this, it was a small leap to understanding the Earth as a system, the goal of NASA's Earth Observing System (EOS).)

The report of this Committee, *Oceanography from Space: A Research Strategy for the Decade, 1985–1995*, linked altimetry, scatterometry, and ocean color with the major global ocean research programs being planned at that time – the World Ocean Circulation Experiment (WOCE), Tropical Ocean Global Atmosphere (TOGA) program, and Joint Global Ocean Flux Study (JGOFS). This strategy, still being followed today, served as a catalyst to engage the greater community, to identify the most important missions, and to develop an approach for their prioritization. Altimetry, scatterometry, and ocean color emerged from this process as national priorities.

Promotion and Advocacy

The *Research Strategy* also provided a basis for promoting and building an advocacy for the NASA program. If requisite funding was to be secured to pay for proposed missions, it was critical that government policymakers, the Congress, the greater oceanographic community, and the public had a good understanding of oceanography from space and its potential benefits. In response to this need, a set of posters, brochures, folders, and slide sets was designed by Payson Stevens of Internetwork Incorporated and distributed to a mailing list which grew to exceed 3000. These award-winning materials – sharing a common recognizable identity – were both scientifically accurate and esthetically pleasing.

At the same time, dedicated issues of magazines and journals were prepared by the community of involved researchers. The first example was the issue of *Oceanus* (1981), which presented results from the second-generation missions and represented a first step toward educating the greater oceanographic community in a scientifically useful and balanced way about realistic prospects for satellite oceanography.

Implementation Studies

Given the SWG reports taken in the context of the *Research Strategy*, the NASA effort focused on the following sensor systems (listed with each are the various flight opportunities which were studied):

- altimetry – the flight of a dedicated altimeter mission, first *TOPEX* as a NASA mission, and then *TOPEX/Poseidon* jointly with the French Centre Nationale d'Etudes Spatiales (CNES);
- scatterometry – the flight of a NASA scatterometer (NSCAT), first on NOSS, then on the *Navy Remote Ocean Observing Satellite* (*NROSS*), and finally on the *Advanced Earth Observing Satellite* (*ADEOS*) of the Japanese National Space Development Agency (NASDA);
- visible radiometry – the flight of a NASA color scanner on a succession of missions (NOSS, *NOAA-H/-I*, *SPOT-3* (*Système Pour l'Observation de la Terre*), and *Landsat-6*) and finally the purchase of ocean color data from the Sea-viewing Wide Field-of-view Sensor (SeaWiFS) to be flown by the Orbital Sciences Corporation (OSC);
- microwave radiometry – a system to utilize data from the series of special sensor microwave imager (SSMI) radiometers to fly on the Defense Meteorological Satellite Program satellites;
- SAR – a NASA ground station, the Alaska SAR Facility, to enable direct reception of SAR data from the *ERS-1/-2*, *JERS-1*, and *Radarsat* satellites of the ESA, NASDA, and the Canadian Space Agency, respectively.

New Starts

Using the results of the studies listed above, the Oceans Program entered the new start process at NASA headquarters numerous times attempting to secure funds for implementation of elements of the third generation. *TOPEX* was first proposed as a NASA mission in 1980. However, considering limited prospects for success, partnerships were sought and the most promising one was with the French. CNES initially proposed a mission using a *SPOT* bus with a US launch. However, NASA rejected this because *SPOT*, constrained to be Sun-synchronous, would alias solar tidal components. NASA proposed instead a mission using a US bus capable of flying in a non-Sun-synchronous orbit with CNES providing an *Ariane* launch. The NASA proposal was accepted for study in fiscal year (FY) 1983, and a new start was finally secured for the combined *TOPEX/Poseidon* in FY 1987.

In 1982 when the US Navy first proposed *NROSS*, NASA offered to be a partner and provide a scatterometer. The US Navy and NASA obtained new starts for both *NROSS* and NSCAT in FY 1985. However, *NROSS* suffered from a lack of strong support within the navy, experienced a number of delays, and was finally terminated in 1987. Even with this termination, NASA was able to keep NSCAT alive until establishing the partnership with NASDA for its flight on their *ADEOS* mission.

Securing a means to obtain ocean color observations as a follow-on to the coastal zone color scanner (CZCS) was a long and arduous process, finally coming to fruition in 1991 when a contract was signed with the OSC to purchase data from the flight of their SeaWiFS sensor. By that time, a new start had already been secured for NASA's EOS, and ample funds were available in that program for the SeaWiFS data purchase.

Finally, securing support for the Alaska SAR Facility (now the Alaska Satellite Facility to reflect its broader mission) was straightforward; being small in comparison with the cost of flying space hardware, its funding had simply been included in the new start that NSCAT obtained in FY 1985. Also, funding for utilization of SSMI data was small enough to be covered by the Oceans Program itself.

Implementing the Third Generation

With the exception of the US Navy's *Geosat*, these third-generation missions would take a very long time to come into being. As seen in **Table 1**, *TOPEX/Poseidon* was launched in 1992 – 14 years after *Seasat*; NSCAT was launched on *ADEOS* in 1996 – 18 years after *Seasat*; and SeaWiFS was launched in 1997 – 19 years after *Nimbus-7*. (In addition to the missions mentioned in **Table 1**, the Japanese *ADEOS-1* included the US NSCAT in its sensor complement, and the US *Aqua* included the Japanese advanced microwave scanning radiometer (AMSR); the United States provided a launch for the Canadian *RADARSAT-1*.) In fact, these missions came so late that they had limited overlap with the field phases of the major ocean research programs (WOCE, TOGA, and JGOFS) they were to complement. Why did it take so long?

Understanding and Consensus

First, it took time to develop a physically unambiguous understanding of how well the satellite sensors actually performed, and this involved learning to cope with the data – satellite data rates being orders of magnitude larger than those encountered in traditional oceanography. For example, it was not until 3 years after the launch of *Nimbus-7* that CZCS data could be processed as fast as collected by the satellite. And even with only a 3-month data set from *Seasat*, it took 4 years to produce the first global maps of variables such as those shown in **Figure 4**.

In evaluating the performance of both *Seasat* and *Nimbus-7*, it was necessary to have access to the data. *Seasat* had a free and open data policy; and after a very slow start, the experiment team concept (where team members had a lengthy period of exclusive access to the data) for the *Nimbus-7* CZCS was replaced with that same policy. Given access to the data, delays were due to a combination of sorting out the algorithms for converting the satellite observations into variables of interest, as well as being constrained by having limited access to raw computing power.

In addition, the rationale for the third-generation missions represented a major paradigm shift. While earlier missions had been justified largely as demonstrations of remote sensing concepts, the third-generation missions would be justified on the basis of their potential contribution to oceanography. Hence, the long time it took to understand sensor performance translated into a delay in being able to convince traditional oceanographers that satellites were an important observational tool ready to be exploited for ocean science. As this case was made, it was possible to build consensus across the remote sensing and oceanographic communities.

Space Policy

Having such consensus reflected at the highest levels of government was another matter. The *White House*

Table 1 Some major ocean-related missions

Year	USA	Russia	Japan	Europe	Canada	Other
1968		Kosmos 243				
	Nimbus-3					
1970	Nimbus-4	Kosmos 384				
1972	Nimbus-5					
1974	Skylab					
	Nimbus-6, Geos-3					
1976						
1978	Nimbus-7, Seasat					
		Kosmos 1076				
1980		Kosmos 1151				
1982						
		Kosmos 1500				
1984		Kosmos 1602				
	Geosat					
1986		Kosmos 1776				
		Kosmos 1870	MOS 1A			
1988		OKEAN 1				
1990		OKEAN 2	MOS 1B			
		Almaz-1, OKEAN 3		ERS-1		
1992	Topex/Poseidon[a]		JERS-1			
1994		OKEAN 7				
		OKEAN 8		ERS-2	RADARSAT-1	
1996			ADEOS-1			
	SeaWiFS		TRMM[b]			
1998	GFO					
	Terra, QuikSCAT	OKEAN-O #1				OCEANSAT-1[i]
2000				CHAMP[c]		
		Meteor-3 #1		Jason-1[e]		
2002	Aqua; GRACE[d]			ENVISAT		HY-1A[j]
	WINDSAT		ADEOS-2			
2004	ICESat	SICH-1M[f]				
				CRYOSat		
2006			ALOS	GOCE, MetOp-1		HY-1B[j]
		Meteor-3M #2		SMOS	RADARSAT-2	OCEANSAT-2[i]
2008				OSTM/Jason-2[g]		
	NPP; Aquarius[h]			CryoSat-2		HY-1C, HY-2A[j]
2010			GCOM-W	Sentinel-3, MetOp-2		OCEANSAT-3[i]

[a] US/France TOPEX/Poseidon.
[b] Japan/US TRMM.
[c] German CHAMP.
[d] US/German GRACE.
[e] France/US Jason-1.
[f] Russia/Ukraine Sich-1M.
[g] France/US Jason-2/OSTM.
[h] US/Argentina Aquarius.
[i] India OCEANSAT series.
[j] China HY series.

Updated version of similar data in Wilson WS, Fellous JF, Kawamura H, and Mitnik L (2006) A history of oceanography from space. In: Gower JFR (ed.) *Manual of Remote Sensing, Vol. 6: Remote Sensing of the Environment*, pp. 1–31. Bethesda, MD: American Society for Photogrammetry and Remote Sensing.

Fact Sheet on US Civilian Space Policy of 11 October 1978 states, "... emphasizing space applications ... will bring important benefits to our understanding of earth resources, climate, weather, pollution ... and provide for the private sector to take an increasing responsibility in remote sensing and other applications." *Landsat* was commercialized in 1979 as part of this space policy. As Robert Stewart explains, "Clearly the mood at the presidential level was that earth remote sensing, including the oceans, was a practical space application more at home outside the scientific community. It took almost a decade to get an

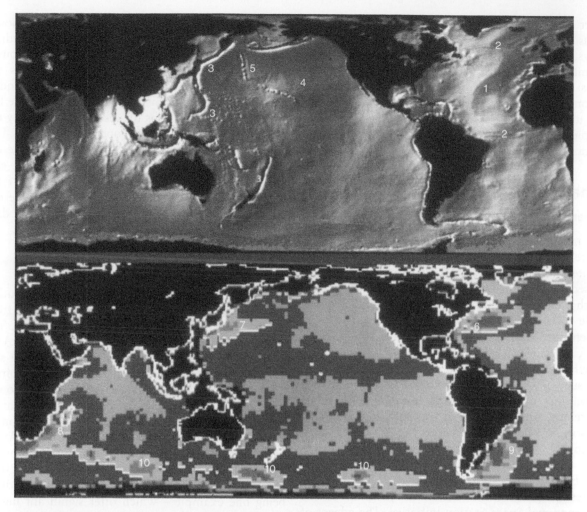

Figure 4 Global sea surface topography *c.* 1983. This figure shows results computed from the 70 days of *Seasat* altimeter data in 1978. Clearly visible in the mean sea surface topography, the marine geoid (upper panel), are the Mid-Atlantic Ridge (1) and associated fracture zones (2), trenches in the western Pacific (3), the Hawaiian Island chain (4), and the Emperor seamount chain (5). Superimposed on the mean surface is the time-varying sea surface topography, the mesoscale variability (lower panel), associated with the variability of the ocean currents. The largest deviations (10–25 cm), yellow and orange, are associated with the western boundary currents: Gulf Stream (6), Kuroshio (7), Agulhas (8), and Brazil/Falkland Confluence (9); large variations also occur in the West Wind Drift (10). Courtesy of NASA.

understanding at the policy level that scientific needs were also important, and that we did not have the scientific understanding necessary to launch an operational system for climate." The failures of NOSS, and later *NROSS*, were examples of an effort to link remote sensing directly with operational applications without the scientific underpinning.

The view in Europe was not dissimilar; governments felt that cost recovery was a viable financial scheme for ocean satellite missions, that is, the data have commercial value and the user would be willing to pay to help defray the cost of the missions.

Joint Satellite Missions

It is relatively straightforward to plan and implement missions within a single agency, as with NASA's space science program. However, implementing a satellite mission across different organizations, countries, and cultures is both challenging and time-consuming. An enormous amount of time and energy was invested in studies of various flight options, many of which fizzled out, but some were implemented. With the exception of the former Soviet Union, NASA's third-generation missions would be joint with each nation having a space program at that time, as well as with a private company.

The *Geosat* Exception

Geosat was the notable US exception, having been implemented so quickly after the second generation. It was approved in 1981 and launched in 1985 in order to address priority operational needs on the part of the US Navy. During the second half of its mission, data would become available within 1–2 days. As will be discussed below, *Geosat* shared a number of attributes with the meteorological satellites: it had a specific focus; it met priority operational needs for its user; experience was available for understanding and using the observations; and its implementation was done in the context of a single organization.

Challenges Ahead

Scientific Justification

As noted earlier, during the decade of the 1980s, there was a dearth of ocean-related missions in the United States, it being difficult to justify a mission based on its contribution to ocean science. Then later in that decade, NASA conceived of the EOS and was able to make the case that Earth science was sufficient justification for a mission. Also noted earlier, ESA initially had no appropriate framework for Earth science missions, and a project like *ERS-1* was pursued under the assumption that it would help develop commercial and/or operational applications of remote sensing of direct societal benefit. Its successor, *ERS-2*, was justified on the basis of needing continuity of SAR coverage for the land surface, rather than the need to monitor ocean currents. And the *ENVISAT* was initially decided by ESA member states as part of the Columbus program of the International Space Station initiative. The advent of an Earth Explorer program in 1999 represented a change in this situation. As a consequence, new Earth science missions – *GOCE*, *CryoSat*, and *SMOS* – all represent significant steps forward.

This ESA program, together with similar efforts at NASA, are leading to three sets of ground-breaking scientific missions which have the potential to significantly impact oceanography. *GOCE* and *GRACE* will contribute to an improved knowledge of the Earth's gravity field, as well as the mass of water on the surface of the Earth. *CryoSat-2* and *ICESat* will contribute to knowledge of the volume of water locked up in polar and terrestrial ice sheets. Finally, *SMOS* and *Aquarius* will contribute to knowledge of the surface salinity field of the global oceans. Together, these will be key ingredients in addressing the global water cycle.

Data Policy

The variety of missions described above show a mix of data policies, from full and open access without any period of exclusive use (e.g., *TOPEX/Poseidon*, *Jason*, and *QuikSCAT*) to commercial distribution

(e.g., real-time SeaWiFS for nonresearch purposes, *RADARSAT*), along with a variety of intermediate cases (e.g., *ERS*, *ALOS*, and *ENVISAT*). From a scientific perspective, full and open access is the preferred route, in order to obtain the best understanding of how systems perform, to achieve the full potential of the missions for research, and to lay the most solid foundation for an operational system. Full and open access is also a means to facilitate the development of a healthy and competitive private sector to provide value-added services. Further, if the international community is to have an effective observing system for climate, a full and open data policy will be needed, at least for that purpose.

In Situ Observations

Satellites have made an enormous contribution enabling the collection of *in situ* observations from *in situ* platforms distributed over global oceans. The Argos (plural spelling; not to be confused with *Argo* profiling floats) data collection and positioning system has flown on the NOAA series of polar-orbiting operational environmental satellites continuously since 1978. It provides one-way communication from data collection platforms, as well as positioning of those platforms. While an improved Argos capability (including two-way communications) is coming with the launch of *MetOp-1* in 2006, oceanographers are looking at alternatives – Iridium being one example – which offer significant higher data rates, as well as two-way communications.

In addition, it is important to note that the Intergovernmental Oceanographic Commission and the World Meteorological Organization have established the Joint Technical Commission for Oceanography and Marine Meteorology (JCOMM) to bring a focus to the collection, formatting, exchange, and archival of data collected at sea, whether they be oceanic or atmospheric. JCOMM has established a center, JCOMMOPS, to serve as the specific institutional focus to harmonize the national contributions of *Argo* floats, surface-drifting buoys, coastal tide gauges, and fixed and moored buoys. JCOMMOPS will play an important role helping contribute to 'integrated observations' described below.

Integrated Observations

To meet the demands of both the research and the broader user community, it will be necessary to focus, not just on satellites, but 'integrated' observing systems. Such systems involve combinations of satellite and *in situ* systems feeding observations into data-processing systems capable of delivering a comprehensive view of one or more geophysical variables (sea level, surface temperature, winds, etc.).

Three examples help illustrate the nature of integrated observing systems. First, consider global sea level rise. The combination of the *Jason-1* altimeter, its precision orbit determination system, and the suite of precision tide gauges around the globe allow scientists to monitor changes in volume of the oceans. The growing global array of *Argo* profiling floats allows scientists to assess the extent to which those changes in sea level are caused by changes in the temperature and salinity structure of the upper ocean. *ICESat* and *CryoSat* will provide estimates of changes in the volume of ice sheets, helping assess the extent to which their melting contributes to global sea level rise. And GRACE and GOCE will provide estimates of the changes in the mass of water on the Earth's surface. Together, systems such as these will enable an improved understanding of global sea level rise and, ultimately, a reduction in the wide range of uncertainty in future projections.

Second, global estimates of vector winds at the sea surface are produced from the scatterometer on *QuikSCAT*, a global array of *in situ* surface buoys, and the Seawinds data processing system. Delivery of this product in real time has significant potential to improve marine weather prediction. The third example concerns the *Jason-1* altimeter together with *Argo*. When combined in a sophisticated data assimilation system – using a state-of-the-art ocean model – these data enable the estimation of the physical state of the ocean as it changes through time. This information – the rudimentary 'weather map' depicting the circulation of the oceans – is a critical component of climate models and provides the fundamental context for addressing a broad range of issues in chemical and biological oceanography.

Transition from Research to Operations

The maturing of the discipline of oceanography includes the development of a suite of global oceanographic services being conducted in a manner similar to what exists for weather services. The delivery of these services and their associated informational products will emerge as the result of the successes in ocean science ('research push'), as well as an increasing demand for ocean analyses and forecasts from a variety of sectors ('user pull').

From the research perspective, it is necessary to 'transition' successfully demonstrated ('experimental') observing techniques into regular, long-term, systematic ('operational') observing systems to meet a broad range of user requirements, while maintaining the capability to collect long-term, 'research-quality'

observations. From the operational perspective, it is necessary to implement proven, scientifically sound, cost-effective observing systems – where the uninterrupted supply of real-time data is critical. This is a big challenge to be met by the space systems because of the demand for higher reliability and redundancy, at the same time calling for stringent calibration and accuracy requirements. Meeting these sometimes competing, but quite complementary demands will be the challenge and legacy of the next generation of ocean remote-sensing satellites.

Meteorological Institutional Experience

With the launch in 1960 of the world's first meteorological satellite, the polar-orbiting *Tiros-1* carrying two TV cameras, the value of the resulting imagery to the operational weather services was recognized immediately. The very next year a National Operational Meteorological System was implemented, with NASA to build and launch the satellites and the Weather Bureau to be the operator. The feasibility of using satellite imagery to locate and track tropical storms was soon demonstrated, and by 1969 this capability had become a regular part of operational weather forecasting. In 1985, Richard Hallgren, former Director of the National Weather Service, stated, "the use of satellite information simply permeates every aspect of the [forecast and warning] process and all this in a mere 25 years." In response to these operational needs, there has been a continuing series of more than 50 operational, polar-orbiting satellites in the United States alone!

The first meteorological satellites had a specific focus on synoptic meteorology and weather forecasting. Initial image interpretation was straightforward (i.e., physically unambiguous), and there was a demonstrated value of resulting observations in meeting societal needs. Indeed, since 1960 satellites have ensured that no hurricane has gone undetected. In addition, the coupling between meteorology and remote sensing started very near the beginning. An 'institutional mechanism' for transition from research to operations was established almost immediately. Finally, recognition of this endeavor extended to the highest levels of government, resulting in the financial commitment needed to ensure success.

Oceanographic Institutional Issue

Unlike meteorology where there is a National Weather Service in each country to provide an institutional focus, ocean-observing systems have multiple performer and user institutions whose interests must be reconciled. For oceanography, this is a significant challenge working across 'institutions', where the *in situ* research is in one or more agencies, the space research and development is in another, and operational activities in yet another – with possibly separate civil and military systems. In the United States, the dozen agencies with ocean-related responsibilities are using the National Oceanographic Partnership Program and its Ocean.US Office to provide a focus for reconciling such interests. In the United Kingdom, there is the Interagency Committee on Marine Science and Technology.

In France, there is the Comité des Directeurs d'Organismes sur l'Océanographie, which gathers the heads of seven institutions interested in the development of operational oceanography, including CNES, meteorological service, ocean research institution, French Research Institute for Exploitation of the Sea (IFREMER), and the navy. This group of agencies has worked effectively over the past 20 years to establish a satellite data processing and distribution system (AVISO), the institutional support for a continuing altimetric satellite series (*TOPEX/Poseidon*, *Jason*), the framework for the French contribution to the *Argo* profiling float program (CORIOLIS), and to create a public corporation devoted to ocean modeling and forecasting, using satellite and *in situ* data assimilation in an operational basis since 2001 (Mercator). This partnership could serve as a model in the effort to develop operational oceanography in other countries. Drawing from this experience working together within France, IFREMER is leading the European integrated project, MERSEA, aimed at establishing a basis for a European center for ocean monitoring and forecasting.

Ocean Climate

If we are to adequately address the issue of global climate change, it is essential that we are able to justify the satellite systems required to collect the global observations 'over the long term'. Whether it be global sea level rise or changes in Arctic sea ice cover or sea surface temperature, we must be able to sustain support for the systems needed to produce climate-quality data records, as well as ensure the continuing involvement of the scientific community.

Koblinsky and Smith have outlined the international consensus for ocean climate research needs and identified the associated observational requirements. In addition to their value for research, we are compelled by competing interests to demonstrate the value of such observations in meeting a broad range of societal needs. Climate observations pose

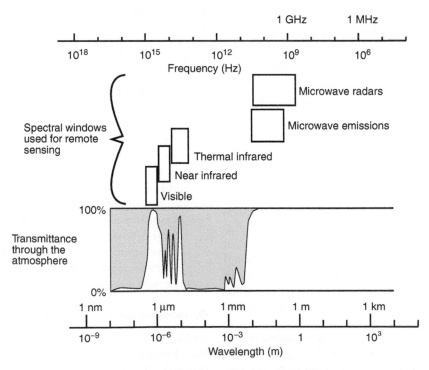

Figure 5 The electromagnetic spectrum showing atmospheric transmitance as a function of frequency and wavelength, along with the spectral windows used for remote sensing. Microwave bands are typically defined by frequency and the visible/infrared by wavelength. Adapted from Robinson IS and Guymer T (1996) Observing oceans from space. In: Summerhayes CP and Thorpe SA (eds.) *Oceanography: An Illustrated Guide*, pp. 69–87. Chichester, UK: Wiley.

challenges, since they require operational discipline to be collected in a long-term systematic manner, yet also require the continuing involvement of the research community to ensure their scientific integrity, and have impacts that may not be known for decades (unlike observations that support weather forecasting whose impact can be assessed within a matter of hours or days). Together, the institutional and observational challenges for ocean climate have been difficult to surmount.

International Integration

The paper by the Ocean Theme Team prepared under the auspices of the Integrated Global Observing Strategy (IGOS) Partnership represents how the space-faring nations are planning for the collection of global ocean observations. IGOS partners include the major global research program sponsors, global observing systems, space agencies, and international organizations.

On 31 July 2003, the First Earth Observations Summit – a high-level meeting involving ministers from over 20 countries – took place in Washington, DC, following a recommendation adopted at the G-8 meeting held in Evian the previous month; this summit proposed to 'plan and implement' a Global Earth Observation System of Systems (GEOSS). Four additional summits have been held, with participation having grown to include 60 nations and 40 international organizations; the GEOSS process provides the political visibility – not only to implement the plans developed within the IGOS Partnership – but to do so in the context of an overall Earth observation framework. This represents a remarkable opportunity to develop an improved understanding of the oceans and their influence on the Earth system, and to contribute to the delivery of improved oceanographic products and services to serve society.

Appendix 1: A Brief Overview of Satellite Remote Sensing

Unlike the severe attenuation in the sea, the atmosphere has 'windows' in which certain electromagnetic (EM) signals are able to propagate. These windows, depicted in **Figure 5**, are defined in terms

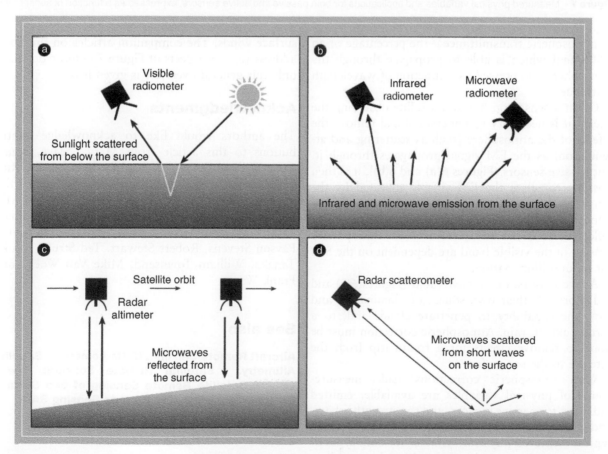

Figure 6 Four techniques for making oceanic observations from satellites: (a) visible radiometry, (b) infrared and microwave radiometry, (c) altimetry, and (d) scatterometry. Adapted from Robinson IS and Guymer T (1996) Observing oceans from space. In: Summerhayes CP and Thorpe SA (eds.) *Oceanography: An Illustrated Guide*, pp. 69–87. Chichester, UK: Wiley.

	Passive sensors (radiometers)			Active sensors (microwave radars)		
Sensor type	Visible	Infrared	Microwave	Altimetry	Scatterometry	SAR
Measured physical variable	Solar radiation backscattered from beneath the sea surface	Infrared emission from the sea surface	Microwave emission from the sea surface	Travel time, shape, and strength of reflected pulse	Strength of return pulse when illuminated from different directions	Strength and phase of return pulse
Applications	Ocean color; chlorophyll; primary production; water clarity; shallow-water bathymetry	Surface temperature; ice cover	Ice cover, age and motion; sea surface temperature; wind speed	Surface topography for geostrophic currents and tides; bathymetry; oceanic geoid; wind and wave conditions	Surface vector winds; ice cover	Surface roughness at fine spatial scales; surface and internal wave patterns; bathymetric patterns; ice cover and motion

Figure 7   Measured physical variables and applications for both passive and active sensors, expressed as a function of sensor type.

of atmospheric transmittance – the percentage of an EM signal which is able to propagate through the atmosphere – expressed as a function of wavelength or frequency.

Given a sensor on board a satellite observing the ocean, it is necessary to understand and remove the effects of the atmosphere (such as scattering and attenuation) as the EM signal propagates through it. For passive sensors (**Figures 6(a)** and **6(b)**), it is then possible to relate the EM signals collected by the sensor to the associated signals at the bottom of the atmosphere, that is, the natural radiation emitted or reflected from the sea surface. Note that passive sensors in the visible band are dependent on the Sun for natural illumination.

Active sensors, microwave radar (**Figures 6(c)** and **6(d)**), provide their own source of illumination and have the capability to penetrate clouds and, to a certain extent, rain. Atmospheric correction must be done to remove effects for a round trip from the satellite to the sea surface.

With atmospheric corrections made, measurements of physical variables are available: emitted radiation for passive sensors, and the strength, phase, and/or travel time for active sensors. **Figure 7** shows typical measured physical variables for both types of sensors in their respective spectral bands, as well as applications or derived variables of interest – ocean color, surface temperature, ice cover, sea level, and

surface winds. The companion articles on this topic address various aspects of **Figure 7** in more detail, so only this general overview is given here.

Acknowledgments

The authors would like to acknowledge contributions to this article from Mary Cleave, Murel Cole, William Emery, Michael Freilich, Lee Fu, Rich Gasparovic, Trevor Guymer, Tony Hollingsworth[†], Hiroshi Kawamura, Michele Lefebvre, Leonid Mitnik, Jean-François Minster, Richard Moore, William Patzert, Willard Pierson[†], Jim Purdom, Keith Raney, Payson Stevens, Robert Stewart, Ted Strub, Tasuku Tanaka, William Townsend, Mike Van Woert, and Frank Wentz.

See also

Aircraft Remote Sensing. IR Radiometers. Satellite Altimetry.   Satellite Remote Sensing: Ocean Colour. Satellite Remote Sensing of Sea Surface Temperatures. Satellite Remote Sensing SAR.

Further Reading

Apel JR (ed.) (1972) Sea surface topography from space. *NOAA Technical Reports: ERL No. 228, AOML No. 7.* Boulder, CO: NOAA.

Cherny IV and Raizer VY (1998) *Passive Microwave Remote Sensing of Oceans*. Chichester, UK: Praxis.

Committee on Earth Sciences (1995) *Earth Observations from Space: History, Promise, and Reality*. Washington, DC: Space Studies Board, National Research Council.

Ewing GC (1965) Oceanography from space. *Proceedings of a Conference held at Woods Hole, 24–28 August 1964. Woods Hole Oceanographic Institution Ref. No. 65-10*. Woods Hole, MA: WHOI.

Fu L-L, Liu WT, and Abbott MR (1990) Satellite remote sensing of the ocean. In: Le Méhauté B (ed.) *The Sea, Vol. 9: Ocean Engineering Science*, pp. 1193–1236. Cambridge, MA: Harvard University Press.

Guymer TH, Challenor PG, and Srokosz MA (2001) Oceanography from space: Past success, future challenge. In: Deacon M (ed.) *Understanding the Oceans: A Century of Ocean Exploration*, pp. 193–211. London: UCL Press.

JOI Satellite Planning Committee (1984) *Oceanography from Space: A Research Strategy for the Decade, 1985–1995*, parts 1 and 2. Washington, DC: Joint Oceanographic Institutions.

Kaula WM (ed.) (1969) The terrestrial environment: solid-earth and ocean physics. *Proceedings of a Conference held at William College, 11–21 August 1969, NASA CR-1579*. Washington, DC: NASA.

Kawamura H (2000) Era of ocean observations using satellites. *Sokko-Jiho* 67: S1–S9 (in Japanese).

Koblinsky CJ and Smith NR (eds.) (2001) *Observing the Oceans in the 21st Century*. Melbourne: Global Ocean Data Assimilation Experiment and the Bureau of Meteorology.

Masson RA (1991) *Satellite Remote Sensing of Polar Regions*. London: Belhaven Press.

Minster JF and Lefebvre M (1997) *TOPEX/Poseidon* satellite altimetry and the circulation of the oceans. In: Minster JF (ed.) *La Machine Océan*, pp. 111–135 (in French). Paris: Flammarion.

Ocean Theme Team (2001) *An Ocean Theme for the IGOS Partnership*. Washington, DC: NASA. http:// www.igospartners.org/docs/theme_reports/IGOS-Oceans-Final-0101.pdf (accessed Mar. 2008).

Purdom JF and Menzel WP (1996) Evolution of satellite observations in the United States and their use in meteorology. In: Fleming JR (ed.) *Historical Essays on Meteorology: 1919–1995*, pp. 99–156. Boston, MA: American Meteorological Society.

Robinson IS and Guymer T (1996) Observing oceans from space. In: Summerhayes CP and Thorpe SA (eds.) *Oceanography: An Illustrated Guide*, pp. 69–87. Chichester, UK: Wiley.

Victorov SV (1996) *Regional Satellite Oceanography*. London: Taylor and Francis.

Wilson WS (ed.) (1981) *Special Issue: Oceanography from Space. Oceanus* 24: 1–76.

Wilson WS, Fellous JF, Kawamura H, and Mitnik L (2006) A history of oceanography from space. In: Gower JFR (ed.) *Manual of Remote Sensing, Vol. 6: Remote Sensing of the Environment*, pp. 1–31. Bethesda, MD: American Society for Photogrammetry and Remote Sensing.

Wilson WS and Withee GW (2003) A question-based approach to the implementation of sustained, systematic observations for the global ocean and climate, using sea level as an example. *MTS Journal* 37: 124–133.

Relevant Websites

http://www.aviso.oceanobs.com
– AVISO.
http://www.coriolis.eu.org
– CORIOLIS.
http://www.eohandbook.com
– Earth Observation Handbook, CEOS.
http://www.igospartners.org
– IGOS.
http://wo.jcommops.org
– JCOMMOPS.
http://www.mercator-ocean.fr
– Mercator Ocean.

SATELLITE PASSIVE-MICROWAVE MEASUREMENTS OF SEA ICE

C. L. Parkinson, NASA Goddard Space Flight Center, Greenbelt, MD, USA

Published by Elsevier Ltd.

Introduction

Satellite passive-microwave measurements of sea ice have provided global or near-global sea ice data for most of the period since the launch of the *Nimbus 5* satellite in December 1972, and have done so with horizontal resolutions on the order of 25–50 km and a frequency of every few days. These data have been used to calculate sea ice concentrations (percent areal coverages), sea ice extents, the length of the sea ice season, sea ice temperatures, and sea ice velocities, and to determine the timing of the seasonal onset of melt as well as aspects of the ice-type composition of the sea ice cover. In each case, the calculations are based on the microwave emission characteristics of sea ice and the important contrasts between the microwave emissions of sea ice and those of the surrounding liquid-water medium.

The passive-microwave record is most complete since the launch of the scanning multichannel microwave radiometer (SMMR) on the *Nimbus 7* satellite in October 1978; and the SMMR data and follow-on data from the special sensor microwave imagers (SSMIs) on satellites of the United States Defense Meteorological Satellite Program (DMSP) have been used to determine trends in the ice covers of both polar regions since the late 1970s. The data have revealed statistically significant decreases in Arctic sea ice coverage and much smaller magnitude increases in Antarctic sea ice coverage.

Background on Satellite Passive-Microwave Sensing of Sea Ice

Rationale

Sea ice is a vital component of the climate of the polar regions, insulating the oceans from the atmosphere, reflecting most of the solar radiation incident on it, transporting cold, relatively fresh water toward equator, and at times assisting overturning in the ocean and even bottom water formation through its rejection of salt to the underlying water. Furthermore, sea ice spreads over vast distances, globally covering an area approximately the size of North America, and it is highly dynamic, experiencing a prominent annual cycle in both polar regions and many short-term fluctuations as it is moved by winds and waves, melted by solar radiation, and augmented by additional freezing. It is a major player in and indicator of the polar climate state and has multiple impacts on all levels of the polar marine ecosystems. Consequently it is highly desirable to monitor the sea ice cover on a routine basis. In view of the vast areal coverage of the ice and the harsh polar conditions, the only feasible means of obtaining routine monitoring is through satellite observations. Visible, infrared, active-microwave, and passive-microwave satellite instruments are all proving useful for examining the sea ice cover, with the passive-microwave instruments providing the longest record of near-complete sea ice monitoring on a daily or near-daily basis.

Theory

The tremendous value of satellite passive-microwave measurements for sea ice studies results from the combination of the following four factors:

1. Microwave emissions of sea ice differ noticeably from those of seawater, making sea ice generally readily distinguishable from liquid water on the basis of the amount of microwave radiation received by the satellite instrument. For example, **Figure 1** presents color-coded images of the data from one channel on a satellite passive-microwave instrument, presented in units (termed 'brightness temperatures') indicative of the intensity of emitted microwave radiation at that channel's frequency, 19.4 GHz. The ice edge, highlighted by the broken white curve, is readily identifiable from the brightness temperatures, with open-ocean values of 172–198 K outside the ice edge and sea ice values considerably higher, predominantly greater than 230 K, within the ice edge.

2. The microwave radiation received by Earth-orbiting satellites derives almost exclusively from the Earth system. Hence, microwave sensing does not require sunlight, and the data can be collected irrespective of the level of daylight or darkness. This is a major advantage in the polar latitudes, where darkness lasts for months at a time, centered on the winter solstice.

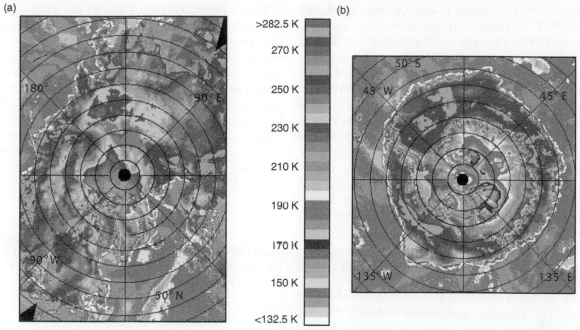

Figure 1 Late-winter brightness temperature images of 19.4-GHz vertically polarized (19.4 V) data from the DMSP SSMI for (a) the north polar region on 15 Mar. 1998 and (b) the south polar region on 15 Sep. 1998, showing near-maximum sea ice coverage in each hemisphere. The broken white curve has been added to indicate the location of the sea ice edge. Black signifies areas of no data; the absence of data poleward of 87.6° latitude results from the satellite's near-polar orbit and is consistent throughout the SSMI data set.

3. Many of the microwave data are largely un-affected by atmospheric conditions, including the presence or absence of clouds. Storm systems can produce atmospheric interference, but, at selected wavelengths, the microwave signal from the ice–ocean surface can pass through most nonprecipitating clouds essentially unhindered. Hence, microwave sensing of the surface does not require cloud-free conditions.

4. Satellite passive-microwave instruments can obtain a global picture of the sea ice cover at least every few days with a resolution of 50 km or better, providing reasonable spatial resolution and extremely good temporal resolution for most large-scale or climate-related studies.

Satellite Passive-Microwave Instruments

The first major satellite passive-microwave imager was the electrically scanning microwave radiometer (ESMR) launched on the *Nimbus 5* satellite of the United States National Aeronautics and Space Administration (NASA) in December 1972, preceded by a nonscanning passive-microwave radiometer launched on the Russian Cosmos satellite in September 1968. The ESMR was a single-channel instrument recording radiation at a wavelength of 1.55 cm and corresponding frequency of 19.35 GHz. It collected good-quality data for much of the 4-year period from January 1973 through December 1976, although with some major data gaps, including one that lasted for 3 months, from June through August 1975. Being a single-channel instrument, it did not allow some of the more advanced studies that have been done with subsequent instruments, but its flight was a highly successful proof-of-concept mission, establishing the value of satellite passive-microwave technology for observing the global sea ice cover and other variables. The ESMR data were used extensively in the determination and analysis of sea ice conditions in both the Arctic and the Antarctic over the 4 years 1973–76. Emphasis centered on the determination of ice concentrations (percent areal coverages of ice) and, based on the ice concentration results, the calculation of ice extents (integrated areas of all grid elements with ice concentration ≥15%). This 4-year data set established key aspects of the annual cycles of the polar sea ice covers, including the non-uniformity of the growth and decay seasons and the marked interannual differences even within a 4-year data set.

The *Nimbus 5* ESMR was followed by a less successful ESMR on the *Nimbus 6* satellite and then by the more advanced 10-channel SMMR on board NASA's *Nimbus 7* satellite and a sequence of seven-channel SSMIs on board satellites of the DMSP.

Nimbus 7 was launched in late October 1978, and the SMMR on board was operational through mid-August 1987. The first of the DMSP SSMIs was operational as of early July 1987, providing a welcome data overlap with the *Nimbus 7* SMMR and thereby allowing intercalibration of the SMMR and SSMI data sets. SSMIs continue to operate into the twenty-first century. There was also an SMMR on board the short-lived *Seasat* satellite in 1978; and there was a two-channel microwave scanning radiometer (MSR) on board the Japanese Marine Observation Satellites starting in February 1987. Each of these successor satellite passive-microwave instruments, after the ESMR, has been multichannel, allowing both an improved accuracy in the ice concentration derivations and the calculation of additional sea ice variables, including ice temperature and the concentrations of separate ice types.

The Japanese developed a 12-channel advanced microwave scanning radiometer (AMSR) for the Earth Observing System's (EOS) *Aqua* satellite (formerly named the *PM-1* satellite), launched by NASA in May 2002, and for the Advanced Earth Observing Satellite II (*ADEOS-II*), launched by the Japanese National Space Development Agency in December 2002. *ADEOS-II* prematurely ceased operations in October 2003, but the *Aqua* AMSR, labeled AMSR-E in recognition of its place in the EOS and to distinguish it from the *ADEOS-II* AMSR, has collected a multiyear data record. The AMSR-E has a major advantage over the SMMR and SSMI instruments in allowing sea ice measurements at a higher spatial resolution (12–25 km vs. 25–50 km for the major derived sea ice products). It furthermore has an additional advantage over the SMMR in having channels at 89 GHz in addition to its lower-frequency channels, at 6.9, 10.7, 18.7, 23.8, and 36.5 GHz.

Sea Ice Determinations from Satellite Passive-Microwave Data

Sea Ice Concentrations

Ice concentration is among the most fundamental and important parameters for describing the sea ice cover. Defined as the percent areal coverage of ice, it is directly critical to how effectively the ice cover restricts exchanges between the ocean and the atmosphere and to how much incoming solar radiation the ice cover reflects. Ice concentration is calculated at each ocean grid element, for whichever grid is being used to map or otherwise display the derived satellite products. A map of ice concentrations presents the areal distribution of the ice cover, to the resolution of the grid.

With a single channel of microwave data, taken at a radiative frequency and polarization combination that provides a clear distinction between ice and water, approximate sea ice concentrations can be calculated by assuming a uniform radiative brightness temperature TB_w for water and a uniform radiative brightness temperature TB_I for ice, with both brightness temperatures being appropriate for the values received at the altitude of the satellite, that is, incorporating an atmospheric contribution. Assuming no other surface types within the field of view, the observed brightness temperature TB is given by

$$TB = C_w TB_w + C_I TB_I \qquad [1]$$

C_w is the percent areal coverage of water and C_I is the ice concentration. With only the two surface types, $C_w + C_I = 1$, and eqn [1] can be expressed as

$$TB = (1 - C_I)TB_w + C_I TB_I \qquad [2]$$

This is readily solved for the ice concentration:

$$C_I = \frac{TB - TB_w}{TB_I - TB_w} \qquad [3]$$

Equation [3] is the standard equation used for the calculation of ice concentrations from a single channel of microwave data, such as the data from the ESMR instrument. A major limitation of the formulation is that the polar ice cover is not uniform in its microwave emission, so that the assumption of a uniform TB_I for all sea ice is only a rough approximation, far less justified than the assumption of a uniform TB_w for seawater, although that also is an approximation.

Multichannel instruments allow more sophisticated, and generally more accurate, calculation of the ice concentrations. They additionally allow many options as to how these calculations can be done. To illustrate the options, two algorithms will be described, both of which assume two distinct ice types, thereby advancing over the assumption of a single ice type made in eqns [1]–[3] and being particularly appropriate for the many instances in which two ice types dominate the sea ice cover. For this approximation, the assumption is that the field of view contains only water and two ice types, type 1 ice and type 2 ice (see the section 'Sea ice types' for more information on ice types), and that the three surface types have identifiable brightness temperatures, TB_w, TB_{I1}, and TB_{I2}, respectively. Labeling the concentrations of the two ice types as C_{I1} and C_{I2}, respectively, the percent coverage of water is $1 - C_{I1} - C_{I2}$, and the integrated observed brightness temperature is

$$TB = (1 - C_{I1} - C_{I2})TB_w + C_{I1}TB_{I1} + C_{I2}TB_{I2} \quad [4]$$

With two channels of information, as long as appropriate values for TB_w, TB_{I1}, and TB_{I2} are known for each of the two channels, eqn [4] can be used individually for each channel, yielding two linear equations in the two unknowns C_{I1} and C_{I2}. These equations are immediately solvable for C_{I1} and C_{I2}, and the total ice concentration C_I is then given by

$$C_I = C_{I1} + C_{I2} \quad [5]$$

Although the scheme described in the preceding paragraph is a marked advance over the use of a single-channel calculation (eqn [3]), most algorithms for sea ice concentrations from multichannel data make use of additional channels and concepts to further improve the ice concentration accuracies. A frequently used algorithm (termed the NASA Team algorithm) for the SMMR data employs three of the 10 SMMR channels, those recording horizontally polarized radiation at a frequency of 18 GHz and vertically polarized radiation at frequencies of 18 and 37 GHz. The algorithm is based on both the polarization ratio (PR) between the 18-GHz vertically polarized data (abbreviated 18 V) and the 18-GHz horizontally polarized data (18 H) and the spectral gradient ratio (GR) between the 37-GHz vertically polarized data (37 V) and the 18-V data. PR and GR are defined as:

$$PR = \frac{TB(18\ V) - TB(18\ H)}{TB(18\ V) + TB(18\ H)} \quad [6]$$

$$GR = \frac{TB(37\ V) - TB(18\ V)}{TB(37\ V) + TB(18\ V)} \quad [7]$$

Substituting into eqns [6] and [7] expanded forms of TB(18 V), TB(18 H), and TB(37 V) obtained from eqn [4], the result yields equations for PR and GR in the two unknowns C_{I1} and C_{I2}. Solving for C_{I1} and C_{I2} yields two algebraically messy but computationally straightforward equations for C_{I1} and C_{I2} based on PR, GR, and numerical coefficients determined exclusively from the brightness temperature values assigned to water, type 1 ice, and type 2 ice for each of the three channels (these assigned values are termed 'tie points' and are determined empirically). These are the equations that are then used for the calculation of the concentrations of type 1 and type 2 ice once the observations are made and are used to calculate PR and GR from eqns [6] and [7]. The total ice concentration C_I is then obtained from eqn [5]. The use of PR and GR in this formulation reduces the impact of ice temperature variations on the ice

concentration calculations. This algorithm is complemented by a weather filter that sets to 0 all ice concentrations at locations and times with a GR value exceeding 0.07. The weather filter eliminates many of the erroneous calculations of sea ice presence arising from the influence of storm systems on the microwave data.

For the SSMI data, the same basic NASA Team algorithm is used, although 18 V and 18 H in eqns [6] and [7] are replaced by 19.4 V and 19.4 H, reflecting the placement on the SSMI of channels at a frequency of 19.4 GHz rather than the 18-GHz channels on the SMMR. Also, because the data from the 19.4-GHz channels tend to be more contaminated by water vapor absorption/emission and other weather effects than the 18-GHz data, the weather filter for the SSMI calculations incorporates a threshold level for the GR calculated from the 22.2-GHz vertically polarized data and 19.4-V data as well as a threshold for the GR calculated from the 37 V and 19.4-V data. To illustrate the results of this ice concentration algorithm, **Figure 2** presents the derived sea ice concentrations for 15 March 1998 in the Northern Hemisphere and for 15 September 1998 in the Southern Hemisphere, the same dates as used in **Figure 1**.

As mentioned, there are several alternative ice concentration algorithms in use. Contrasts from the NASA Team algorithm just described include: use of different microwave channels; use of regional tie points rather than hemispherically applicable tie points; use of cluster analysis on brightness temperature data, without PR and GR formulations; use of iterative techniques whereby an initial ice concentration calculation leads to refined atmospheric temperatures and opacities, which in turn lead to refined ice concentrations; use of iterative techniques involving surface temperature, atmospheric water vapor, cloud liquid water content, and wind speed; incorporation of higher-frequency data to reduce the effects of snow cover on the computations; and use of a Kalman filtering technique in conjunction with an ice model. The various techniques tend to obtain very close overall distributions of where the sea ice is located, although sometimes with noticeable differences (up to 20%, and on occasion even higher) in the individual, local ice concentrations. The differences can often be markedly reduced by adjustment of tunable parameters, such as the algorithm tie points, in one or both of the algorithms being compared. However, the lack of adequate ground data often makes it difficult to know which tuning is most appropriate or which algorithm is yielding the better results. To help resolve the uncertainties, *in situ* and aircraft measurements are being made in both the Arctic and the

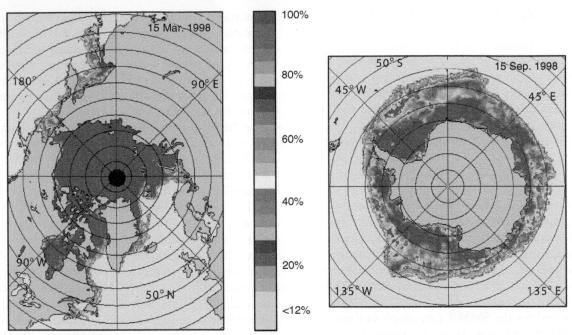

Figure 2 North and south polar sea ice concentration images for 15 Mar. 1998 and 15 Sep. 1998, respectively. The ice concentrations are derived from the data of the DMSP SSMI, including the 19.4-V data depicted in **Figure 1**.

Antarctic to help validate and improve the AMSR-E sea ice products.

Whichever ice concentration algorithm is used, the result provides estimates of ice coverage, not ice thickness. In cases where ice thickness data are also available, the combination of ice concentration and ice thickness allows the calculation of ice volume. Ice thickness data, however, are quite limited both spatially and over time. Furthermore, they are generally not derived from the passive-microwave observations and so are not highlighted in this article. The limited ice thickness data traditionally have come from *in situ* and submarine measurements, although some recent data are also available from satellite radar altimetry and, since the January 2003 launch of the Ice, Cloud and land Elevation Satellite (*ICE-Sat*), from satellite laser altimetry. The laser technique appears promising, although the laser on board ICESat operates only for short periods, and hence the full value of the technique will not be realized until a new laser is launched with a longer operational lifetime. Average sea ice thickness in the Antarctic is estimated to be in the range 0.4–1.5 m, and average sea ice thickness in the Arctic is estimated to be in the range 1.5–3.5 m.

Sea Ice Extents and Trends

Sea ice extent is defined as the total ocean area of all grid cells with sea ice concentration of at least 15% (or, occasionally, an alternative prescribed minimum percentage). Sea ice extents are now routinely calculated for the north polar region as a whole, for the south polar region as a whole, and for each of several subregions within the two polar domains, using ice concentration maps determined from satellite passive-microwave data.

A major early result from the use of satellite passive-microwave data was the detailed determination of the seasonal cycle of ice extents in each hemisphere. Incorporating the interannual variability observed from the 1970s through the early twenty-first century, the Southern Ocean ice extents vary from about $2\text{–}4 \times 10^6\,\text{km}^2$ in February to about $17\text{–}20 \times 10^6\,\text{km}^2$ in September, and the north polar ice extents vary from about $5\text{–}8 \times 10^6\,\text{km}^2$ in September to about $14\text{–}16 \times 10^6\,\text{km}^2$ in March. The exact timing of minimum and maximum ice coverage and the smoothness of the growth from minimum to maximum and the decay from maximum to minimum vary noticeably among the different years.

As the data sets lengthened, a major goal became the determination of trends in the ice extents and the placement of these trends in the context of other climate variables and climate change. Because of the lack of a period of data overlap between the ESMR and the SMMR, matching of the ice extents derived from the ESMR data to those derived from the SMMR and SSMI data has been difficult and uncertain. Consequently, most published results regarding trends found from the SMMR and SSMI data sets do not include the ESMR data.

The SMMR/SSMI record from late 1978 until the early twenty-first century indicates an overall decrease in Arctic sea ice extents of about 3% per decade and an overall increase in Antarctic sea ice extents of about 1% per decade. The Arctic ice decreases have received particular attention because they coincide with a marked warming in the Arctic and likely are tied closely to that warming. Although the satellite data reveal significant interannual variability in the ice cover, even as early as the late 1980s it had become clear from the satellite record that the Arctic as a whole had lost ice since the late 1970s. The picture was mixed regionally, with some regions having lost ice and others having gained ice; and with such a short data record, there was a strong possibility that the decreases through the 1980s were part of an oscillatory pattern and would soon reverse. However, although the picture remained complicated by interannual variability and regional differences, the Arctic decreases overall continued (with fluctuations) through the 1990s and into the twenty-first century, and by the middle of the first decade of the twenty-first century no large-scale regions of the Arctic showed overall increases since late 1978. Moreover, by the early twenty-first century the Arctic sea ice decreases were apparent in all seasons of the year, and the decreases in ice extent found from satellite data were complemented by decreases in ice thickness found from submarine and *in situ* data.

The satellite-derived Arctic sea ice decreases are illustrated in **Figure 3** with 26-year March and September time series for the Northern Hemisphere sea

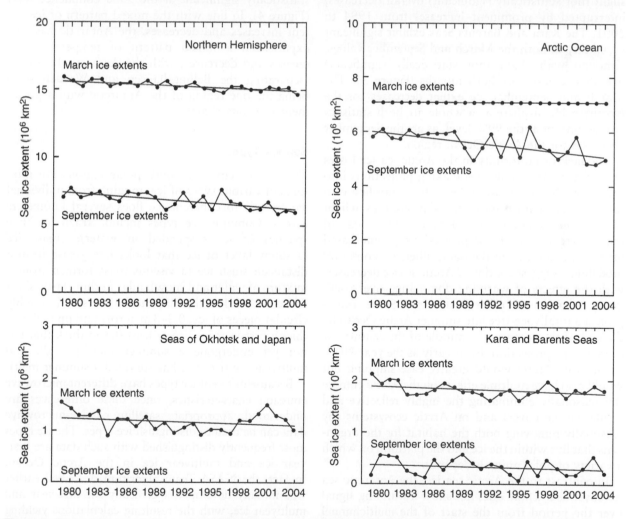

Figure 3 Time series of monthly average 1979–2004 March and September sea ice extents for the Northern Hemisphere and the following three regions within the Northern Hemisphere: the Arctic Ocean, the Seas of Okhotsk and Japan, and the Kara and Barents Seas. All ice extents are derived from the *Nimbus 7* SMMR and DMSP SSMI satellite passive-microwave data. The trend lines are linear least squares fits through the data points, and the slopes of the trend lines for the Northern Hemisphere total are $-29\,500 \pm 5400\,\text{km}^2\,\text{yr}^{-1}$ ($-1.9 \pm 0.35\%$ per decade) for the March values and $-51\,300 \pm 10\,800\,\text{km}^2\,\text{yr}^{-1}$ ($6.9 \pm 1.5\%$ per decade) for the September values. For the Seas of Okhotsk and Japan, all the September ice extents are $0\,\text{km}^2$, as the ice cover fully disappeared from these seas by the end of summer in each of the 26 years.

ice cover as a whole and for three regions within the Northern Hemisphere ice cover, those being the Arctic Ocean, the Seas of Okhotsk and Japan, and the Kara and Barents Seas. March and September are typically the months of maximum and minimum Northern Hemisphere sea ice coverage. Among the regional and seasonal differences visible in the plots, the Arctic Ocean shows little or no variation in March ice extents because of consistently being fully covered with ice in March, to at least 15% ice coverage in each grid square, but shows noticeable fluctuations and overall decreases in September ice coverage. In contrast, the region of the Seas of Okhotsk and Japan has no variability in September, because of having no ice in any of the Septembers, but for March shows marked fluctuations and slight (not statistically significant) overall decreases, interrupted by prominent increases from 1994 to 2001. The Kara and Barents Seas exhibit significant variability in both the March and September values, although with slight (not statistically significant) overall decreases for both months (**Figure 3**). The March and September ice extent decreases for the Northern Hemisphere as a whole are both statistically significant at the 99% level, as are the September decreases for the Arctic Ocean region.

The lack of uniformity in the Arctic sea ice losses (e.g., **Figure 3**) complicates making projections into the future. Nonetheless, in the early twenty-first century, several scientists have offered projections in light of the expectation of continued warming of the climate system. These projections – some based on extrapolation from the data, others on computer modeling – suggest continued Arctic sea ice decreases, with the effects of warming dominating over oscillatory behavior and other fluctuations. Some studies project a totally ice-free late summer Arctic Ocean by the end of the century, the middle of the century, or even, in one projection, by as early as the year 2013. An ice-free Arctic would greatly ease shipping but would also have multiple effects on the Arctic climate (by seasonally eliminating the highly reflective and insulating ice cover) and on Arctic ecosystems (by seasonally removing both the habitat for the organisms that live within the ice and the platform on which a variety of polar animals depend).

In contrast to the Arctic sea ice, the Antarctic sea ice cover as a whole does not show a warming signal over the period from the start of the multichannel satellite passive-microwave record, in late 1978, through the end of 2004. Over this period, some regions in the Antarctic experienced overall ice cover increases and other regions experienced overall ice cover decreases, with the hemispheric 1% per decade ice extent increases incorporating contrasting regional conditions. In **Figure 4**, the 26-year February (generally the month of Antarctic sea ice minimum) and September (generally the month of Antarctic sea ice maximum) ice extent time series are plotted for the Southern Hemisphere total, the Weddell Sea, the Bellingshausen and Amundsen Seas, and the Ross Sea. The region of the Bellingshausen and Amundsen Seas shows statistically significant (99% confidence level) February sea ice decreases but shows very slight (not statistically significant) September sea ice increases. The Weddell Sea instead shows statistically significant (95% confidence level) increases in February ice coverage and slight (not statistically significant) decreases in September ice coverage, while the Ross Sea shows ice increases in both months, with the February increases being statistically significant at the 99% confidence level (**Figure 4**). In line with the mixed pattern of ice extent increases and decreases, the Antarctic has also experienced a mixed pattern of temperature increases and decreases, with the Antarctic Peninsula (separating the Bellingshausen and Weddell seas) being the one region of the Antarctic with a prominent warming signal.

Sea Ice Types

The sea ice covers in both polar regions are mixtures of various types of ice, ranging from individual ice crystals to coherent ice floes several kilometers across. Common ice types include frazil ice (fine spicules of ice suspended in water); grease ice (a soupy layer of ice that looks like grease from a distance); slush ice (a viscous mass formed from a mixture of water and snow); nilas (a thin elastic sheet of ice 0.01–0.1-m thick); pancake ice (small, roughly circular pieces of ice, 0.3–3 m across and up to 0.1-m thick); first-year ice (ice at least 0.3-m thick that has not yet undergone a summer melt period); and multiyear ice (ice that has survived a summer melt).

Because different ice types have different microwave emission characteristics, once these differences are understood, appropriate satellite passive-microwave data can be used to distinguish ice types. The ice types most frequently distinguished with such data are first-year ice and multiyear ice in the Arctic Ocean. In fact, the NASA Team algorithm described earlier was initially developed specifically for first-year and multiyear ice, with the resulting calculations yielding the concentrations, C_{I1} and C_{I2}, of those two ice types. First-year and multiyear ice are distinguishable in their microwave signals because the summer melt process drains some of the salt content downward through the ice, reducing the salinity of the upper layers of the ice and thereby changing the microwave emissions; these

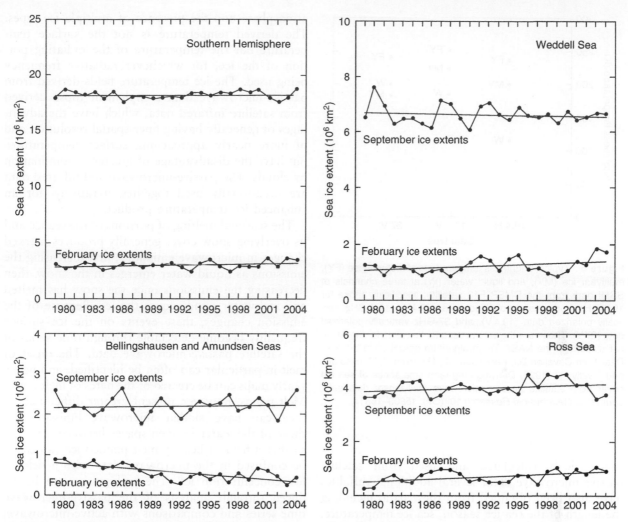

Figure 4   Time series of monthly average 1979–2004 February and September sea ice extents for the Southern Hemisphere and the following three regions within the Southern Hemisphere: the Weddell Sea, the Bellingshausen and Amundsen Seas, and the Ross Sea. All ice extents are derived from the *Nimbus 7* SMMR and DMSP SSMI satellite passive-microwave data. The trend lines are linear least squares fits through the data points, and the slopes of the trend lines for the Southern Hemisphere total are $14\,100\pm8000\,km^2\,yr^{-1}$ ($5.0\pm2.8\%$ per decade) for the February values and $7000\pm8200\,km^2\,yr^{-1}$ ($0.4\pm0.5\%$ per decade) for the September values.

changes are dependent on the frequency and polarization of the radiation. To illustrate the differences, **Figure 5** presents a plot of the tie points employed in the NASA Team algorithm for the Arctic ice. Tie points were determined empirically and are included for each of the three SSMI channels used in the calculation of ice concentrations prior to the application of the weather filter. The plot shows that while the transition from first-year to multiyear ice lowers the brightness temperatures for each of the three channels, the reduction is greatest for the 37-V data and least for the 19.4-V data. The plot further reveals that the polarization PR (eqn [6], revised for 19.4 GHz rather than 18-GHz data) is larger for multiyear ice than for first-year ice, and larger for

water than for either ice type. Furthermore, the GR (eqn [7], revised for 19.4-GHz data) is positive for water, slightly negative for first-year ice, and considerably more negative for multiyear ice (**Figure 5**). The differences allow the sorting out, either through the calculation of C_{I1} and C_{I2} as described earlier or through alternative algorithms, of the first-year ice and multiyear ice percentages in the satellite field of view.

Other Sea Ice Variables: Season Length, Temperature, Melt, Velocity

Although ice concentrations, ice extents, and, to a lesser degree, ice types have been the sea ice variables

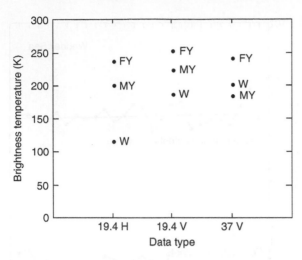

Figure 5 Typical brightness temperatures of first-year ice (FY), multiyear ice (MY), and liquid water (W) at three channels of SSMI data from the DMSP F13 satellite, specifically those for 19.4-GHz horizontally polarized data (19.4 H), 19.4-GHz vertically polarized data (19.4 V), and 37-GHz vertically polarized data (37 V). These are the values used as tie points for the Arctic calculations in the NASA Team algorithm described in the text. Data from Cavalieri DJ, Parkinson CL, Gloersen P, Comiso JC, and Zwally HJ (1999) Deriving long-term time series of sea ice cover from satellite passive-microwave multisensor data sets. *Journal of Geophysical Research* 104(C7): 15803–15814.

atmosphere and the presence of multiple ice types. The derived temperature is not the surface temperature but the temperature of the radiating portion of the ice, for whichever radiative frequency being used. The ice temperature fields derived from passive-microwave data complement those derived from satellite infrared data, which have the advantages of generally having finer spatial resolution and of more nearly approaching surface temperatures but have the disadvantage of greater contamination by clouds. The passive-microwave and infrared data are occasionally used together, iteratively, for an enhanced ice temperature product.

The seasonal melting of portions of the sea ice and its overlying snow cover generally produces marked changes in microwave emissions, first increasing the emissions as liquid water emerges in the snow, then decreasing the emissions once the snow has melted and meltwater ponds cover the ice. Because of the emission changes, these events on the ice surface frequently become detectable through time series of the satellite passive-microwave data. The onset of melt in particular can often be identified, and hence yearly maps can be created of the dates of melt onset. Melt ponds, however, present greater difficulties, as they can have similar microwave emissions to those of the water in open spaces between ice floes, so that a field of heavily melt-ponded ice can easily be confused in the microwave data with a field of low-concentration ice. The ambiguities can be reduced through analysis of the passive-microwave time series and comparisons with active-microwave, visible, and infrared data. Still, because of these complications under melt conditions, the passive-microwave-derived ice concentrations tend to have larger uncertainties for summertime ice than for wintertime ice.

The calculation of sea ice velocities from satellite data has in general relied upon data with fine enough resolution to distinguish individual medium-sized ice floes, such as visible data and active-microwave data rather than the much coarser resolution passive-microwave data. However, in the 1990s, several groups devised methods of determining ice velocity fields from passive-microwave data, some using techniques based on cross-correlation of brightness temperature fields and others using wavelet analysis. These techniques have yielded ice velocity maps on individual dates for the entire north and south polar sea ice covers. Comparisons with buoy and other data have been quite encouraging regarding the potential of using the passive-microwave satellite data for long-term records and monitoring of ice motions.

most widely calculated and used from satellite passive-microwave data, several additional variables have also been obtained from these data, including the length of the sea ice season, sea ice temperature, sea ice melt, and sea ice velocity. The length of the sea ice season for any particular year is calculated directly from that year's daily maps of sea ice concentrations, by counting, at each grid element, the number of days with ice coverage of at least some predetermined (generally 15% or 30%) ice concentration. Trends in the length of the sea ice season from the late 1970s to the early twenty-first century show coherent spatial patterns in both hemispheres, with a predominance of negative values (shortening of the sea ice season) in the Northern Hemisphere and in the vicinity of the Antarctic Peninsula and a much lesser predominance of positive values in the rest of the Southern Hemisphere's sea ice region, consistent with the respective hemispheric trends in sea ice extents.

The passive-microwave-based ice temperature calculations generally depend on the calculated sea ice concentrations, empirically determined ice emissivities, a weighting of the water and ice temperatures within the field of view, and varying levels of sophistication in incorporating effects of the polar

Looking toward the Future

Monitoring of the polar sea ice covers through satellite passive-microwave technology is ongoing with the operational SSMI instruments on the DMSP satellites and the Japanese AMSR-E instrument on NASA's Aqua satellite. Both Japan and the United States anticipate launching additional passive-microwave instruments to maintain an uninterrupted satellite passive-microwave data record. The resulting lengthening sea ice records should continue to provide an improved basis with which scientists can examine trends in the sea ice cover and interactions between the sea ice and other elements of the climate system. For instance, the lengthened records will be essential to answering many of the questions raised concerning whether the negative overall trends found in Arctic sea ice extents for the first quarter century of the SMMR/SSMI record will continue and how these trends relate to temperature trends, in particular to climate warming, and to oscillations within the climate system, in particular the North Atlantic Oscillation, the Arctic Oscillation, and the Southern Oscillation. In addition to covering a longer period, other expected improvements in the satellite passive-microwave record of sea ice include further algorithm developments, following additional analyses of the microwave properties of sea ice and liquid water. Such analyses are likely to lead both to improved algorithms for the variables already examined and to the development of techniques for calculating additional sea ice variables from the satellite data.

Glossary

brightness temperature Unit used to express the intensity of emitted microwave radiation received by the satellite, presented in temperature units (K) following the Rayleigh–Jeans approximation to Planck's law, whereby the radiation emitted from a perfect emitter at microwave wavelengths is proportional to the emitter's physical temperature.

sea ice concentration Percent areal coverage of sea ice.

sea ice extent Integrated area of all grid elements with sea ice concentration of at least 15%.

See also

Acoustics, Arctic. Satellite Oceanography, History and Introductory Concepts. Satellite Remote Sensing SAR.

Further Reading

Barry RG, Maslanik J, Steffen K, et al. (1993) Advances in sea-ice research based on remotely sensed passive microwave data. Oceanography 6(1): 4–12.

Carsey FD (ed.) (1992) Microwave Remote Sensing of Sea Ice. Washington, DC: American Geophysical Union.

Cavalieri DJ, Parkinson CL, Gloersen P, Comiso JC, and Zwally HJ (1999) Deriving long-term time series of sea ice cover from satellite passive-microwave multisensor data sets. Journal of Geophysical Research 104(C7): 15803–15814.

Comiso JC, Yang J, Honjo S, and Krishfield RA (2003) Detection of change in the Arctic using satellite and in situ data. Journal of Geophysical Research 108(C12): 3384 (doi:10.1029/2002JC001347).

Gloersen P, Campbell WJ, Cavalieri DJ, et al. (1992) Arctic and Antarctic Sea Ice, 1978–1987: Satellite Passive Microwave Observations and Analysis. Washington, DC: National Aeronautics and Space Administration.

Jeffries MO (ed.) (1998) Antarctic Sea Ice: Physical Processes, Interactions and Variability. Washington, DC: American Geophysical Union.

Johannessen OM, Bengtsson L, Miles MW, et al. (2004) Arctic climate change: Observed and modelled temperature and sea-ice variability. Tellus 56A(4): 328–341.

Kramer HJ (2002) Observation of the Earth and Its Environment, 4th edn. Berlin: Springer.

Lubin D and Massom R (2006) Polar Remote Sensing, Vol. 1: Atmosphere and Oceans. Berlin: Springer-Praxis.

Parkinson CL (1997) Earth from Above: Using Color-Coded Satellite Images to Examine the Global Environment. Sausalito, CA: University Science Books.

Parkinson CL (2004) Southern Ocean sea ice and its wider linkages: Insights revealed from models and observations. Antarctic Science 16(4): 387–400.

Smith WO, Jr. and Grebmeier JM (eds.) (1995) Arctic Oceanography: Marginal Ice Zones and Continental Shelves. Washington, DC: American Geophysical Union.

Thomas DN and Dieckmann GS (eds.) (2003) Sea Ice: An Introduction to Its Physics, Chemistry, Biology and Geology. Oxford, UK: Blackwell Science.

Ulaby FT, Moore RK, and Fung AK (eds.) (1986) Monitoring sea ice. In: Microwave Remote Sensing: Active and Passive, Vol. III: From Theory to Applications, pp. 1478–1521. Dedham, MA: Artech House.

Walsh JE, Anisimov O, Hagen JOM, et al. (2005) Cryosphere and hydrology. In: Symon C, Arris L, and Heal B (eds.) Arctic Climate Impact Assessment, pp. 183–242. Cambridge, UK: Cambridge University Press.

Relevant Websites

http://www.awi-bremerhaven.de
– Alfred Wegener Institute for Polar and Marine Research.

http://www.antarctica.ac.uk
  – Antarctica, British Antarctic Survey.
http://www.arctic.noaa.gov
  – Arctic Change, NOAA Arctic Theme Page.
http://www.aad.gov.au
  – Australian Antarctic Division.
http://www.dcrs.dtu.dk
  – Danish Center for Remote Sensing.
http://www.jaxa.jp
  – Japan Aerospace Exploration Agency.

http://www.spri.cam.ac.uk
  – Scott Polar Research Institute.
http://www.nasa.gov
  – US National Aeronautics and Space Administration.
http://www.natice.noaa.gov
  – US National Ice Center.
http://nsidc.org
  – US National Snow and Ice Data Center.

SATELLITE REMOTE SENSING MICROWAVE SCATTEROMETERS

W. J. Plant, Applied Physics Laboratory, University of Washington, Seattle, WA, USA

Introduction

Microwave scatterometers are instruments that transmit low-power pulses of radiation toward the Earth's surface at intermediate incidence angles and measure the intensity of the signals scattered back at the same angles from surface areas a few kilometers on a side. Satellite scatterometers operate continuously and therefore scatter from land and ice as well as the ocean. Useful information is available in the signals from land and ice, but will not be discussed here. This article will concentrate on the primary goal of satellite scatterometers: the measurement of near-surface wind speed and direction over the ocean.

Scatterometers achieve this goal by measuring the intensity, or cross-section, of microwave signals backscattered from the ocean surface. Common frequencies of the transmitted signals for satellite scatterometers are near 5.3 GHz (C-band) on European instruments and 14 GHz (Ku-band) on US ones. At these frequencies, microwaves penetrate only a few millimeters into sea water, so all backscatter originates at the surface and is caused by the roughness of the surface; a perfectly calm sea surface produces no detectable scattering in the direction of the incident radiation. Changes in the average roughness of the ocean surface over scales of several kilometers are caused primarily, but not exclusively, by changes in the wind speed or direction at the ocean surface. Standard assumptions of scatterometry are that the backscatter cross-section, usually called σ_0, over such scales depends only on parameters of the scatterometer and on the mean wind, increases with wind speed, is a maximum when the antenna looks upwind, and is a minimum when the antenna looks nearly perpendicular to the wind, or crosswind. These assumptions allow the wind speed and direction to be determined from cross-sections measured for the same patch of the ocean, but with the antenna directed at several different azimuth angles. For satellite scatterometry a given patch of ocean can be viewed from several different directions only by allowing the scatterometer to sweep its antenna beams across the patch, a process that requires as much as 4 min. Thus an additional assumption of scatterometry is that average winds over kilometer-scale patches of ocean surface are stationary for several minutes.

With these assumptions and an adequate definition of the wind being measured (discussed below), satellite scatterometers have proven to be able to measure winds over the ocean with accuracies as good as or better than *in situ* measurement techniques. Because oceans cover most of the Earth, this means that microwave scatterometers carried on satellites can monitor the wind field over most of the globe every few days. **Table 1** gives typical specifications that a satellite scatterometer can be expected to meet.

The spatial coverage offered by satellite scatterometry is far better than can be achieved by *in situ* measurements. It allows scatterometers to provide data to study global weather patterns, monitor storm intensities, improve meteorological forecasts, impact global ocean circulation models, facilitate climate prediction, and much more. In addition to introducing the basics of scatterometry, this article will provide examples of these benefits of satellite scatterometry and indicate how continued improvements in the technique may be expected to provide even better results in the future.

Satellite Scatterometers

Other instruments such as radar altimeters that look straight down and real and synthetic aperture radars (RARs and SARs) that image surface scenes at resolutions of meters to kilometers have been operated in space and are capable of measuring wind

Table 1 Expected specifications of the NSCAT satellite scatterometer

Parameter	Value	Accuracy comment
Wind speed	3–30 m s^{-1}	2 m s^{-1} or 10%
Wind direction	3–30 m s^{-1}	20°
Spatial resolution	50 km	Wind cells
Location accuracy	25/10 km	Absolute/relative
Coverage	90% of ice-free ocean	Every 2 d
Mission duration	3 y	Includes check out

(Data from *Naderi et al.* 1991.)

speed or direction, but not both simultaneously and routinely. Microwave radiometers, passive instruments that measure the naturally occurring radiation from the ocean surface, are presently being developed as spaceborne anemometers capable of measuring wind speed and direction simultaneously. However, only microwave scatterometers, active instruments that both transmit and receive radiation, have a history of wind vector measurement from space. Because they are microwave, scatterometers can make their measurements both day and night and in most kinds of weather. Only very heavy rainfall, as discussed below, can hinder a scatterometer's view of the surface.

For these reasons, microwave scatterometers have been the instruments of choice for measuring near-surface winds from space. **Table 2** lists the scatterometers that have been in space to date and those that are planned for the future. As the table shows, the first scatterometer in space specifically designed to measure winds was the one on Seasat in 1978. An earlier microwave radar on Skylab viewed the ocean surface at intermediate incidence angles from space in 1973, but did not produce multiple looks at a single ocean patch from which wind vector information could be obtained. With the launch of the C-band scatterometer on the first European Remote Sensing Satellite (ERS-1) in 1991, a continuous series of global wind vector measurements was begun. As **Table 2** shows, this series has continued to the present through the launch of ERS-2 and the subsequent decommissioning of ERS-1. If present plans are carried out, at least two microwave scatterometers (and one or more microwave radiometers) will continue to produce global wind vector information into the foreseeable future.

While most scatterometers in space have consisted of multiple, fixed waveguide (stick) antennas, more recent US scatterometers have used rotating parabolic antennas. All satellite scatterometers to date have been in orbits with inclinations near 98.5°, except for Seasat which was 108°. Spatial resolutions have usually been 50×50 km, although 25×25 km is becoming more common. **Table 3** addresses the primary characteristics of the scatterometers listed in **Table 2**. As **Table 3** shows, ERS scatterometers have had about half the spatial coverage of their NASA counterparts because their antennas view only one side of the subsatellite path. As discussed below, however, uniformly reliable wind vectors are not obtained over the whole swath of the SeaWinds instruments. Also a gap is present between the two swaths on either side of the subsatellite path of the NASA stick-type scatterometers because the response

Table 2 Scatterometers in space to date and planned for the future

Satellite	Country/agency	Scatterometer	Launch date	Status
Seasat	USA/NASA	SASS	June 1978	Failed, October 1978
ERS-1	Europe/ESA	CSCAT(AMI)	July 1991	Standby, June 1996
ERS-2	Europe/ESA	CSCAT(AMI)	April 1995	Operational
ADEOS-I	USA/NASA Japan/NASDA	NSCAT	August 1996	ADEOS failed June 1997
QuikSCAT	USA/NASA	SeaWinds-1	June 1999	Operational
ADEOS-II	USA/NASA Japan/NASDA	SeaWinds-2	November 2001	Approved
ASCAT	Europe/ESA	Adv.CSCAT	2003	Proposed

(Adapted from Patzert and Van Woert, private communication.)

Table 3 Principal characteristics of satellite scatterometers

Satellite	Type	Frequency (GHz)	Polarization	Incidence angle (°)	Swath (km)	Altitude
Seasat	4 stick	14.6	4 VV, 4 HH	25–55 0–4	475, 475 140	800
ERS-1	3 stick	5.3	3 VV	18–57	500	785
ERS-2	3 stick	5.3	3 VV	18–57	500	785
ADEOS-I	6 stick	14.0	6 VV, 2 HH	15–63	600, 600	797
QuikSCAT	1 rotating	13.4	1 VV, 1 HH	46(H), 54(V)	1800	803
ADEOS-II	1 rotating	13.4	1 VV, 1 HH	46(H), 54(V)	1800	803
METOP	3 stick	–	–	–	–	835

of the cross-section to the wind vector is weak at low incidence angles. The Seasat scatterometer did have a mode that observed the surface at very low incidence angles, but could not get wind direction in this swath without extrapolation from the wider swaths. **Figure 1** shows the fixed-stick ERS-1 scatterometer (**Figure 1A**) and the rotating-antenna QuickSCAT scatterometer (**Figure 1B**).

The Normalized Radar Cross-section of the Sea

The basis of a scatterometer's ability to measure the near-surface wind vector is the dependence of the normalized radar cross-section of the sea, σ_o, on this vector. This cross-section is defined through the radar equation as follows:

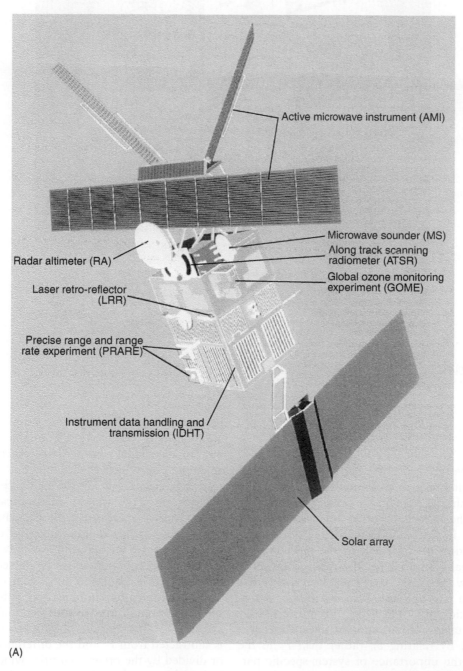

(A)

Figure 1   (A) ERS-1 satellite with two scatterometer waveguide antennas directed upwards. (B) QuikSCAT satellite with parabolic antenna with dual beams pointed downwards.

(B)

Figure 1   *continued*

$$P_r = \frac{P_t G^2 \lambda^2 A \sigma_o}{(4\pi)^3 R^4} + P_n \qquad [1]$$

where P_r is the received power, P_t is the transmitted power, G is antenna gain, λ is microwave length, A is the total illuminated area on the sea surface, and R is the range to the surface. P_n is the noise signal due to thermal noise in the components of the system and natural radiation of the earth to the receiving antenna. In a real scatterometer, the received power is reduced below the above level by system losses that must be taken into account. Also, for greater accuracy, the variation of G and R over the scatterometer footprint are taken into account by integration over the footprint. By relating σ_o rather than P_r to the wind vector, the importance of system-specific parameters, P_t, G, R, and λ, are greatly reduced. In fact, σ_o is independent of the first three of these parameters and depends on λ only because the mechanism of backscatter from the sea surface is weakly dependent on λ. Two other system parameters upon which σ_o depends due to the surface scattering mechanism are the incidence angle and the polarization. The strength of backscatter from the ocean surface increases with decreasing incidence angle and depends on the direction of the electric field of the incident radiation, its polarization.

The radar equation must be solved in order to obtain σ_o. This requires a knowledge of P_n, which is usually obtained by sampling only noise signals on a small fraction of the transmitted pulses, typically about 15%. The noise level obtained in this manner is subtracted from P_r and the difference is multiplied or divided by the other quantities which are known. This yields σ_o values that are accurate down to very low signal-to-noise ratios, but that can become

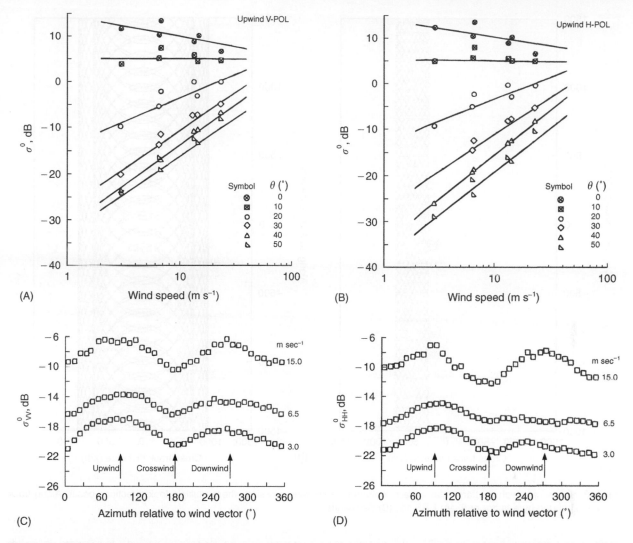

Figure 2 (A) Normalized radar cross section, σ_0 versus wind speed for VV polarization and an upwind look direction (θ is incidence angle); (B) same as (A) for HH polarization; (C) azimuthal dependence of σ_0 for VV polarization and a 30° incidence angle; (D) same as (C) for HH polarization. (Reproduced with permission from Jones WL *et al.* Aircraft measurements of the microwave scattering signature of the ocean. *IEEE Journal of Oceanic Engineering* © 1977 IEEE.)

negative due to sampling variability. This is an important consideration in the measurement of very low wind speeds.

Figure 2 shows the measured dependance of σ_o on polarization and incidence angle as well as on the wind vector. Two polarizations are indicated in the figure: the electric field vertical on both transmission and reception (VV) and the electric field horizontal on both (HH). These are the two polarizations of importance in scatterometry. The characteristics of σ_o that make it useful for wind vector measurement are obvious from this figure: for any given incidence angle and polarization, it depends on both wind speed and direction. In general the slope of σ_o versus wind speed is smaller at lower incidence angles. The dependence of σ_o on azimuth angle, χ, defined as the angle between the horizontal direction the antenna is

pointing and the direction from which the wind comes, has generally been found to fit a three-term Fourier cosine series in χ very well. The relationship between σ_o and incidence angle, θ_i, polarization, p, wind speed, U, and wind direction, χ, is called the geophysical model function. Its general form is taken to be the following:

$$\sigma_o = A_o(U, \theta_i, p) + A_1(U, \theta_i, p)\cos\chi \\ + A_2(U, \theta_i, p)\cos2\chi \qquad [2]$$

Usually the A coefficients are specified in tabular form.

Some characteristics of the dependence of σ_o on χ can be easily discerned in **Figure 2**. Cross-sections measured with the antenna looking nearly perpendicular to the wind direction ($\chi = 90°$ or $270°$)

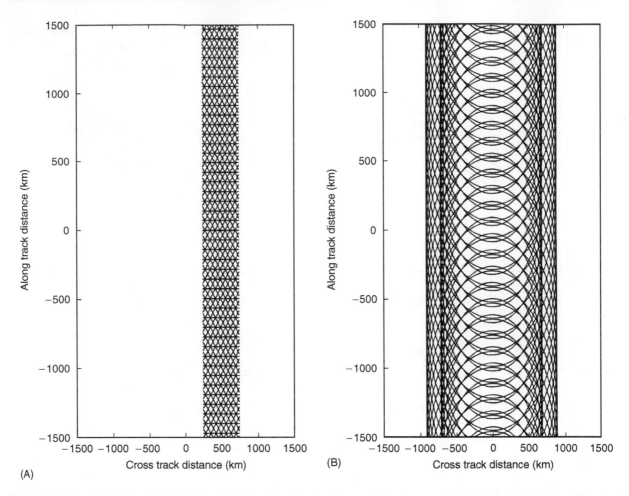

(A)

(B)

Figure 3 Lines on which scatterometer surface footprints lie showing crossing where multiple look directions occur. Along track distance expanded by a factor of 5. (A) ERS-1/2, (B) SeaWinds.

are always lower than those measured with the antenna directed upwind $(\chi = 0°)$ or downwind $(\chi = 180°)$. Furthermore, σ_o is larger when the antenna looks upwind than when it looks downwind, except perhaps for VV polarization at small incidence angles.

Spaceborne Scatterometer Wind Measurements

Given the behavior of σ_o shown in **Figure 2**, a scatterometer fixed on Earth can easily yield the wind vector. If the antenna is rotated and the direction of maximum σ_o is determined, then the level of σ_o in this direction yields the wind speed. In most cases, the direction of maximum σ_o would be the wind direction. The exception to this might be at VV polarization for some incidence angles where the downwind look direction could yield the maximum σ_o. Thus a $180°$ ambiguity might exist in the wind direction.

Unfortunately this simple technique will not work from satellites because of their high speeds. By the time the antenna rotates one revolution, the satellite has travelled several kilometers, so the footprint on the surface samples many different, widely separated areas. The solution to this problem is to use σ_o values measured in different directions at different times so that nearly the same surface area is illuminated in each direction. **Figure 3** indicates how this might be accomplished in the case of three, fixed stick antennas (**Figure 3A**) and in the case of a rotating antenna with two beams (**Figure 3B**). The first case corresponds to the ERS scatterometers while the latter corresponds to SeaWinds whose single rotating antenna transmits beams at two different incidence angles. **Figure 3** shows the lines on which the surface resolution cells are located. For clarity, the vertical distance traveled by the satellite has been expanded by a factor of 5 in **Figure 3**. Thus crossings of the lines, where cross-sections can be measured from different directions, are more frequent than indicated in the figure. The two parts of **Figure 3** have the same

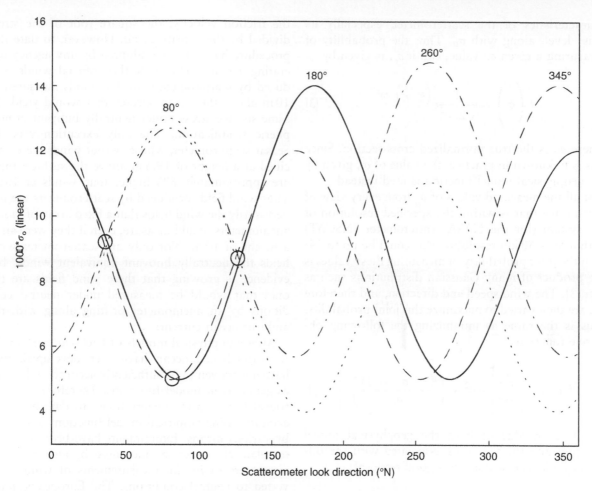

Figure 4 Angular dependence of σ_0 for various wind directions (indicated in the figure). Wind speeds vary little. Circles show possible data points; curves crossing at data points produce ambiguities in wind direction.

horizontal scale to emphasize the much wider swath of the SeaWinds scatterometer compared with ERS. For regions of the swath near the center, however, the azimuth angles at which the SeaWinds beam views the surface are nearly opposite or parallel. Furthermore, near the edges, only two nearly parallel looks at a given surface area can be obtained. Estimation of wind speed and direction is consequently more difficult near the center and edge of the swath due to the ambiguous nature of the model function.

The ambiguities inherent in the form of the model function are illustrated in **Figure 4**. The figure shows the angular dependence of σ_o for winds coming from 80°, 180°, 260°, and 345° with respect to north. Each curve maximizes when the antenna is directed into the wind direction. The circles in the figure represent measurements of the cross-section made by an ERS-type scatterometer where the three beams are 45° apart. As the figure shows, winds from 180° and 345° fit the measurements equally well. If the scatterometer had only two beams that were 90° apart,

like the Seasat scatterometer, then the other two wind directions, 80° and 260° would also fit the data, provided that the wind speeds were slightly higher for the 260° direction and slightly lower for the 80° direction. Ambiguous wind vectors always exist in the output of scatterometers. Attempts to resolve these ambiguities have ranged from using human analysts to insure consistency of the wind fields to using median filters in which a given wind vector is forced to be in the direction of the median of its surrounding vectors. While automatic techniques such as median filtering are now producing skills upwards of 90% in selecting the correct ambiguity, incorrect assignment of vector directions in some locations is still a problem for scatterometry.

Retrieval of the ambiguous wind vectors is accomplished in practice by noting that any single measurement of σ_o is an average of measurements from several scatterometer pulses and therefore has a probability distribution near Gaussian. The variance, δ^2, of this distribution may be determined from the characteristics of the scatterometer, especially its noise level, along with σ_o. Thus the probability of measuring a given σ_o value, call it $\hat{\sigma}_o$, is given by

$$P\left(\hat{\sigma}_o\right) = \frac{1}{\sqrt{2\pi\delta^2}}e^{\left(\hat{\sigma}_o - \sigma_o^2\right)/2\delta^2} \qquad [3]$$

where σ_o is the true normalized cross-section. Since this is unknown in practice, the value of σ_o given by the geophysical model function is used instead.

If all the measured values of $\hat{\sigma}_o$ from every view of the surface that is within the specified resolution of the system (for the NASA scatterometer (NSCAT) with a 50×50 km resolution this could be up to 24) then the joint probability of measuring these values is the product of many Gaussian distributions such as eqn [3]. The wind speed and direction, and therefore σ_o, are then varied to maximize this joint probability. This is the same as minimizing the following objective function:

$$J(U, \chi) = \sum_{i=1}^{N}\left[\ln\delta_i^2 + \left(\frac{\hat{\sigma}_{oi} - \sigma_{oi}(U, \chi)}{\delta_i}\right)^2\right] \qquad [4]$$

Good knowledge of both the geophysical model function and the variances associated with the different looks is essential for a good retrieval.

Calibration

In principle, σ_o values obtained by scatterometers need not be calibrated in an absolute sense; they could simply be related to the surface wind that produced them. In order to cross-check the operation of different scatterometers and to further the development of rough surface scattering theories, however, all scatterometers flown in space have produced normalized radar cross-sections that have been calibrated against a standard target. Usually signals from a geophysical target that has been calibrated by airborne measurements and found to be isotropic are used for calibration of σ_o from spaceborne scatterometers. The Amazon Rain Forest is the most common geophysical target chosen, although areas of Antarctica have also been used.

Calibration and verification of scatterometer winds require an exact definition of the wind that a scatterometer measures. Since a scatterometer really responds to surface roughness, the suggestion has often been made that a scatterometer measures the stress of the wind on the ocean surface rather than the wind vector itself. In fact, a geophysical model function has been developed to relate σ_o directly to

the friction velocity, the square root of the stress divided by the density of air. However, to date this procedure has not been adopted by any agency operating a scatterometer. Rather, official winds produced by scatterometers are the winds measured at 10 m above the ocean surface that would yield the same surface stress under neutrally buoyant atmospheric stratification. The only exception was the Seasat scatterometer, whose output winds were specified at a height of 19.5 m above the surface; these are approximately 6% higher than winds at 10 m. Thus wind fields produced by scatterometers are not necessarily the wind fields that a fixed array of *in situ* anemometers would measure, even if they were all at a height of 10 m. Not only are scatterometer wind fields the neutrally buoyant equivalent winds, but evidence is growing that these wind fields are the ones that would be measured under neutral conditions by an anemometer drifting along with the ambient ocean current.

Although physical models of backscatter from the wind-roughened ocean have been developed, they have not proven to be sufficiently accurate to be used as geophysical model functions. Therefore wind retrieval from scatterometers flown to date has been achieved using empirical model functions developed by various means. Experiments have been mounted to relate cross-sections measured by airborne scatterometers to *in situ* measurements of winds converted to neutral conditions. The Europeans noted that if σ_o depends on no other environmental variables than U and χ, then σ_o values measured by three antennas looking in three different directions must fall on a well-defined surface when σ_o values from the three separate antennas are plotted on three orthogonal axes. This observation has allowed them to determine the properties of the geophysical model function to within a constant calibration factor, which was obtained from comparisons with buoy measurements. US geophysical model functions have also been developed by comparison of cross-sections with buoy measurements and with winds measured by other remote sensing instruments, such as microwave radiometers. Most recently, however, US geophysical model functions have been developed by binning satellite scatterometer σ_o values according to U and χ values produced by numerical weather prediction models. The idea is that errors in the numerical model will cancel out in the mean so that the correct model function can be obtained even in the presence of these errors.

After a model function has been developed by some means, a period of validation of the scatterometer wind fields always follows. Generally scatterometer wind fields are compared with those measured by buoys moored in fixed locations whose anemometer readings have been corrected to 10 m and neutral conditions. As an indication of the accuracy that can be obtained by scatterometers, **Figure 5** shows a comparison of wind speeds and

directions produced by NSCAT with winds measured by anemometers on buoys operated by the US National Data Buoy Center (NDBC). To produce the wind speed comparison in **Figure 5A,** buoy wind speeds have been binned into $0.5\,\mathrm{m\,s^{-1}}$ bins and the

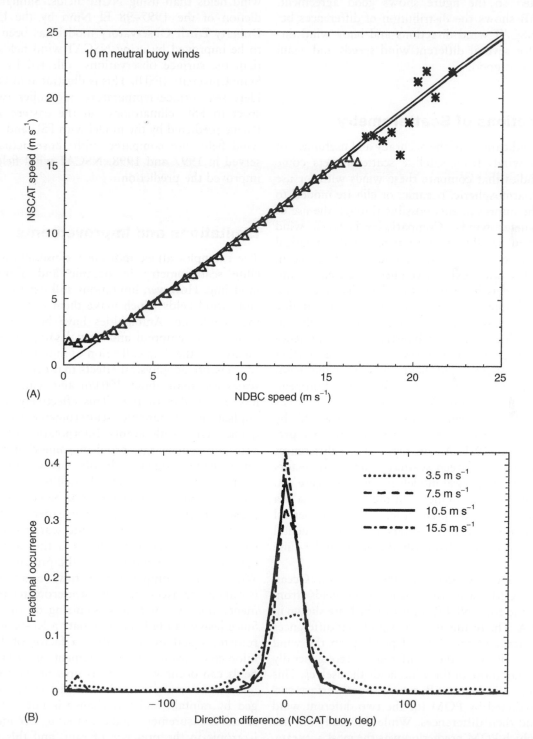

Figure 5 (A) Sample mean NSCAT wind speeds in $0.5\,\mathrm{m\,s^{-1}}$ buoy wind speed bins. All NSCAT measurements were located within 50 km and 30 min of buoy measurements. Triangles denote >100 NSCAT measurements in the mean; asterisks denote 5–99 samples. (B) Distributions of directional differences (NSCAT buoy) for $1\,\mathrm{m\,s^{-1}}$ buoy wind speed bins centered on the indicated wind speeds. (Reproduced with permission from Freilich and Dunbar, 1999.)

corresponding NSCAT winds measured near the buoy have been averaged for each bin. The nonzero values of NSCAT wind speed at zero buoy wind speed are known to be a result of comparing magnitudes of vectors whose components are Gaussian distributed so the figure shows good agreement. **Figure 5B** shows the distribution of differences between NSCAT wind directions and buoy wind directions for several different wind speeds and again agreement is good.

Applications of Scatterometry

Further indications of the accuracy and usefulness of data on winds from satellite scatterometers come from studies that compare these winds with, or use them in, atmospheric, oceanic, or climate models to assess the improvements possible through the use of scatterometer winds. Comparisons between wind fields from satellite scatterometers and numerical models have shown that significant differences in these fields often exist. Cyclones predicted by numerical models have been found to disagree in location and intensity with those observed in satellite scatterometry, sometimes by 300 km and 10 mbar or more. The location of the Intertropical Convergence Zone (ITCZ) has been shown to be 1°–2° farther south in NSCAT wind fields than in wind fields predicted by the numerical model of the European Centre for Medium-Range Weather Forecasts (ECMWF). Furthermore, the ITCZ observed by NSCAT was stronger and narrower than that predicted by ECMWF. An example of the differences that can occur between scatterometer wind vectors and those of numerical models is given in **Figure 6A**. Here NSCAT measurements in the South China Sea are compared with predictions of the numerical model of the National Centers for Environmental Prediction (NCEP). Obviously the fields can be quite different.

In the South China Sea, the Princeton Ocean Model (POM) was run using wind fields from NSCAT and from NCEP, some of which are shown in **Figure 6A**. The results indicate significant differences in the output of the POM depending on the wind field used, although these differences were generally smaller than those of the wind fields themselves. This is illustrated in **Figure 6B**, which shows surface currents predicted by POM for the two different wind fields and their differences. While this study did not assess which POM prediction was the most accurate, other studies have indicated that the predictions of ocean models and coupled atmosphere–ocean models using scatterometer wind fields agreed with observations better than those using other wind fields. Comparisons of sea levels predicted by the modular ocean model (MOM) with those of the TOPEX/Poseidon radar altimeter have indicated that more accurate predictions are achieved using ERS-1 wind fields than using NCEP fields. Similarly, prediction of the 1997–98 El Niño by the Lamont-Doherty Earth Observatory model has been shown to be improved by using NSCAT wind fields rather than the surface observations collected by Florida State University (FSU). This is illustrated in **Figure 7**. Here sea surface temperature anomalies (with respect to FSU climatology) in the eastern tropical Pacific predicted by the model with FSU and NSCAT wind fields are compared with those actually observed in 1997 and 1998. NSCAT wind fields have improved the prediction.

Limitations and Improvements

The examples above indicate the usefulness of satellite scatterometry in oceanic and atmospheric modeling. However, limitations still exist in the satellite wind fields which make them less useful than they could be. Ambiguities have been mentioned earlier. The temporal and spatial sampling patterns inherent in the data collection of individual satellites can obscure geophysical effects that occur on spatial scales less than about 200 km and temporal scales less than a day or two. This effectively limits the application of satellite scatterometry to low-frequency, large-scale events. Interpolation of the raw data can reduce this problem somewhat but may introduce false signals. The obvious way to alleviate the problem is to put fleets of satellite scatterometers into orbit, an idea that may not be completely impossible in this day of smaller, cheaper satellites.

Another limitation on satellite scatterometry that has been observed many times is the mismeasurement of surface wind fields in the presence of rain. Without an accompanying microwave radiometer, it is difficult to ascertain from scatterometer measurements alone when rain is occurring on the surface. Since heavy rainfall can attenuate a Ku-band scatterometer signal by more than a factor of 10 every kilometer, serious mismeasurement of surface wind fields can occur when rain is present. Furthermore, microwave return from the sea surface is also changed by rainfall, and this change is not well understood. Measurements indicate that σ_o becomes more isotropic in the presence of rain, and this effect is being utilized in an attempt to develop a rain flag from scatterometer data alone. Such a flag would allow users to determine when rain effects might be

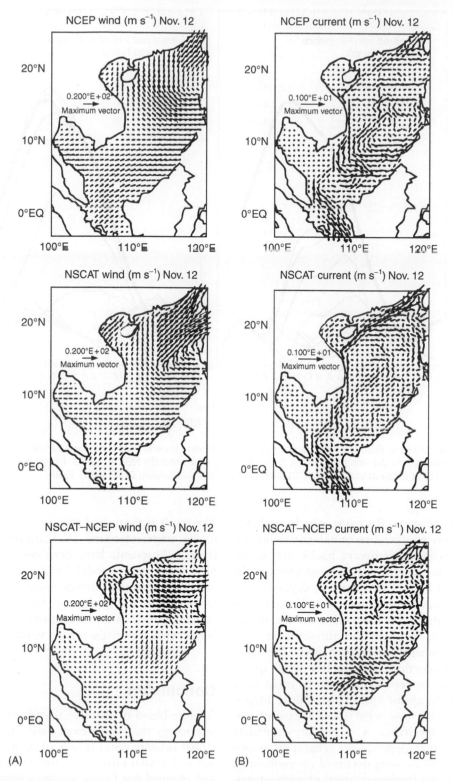

Figure 6 (A) NCEP and NSCAT wind vectors in the South China Sea and their differences. (B) Surface currents in the South China Sea from NCEP model under NCEP and NSCAT wind forcing. (Reproduced with permission from Chu *et al.*, 1999.)

present in scatterometer data, but not what the proper wind vectors are.

Winds below $3\,\mathrm{m\,s^{-1}}$ are difficult to measure accurately using satellite scatterometry. Present

geophysical model functions are smoothly decreasing functions of wind speed that are nonzero even at zero wind speeds. The Europeans have chosen not to provide wind speed estimates below $3\,\mathrm{m\,s^{-1}}$ using

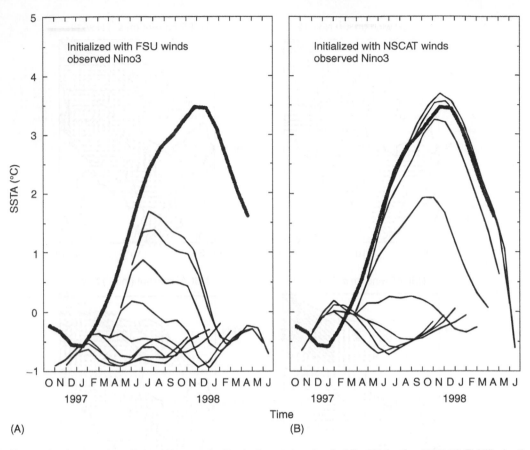

Figure 7 Forecasts of sea surface temperature anomalies in the eastern tropical Pacific for the 1997–98 El Niño by the Lamont-Doherty Earth Observatory model using (A) Florida State University winds and (B) NSCAT winds. Heavy curves are observations; other lines are predictions made at various times (Reproduced with permission from Chen *et al.*, 1999.)

their ERS scatterometers. Recent studies have shown that, in addition to the directional variability inherent in low winds, microwave backscatter is virtually zero below some threshold wind speed in the neighborhood of $2 \, \mathrm{m \, s}^{-1}$ to $4.5 \, \mathrm{m \, s}^{-1}$, depending on incidence angle and water temperature. Variability of the wind over the surface footprint of the scatterometer, however, obscures this threshold in most satellite scatterometry measurements. The geophysical model function at these low wind speeds therefore appears to depend on both the mean wind vector over the footprint and its variability. Suggestions for improving low wind speed measurements have included specifying a variability-dependent model function that depends on geographic region and season, as well as using details of the probability distribution on σ_o. Implementation of such suggestions in future retrieval schemes offers hope of improving low wind speed measurements.

Finally, high wind speeds have also proven to be a problem for scatterometry. Recent studies indicate that σ_o increases less rapidly with wind speed above about $25 \, \mathrm{m \, s}^{-1}$ than current model functions predict.

The result is that wind speeds above this value tend to be underestimated in scatterometry wind retrievals. Proposals have been offered for improved high wind speed model functions based on simultaneous scatterometry/radiometry measurements from aircraft flying through hurricanes. Implementation of such model functions promises to yield better wind retrievals at high wind speeds.

Conclusion

Satellite-based microwave scatterometry is a mature technology that has proven itself capable of yielding global oceanic wind speeds of unprecedented accuracy and spatial coverage. Satellites currently in orbit and planned for future missions promise a continuous long-term series of global wind measurements that can aid in climate studies. Based on the time-series presently available, satellite scatterometer measurements have proven themself capable of improving present oceanographic and atmospheric models. Future improvements in scatterometry promise to make this technology even more valuable in studying the dynamics of the atmosphere and oceans.

See also

Aircraft Remote Sensing. Satellite Altimetry. Satellite Remote Sensing SAR. Satellite Remote Sensing of Sea Surface Temperatures. Sensors for Mean Meteorology. Sensors for Micrometeorological and Flux Measurements.

Further Reading

Brown RA and Zeng L (1994) Estimating central pressures of oceanic midlatitude cyclones. *Journal of Applied Meteorology* 33: 1088–1095.

Chen D, Cane MA, and Zebiak SE (1999) The impact of NSCAT winds on predicting the 1997/1998 El Niño: A case study with the Lamont-Doherty Earth Observatory model. *Journal of Geophysical Research* 104: 11322–11327.

Chu PC, Lu S, and Liu WT (1999) Uncertainty of South China Sea prediction using NSCAT and National Centers for Environmental Prediction winds during tropical storm Ernie, 1996. *Journal of Geophysical Research* 104: 11273–11289.

Donelan MA and Pierson WJ (1987) Radar scattering and equilibrium ranges in wind-generated waves with application to scatterometry. *Journal of Geophysical Research* 92: 4971–5029.

Fisher RE (1972) Standard deviation of scatterometer measurements from space. *IEEE Transactions Geoscience Electronics* GE-10: 106–113.

Freilich MH (1997) Validation of vector magnitude datasets: effects of random component errors. *Journal of Atmospheric and Oceanic Technology* 14: 695–703.

Freilich MH and Dunbar RS (1999) The accuracy of the NSCAT 1 vector winds: Comparisons with National Data Buoy Center buoys. *Journal of Geophysical Research* 104: 11231–11246.

Fu Lee-Lueng and Yi Chao (1997) The sensitivity of a global ocean model to wind forcing. A test using sea level and wind observations from satellites and operational wind analysis. *Geophysical Research Letters* 24: 1783–1786.

Graf J, Sasaki C, Winn C, *et al.* (1998) NASA scatterometer experiment. *Acta Astronautica* 43: 397–407.

Jones WL, Schroeder LC, and Mitchell JL (1977) Aircraft measurements of the microwave scattering signature of the ocean. *IEEE Journal of Oceanic Engineering* OE-2: 52–61.

Kelly KA, Dickinson S, and Yu Z (1999) NSCAT tropical wind stress maps: implications for improving ocean modeling. *Journal of Geophysical Research* 104: 11291–11310.

Moore RK and Pierson WJ (1996) *Measuring Sea State and Estimating Surface Winds from a Polar Orbiting Satellite.* In: Proceedings of the International Symposium on Electromagnetic Sensing of Earth from Satellites, Miami Beach, FL, 1966.

Moore RK and Fung AK (1979) Radar determination of winds at sea. *Proceedings of the IEEE* 67: 1504–1521.

Naderi FM, Freilich MH, and Long DG (1991) Spaceborne radar measurement of wind velocity over the ocean – An overview of the NSCAT Scatterometer system. *Proceedings of the IEEE* 79: 850–866.

Pierson WJ Jr (1983) The measurement of the synoptic scale wind over the ocean. *Journal of Geophysical Research* 88: 1682–1780.

SATELLITE REMOTE SENSING OF SEA SURFACE TEMPERATURES

P. J. Minnett, University of Miami, Miami, FL, USA

Introduction

The ocean surface is the interface between the two dominant, fluid components of the Earth's climate system: the oceans and atmosphere. The heat moved around the planet by the oceans and atmosphere helps make much of the Earth's surface habitable, and the interactions between the two, that take place through the interface, are important in shaping the climate system. The exchange between the ocean and atmosphere of heat, moisture, and gases (such as CO_2) are determined, at least in part, by the sea surface temperature (SST). Unlike many other critical variables of the climate system, such as cloud cover, temperature is a well-defined physical variable that can be measured with relative ease. It can also be measured to useful accuracy by instruments on observation satellites.

The major advantage of satellite remote sensing of SST is the high-resolution global coverage provided by a single sensor, or suite of sensors on similar satellites, that produces a consistent data set. By the use of onboard calibration, the accuracy of the time-series of measurements can be maintained over years, even decades, to provide data sets of relevance to research into the global climate system. The rapid processing of satellite data permits the use of the global-scale SST fields in applications where the immediacy of the data is of prime importance, such as weather forecasting – particularly the prediction of the intensification of tropical storms and hurricanes.

Measurement Principle

The determination of the SST from space is based on measuring the thermal emission of electromagnetic radiation from the sea surface. The instruments, called radiometers, determine the radiant energy flux, B_λ, within distinct intervals of the electromagnetic spectrum. From these the brightness temperature (the temperature of a perfectly emitting 'black-body' source that would emit the same radiant flux) can be calculated by the Planck equation:

$$B_\lambda(T) = 2hc^2\lambda^{-5}\left(e^{hc/(\lambda kT)} - 1\right)^{-1} \qquad [1]$$

where h is Planck's constant, c is the speed of light in a vacuum, k is Boltzmann's constant, λ is the wavelength and T is the temperature. The spectral intervals (wavelengths) are chosen where three conditions are met: (1) the sea emits a measurable amount of radiant energy, (2) the atmosphere is sufficiently transparent to allow the energy to propagate to the spacecraft, and (3) current technology exists to build radiometers that can measure the energy to the required level of accuracy within the bounds of size, weight, and power consumption imposed by the spacecraft. In reality these constrain the instruments to two relatively narrow regions of the infrared part of the spectrum and to low-frequency microwaves. The infrared regions, the so-called atmospheric windows, are situated between wavelengths of $3.5-4.1\mu m$ and $8-12\mu m$ (**Figure 1**); the microwave measurements are made at frequencies of 6–12 GHz.

As the electromagnetic radiation propagates through the atmosphere, some of it is absorbed and scattered out of the field of view of the radiometer, thereby attenuating the original signal. If the attenuation is sufficiently strong none of the radiation from the sea reaches the height of the satellite, and such is the case when clouds are present in the field of view of infrared radiometers. Even in clear-sky conditions a significant fraction of the sea surface emission is absorbed in the infrared windows. This energy is re-emitted, but at a temperature characteristic of that height in the atmosphere. Consequently the brightness temperatures measured through the clear atmosphere by a spacecraft radiometer are cooler than would be measured by a similar device just above the surface. This atmospheric effect, frequently referred to as the temperature deficit, must be corrected accurately if the derived sea surface temperatures are to be used quantitatively.

Infrared Atmospheric Correction Algorithms

The peak of the Planck function for temperatures typical of the sea surface is close to the longer wavelength infrared window, which is therefore well suited to SST measurement (**Figure 1**). However, the

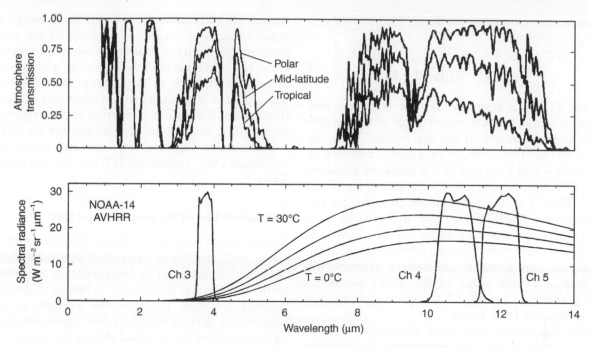

Figure 1 Spectra of atmospheric transmission in the infrared (wavelengths 1–14 μm) calculated for three typical atmospheres from diverse parts of the ocean; polar, mid-latitude and tropical with integrated water vapor content of 7 kg m^{-2} (polar), 29 kg m^{-2} (mid-latitude) and 54 kg m^{-2} (tropical). Regions where the transmission is high are well suited to satellite remote sensing of SST. The lower panel shows the electromagnetic radiative flux for four sea surface temperatures (0, 10, 20, and 30°C) with the relative spectral response functions for channels 3, 4, and 5 of the AVHRR on the NOAA-14 satellite. The so-called 'split-window' channels, 4 and 5, are situated where the sea surface emission is high, and where the atmosphere is comparatively clear but exhibits a strong dependence on atmospheric water vapor content.

main atmospheric constituent in this spectral interval that contributes to the temperature deficit is water vapor, which is very variable both in space and time. Other molecular species that contribute to the temperature deficit are quite well mixed throughout the atmosphere, and therefore inflict a relatively constant temperature deficit that is simple to correct.

The variability of water vapor requires an atmospheric correction algorithm based on the information contained in the measurements themselves. This is achieved by making measurements at distinct spectral intervals in the windows when the water vapor attenuation is different. These spectral intervals are defined by the characteristics of the radiometer and are usually referred to as bands or channels (**Figure 1**). By invoking the hypothesis that the difference in the brightness temperatures measured in two channels, i and j, is related to the temperature deficit in one of them, the atmospheric correction algorithm can be formulated thus:

$$SST_{ij} - T_i = f(T_i - T_j) \qquad [2]$$

where SST_{ij} is the derived SST and T_i, T_j are the brightness temperatures in channels i, j.

Further, by assuming that the atmospheric attenuation is small in these channels, so that the radiative transfer can be linearized, and that the channels are spectrally close so that Planck's function can be linearized, the algorithm can be expressed in the very simple form:

$$SST_{ij} = a_o + a_i T_i + a_j T_j \qquad [3]$$

where are a_o, a_i, and a_j are coefficients. These are determined by regression analysis of either coincident satellite and *in situ* measurements, mainly from buoys, or of simulated satellite measurements derived by radiative transfer modeling of the propagation of the infrared radiation from the sea surface through a representative set of atmospheric profiles.

The simple algorithm has been applied for many years in the operational derivation of the sea surface from measurements of the Advanced Very High Resolution Radiometer (AVHRR, see below), the product of which is called the multi-channel SST (MCSST), where i refers to channel 4 and j to channel 5.

More complex forms of the algorithms have been developed to compensate for some of the shortcomings of the linearization. One such widely applied algorithm takes the form:

$$SST_{ij} = b_o + b_1 T_i + b_2(T_i - T_j)SST_r$$
$$+ b_3(T_i - T_j)(\sec\theta - 1) \quad [4]$$

where SST_r is a reference SST (or first-guess temperature), and θ is the zenith angle to the satellite radiometer measured at the sea surface. When applied to AVHRR data, with i and j referring to channels 4 and 5 derived SST is called the nonlinear SST (NLSST). A refinement is called the Pathfinder SST (PFSST) in the research program designed to post-process AVHRR data over a decade or so to provide a consistent data set for climate research. In the PFSST, the coefficients are derived on a monthly basis for two different atmospheric regimes, distinguished by the value of the T_4–T_5 differences being above or below 0.7 K, by comparison with measurements from buoys.

The atmospheric correction algorithms work effectively only in the clear atmosphere. The presence of clouds in the field of view of the infrared radiometer contaminates the measurement so that such pixels must be identified and removed from the SST retrieval process. It is not necessary for the entire pixel to be obscured, even a portion as small as 3–5%, dependent on cloud type and height, can produce unacceptable errors in the SST measurement. Thin, semi-transparent cloud, such as cirrus, can have similar effects to subpixel geometric obscuration by optically thick clouds. Consequently, great attention must be paid in the SST derivation to the identification of measurements contaminated by even a small amount of clouds. This is the principle disadvantage to SST measurement by spaceborne infrared radiometry. Since there are large areas of cloud cover over the open ocean, it may be necessary to composite the cloud-free parts of many images to obtain a complete picture of the SST over an ocean basin.

Similarly, aerosols in the atmosphere can introduce significant errors in SST measurement. Volcanic aerosols injected into the cold stratosphere by violent eruptions produce unmistakable signals that can bias the SST too cold by several degrees. A more insidious problem is caused by less readily identified aerosols at lower, warmer levels of the atmosphere that can introduce systematic errors of a much smaller amplitude.

Microwave Measurements

Microwave radiometers use a similar measurement principle to infrared radiometers, having several spectral channels to provide the information to correct for extraneous effects, and black-body calibration targets to ensure the accuracy of the measurements. The suite of channels is selected to include sensitivity to the parameters interfering with the SST measurements, such as cloud droplets and surface wind speed, which occurs with microwaves at higher frequencies. A simple combination of the brightness temperature, such as eqn [2], can retrieve the SST.

The relative merits of infrared and microwave radiometers for measuring SST are summarized in **Table 1**.

Characteristics of Satellite-derived SST

Because of the very limited penetration depth of infrared and microwave electromagnetic radiation in sea water the temperature measurements are limited to the sea surface. Indeed, the penetration depth is typically less than 1 mm in the infrared, so that temperature derived from infrared measurements is characteristic of the so-called skin of the ocean. The skin temperature is generally several tenths of a degree cooler than the temperature measured just below, as a result of heat loss from the ocean to atmosphere. On days of high insolation and low wind speed, the absorption of sunlight causes an increase in near surface temperature so that the water just below the skin layer is up to a few degrees warmer than that measured a few meters deeper, beyond the influence of the diurnal heating. For those people interested in a temperature characteristic of a depth of a few meters or more, the decoupling of the skin and deeper, bulk temperatures is perceived as a disadvantage of using satellite SST. However, algorithms generated by comparisons between satellite

Table 1 Relative merits of infrared and microwave radiometers for sea surface temperature measurement

Infrared	Microwave
Good spatial resolution ($\sim$1 km)	Poor spatial resolution ($\sim$50 km)
Surface obscured by clouds	Clouds largely transparent, but measurement perturbed by heavy rain
No side-lobe contamination	Side-lobe contamination prevents measurements close to coasts or ice
Aperture is reasonably small; instrument can be compact for spacecraft use	Antenna is large to achieve spatial resolution from polar orbit heights ($\sim$800 km above the sea surface)
4 km resolution possible from geosynchronous orbit; can provide rapid sampling data	Distance to geosynchronous orbit too large to permit useful spatial resolution with current antenna sizes

Table 2 Spectral characteristics of current and planned satellite-borne infrared radiometers

AVHRR		ATSR		MODIS		OCTS		GLI	
λ (μm)	NEΔT (K)	λ (μm)	NEΔT (K)	λ (μm)	NEΔT (K)	λ (μm)	NEΔT (K)	λ (μm)	NEΔT (K)
3.75	0.12	3.7	0.019	3.75	0.05	3.7	0.15	3.715	<0.15
				3.96	0.05				
				4.05	0.05				
				8.55	0.05	8.52	0.15	8.3	<0.1
10.5	0.12	10.8	0.028	11.03	0.04	10.8	0.15	10.8	<0.1
11.5	0.12	12.0	0.025	12.02	0.04	11.9	0.15	12	<0.1

and *in situ* measurements from buoys include a mean skin effect masquerading as part of the atmospheric effect, and so the application of these results in an estimate of bulk temperatures.

The greatest advantage offered by satellite remote sensing is, of course, coverage. A single, broad-swath, imaging radiometer on a polar-orbiting satellite can provide global coverage twice per day. An imaging radiometer on a geosynchronous satellite can sample much more frequently, once per half-hour for the Earth's disk, or smaller segments every few minutes, but the spatial extent of the data is limited to that part of the globe visible from the satellite.

The satellite measurements of SST are also reasonably accurate. Current estimates for routine measurements show absolute accuracies of ±0.3 to ±0.5 K when compared to similar measurements from ships, aircraft, and buoys.

Spacecraft Instruments

All successful instruments have several attributes in common: a mechanism for scanning the Earth's surface to generate imagery, good detectors, and a mechanism for real-time, in-flight calibration. Calibration involves the use of one or more black-body calibration targets, the temperatures of which are accurately measured and telemetered along with the imagery. If only one black-body is available a measurement of cold dark space provides the equivalent of a very cold calibration target. Two calibration points are needed to provide in-flight calibration; nonlinear behavior of the detectors is accounted for by means of pre-launch instrument characterization measurements.

The detectors themselves inject noise into the data stream, at a level that is strongly dependent on their temperature. Therefore, infrared radiometers require cooled detectors, typically operating from 78 K (−195°C) to 105 K (−168°C) to reduce the noise

equivalent temperature difference (NEΔT) to the levels shown in **Table 2**.

The Advanced Very High Resolution Radiometer (AVHRR)

The satellite instrument that has contributed the most to the study of the temperature of the ocean surface is the AVHRR that first flew on TIROS-N launched in late 1978. AVHRRs have flown on successive satellites of the NOAA series from NOAA-6 to NOAA-14, with generally two operational at any given time. The NOAA satellites are in a near-polar, sun-synchronous orbit at a height of about 780 km above the Earth's surface and with an orbital period of about 100 min. The overpass times of the two NOAA satellites are about 2.30 a.m. and p.m. and about 7.30 a.m. and p.m. local time. The AVHRR has five channels: 1 and 2 at ∼0.65 and ∼0.85 μm are responsive to reflected sunlight and are used to detect clouds and identify coastlines in the images from the daytime part of each orbit. Channels 4 and 5 (**Table 2** and **Figure 1**) are in the atmospheric window close to the peak of the thermal emission from the sea surface and are used primarily for the measurement of sea surface temperature. Channel 3, positioned at the shorter wavelength atmospheric window, is responsive to both surface emission and reflected sunlight. During the nighttime part of each orbit, measurements of channel 3 brightness temperatures can be used with those from channels 4 and 5 in variants of the atmospheric correction algorithm to determine SST. The presence of reflected sunlight during the daytime part of the orbit prevents much of these data from being used for SST measurement. Because of the tilting of the sea surface by waves, the area contaminated by reflected sunlight (sun glitter) can be quite extensive, and is dependent on the local surface wind speed. It is limited to the point of specular reflection only in very calm seas.

The images in each channel are constructed by scanning the field of view of the AVHRR across the

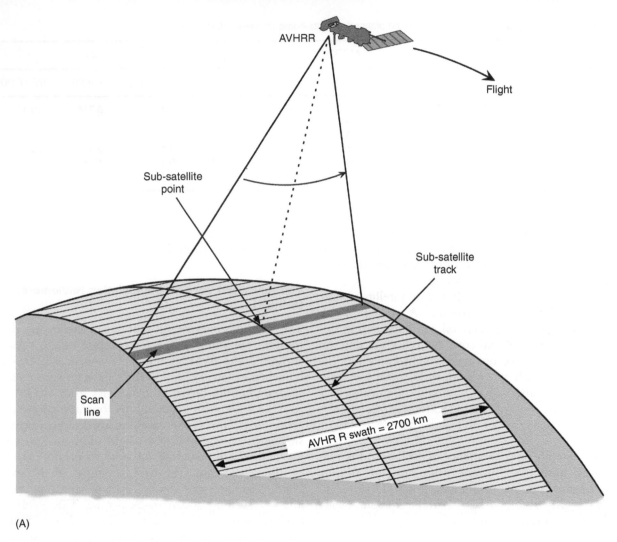

(A)

Figure 2 Scan geometries of AVHRR (A) and ATSR (B). The continuous wide swath of the AVHRR is constructed by linear scan lines aligned across the direction of motion of the subsatellite point. The swaths of the ATSR are generated by an inclined conical scan, which covers the same swath through two different atmospheric path lengths. The swath is limited to 512 km by geometrical constraints. Both radiometers are on sun-synchronous, polar-orbiting satellites.

Earth's surface by a mirror inclined at 45° to the direction of flight (**Figure 2A**). The rate of rotation, 6.67 Hz, is such that successive scan lines are contiguous at the surface directly below the satellite. The width of the swath (∼2700 km) means that the swaths from successive orbits overlap so that the whole Earth's surface is covered without gaps each day.

The Along-Track Scanning Radiometer (ATSR)

An alternative approach to correcting the effects of the intervening atmosphere is to make a brightness temperature measurement of the same area of sea surface through two different atmospheric path lengths. The pairs of such measurements must be made in quick succession, so that the SST and atmospheric conditions do not change in the time interval. This approach is that used by the ATSR, two of which have flown on the European satellites ERS-1 and ERS-2.

The ATSR has infrared channels in the atmospheric windows comparable to those of AVHRR, but the rotating scan mirror sweeps out a cone inclined from the vertical by its half-angle (**Figure 2B**). The field of view of the ATSR sweeps out a curved path on the sea surface, beginning at the point directly below the satellite, moving out sideways and forwards. Half a mirror revolution later, the field of view is about 900 km ahead of the sub-satellite track in the center of the 'forward view'. The path of the field of view returns to the sub-satellite point, which,

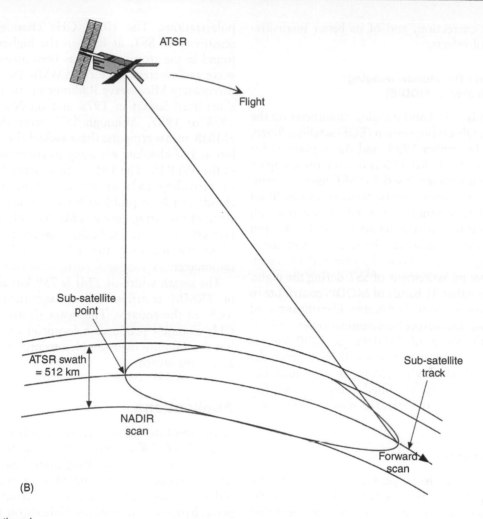

(B)

Figure 2 continued

during the period of the mirror rotation, has moved 1 km ahead of the starting point. Thus the pixels forming the successive swaths through the nadir point are contiguous. The orbital motion of the satellite means that the nadir point overlays the center of the forward view after about 2 min. The atmospheric path length of the measurement at nadir is simply the thickness of the atmosphere, whereas the slant path to the center of the forward view is almost double that, resulting in colder brightness temperatures. The differences in the brightness temperatures between the forward and nadir swaths are a direct measurement of the effect of the atmosphere and permit a more accurate determination of the sea surface temperature. The atmospheric correction algorithm takes the form:

$$SST = c_o + \sum_i c_{n,i} T_{n,i} + \sum_i c_{f,i} T_{f,i} \qquad [5]$$

where the subscripts n and f refer to measurements from the nadir and forward views, i indicates two or three atmospheric window channels and the set of c are coefficients. The coefficients, derived by radiative transfer simulations, have an explicit latitudinal dependence.

Accurate calibration of the brightness temperatures is achieved by using two onboard black-body cavities, situated between the apertures for the nadir and forward views such that they are scanned each rotation of the mirror. One calibration target is at the spacecraft ambient temperature while the other is heated, so that the measured brightness temperatures of the sea surface are straddled by the calibration temperatures.

The limitation of the simple scanning geometry of the ATSR is a relatively narrow swath width of 512 km. The ERS satellites have at various times in their missions been placed in orbits with repeat patterns of 3, 35, and 168 days, and given the narrow ATSR swath, complete coverage of the globe has been possible only for the 35 and 186 day cycles. This disadvantage is offset by the intended improvement in absolute accuracy of the atmospheric correction, and of its better insensitivity to aerosol effects.

The Moderate Resolution Imaging Spectroradiometer (MODIS)

The MODIS is a 36-band imaging radiometer on the NASA Earth Observing System (EOS) satellites *Terra*, launched in December 1999, and *Aqua*, planned for launch by late 2001. MODIS is much more complex than other radiometers used for SST measurement, but uses the same atmospheric windows. In addition to the usual two bands in the 10–12 μm interval, MODIS has three narrow bands in the 3.7–4.1 μm windows, which, although limited by sun-glitter effects during the day, hold the potential for much more accurate measurement of SST during the night. Several of the other 31 bands of MODIS contribute to the SST measurement by better identification of residual cloud and aerosol contamination.

The swath width of MODIS, at 2330 km, is somewhat narrower than that of AVHRR, with the result that a single day's coverage is not entire, but the gaps from one day are filled in on the next. The spatial resolution of the infrared window bands is 1 km at nadir.

The GOES Imager

SST measurements from geosynchronous orbit are made using the infrared window channels of the GOES Imager. This is a five-channel instrument that remains above a given point on the Equator. The image of the Earth's disk is constructed by scanning the field of view along horizontal lines by an oscillating mirror. The latitudinal increments of the scan line are done by tilting the axis of the scan mirror. The spatial resolution of the infrared channels is 2.3 km (east–west) by 4 km (north–south) at the subsatellite point. There are two imagers in orbit at the same time on the two GOES satellites, covering the western Atlantic Ocean (GOES-East) and the eastern Pacific Ocean (GOES-West). The other parts of the global oceans visible from geosynchronous orbit are covered by three other satellites operated by Japan, India, and the European Meteorological Satellite organization (Eumetsat). Each carries an infrared imager, but with lesser capabilities than the GOES Imager.

TRMM Microwave Imager (TMI)

The TMI is a nine-channel microwave radiometer on the Tropical Rainfall Measuring Mission satellite, launched in 1997. The nine channels are centered at five frequencies: 10.65, 19.35, 21.3, 37.0, and 85.5 GHz, with four of them being measured at two polarizations. The 10.65 GHz channels confer a sensitivity to SST, at least in the higher SST range found in the tropics, that has been absent in microwave radiometers since the SMMR (Scanning Multifrequency Microwave Radiometer) that flew on the short-lived Seasat in 1978 and on Nimbus-7 from 1978 to 1987. Although SSTs were derived from SMMR measurements, these lacked the spatial resolution and absolute accuracy to compete with those of the AVHHR. The TMI complements AVHRR data by providing SSTs in the tropics where persistent clouds can be a problem for infrared retrievals. Instead of a rotating mirror, TMI, like other microwave imagers, uses an oscillating parabolic antenna to direct the radiation through a feed-horn into the radiometer.

The swath width of TMI is 759 km and the orbit of TRMM restricts SST measurements to within 38.5° of the equator. The beam width of the 10.65 GHz channels produces a footprint of 37×63 km, but the data are over-sampled to produce 104 pixels across the swath.

Applications

With absolute accuracies of satellite-derived SST fields of ~ 0.5 K or better, and even smaller relative uncertainties, many oceanographic features are resolved. These can be studied in a way that was hitherto impossible. They range from basin-scale perturbations to frontal instabilities on the scales of tens of kilometers. SST images have revealed the great complexity of ocean surface currents; this complexity was suspected from shipboard and aircraft measurements, and by acoustically tacking neutrally buoyant floats. However, before the advent of infrared imagery the synoptic view of oceanic variability was elusive, if not impossible.

El Niño

The El Niño Southern Oscillation (ENSO) phenomenon has become a well-known feature of the coupled ocean–atmosphere system in terms of perturbations that have a direct influence on people's lives, mainly by altering the normal rainfall patterns causing draughts or deluges – both of which imperil lives, livestock, and property.

The normal SST distribution in the topical Pacific Ocean is a region of very warm surface waters in the west, with a zonal gradient to cooler water in the east; superimposed on this is a tongue of cool surface water extending westward along the Equator. This situation is associated with heavy rainfall over the western tropical Pacific, which is in turn associated with lower level atmospheric convergence and deep atmospheric convection. The atmospheric convergence and convection are part of the large-scale global circulation. The warm area of surface water, enclosed by the 28°C isotherm, is commonly referred to as the 'Warm Pool' and in the normal situation is confined to the western part of the tropical Pacific. During an El Niño event the warm surface water, and associated convection and rainfall, migrate eastward perturbing the global atmospheric circulation. El Niño events occur up to a few times per decade and are of very variable intensity. Detailed knowledge of

the shape, area, position, and movement of the Warm Pool can be provided from satellite-derived SST to help study the phenomenon and forecast its consequences.

Figure 3 shows part of the global SST fields derived from the Pathfinder SST algorithm applied to AVHRR measurements. The tropical Pacific SST field in the normal situation (December 1993) is shown in the upper panel, while the lower panel shows the anomalous field during the El Niño event of 1997–98. This was one of the strongest El Niños on record, but also the best documented and forecast. Seasonal

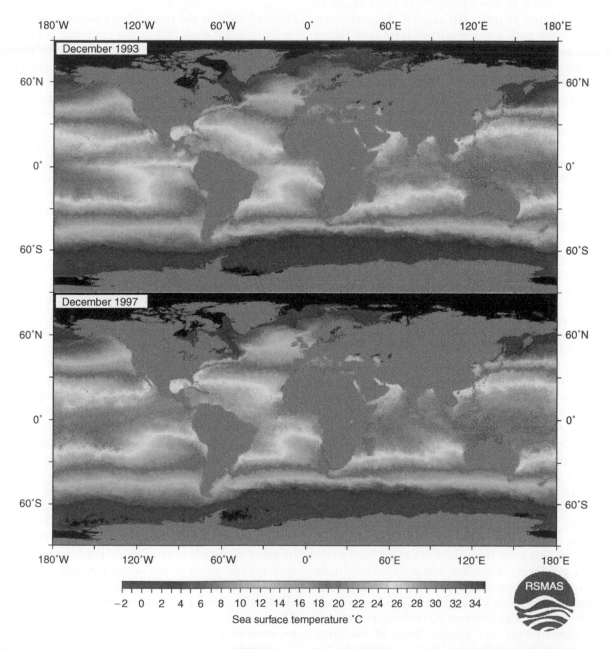

Figure 3 Global maps of SST derived from the AVHRR Pathfinder data sets. These are monthly composites of cloud-free pixels and show the normal situation in the tropical Pacific Ocean (above) and the perturbed state during an El Niño event (below).

predictions of disturbed patterns of winds and rainfall had an unprecedented level of accuracy and provided improved useful forecasts for agriculture in many affected areas. Milder than usual hurricane and tropical cyclone seasons were successfully forecast, as were much wetter winters and severe coastal erosion on the Pacific coasts of the Americas.

Hurricane Intensification

The Atlantic hurricane season in 1999 was one of the most damaging on record in terms of land-falling storms in the eastern USA, Caribbean, and Central America. Much of the damage was not a result of high winds, but of torrential rainfall. Accurate forecasting of the path and intensity of these land-falling storms is very important, and a vital component of this forecasting is detailed knowledge of SST patterns in the path of the hurricanes. The SST is indicative of the heat stored in the upper ocean that is available to intensify the storms, and SSTs of $>26°C$ appear to be necessary to trigger the intensification of the hurricanes. Satellite-derived SST maps are used in the prediction of the development of storm propagation across the Atlantic Ocean from the area off Cape Verde where atmospheric disturbances spawn the nascent storms. Closer to the USA and Caribbean, the SST field is important in determining the sudden intensification that can occur just before landfall. After the hurricane has passed, they sometimes leave a wake of cooler water in the surface that is readily identifiable in the satellite-derived SST fields.

Frontal Positions

One of the earliest features identified in infrared images of SST were the positions of ocean fronts, which delineate the boundaries between dissimilar surface water masses. Obvious examples are western boundary currents, such as the Gulf Stream in the Atlantic Ocean (**Figure 4**) and the Kuroshio in the Pacific Ocean, both of which transport warm surface water poleward and away from the western coastlines. In the Atlantic, the path of the warm surface water of the Gulf Stream can be followed in SST images across the ocean, into the Norwegian Sea, and into the Arctic Ocean. The surface water loses heat to the atmosphere, and to adjacent cooler waters on this path from the Gulf of Mexico to the Arctic, producing a marked zonal difference in the climates of the opposite coasts of the Atlantic and Greenland-Norwegian Seas. Instabilities in the fronts at the sides of the currents have been revealed in great detail in the SST images. Some of the large-scale instabilities can lead to loops on scales of a few tens to hundreds of kilometers that can become

'pinched off' from the flow and evolve as independent features, migrating away from the currents. When these occur on the equator side of the current these are called 'Warm Core Rings' and can exist for many months; in the case of the Gulf Stream these can propagate into the Sargasso Sea.

Figure 5 shows a series of instabilities along the boundaries of the Equatorial current system in the Pacific Ocean. The extent and structure of these features were first described by analysis of satellite SST images.

Coral Bleaching

Elevated SSTs in the tropics have adverse influences on living coral reefs. When the temperatures exceed the local average summertime maximum for several days, the symbiotic relationship between the coral polyps and their algae breaks down and the reef-building animals die. The result is extensive areas where the coral reef is reduced to the skeletal structure without the living and growing tissue, giving the reef a white appearance. Time-series of AVHRR-derived SST have been shown to be valuable predictors of reef areas around the globe that are threatened by warmer than usual water temperatures. Although it is not possible to alter the outcome, SST maps have been useful in determining the scale of the problem and identifying threatened, or vulnerable reefs.

The 'Global Thermometer'

Some of the most pressing problems facing environmental scientists are associated with the issue of global climate change: whether such changes are natural or anthropogenic, whether they can be forecast accurately on regional scales over decades, and whether undesirable consequences can be avoided. The past decade has seen many air temperature records being surpassed and indeed the planet appears to be warming on a global scale. However, the air temperature record is rather patchy in its distribution, with most weather stations clustered on Northern Hemisphere continents.

Global SST maps derived from satellites provide an alternative approach to measuring the Earth's temperature in a more consistent fashion. However, because of the very large thermal inertia of the ocean (it takes as much heat to raise the temperature of only the top meter of the ocean through one degree as it does for the whole atmosphere), the SST changes indicative of global warming are small. Climate change forecast models indicate a rate of temperature increase of only a few tenths of a degree per decade, and this is far from certain because of our incomplete understanding of how the climate

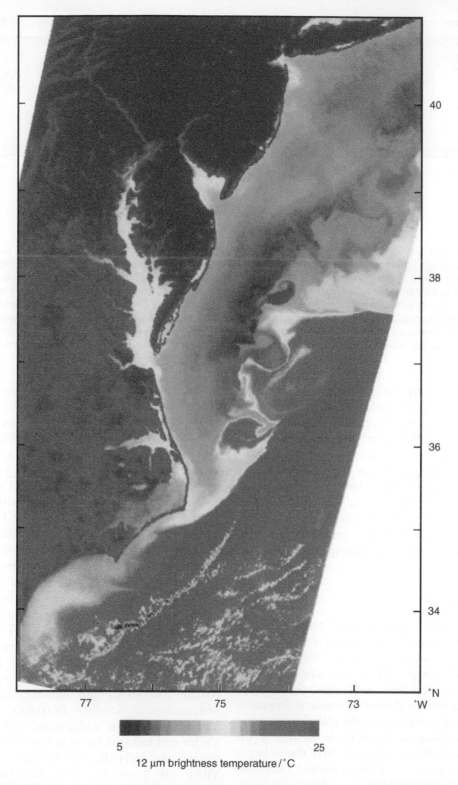

5   12 µm brightness temperature / °C   25

Figure 4 Brightness temperature image derived from the measurements of the ATSR on a nearly cloud-free day over the eastern coast of the USA. The warm core of the Gulf Stream is very apparent; it departs from the coast at Cape Hatteras. The cool, shelf water from the north entrains the warmer outflows from the Chesapeake and Delaware Bays. The north wall of the Gulf Stream reveals very complex structure associated with frontal instabilities that lead to exchanges between the Gulf Stream and inshore waters. The small-scale multicolored patterns over the warm Gulf Stream waters to the south indicate the presence of cloud. This image was taken at 15.18 UTC on 21 May 1992, and is derived from nadir view data from the 12 µm channel. (Generated from data © NERC/ESA/RAL/BNSC, 1992)

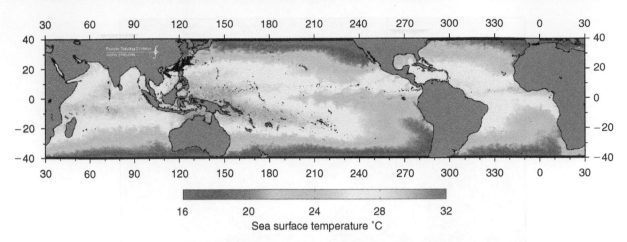

Figure 5 Tropical SSTs produced by microwave radiometer measurements from the TRMM (Tropical Rainfall Measuring Mission) Microwave Imager (TMI). This is a composite generated from data taken during the week ending December 22, 1999. The latitudinal extent of the data is limited by the orbital geometry of the TRMM satellite. The measurement is much less influenced by clouds than those in the infrared, but the black pixels in parts of the oceans where there are no islands indicate areas of heavy rainfall. The image reveals the cold tongue of surface water along the Equator in the Pacific Ocean and cold water off the Pacific coast of South America, indicating a non-El Niño situation. Note that the color scale is different from that used in **Figure 3**. The image was produced by Remote Sensing Systems, sponsored in part by NASA's Earth Science Information Partnerships (ESIP) (a federation of information sites for Earth science); and by the NOAA/NASA Pathfinder Program for early EOS products; principal investigator: Frank Wentz.

system functions, especially in terms of various feedback factors such as those involving changes in cloud and aerosol properties. Such a rate of temperature increase will require SST records of several decades length before the signal, if present, can be unequivocally identified above the uncertainties in the accuracy of the satellite-derived SSTs. Furthermore, the inherent natural variability of the global SST fields tends to mask any small, slow changes. Difficult though this task may be, global satellite-derived SSTs are an important component in climate change research.

Air-sea Exchanges

The SST fields play further indirect roles in the climate system in terms of modulating the exchanges of heat and greenhouse gases between the ocean and atmosphere. Although SST is only one of several variables that control these exchanges, the SST distributions, and their evolution on seasonal timescales can help provide insight into the global patterns of the air–sea exchanges. An example of this is the study of tropical cloud formation over the ocean, a consequence of air–sea heat and moisture exchange, in terms of SST distributions.

Future Developments

Over the next several years continuing improvement of the atmospheric correction algorithms can be anticipated to achieve better accuracies in the derived SST fields, particularly in the presence of atmospheric aerosols. This will involve the incorporation of information from additional spectral channels, such as those on MODIS or other EOS era satellite instruments. Improvements in SST coverage, at least in the tropics, can be expected in areas of heavy, persistent cloud cover by melding SST retrievals from high-resolution infrared sensors with those from microwave radiometers, such as the TMI.

Continuing improvements in methods of validating the SST retrieval algorithms will improve our understanding of the error characteristics of the SST fields, guiding both the appropriate applications of the data and also improvements to the algorithms.

On the hardware front, a new generation of infrared radiometers designed for SST measurements will be introduced on the new operational satellite series, the National Polar-Orbiting Environmental Satellite System (NPOESS) that will replace both the civilian (NOAA-n) and military (DMSP, Defense Meteorological Satellite Program) meteorological satellites. The new radiometer, called VIIRS (the Visible and Infrared Imaging Radiometer Suite), will replace the AVHRR and MODIS. The prototype VIIRS will fly on the NPP (NPOESS Preparatory Program) satellite scheduled for launch in late 2005. At present, the design details of the VIIRS are not finalized, but the physics of the measurement constrains the instrument to use the same atmospheric window channels as previous and current instruments, and have comparable, or better, measurement accuracies.

The ATSR series will continue with at least one more model, called the Advanced ATSR (AATSR) to fly on Envisat to be launched in 2001. The SST capability of this will be comparable to that of its predecessors.

Thus, the time-series of global SSTs that now extends for two decades will continue into the future to provide invaluable information for climate and oceanographic research.

See also

IR Radiometers. Radiative Transfer in the Ocean. Satellite Altimetry. Satellite Remote Sensing SAR.

Further Reading

Barton IJ (1995) Satellite-derived sea surface temperatures: Current status. *Journal of Geophysical Research* 100: 8777–8790.

Gurney RJ, Foster JL, and Parkinson CL (eds.) (1993) *Atlas of Satellite Observations Related to Global Change*. Cambridge: Cambridge University Press.

Ikeda M and Dobson FW (1995) *Oceanographic Applications of Remote Sensing*. London: CRC Press.

Kearns EJ, Hanafin JA, Evans RH, Minnett PJ, and Brown OB (2000) An independent assessment of Pathfinder AVHRR sea surface temperature accuracy using the Marine-Atmosphere Emitted Radiance Interferometer (M-AERI). Bulletin of the American Meteorological Society. 81: 1525–1536.

Kidder SQ and Vonder Haar TH (1995) *Satellite Meteorology: An Introduction*. London: Academic Press.

Legeckis R and Zhu T (1997) Sea surface temperature from the GEOS-8 geostationary satellite. *Bulletin of the American Meteorological Society* 78: 1971–1983.

May DA, Parmeter MM, Olszewski DS, and Mckenzie BD (1998) Operational processing of satellite sea surface temperature retrievals at the Naval Oceanographic Office. *Bulletin of the American Meteorological Society*, 79: 397–407.

Robinson IS (1985) *Satellite Oceanography: An Introduction for Oceanographers and Remote-sensing Scientists*. Chichester: Ellis Horwood.

Stewart RH (1985) *Methods of Satellite Oceanography*. Berkeley, CA: University of California Press.

Victorov S (1996) *Regional Satellite Oceanography*. London: Taylor and Francis.

SATELLITE REMOTE SENSING SAR

A. K. Liu and S. Y. Wu, NASA Goddard
Space Flight Center, Greenbelt, MD, USA

Introduction

Synthetic aperture radar (SAR) is a side-looking imaging radar usually operating on either an aircraft or a spacecraft. The radar transmits a series of short, coherent pulses to the ground producing a footprint whose size is inversely proportional to the antenna size, its aperture. Because the antenna size is generally small, the footprint is large and any particular target is illuminated by several hundred radar pulses. Intensive signal processing involving the detection of small Doppler shifts in the reflected signals from targets to the moving radar produces a high resolution image that is equivalent to one that would have been collected by a radar with a much larger aperture. The resulting larger aperture is the 'synthetic aperture' and is equal to the distance traveled by the spacecraft while the radar antenna is collecting information about the target. SAR techniques depend on precise determination of the relative position and velocity of the radar with respect to the target, and on how well the return signal is processed.

SAR instruments transmit radar signals, thus providing their own illumination, and then measure the strength and phase of the signals scattered back to the instrument. Radar waves have much longer wavelengths compared with light, allowing them to penetrate clouds with little distortion. In effect, radar's longer wavelengths average the properties of air with the properties and shapes of many individual water droplets, and are only affected while entering and exiting the cloud. Therefore, microwave radar can 'see' through clouds.

SAR images of the ocean surface are used to detect a variety of ocean features, such as refracting surface gravity waves, oceanic internal waves, wind fields, oceanic fronts, coastal eddies, and intense low pressure systems (i.e. hurricanes and polar lows), since they all influence the short wind waves responsible for radar backscatter. In addition, SAR is the only sensor that provides measurements of the directional wave spectrum from space. Reliable coastal wind vectors may be estimated from calibrated SAR images using the radar cross-section. The ability of a SAR to provide valuable information on the type, condition, and motion of the sea ice and surface signatures of swells, wind fronts, and eddies near the ice edge has also been amply demonstrated.

With all-weather, day/night imaging capability, SAR penetrates clouds, smoke, haze, and darkness to acquire high quality images of the Earth's surface. This makes SAR the frequent sensor of choice for cloudy coastal regions. Space agencies from the USA, Canada, and Europe use SAR imagery on an operational basis for sea ice monitoring, and for the detection of icebergs, ships, and oil spills. However, there can be considerable ambiguity in the interpretation of physical processes responsible for the observed ocean features. Therefore, the SAR imaging mechanisms of ocean features are briefly described here to illustrate how SAR imaging is used operationally in applications such as environmental monitoring, fishery support, and marine surveillance.

History

The first spaceborne SAR was flown on the US satellite Seasat in 1978. Although Seasat only lasted 3 months, analysis of its data confirmed the sensitivity of SAR to the geometry of surface features. On March 31, 1991 the Soviet Union became the next country to operate an earth-orbiting SAR with the launch of Almaz-1. Almaz-1 returned to earth in 1992 after operating for about 18 months. The European Space Agency (ESA) launched its first remote sensing satellite, ERS-1, with a C-band SAR on July 17, 1991. Shortly thereafter, the JERS-1 satellite, developed by the National Space Development Agency of Japan (NASDA), was launched on February 11, 1992 with an L-band SAR. This was followed a few years later by Radarsat-1, the first Canadian remote sensing satellite, launched in November 1995. Radarsat-1 has a ScanSAR mode with a 500 km swath and a 100 m resolution, an innovative variation of the conventional SAR (with a swath of 100 km and a resolution of 25 m). ERS-2 was launched in April 1995 by ESA, and Envisat-1 with an Advanced SAR is underway with a scheduled launch date in July 2001. The Canadian Space Agency (CSA) has Radarsat-2 planned for 2002, and NASDA has Advanced Land Observing Satellite (ALOS) approved for 2003. **Table 1** shows all major ocean-oriented spaceborne SAR missions worldwide from 1978 to 2003.

Table 1 Major ocean-oriented spaceborne SAR missions

Platform	Nation	Launch	Band[a]	Status
Seasat	USA	1978	L	Ended
Almaz-1	USSR	1991	S	Ended
ERS-1	Europe	1991	C	Standby
JERS-1	Japan	1992	L	Ended
ERS-2	Europe	1995	C	Operational
Radarsat-1	Canada	1995	C	Operational
Envisat-1	Europe	2001	C	Launch scheduled
Radarsat-2	Canada	2002	C	Approved
ALOS	Japan	2003	L	Approved

[a]Some frequently used radar wavelengths are: 3.1 cm for X-band, 5.66 cm for C-band, 10.0 cm for S-band, and 23.5 cm for L-band.

Aside from these free-flying missions, a number of early spaceborne SAR experiments in the USA were conducted using shuttle imaging radar (SIR) systems flown on NASA's Space Shuttle. The SIR-A and SIR-B experiments, in November 1981 and October 1984, respectively, were designed to study radar system performance and obtain sample data of the land using various incidence angles. The SIR-B experiment provided a unique opportunity for studying ocean wave spectra due to the relatively lower orbit of the Shuttle as compared with satellites. The low orbital altitude increases the frequency range of ocean waves that could be reliably imaged, because blurring of the detected waves caused by the motion of ocean surface during the imaging process is reduced. The final SIR mission, SIR-C in April and October 1994, simultaneously recorded SAR data at three wavelengths (L-, C-, and X-bands). These multiple-frequency data from SIR-C improved our understanding of the radar scattering properties of the ocean surface.

Imaging Mechanism of Ocean Features

For a radar with an incidence angle of 20°–50°, such as all spaceborne SARs, backscatter from the ocean

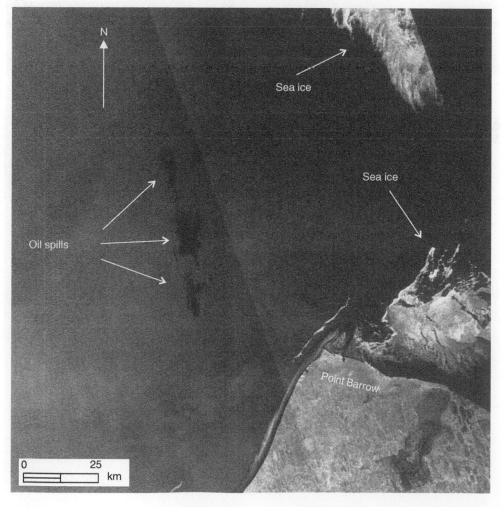

Figure 1 Radarsat ScanSAR image of oil spills off Point Barrow, Alaska, collected on November 2, 1997. (© CSA 1997.)

surface is produced primarily by the Bragg resonant scattering mechanism. That is, surface waves traveling in the radar range (across-track) direction with a wavelength of $\lambda/(2 \sin \theta)$, called the Bragg resonant waves, account for most of the backscattering. In this formula, λ is the radar wavelength, and θ is the incidence angle. In general, the Bragg resonant waves are short gravity waves with wavelengths in the range of 3–30 cm, depending on the radar wavelength or band, as shown in **Table 1**. Because SAR is most sensitive to waves of this wavelength, or roughness of this scale, any ocean phenomenon or process that produces modulation in these particular wavelengths is theoretically detectable by SAR. The radar cross-section of the ocean surface is affected by any geophysical variable, such as wind stress, current shear, or surface slicks, that can modulate the ocean surface roughness at

Bragg-scattering scales. Thus, SARs have proven to be an excellent means of mapping ocean features.

For ocean current features, the essential element of the surface manifestation is the interaction between the current field and the wind-driven ocean surface waves. The effect of the surface current is to alter the short-wave spectrum from its equilibrium value, while the natural processes of wave energy input from the wind restores the ambient equilibrium spectrum. A linear SAR system is one for which the variation of the SAR image intensity is proportional to the gradient of the surface velocity. The proportionality depends on radar wavelength, radar incidence angle, angle between the radar look direction and the current direction, azimuth angle, and the wind velocity. Under high wind condition, large wind waves may overwhelm the weaker current feature. When current flows in the cross-wind direction, the

Figure 2   ERS-1 SAR image of shallow water bathymetry at Taiwan Tan acquired on July 27, 1994. (© ESA 1994.)

wave–current interaction is relatively weak, causing a weak radar backscattering signal.

For ocean frontal features, the change in surface brightness across a front in a SAR image is caused by the change in wind stress exerted onto the ocean surface. The wind stress in turn depends on wind speed and direction, air–sea temperature difference, and surface contamination. The effects of wind stress upon surface ripples and therefore upon radar cross-section, have been modeled and demonstrated as shown in the example below. In the high wind stress area, the ocean surface is rougher and appears as a brighter area in a SAR image. On the other side of the front where the wind is lower, the surface is smoother and appears as a darker area in a SAR image.

The reason why surface films are detectable on radar images is that oil films have a dampening effect on short surface waves. Radar is remarkably sensitive to small changes in the roughness of sea surface. Oil slicks also have a dark appearance in radar images

and are thus similar to the appearance of areas of low winds. The distinctive shape and sharp boundary of localized surface films allows them to be distinguished from the relatively large regions of low wind.

Examples of Ocean Features from SAR Applications

A number of important SAR applications have emerged recently, particularly since ERS-1/2, and Radarsat-1 data became available and the ability to process SAR data has improved. In the USA, the National Oceanic and Atmospheric Administration (NOAA) and the National Ice Center use SAR imagery on an operational basis for sea ice monitoring, iceberg detection, fishing enforcement, oil spill detection, wind and storm information. In Canada, sea ice surveillance is now a proven near-real-time operation, and new marine and coastal applications for

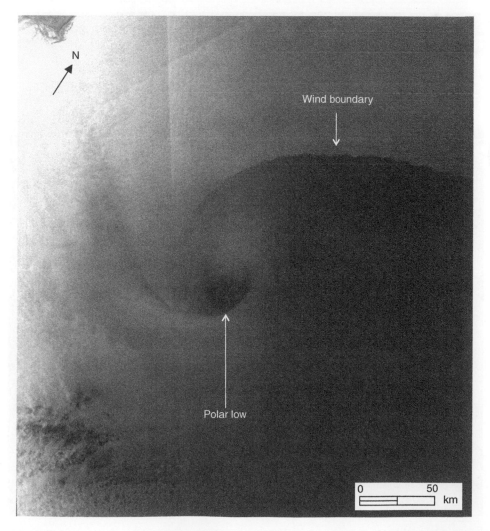

Figure 3 Radarsat ScanSAR image of a polar low in the Bering Sea collected on February 5, 1998. (© CSA 1998.)

SAR imagery are still emerging. In Europe, research on SAR imaging of ocean waves has received great attention in the past 10 years, and has contributed to better global ocean wave forecasts. However, the role of SAR in the coastal observing system still remains at the research and development stage. For reference, examples of some typical ocean features from SAR applications are provided below.

For marine environmental monitoring, features such as oil spills, bathymetry, and polar lows are important for tracking and can often be identified easily with SAR. In early November 1997, Radarsat's SAR sensor captured an oil spill off Point Barrow, Alaska. The oil slicks showed up clearly on the ScanSAR imagery on November 2, 3 and 9. The oil spill is suspected to be associated with the Alaskan Oil Pipeline. **Figure 1** shows a scene containing the oil slicks cropped out from the original ScanSAR image for a closer look. Tracking oil spills using SAR is useful for planning clean-up activities. Early

detection, monitoring, containment, and clean up of oil spills are crucial to the protection of the environment.

Under favorable wind conditions with strong tidal current, the surface signature of bottom topography in shallow water has often been observed in SAR images. **Figure 2**, showing an ERS-1 SAR image of the shallow water bathymetry of Taiwan Tan collected on July 27, 1994, is such an example. Taiwan Tan is located south west of Taiwan in the Taiwan Strait. Typical water depth there is around 30 m with a valley in the middle and the continental shelf break to the south. An extensive sand wave field (**Figure 2**) is developed at Taiwan Tan regularly by wind, tidal current, and surface waves. Monitoring the changes of bathymetry is critical to ship navigation, especially in the areas where water is shallow and ship traffic is heavy.

Intense low pressure systems in polar regions, often referred to as polar lows, may develop above

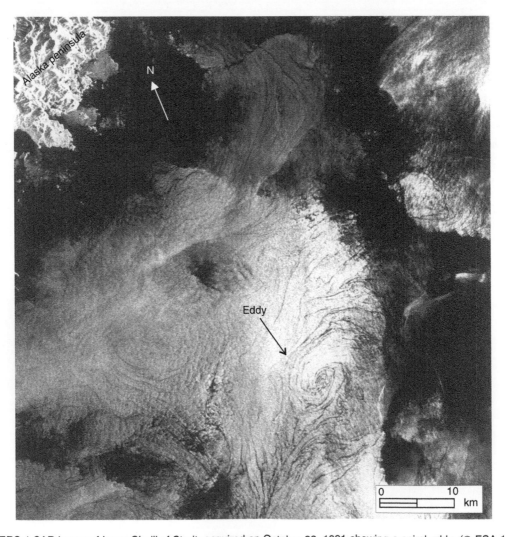

Figure 4   ERS-1 SAR image of lower Shelikof Strait, acquired on October 23, 1991 showing a spiral eddy. (© ESA 1991.)

regions between colder ice/land and warmer ocean during cold air outbreaks. These intense polar lows are formed off major jet streams in cold air masses. Since they usually occur near polar regions where data are sparse, SAR images have been a useful tool for studying these phenomena. **Figure 3** shows a Radarsat ScanSAR image of a polar low in the Bering Sea (centered at 58.0°N, 174.9°E) collected on February 5, 1998. It has a wind boundary to the north spiraling all the way to the center of the storm that separates the high wind (bright) area from the low wind (dark) area. The rippled character along the wind boundary indicates the presence of an instability disturbance induced by the shear flow, which in turn is caused by the substantial difference in wind speed across the boundary.

Ocean features such as eddies, fronts, and ice edges can result in changes in water temperature, turbulence, or transport and may be the primary determinant of recruitment to fisheries. The survival of larvae is enhanced if they remain on the continental shelf and ultimately recruit to nearshore nursery areas. Features such as fronts and eddies can retain larval patches within the shelf zone. **Figure 4** shows an ERS-1 SAR image acquired on October 23, 1991 (centered at 56.69°N, 156.07°W) in lower Shelikof Strait, the Gulf of Alaska. In this image, an eddy with a diameter of approximately 20 km is visible due to low wind conditions. The eddy is characterized by spiraling curvilinear lines which are most likely associated with current shears, surface films, and to a lesser extent temperature contrasts. SAR has the potential to locate these eddies over extensive areas in coastal oceans.

A Radarsat ScanSAR image over the Gulf of Mexico taken on November 23, 1997 (**Figure 5**) shows a distinct, nearly straight front stretching at least 300 km in length. The center of the front in this scene is in the Gulf of Mexico some 400 km south west of New Orleans. The frontal orientation is about 76° east of north. Closer inspection reveals that there are many surface film-like filaments on the

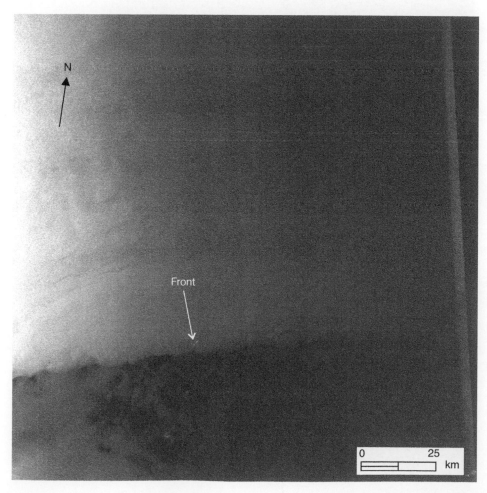

Figure 5 Radarsat ScanSAR image collected over the Gulf of Mexico on November 23, 1997 showing a frontal boundary. (© CSA 1997.)

south side of the front. Concurrent wind data suggest that surface currents converge along the front. Therefore, the formation of this front is probably caused by the accumulation of natural surface films brought about by the convergence around the front. This example highlights SAR's sensitivity to the changes of wind speed and the presence of surface films.

The edge of the sea ice has been found to be highly productive for plankton spring bloom and fishery feeding. In the Bering Sea, fish abundance is highly correlated with yearly ice extent because for their survival many species of fish prefer the cold pools left behind after ice retreat. SAR images are very useful for tracking the movement of the ice edge. **Figure 6** shows a Radarsat ScanSAR image near the ice edge in the Bering Sea collected on February 29, 2000 (centered at 59.6°N and 177.3°W). The sea ice pack with ice bands extending from the ice edge can be clearly seen as the bright area because sea ice surface is rougher than ocean surface (dark area). In the same image, a front is also visible and may or may not be associated with the ice edge to the north. The cold water near the ice edge dampens wave action and appears as a darker area compared with the other side of the front, where it shows up as brighter area due to higher wind and higher sea states.

Information on surface and internal waves, as well as ship wakes, are very important and valuable for marine surveillance and ship navigation. The principal use of SAR for oceanographic studies has been for the detection of ocean waves. The wave direction and height derived from SAR data can be incorporated into models of wind–wave forecast and other applications such as wave–current interaction. **Figure 7** shows an ERS-1 SAR image of long surface waves (or swells) in the lower Shelikof Strait collected on October 17, 1991 (centered at 56.11°N,

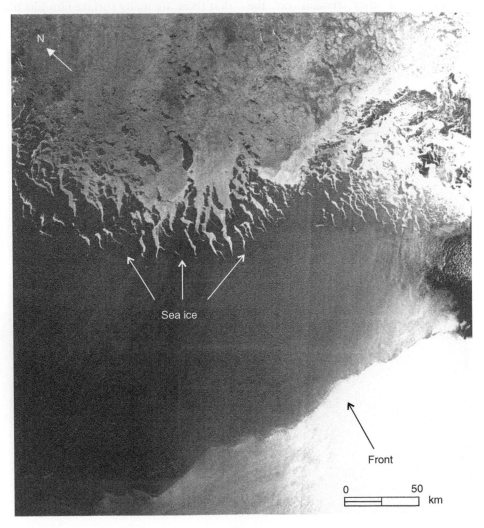

Figure 6   Radarsat ScanSAR image collected over the Bering Sea on February 29, 2000 showing a front near the ice edge. (© CSA 2000.)

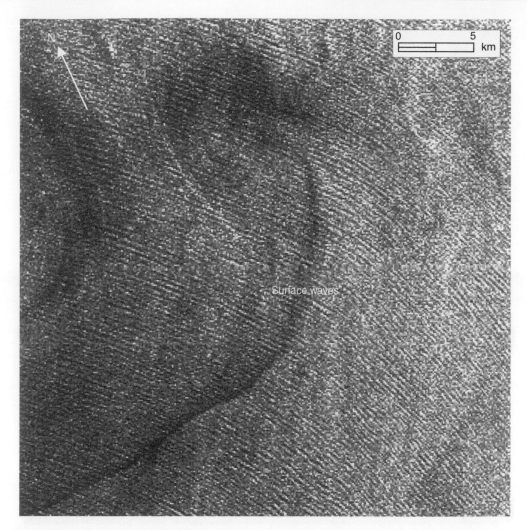

Figure 7   ERS-1 SAR image of lower Shelikof Strait, obtained on October 17, 1991 showing long surface gravity waves refracted by current. (© ESA 1991.)

156.36°W). The location is close to that of the spiral eddy shown in **Figure 4**. Because of the higher winds and higher sea states at the time the image was acquired, the eddy is less conspicuous in this SAR image taken 6 days earlier. Although the direct surface signature of the eddy cannot be discerned clearly in this image, the wave refraction in the eddy area can still be observed. The rays of the wave field can be traced out directly from the SAR image. The ray pattern provides information on the wave refraction pattern and on the relative variation of wave energy along a ray through wave–current interaction.

Tidal currents flowing over submarine topographic features such as a sill or continental shelf in a stratified ocean can generate nonlinear internal waves of tidal frequency. This phenomenon has been studied by many investigators. Direct observations have lent valuable insight into the internal wave generation process and explained the role they play in the transfer of energy from tides to ocean mixing. These nonlinear internal waves are apparently generated by internal mixing as tidal currents flow over bottom features and propagate in the open ocean. In the South China Sea near DongSha Island, enormous westward propagating internal waves from the open ocean are often confronted by coral reefs on the continental shelf. As a result, the waves are diffracted upon passing the reefs. **Figure 8** shows a Radarsat ScanSAR image collected over the northern South China Sea on April 26, 1998, showing at least three packets of internal waves. Each packet consists of a series of internal waves, and the pattern of each wave is characterized by a bright band followed immediately by a dark band. The bright/dark bands indicate the contrast in ocean rough/smooth surfaces caused by convergence/divergence areas induced by the internal waves. At times, the wave 'crest' as observed by SAR from the length of bands can be over 200 km

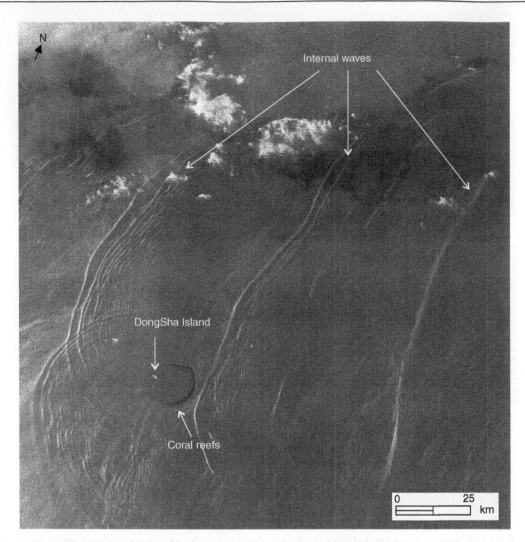

Figure 8 Radarsat ScanSAR image collected over the South China Sea on April 26, 1998 showing three internal wave packets. (© CSA 1998.)

long. After passing the DongSha coral reefs, the waves regroup themselves into two separate packets of internal waves. Later, they interact with each other and emerge as a single wave packet again. SAR can be a very useful tool for studying these shelf processes and the effect of the internal waves on oil drilling platforms, nutrient mixing, and sediment transport.

Ships and their wakes are commonly observable in high-resolution satellite SAR imagery. Detection of ships and ship wakes by means of remote sensing can be useful in the areas of national defense intelligence, shipping traffic, and fishing enforcement. **Figure 9** is an ERS-1 SAR image collected on May 31, 1995 near the northern coast of Taiwan. The image is centered at 25.62°N and 121.15°E, approximately 30 km offshore in the East China Sea. A surface ship heading north east, represented by a bright spot, can be easily identified. Behind this ship, a long dark turbulence

wake is clearly visible. The turbulent wake dampens any short waves, resulting in an area with low back-scattering as indicated by the arrow A in **Figure 9**. Near the ship, the dark wake is accompanied by a bright line which may be caused by the vortex shed by the ship into its wake. The ship track follows the busy shipping lane between Hong Kong, Taiwan, and Japan. The ambient dark slicks are natural surface films induced by upwelling on the continental shelf. In the lower part of the image, another ship turbulent wake (long and dark linear feature oriented east–west) can be identified near the location B in **Figure 9**. A faint bright line connects to the end of this turbulent wake, forming a V-shaped wake in the box B. The faintness of this second ship may be caused by very low backscattering of the ship configuration or the wake could have been formed by a submarine. In the latter case, it must have been operating very close to the ocean surface, since the surface wake is

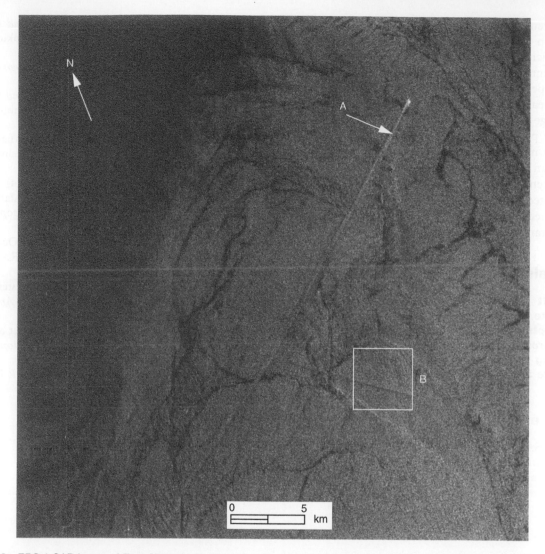

Figure 9 ERS-1 SAR image of East China Sea, obtained on May 31, 1995 showing a surface ship and its wake (arrow A) and a V-shaped wake in box B. (© ESA 1995.)

observable. The ship wake is pointing to the east, indicating that the faint ship was moving from mainland China toward the open ocean.

Discussion

As mentioned earlier, SAR has the unique capability of operating during the day or night and under all weather conditions. With repeated coverage, spaceborne SAR instruments provide the most efficient means to monitor and study the changes in important elements of the marine environment. As demonstrated by the above examples, the use of SAR-derived observations to track eddies, fronts, ice edges, and oil slicks can supply valuable information and can aid in the management of the fishing industry and the protection of the environment. In overcast coastal

areas at high latitudes, the uniformly cold sea surface temperature and persistent cloud cover preclude optical and infrared measurement of surface temperature features, and obscure ocean color observations. The mapping of ocean features by SAR in these challenging coastal regions is, therefore, a potentially major application for satellite-based SAR, particularly for the wider swath ScanSAR mode. Furthermore, SAR data provide unique information for studying the health of the Earth system, as well as critical data for natural hazards and resource assessments.

The prospect of SAR data collection extending well into the twenty-first century gives impetus to current research in SAR applications in ocean science and opens the doors to change detection studies on decadal timescales. The next step is to move into the operational use of SAR data to complement ground measurements. The challenge is to increase cooperation in the scheduling, processing, dissemination, and pricing of SAR data from all SAR satellites between international space agencies. Such cooperation might permit near-real-time high-resolution coastal SAR measurements of sufficient temporal and spatial coverage to impact weather forecasting for selected heavily populated coastal regions. It is necessary to bear in mind that each satellite image is a snapshot and can be complemented with buoy and ship measurements. Ultimately, these data sets should be integrated by numerical models. Such validated and calibrated models will prove extremely useful in understanding a wide variety of oceanic processes.

See also

Aircraft Remote Sensing. Satellite Altimetry. Satellite Oceanography, History and Introductory Concepts. Satellite Passive-Microwave Measurements of Sea Ice. Satellite Remote Sensing Microwave Scatterometers. Satellite Remote Sensing of Sea Surface Temperatures.

Further Reading

Alaska SAR Facility User Working Group (1999) *The Critical Role of SAR in Earth System Science.* (http://www.asf.alaska.edu/)

Beal RC and Pichel WG (eds) (2000) *Coastal and Marine Applications of Wide Swath SAR.* Johns Hopkins APL Technical Digest, 21.

European Space Agency (1995) *Scientific Achievements of ERS-1.* ESA SP-1176/I.

Fu L and Holt B (1982) *Seasat Views Oceans and Sea Ice with Synthetic Aperture Radar,* JPL Publication, pp. 81–120. Pasadena, CA: NASA, JPL/CIT.

Hsu MK, Liu AK, and Liu C (2000) An internal wave study in the China Seas and Yellow Sea by SAR. *Continental Shelf Research* 20: 389–410.

Liu AK, Peng CY, and Schumacher JD (1994) Wave–current interaction study in the Gulf of Alaska for detection of eddies by SAR. *Journal of Geophysical Research* 99: 10075–10085.

Liu AK, Peng CY, and Weingartner TJ (1994) Ocean–ice interaction in the marginal ice zone using SAR. *Journal of Geophysical Research* 99: 22391–22400.

Liu AK, Peng CY, and Chang YS (1996) Mystery ship detected in SAR image. *EOS, Transactions, American Geophysical Union* 77: 17–18.

Liu AK, Peng CY, and Chang YS (1997) Wavelet analysis of satellite images for coastal watch. *IEEE Journal of Oceanic Engineering* 22: 9–17.

Tsatsoulis C and Kwok R (1998) *Analysis of SAR Data of the Polar Oceans.* Berlin: Springer-Verlag.

SATELLITE REMOTE SENSING: OCEAN COLOUR

C. R. McClain, NASA Goddard Space Flight Center, Greenbelt, MD, USA

Introduction

The term 'ocean color' refers to the spectral composition of the visible light field that emanates from the ocean. The color of the ocean depends on the solar irradiance spectra, atmospheric conditions, solar and viewing geometries, and the absorption and scattering properties of water and the substances that are dissolved and suspended in the water column, for example, phytoplankton and suspended sediments. Water masses whose reflectance is determined primarily by absorption due to water and phytoplankton are generally referred to as 'case 1' waters. In other situations, where scattering is the dominant process or where absorption is dominated by substances other than phytoplankton or their derivatives, the term 'case 2' is applied.

The primary optical variable of interest for remote sensing purposes is the water-leaving radiance, L_w, that is, the subsurface upwelled radiance (light moving upward in the water column) propagating through the air–sea interface, but not including the downwelling irradiance (light moving downward through the atmosphere) reflected at the interface. To simplify the interpretation of ocean color, measurements of the water-leaving radiances are normalized by the surface downwelling irradiances to produce 'remote sensing' reflectances, which provide an unambiguous measure of the ocean's subsurface optical signature. Clear open-ocean reflectances have a spectral peak at blue wavelengths because water absorbs strongly in the near-infrared (NIR) and scatters blue light more effectively than at longer wavelengths. As the concentrations of microscopic green plants (phytoplankton) and suspended materials increase, absorption and scattering reduce the reflectance at blue wavelengths and increase the reflectance at green wavelengths, that is, the color shifts from blue to green and brown. This spectral shift in reflectance can be quantified and used to estimate concentrations of optically active components such as chlorophyll a.

The goal of satellite ocean color analysis is to accurately estimate the water-leaving radiance spectra in order to derive other geophysical quantities, for example, chlorophyll a concentration and the diffuse attenuation coefficient. The motivation for spaceborne observations of this kind lies in the need for frequent high-resolution spatial measurements of these geophysical parameters on regional and global scales for addressing both research and operational requirements associated with marine primary production, ecosystem dynamics, fisheries management, ocean dynamics, and coastal sedimentation and pollution, to name a few. The first proof-of-concept satellite ocean color mission was the Coastal Zone Color Scanner (CZCS) on the *Nimbus-7* spacecraft which was launched in the summer of 1978. The CZCS was intended to be a 1-year demonstration with very limited data collection, ground processing, and data validation requirements. However, because of the extraordinary quality and unexpected utility of the data for both coastal and open-ocean research, data collection continued until June 1986 when the sensor ceased operating. The entire CZCS data set was processed, archived, and released to the research community by 1990. As a result of the CZCS experience, a number of other ocean color missions have been launched, for example, the Ocean Color and Temperature Sensor (OCTS; Japan; 1996–97), the Sea-viewing Wide Field-of-view Sensor (SeaWiFS, 1997 and continuing; US), and two Moderate Resolution Imaging Spectroradiometer (MODIS on the *Terra* spacecraft, 2000 and continuing; MODIS on the *Aqua* spacecraft, 2002 and continuing; US), with the expectation that continuous global observations will be maintained in the future.

Ocean Color Theoretical and Observational Basis

Reflectance can be defined in a number of ways and the relationship between the various quantities can be confusing. The most common definition is irradiance reflectance, R, just below the surface, as given in eqn [1], where E_u and E_d are the upwelling and downwelling irradiances, respectively, and the superscript minus sign implies the value just beneath the surface:

$$R(\lambda) = \frac{E_u(\lambda, 0^-)}{E_d(\lambda, 0^-)} \qquad [1]$$

In general, irradiance and radiance are functions of depth (or altitude in the atmosphere) and viewing geometry with respect to the sun. R has been theoretically related to the absorption and scattering properties of the ocean as in eqn [2], where Q is the ratio of $E_u(\lambda, 0^-)$ divided by the upwelling radiance, $L_u(\lambda, 0^-)$, $\ell_1 = 0.0949$, $\ell_2 = 0.0794$, $b_b(\lambda)$ is the backscattering coefficient, and $a(\lambda)$ is the absorption coefficient.

$$R(\lambda) = Q(\lambda) \sum_{i=1}^{2} \ell_i \left[\frac{b_b(\lambda)}{a(\lambda) + b_b(\lambda)} \right]^i \qquad [2]$$

Both $b_b(\lambda)$ and $a(\lambda)$ represent the sum of the contributions of the various optical components (water, inorganic particulates, dissolved substances, phytoplankton, etc.) which are often specified explicitly. If the angular distribution of E_u were directionally uniform, that is, Lambertian, Q would equal π. However, the irradiance distribution is not uniform and is dependent on a number of variables. Some experimental results indicate that Q is roughly 4.5 and, to a first approximation, independent of wavelength. In case 1 water, the approximation, $R(\lambda) \sim f[b_b(\lambda)/a(\lambda)]$ is often used and f is assigned a constant value of 0.33. In reality, f is wavelength-dependent. f and Q are also functions of the solar zenith and viewing angles.

When concentrations of substances, for example, chlorophyll a, are measured, the coefficients can be specified in terms of the concentrations, for example, $a_\varphi(\lambda) = a_\varphi^*(\lambda)[\text{chl } a]$ where $a_\varphi(\lambda)$ and $a_\varphi^*(\lambda)$ are the phytoplankton absorption coefficient and specific absorption coefficient, respectively, and $[\text{chl } a]$ is the chlorophyll a concentration. The absorption coefficients are designated for phytoplankton rather than chlorophyll a because the actual absorption by living cells can vary substantially for a fixed amount of chlorophyll a. Thus, $R(\lambda)$ can be expressed in terms of the specific absorption and scattering coefficients and pigment and particle concentrations. When $R(\lambda)$ is observed at a sufficient number of wavelengths across the spectrum, the $R(\lambda)$ values can be inverted to provide estimates of the scattering and absorption coefficients and pigment concentrations.

For satellite applications, the reflectances and radiances just above the surface are more appropriate to use than $R(\lambda)$. Therefore, water-leaving radiance, remote sensing reflectance, $R_{rs}(\lambda)$, and normalized water-leaving radiance, $L_{wn}(\lambda)$, are commonly used. These are defined in eqns [3]–[5], respectively, where ρ is the Fresnel reflectance of the air–sea interface, n is the index of refraction of seawater, $F_o(\lambda)$ is the extraterrestial solar irradiance, and the plus sign denotes the value just above the surface:

$$L_w(\lambda) = \left[\frac{1-\rho}{n^2} \right] L_u(\lambda, 0^-) \qquad [3]$$

$$R_{rs}(\lambda) = \frac{L_w(\lambda)}{E_d(\lambda, 0^+)} \qquad [4]$$

and

$$L_{wn}(\lambda) = F_o(\lambda) \frac{L_w(\lambda)}{E_d(\lambda, 0^+)} \qquad [5]$$

The term $(1-\rho)/n^2$ is approximately equal to 0.54. Furthermore, it can be shown that $L_{wn}(\lambda)$ is related to $R(\lambda)$ by eqn [6] where the $\langle \rho \rangle$ and $\langle r \rangle$ denote the angular mean Fresnel reflectances for downwelling irradiance above the surface and irradiance below the surface, respectively, and depend on the angular distributions of those irradiances:

$$L_{wn}(\lambda) = F_o(\lambda) \left[\frac{1-\rho}{n^2} \right] \left[\frac{1-\langle \rho \rangle}{1 - R\langle r \rangle} \right] \frac{R(\lambda)}{Q(\lambda)} \qquad [6]$$

Combining eqns [2] and [6] for case 1 waters yields

$$L_{wn}(\lambda) \approx F_o(\lambda) \mathscr{R}(\lambda) \left[\frac{f(\lambda)}{Q(\lambda)} \right] \left[\frac{b_b(\lambda)}{a(\lambda)} \right] \qquad [7]$$

$\mathscr{R}$ is the product of the two bracketed terms in eqn [6] and is largely dependent on surface roughness (wind speed). f/Q is usually given in tables derived from Monte Carlo simulations.

The actual surface reflectance includes not only L_w, but also the Fresnel reflection of photons not scattered by the atmosphere (the direct component), photons that are scattered by the atmosphere (skylight or the indirect component), and light reflected off whitecaps. The angular distribution of total reflection broadens as wind speed increases and the sea surface roughens, and the total surface whitecap coverage also increases with wind speed. For clear case 1 waters, that is, areas where the chlorophyll a is less than about 0.25 mg chl a/m^3, the L_{wn} spectrum is fairly constant, for example, $L_{wn}(550) \sim 0.28$ mW/ cm$^2 \cdot \mu$m $\cdot$ sr, and the values are referred to as 'clear water normalized water-leaving radiances'.

Another optical parameter of interest is the diffuse attenuation coefficient for upwelled radiance, K_{L_u}, (eqn [8]):

$$K_{L_u}(\lambda, z) = -\left(\frac{1}{L_u(\lambda, z)} \right) \left(\frac{dL_u(\lambda, z)}{dz} \right) \qquad [8]$$

Near the surface, where optical constituents are relatively uniform and K_{L_u} is constant,

$$L_u(\lambda, z) = L_u(\lambda, 0^-) e^{-K_{L_u}(\lambda)z} \qquad [9]$$

Similar K relationships hold for irradiance. For convenience, $K(\lambda)$ will be used to denote $K_{L_u}(\lambda)$. The various upwelling and downwelling diffuse

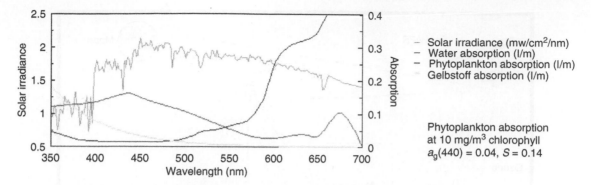

Figure 1 Solar irradiance, chlorophyll specific absorption, yellow substance, and water absorption spectra.

attenuation coefficients for radiance and irradiance are commonly interchanged, but, strictly speaking, they are different quantities and are not equal.

Key to the ocean color measurement technique are the relationships between the solar irradiance, water absorption, and chlorophyll a absorption spectra, chlorophyll a being the primary chemical associated with photosynthesis. The solar spectrum peaks at blue wavelengths which correspond to the maximum transparency of water and the peak in chlorophyll a absorption. Thus, phytoplankton photosynthesis is tuned to the spectral range of maximum light. Figure 1 provides spectra for $F_o(\lambda)$, $a_\varphi^*(\lambda)$, ocean water absorption ($a_w(\lambda)$), and colored dissolved organic matter ($a_g(\lambda)$; CDOM, also known as yellow substance, gelbstoff, and gilvin). The spectrum for CDOM has the form $a_g(\lambda) = a_g(440) \exp[S(\lambda-440)]$, where S ranges from 0.01 to 0.02 nm^{-1}. There are other optical constituents besides phytoplankton and CDOM, such as sediments, which are particularly important in coastal regions.

Satellite Ocean Color Methodology

In order to obtain accurate estimates of geophysical quantities such as chlorophyll a and $K(490)$ from satellite measurements, a number of radiometric issues must be addressed including (1) sensor design and performance; (2) postlaunch sensor calibration stability; (3) atmospheric correction, that is, the removal of light due to atmospheric scattering, atmospheric absorption, and surface reflection; and (4) bio-optical algorithms, that is, the transformation of $R_{rs}(\lambda)$ or $L_{wn}(\lambda)$ values into geophysical parameter values. Items 2–4 represent developments that progress over time during a mission. For example, over the first 9 years of operation, the radiometric sensitivity of *SeaWiFS* degraded by as much as 18% in one of the NIR bands. Also, as radiative transfer theory develops and additional optical data are

obtained, atmospheric correction and bio-optical algorithms will improve and replace previous versions, and new data products will be defined by the research community. Therefore, flight projects, such as *SeaWiFS*, are prepared to periodically reprocess their entire data set. In fact, the *SeaWiFS* data was reprocessed 5 times in the first 8 years. **Figure 2** provides a graphical depiction of all components of the satellite measurement scenario, each aspect of which is discussed below.

Sensor Design and Performance

Sensor design and performance characteristics encompass many considerations which cannot be elaborated on here, but are essential to meeting the overall measurement accuracy requirements. Radiometric factors include wavelength selection, bandwidth, saturation radiances, signal-to-noise ratios (SNRs), polarization sensitivity, temperature sensitivity, scan angle dependencies (response vs. scan, RVS), stray-light rejection, out-of-band contamination, field of view (spatial resolution), band co-registration, and a number of others, all of which must be accurately quantified (characterized) prior to launch and incorporated into the data-processing algorithms. Other design features may include sensor tilting for sunglint avoidance and capacities for tracking the sensor stability on-orbit (e.g., internal lamps, solar diffusers, and lunar views), as instruments generally lose sensitivity over time due to contamination of optical components, filter and detector degradation, etc. Also, there are a variety of spacecraft design criteria including attitude control for accurate navigation, power (solar panel and battery capacities), onboard data storage capacity, telemetry bandwidth (command uplink and downlink data volumes, transmission frequencies, and ground station compatibility and contact constraints), and real-time data broadcast (high-resolution picture transmission (HRPT) station compatibility). Depending on the specifications,

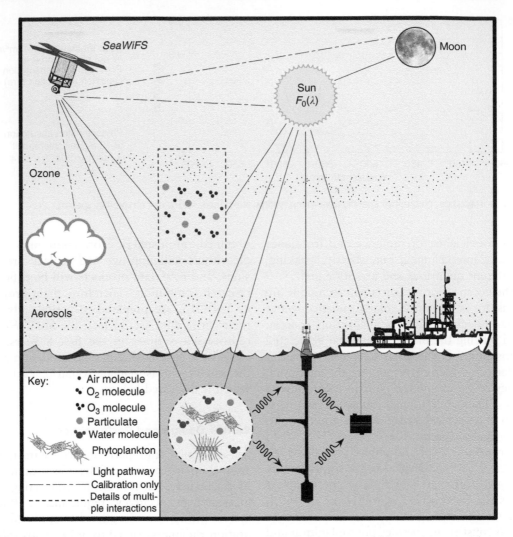

Figure 2 Depiction of the various sensor, atmospheric, and oceanic optical processes which must be accurately accounted for in satellite ocean color data processing.

which must be accurately described and linked to the final data product accuracy requirements, instruments can be built in a variety of ways to optimize performance. For example, the CZCS used a grating for spectral separation, while *SeaWiFS* and *MODIS* use filters. Along with the specifications, the testing procedures and test equipment must be carefully designed to ensure that the characterization data adequately reflect the true sensor performance. Sensor characterization can be a technically difficult, time-consuming, and expensive process. **Table 1** provides a summary of past, present, and planned ocean color sensors and certain observational attributes. It is important to note that only the two *MODIS* instruments have a common design, although their behaviors on orbit were significantly different.

To date, all ocean color missions, both previous and currently approved, have been designed for low-altitude Sun-synchronous orbits, although sensors on high-altitude geostationary platforms are being considered. Sun-synchronous orbits provide global coverage as the Earth rotates so the data is collected at about the same local time, for example, local noon. Typically, the satellite circles the Earth about every 90–100 min (14–15 orbits/day). Multiple views for a given day are possible only at high latitudes where the orbit tracks converge. **Figure 3** shows the daily global area coverage (GAC) of *SeaWiFS*. The gaps between the swaths are filled on the following day as the ground track pattern progressively shifts. The *SeaWiFS* scan is $\pm 57°$, but the GAC is truncated to $\pm 45°$ and subsampled (every fourth pixel and line) on the spacecraft. The full-resolution local area coverage (LAC, 1.1 km at nadir) is broadcast as HRPT data and includes the entire swath at full resolution. Note that as the scan angle increases, pixels (the area on the surface being viewed) become much larger and the atmospheric corrections are less

Table 1 Summary of satellite ocean color sensor and orbit characteristics

Mission instrument (launch yr)	Visible bands	Resolution (km)	Swath (km)	Tilt (deg)	Onboard calibration	Orbit node (inclination)
Nimbus-7 CZCS (1978)	443, 520, 550, 670, 750 11.5 µm	0.8	1600	0, ±20	Internal lamps	12:00 noon (ascending)
ADEOS-I OCTS (1996)	412, 443, 490, 520, 565, 670, 765, 865 nm	0.7	1400	0, ±20	Solar	10:30 am (descending)
					Internal lamps	
ADEOS-I (1996) ADEOS-II (2002) POLDER	443, 490, 565, 670, 763, 765, 865, 910 nm	6.2	2471	N/A	None	10:30 am (descending)
OrbView-2 SeaWiFS (1997)	412, 443, 490, 510, 555, 670, 765, 865 nm	1.1 (LAC) 4.5 (GAC; subsampled)	2800 (LAC) 1500 (GAC)	0, ±20	Solar diffuser Lunar imaging	12:00 noon (descending)
Terra (2000) Aqua (2002) MODIS	412, 443, 488, 531, 551, 667, 678, 748, 869 nm	1.0	1500	None	Solar diffuser (with stability monitor) Spectral radiometric calibration assembly	10:30 (ascending)
Envisat-1 MERIS (2002)	412, 443, 490, 510, 560, 620, 665, 681, 709, 754, 760, 779, 870, 890, 900 nm	0.3 (LAC) 1.2 (GAC)	1150	None	Solar diffusers (3)	10:30 (descending)
ADEOS-II GLI (2002)	400, 412, 443, 460, 490, 520, 545, 565, 625, 666, 680, 710, 749, 865 nm	1.0	1600	0, ±20	Solar diffuser Internal lamps	10:30 (descending)

Some sensors have more channels than are indicated which are used for other applications. There are a number of other ocean color missions not listed because the missions are not designed to routinely generate global data sets. Instruments not mentioned in the text are the Polarization and Directionality of the Earth's Reflectances (POLDER) and the Global Imager (GLI).

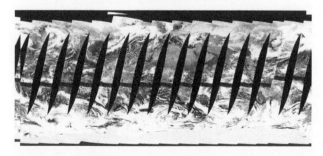

Figure 3 An example of the daily global area coverage (GAC) from *SeaWiFS*. The *SeaWiFS* scan extends to ±57° and samples at about a 1-km resolution. This data is the LAC and is continuously broadcast for HRPT station reception. The GAC data are subsampled at every fourth pixel and line over only the center ±45° of the scan. GAC data are stored on the spacecraft and downlinked to the NASA and Orbimage Corporation (the company that owns *SeaWiFS*) ground stations twice per day.

reliable. *MODIS* has a swath similar to *SeaWiFS* LAC, all of which is recorded and broadcast real-time at full resolution. The data gaps in each swath about the subsolar point are where *SeaWiFS* is tilted from −20° to +20° to avoid viewing into the sunglint. The tilt operation is staggered on successive days in order to ensure every-other-day coverage of the gap. The *MODIS* and Medium Resolution Imaging Spectrometer (MERIS) do not tilt.

Geostationary orbits, that is, orbits having a fixed subsatellite (nadir) point on the equator, only allow hemispheric coverage with decreased spatial resolution away from nadir, but can provide multiple views each day. Multiple views per day allow for the evaluation of tidal and other diurnal time-dependent biases in sampling to be evaluated, and also provide more complete sampling of a given location as cloud patterns change.

Postlaunch Sensor Calibration Stability

The CZCS sensitivity at 443 nm changed by about 40% during its 7.7 years of operation. Even the relatively small changes noted above for *SeaWiFS* undermine the mission objectives for global change research because they introduce spurious trends in the derived products. Quantifying changes in the sensor can be very difficult, especially if the changes are gradual. In the case of the CZCS, there was no ongoing comprehensive validation program after its first year of operation, because the mission was a proof-of-concept. Subsequent missions have some level of continuous validation. In the case of *SeaWiFS*, a combination of solar, lunar, and field observations (oceanic and atmospheric) are used. The solar measurements are made daily using a solar diffuser to detect sudden changes in the sensor. The solar measurements cannot be used as an absolute calibration because the diffuser reflectance gradually changes over time. *MODIS* has a solar diffuser stability monitor which accounts for diffuser degradation. *SeaWiFS* lunar measurements are made once a month at a fixed lunar phase angle (7°) using a spacecraft pitch maneuver that allows the sensor to image the Moon through the Earth-viewing optics. This process provides an accurate estimate of the sensor stability relative to the first lunar measurement (**Figure 4**). The data do require a number of corrections, for example, Sun–Moon distance, satellite–Moon distance, and lunar libration variations. The lunar measurements are not used for an absolute calibration because the moon's surface reflectance is not known to a sufficient accuracy.

Once sensor degradation is removed, field measurements can be used to adjust the calibration gain factors so the satellite retrievals of L_{wn} match independent radiometric field observations, the so-called 'vicarious' calibration (meaning that the vicarious calibration replaces the original laboratory-based prelaunch calibration), and involves the atmospheric correction (discussed below). This adjustment is necessary because the prelaunch calibration is only accurate to within about 3%, the sensor can change during launch and orbit raising, and any biases in the atmospheric correction can be removed in this way. The Marine Optical Buoy (MOBY) was developed and located off Lanai, Hawaii, to provide the *SeaWiFS* and *MODIS* vicarious calibration data. Given cloud cover, sun glint contamination, and satellite sampling frequency at Hawaii, it can take up to 3 years to collect enough high-quality comparisons (25–40 depending on the sensor and the wavelength), to derive statistically stable gain correction factors which should be independent of the sensor-viewing geometry and the solar zenith angle. It is important to note that in-water measurements cannot help in accessing biases in the NIR band calibrations. Atmospheric measurements of optical depth and other parameters are needed to evaluate the calibration of these wavelengths.

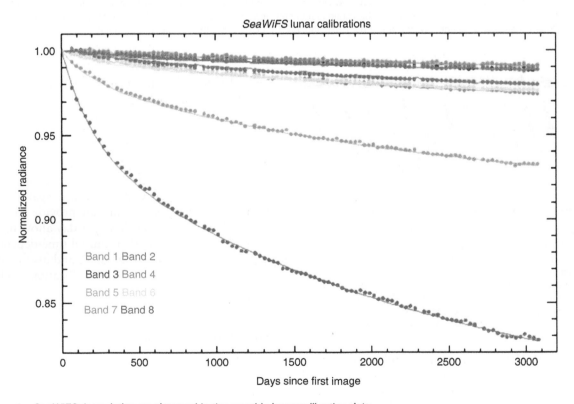

Figure 4 *SeaWiFS* degradation as observed in the monthly lunar calibration data.

Atmospheric Correction

Solar irradiance propagates through the atmosphere where it is attenuated by molecular (Rayleigh) and aerosol scattering and absorption. Rayleigh scattering can be calculated theoretically with a high degree of accuracy. Aerosol scattering and absorption are much more difficult to estimate because their horizontal and vertical distributions are highly variable, as are their absorption and scattering properties. The estimation of the aerosol effects on the upwelling radiance at the top of the atmosphere is one of the most difficult aspects of satellite remote sensing. Ozone (O_3) is the primary absorbing gas that must be considered. Fortunately, ozone is concentrated in a thin band near the top of the atmosphere and its global distribution is mapped daily by sensors such as the Total Ozone Mapping Spectrometer (TOMS), Total Ozone Vertical Sounder (TOVS), and the Ozone Monitoring Instrument (OMI) which have been deployed separately on various spacecraft. Continuous global satellite ozone measurements have been made since 1978 when the first TOMS was flown with the CZCS on *Nimbus-7*. The other absorbing gases of interest are O_2 which has a strong absorption band (the A-band) between 750 and 770 nm and NO_2 which absorbs ultraviolet (UV) and blue light. O_2 is well mixed in the atmosphere and only requires atmospheric pressure data for correction. NO_2 is highly variable and requires global concentration data from satellite sensors such as OMI.

Light that reaches the surface is either reflected or penetrates through the air–sea interface. Simple Fresnel reflection off a flat interface is easily computed theoretically. However, once the surface is windroughened and includes foam (whitecaps), the estimation of the reflected light is more complex and empirical relationships must be invoked. Only a small percentage of the light that enters the water column is reflected upward back through the air–sea interface in the general direction of the satellite sensor. Of that light, only a fraction makes its way back through the atmosphere into the sensor. L_w accounts for no more than about 15% of the top-of-the-atmosphere radiance, L_t. Therefore, each process must be accurately accounted for in estimating L_w. The radiances associated with each process are additive, to the first order, and can be expressed as eqn [10] where the subscripts denote 'total' (t), 'Rayleigh' (r), 'aerosol' (a), 'Rayleigh–aerosol interaction' (ra), 'sun glint' (g), and 'foam' (f), $T(\lambda)$ is the direct transmittance, and $t(\lambda)$ is the diffuse transmittance:

$$L_t(\lambda) = L_r(\lambda) + L_a(\lambda) + L_{ra}(\lambda) + T(\lambda)L_g(\lambda) + t(\lambda)[L_f(\lambda) + L_w(\lambda)] \qquad [10]$$

Note that radiance lost to each absorbing gas is estimated separately and added to the satellite radiance to yield L_t. The Rayleigh radiances can be estimated theoretically and the aerosol radiances are usually inferred from NIR wavelengths using aerosol models and the assumption that L_w is negligible. Determining values for all terms on the right side of eqn [10], with the exception of L_w, constitutes the 'atmospheric correction' which allows eqn [10] to be solved for L_w. As mentioned earlier, if L_w is known, then L_t can be adjusted to balance eqn [10] to derive a 'vicarious' calibration.

Bio-Optical Algorithms

Bio-optical algorithms are used to define relationships between the water-leaving radiances or reflectances and constituents in the water. The basis for the chlorophyll *a* algorithm is the change in reflectance spectral slope with increasing concentration, that is, as chlorophyll *a* increases, the blue end of the spectrum is depressed by pigment absorption and the red end is elevated by increased particle scattering. In case 1 waters, $R(510)$ shows little variation with chlorophyll *a* and is alluded to as the 'hinge point'. Algorithms can be strictly empirical (statistical regressions) or semi-empirical relationships, that is, based on theoretical expressions using measured values of certain optical variables. For example, the empirical chlorophyll *a* (chl *a*) relationship, OC4v4, being used by the *SeaWiFS* mission is depicted in **Figure 5** and is expressed in eqn [11], where $R = \log_{10}[\max(R_{rs}(443),\ R_{rs}(490),\ R_{rs}(510)/R_{rs}(555))]$:

$$\text{chl } a = 10^{0.366 - 3.067R + 1.930R^2 + 0.649R^3 - 1.532R^4} \qquad [11]$$

This relationship was based on observed R_{rs} and chlorophyll *a* data from many locations and, therefore, is not optimized to a particular biological or optical regime, that is, a bio-optical province.

Product Validation

Validation of the derived products can be approached in several ways. The most straightforward approach is to compare simultaneous field and satellite data. Another approach is to make statistical comparisons, for example, frequency distributions, of large *in situ* and satellite data sets. Simultaneous comparisons can provide accurate error estimates, but typically only about 15% of the observations result in valid matchups mainly because of cloud cover, sun glint, spatial inhomogeneities, and time differences. **Figure 6** represents all comparisons for the first 9 years of *SeaWiFS*. Statistical comparisons

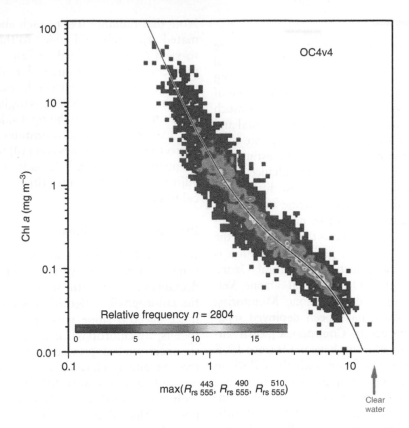

Figure 5   An empirical chlorophyll *a* algorithm, OC4v4, based on a regression of *in situ* chlorophyll *a* vs. R_{rs} ratio observations. In this algorithm, the exponent numerator is the R_{rs} that has the greatest value of the three shown.

of cumulative data sets allow utilization of much more data, but can be subject to sampling biases. Differences in field measurements and satellite estimates can be due to a number of sources including erroneous satellite estimations of L_w, and inaccurate *in situ* values. In the early 1990s, to minimize *in situ* measurement errors, the *SeaWiFS* project initiated a calibration comparison program, the development and documentation of *in situ* measurement protocols, and a number of technology development activities. These were continued under the Sensor Intercomparison and Merger for Biological and Interdisciplinary Oceanic Studies (SIMBIOS) program until 2003. Much of the documentation from *SeaWiFS* and SIMBIOS is available online at http://oceancolor.gsfc.nasa.gov/DOCS/.

An important aspect of satellite data validation is the comparison of data from different missions to assess the level of agreement in the derived products, especially the L_{wn} and chlorophyll *a* values. Because every sensor has a different design and, therefore, different sensitivities to the host of attributes listed earlier, identifying the causes for any differences in the L_{wn} values can be difficult. Good examples are *SeaWiFS* and *MODIS*, which have very different designs. The only design feature these sensors have in common is the use of filters, rather than prisms or gratings, to define the spectral bands. In some situations, the design differences can be used to identify problems in the data products. This point is illustrated by two examples, polarization sensitivity and noise characteristics.

SeaWiFS has a polarization scrambler, which reduces the sensor polarization sensitivity to $\sim 0.25\%$. *SeaWiFS* also incorporates a rotating telescope rather than a mirror which reduces the amount of polarization introduced by the optical surfaces. *MODIS* has a rotating mirror and a polarization sensitivity of several percent. Because the Rayleigh radiance is highly polarized, errors in the polarization characterization can introduce substantial errors in the L_{wn} values. In fact, an error in the *MODIS* polarization tables did initially cause large regional differences ($>50\%$) between the *SeaWiFS* and *MODIS* L_{wn} values. This difference prompted a review of the *MODIS* calibration inputs resulting in the error being corrected, greatly reducing the regional and seasonal differences in the *SeaWiFS* and *MODIS* products.

While the *MODIS* and *SeaWiFS* mean L_{wn} and chlorophyll values are quite comparable, the *MODIS* products have much lower variances (less

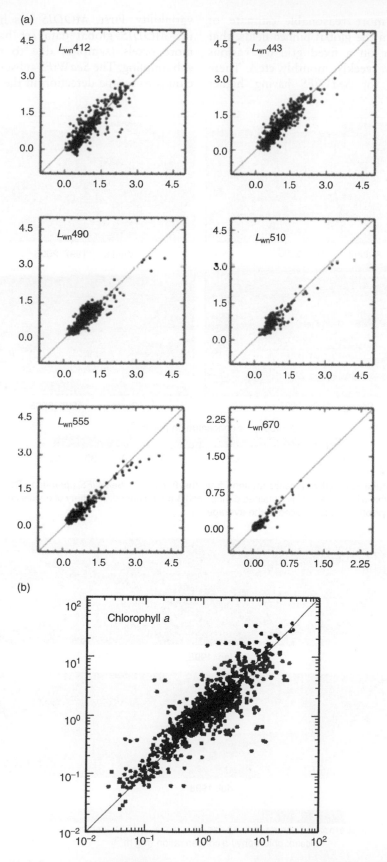

Figure 6 A comparison of simultaneous *SeaWiFS*-derived and *in situ* (a) L_{wn} and (b) chlorophyll *a* data. Only about 15% of the field data collected are used in the comparisons, because many of the data are excluded due to cloud cover, sun glint, time difference, and other rejection criteria. The dashed line is the least squares fit to the data.

noise) providing a more reasonable estimate of sampling variability in the global binned fields, that is, data merged over on a fixed grid for various lengths of time (daily, weekly, monthly, etc.). There are several reasons for *SeaWiFS* having higher

variability. First, *MODIS* has higher SNRs. Also, the *MODIS* global data set has about 20 times more pixels (samples) due to the *SeaWiFS* GAC subsampling. The *SeaWiFS* subsampling allows small clouds to escape detection in the GAC processing in

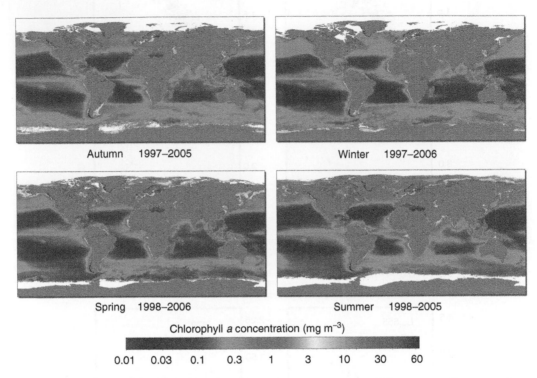

Chlorophyll *a* concentration (mg m^{-3})

0.01   0.03   0.1   0.3   1   3   10   30   60

Figure 7   Seasonal average chlorophyll *a* concentrations from the 9 years of *SeaWiFS* operation. The composites combine all chlorophyll *a* estimates within 9 km square 'bins' obtained during each 3-month period. A variety of quality control exclusion criteria are applied before a sample (pixel) value is included in the average.

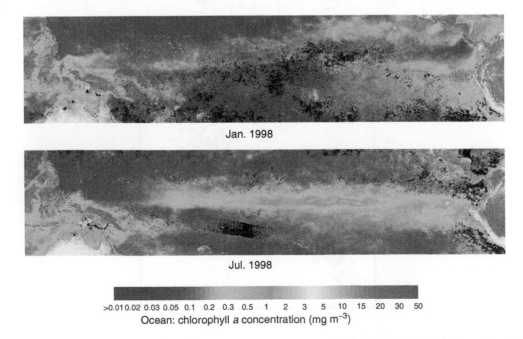

Jul. 1998

>0.01 0.02 0.03 0.05 0.1   0.2   0.3   0.5   1   2   3   5   10   15   20   30   50
Ocean: chlorophyll *a* concentration (mg m^{-3})

Figure 8   A comparison of the monthly average chlorophyll *a* concentrations at the peaks of the 1997 El Niño and the 1998 La Niña in the equatorial Pacific Ocean.

which case stray light is uncorrected (stray light is scattered light within the instrument that contaminates measurements in adjacent pixels), thereby elevating the L_t values. Finally, the *SeaWiFS* data is truncated from 12 to 10 bits on the data recorder resulting in coarser digitization, especially in the NIR bands where the SNRs are relatively low. Noise can cause 'jitter' in the aerosol model selection which amplifies the variability in visible L_{wn} values via the aerosol correction. Undetected clouds in the GAC data and digitization truncation are thought to be the primary reasons for 'speckling' in the *SeaWiFS*-derived products. One strength of the *SeaWiFS* detector array or focal plane design is the bilinear gain which prevents bright pixels from saturating any band. The prelaunch characterization data provided enough information for a stray light correction algorithm to be derived. This correction works well in

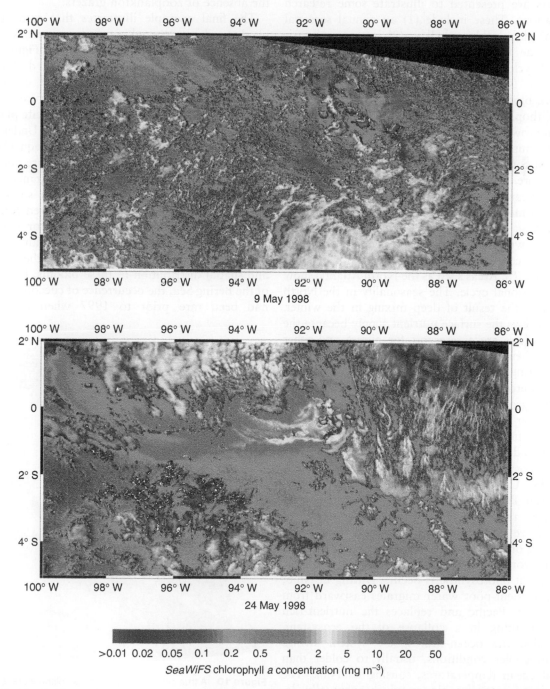

Figure 9 Mesoscale temporal and spatial chlorophyll *a* variability around the Galapagos Islands in the eastern equatorial Pacific Ocean before and after the sudden onset of the 1998 La Niña. The sea surface temperature near the islands dropped nearly 10 °C between 9 and 24 May.

the LAC data processing and for correcting the effects of large bright targets in the GAC. On the other hand, the *MODIS* NIR bands saturate over bright targets including the Moon.

Satellite Ocean Color Data Sets and Applications

In this section, examples of satellite ocean color data products are presented to illustrate some research applications. These include (1) an annual seasonal cycle of global chlorophyll *a*, (2) interannual variability due to the El Niño–Southern Oscillation (ENSO) cycle in the equatorial Pacific, (3) mesoscale variability (scenes from the Galápagos Islands), and (4) blooms of special phytoplankton groups (coccolithophores).

Plant growth in the ocean is regulated by the supply of macronutrients (e.g., nitrate, phosphate, and silicate), micronutrients (iron, in particular), light, and temperature. Light is modulated by cloud cover and time of year (solar zenith angle). Nutrient supply and temperature are determined by ocean circulation and mixing, especially the vertical fluxes, and heat exchange with the atmosphere. **Figure 7** provides seasonal average chlorophyll *a* concentrations derived from *SeaWiFS*. Areas such as the North Atlantic show a clear seasonal cycle. The seasonality in the North Atlantic is the result of deep mixing in the winter, which renews the surface nutrient supply because the deeper waters are a reservoir for nitrate and other macronutrients. Once illumination begins to increase in the spring, the mixed layer shallows providing a well-lit, nutrient-rich surface layer ideal for phytoplankton growth. A bloom results and persists into the summer until zooplankton grazing and nutrient depletion curtail the bloom.

Figure 8 depicts the effects of El Niño and La Niña on the ecosystems of the equatorial Pacific during 1997–98. Under normal conditions, the eastern equatorial Pacific is one of the most biologically productive regions in the world ocean as westward winds force a divergent surface flow resulting in upwelling of nutrient-rich subsurface water into the euphotic zone, that is, the shallow illuminated layer where plant photosynthesis occurs. During El Niño, warm, nutrient-poor water migrates eastward from the western Pacific and replaces the nutrient-rich water, resulting in a collapse of the ecosystem. Eventually, the ocean–atmosphere system swings back to cooler conditions, usually to colder than normal ocean temperatures, causing La Niña. The result is an extensive bloom which eventually declines to more typical concentrations as the atmosphere–ocean system returns to a more normal state.

The 1998 transition from El Niño to La Niña occurred very rapidly. **Figure 9** compares the chlorophyll *a* concentrations around the Galapagos Islands in early and late May. During the time between these two high-resolution scenes, ocean temperatures around the islands dropped by about $10\,°C$. The chlorophyll *a* concentrations jumped dramatically as nutrient-rich waters returned and phytoplankton populations could grow unabated in the absence of zooplankton grazers.

The final example illustrates that some phytoplankton have special optical properties which allows them to be uniquely identified. **Figure 10**, a composite of three Rayleigh-corrected *SeaWiFS* bands, shows an extensive bloom of coccolithophores in the Bering Sea. In their mature stage of development, coccolithophores shed calcite platelets, which turn the water a milky white. Under these conditions, algorithms for reflectance (eqn [2]) and chlorophyll *a* concentration (eqn [10]) are not valid. However, the anomalously high reflectance allows for their detection and removal from spatial and temporal averages of the satellite-derived products. Since coccolithophores are of interest for a number of ecological and biogeochemistry pursuits, satellite ocean color data can be used to map the temporal and spatial distribution of these blooms. In the case of the Bering Sea, the occurrence of coccolithophores had been rare prior to 1997 when the bloom

Figure 10 A true color depiction of a coccolithophore, *Emiliania huxleyi*, bloom in the Bering Sea. The true color composite is formed by summing the Rayleigh radiance-corrected 412-, 555-, and 670-nm *SeaWiFS* images.

persisted for *c*. 6 months. The ecological impact of the blooms in 1997 and 1998, which encompassed the entire western Alaska continental shelf, was dramatic and caused fish to avoid the bloom region resulting in extensive starvation of certain marine mammals and seabirds (those with limited foraging range), and formed a barrier for salmon preventing them from spawning in the rivers along that coast. Research is presently being conducted on methods of identifying other types of blooms such as trichodesmium, important for understanding nitrification in the ocean, and red tides (toxic algal blooms).

Conclusions

Satellite ocean color remote sensing combines a broad spectrum of science and technology. In many ways, the CZCS demonstrated that the technique could work. However, to advance the ocean biology and biogeochemistry on a global scale and conduct research on the effects of global warming, many improvements in satellite sensor technology, atmospheric and oceanic radiative transfer modeling, field observation methodologies, calibration metrology, and other areas have been necessary and are continuing to evolve. As these develop, new products and applications will become feasible and will require periodic reprocessing of the satellite data to incorporate these advances. Ultimately, the goal of the international ocean science community, working with the various space agencies, is to develop a continuous long-term global time series of highly accurate and well-documented satellite ocean color observations.

Nomenclature

a, a_φ	absorption coefficents (m^{-1})
a^*_φ	specific absorption coefficient (m^2/mg Chl a)
b_b	backscattering coefficent (m^{-1})
chl a	chlorophyll a concentration (mg (Chl a)/m^3)
E_d, E_u, F_o	irradiances (mW/cm^2 μm)
K	diffuse attenuation coefficient (m^{-1})
L_a, L_f, L_g, L_p $L_{ra}, L_p L_u, L_w$	radiances (mW/cm^2 μm sr)
L_{wn}	normalized water-leaving radiance (mW/cm^2 μm sr)
n	index of refraction (dimensionless)
ρ, r	Fresnel reflectances (dimensionless)
Q	E_u/L_u ratio (sr)
R	irradiance reflectance (dimensionless)
R_{rs}	remote sensing reflectance (sr^{-1})
S	gelvin absorption spectra parameter (nm^{-1})
t	diffuse transmittance (dimensionless)
T	direct transmittance (dimensionless)

See also

Bio-Optical Models. Inherent Optical Properties and Irradiance. Remote Sensing of Coastal Waters. Satellite Oceanography, History and Introductory Concepts.

Further Reading

Hooker SB and Firestone ER (eds.) (1996) *SeaWiFS Technical Report Series, NASA Technical Memorandum 104566*, vols. 1–43. Greenbelt, MD: NASA Goddard Space Flight Center.

Hooker SB and Firestone ER (eds.) (2000) *SeaWiFS Postlaunch Technical Report Series, NASA Technical Memorandum Year-206892*, vols. 1–29. Greenbelt, MD: NASA Goddard Space Flight Center.

Jerlov NG (1976) *Marine Optics*, 231pp. New York: Elsevier.

Kirk JTO (1994) *Light and Photosynthesis in Aquatic Ecosystems*, 509pp. Cambridge, UK: Cambridge University Press.

Martin S (2004) *An Introduction to Ocean Remote Sensing*, 426pp. New York: Cambridge University Press.

McClain CR, Feldman GC, and Hooker SB (2004) An overview of the SeaWiFS project and strategies for producing a climate research quality global ocean bio-optical time series. *Deep-Sea Research II* 51(1–3): 5–42.

Mobley CD (1994) *Light and Water, Radiative Transfer in Natural Waters*, 592 pp. New York: Academic Press.

Robinson IS (1985) *Satellite Oceanography*, 455pp. Chichester, UK: Wiley.

Shifrin KS (1988) *Physical Optics of Ocean Water*, 285pp. New York: American Institute of Physics.

Stewart RH (1985) *Methods of Satellite Oceanography*, 360pp. Los Angeles: University of California Press.

Relevant Website

http://oceancolor.gsfc.nasa.gov
– OceanColor Documentation, OceanColor Home Page.

SATELLITE REMOTE SENSING: SALINITY MEASUREMENTS

G. S. E. Lagerloef, Earth and Space Research, Seattle, WA, USA

Introduction

Surface salinity is an ocean state variable which controls, along with temperature, the density of seawater and influences surface circulation and formation of dense surface waters in the higher latitudes which sink into the deep ocean and drive the thermohaline convection. Although no satellite measurements are made at present, emerging new technology and a growing scientific need for global measurements have stimulated programs now underway to launch salinity-observing satellite sensors within the present decade. Salinity remote sensing is possible because the dielectric properties of seawater which depend on salinity also affect the surface emission at certain microwave frequencies. Experimental heritage extends more than 35 years in the past, including laboratory studies, airborne sensors, and one instrument flown briefly in space on Skylab. Requirements for very low noise microwave radiometers and large antenna structures have limited the advance of satellite systems, and are now being addressed. Science needs, primarily for climate studies, dictate a resolution requirement of approximately 100 km spatial grid, observed monthly, with approximately 0.1 error on the Practical Salinity Scale (or 1 part in 10 000), which demand very precise radiometers and that several ancillary errors be accurately corrected. Measurements will be made in the 1.413 GHz astronomical hydrogen absorption band to avoid radio interference.

Definition and Theory

How Salinity Is Defined and Measured

Salinity represents the concentration of dissolved inorganic salts in seawater (grams salt per kilogram seawater, or parts per thousand, and historically given by the symbol ‰). Oceanographers have developed modern methods based on the electrical conductivity of seawater which permit accurate measurement by use of automated electronic *in situ* sensors. Salinity is derived from conductivity, temperature, and pressure with an international standard set of empirical equations known as the Practical Salinity Scale, established in 1978 (PSS-78), which is much easier to standardize and more precise than previous chemical methods and which numerically represents grams per kilogram. Accordingly, the modern literature often quotes salinity measurements in practical salinity units (psu), or refers to PSS-78. Salinity ranges from near-zero adjacent to the mouths of major rivers to more than 40 in the Red Sea. Aside from such extremes, open ocean surface values away from coastlines generally fall between 32 and 37 (**Figure 1**).

This global mean surface salinity field has been compiled from all available oceanographic observations. A significant fraction of the 1° latitude and longitude cells have no observations, requiring such maps to be interpolated and smoothed over scales of several hundred kilometers. Seasonal to interannual salinity variations can only be resolved in very limited geographical regions where the sampling density is suitable. Data are most sparse over large regions of the Southern Hemisphere. Remote sensing from satellite will be able to fill this void and monitor multiyear variations globally.

Remote sensing theory Salinity remote sensing with microwave radiometry is likewise possible through the electrically conductive properties of seawater. A radiometric measurement of an emitting surface is given in terms of a 'brightness temperature' (T_B), measured in kelvin (K). T_B is related to the true absolute surface temperature (T) through the emissivity coefficient (e):

$$T_B = eT$$

For seawater, e depends on the complex dielectric constant (ε), the viewing angle (Fresnel laws), and surface roughness (due to wind waves). The complex dielectric constant is governed by the Debye equation

$$\varepsilon = \varepsilon_\infty + \frac{\varepsilon_s(S, T) - \varepsilon_\infty}{1 + i2\pi f\tau(S, T)} - \frac{iC(S, T)}{2\pi f\varepsilon_0}$$

and includes electrical conductivity (C), the static dielectric constant ε_s, and the relaxation time τ, which are all sensitive to salinity and temperature (S, T). The equation also includes radio frequency (f)

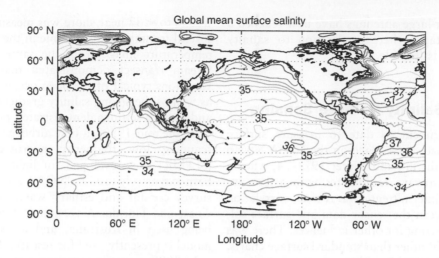

Figure 1 Contour map of the mean global surface salinity field (contour interval 0.5) based on the World Ocean Atlas, 1998 (WOA98). Arctic Ocean salinities <32 are not contoured. Elevated mid-ocean salinities, especially the subtropical Atlantic ones, are caused by excess evaporation. Data were obtained from the US Department of Commerce, NOAA, National Oceanographic Data Center.

and terms for permittivity at infinite frequency (ε_∞) which may vary weakly with T, and permittivity of free space (ε_0, a constant). The relation of electrical conductivity to salinity and temperature is derived from the Practical Salinity Scale. The static dielectric and time constants have been modeled by making laboratory measurements of ε at various frequencies, temperatures, and salinities, and fitting ε_s and τ to polynomial expressions of (S, T) to match the ε data. Different models in the literature show similar variations with respect to (f, S, T).

Emissivity for the horizontal (H) and vertical (V) polarization state is related to ε by Fresnel reflection:

$$e_H = 1 - \left[\frac{\cos\theta - (\varepsilon - \sin^2\theta)^{1/2}}{\cos\theta + (\varepsilon - \sin^2\theta)^{1/2}} \right]^2$$

$$e_V = 1 - \left[\frac{\varepsilon\cos\theta - (\varepsilon - \sin^2\theta)^{1/2}}{\varepsilon\cos\theta + (\varepsilon - \sin^2\theta)^{1/2}} \right]^2$$

where θ is the vertical incidence angle from which the radiometer views the surface, and $e_H = e_V$ when $\theta = 0$. The above set of equations provides a physically based model function relating T_B to surface S, T, θ, and H or V polarization state for smooth water (no wind roughness). This can be inverted to retrieve salinity from radiometric T_B measurements provided the remaining parameters are known. The microwave optical depth is such that the measured emission originates in the top 1 cm of the ocean, approximately.

The rate at which T_B varies with salinity is sensitive to microwave frequency, achieving levels practical for salinity remote sensing at frequencies below

c. 3 GHz. Considerations for selecting a measurement radio frequency include salinity sensitivity, requisite antenna size (see below), and radio interference from other (mostly man-made) sources. A compromise of these factors, dominated by the interference issue, dictates a choice of about a 27-MHz-wide frequency band centered at 1.413 GHz, which is the hydrogen absorption band protected by international treaty for radio astronomy research. This falls within a frequency range known as L-band. Atmospheric clouds have a negligible effect, allowing observations in all weather except possibly heavy rain. Accompanying illustrations are based on applying $f = 1.413$ GHz in the Debye equation and using a model that included laboratory dielectric constant measurements at the nearby frequency 1.43 GHz. Features of this model function and their influence on measurement accuracy are discussed in the section on resolution and error sources.

Antennas Unusually large radiometer antennas will be required to be deployed on satellites for measuring salinity. Radiometer antenna beam width varies inversely with both antenna aperture and radio frequency. 1.413 GHz is a significantly lower frequency than found on conventional satellite microwave radiometers, and large antenna structures are necessary to avoid excessive beam width and accordingly large footprint size. For example, a 50 km footprint requires about a 6-m-aperture antenna whereas conventional radiometer antennas are around 1–2 m. To decrease the footprint by a factor of 2 requires doubling the antenna size. Various filled and thinned array technologies for large antennas have now reached a development stage where application to salinity remote sensing is feasible.

History of Salinity Remote Sensing

The only experiment to date to measure surface salinity from space took place on the NASA Skylab mission during the fall and winter 1973–74, when a 1.413 GHz microwave radiometer with a 1 m antenna collected intermittent data. A weak correlation was found between the sensor data and surface salinity, after correcting for other influences. There was no 'ground truth' other than standard surface charts, and many of the ambient corrections were not as well modeled then as they could be today. Research leading up to the Skylab experiment began with several efforts during the late 1940s and early 1950s to measure the complex dielectric constant of saline solutions for various salinities, temperatures, and microwave frequencies. These relationships provide the physical basis for microwave remote sensing of the ocean as described above.

The first airborne salinity measurements were demonstrated in the Mississippi River outflow and published in 1970. This led to renewed efforts during the 1970s to refine the dielectric constants and governing equations. Meanwhile, a series of airborne experiments in the 1970s mapped coastal salinity patterns in the Chesapeake and Savannah river plumes and freshwater sources along the Puerto Rico shoreline. In the early 1980s, a satellite concept was suggested that might achieve an ideal precision of about 0.25 and spatial resolution of about 100 km. At that time, space agencies were establishing the oceanic processes remote sensing program around missions and sensors for measuring surface dynamic topography, wind stress, ocean color, surface temperature, and sea ice. For various reasons, salinity remote sensing was then considered only marginally feasible from satellite and lacked a strongly defined scientific need.

Interest in salinity remote sensing revived in the late 1980s with the development of a 1.4 GHz airborne electrically scanning thinned array radiometer (ESTAR) designed primarily for soil moisture measurements. ESTAR imaging is done electronically with no moving antenna parts, thus making large antenna structures more feasible. The airborne version was developed as an engineering prototype and to provide the proof of concept that aperture synthesis can be extrapolated to a satellite design. The initial experiment to collect ocean data with this sensor consisted of a flight across the Gulf Stream in 1991 near Cape Hatteras. The change from 36 in the offshore

waters to <32 near shore was measured, along with several frontal features visible in the satellite surface temperature image from the same day. This Gulf Stream transect demonstrated that small salinity variations typical of the open ocean can be detected as well as the strong salinity gradients in the coastal and estuary settings demonstrated previously.

By the mid-1990s, a new airborne salinity mapper scanning low frequency microwave radiometer (SLFMR) was developed for light aircraft and has been extensively used by NOAA and the US Navy to survey coastal and estuary waters on the US East Coast and Florida. A version of this sensor is now being used in Australia, and a second-generation model is presently used for research by the US Navy.

In 1999 a satellite project was approved by the European Space Agency for the measurement of Soil Moisture and Ocean Salinity (SMOS), now with projected launch in 2009. The SMOS mission design emphasizes soil moisture measurement requirements, which is done at the same microwave frequency for many of the same reasons as salinity. The T_B dynamic range is about 70–80 K for varying soil moisture conditions and the precision requirement is therefore much less rigid than for salinity. SMOS will employ a large two-dimensional phased array antenna system that will yield 40–90 km resolution across the measurement swath. The Aquarius/SAC-D mission is being jointly developed by the US (NASA) and Argentina (CONAE) and is due to be launched in 2010. The Aquarius/SAC-D mission design puts primary emphasis on ocean salinity rather than soil moisture, with the focus on optimizing salinity accuracy with a very precisely calibrated microwave radiometer and key ancillary measurements for addressing the most significant error sources.

Requirements for Observing Salinity from Satellite

Scientific Issues

Three broad scientific themes have been identified for a satellite salinity remote sensing program. These themes relate directly to the international climate research and global environmental observing program goals.

Improving seasonal to interannual climate predictions This focuses primarily on El Niño forecasting and involves the effective use of surface salinity data (1) to initialize and improve the coupled climate forecast models, and (2) to study and model the role of freshwater flux in the formation and maintenance of barrier layers and mixed layer heat budgets in the Tropics. Climate prediction models in which satellite altimeter sea level data are assimilated must be adjusted for steric height (sea level change due to ocean density) caused by the variations in upper layer salinity. If not, the adjustment for model heat content is incorrect and the prediction skill is degraded. Barrier layer formation occurs when excessive rainfall creates a shallow, freshwater-stratified, surface layer which effectively isolates the deeper thermocline from exchanging heat with the atmosphere with consequences on the air–sea coupling processes that govern El Niño dynamics.

Improving ocean rainfall estimates and global hydrologic budgets Precipitation over the ocean is still poorly known and relates to both the hydrologic budget and to latent heating of the overlying atmosphere. Using the ocean as a rain gauge is feasible with precise surface salinity observations coupled with ocean surface current velocity data and mixed layer modeling. Such calculations will reduce uncertainties in the surface freshwater flux on climate timescales and will complement satellite precipitation and evaporation observations to improve estimates of the global water and energy cycles.

Monitoring large-scale salinity events and thermohaline convection Studying interannual surface salinity variations in the subpolar regions, particularly the North Atlantic and Southern Oceans, is essential to long-time-scale climate prediction and modeling. These variations influence the rate of oceanic convection and poleward heat transport (thermohaline circulation) which are known to have been coupled to extreme global climate changes in the geologic record. Outside of the polar regions, salinity signals are stronger in the coastal ocean and marginal seas than in the open ocean in general, but large footprint size will limit near-shore applications of the data. Many of the larger marginal seas which have strong salinity signals might be adequately resolved nonetheless, such as the East China Sea, Bay of Bengal, Gulf of Mexico, Coral Sea/Gulf of Papua, and the Mediterranean.

Science requirements From the above science themes, preliminary accuracy and spatial and temporal resolution requirements from satellite observations have been suggested as the minimum to study the following ocean processes: (1) barrier layer effects on tropical Pacific heat flux: 0.2 (PSS-78), 100 km, 30 days; (2) steric adjustment of heat storage from sea level: 0.2, 200 km, 7 days; (3) North Atlantic thermohaline circulation: 0.1, 100 km, 30 days; and (4) surface freshwater flux balance: 0.1, 300 km, 30 days. Thermohaline circulation and convection in the subpolar seas has the most demanding requirement, and is the most technically challenging because of the reduced T_B/salinity ratio at low seawater temperatures (see below). This can serve as a prime satellite mission requirement, allowing for the others to be met by reduced mission requirements as appropriate. Aquarius/SAC-D is a pathfinder mission capable of meeting the majority of these requirements with a grid scale of 150 km and salinity error less than 0.2 observed monthly.

Resolution and Error Sources

Model function Figure 2 shows that the dynamic range of T_B is about 4 K over the range of typical open ocean surface salinity and temperature conditions. T_B gradients are greater with respect to salinity than to temperature. At a given temperature, T_B decreases as salinity increases, whereas the tendency with respect to temperature changes sign. The differential of T_B with respect to salinity ranges from -0.2 to -0.7 K per salinity unit. Corrected T_B will need to be measured to 0.02–0.07 K precision to achieve 0.1 salinity resolution. The sensitivity is strongly affected by temperature, being largest at the highest temperatures and yielding better measurement precision in warm versus cold ocean conditions. Random error can be reduced by temporal and spatial averaging. The degraded measurement precision in higher latitudes will be somewhat compensated by the greater sampling frequency from a polar orbiting satellite.

The T_B variation with respect to temperature falls generally between ± 0.15 K $°C^{-1}$ and near zero over a broad S and T range. Knowledge of the surface temperature to within a few tenths of degrees Celsius will be adequate to correct T_B for temperature effects and can be obtained using data from other satellite systems. The optical depth for this microwave frequency in seawater is about 1–2 cm, and the remotely sensed measurement depends on the T and S in that surface layer thickness. T_B for the H and V polarizations have large variations with incidence angle and spacecraft attitude will need to be monitored very precisely.

Other errors Several other error sources will bias T_B measurements and must be either corrected or avoided. These include ionosphere and atmosphere effects, cosmic and galactic background radiations, surface roughness from winds, sun glint, solar flux, and rain effects. Cosmic background and lower atmospheric adsorption are nearly constant biases

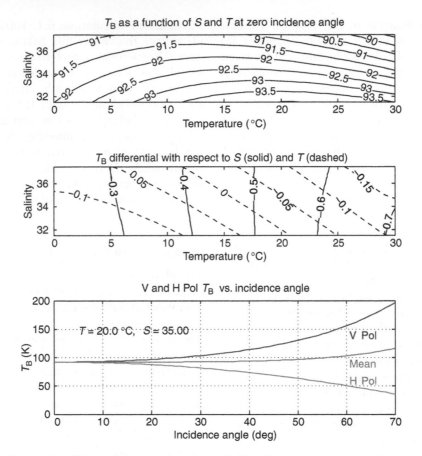

Figure 2 Brightness temperature (T_B) properties as a function of S, T, and incidence angle for typical ocean surface conditions. Top: T_B contours. Middle: T_B derivatives relative to S (solid curves) and T (dashed curves). Bottom: T_B variation vs. incidence angle for H and V polarization. Calculations based on formulas in Klein LA and Swift CT (1977). An improved model for the dielectric constant of sea water at microwave frequencies. *IEEE Transactions on Antennas and Propagation* AP-25(1): 104–111.

easily corrected. An additional correction will be needed when the reflected radiation from the galactic core is in the field of view. The ionosphere and surface winds (roughness) have wide spatial and temporal variations and require ancillary data and careful treatment to avoid T_B errors of several kelvins.

The ionosphere affects the measurement though attenuation and through Faraday rotation of the H and V polarized signal. There is no Faraday effect when viewing at nadir ($\theta = 0$) because H and V emissivities are identical, whereas off-nadir corrections will be needed to preserve the polarization signal. Correction data can be obtained from ionosphere models and analyses but may be limited by unpredictable short-term ionosphere variations. Onboard correction techniques have been developed that require fully polarimetric radiometer measurements from which the Faraday rotation may be derived. Sun-synchronous orbits can be selected that minimize the daytime peak in ionosphere activity as well as solar effects.

The magnitude of the wind roughness correction varies with incidence angle and polarization, and

ranges between 0.1 and 0.4 K/(ms^{-1}). Sea state conditions can change significantly within the few hours that may elapse until ancillary measurements are obtained from another satellite. Simultaneous wind roughness measurement can be made with an onboard radar backscatter sensor. A more accurate correction can be applied using a direct relationship between the radar backscatter and the T_B response rather than rely on wind or sea state information from other sensors.

Microwave attenuation by rain depends on rain rate and the thickness of the rain layer in the atmosphere. The effect is small at the intended microwave frequency, but for the required accuracy the effect must be either modeled and corrected with ancillary data, or the contaminated data discarded. For the accumulation of all the errors described here, it is anticipated that the root sum square salinity error will be reduced to less than 0.2 with adequate radiometer engineering, correction models, onboard measurements, ancillary data, and spatiotemporal filtering with methods now in development.

See also

Satellite Altimetry. Satellite Remote Sensing of Sea Surface Temperatures.

Further Reading

Blume HC and Kendall BNM (1982) Passive microwave measurements of temperature and salinity in coastal zones. *IEEE Transactions on Geoscience and Remote Sensing* GE 20: 394–404.

Blume H-JC, Kendall BM, and Fedors JC (1978) Measurement of ocean temperature and salinity via microwave radiometry. *Boundary-Layer Meteorology* 13: 295–380.

Blume H-JC, Kendall BM, and Fedors JC (1981) Multifrequency radiometer detection of submarine freshwater sources along the Puerto Rican coastline. *Journal of Geophysical Research* 86: 5283–5291.

Broecker WS (1991) The great ocean conveyer. *Oceanography* 4: 79–89.

Delcroix T and Henin C (1991) Seasonal and interannual variations of sea surface salinity in the tropical Pacific Ocean. *Journal of Geophysical Research* 96: 22135–22150.

Delworth T, Manabe S, and Stouffer RJ (1993) Interdecadal variations of the thermohaline circulation in a coupled ocean–atmosphere model. *Journal of Climate* 6: 1993–2011.

Dickson RR, Meincke R, Malmberg S-A, and Lee JJ (1988) The 'great salinity anomaly' in the northern North Atlantic, 1968–1982. *Progress in Oceanography* 20: 103–151.

Droppelman JD, Mennella RA, and Evans DE (1970) An airborne measurement of the salinity variations of the Mississippi River outflow. *Journal of Geophysical Research* 75: 5909–5913.

Kendall BM and Blanton JO (1981) Microwave radiometer measurement of tidally induced salinity changes off the Georgia coast. *Journal of Geophysical Research* 86: 6435–6441.

Klein LA and Swift CT (1977) An improved model for the dielectric constant of sea water at microwave frequencies. *IEEE Transactions on Antennas and Propagation* AP-25(1): 104–111.

Lagerloef G, Swift C, and LeVine D (1995) Sea surface salinity: The next remote sensing challenge. *Oceanography* 8: 44–50.

Lagerloef GSE (2000) Recent progress toward satellite measurements of the global sea surface salinity field. In: Halpern D (ed.) *Elsevier Oceanography Series, No. 63: Satellites, Oceanography and Society*, pp. 309–319. Amsterdam: Elsevier.

Lerner RM and Hollinger JP (1977) Analysis of 1.4 GHz radiometric measurements from Skylab. *Remote Sensing Environment* 6: 251–269.

Le Vine DM, Kao M, Garvine RW, and Sanders T (1998) Remote sensing of ocean salinity: Results from the Delaware coastal current experiment. *Journal of Atmospheric and Ocean Technology* 15: 1478–1484.

Le Vine DM, Kao M, Tanner AB, Swift CT, and Griffis A (1990) Initial results in the development of a synthetic aperture radiometer. *IEEE Transactions on Geoscience and Remote Sensing* 28(4): 614–619.

Lewis EL (1980) The Practical Salinity Scale 1978 (PSS-78) and its antecedents. *IEEE Journal of Oceanic Engineering* OE-5: 3–8.

Miller J, Goodberlet M, and Zaitzeff J (1998) Airborne salinity mapper makes debut in coastal zone. *EOS Transactions, American Geophysical Union* 79(173): 176–177.

Reynolds R, Ji M, and Leetmaa A (1998) Use of salinity to improve ocean modeling. *Physics and Chemistry of the Earth* 23: 545–555.

Swift CT and McIntosh RE (1983) Considerations for microwave remote sensing of ocean-surface salinity. *IEEE Transactions on Geoscience and Remote Sensing* GE-21: 480–491.

UNESCO (1981) The Practical Salinity Scale 1978 and the International Equation of State of Seawater 1980. *Technical Papers in Marine Science* 36, 25pp.

Webster P (1994) The role of hydrological processes in ocean–atmosphere interactions. *Reviews of Geophysics* 32(4): 427–476.

Yueh SH, West R, Wilson WJ, Li FK, Njoku EG, and Rahmat-Samii Y (2001) Error sources and feasibility for microwave remote sensing of ocean surface salinity. *IEEE Transactions on Geoscience and Remote Sensing* 39: 1049–1060.

SEA SURFACE EXCHANGES OF MOMENTUM, HEAT, AND FRESH WATER DETERMINED BY SATELLITE REMOTE SENSING

L. Yu, Woods Hole Oceanographic Institution, Woods Hole, MA, USA

Introduction

The ocean and the atmosphere communicate through the interfacial exchanges of heat, fresh water, and momentum. While the transfer of the momentum from the atmosphere to the ocean by wind stress is the most important forcing of the ocean circulation, the heat and water exchanges affect the horizontal and vertical temperature gradients of the lower atmosphere and the upper ocean, which, in turn, modify wind and ocean currents and maintain the equilibrium of the climate system. The sea surface exchanges are the fundamental processes of the coupled atmosphere–ocean system. An accurate knowledge of the flux variability is critical to our understanding and prediction of the changes of global weather and climate.

The heat exchanges include four processes: the short-wave radiation (Q_{SW}) from the sun, the outgoing long-wave radiation (Q_{LW}) from the sea surface, the sensible heat transfer (Q_{SH}) resulting from air–sea temperature differences, and the latent heat transfer (Q_{LH}) carried by evaporation of sea surface water. Evaporation releases both energy and water vapor to the atmosphere, and thus links the global energy cycle to the global water cycle. The oceans are the key element of the water cycle, because the oceans contain 96% of the Earth's water, experience 86% of planetary evaporation, and receive 78% of planetary precipitation.

The amount of air–sea exchange is called sea surface (or air–sea) flux. Direct flux measurements by ships and buoys are very limited. Our present knowledge of the global sea surface flux distribution stems primarily from bulk parametrizations of the fluxes as functions of surface meteorological variables that can be more easily measured (e.g., wind speed, temperature, humidity, cloud cover, precipitation, etc.). Before the advent of satellite remote sensing, marine surface weather reports collected from voluntary observing ships (VOSs) were the backbone for constructing the climatological state of the global flux fields. Over the past two decades, satellite remote sensing has become a mature technology for remotely sensing key air–sea variables. With continuous global spatial coverage, consistent quality, and high temporal sampling, satellite measurements not only allow the construction of air–sea fluxes at near-real time with unprecedented quality but most importantly, also offer the unique opportunity to view the global ocean synoptically as an entity.

Flux Estimation Using Satellite Observations

Sea Surface Wind Stress

The *Seasat-A* satellite scatterometer, launched in June 1978, was the first mission to demonstrate that ocean surface wind vectors (both speed and direction) could be remotely sensed by active radar backscatter from measuring surface roughness. Scatterometer detects the loss of intensity of transmitted microwave energy from that returned by the ocean surface. Microwaves are scattered by wind-driven capillary waves on the ocean surface, and the fraction of energy returned to the satellite (backscatter) depends on both the magnitude of the wind stress and the wind direction relative to the direction of the radar beam (azimuth angle). By using a transfer function or an empirical algorithm, the backscatter measurements are converted to wind vectors. It is true that scatterometers measure the effects of small-scale roughness caused by surface stress, but the retrieval algorithms produce surface wind, not wind stress, because there are no adequate surface-stress 'ground truths' to calibrate the algorithms. The wind retrievals are calibrated to the equivalent neutral-stability wind at a reference height of 10 m above the local-mean sea surface. This is the 10-m wind that would be associated with the observed surface stress if the atmospheric boundary layer were neutrally stratified. The 10-m equivalent neutral wind speeds differ from the 10-m wind speeds measured by anemometers, and these differences are a function of atmospheric stratification and are normally in the order of $0.2\,\mathrm{m\,s^{-1}}$. To compute the surface wind stress, τ, the conventional bulk formulation is then employed:

$$\tau = (\tau_x, \tau_y) = \rho c_d W(u, v) \qquad [1]$$

where τ_x and τ_y are the zonal and meridional components of the wind stress; W, u, and v are the scatterometer-estimated wind speed at 10 m and its zonal component (eastward) and meridional component (northward), respectively. The density of surface air is given by ρ and is approximately equal to $1.225 \, kg \, m^{-3}$, and c_d is a neutral 10-m drag coefficient.

Scatterometer instruments are typically deployed on sun-synchronous near-polar-orbiting satellites that pass over the equator at approximately the same local times each day. These satellites orbit at an altitude of approximately 800 km and are commonly known as Polar Orbiting Environmental Satellites (POES). There have been six scatterometer sensors aboard POES since the early 1990s. The major characteristics of all scatterometers are summarized in **Table 1**. The first European Remote Sensing (*ERS-1*) satellite was launched by the European Space Agency (ESA) in August 1991. An identical instrument aboard the successor *ERS-2* became operational in 1995, but failed in 2001. In August 1996, the National Aeronautics and Space Administration (NASA) began a joint mission with the National Space Development Agency (NASDA) of Japan to maintain continuous scatterometer missions beyond ERS satellites. The joint effort led to the launch of the NASA scatterometer (NSCAT) aboard the first Japanese Advanced Earth Observing Satellite (*ADEOS-I*). The ERA scatterometers differ from the NASA scatterometers in that the former operate on the C band ($\sim 5 \, GHz$), while the latter use the Ku band ($\sim 14 \, GHz$). For radio frequency band, rain attenuation increases as the signal frequency increases. Compared to C-band satellites, the higher frequencies of Ku band are more vulnerable to signal quality problems caused by rainfall. However, Ku-band satellites have the advantage of being more sensitive to wind variation at low winds and of covering more area.

Rain has three effects on backscatter measurements. It attenuates the radar signal, introduces volume scattering, and changes the properties of the sea surface and consequently the properties of microwave signals scattered from the sea surface. When the backscatter from the sea surface is low, the additional volume scattering from rain will lead to an overestimation of the low wind speed actually present. Conversely, when the backscatter is high,

attenuation by rain will reduce the signal causing an underestimation of the wind speed.

Under rain-free conditions, scatterometer-derived wind estimates are accurate within $1 \, m \, s^{-1}$ for speed and $20°$ for direction. For low (less than $3 \, m \, s^{-1}$) and high winds (greater than $20 \, m \, s^{-1}$), the uncertainties are generally larger. Most problems with low wind retrievals are due to the weak backscatter signal that is easily confounded by noise. The low signal/noise ratio complicates the ambiguity removal processing in selecting the best wind vector from the set of ambiguous wind vectors. Ambiguity removal is over 99% effective for wind speed of $8 \, m \, s^{-1}$ and higher. Extreme high winds are mostly associated with storm events. Scatterometer-derived high winds are found to be underestimated due largely to deficiencies of the empirical scatterometer algorithms. These algorithms are calibrated against a subset of ocean buoys – although the buoy winds are accurate and serve as surface wind truth, few of them have high-wind observations.

NSCAT worked flawlessly, but the spacecraft (*ADEOS-I*) that hosted it demised prematurely in June 1997 after only 9 months of operation. A replacement mission called *QuikSCAT* was rapidly developed and launched in July 1999. To date, *QuikSCAT* remains in operation, far outlasting the expected 2–3-year mission life expectancy. *QuikSCAT* carries a Ku-band scatterometer named SeaWinds, which has accuracy characteristics similar to *NSCAT* but with improved coverage. The instrument measures vector winds over a swath of 1800 km with a nominal spatial resolution of 25 km. The improved sampling size allows approximately 93% of the ocean surface to be sampled on a daily basis as opposed to 2 days by NSCAT and 4 days by the ERS instruments. A second similar-version SeaWinds instrument was placed on the ADEOS-II mission in December 2002. However, after only a few months of operation, it followed the unfortunate path of *NSCAT* and failed in October 2003 due – once again – to power loss.

The Advanced Scatterometer (*ASCAT*) launched by ESA/EUMETSAT in March 2007 is the most recent satellite designed primarily for the global measurement of sea surface wind vectors. *ASCAT* is flown on the first of three METOP satellites. Each METOP has a design lifetime of 5 years and thus, with overlap, the series has a planned duration of 14 years. *ASCAT* is similar to *ERS-1/2* in configuration except that it has increased coverage, with two 500-km swaths (one on each side of the spacecraft nadir track).

The data collected by scatterometers on various missions have constituted a record of ocean vector winds for more than a decade, starting in August 1992. These satellite winds provide synoptic global

Table 1 Major characteristics of the spaceborne scatterometers

Characteristics	Scatterometer						
	SeaSat-A	ERS-1	ERS-2	NSCAT	SeaWinds on QuikSCAT	SeaWinds on ADEOS II	ASCAT
Operational frequency	Ku band	C band	C band	Ku band	Ku band	Ku band	C band
Spatial resolution	14.6 GHz 50 km × 50 km with 100-km spacing	5.255 GHz 50 km × 50 km	5.255 GHz 50 km × 50 km	13.995 GHz 25 km × 25 km	13.402 GHz 25 km × 25 km	13.402 GHz 25 km × 6 km	5.255 GHz 25 km × 25 km
Scan characteristics	Two-sided, double 500 km swaths separated by a 450 km nadir gap	One-sided, single 500-km swath	One-sided, single 500-km swath	Two-sided, double 600-km swaths separated by a 329-km nadir gap	Conical scan, one wide swath of 1800 km	Conical scan, one wide swath of 1800 km	Two-sided, double 500-km swaths separated by a 700-km nadir gap
Daily coverage	Variable	41%	41%	77%	93%	93%	60%
Period in service	Jul. 1978–Oct. 1978	Aug. 1991–May. 1997	May. 1995–Jan. 2001	Sep. 1996–Jun. 1997	Jun. 1999–current	Dec. 2002–Oct. 2003	Mar. 2007–current

view from the vantage point of space, and provide excellent coverage in regions, such as the southern oceans, that are poorly sampled by the conventional observing network. Scatterometers have been shown to be the only means of delivering observations at adequate ranges of temporal and spatial scales and at adequate accuracy for understanding ocean–atmosphere interactions and global climate changes, and for improving climate predictions on synoptic, seasonal, and interannual timescales.

Surface Radiative Fluxes

Direct estimates of surface short-wave (SW) and long-wave (LW) fluxes that resolve synoptic to regional variability over the globe have only become possible with the advent of satellite in the past two decades. The surface radiation is a strong function of clouds. Low, thick clouds reflect large amounts of solar radiation and tend to cool the surface of the Earth. High, thin clouds transmit incoming solar radiation, but at the same time, they absorb the outgoing LW radiation emitted by the Earth and radiate it back downward. The portion of radiation, acting as an effective 'greenhouse gas', adds to the SW energy from the sun and causes an additional warming of the surface of the Earth. For a given cloud, its effect on the surface radiation depends on several factors, including the cloud's altitude, size, and the particles that form the cloud. At present, the radiative heat fluxes at the Earth's surface are estimated from top-of-the-atmosphere (TOA) SW and LW radiance measurements in conjunction with radiative transfer models.

Satellite radiance measurements are provided by two types of radiometers: scanning radiometers and nonscanning wide-field-of-view radiometers. Scanning radiometers view radiance from a single direction and must estimate the hemispheric emission or reflection. Nonscanning radiometers view the entire hemisphere of radiation with a roughly 1000-km field of view. The first flight of an Earth Radiation Budget Experiment (ERBE) instrument in 1984 included both a scanning radiometer and a set of nonscanning radiometers. These instruments obtain good measurements of TOA radiative variables including insolation, albedo, and absorbed radiation. To estimate surface radiation fluxes, however, more accurate information on clouds is needed.

To determine the physical properties of clouds from satellite measurements, the International Satellite Cloud Climatology Project (ISCCP) was established in 1983. ISCCP pioneered the cross-calibration, analysis, and merger of measurements from the international constellation of operational weather satellites. Using geostationary satellite measurements with polar orbiter measurements as supplemental when there are no geostationary measurements, the ISCCP cloud-retrieval algorithm includes the conversion of radiance measurements to cloud scenes and the inference of cloud properties from the radiance values. Radiance thresholds are applied to obtain cloud fractions for low, middle, and high clouds based on radiance computed from models using observed temperature and climatological lapse rates.

In addition to the global cloud analysis, ISCCP also produces radiative fluxes (up, down, and net) at the Earth's surface that parallels the effort undertaken by the Global Energy and Water Cycle Experiment – Surface Radiation Budget (GEWEX-SRB) project. The two projects use the same ISCCP cloud information but different ancillary data sources and different radiative transfer codes. They both compute the radiation fluxes for clear and cloudy skies to estimate the cloud effect on radiative energy transfer. Both have a 3-h resolution, but ISCCP fluxes are produced on a 280-km equal-area (EQ) global grid while GEWEX-SRB fluxes are on a $1° \times 1°$ global grid. The two sets of fluxes have reasonable agreement with each other on the long-term mean basis, as suggested by the comparison of the global annual surface radiation budget in **Table 2**. The total net radiation differs by about $5\,W\,m^{-2}$, due mostly to the SW component. However, when compared with ground-based observations, the uncertainty of these fluxes is about $10–15\,W\,m^{-2}$. The main cause is the uncertainties in surface and near-surface atmospheric properties such as surface skin temperature, surface air and near-surface-layer temperatures and humidity, aerosols, etc. Further improvement requires improved retrievals of these properties.

In the late 1990s, the Clouds and the Earth's Radiant Energy System (CERES) experiment was

Table 2 Annual surface radiation budget (in $W\,m^{-2}$) over global oceans. Uncertainty estimates are based on the standard error of monthly anomalies

Data 21-year mean 1984–2004	Parameter		
	SW Net downward	LW Net downward	SW+LW Net downward
ISCCP (Zhang et al., 2004)	173.2±9.2	−46.9±9.2	126.3±11.0
GEWEX-SRB (Gupta et al., 2006)	167.2±13.9	−46.3±5.5	120.9±11.9

developed by NASA's Earth Observing System (EOS) not only to measure TOA radiative fluxes but also to determine radiative fluxes within the atmosphere and at the surface, by using simultaneous measurements of complete cloud properties from other EOS instruments such as the moderate-resolution imaging spectroradiometer (MODIS). CERES instruments were launched aboard the Tropical Rainfall Measuring Mission (TRMM) in November 1997, on the EOS *Terra* satellite in December 1999, and on the EOS *Aqua* spacecraft in 2002. There is no doubt that the EOS era satellite observations will lead to great improvement in estimating cloud properties and surface radiation budget with sufficient simultaneity and accuracy.

Sea Surface Turbulent Heat Fluxes

Latent and sensible heat fluxes are the primary mechanism by which the ocean transfers much of the absorbed solar radiation back to the atmosphere. The two fluxes cannot be directly observed by space sensors, but can be estimated from wind speed and sea–air humidity/temperature differences using the following bulk parametrizations:

$$Q_{LH} = \rho L_e c_e W (q_s - q_a) \qquad [2]$$

$$Q_{SH} = \rho c_p c_h W (T_s - T_a) \qquad [3]$$

where L_e is the latent heat of vaporization and is a function of sea surface temperature (SST, T_s) expressed as $L_e = (2.501 - 0.002\,37 \times T_s) \times 1.0^6$. c_p is the specific heat capacity of air at constant pressure; c_e and c_h are the stability- and height-dependent turbulent exchange coefficients for latent and sensible heat, respectively. T_a/q_a are the temperature/specific humidity at a reference height of 2 m above the sea surface. q_s is the saturation humidity at T_s, and is multiplied by 0.98 to take into account the reduction in vapor pressure caused by salt water.

The two variables, T_s and W, in eqns [2] and [3] are retrieved from satellites, and so q_s is known. The remote sensing of T_s is based on techniques by which spaceborne infrared and microwave radiometers detect thermally emitted radiation from the ocean surface. Infrared radiometers like the five-channel advanced very high resolution radiometer (AVHRR) utilize the wavelength bands at 3.5–4 and 10–12 μm that have a high transmission of the cloud-free atmosphere. The disadvantage is that clouds are opaque to infrared radiation and can effectively mask radiation from the ocean surface, and this affects temporal resolution. Although the AVHRR satellite orbits the Earth 14 times each day from 833 km

above its surface and each pass of the satellite provides a 2399-km-wide swath, it usually takes 1 or 2 weeks, depending on the actual cloud coverage, to obtain a complete global coverage. Clouds, on the other hand, have little effect on the microwave radiometers so that microwave T_s retrievals can be made under complete cloud cover except for raining conditions. The TRMM microwave imager (TMI) launched in 1997 has a full suite of channels ranging from 10.7 to 85 GHz and was the first satellite sensor capable of accurately measuring SST through clouds. The low-inclination equitorial orbit, however, limits the TMI's coverage only up to c. 38° latitude. Following TMI, the first polar-orbiting microwave radiometer capable of measuring global through-cloud SST was made possible by the NASDA's advanced microwave scanning radiometer (AMSR) flown aboard the NASA's EOS Aqua mission in 2002.

While SST can be measured in both infrared and microwave regions, the near-surface wind speed can only be retrieved in the microwave region. The reason is that the emissivity of the ocean's surface at wavelengths of around 11 μm is so high that it is not sensitive to changes in the wind-induced sea surface roughness or humidity fluctuations in the lower atmosphere. Microwave wind speed retrievals are provided by the special sensor microwave/imager (SSM/I) that has been flown on a series of polar-orbiting operational spacecrafts of the Defense Meteorological Space Program (DMSP) since July 1987. SSM/I has a wide swath ($\sim$ 1400 km) and a coverage of 82% of the Earth's surface within 1 day. But unlike scatterometers, SSM/I is a passive microwave sensor and cannot provide information on the wind direction. This is not a problem for the computation in eqns [2] and [3] that requires only wind speed observations. In fact, the high space-time resolution and good global coverage of SSM/I has made it serving as a primary database for computing the climate mean and variability of the oceanic latent and sensible heat fluxes over the past $\sim$20-year period. At present, wind speed measurements with good accuracy are also available from several NASA satellite platforms, including TMI and AMSR.

The most difficult problem for the satellite-based flux estimation is the retrieval of the air humidity and temperature, q_a and T_a, at a level of several meters above the surface. This problem is inherent to all spaceborne passive radiometers, because the measured radiation emanates from relatively thick atmospheric layers rather than from single levels. One common practice to extract satellite q_a is to relate q_a to the observed column integrated water vapor (IWV, also referred to as the total precipitable water) from

SSM/I. Using IWV as a proxy for q_a is based on several observational findings that on monthly timescales the vertical distribution of water vapor is coherent throughout the entire atmospheric column. The approach, however, produces large systematic biases of over $2\,g\,kg^{-1}$ in the Tropics, as well as in the mid- and high latitudes during summertime. This is caused by the effect of the water vapor convergence that is difficult to assess in regions where the surface air is nearly saturated but the total IWV is small. Under such situations, the IWV cannot reflect the actual vertical and horizontal humidity variations in the atmosphere. Various remedies have been proposed to improve the q_a–IWV relation and to make it applicable on synoptic and shorter timescales. There are methods of including additional geophysical variables, replacing IWV with the IWV in the lower 500 m of the planetary layer, and/or using empirical orthogonal functions (EOFs). Although overall improvements were achieved, the accuracy remains poor due to the lack of detailed information on the atmospheric humidity profiles.

Retrieving T_a from satellite observations is even more challenging. Unlike humidity, there is no coherent vertical structure of temperature in the atmosphere. Satellite temperature sounding radiometers offer little help, as they generally are designed for retrieval in broad vertical layers. The sounder's low information content in the lower atmosphere does not enable the retrieval of near-surface air temperature with sufficient accuracy. Different methods have been tested to derive T_a from the inferred q_a, but all showed limited success. Because of the difficulties in determining q_a and T_a, latent and sensible fluxes estimated from satellite measurements have large uncertainties.

Three methods have been tested for obtaining better q_a and T_a to improve the estimates of latent and sensible fluxes. The first approach is to enhance the information on the temperature and moisture in the lower troposphere. This is achieved by combining SSM/I data with additional microwave sounder data that come from the instruments like the advanced microwave sounding unit (AMSU-A) and microwave humidity sounder (MHS) flown aboard the National Oceanic and Atmospheric Administration (NOAA) polar-orbiting satellites, and the special sensor microwave temperature sounder (SSM/T) and (SSM/T-2) on the DMSP satellites. Although the sounders do not directly provide shallow surface measurements, detailed profile information provided by the sounders can help to remove variability in total column measurements not associated with the surface. The second approach is to capitalize on the progress made in numerical weather prediction models that assimilate sounder observations into

the physically based system. The q_a and T_a estimates from the models contain less ambiguity associated with the vertical integration and large spatial averaging of the various parameters, though they are subject to systematic bias due to model's subgrid parametrizations. The third approach is to obtain a better estimation of q_a and T_a through an optimal combination of satellite retrievals with the model outputs, which has been experimented by the Objectively Analyzed air–sea Fluxes (OAFlux) project at the Woods Hole Oceanographic Institution (WHOI). The effort has led to improved daily estimates of global air–sea latent and sensible fluxes.

Freshwater Flux

The freshwater flux is the difference between precipitation (rain) and evaporation. Evaporation releases both water vapor and latent heat to the atmosphere. Once latent heat fluxes are estimated, the sea surface evaporation (E) can be computed using the following relation:

$$E = Q_{LH}/\rho_w L_e \qquad [4]$$

where Q_{LH} denotes latent heat flux and ρ_w is the density of seawater.

Spaceborne sensors cannot directly observe the actual precipitation reaching the Earth's surface, but they can measure other variables that may be highly correlated with surface rainfall. These include variations in infrared and microwave brightness temperatures, as well as visible and near-infrared albedo. Infrared techniques are based on the premise that rainfall at the surface is related to cloud-top properties observed from space. Visible/infrared observations supplement the infrared imagery with visible imagery during daytime to help eliminate thin cirrus clouds, which are cold in the infrared imagery and are sometimes misinterpreted as raining using infrared data alone. Visible/infrared sensors have the advantage of providing good space and time sampling, but have difficulty capturing the rain from warm-topped clouds. By comparison, microwave (MW) estimates are more physically based and more accurate although time and space resolutions are not as good. The principle of MW techniques is that rainfall at the surface is related to microwave emission from rain drops (low-frequency channels) and microwave scattering from ice (high-frequency channels). While the primary visible/infrared data sources are the operational geostationary satellites, microwave observations are available from SSM/I, the NOAA AMSU-B, and the TRMM spacecraft.

TRMM opened up a new era of estimating not only surface rainfall but also rain profiles. TRMM is equipped with the first spaceborne precipitation radar (PR) along with a microwave radiometer (TMI) and a visible/infrared radiometer (VIRS). Coincident measurements from the three sensors are complementary. PR provides detailed vertical rain profiles across a 215-km-wide strip. TMI (a five-frequency conical scanning radiometer) though has less vertical and horizontal fidelity in rain-resolving capability, and it features a swath width of 760 km. The VIRS on TRMM adds cloud-top temperatures and structures to complement the description of the two microwave sensors. While direct precipitation information from VIRS is less reliable than that obtained by the microwave sensors, VIRS serves an important role as a bridge between the high-quality but infrequent observations from TMI and PR and the more available data and longer time series data available from the geostationary visible/infrared satellite platforms.

The TRMM satellite focuses on the rain variability over the tropical and subtropical regions due to the low inclination. An improved instrument, AMSR, has extended TRMM rainfall measurements to higher latitudes. AMSR is currently aboard the *Aqua* satellite and is planned by the Global Precipitation Measurement (GPM) mission to be launched in 2009. Combining rainfall estimates from visible/infrared with microwave measurements is being undertaken by the Global Climatology Project (GPCP) to produce global precipitation analyses from 1979 and continuing.

Summary and Applications

The satellite sensor systems developed in the past two decades have provided unprecedented observations of geophysical parameters in the lower atmosphere and upper oceans. The combination of measurements from multiple satellite platforms has demonstrated the capability of estimating sea surface heat, fresh water, and momentum fluxes with sufficient accuracy and resolution. These air–sea flux data sets, together with satellite retrievals of ocean surface topography, temperature, and salinity (**Figure 1**), establish a complete satellite-based observational infrastructure for fully monitoring the ocean's response to the changes in air–sea physical forcing.

Atmosphere and the ocean are nonlinear turbulent fluids, and their interactions are nonlinear scale-dependent, with processes at one scale affecting processes at other scales. The synergy of various satellite-based products makes it especially advantageous to study the complex scale interactions between the atmosphere and the ocean. One clear example is the satellite monitoring of the development of the El Niño–Southern Oscillation (ENSO) in 1997–98. ENSO is the largest source of interannual variability in the global climate system. The phenomenon is characterized by the appearance of extensive warm surface water over the central and eastern tropical

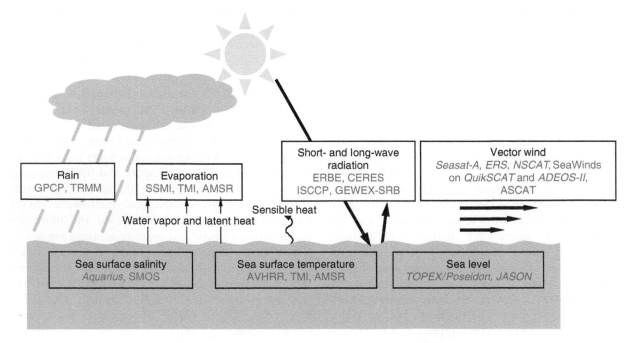

Figure 1 Schematic diagram of the physical exchange processes at the air–sea interface and the upper ocean responses, with corresponding sensor names shown in red.

Pacific Ocean at a frequency of *c.* 3–7 years. The 1997–98 El Niño was one of the most severe events experienced during the twentieth century. During the peak of the event in December 1997 (**Figure 2**), the SST in the eastern equatorial Pacific was more than 5 °C above normal, and the warming was accompanied by excessive precipitation and large net heat transfer from the ocean to the atmosphere.

The 1997–98 event was also the best observed thanks largely to the expanded satellite-observing capability. One of the major observational findings was the role of synoptic westerly wind bursts (WWBs) in the onset of El Niño. **Figure 3** presents the evolution of zonal wind from NSCAT scatterometer combined with SSM/I-derived wind product, sea surface height (SSH) from TOPEX altimetry, and SST from AVHRR imagery in 1996–98. The appearance of the anomalous warming in the eastern basin in February 1997 coincided with the arrival of the downwelling Kelvin waves generated by the

WWB of December 1996 in the western Pacific. A series of subsequent WWB-induced Kelvin waves further enhanced the eastern warming, and fueled the El Niño development. The positive feedback between synoptic WWB and the interannual SST warming in making an El Niño is clearly indicated by satellite observations. On the other hand, the synoptic WWB events were the result of the development of equatorial twin cyclones under the influence of northerly cold surges from East Asia/western North Pacific. NSCAT made the first complete recording of the compelling connection between near-equatorial wind events and mid-latitude atmospheric transient forcing.

Clearly, the synergy of various satellite products offers consistent global patterns that facilitate the mapping of the correlations between various processes and the construction of the teleconnection pattern between weather and climate anomalies in one region and those in another. The satellite

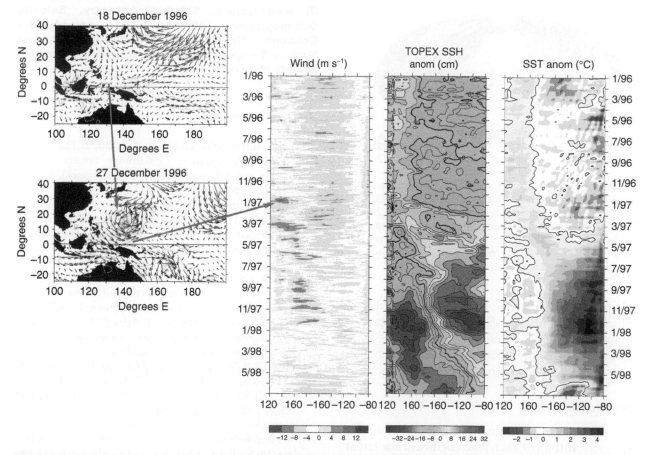

Figure 2 (First column) An example of the scatterometer observations of the generation of the tropical cyclones in the western tropical Pacific under the influence of northerly cold surges from East Asia/western North Pacific. The effect of westerly wind bursts on the development of El Niño is illustrated in the evolution of the equatorial sea level observed from *TOPEX/Poseidon* altimetry and SST from AVHRR. The second to fourth columns show longitude (horizontally) and time (vertically, increasing downwards). The series of westerly wind bursts (second column, SSM/I wind analysis by Atlas *et al.* (1996)) excited a series of downwelling Kelvin waves that propagated eastward along the equator (third column), suppressed the thermocline, and led to the sea surface warming in the eastern equatorial Pacific (fourth column).

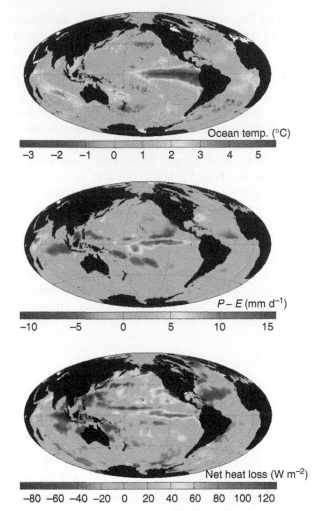

Figure 3 Satellite-derived global ocean temperature from AVHRR (top), precipitation minus evaporation from GPCP and WHOI OAFlux, respectively (middle), and net heat loss $(Q_{LH} + Q_{SH} + Q_{LW} - Q_{SW})$ from the ocean (bottom) during the El Niño in Dec. 1997. The latent and sensible heat fluxes $Q_{LH} + Q_{SH}$ are provided by WHOI OAFlux, and the short- and long-wave radiative fluxes are by ISCCP.

observing system will complement the *in situ* ground observations and play an increasingly important role in understanding the cause of global climate changes and in improving the model skills on predicting weather and climate variability.

Nomenclature

c_d	drag coefficient
c_e	turbulent exchange coefficient for latent heat
c_h	turbulent exchange coefficient for sensible heat
c_p	specific heat capacity of air at constant pressure
E	evaporation
L_e	latent heat of vaporization
q_a	specific humidity at a reference height above the sea surface
q_s	specific humidity at the sea surface
Q_{LH}	latent heat flux
Q_{SH}	sensible heat flux
T_a	temperature at a reference height above the sea surface
T_s	temperature at the sea surface
u	zonal component of the wind speed
v	meridional component of the wind speed
W	wind speed
ρ	density of surface air
ρ_w	density of sea water
τ	wind stress
τ_x	zonal component of the wind stress
τ_y	meridional component of the wind stress

See also

IR Radiometers. Satellite Altimetry. Satellite Oceanography, History and Introductory Concepts. Satellite Remote Sensing of Sea Surface Temperatures. Satellite Remote Sensing: Salinity Measurements.

Further Reading

Adler RF, Huffman GJ, Chang A, *et al.* (2003) The Version 2 Global Precipitation Climatology Project (GPCP) monthly precipitation analysis (1979–present). *Journal of Hydrometeorology* 4: 1147–1167.

Atlas R, Hoffman RN, Bloom SC, Jusem JC, and Ardizzone J (1996) A multiyear global surface wind velocity dataset using SSM/I wind observations. *Bulletin of the American Meteorological Society* 77: 869–882.

Bentamy A, Katsaros KB, Mestas-Nuñez AM, *et al.* (2003) Satellite estimates of wind speed and latent heat flux over the global oceans. *Journal of Climate* 16: 637–656.

Chou S-H, Nelkin E, Ardizzone J, Atlas RM, and Shie C-L (2003) Surface turbulent heat and momentum fluxes over global oceans based on the Goddard satellite retrievals, version 2 (GSSTF2). *Journal of Climate* 16: 3256–3273.

Gupta SK, Ritchey NA, Wilber AC, Whitlock CH, Gibson GG, and Stackhouse RW, Jr. (1999) A climatology of surface radiation budget derived from satellite data. *Journal of Climate* 12: 2691–2710.

Liu WT and Katsaros KB (2001) Air–sea flux from satellite data. In: Siedler G, Church J, and Gould J (eds.) *Ocean Circulation and Climate*, pp. 173–179. New York: Academic Press.

Kubota M, Iwasaka N, Kizu S, Konda M, and Kutsuwada K (2002) Japanese Ocean Flux Data Sets with Use of Remote Sensing Observations (J-OFURO). *Journal of Oceanography* 58: 213–225.

Wentz FJ, Gentemann C, Smith D, and Chelton D (2000) Satellite measurements of sea surface temperature through clouds. *Science* 288: 847–850.

Yu L and Weller RA (2007) Objectively analyzed air–sea heat fluxes (OAFlux) for the global oceans. *Bulletin of the American Meteorological Society* 88: 527–539.

Zhang Y-C, Rossow WB, Lacis AA, Oinas V, and Mishchenko MI (2004) Calculation of radiative fluxes from the surface to top of atmosphere based on ISCCP and other global data sets: Refinements of the radiative transfer model and the input data. *Journal of Geophysical Research* 109: D19105 (doi:10.1029/2003JD004457).

Relevant Websites

http://winds.jpl.nasa.gov
– Measuring Ocean Winds from Space.

http://www.gewex.org
– The Global Energy and Water Cycle Experiment (GEWEX).

http://precip.gsfc.nasa.gov
– The Global Precipitation Climatology Project.

http://isccp.giss.nasa.gov
– The International Cloud Climatology Project.

http://oaflux.whoi.edu
– The Objectively Analyzed air–sea Fluxes project.

http://www.ssmi.com
– The Remote Sensing Systems Research Company

http://www.gfdi.fsu.edu
– The SEAFLUX Project, Geophysical Fluid Dynamics Institute.

http://eosweb.larc.nasa.gov
– The Surface Radiation Budget Data, Atmospheric Science Data Center.

GRAVITY

M. McNutt, MBARI, Moss Landing, CA, USA

Introduction

The gravity field varies over the oceans on account of lateral variations in density beneath the ocean surface. The most prominent anomalies arise from undulations on density interfaces, such as occur at the water–rock interface at the seafloor or at the crust–mantle interface, also known as the Moho discontinuity. Because marine gravity is relatively easy to measure, it serves as a remote sensing tool for exploring the earth beneath the oceans. The interpretation of marine gravity anomalies in terms of the Earth's structure is highly nonunique, however, and thus requires simultaneous consideration of other geophysically observed quantities. The most useful auxiliary measurements include depth of the ocean from echo sounders, the shape of buried reflectors from marine seismic reflection data, and/or the density of ocean rocks as determined from dredge samples or inferred from seismic velocities.

Depending on the spatial wavelength of the observed variation in the gravity field, marine gravity observations are applied to the solution of a number of important problems in earth structure and dynamics. At the very longest wavelengths of 1000 to 10 000 km, the marine gravity field is usually combined with anomalies over land to infer the dynamics of the entire planet. At medium wavelengths of several tens to hundreds of kilometers, the gravity field contains important information on the thermal and mechanical properties of the lithospheric plates and on the thickness of their sedimentary cover. At even shorter wavelengths, the field reflects local irregularities in density, such as produced by seafloor bathymetric features, magma chambers, and buried ore bodies. On account of the large number of potential contributions to the marine gravity field, modern methods of analysis include spectral filtering to remove signals outside of the waveband of interest and interpretation within the context of models that obey the laws of phyics.

Units

Gravity is an acceleration. The acceleration of gravity on the earth's surface is about 9.81 m s^{-2}.

Gravity anomalies (observed gravitational acceleration minus an expected value) are typically much smaller, about 0.5% of the total field. The SI-compatible unit for gravity anomaly is the gravity unit (gu): 1 gu = 10^{-6} m s^{-2}. However, the older c.g.s unit for gravity anomaly, the milligal (mGal), is still in very wide use: 1 mGal = 10 gu. Typical small-scale variations in gravity over the ocean range from a few tens to a few hundreds of gravity units. Lateral variations in gravitational acceleration (gravity gradients) are measured in Eötvös units (E): 1 E = 10^{-9} s^{-2}. Another quantity useful in gravity interpretation is the density of earth materials, measured in kg m^{-3}. In the marine realm, relevant densities range from about 1000 kg m^{-3} for water to more than 3300 kg m^{-3} for mantle rocks.

A close relative of the marine gravity field is the marine geoid. The geoid, measured in units of height, is the elevation of the sea level equipotential surface. Geoid anomalies are measured in meters and are the departure of the true equipotential surface from that predicted for an idealized spheroidal Earth whose density structure varies only with radius. Geoid anomalies range from 0 to more than ± 100 m. The direction of the force of gravity is everywhere perpendicular to the geoid surface, and the magnitude of the gravitational attraction is the vertical derivative of the geopotential U (eqn [1]).

$$g = -\frac{\partial U}{\partial z} \qquad [1]$$

Geoid height N is related to the same equipotential U via Brun's formula eqn ([2]).

$$N = -\frac{U}{g_0} \qquad [2]$$

in which g_0 is the acceleration of gravity on the spheroid.

For a ship sailing on the sea surface (the equipotential), it is easier to measure gravity. From a satellite in free-fall orbit high over the Earth's surface, radar altimeters can measure with centimeter precision variations in the elevation of sea level, an excellent approximation to the true geoid that would follow the surface of a motionless ocean. Regardless of whether geoid or gravity is the quantity measured directly, simple formulas in the wavenumber domain allow gravity to be computed from geoid and vice versa. Given the same equipotential, the gravity representation emphasizes the power in the high-frequency (short-wavelength) part of the spectrum, whereas the geoid representation emphasizes the longer wavelengths. Therefore, for investigations of high-frequency phenomena, gravity is generally the quantity interpreted even if geoid is what was measured. The opposite is true for long-wavelength phenomena.

Measurement of Marine Gravity

Marine gravity measurements can be and have been acquired with several different sorts of sensors and from a variety of platforms, including ships, submarines, airplanes, and satellites. The ideal combination of sensor system and platform depends upon the needed accuracy, spatial coverage, and available time and funds.

Gravimeters

The design for most marine gravimeters is borrowed from their terrestrial counterparts and are either absolute or relative in their measurements. Absolute gravimeters measure the full acceleration of gravity g at the survey site along the direction of the local vertical. Modern marine absolute gravimeters measure precisely the vertical position z of a falling mass (e.g., a corner cube reflector) as a function of time t in a vacuum cylinder using laser interferometry and an atomic clock. The acceleration is then calculated as the second derivative of the position of the falling mass as a function of time (eqn [3]).

$$g = \frac{\mathrm{d}^2 z(t)}{\mathrm{d}t^2} \qquad [3]$$

Absolute gravimeters tend to be larger, more difficult to deploy, costlier to build, and more expensive to run than relative gravimeters, and thus are only used when relative gravimeters are inadequate for the problem being addressed.

Most gravity measurements at sea are relative measurements, Δg: the instrument measures the difference between gravity at the study site and at another site where absolute gravity is known (e.g., the dock where the expedition originates). Modern relative gravimeters are based on Hooke's law for the force F required to extend a spring a distance x (eqn [4]), where k_s is the spring constant, calculated by extending the spring under a known force.

$$F = -k_s x = mg \qquad [4]$$

If a mass m is suspended from this spring at a site where gravity g is known (e.g., by deploying an absolute gravimeter at that base station), then gravity

at other locations can be calculated by observing how much more or less that same spring is stretched at other locations by the same mass. Although such systems are relatively inexpensive to build and easy to deploy, they suffer from drift: in effect, the spring constant changes with time because no physical spring is perfectly elastic. To first order, the drift can be corrected by returning to the same or another base station with the same instrument, and assuming that the drift was linear with time in between. The accuracy of this linear drift assumption improves with more frequent visits to the base station, but this is usually impractical for marine surveys. Through clever design, the latest generation of marine gravimeters has greatly reduced the drift problem as compared with earlier instruments.

The measurement of the gravity gradient tensor was widely used early in the twentieth century for oil exploration, but fell into disfavor in the 1930s as scalar gravimeters became more reliable and easy to use. Gravity gradiometry at sea is currently making a comeback as the result of declassification of military gradiometer technology developed for use in submarines during the Cold War. Gravity gradiometers measure the three-dimensional gradient in the gravity vector using six pairs of aligned gravimeters, with accuracies reaching better than 1 Eotvos. In comparison with measurements of gravity, the gravity gradient has more sensitivity to variations at short wavelenghts (~5 km or less), making it useful for delimiting shallow structures buried beneath the seafloor.

Geoid anomalies can be directly measured from orbiting satellites carrying radar altimeters. The altimeters measure the travel time of a radar pulse from the satellite to the ocean surface, from which it is reflected and bounced back to the satellite. Tracking stations on Earth solve for the position of the satellite with respect to the center of the Earth. These two types of information are then combined to calculate the height of the sea level equipotential surface above the center of the Earth. Because the solid land surface does not follow an equipotential, altimeters cannot be used to constrain the terrestrial geoid. Furthermore, it is difficult to extract geoid from ocean areas covered by sea ice. However, in the near future, laser altimeters deployed from satellites hold the promise of extracting geoid information even over ocean surfaces marred with sea ice, on account of their enhanced resolution.

Platforms

Marine gravity data can be acquired either from moving or from stationary platforms. Because the gravity field from variations in the depth of the sea floor is such a large component of the observed signal, most marine gravity surveys have relied on ships or submarines that enable the simultaneous acquisition of depth observations. However, airborne gravity measurements have been acquired successfully over ice-covered areas of the polar oceans, and orbiting satellites have measured the marine geoid from space.

A major challenge in acquiring gravity data from a floating platform at the sea surface is in separating the acceleration of the platform in the dynamic ocean from the acceleration of gravity. This problem is overcome by mounting marine gravimeters deployed from ships on inertially stabilized tables. These tables employ gyroscopes to maintain a constant attitude despite the pitching and rolling of the ship beneath the table. The nongravitational acceleration is somewhat mitigated by mounting marine gravimeters deep in the hold and as close to the ship's center of motion as possible. Special damping mechanisms also prevent the spring in the gravimeter from responding to extremely high-frequency changes in the force on the suspended mass.

Instruments deployed in submersibles resting on the bottom of the ocean or in instrument packages lowered to the bottom of the ocean do not suffer from the dynamic accelerations of the moving ocean surface, but bottom currents can also be an important source of noise in submarine gravimetry. Installing instruments in boreholes is the most effective way to counter this problem, but it is also an expensive solution.

Reduction of Marine Data

A number of standard corrections must be applied to the raw gravity data (either g or Δg) prior to interpretation. In addition to any drift correction, as mentioned above for relative gravity measurements, a latitude correction is immediately applied to account for the large change in gravity between the poles and the Equator caused by Earth's rotation. Near the Equator, the centrifugal acceleration from the Earth's spin is large, and gravity is about 50 000 gu less, on average, than at the poles. Because this effect is 5000 times larger than typical regional gravity signals of interest, it must be removed from the data using a standard formula for the variation of gravity g_0 on a spheroid of revolution best fitting the shape of the Earth (eqn [5]; Θ = latitude).

$$g_0(\Theta) = 9.7803185\big(1 + 5.278895 \times 10^{-3}\sin^2\Theta$$
$$+ 2.3462 \times 10^{-5}\sin^4\Theta\big)\mathrm{ms}^{-2} \quad [5]$$

A second correction that must be made if the gravity is measured from a moving vehicle, such as a ship or airplane, accounts for the effect on gravity of the motion of the vehicle with respect to the Earth's spin. A ship steaming to the east is, in effect, rotating faster than the Earth. The centrifugal effect of this increased rate of rotation causes gravity to be less than it would be if the ship were stationary. The opposite effect occurs for a ship steaming to the west. This term, called the Eötvös correction, is largest near the Equator and involves only the east–west component v_{EW} of the ship's velocity vector (eqn [6]), in which ω is the angular velocity of the Earth's rotation.

$$g_{EOT} = 2\omega v_{EW}\cos\Theta \quad [6]$$

The free air gravity correction, which accounts for the elevation of the measurement above the Earth's sea level equipotential surface, is obviously not needed if the measurement is made on the sea surface. The free air correction g_{FA} is required if the measurement is made from a submersible or an airplane: eqn [7], where h is elevation above sea level in meters.

$$g_{FA} = 3.1h \quad \mathrm{gu} \quad [7]$$

This correction is added to the observation if the sensor is deployed above the Earth's surface, and subtracted for stations below sea level.

For land surveys, the Bouguer correction accounts for the extra mass of the topography between the observation and sea level. For its marine equivalent, it adds in the extra gravitational attraction that would be present if rock rather than water existed between sea level and the bottom of the ocean. Except in areas of rugged bathymetry, the Bouguer correction g_B is calculated using the slab formula (eqn [8]).

$$g_B = -2\pi\Delta\rho Gz \quad [8]$$

Here $\Delta\rho$ is the density difference between oceanic crust and sea water, G is Newton's constant, and z is the depth of the sea floor. This correction is seldom used because it produces very large positive gravity anomalies. Furthermore, there are more accurate corrections for the effect of bathymetry that do not make the unrealistic assumption that the expected state for the oceans should be that the entire depth is filled with crustal rocks displacing the water. The Bouguer correction is necessary, however, when gravity measurements are made from a submarine, in order to combine those data with more conventional observations from the sea surface. In this case, the Bouguer correction is applied twice: once to remove the upward attraction of the layer of water above the submarine, and once more to add in that layer's gravitation field below the sensor.

Satellite measurements of sea surface height go through a different processing sequence to recover marine geoid anomalies. The most important step is in calculating precise orbits. Information from tracking stations is supplemented with a 'crossover analysis' that removes long-wavelength bias in orbit elevation by forcing the height valves to agree wherever orbits cross. Corrections are then made for known physical oceanographic effects such as tides, and wave action is averaged out. The height of the sea level geoid above the Earth's center, assuming the standard spheroid, is subtracted from the data to create geoid anomalies.

History

A principal impediment to the acquisition of useful gravity observations at sea was the difficulty in separating the desired acceleration of gravity from the acceleration of the platform floating on the surface of the moving ocean. For this reason, the first successful gravity measurements to be acquired at sea were taken from a submarine by the Dutch pioneer, Vening Meinesz, in 1923. He used a pendulum gravimeter, which was the state of the art for measuring absolute gravity at that time. By accurately timing the period, T, of the swinging pendulum, the acceleration of gravity, g, can be recovered according to eqn [9], in which l is the length of the pendulum arm.

$$T = 2\pi\sqrt{\frac{l}{g}} \qquad [9]$$

By 1959, five thousand gravity measurements had been acquired from submarines globally. These measurements were instrumental in revealing the large gravity anomalies associated with the great trenches along the western margin of the Pacific. However, these gravity observations were very time-consuming to acquire because of the long integration times needed to achieve a high-precision estimate of the pendulum's period, and could not be adapted for use on a surface ship.

Gravity measurements at sea became routine and reliable in the late 1950s with the development of gyroscopically stabilized platforms and heavily damped mass-and-spring systems constrained to move only vertically. The new platforms compensated for the pitch and roll of the ship such that simple mass-and-spring gravimeters could collect time series of variations in gravity over the oceans from vessels under way. Without any need to stop the ship on station, a time series of gravity measurements could be obtained at only small incremental cost to ship operations. With the advent of the new instrumentation, the catalogue of marine gravity values has grown in the past 40 years to more than 2.5 million measurements.

A new era of precision in marine gravity began with the advent of the Global Positioning System (GPS) in the late 1980s. Prior to this time, the largest source of uncertainty in marine gravity lay in the Eötvös correction. Older navigation systems (dead reckoning, celestial, and even the TRANSIT satellite system) were too imprecise in the absolute position of the ship and too infrequently available to allow accurate velocity estimation from minute to minute, especially if the ship was maneuvering. Typically, gravity data had to be discarded for an hour or so near the time of any change in course. The high positioning accuracy and frequency of GPS fixes now allows such precise calculation of the Eötvös correction that it is no longer the limiting factor in the accuracy of marine gravity data.

A breakthough in determining the global marine gravity field was achieved with the launching of the GEOS-3 (1975–1977) and Seasat (1978) satellites, which carried radar altimeters. Altimeters were deployed for the purpose of measuring dynamic sea surface elevation associated with physical oceanographic effects. The Seasat satellite carried a new, high-precision altimeter that characterized the variations in sea surface elevation with unprecedented detail. The satellite failed prematurely, but not before it returned a wealth of data on the marine gravity field from its observations of the marine geoid. The geoid variations at mid- and short-wavelength were so large that the dynamic oceanographic effects motivating the mission could be considered a much smaller noise term. The success of the Seasat mission led to the launch of Geosat, which measured the geoid at even higher precision and resolution. Unfortunately, most of that data remained classified by the US military until the results from a similar European mission were about to be released into the public domain. The declassification of the Geosat data in 1995 fueled a major revolution in our understanding of the deep seafloor (**Figure 1**).

The latest developments in marine gravity stem from the desire to detect the shortest spatial wavelengths of gravity variations by taking gravimeters to the bottom of the ocean. Gravity is one example of a potential field, and as such the amplitude, A, of the signal of interest decays with distance, z, between source and detector as in eqn [10], where k is the

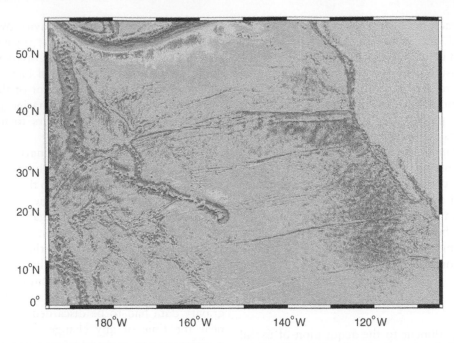

Figure 1 Gravity field over the North Pacific. This view is based on satellite altimetry data from the Geosat and other missions. (Data from Sandwell and Smith (1997).)

modulus of the spatial wavenumber, the reciprocal of the spatial wavelength.

$$A \sim e^{-2\pi kz} \qquad [10]$$

For sensors located on a ship at the sea surface in average ocean depths of 4.5 km, it is extremely difficult to detect short-wavelength variations in gravity of a few kilometers or less. Even lowering the gravimeter to the cruising depth of most submarines (a few hundred meters) does little to overcome the upward attenuation of the signal from localized sources on and beneath the seafloor. The solution to this problem recently has been to take gravimeters to the bottom of the ocean, either in a deep-diving submersible such as *Alvin*, or as an instrument package lowered on a cable. Most gravity measurements at sea are relative measurements. However, recent advances in instrumentation now allow absolute gravimeters to be deployed on the bottom of the ocean, avoiding the problem of instrument drift that adds error to relative gravity measurements. However, noise associated with the short baselines required for operation in the deep sea remains problematic.

Interpretation of Marine Gravity

Short-Wavelength Anomalies

The shortest-wavelength gravity anomalies over the oceans (less than a few tens of kilometers) are the least ambiguous to interpret since they invariably are of shallow origin. The upward continuation factor guarantees that any spatially localized anomalies with deep sources will be undetectable at the ocean surface. Near-bottom gravity measurements are able to improve somewhat the detection of concentrated density anomalies buried at deeper levels, but most are assumed to lie within the oceanic crust.

One of the most useful applications of short-wavelength gravity anomalies has been to predict ocean bathymetry (**Figure 2**). Radar altimeters deployed on the Seasat and Geosat missions measured with centimeter accuracy the height of the underlying sea surface, an excellent approximation to the marine geoid, over all ice-free marine regions. The accuracy and spatial coverage was far better than had been provided from more than a century of marine surveys from ships. At short wavelengths, undulations of the rock–water interface are the largest contribution to the short-wavelength portion of the geoid spectrum, which opened up the possibility of predicting ocean depth from the excellent geoid data. For example, an undersea volcano, or seamount, represents a mass excess over the water it displaces. The extra mass locally raises the equipotential surface, such that positive geoid anomalies are seen over volcanoes and ridges while geoid lows are seen over narrow deeps and trenches. The prediction of bathymetry from marine geoid or gravity data is tricky: the highest frequencies in the bathymetry cannot be estimated because of the upward attenuation

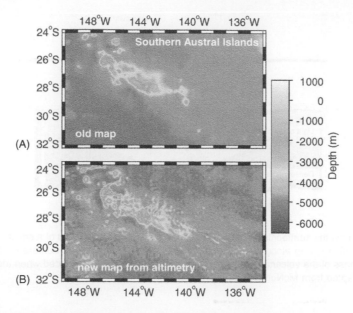

Figure 2 Example of bathymetric prediction from gravity anomalies in a largely unexplored region of the South Pacific. (A) The best available bathymetry from sparse echo soundings available in the early 1990s. (B) A diagram shows a dramatic improvement in definition of the bathymetry when satellite gravity observations are used to constrain the short-wavelength component of the bathymetry. (Adapted from McNutt and Bonneville (1996).)

problem, and the longer wavelengths are canceled out in the geoid by their isostatic compensation (see following section). These longer wavelengths in the bathymetry must be introduced into the solution using traditional echo soundings from sparse ship tracks. Nevertheless, the best map we currently have of the depth of the global ocean is courtesy of satellite altimetry.

Mid-Wavelength Anomalies

The mid-wavelength part of the gravity spectrum (tens to hundreds of kilometers) is dominated by the effects of isostatic compensation. Isostasy is the process by which the Earth supports variations in topography or bathymetry in order to bring about a condition of hydrostatic equilibrium at depth. The definition of isostasy can be extended to include both static and dynamic compensation mechanisms, but at these wavelengths the static mechanisms are most important. There are a number of different types of isostatic compensation at work in the oceans, and the details of the gravity field can be used to distinguish them and to estimate the thermomechanical behavior of oceanic plates.

One of the simplest mechanisms for isostatic compensation is Airy isostasy: the oceanic crust is thickened beneath areas of shallow bathymetry. The thick crustal roots displace denser mantle material, such that the elevated features float on the mantle much like icebergs float in the ocean. Of the various

methods of isostatic compensation, this mechanism predicts the smallest gravity anomalies over a given feature. From analysis of marine gravity, we now know that this sort of compensation mechanism is only found where the oceanic crust is extremely weak, such as on very young lithosphere near a midocean ridge. For example, large plateaus formed when hotspots intersect midocean ridges are largely supported by Airy-type isostasy. Elsewhere the oceanic lithosphere is strong enough to exhibit some lateral strength in supporting superimposed volcanoes and other surface loads.

An extremely common form for support of bathymetric features in the oceans is elastic flexure. Oceanic lithosphere has sufficient strength to bend elastically, thus distributing the weight of a topographic feature over an area broader than that of the feature itself (**Figure 3**). Analysis of marine gravity has been instrumental in establishing that the elastic strength of the oceanic lithosphere increases with increasing age. Young lithosphere near the midocean ridge is quite weak, in some cases hardly distinguished from Airy-type isostasy. The oldest oceanic lithosphere displays an effective thickness equivalent to that of a perfectly elastic plate 40 km thick. The fact that this thickness is less than that of the commonly accepted value for the thickness of the mechanical plate that drifts over the asthenosphere indicates that the base of the oceanic lithosphere is not capable of sustaining large deviatoric stresses (of the order of 100 MPa or more) over million-year timescales.

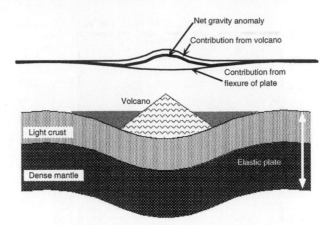

Figure 3 Cartoon showing how the seafloor is warped as an elastic plate under the weight of a small volcano. The gravity anomaly that would be detected by a ship sailing along the sea surface over this feature is the net difference between the positive gravity perturbation from the extra mass of the volcano and the negative gravity perturbation produced when the elastically flexed light crust replaces denser mantle. (Adapted from McNutt and Bonneville (1996).)

Another very important method of isostatic compensation in the ocean is Pratt isostasy. This method of support supposes that the height of a vertical column of bathymetry is inversely proportional to its density. Low-density columns can be higher because they are lighter, whereas heavy columns must be short in order to produce the same integrated mass at some assumed depth of compensation. In the oceans, variations in the temperature of the lithosphere produce elevation changes in the manner of Pratt isostasy. For example, ridges stand 4 km above the deep ocean basins because the underlying lithosphere is hotter when the plate is young. The bathymetric swells around young hotspot volcanoes may also be supported by Pratt-type isostasy, although some combination of crustal thickening and dynamic isostasy may be operating as well. Again, gravity and geoid anomalies have been principal constraints in arguing for the mechanism of support for bathymetric swells.

Long-Wavelength Anomalies

At wavelengths from 1000 to several thousand kilometers, gravity anomalies are usually derived from satellite observations and interpreted using equations appropriate for a spherical earth. Geoid is interpreted more commonly than gravity directly, as it emphasizes the longer wavelengths in the geopotential field. Isostatic compensation for smaller-scale bathymetric features, such as seamounts, can usually be ignored in that the gravity anomaly from bathymetry is canceled out by that from its compensation when spatially averaged over longer wavelengths.

The principal signal at these wavelengths arises from the subduction of lithospheric slabs and other sorts of convective overturn within the mantle. Three sorts of gravitational contributions must be considered: (1) the direct effect of mass anomalies within the mantle, either buoyant risers or dense sinkers which drive convection; (2) the warping of the surface caused by viscous coupling of the risers or sinkers to the earth's surface; and (3) the warping of any deeper density discontinuities (such as the core–mantle boundary) also caused by viscous coupling.

In the 1980s, estimates of the locations and densities of mass anomalies in the mantle responsible for the first contribution above began to become available courtesy of seismic tomography. Travel times of earthquake waves constrained the locations of seismically fast and slow regions in the mantle. By assuming that the seismic velocity variations were caused by temperature differences between hot, rising material and cold, sinking material, it was possible to convert velocity to density using standard relations. Knowing the locations of the mass anomalies driving convection inside the Earth led to a breakthrough in understanding the long-wavelength gravity and geoid fields.

The amount of deformation on density interfaces above and below the mass anomalies inferred from tomography (contributions (2) and (3) above) depends upon the viscosity structure of Earth's mantle. Coupling is more efficient with a more viscous mantle, whereas a weaker mantle is able to soften the transmission of the viscous stresses from the risers and sinkers. Therefore, one of the principal uses of marine gravity anomalies at long wavelengths has been to calibrate the viscosity structure of the oceanic upper mantle. This interpretation must be constrained by estimates of the dynamic surface topography over the oceans, which is actually easier to estimate than over the continents because of the relatively uniform thickness of oceanic crust.

A fairly common result from this sort of analysis is that the oceanic upper mantle must be relatively inviscid. The geoid shows that there are large mass anomalies within the mantle driving convection that are poorly coupled to variations in the depth of the seafloor. If the upper mantle were more viscous, there should be a stronger positive correlation between marine geoid and depth of the seafloor at long wavelengths.

See also

Manned Submersibles, Deep Water. Satellite Altimetry.

Further Reading

Bell RE, Anderson R, and Pratson L (1997) Gravity gradiometry resurfaces. *The Leading Edge* 16: 55–59.

Garland GD (1965) *The Earth's Shape and Gravity.* New York: Pergamon Press.

McNutt MK and Bonneville A (1996) Mapping the seafloor from space. *Endeavour* 20: 157–161.

McNutt MK and Bonneville A (2000) Chemical origin for the Marquesas swell. *Geochemistry, Geophysics and Geosystems* 1.

Sandwell DT and Smith WHF (1997) Marine gravity for Geosat and ERS-1 altimetry. *Journal of Geophysical Research* 102: 10039–10054.

Smith WHF and Sandwell DT (1997) Global seafloor topography from satellite altimetry and ship depth soundings. *Science* 277: 1956–1962.

Turcotte DL and Schubert G (1982) *Geodynamics: Applications of Continuum Physics to Geological Problems.* New York: Wiley.

Watts AB, Bodine JH, and Ribe NM (1980) Observations of flexure and the geological evolution of the Pacific Ocean basin. *Nature* 283: 532–537.

Wessel P and Watts AB (1988) On the accuracy of marine gravity measurements. *Journal of Geophysical Research* 93: 393–413.

Zumberg MA, Hildebrand JA, Stevenson JM, *et al.* (1991) Submarine measurement of the Newtonian gravitational constant. *Physical Review Letters* 67: 3051–3054.

Zumberg MA, Ridgeway JR, and Hildebrand JA (1997) A towed marine gravity meter for near-bottom surveys. *Geophysics* 62: 1386–1393.

APPENDICES

APPENDICES

APPENDIX 1. SI UNITS AND SOME EQUIVALENCES

Wherever possible the units used are those of the International System of Units (SI). Other "conventional" units (such as the liter or calorie) are frequently used, especially in reporting data from earlier work. Recommendations on standardized scientific terminology and units are published periodically by international committees, but adherence to these remains poor in practice. Conversion between units often requires great care.

The base SI units

Quantity	Unit	Symbol
Length	meter	m
Mass	kilogram	kg
Time	second	s
Electric current	ampere	A
Thermodynamic temperature	kelvin	K
Amount of substance	mole	mol
Luminous intensity	candela	cd

Some SI derived and supplementary units

Quantity	Unit	Symbol	Unit expressed in base or other derived units
Frequency	hertz	Hz	s^{-1}
Force	newton	N	$kg\,m\,s^{-2}$
Pressure, stress	pascal	Pa	$N\,m^{-2}$
Energy, work, quantity of heat	joule	J	$N\,m$
Power	watt	W	$J\,s^{-1}$
Electric charge, quantity of electricity	coulomb	C	$A\,s$
Electric potential, potential difference, electromotive force	volt	V	$J\,C^{-1}$
Electric capacitance	farad	F	$C\,V^{-1}$
Electric resistance	ohm	ohm (Ω)	$V\,A^{-1}$
Electric conductance	Siemens	S	Ω^{-1}
Magnetic flux	weber	Wb	$V\,s$
Magnetic flux density	tesla	T	$Wb\,m^{-2}$
Inductance	henry	H	$Wb\,A^{-1}$
Luminous flux	lumen	lm	$cd\,sr$
Illuminance	lux	lx	$lm\,m^{-2}$
Activity (of a radionuclide)	becquerel	Bq	s^{-1}
Absorbed dose, specific energy	gray	Gy	$J\,kg^{-1}$
Dose equivalent	sievert	Sv*	$J\,kg^{-1}$
Plane angle	radian	rad	
Solid angle	steradian	sr	

*Not to be confused with Sverdrup conventionally used in oceanography: see SI Equivalences of Other Units.

SI base units and derived units may be used with multiplying prefixes (with the exception of kg, though prefixes may be applied to gram $= 10^{-3}$kg; for example, $1\,Mg = 10^6\,g = 10^6$kg)

Prefixes used with SI units

Prefix	Symbol	Factor
yotta	Y	10^{24}
zetta	Z	10^{21}
exa	E	10^{18}
peta	P	10^{15}
tera	T	10^{12}
giga	G	10^{9}
mega	M	10^{6}
kilo	k	10^{3}
hecto	h	10^{2}
deca	da	10
deci	d	10^{-1}
centi	c	10^{-2}
milli	m	10^{-3}
micro	μ	10^{-6}
nano	n	10^{-9}
pico	p	10^{-12}
femto	f	10^{-15}
atto	a	10^{-18}
zepto	z	10^{-21}
yocto	y	10^{-24}

SI Equivalences of Other Units

Physical quantity	Unit	Equivalent	Reciprocal
Length	nautical mile (nm)	$1.85318\,km$	$km = 0.5396\,nm$
Mass	tonne (t)	$10^3\,kg = 1\,Mg$	
Time	min	$60\,s$	
	h	$3600\,s$	
	day or d	$86\,400\,s$	$s = 1.1574 \times 10^{-5}\,day$
	y	$3.1558 \times 10^7\,s$	$s = 3.1688 \times 10^{-8}\,y$
Temperature	°C	$°C = K - 273.15$	
Velocity	knot ($1\,nm\,h^{-1}$)	$0.51477\,m\,s^{-1}$	$m\,s^{-1} = 1.9426\,knot$
		$44.5\,km\,d^{-1}$	
		$16\,234\,km\,y^{-1}$	
Density	$gm\,cm^{-3}$	$tonne\,m^{-3} = 10^3\,kg\,m^{-3}$	
Force	dyn	$10^{-5}\,N$	
Pressure	$dyn\,cm^{-2}$	$10^{-1}\,N\,m^{-2} = 10^{-1}\,Pa$	
	bar	$10^5\,N\,m^{-2} = 10^5\,Pa$	
	atm (standard atmosphere)	$101\,325\,N\,m^{-2}$ $= 101.325\,kPa$	
Energy	erg	$10^{-7}\,J$	
	cal (I.T.)	$4.1868\,J$	
	cal (15°C)	$4.1855\,J$	
	cal (thermochemical)	$4.184\,J$	$J = 0.239\,cal$

(*Note*: The last value is the one used for subsequent conversions involving calories.)

Energy flux	langley (ly) min^{-1} = cal cm^{-2} min^{-1}	697 W m^{-2}	W m^{-2} = 1.434 × 10^{-3} ly min^{-1}
	ly h^{-1}	11.6 W m^{-2}	W m^{-2} = 0.0860 ly h^{-1}
	ly d^{-1}	0.484 W m^{-2}	W m^{-2} = 2.065 ly d^{-1}
	kcal cm^{-2} y^{-1}	1.326 W m^{-2}	W m^{-2} = 0.754 kly y^{-1}
Volume flux	Sverdrup	10^6 m^3 s^{-1}	
		3.6 km^3 h^{-1}	
Latent heat	cal g^{-1}	4184 J kg^{-1}	J kg^{-1} = 2.39 × 10^{-4} cal g^{-1}
Irradiance	Einstein m^{-2} s^{-1} (mol photons m^{-2} s^{-1})		

*Most values are taken from or derived from *The Royal Society Conference of Editors Metrication in Scientific Journals*, 1968, The Royal Society, London.

The SI units for pressure is the pascal (1 Pa = 1 N m^{-2}). Although the bar (1 bar = 10^5 Pa) is also retained for the time being, it does not belong to the SI system. Various texts and scientific papers still refer to gas pressure in units of the torr (symbol: Torr), the bar, the conventional millimetre of mercury (symbol: mmHg), atmospheres (symbol: atm), and pounds per square inch (symbol: psi) – although these units will gradually disappear (see Conversions between Pressure Units).

Irradiance is also measured in W m^{-2}. Note: 1 mol photons = 6.02 × 10^{23} photons.

The SI unit used for the amount of substance is the mole (symbol: mol), and for volume the SI unit is the cubic metre (symbol: m^3). It is technically correct, therefore, to refer to concentration in units of mol m^3. However, because of the volumetric change that sea water experiences with depth, marine chemists prefer to express sea water concentrations in molal units, mol kg^{-1}.

Conversions between Pressure Units

	Pa	kPa	bar	atm	Torr	psi
1 Pa =	1	10^{-3}	10^{-5}	9.869 23 × 10^{-6}	7.500 62 × 10^{-3}	1.450 38 × 10^{-4}
1 kPa =	10^3	1	10^{-2}	9.869 23 × 10^{-3}	7.500 62	0.145 038
1 bar =	10^5	10^2	1	0.986 923	750.062	145.038
1 atm =	101 325	101.325	1.013 25	1	760	14.6959
1 Torr =	133.322	0.133 322	1.333 22 × 10^{-3}	1.315 79 × 10^{-3}	1	1.933 67 × 10^{-2}
1 psi	6894.76	6.894 76	6.894 76 × 10^{-2}	6.804 60 × 10^{-2}	51.715 07	1

psi = pounds force per square inch.
1 mmHg = 1 Torr to better than 2 × 10^{-7} Torr.

APPENDIX 5. PROPERTIES OF SEAWATER

A5.1 The Equation of State

It is necessary to know the equation of state for the ocean very accurately to determine stability properties, particularly in the deep ocean. The equation of state defined by the Joint Panel on Oceanographic Tables and Standards fits available measurements with a standard error of 3.5 ppm for pressure up to 1000 bar, for temperatures between freezing and 40 °C, and for salinities between 0 and 42. The density ρ (kg m^{-3}) is expressed in terms of pressure p (bar), temperature t (°C), and practical salinity S. The last quantity is defined in such a way that its value (in practical salinity units or psu) is very close to the old value expressed in parts per thousand (‰ or ppt). Its relation to previously defined measures of salinity is given by Lewis and Perkin.

The equation for ρ is obtained in a sequence of steps. First, the density ρ_w of pure water ($S = 0$) is given by

$$\rho_w = 999.842\ 594 + 6.793\ 952 \times 10^{-2}t$$
$$- 9.095\ 290 \times 10^{-3}t^2 + 1.001\ 685 \times 10^{-4}t^3$$
$$- 1.120\ 083 \times 10^{-6}t^4$$
$$+ 6.536\ 332 \times 10^{-9}t^5 \qquad \text{[A5.1]}$$

Second, the density at one standard atmosphere (effectively $p = 0$) is given by

$$\rho(S, t, 0) = \rho_w + S(0.824\ 493 - 4.089\ 9 \times 10^{-3}t$$
$$+ 7.643\ 8 \times 10^{-5}t^2 - 8.246\ 7 \times 10^{-7}t^3$$
$$+ 5.387\ 5 \times 10^{-9}t^4) + S^{3/2}(-5.724\ 66$$
$$\times 10^{-3} + 1.022\ 7 \times 10^{-4}t - 1.654\ 6$$
$$\times 10^{-6}t^2) + 4.831\ 4 \times 10^{-4}S^2 \qquad \text{[A5.2]}$$

Finally, the density at pressure p is given by

$$\rho(S, t, p) = \frac{\rho(S, t, 0)}{1 - p/K(S, t, p)} \qquad \text{[A5.3]}$$

where K is the secant bulk modulus. The pure water value K_w is given by

$$K_w = 19\ 652.21 + 148.420\ 6t - 2.327\ 105t^2$$
$$+ 1.360\ 477 \times 10^{-2}t^3 - 5.155\ 288$$
$$\times 10^{-5}t^4 \qquad \text{[A5.4]}$$

The value at one standard atmosphere ($p = 0$) is given by

$$K(S, t, 0) = K_w + S(54.674\ 6 - 0.603\ 459t$$
$$+ 1.099\ 87 \times 10^{-2}t^2 - 6.167\ 0$$
$$\times 10^{-5}t^3) + S^{3/2}(7.944 \times 10^{-2}$$
$$+ 1.648\ 3 \times 10^{-2}t$$
$$- 5.300\ 9 \times 10^{-4}t^2) \qquad \text{[A5.5]}$$

and the value at pressure p by

$$K(S, t, p) = K(S, t, 0) + p(3.239\ 908 + 1.437\ 13$$
$$\times 10^{-3}t + 1.160\ 92 \times 10^{-4}t^2$$
$$- 5.779\ 05 \times 10^{-7}t^3)$$
$$+ pS(2.283\ 8 \times 10^{-3} - 1.098\ 1$$
$$\times 10^{-5}t - 1.607\ 8 \times 10^{-6}t^2)$$
$$+ 1.910\ 75 \times 10^{-4}pS^{3/2}$$
$$+ p^2(8.509\ 35 \times 10^{-5} - 6.122\ 93$$
$$\times 10^{-6}t + 5.278\ 7 \times 10^{-8}t^2)$$
$$+ p^2S(-9.934\ 8 \times 10^{-7} + 2.081\ 6$$
$$\times 10^{-8}t + 9.169\ 7 \times 10^{-10}t^2) \qquad \text{[A5.6]}$$

Values for checking the formula are $\rho(0, 5, 0) = 999.966\ 75$, $\rho(35, 5, 0) = 1027.675\ 47$, and $\rho(35, 25, 1000) = 1062.538\ 17$.

Since ρ is always close to 1000 kg m^{-3}, values quoted are usually those of the difference $(\rho - 1000)$ in kg m^{-3} as is done in **Table A5.1**. The table is constructed so that values can be calculated for 98% of the ocean (see **Figure A5.1**). The maximum errors in density on straight linear interpolation are 0.013 kg m^{-3} for both temperature and pressure interpolation and only 0.006 for salinity interpolation in the range of salinities between 30 and 40. The error when combining all types of interpolation for the 98% range of values is less than 0.03 kg m^{-3}.

A5.2 Other Quantities Related to Density

Older versions of the equation of state usually gave formulas not for calculating the absolute density ρ, but for the 'specific gravity' ρ/ρ_m, where ρ_m is the maximum density of pure water. Since this is always

Table A5.1

p (bar)	S	t (°C)	$\rho-1000$ (kg m⁻³)	$\partial\rho/\partial S$	α (10⁻⁷ K⁻¹)	$\partial\alpha/\partial S$	c_p (J kg⁻¹ K⁻¹)	$\partial c_p/\partial S$	θ (10⁻³ °C)	$\partial\theta/\partial S$	c_s (m s⁻¹)	$\partial c_s/\partial S$
0	35	−2	28.187	0.814	254	33	3989	−6.2	−2000	0	1439.7	1.37
0	35	0	28.106	0.808	526	31	3987	−6.1	0	0	1449.1	1.34
0	35	2	27.972	0.801	781	28	3985	−5.9	2000	0	1458.1	1.31
0	35	4	27.786	0.796	1021	26	3985	−5.8	4000	0	1466.6	1.29
0	35	7	27.419	0.788	1357	23	3985	−5.6	7000	0	1478.7	1.25
0	35	10	26.952	0.781	1668	20	3986	−5.5	10000	0	1489.8	1.22
0	35	13	26.394	0.775	1958	17	3988	−5.3	13000	0	1500.2	1.19
0	35	16	25.748	0.769	2230	15	3991	−5.2	16000	0	1509.8	1.16
0	35	19	25.022	0.764	2489	14	3993	−5.1	19000	0	1518.7	1.13
0	35	22	24.219	0.760	2734	12	3996	−4.9	22000	0	1526.8	1.10
0	35	25	23.343	0.756	2970	11	3998	−4.9	25000	0	1534.4	1.08
0	35	28	22.397	0.752	3196	9	4000	−4.8	28000	0	1541.3	1.06
0	35	31	21.384	0.749	3413	8	4002	−4.7	31000	0	1547.6	1.03
100	35	−2	32.958	0.805	552	31	3953	−5.8	−2029	−2	1456.1	1.38
100	35	0	32.818	0.799	799	28	3953	−5.7	−45	−2	1465.5	1.35
100	35	2	32.629	0.793	1031	26	3954	−5.6	1939	−2	1474.5	1.33
100	35	4	32.393	0.788	1251	24	3955	−5.5	3923	−2	1483.1	1.30
100	35	7	31.958	0.781	1559	21	3957	−5.3	6901	−1	1495.1	1.26
100	35	10	31.431	0.774	1844	18	3960	−5.2	9879	−1	1506.3	1.22
100	35	13	30.818	0.769	2111	16	3963	−5.1	12858	−1	1516.7	1.19
100	35	16	30.126	0.763	2363	14	3967	−5.0	15838	−1	1526.4	1.16
100	35	19	29.359	0.759	2603	13	3970	−4.9	18819	−1	1535.3	1.13
200	35	−2	37.626	0.797	834	28	3922	−5.5	−2076	−3	1472.8	1.39
200	35	0	37.429	0.791	1058	26	3923	−5.4	−107	−3	1482.3	1.36
200	35	2	37.187	0.786	1269	24	3925	−5.3	1862	−3	1491.2	1.33
200	35	4	36.903	0.781	1469	22	3927	−5.2	3832	−3	1499.8	1.30
200	35	7	36.402	0.774	1750	19	3931	−5.1	6789	−3	1511.8	1.26
300	35	−2	42.191	0.789	1101	26	3893	−5.2	−2140	−5	1489.9	1.39
300	35	0	41.941	0.783	1303	24	3896	−5.1	−186	−5	1499.3	1.36
300	35	2	41.649	0.778	1494	22	3899	−5.0	1771	−5	1508.2	1.33
300	35	4	41.319	0.774	1676	20	3903	−5.0	3728	−5	1516.6	1.30
400	35	−2	46.658	0.781	1351	24	3867	−4.9	−2221	−7	1507.2	1.39
400	35	0	46.356	0.776	1534	22	3871	−4.8	−279	−6	1516.5	1.36
400	35	2	46.017	0.771	1707	20	3876	−4.8	1665	−6	1525.3	1.33
400	35	4	45.643	0.767	1872	19	3880	−4.7	3610	−6	1533.7	1.30
500	35	−2	51.029	0.773	1587	22	3844	−4.7	−2316	−8	1524.8	1.38
500	35	0	50.678	0.769	1751	20	3849	−4.6	−386	−8	1534.0	1.35
500	35	2	50.293	0.764	1907	19	3854	−4.6	1546	−7	1542.7	1.32
600	35	−2	55.305	0.766	1807	20	3824	−4.4	−2426	−9	1542.6	1.37
600	35	0	54.908	0.762	1954	18	3829	−4.4	−506	−9	1551.6	1.34
600	35	2	54.481	0.758	2094	17	3835	−4.4	1416	−9	1560.2	1.31

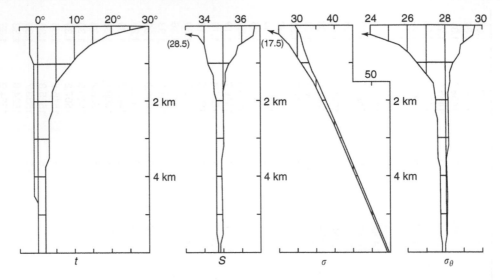

Figure A5.1 The ranges of temperature t (in °C) and salinity S for 98% of the ocean as a function of depth and the corresponding ranges of density σ and potential density σ_θ. From Bryan K and Cox MD (1972) An approximate equation of state for numerical models of ocean circulation. *Journal of Physical Oceanography* 2: 510–514.

close to unity, a quantity called σ was defined by

$$1000\left(\frac{\rho}{\rho_m} - 1\right) = \frac{1000}{\rho_m}(\rho - \rho_m) \quad [A5.7]$$

Since

$$\rho_m = 999.975 \text{ kg m}^{-3} \quad [A5.8]$$

it follows that σ, as defined above, is related to the $(\rho - 1000)$ values by

$$\sigma = (\rho - 1000) + 0.025 \quad [A5.9]$$

that is, 0.025 must be added to the values of $(\rho - 1000)$ on the table to obtain the old σ value. The notation σ, (sigma tau) was used for the value of σ calculated at zero pressure, and σ_θ (sigma theta) for the quantity corresponding to potential density. Another quantity commonly used in oceanography is the specific volume (or steric) 'anomaly' δ defined by

$$\delta = v_s(S, t, p) - v_s(35, 0, p) \quad [A5.10]$$

and usually reported in units of $10^{-8} \text{ m}^3 \text{ kg}^{-1}$.

A5.3   Expansion Coefficients

The thermal expansion coefficient α is given in **Table A5.1** in units of 10^{-7} K^{-1} along with its S derivative. The maximum error from pressure interpolation is 2 units, that from temperature interpolation is 3 units, and that for salinity interpolation ($30 < S < 40$) is 2 units plus a possible round-off error of 2 units.

The salinity expansion coefficient β can be calculated by using the given values of $\partial\rho/\partial S$.

A5.4   Specific Heat

The specific heat at surface pressure is given by Millero *et al.* and can be calculated in two stages. First, the value in $\text{J kg}^{-1} - \text{K}^{-1}$ for fresh water is given by

$$\begin{aligned}c_p(0, t, 0) = {} & 4217.4 - 3.720\ 283t + 0.141\ 285\ 5t^2 \\ & - 2.654\ 387 \times 10^{-3}t^3 \\ & + 2.093\ 236 \times 10^{-5}t^4 \quad [A5.11]\end{aligned}$$

Second,

$$\begin{aligned}c_p(S, t, 0) = {} & c_p(0, t, 0) + S(-7.644\ 4 \\ & + 0.107\ 276t - 1.383\ 9 \times 10^{-3}t^2) \\ & + S^{3/2}(0.177\ 09 - 4.077\ 2 \times 10^{-3}t \\ & + 5.353\ 9 \times 10^{-5}t^2) \quad [A5.12]\end{aligned}$$

The formula can be checked against the result $c_p(40, 40, 0) = 3981.050$. The standard deviation of the algorithm fit is 0.074. Values at nonzero pressures can be calculated by using eqn [A5.13] and the equation of state:

$$\left(\frac{\partial c_p}{\partial p}\right)_T = -T\left(\frac{\partial^2 v_s}{\partial T^2}\right)_p \quad [A5.13]$$

The values in **Table A5.1** are based on the above formula and a polynomial fit for higher pressures derived from the equation of state by N.P. Fofonoff.

The intrinsic interpolation errors in the table are 0.4, 0.1, and 0.3 J kg^{-1} K^{-1} for pressure, temperature, and salinity interpolation, respectively, and there are additional obvious round-off errors.

A5.5 Potential Temperature

The 'adiabatic lapse rate' Γ is given by

$$\Gamma = \frac{g\alpha T}{c_p} \qquad [A5.14]$$

and therefore can be calculated from the above formulas. The definition of potential temperature

$$\frac{\theta}{T} = \left(\frac{p_r}{p}\right)^{\kappa} \qquad [A5.15]$$

where p_r is a reference pressure level (usually 1 bar) and $\kappa = (\gamma - 1)/\gamma$, where γ is the ratio of specific heats at constant pressure and at constant volume, can then be used to obtain θ. The following algorithm, however, was derived by Bryden, using experimental compressibility data, to give θ (°C) as a function of salinity S, temperature t (°C), and pressure p (bar) for $30 < S < 40$, $2 < t < 30$, and $0 < p < 1000$:

$$
\begin{aligned}
\theta(S,t,p) = {} & t - p(3.650\ 4 \times 10^{-4} + 8.319\ 8 \times 10^{-5}t \\
& - 5.406\ 5 \times 10^{-7}t^2 + 4.027\ 4 \times 10^{-9}t^3) \\
& - p(S - 35)(1.743\ 9 \times 10^{-5} - 2.977\ 8 \\
& \times 10^{-7}t) - p^2(8.930\ 9 \times 10^{-7} - 3.162\ 8 \\
& \times 10^{-8}t + 2.198\ 7 \times 10^{-10}t^2) + 4.105\ 7 \\
& \times 10^{-9}(S - 35)p^2 - p^3(-1.605\ 6 \times 10^{-10} \\
& + 5.048\ 4 \times 10^{-12}t) \qquad [A5.16]
\end{aligned}
$$

A check value is $\theta(25, 10, 1000) = 8.467\ 851\ 6$, and the standard deviation of Bryden's polynomial fit was 0.001 K. Values in **Table A5.1** are given in millidegrees, the intrinsic interpolation errors being 2, 0.3, and 0 millidegrees for pressure, temperature, and salinity interpolation, respectively (**Figure A5.2**).

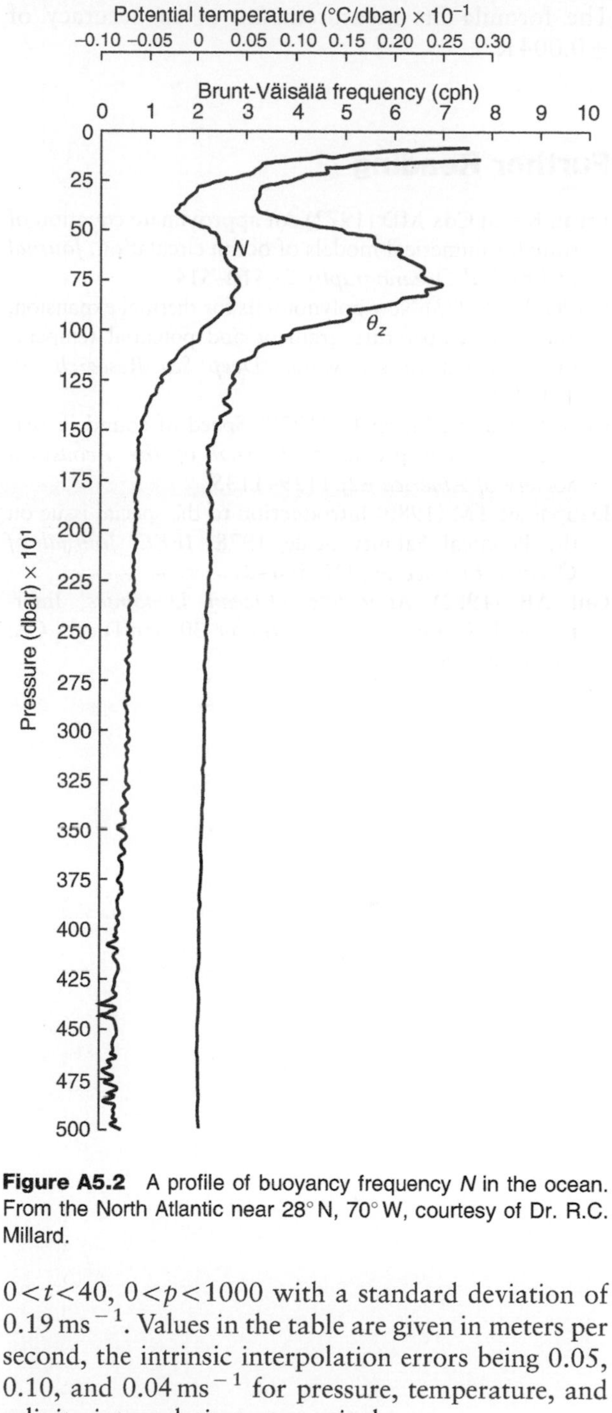

A5.6 Speed of Sound

The speed of sound c_s can be calculated from the equation of state, using eqn [A5.17]

$$c_s^2 = \left(\frac{\partial p}{\partial \rho}\right)_{\theta, S} \qquad [A5.17]$$

Values given in **Table A5.1** use algorithms derived by Chen and Millero on the basis of direct measurements. The formula applies for $0 < S < 40$,

Figure A5.2 A profile of buoyancy frequency N in the ocean. From the North Atlantic near 28° N, 70° W, courtesy of Dr. R.C. Millard.

$0 < t < 40$, $0 < p < 1000$ with a standard deviation of 0.19 ms^{-1}. Values in the table are given in meters per second, the intrinsic interpolation errors being 0.05, 0.10, and 0.04 ms^{-1} for pressure, temperature, and salinity interpolation, respectively.

A5.7 Freezing Point of Sea Water

The freezing point t_f of sea water (°C) is given by

$$
\begin{aligned}
t_f(S,p) = {} & -0.057\ 5S + 1.710\ 523 \\
& \times 10^{-3}S^{3/2} - 2.154\ 996 \\
& \times 10^{-4}S^2 - 7.53 \times 10^{-3}p \qquad [A5.18]
\end{aligned}
$$

The formula fits measurements to an accuracy of $\pm 0.004\,\text{K}$.

Further Reading

Bryan K and Cox MD (1972) An approximate equation of state for numerical models of ocean circulation. *Journal of Physical Oceanography* 2: 510–514.

Bryden HL (1973) New polynomials for thermal expansion, adiabatic temperature gradient and potential temperature gradient of sea water. *Deep Sea Research* 20: 401–408.

Chen C-T and Millero FJ (1977) Speed of sound in seawater at high pressures. *Journal of the Acoustical Society of America* 62: 1129–1135.

Dauphinee TM (1980) Introduction to the special issue on the Practical Salinity Scale 1978. *IEEE, Journal of Oceanic Engineering* OE 5: 1–2.

Gill AE (1982) *Atmosphere–Ocean Dynamics*, International Geophysics Series Volume 30. San Diego, CA: Academic Press.

Kraus EB (1972) *Atmosphere–Ocean Interaction*. London: Oxford University Press.

Lewis EL and Perkin RG (1981) The Practical Salinity Scale 1978: Conversion of existing data. *Deep Sea Research* 28A: 307–328.

Millero FJ (1978) Freezing point of seawater. In: *Eighth Report of the Joint Panel on Oceanographic Tables and Standards*, UNESCO Technical Papers in Marine Science No. 28, Annex 6. Paris: UNESCO.

Millero FJ, Chen C-T, Bradshaw A, and Schleicher K (1980) A new high pressure equation of state for seawater. *Deep Sea Research* 27A: 255–264.

Millero FJ and Poisson A (1981) International one-atmosphere equation of state for seawater. *Deep Sea Research* 28A: 625–629.

Millero FJ, Perron G, and Desnoyers JE (1973) Heat capacity of seawater solutions from 5 to 25 °C and 0.5 to 22% chlorinity. *Journal of Geophysical Research* 78: 4499–4507.

UNESCO (1981) *Tenth Report of the Joint Panel on Oceanographic Tables and Standards*, UNESCO Technical Papers in Marine Science No. 36. Paris: UNESCO.

INDEX

Notes

Cross-reference terms in italics are general cross-references, or refer to subentry terms within the main entry (the main entry is not repeated to save space). Readers are also advised to refer to the end of each article for additional cross-references - not all of these cross-references have been included in the index cross-references.

The index is arranged in set-out style with a maximum of three levels of heading. Major discussion of a subject is indicated by bold page numbers. Page numbers suffixed by T and F refer to Tables and Figures respectively. vs. indicates a comparison.

This index is in letter-by-letter order, whereby hyphens and spaces within index headings are ignored in the alphabetization. For example, 'oceanography' is alphabetized before 'ocean optics.' Prefixes and terms in parentheses are excluded from the initial alphabetization.

Where index subentries and sub-subentries pertaining to a subject have the same page number, they have been listed to indicate the comprehensiveness of the text.

Abbreviations used in subentries

AUV - autonomous underwater vehicle
CPR - continuous plankton recorder
DOC - dissolved organic carbon
ENSO - El Niño Southern Oscillation
MOR - mid-ocean ridge
ROV - remotely operated vehicle
SAR - synthetic aperture radar
SST - sea surface temperature

Additional abbreviations are to be found within the index.

A

Aanderaa RCM4 deep ocean rotor-vane instrument, 116F
Aanderaa RCM8 current meter, 117T
Aanderaa RCM9 current meter, 116F, 117T
ABE *see* Autonomous Benthic Explorer
ABE Autonomous Benthic Explorer (AUV), 129F, 137, 138F, 163F
ABS *see* Acoustic backscatter system (ABS)
Absolute surface temperature, 584
Absorbance (of EM radiation by a compound), 239–240
see also Absorbance spectroscopy
Absorbance spectroscopy, chemical sensors, **239–246**
   background subtraction, 244
   basic principles, 239–240
   molecular recognition element, 242–243
   optical fibers for, 240–241
   optoelectronics, 241–242
   reagent support material, 242

response characteristics, 242–243
sensor design, 243
   see also Absorption spectra; Optical fibers; Reflectance spectroscopy
Absorption (optical) coefficient, 272–273, 323–324, 493
   components, 494
   definition, 315–316
   detritus, 273
   gelbstoff, 273
   Hydrolight simulation, 319–320, 319F
   oceanic water, 251F, 253
      as bio-optical model quantity, 248T, 249
   phytoplankton, 273
   pure sea water, 273
   sewage plume waters, 276–279, 279F
      see also Ocean optics
Absorption spectra
   chlorophyll, 573F
   compared with fluorescence spectra, 258F
   gelbstoff, 494, 573F

Accelerometer, expendable bottom penetrometer probe (XBP), 46
Accessory pigments, 258
Acetylcholine esterase, pesticide sensors and, 245
Ac-meters, 274
Acoustical systems *see* Acoustic systems
Acoustic backscatter, diffuse seafloor flows, 376–377
   temporal correlation, 377–378
Acoustic backscatter system (ABS), 344, 344F
   STABLE II, 351, 352
Acoustic Doppler current meters, 116–118, 116F, 117T, 234
Acoustic Doppler current profiler (ADCP), 204, 347
   see also Moorings; Single Point Current Meters
Acoustic Doppler velocimeter (ADV), 347
Acoustic floats, 92, 94–95
Acoustic imaging, high-frequency, 433
Acoustic measurement, sediment transport, **343–356**

U

V

Printed and bound by CPI Group (UK) Ltd, Croydon, CR0 4YY

03/10/2024

01040315-0010